Drugs of Natural Origin

A Treatise of Pharmacognosy

6th revised edition

GUNNAR SAMUELSSON

LARS BOHLIN

Drugs of Natural Origin

A Treatise of Pharmacognosy

6th revised edition

APOTEKARSOCIETETEN

SWEDISH ACADEMY OF
PHARMACEUTICAL SCIENCES

The cover shows
Viola odorata L., a rich source of a novel family of macrocyclic peptides (cyclotides), represented by cycloviolacin O2. The figure in the background is Carl von Linné, who was professor at Uppsala University in the 18th century. His system for nomenclature and classification has been of profound importance for scientific communication in life sciences, including pharmacognosy.

Concept of cover: Lars Bohlin
Design: Matador kommunikation AB and HKH Media AB, Uppsala, Sweden

Drugs of Natural Origin, by Gunnar Samuelsson and Lars Bohlin
©2009 Gunnar Samuelsson, Lars Bohlin and Apotekarsocieteten –
Swedish Pharmaceutical Society, Swedish Pharmaceutical Press,
P.O. Box 1136, S-111 81 Stockholm, Sweden

ISBN 978 91 976510 5 9

Kristianstads Boktryckeri AB, Kristianstad, Sweden, 2009

Preface to the 3rd edition

Pharmacognosy is a multidisciplinary subject which comprises parts of botany, organic chemistry, biochemistry, and pharmacology. The importance of the different parts of the subject has varied with time. This is reflected in how the material in the textbooks on the subject have been arranged. Since the constituents of the crude drugs are the reason for their use as drugs or as raw materials for drugs, most modern textbooks arrange the material according to the chemistry of the active principles. The constituents are usually grouped together as carbohydrates, glycosides, fixed oils, alkaloids etc. However, today we have a relatively good knowledge of the biosynthetic processes occurring in the plants which lead to secondary metabolites that are used as drugs. Knowledge of these biosynthetic pathways is an important part of the subject of pharmacognosy. In a system where the constituents are grouped together according to their structures, it is rather difficult to appreciate the underlying processes involved. This textbook therefore uses the biosynthetic pathways as the basis on which to arrange the material. This has the advantage that it is easier for students to learn and understand the often complicated structures involved.

The two previous editions of this textbook were published in Swedish, in 1980 and 1982, and were used extensively in the Scandinavian countries. The present edition has been revised and enlarged. The chapter on photosynthesis has been completely rewritten. A chapter on plant tissue and cell culture has been added. Antibiotics have been included and the treatment is mainly of the biosynthesis and the industrial production. Several biosynthetic schemes, e.g. opium alkaloids and tropane alkaloids, have been brought up to date in accordance with recent findings

As in the previous editions, no descriptions of the macro- and micromorphology of the crude drugs are included, as there are excellent books already available; see list of references.

I wish to express my sincere gratitude to Dr. Gunhild Aulin-Erdtman, who has translated the original text from Swedish and checked the chemical nomenclature; she has also helped in many other ways during the preparation of the manuscript. My thanks are also due to Prof. Norman Bisset for improving the english as well as providing useful information and suggesting a number of improvements. Dr. Aulin Erdtman has given valuable help in the proofreading. In the final analysis, it is of course, the author who has to take the responsibility for any errors, of whatever kind, that may remain.

It is a pleasure to thank Prof. Finn Sandberg, who kindly made available a number of slides from his collection which are reproduced in the coloured illustrations. Prof. Xorge A. Dominguez kindly provided a slide of the peyote cactus, *Lophophora williamsii*.

5

Finally, I would like to thank the publishers, the Swedish Pharmaceutical Press, for the excellent way in which the book has been produced.

Rimforsa, November 1991
Gunnar Samuelsson

Preface to the 4th edition

The third edition of this book, which was the first edition in english, was very well received and has been translated into Italian and Greek. During the six years since it's publication, our knowledge of the biosynthesis of natural products has increased tremendously, particularly concerning the enzymes that catalyse the reactions. This knowledge has in many cases also been extended to the genes that code for the enzymes. It therefore seems appropriate to include also these areas in the book. The chapter on crude drugs has been enlarged and now also comprises preparation of extracts of crude drugs as well as the quality control of these. Plant tissue and cell culture is subject to intensive research and development and consequently the chapter has been enlarged and now also deals with the possibilities of using transgenic plants for production of drugs, i.e. antibiotics. Also in the field of photosynthesis research is progressing and this has been reflected in a revision of the part of the chapter dealing with the light reactions and the reaction centers of the two photosystems. Several drugs are polyketides and our knowledge about the enzymes and genes that are responsible for their biosynthesis has been extensively enlarged. This opens up a new field for drug discovery as it is possible to manipulate the multifunctional enzymes involved, thus making them produce "unnatural" metabolites with other properties than those of the natural ones. This new knowledge has been included in the corresponding chapter. A similar development is in progress for non-ribosomal polypeptides and this chapter has consequently also been subjected to revision and enlargement. The information in these two chapters represents the situation at the end of 1997 and beginning of 1998 but as these fields are rapidly expanding much more information might be available in the near future.

As in the third edition, the chapter (8) on amino acids is rather brief and the reason for including the chapter at all is to form a bridge to the following chapters. More extensive information on amino acids is available in textbooks on biochemistry. Finally the exciting biosynthesis of cyanocobalamin has been included in chapter 12 as an example of a remarkable achievement in the study of biologically and pharmacologically active natural products.

I am indebted to Dr Steven Lucas for valuable linguistic corrections of the revised and added text and to Dr Alice Sparks for excellent proof reading. Any remaining errors and mistakes, of whatever kind, are, however, the sole responsibility of the author. As for the previous editions Prof. Finn Sandberg has kindly made available a number of coloured slides from his collection. This valuable contribution is gratefully acknowledged. Dr Kirsi-Marja Oksman-Caldentey (Helsinki, Finland) has made important suggestions for the revision of the chapter on plant biotechnology. It is a pleasure for me to acknowledge this valuable contribution. My thanks are also due to Prof. Lars Bohlin, Dr Jan Bruhn and Dr. Per Claeson for valuable discussions about the manuscript.

Finally I would like to thank the publishers – the Swedish Pharmaceutical Press – for excellent production of the book and for a very good cooperation in the process of publishing.

Rimforsa, September 1998
Gunnar Samuelsson

Preface to the 5th edition

Although only 4 years has passed since the the 4th edition of this book was published, an extensive revision was called upon by the fast development in the field. This is particularly true for polyketide antibiotics which have been subject to intensive research activities resulting in the elucidation of biosynthetic pathways and characterization of the large enzymes involved. Modern molecular biology techniques have played an extensive role in this development. Based on these findings it is now possible to construct new molecules with different pharmacological activities which can be developed into new and better drugs.

The non-mevalonate pathway for the biosynthesis of isoprenoids has turned out to be of much greater importance than originally realized. This has necessitated rewriting of the corresponding parts of the book. Also newly published information on the biosynthesis of artemisinin, ginkgolides and paclitaxel has been included.

The biosynthesis of the important antibiotic vancomycin has now been elucidated at the genetic and enzymatic level and the information has been included in the book as it is a very important development in the field of antibiotics which opens up new possibilities for construction of new compounds, greatly needed in the never ceasing battle against antibiotic-resistant strains of bacteria.

The giant work on the biosynthesis of coenzyme B12 has continued and the new findings have been included in chapter 12.

Finally, a long overdue revision of the text concerning the light and the dark reactions in photosynthesis (chapter 4) has been performed. Also the biosynthesis of sucrose has been revised and a paragraph on the biosynthesis of starch has been added.

Dr Anders Backlund (Dept. of Medicinal Chemistry, Div. of Pharmacognosy, Uppsala University) has revised all scientific plant names and has made many suggestions for improvement of the text, particularly concerning the botanical parts of the book. This valuable contribution is gratefully acknowledged.

I am indebted to Dr Alice Sparks for valuable linguistic revision of the manuscript for the parts of the text which have been added in this edition.

As for the previous editions I am indebted to the publishers – the Swedish Pharmaceutical Press – for excellent production of the book and for a very good cooperation in the process of publishing.

Rimforsa, March 2004
Gunnar Samuelsson

Preface to the 6th edition

The development in the field of natural products of medical importance continues to expand. Since the publication of the 5th edition of this book, new trends of importance for research aiming at development of new drugs from natural products have appeared. Thus new concepts such as chemo- and bioinformatics, phylogenetics and systems biology with the new "omics" techniques (genomics, proteomics, metabolomics) have entered the field. These are introduced in the largely expanded first chapter of the book. The explanatory model of pharmacognostic research, which was first presented in the 4th edition, has been revised with reference to these developments. This chapter now also contains a section on natural products from marine organisms, an area of increasing importance in drug discovery.

Herbal remedies have been subjected to more rigid legislation aiming at improvement of the quality and the laying of a better foundation for their use, particularly as over-the-counter preparations for self-medication but also as constituents of conventional drugs. This development is described in chapter 2.

In chapter 3 a section on metabolic engineering of secondary metabolites has been included. In previous editions, the enzymes which operate in biosynthetic pathways have been denoted by the names used by the authors of the publications which constitute the sources for the text in the book. In this edition, however, the EC numbers and the names approved by the Nomenclature Committee of the International Union of Biochemistry have been introduced wherever applicable. A description of the classification of enzymes as represented by the EC code is given in chapter 4, which now also contains sections on glycolysis, the citric acid cycle and transporters of secondary metabolites.

In chapter 5 the section on biosynthesis of starch has been revised and an enlarged section on aminoglycoside antibiotics has been introduced. In chapter 6 the concept of adaptogens has been added as well as desciption of several adaptogenic crude drugs.

Chapter 7a contains a revision of the description of fatty acid syntheses and sections on the antiobesity drug Orlistat, the antibiotics mupirocin and tigecycline as well as the structurally unique enediyne antibiotics have been added. The isoflavones genistein and daidzein from soybean seeds, which are of interest as phytoestrogens, have also been included.

Chapter 7b includes a new section on the endocannabinoid system. A description of the biosysnthesis of vitamins E and K as well as new information on the biosynthesis of paclitaxel has been included. Chapter 8 has been thorougly revised, particularly with reference to the enzyme nomenclature.

Chapter 9 contains new sections on antidiabetic lizard toxins, the analgetic ziconotide from a marine cone snail and the new peptide family of cyclotides which currently are of great interest for pharmaceutical and agricultural applications. Sections have been added on the antibiotic daptomycin as well as on the antifungal echinocandins and on streptogramin antibiotics, which are particularly interesting because of their synergistic activities. This chapter also contains new sections on pantothenic acid, vitamins B6, and vitamin H. The section on vitamin B12, which was contained in chapter 12 in previous editions, has been moved to chapter 9.

Chapter 10 contains further information on the biosynthesis of hyoscyamine and the pyrrolizidine alkaloids as well as a new section on the biosynthesis of ajmaline.

Chapter 11 has been entirely revised and now deals extensively with purines and pyrimidines including biosynthesis of ATP, GTP, NAD+, NADP+, riboflavin, FMN, FAD and THF.

The section on caffeine, theobromine and theophylline now also includes a description of their biosynthesis. The biosynthesis of the pyrimidine derivatives UTP and CTP as well as of thiamin is also included. Finally a section dealing with the leukaemia remedy cytarabine has been added.

The former chapters 12 and 13 have been removed and the section on vitamin B12 moved to chapter 9 as mentioned above.

This edition of *Drugs of Natural Origin* also welcomes a new co-author – Professor Lars Bohlin – who succeeded Gunnar Samuelsson as professor of Pharmacognosy at Uppsala University in 1991. We are indebted to Dr Alice Sparks for excellent linguistic revision and proof-reading of the text. Our thanks are also due to the Swedish Pharmaceutical Press for excellent production of the book and for very good cooperation in the process of publishing.

Uppsala, December 2009
Gunnar Samuelsson
Lars Bohlin

Contents

1. Introduction

Definitions

The subject of pharmacognosy deals with *natural products used as drugs or for the production and discovery of drugs.*

Natural product   A *natural product* can be an entire organism such as a plant, an animal or a microorganism, which has not been subjected to any treatment except, perhaps, to a simple process of preservation such as drying. A natural product can also be part of an organism, e.g. a leaf or flower of a plant, or an isolated gland or other organ of an animal. An extract of an organism or an exudate can also be regarded as a natural product, as can a pure compound isolated from a plant, an animal or a microorganism. In pharmacognosy, the term

Crude drug   *crude drug* is used for those natural products such as plants or parts of plants, extracts and exudates which are not pure compounds. Most of the crude drugs are obtained from plants but some are of animal or insect origin such as lard and beeswax. Also microbial products such as dried yeast (page 577) and lyophilized bacterial cultures are regarded as crude drugs.

The history of natural products in medicine

A great proportion of the natural products used as drugs are derived from plants. The healing or poisonous properties of plants and other organisms were early explored by man. Some plants were found which had very dramatic effects on the body and some were found to cure certain diseases. The knowledge of these plants was passed on through the generations and thus man gathered considerable experience of drugs which could be obtained from the plants in his surroundings. It became the task of the medicine man or shaman to maintain this knowledge and pass it on to his successor. The medicine men were often also priests and thus the actual knowledge in many cases became enmeshed in a veil of myth and magic. This process can still be observed in developing countries and the study of drugs used by traditional healers is an important object of pharmacognostical research.

The oldest information about plants used as drugs originates from the Sumerians and Akkadians (3rd millenium BC). The Egyptians had extensive knowledge of the technique of embalming, derived from their knowledge of plants. The famous *Ebers Papyrus,* which dates from about 1550 BC, presents a large number of crude

17

drugs that are still of great importance, such as castor seeds and gum arabic. Many authors of antiquity described plants and animals that could be used as drugs. Among them were Hippocrates (ca. 460–377 BC) "The Father of Medicine", Theophrastus (372–287 BC), Pliny the Elder (AD 23–79), Dioscorides (AD ca. 40–90) and Galenos (AD 129–199). Around AD 77 Dioscorides wrote *Peri hyales iatrikes* in which he described more than 600 medicinal plants. The book was later translated to many languages among them Latin. The title of the Latin translation is *De Materia Medica* and the term "Materia Medica" was used to define the knowledge of drugs for many hundreds of years. During the Middle Ages very little progress was made in the development of the subject. Following the introduction of the art of printing in the 15th century several so-called herbals (in German: *Kräuterbuch*) were published containing information, often with pictures, on the medicinal plants. These books are today in very high demand among book-collectors and command very high prices on the rare occasions when they are available. During the 16th and 17th centuries, the era of European exploration overseas, many new crude drugs were brought to Europe, e.g. coffee, tea, cocoa seeds, ipecacuanha root and *Cinchona* bark.

The term *pharmacognosy* appears to have been coined in the 18th century by Johann Adam Schmidt (1759–1809), who was Professor of General Pathology, Therapeutics, Materia Medica and Prescription of Medicines at the Medico-Surgical Joseph Academy in Vienna, founded in 1785. The term is to be found in his *Lehrbuch der Materia Medica* which was published posthumously in Vienna in 1811. The word pharmacognosy is derived from the two Greek words *pharmakon* = drug and *gnosis* = knowledge and thus defines the subject as the knowledge of drugs. At that time the knowledge of drugs was limited and could easily be contained in one subject.

In the 18th century, Carl von Linné, who was a professor at Uppsala University, made an important contribution to the development of pharmacognosy through the introduction of his new system for naming and classifying plants. This system had profound impact on scientific communication, which could be performed more easily and with higher accuracy. Linnaeus also practised medicine which is reflected in his publications *Materia Medica* and *Inebriantia*. Materia Medica has its focus on the multidisciplinary approach of combining botany with medicine. In *Inebriantia* a number of psychoactive plants and vegetal products are discussed together with the effects of various alcoholic beverages.

At the end of the 18th century, crude drugs were still being used as powders, simple extracts, or tinctures. However, in the beginning of the 19th century Sertürner isolated morphine from opium, and this discovery ushered in a new era in the history of medicine, characterized by the isolation and chemical identification of pharmacologically active compounds from crude drugs. The discovery of morphine was soon followed by the isolation of many other important compounds, such as strychnine (1817), quinine and caffeine (1820), nicotine (1828), atropine (1833), cocaine (1855) and the mixture of cardioactive glycosides, digitaline, from foxglove leaves in 1868. With the development of organic chemistry during the 19th century, the chemical structures of many of the isolated compounds were determined. This development has continued with ever-increasing speed in the 20th century, and today the main constituents of all important crude drugs have been isolated and their structures determined.

The 20th century has also witnessed the discovery of important drugs from the animal kingdom, particularly hormones and vitamins. Although these compounds were detected in animals, they are often present in amounts too small to permit isolation of large enough quantities for use as drugs. Some of these, e.g. the steroidal hormones, are prepared from plant-derived compounds, the chemical structures of which permit their conversion to the desired drugs. Others have to be manufactured by biotechnological methods.

Following the discovery of penicillin (page 561) during the Second World War, natural products produced by microorganisms have become a very important source of drugs yielding above all antibiotics but also other important remedies such as immunosuppressants and blood-cholesterol lowering drugs.

Natural products as modern drugs

As discussed above, natural products were for a long time the only drugs available, and among the modern drugs in use today about 40 % are of natural origin. Among some groups of drugs the figure is even higher. Thus approximately 60 % of anticancer remedies and 75 % of drugs for infectious diseases are either natural products or derivatives of natural products. For some of our most important drugs, like morphine, cardiac glycosides, penicillin and other antibiotics no synthetic substitutes are available. Many of the natural compounds are isolated from the producing organisms, purified and compounded into tablets, injectables etc. for direct use as drugs. Other drugs are derivatives of natural products where the chemical structure has been modified to yield a product with the desired pharmacological properties. An example is the large group of steroids – sex hormones, corticosteroids, contraceptive drugs, etc. – which are made from plant steroids such as sapogenins (diosgenin, hecogenin), alkaloids (solasodine) or plant sterols (stigmasterol).

Some drugs produced by total chemical synthesis can also be considered to be of natural origin. This is the case when the structure of a pharmacologically active, naturally occurring, substance has served as a model for the synthetic compound but where it has been possible to simplify the structure and retain or improve the pharmacological properties. An example is the alkaloid tubocu-

Lidocaine

rarine (page 662) which was introduced in surgery (and is still used) as a muscle relaxant. Studies on the structure–activity relationship showed that the important part of the structure was two supposedly quaternary nitrogen atoms positioned at a certain distance from each other. Based on that concept, simple synthetic compounds were produced which proved to have even better properties than the natural compound. Such compounds have now to a great extent been substituted for the natural compound. Ironically enough, it was later found that only one nitrogen atom in the structure of tubocurarine from the South American liana *Chondodendron tomentosum* Ruiz et Pav. is quaternary but at physiological pH the other nitrogen is positively charged. The good result thus eminated from a partly false structure–activity relationship. The history of the local anaesthetic lidocaine goes back to studies on the structure of natural products. In 1935 Holger Erdtman, a famous Swedish chemist, was engaged in elucidation of the structure of gramine, an alkaloid present in a chlorophyll-deficient mutant of barley (*Hordeum vulgare* L.). During this work he synthesized 2-(dimethylamino-methyl)-indole, a compound which, however, was not identical with the natural gramine (which is 3-[dimethylamino-methyl]-indole). The synthesized compound was called *iso*-gramine and when tasting it Erdtman found that it numbed his tongue. Also, the starting material for the synthesis of *iso*-gramine – dimethylamino-acet-*o*-toluidide – had this property. This initiated synthesis of a series of α-*N*-dialkylaminoacid anilides and this eventually led to the marketing of lidocaine by the Swedish pharmaceutical company Astra in 1948. This drug became a great success and is perhaps the most widely used local anaesthetic in the world.

Production of drugs based on natural products

The production of plant-derived crude drugs for use as herbal remedies (page 78) or as raw materials in the pharmaceutical industry is described in chapter 2. The majority of the drugs of natural origin used in modern medicine are pure, chemically well defined compounds. These are obtained from the organism which produces them by extraction followed by various isolation and purification procedures (page 81). The isolated material can be used directly for preparation of a drug, e.g. cardiac glycosides, page 459, morphine, page 648 and benzyl penicillin, page 561. It might also be necessary to subject the isolated substance to modification by chemical or microbiological methods before it can be used as a drug. Examples are the production of steroids from diosgenin and sitosterol (page 452).

Production of pure compounds by isolation from crude plant drugs, fermentation broths or from cell- and tissue cultures is expensive and tedious. Also, other difficulties might be encountered

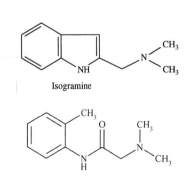

Isogramine

Dimethylamino-acet-*o*-toluidide

such as limited access to raw materials, e.g. bark of the pacific yew tree (*Taxus brevifolia* Nutt.), for production of paclitaxel (page 432). From an industrial point of view production by total synthesis is preferable in most cases. The structures of most of the natural products are, however, very complicated and although many of them have been synthesized in the laboratory it has not been possible to develop commercially profitable methods of synthesis. This problem may be at least partially solved by the combination of enzymatic and chemical synthesis which has been made possible by the techniques of molecular biology, by which large amounts of biosynthetically active enzymes are obtainable. A striking example of this technique is the single-pot complete synthesis of the corrin hydrogenobyrinic acid (a precursor of vitamin B_{12}) from the starting substances 5-aminolevulinic acid (ALA) and S-adenosylmethionine (SAM). The 12 enzymes required for this synthesis were all produced in *Escherichia coli* by genetic engineering. The overall yield, based on ALA, was 20 % which indicates an average of at least 90 % conversion at each of the 17 individual steps in the pathway. Hydrogenobyrinic acid can be converted to vitamin B_{12} by chemical methods and the complete *in vitro* – chemo-enzymatic synthesis of this important vitamin is now possible (see also page 587).

Another possibility for drug production which might be of increased importance in the future is the use of transgenic plants for production of compounds, e.g. antibodies which are normally not present in plants and which are more difficult and expensive to obtain by other means (page 104).

The role of natural products in drug discovery

Natural organisms are a rich source of secondary metabolites. These are produced by the organisms to cope with the challenges of their local environments and to gain advantages in the competition for living space and nutrients. During evolution mutations and recombinations of existing genetic material have resulted in new or modified enzymatic machineries, which under selective pressures have become fixed and inherited. This has resulted in development of extremely complex chemical structures, rich in stereochemistry, concatenated rings and reactive functional groups. The number of plant species has been estimated to be more than 300 000 and fungi, bacteria, marine invertebrates and insects add another million species or more. The number of drug-like molecules possibly present in this vast amount of species has been estimated to exceed 10^{60}. Most of this biological diversity is still unexplored and Nature is thus by far the richest source of novel compound classes to be explored for leads and targets for the development of new drugs

(See Larsson 2005 in *Further reading* (page 42)). There are, however, many obstacles to be overcome in this task. Looking for a new compound, suitable for development as a drug, in this immense amount of species and chemical structures is like trying to find a needle in a haystack. Several approaches to utilize natural products in the search for new drugs are descibed below.

Plants used in traditional medicine. Ethnopharmacology

New drugs can be discovered as a result of research programmes aimed at the study of medicinal plants used in traditional medicine. These have often been used for thousands of years and are therefore likely to contain compounds with pharmacological activities. They are also less likely to contain toxic compounds as such plants would have been sorted out at an early stage. This branch of drug research has been termed *ethnopharmacology* and can be applied both to so called "herbal remedies" (page 78) and to plants used by traditional healers (e.g. medicine men or shamans) in developing countries. An example where this approach has been successful is the antimalarial drug artemisinin (page 408) which was isolated from a plant – *Artemisia annua* L. – that for at least 1 700 years has been used in traditional Chinese medicine for treatment of fevers, including malaria. That clinical studies of an extract from a herbal remedy can be rewarding is illustrated by results of studies of the crude drug Hyperici herba (page 325) which is used as a herbal remedy against depression. Controlled clinical trials of an ethanol–water extract on a fairly large number of patients has verified that the extract has antidepressive activity. The pharmacologically active compound(s) has, however, not yet been identified with certainty.

Traditional medicine in developing countries is very complicated and different types of healers differ widely in their beliefs of the action of the plants. One group – magicians – use the plants as adjuvants to a treatment that mainly aims at soothing spirits which are believed to cause the disease, while others – herbalists – believe that the plants they use contain something which helps alleviate the illness, i.e. they use the plants as drugs. Clearly it is primarily in the plants used by the latter group that one can expect to find pharmacologically active compounds. An ethnopharmacological drug research programme in a developing country could start with observation by a trained clinical pharmacologist of the practice of a traditional healer. During this process some plants might be identified which seem to have an effect on a properly diagnosed disease. During this study any adverse effects of the treatment might also be discovered. An extract of the plant could then be prepared in agreement with the way the healer prepares his remedy (in most cases this implies an aqueous extract). This extract could then be subjected to a placebo-controlled clinical trial in the country where the traditional remedy is used, employing only patients who prefer treatment with the traditional remedy to treatment with a modern drug. An extract which gives a positive result in such a trial would then be an excellent subject for isolation of a pharmacologically active compound. Further chemical, pharmacological and clinical studies of this compound could result in the discovery of a new drug.

Intellectual property rights
For some time it has been generally accepted that countries own the right to the biological property present within their borders. This includes both terrestrial and marine organisms. Also, indigenous people have the right to the knowledge which they have developed over a long period and which is based on organisms in their local environment. This can cause problems with respect to access and sharing of the benefits of genetic material and is particularly pronounced within the field of pharmaceutical products based on natural resources. Most of the world's biodiversity exists in developing countries. This has encouraged more developed countries to organize and finance expeditions to search for organisms from which drugs could be developed or which could be used for other commercial purposes, such as plant breeding. Sometimes this resulted in discoveries of products which could be patented, produced and sold with a large profit. Often, however, the countries from which the organisms originated received little or no compensation.

The *Convention on Biological Diversity (CBD)*, which was adopted at the Earth Summit in Rio de Janeiro in 1992, codified many of these intellectual property concerns. Thus Article 15 recognizes "the sovereign rights of States over their natural resources" and that "authority to determine access to genetic resources rests with the national governments and is subject to national legislation". In the same article it is also stated that "each contracting party shall endeavour to create conditions to facilitate access to genetic resources for environmentally sound uses" and that "access, where granted, shall be on mutually agreed terms and subject to the provisions of this Article". When a microorganism, plant or animal is used for commercial application, the country from which it came has the right to benefit. Such benefits can include cash, samples of what is collected, the participation or training of national researchers, the transfer of biotechnology equipment and know-how, and shares of any profits from the use of such resources.

The adoption of *CBD* has encouraged some developing countries to put extreme restrictions on the cooperation with researchers from the Western World. In most cases these countries do not themselves have the necessary resources for development of drugs from natural products. This attitude will therefore seriously hamper the ethnopharmacological approach to drug development and the result may well be that these countries will lag behind in the development of their pharmaceutical systems, eventually resulting in increased costs for pharmaceutical products available to their people. Clearly some balance must be achieved regarding the way in which *CBD* is applied. On one side indigenous groups must feel that their knowledge is not being stolen from them or their legal rights violated. On the other side, those organizations or companies which are making substantial investments in the development of products, and thus also infrastructure, must feel that they can protect their investments. Sharing of resources and knowledge therefore becomes critical. Countries that have the foresight to provide access under reasonable terms of immediate and longer term compensation will find satisfactory economic reward and pharmaceutical development. Pharmaceutical companies which can offer and negotiate compensation packages in exchange for long-term access and provision of materials will be welcomed as global contrtibutors to local economic growth and health care.

Natural products from marine organisms

Some natural products of marine origin, such as alginic acid from seaweeds (page 170) and cod-liver oil (page 474) have been used as drugs for a long time. Compared with natural product research based on terrestrial organisms, which has a long history, the marine environment has only recently been explored. In the book *Chemistry of Marine Natural Products*, written by the late professor Paul Scheuer 1973, the term marine natural products was coined. Since then marine natural products have been established as an important group of diverse compounds with biomedical and pharmacological potential. The reported number of novel compounds isolated from marine organisms has steadily increased since the beginning of the 1960s. Recently a yearly number of about 800 new compounds have been discovered. Up to now more than 17 000 compounds have been reported in the literature from marine sources.

The marine ecosystem   The marine ecosystems represent a great biodiversity where 34 out of 36 phyla of life are represented. For life in the oceans the physical conditions, such as high salinity, extreme pressure, lack of light and large differences in temperature, differ substantially compared with the conditions for terrestrial organisms. The vast majority of marine organisms studied originate from tropical and temperate waters. However, organisms in cold-water habitats have so far not been fully exploited. In recent years microorganisms have attracted an increased interest as the source for many novel compounds, and have also turned out to be the true producers of previously isolated compounds reported from, e.g., sponges. The predominant classes of compounds found in brown algae are terpenes and polyphenolic compounds, while red algae contain many halogenated substances. However, in green algae relatively few metabolites have been reported so far. Sponges have attracted a great interest as a rich source of many unique marine natural products, e.g. unusual fatty acid derived metabolites, and are also a rich source of bioactive nitrogen-containing compounds. Another rich source of secondary metabolites, e.g. terpenoids, are the *coelenterates* (e.g. soft corals and sea anemones). *Echinoderms* including starfish and sea cucumbers are a well-documented source of glycolipids and saponins. *Bryozoans* (encrusting colonial marine forms, also called moss animals) have yielded a number of alkaloids while *tunicates* in the phylum *Chordata* are known for metabolites derived from amino acids.

Research on secondary metabolites from marine organisms

The production of secondary metabolites in marine organisms is largely driven by ecological requirements of the producing organism, such as predator–prey interactions, competition for food and space, reproduction and for communication. The understanding of the diversity of ecological functions for secondary metabolites is increasing. However, many ecological questions remain, such as where in an organism the metabolites are produced and how the biosynthesis is regulated. Perhaps recent advances in the discovery of biosynthetic genes, responsible for production of secondary metabolites, and development of techniques for chemical characterization of small amounts of bioactive compounds, involved in behavioural processes, will answer some of these questions.

In the beginning of this century molecular genetic techniques were introduced in a higher degree in the field of marine natural products research. Bioinformatics and recombinant techniques have been instrumental for the understanding of in-depth complex marine microbial pathways, while earlier feeding experiments, using isotopically labelled precursors, had been the main analytical technique to study biosynthesis. Biosynthetic pathways at the molecular level have been elucidated mainly in marine microbes such as bacteria and cyanobacteria. A novel polyketide biosynthetic pathway producing long chain polyunsaturated fatty acids (PUFAs), such as eicosapentanoic acid, was discovered in a marine bacterium of the genus *Shewanella*. This pathway is anaerobic and differs from the well characterized aerobic pathway in plants and animals.

Exploration of the marine environment for useful molecules

Marine bioprospecting is a rapidly expanding enterprise that entails the investigation of the marine environment in search of novel biomolecules. There is potential not only for pharmaceuticals but also for fine chemicals, paint ingredients for anti-fouling, nutritional supplements and agrochemicals. In the search for lead candidates for drug development, biological testing has so far mainly been focused on anti-tumour and antibiotic activity.

The first marine derived drugs were based on discoveries more than 50 years ago. The antibiotic cephalosporin (page 562) was isolated from the marine fungus *Acremonium chrysogenum* in 1948 by an Italian scientist. This was followed by the isolation of the novel unusual natural nucleosides spongothymidine and spongouridine from the Caribbean sponge *Cryptotethya crypta*, which led to commercially available synthetic anticancer and antiviral drugs (page 740).

In the last decade an increase is seen in the number of lead compounds from diverse marine life that enter preclinical and clinical trials. However, so far very few marine biomolecules have been commercialized. Some examples are briefly described here to show the great diversity of organisms, chemical structures and bioactivity in the marine environment and to indicate the potential for drug development.

Didemnin B, a cyclic depsipeptide isolated from the Caribbean tunicate *Trididemnum solidum*, was shown to be an effective inhibitor of nucleotide and protein synthesis. Preclinical anti-tumour activity caused didemnin-B to be selected for human clinical trials as the first marine natural product of this group of compounds. Clinical development was later halted but the related compound aplidine (Aplidin®, dehydrodidemnin B), isolated from another tunicate and later synthesized, is in advanced clinical trials against the treatment of solid tumours.

Two other promising lead candidates with strong potential as anti-tumour agents are (+)-*discodermolide* and *bryostatin 1*. (+)-Discodermolide, a polyketide isolated from the marine sponge *Discodermia dissolute*, was further developed as it had a better microtubuli binding effect than paclitaxel. Bryostatin 1 is a macrocyclic lactone belonging to the chemical class polyketides. It was first isolated from the bryozoan *Bugula neritina* but was later shown to originate from a microbial symbiont. Both compounds have limited effects in clinical trials. However, in combination with existing chemotherapies, e.g. paclitaxel, synergistic effects have been seen which can lead to more efficient combination therapies.

Another fascinating discovery is that a defined standardized extract derived from shark cartilage, *Neovastat®* (AE-941), has shown effect on blood vessel formation (anti-angiogenic) and is in advanced clinical trials for several types of cancer. That an extract, containing a mixture of compounds, can be a lead candidate for drug development is extraordinary.

A recent development is the commercialization of *Ziconotide* (Prialt®), a synthetic form of an isolated polypeptide from the tropical cone snail *Conus magus*, which is used against severe chronic pain (page 526).

A general development in the marine natural product research is exploration of new geographical areas and alternative sources of marine organisms. The field also attracts new research groups. However, in the exploitation of marine resources it is also a growing awareness of sustainable use and obtaining sufficient biomass for further development, without disturbing the ecosystem.

Differences between marine and terrestrial organisms in drug research

The search for lead candidates for drug development from marine organisms is more complicated than research using terrestrial organisms. The collection of organisms is done in water and often requires SCUBA equipment, depending on the temperature and depth. It is often difficult to obtain sufficient amounts of material due to the habitats and living conditions of marine organisms. Furthermore, the harvesting of wild populations of some organisms, e.g. marine sponges, is restricted because of biodiversity conservation. Due to a limited knowledge of the taxonomy of marine organisms, the identification of the collected material is in many cases uncertain. Usually the yield of secondary metabolites turns out to be very low and many bioactive compounds with drug potential have complex chemical structures and are also highly toxic. However, symbiotic microorganisms have recently been shown to be the producers of many secondary metabolites. Therefore, if the right conditions for cultivation of symbionts can be found, which is often difficult, cultured organisms can provide a renewable resource and at the same time preserve the natural population of the host organisms. The extraction, purification and isolation procedures, used for marine organisms, are in general the same as those used for natural product research based on terrestrial organisms.

The development of drugs from bioactive compounds found in plants has often been based on ethnopharmacological data. Reported use of marine organisms for medical purposes is very limited in comparison with plants on land. The selection of marine organisms in the search for potential lead candidates for drug development is therefore mostly based on ecological observations or results of high throughput screening of a large number of organisms.

Combinatorial biosynthesis

Elucidation of the biosynthetic pathways for natural products has now in several cases progressed to isolation and characterization of the enzymes involved and even to the recognition and cloning of the genes which code for these enzymes. This opens up new avenues for the discovery of bioactive molecules as in the case of some polyketides, e.g. the macrolides (page 271), that are synthesized by very large enzymes (PKSs) which carry out all synthetic steps in a predetermined order. It is now possible to characterize the genes coding for the different subunits of the PKS which catalyse these steps. This opens up the field for a new strategy termed "combinatorial biosynthesis". In this concept mixing, matching and shuffling of PKS gene segments might lead to entirely new classes of polyketides that might have valuable pharmacological properties. The same approach can also be applied to the non-ribosomal polypeptides (page 531), while ribosomal polypeptides may be manipulated directly via site-directed mutagenesis of the corresponding genes. However, the study of biosynthesis of natural products of importance for the production of drugs is only at the beginning. In higher plants only one pathway, that leading to the alkaloids berberine and morphine, has been completely elucidated at the enzyme level (page 645 ff]) and is now open for characterization of the genes coding for the enzymes involved. This has been made possible by employment of the techniques of cell and tissue culture and these will certainly be of great use also in the future (page 109).

Screening of randomly chosen organisms

This is the classical method for detection of antimicrobial activity for development of antibiotics and has been applied mainly to bacteria and microscopic fungi. Screening for antimicrobial activity is fairly simple and can be applied to a large number of samples. Screening for other pharmacological activities is in most cases more difficult and time-consuming. An example is the programme for screening of extracts of thousands of organisms for antitumour activity performed by The National Institutes of Health (NIH) in the USA. This programme has been in operation for many years at very high cost but has so far resulted in few drugs. The best known are paclitaxel (= taxol, page 426) from the Western yew tree *Taxus brevifolia* Nutt. and camptothecin from the tree *Camptotheca acuminata* Decne (page 696).

The pharmaceutical industry has developed a technique called *high-throughput screening* (HTS, page 86), originally aimed at screening thousands of synthetic compounds produced by combinatorial chemistry. This technique has also been applied to extracts of plants and other organisms as well as to libraries of isolated compounds of natural origin. The approach has so far met with limited success and many big companies have terminated their activities in natural product discovery and sold or disposed of their collections of extracts for screening. A few companies such as Novartis, Bayer and Wyeth are likely to continue natural product research and a few small companies or institutions have developed the unique set of skills to conduct bioprospecting efforts on behalf of a number of collaborating institutions. Some well-known contract research organizations such as Albany Molecular Research (www.albmolecular.com), Analyticon Discovery (www.ac-discovery.com) and Magellan Biosciences (www.magellanbioscience.com) offer natural product services (see Baker 2007 in *Further reading* (page 41)).

This development is unfortunate as there is a great need of many new drugs, particularly in the antibiotics area, where natural products so far have been much more important than synthetic compounds. Clearly more basic research is needed to simplify the discovery of new natural products that are suitable for development of drugs. The solution might be found in two new research areas: Chemoinformatics combined with Phylogenetics and Systems Biology.

Chemoinformatics and phylogenetics

Chemography   Chemical compounds can be characterized by a wide range of descriptors such as their molecular mass, lipophilicity as well as by topological features. Chemical space can be a region defined by a particular choice of descriptors and the limits placed on them. A drug-like molecule is characterized by descriptors found to be of importance in chemical compounds that are used as drugs, such as size, lipophilicity, polarizability, charge, flexibility, rigidity and hydrogen bond capacity. Combinatorial chemistry has evolved in the pharmaceutical industry as an efficient technology for synthesis of large libraries of compounds for screening by HTS. The range of structural possibilities is, however, so large that a selection of the most promising compound classes must be performed, instead of just synthesizing and screening enormous chemical libraries. There is thus a need for tools for navigation in the drug-like chemical space. This has been termed *chemography* and a method called *ChemGPS* (CHEMical Global Positioning System) has been developed for this purpose, which allows navigation in the drug-like chemical space in a similar way as geographical navigation. In conventional mapping systems (e.g. Mercator), which allow projection on the same plane of objects located on a geosphere, there is a set of rules (e.g. meridians and parallels) and a set of objects (e.g. mountains, cities, countries etc.). In chemography the rules are provided by the principal physico-chemical properties and the objects are the molecular structures. The original ChemGPS data set consists of 531 structures selected to provide a balanced chemical space largely defined by Lipinski's rule of five and general drug-like properties, defined here by 60 descriptors. ChemGPS also involves structures intentionally located outside the drug-like chemical space in analogy with satellites in the Navstar global positioning system. Satellite molecules are chemical structures that have at least one property value located outside the property range defined by the known drugspace, but still have drug-like fragments. In the ChemGPS model chemographic map, coordinates for compounds are predicted by *principal component analysis* (PCA) scores prediction. Principal component analysis is a mathematical procedure that transforms a number of (possibly) correlated variables into a smaller number of uncorrelated variables called *principal components* (PC). The first component (PC1) accounts for as much of the variability in the data as possible and each succeeding PC accounts for as much of the remaining variability as possible. In ChemGPS 6 PCs are required to reach an acceptable level of explanation. The results can be presented as observation- or variable-related projections visualized in plots called score (related to the objects) and loading plots (related to the variables).

When applied to natural products the original ChemGPS was found to produce numerous outliers, i.e. extrapolations predicted outside the model. A modified model called *ChemGPS-NP* was therefore constructed, which provides a tool for charting the biologically relevant chemical space and is an efficient mapping device for selection of high-probability hits and prediction of their properties and activities. It also has the capacity of serving as a reference system enabling characterization and comparison of molecules from various research groups. ChemGPS-NP is based on evaluation of more than 1.2 million compounds. The ChemGPS-NP model consists of 35 descriptors chosen on the basis of a set of criteria including chemical intuitiveness. In contrast to the original ChemGPS, which requires 6 PCs for an acceptable level of explanation, eight PCs are required in ChemGPS-NP, due to the more complex data. For a detailed description of ChemGPS and ChemGPS-NP the reader is referred to the papers by Oprea (2001), and Larsson (2005 and 2007a) in *Further reading* (page 42).

An example of the application of ChemGPS and ChemGPS-NP to natural products is given in the above cited papers by Larsson *et al.* They applied the models to natural products reported as inhibitors of the isoforms (COX-1 and COX-2) of the enzyme which mediates the two first steps in the biosynthesis of prostaglandins (page 259). Of the compounds studied, 30 % were flavonoids, 20 % terpenoids, 10 % phenylpropanoids and 5 % alkaloids. The study showed that relatively large, rigid structures with at least one ring were structural features of importance for inhibitors of COX-2. All of these compounds consisted of 11 or more carbons and had a molecular weight of 270 or more. Structural features of importance for COX-1 inhibitors were: large distance between a donor for hydrogen bonds and an acceptor, low flexibility and small number of rings. Few negatively charged substances were found, strong H-bond donors were uncommon and nitrogen was not preferred among inhibitors of COX-1. As mentioned above, 30 % of the studied compounds were flavonoids, which shows that this group of natural products might be of importance for further studies, as there are over 6 000 naturally occurring flavonoids found in a wide variety of different plants and most of them are unexplored in terms of COX-inhibiting activity. Fig. 1 illustrates the plots obtained by ChemGPS and ChemGPS-NP.

Phylogenetics   Chemoinformatics is used to describe the chemical space and phylogenetics applies to the evolutionary space. Chemosystematics, sometimes referred to by the older term chemotaxonomy, is the attempt to classify or organize organisms based their content of chemical compounds. This process is based on the supposition that closely related organisms share a biogenic pool of common core structures. If a pharmacologically active, clinically useful compound has been found in one organism it could therefore be profitable to search for similar – and also evolutionary pre-validated – structures in closely related organisms. What is considered to constitute "closely related organisms" may, however, vary greatly among different scientists and between applications. In the case of the flowering plants, molecular biology and molecular systematics brought about a classificatory revolution at the end of the 1990s. Previously well known classification systems, such as those of Dahlgren, Cronquist, Takhtajan and Thorne, only to a limited extent reflected the evolution of angiosperms. They were anyhow used in attempts to draw chemosystematic conclusions, and in some cases, such as Dahlgren, were also based on chemical observations. These older systems were subsequently replaced with that of Chase (1993)

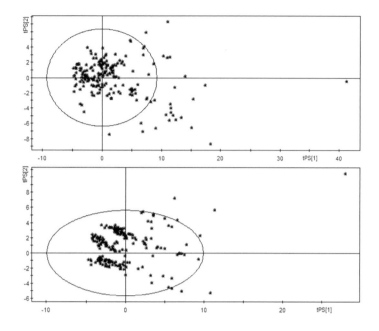

Fig. 1. Comparison of the first two dimensions of the predicted score plots (tPS1 and tPS2) for a sample of natural COX inhibitors, charted using ChemGPS (upper plot) and ChemGPS-NP (lower plot). More distinct clustering and fewer outliers are obtained with ChemGPS-NP. The circle and ellipse define confidence limits on this view of chemical space. (From Larsson et al. (2007a), reprinted with permission from the publisher. Copyright 2007, American Chemical Society and the American Society of Pharmacognosy)

and later the Angiosperm Phylogeny Group (APG 2003), based on a broad cooperation of a large number of molecular systematics laboratories. The new system has its base in the phylogenies (the evolutionary history) and may therefore be a more reliable tool in the search for structurally related, pharmacologically active, chemical compounds. See further the paper by Larsson (2007b) in *Further reading* page 43.

Systems biology

A number of new technologies for the study of biology were developed in the last part of the previous century and are now gradually entering pharmacognosy research programmes. These technologies are *genomics* with *transcriptomics, proteomics* and *metabolomics*. Systems biology is defined as the integrated approach to study biological systems – intracellular networks, cells, organs and any biological entity – by measuring and integrating genetic, proteomic and metabolic data (Wang 2005a). Locate this and following refer-

31

ences in *Further reading* (page 41). Systems biology can thus be said to be in line with the concept molecular pharmacognosy described below (page 38). Before looking at the possible impact of systems biology for research on natural products and medicinal plants the "-omic" technologies will be briefly described.

Genomics and transcriptomics

The *genome* is the total amount of DNA in a cell. One can also speak of the genome of an individual or of a species. The definition of *genomics* is not precise. Originally it meant analysis of the whole genome, but now it commonly refers to large-scale, high-through-put molecular analysis of multiple genes, gene products or regions of genes. Transcriptomics, which is often included in the term genomics depicts the expression level of genes (Ulrich-Merzenich 2007).

As complete human and mouse DNA maps are now available it is possible, by using molecular biology techniques, particularly microarrays, to study the influence of phytochemicals, and even crude extracts, on the expression of the genes that are involved in the particular pharmacological events that are being studied. A microarray or biochip is the arrangement of a high number of spots of DNA, cDNA or oligonucleotides (up to 200 000 spots/cm^2) on an extremely small space, commonly a glass surface. A hybridization solution containing a mixture of fluorescently labelled cDNAs or RNAs from the system to be analysed is prepared and incubated with the DNA chip. The cDNA or RNA which binds to the chip is detected with laser technology and the data are stored in a computer and subsequently analysed. A good desciption of the microarrays technique is available from The National Center for Biotechnology (USA) at the site http://www.ncbi.nlm.nih.gov/About/primer/microarrays.html.

Proteomics

The *proteome* is the entire complement of proteins expressed by a genome, cell, tissue or organism. *Proteomics* is the large-scale study of proteins, particularly their structures, function and interaction (Ulrich-Merzenich 2007). There are fewer protein-coding genes in the human genome than there are proteins in the human proteome (~ 22 000 genes versus ~ 400 000 proteins), implying that protein diversity cannot be fully characterized by gene expression analysis alone. Proteomics is therefore a useful tool for characterizing cells and tissues of interest (Ulrich-Merzenich 2007). Proteomics depends on access to automated methods capable of handling large numbers of samples. Important techniques are automated gel electrophoresis for separation of proteins and advanced mass spectroscopy techniques for determination of molecular size and structure analysis such as matrix-assisted laser desorption ionization (MALDI) and electrospray ionization (ESI). The microarray technique has also entered into proteomics for the study of the function of proteins and for the detection of proteins. In this technique a great number of microscopic samples of proteins are immobilized on surfaces such as glass, membranes and beads and exposed to labelled molecules that might bind to the proteins. The binding is then detected and analysed by an automatic system. Also determination of protein 3-D quarternary structures can be performed by automated systems (Alterovitz 2006, Bradshaw 2005).

Metabolomics

The term *metabolome* indicates the totality of small molecules that are formed by a cell, a tissue or an organism under certain conditions and under consideration of the concentration. *Metabolomics* aims at measuring all metabolites in an organism qualitatively and quantitatively. Subdivisions of metabolomics are *metabolic profiling* which aims at measuring a selected group of metabolites in an organism and *metabolic fingerprinting,* the aim of which is measuring a fingerprint of the metabolites in an organism, without identification of all compounds. *Metabonomics* is a term used in medicine/pharmacology and describes the measurement of the metabolic reaction towards medication, environment or diseases. The term metabonomics is often replaced by the term metabolomics (Ulrich-Merzenich 2007, Verpoorte 2005). The main tools in metabolomics are gas chromatography (GC), high-performance liquid chromatography (LC), nuclear magnetic resonance spectroscopy (NMR) and mass spectroscopy (MS). There are also techniques available to connect these methods on line, e.g. GC/MS, LC/MS and LC/NMR. Without any preceeding separation steps crude extracts can be analysed by NMR or MS in combination with post-analysis chemometric methods.

Bioinformatics

Genomics, proteomics and metabolomics all generate huge amounts of data, the transformation of which into biological information requires informatic tools. Bioinformatics is a multidisciplinary area of research aiming at the creation and advancement of algorithms, computational and statistical techniques, and theory to solve formal and practical problems arising from the management and analysis of biological data.

Application of the systemic biology approach to the study of natural products used in medicine

Genomics

Examples of the use of DNA microarrays in herbal drug research are presented in an excellent review by Chavan (2006). One striking example is the study of the gum resin of the plant *Boswellia serrata* Roxb. ex Colebr. (*Burseraceae*), which has been used in traditional medicine for treatment of inflammatory diseases since antiquity. An extract of the resin (BE) has proven to be anti-inflammatory in clinical trials. The genetic basis of the anti-inflammatory effect of this extract was tested in a system of TNFα-induced gene expression in human microvascular endothelial cells. TNFα is one of the most widely recognized mediators of inflammation. In the system used, TNFα induced 522 genes and downregulated 141 genes in nine out of nine pairwise comparisons; 113 of the induced genes were sensitive to treatment with BE and are directly related to inflammation, cell adhesion and proteolysis. The sensitive genes were then subjected to further processing for the identification of sensitive signalling pathways. One mechanism by which TNFα causes inflammation is by potently inducing expression of adhesion molecules such as VCAM-1 and ICAM-1. The *Boswellia* extract completely prevented the expression of VCAM-1 and also the expression of ICAM-1 was found to be sensitive. This work represents the first effort to conduct a whole genome screen to delineate the molecular basis of the anti-inflammatory activity of a medicinal plant extract (Roy 2005).

Identification of crude drugs has hitherto been based on macroscopic and microscopic examination, coupled with chemical identification and quantitative determination of pharmacologically active compounds or of suitable markers. This is for the most part sufficient but there is always a chance that the macro- and micromorphological characters of material from another plant species are so similar that the identity is questionable. When it comes to identification of powdered plant material or mixtures of powdered crude drugs the macro- and micromorphological methods are very difficult to use. As regards the chemical profile this can be affected by growing conditions, harvesting periods, post-harvest processes and storage. The application of genomics for identification purposes has entered the stage in recent years. Particularly the application of microarrays for DNA sequence-based identification seems promising (Chavan 2006, Chiou 2007, Zhang 2007).

Proteomics   Proteomics is of importance for the study of the enzymes involved in the biosynthesis of the pharmacologically active compounds which are present in plants and microorganisms. In this book examples will be found on the application of genomics and proteomics for the elucidation of the structure and function of macromolecular enzymes involved in the biosynthesis of polyketides (page 270) and non-ribosomal peptides (page 531).

Metabolomics   A review on the application of metabolomics for phytochemical analysis of plants has been presented by Sumner (2003). Another review, covering the application of metabolomics for phytomedicine research and drug development was written by Shyur (2008).

Two-dimensional ^{1}H-NMR has been applied for control of Ginseng commercial products. A metabolic fingerprint of a crude extract, comprising amino acids, carbohydrates, organic acids, phenolics and terpenoids was obtained by this technique in combination with principal component analysis of the spectra (Yang 2006).

In combination with multivariate analysis tecchniques, ^{1}H-NMR spectra of methanol-d_4 extracts of plant parts from three *Strychnos* species gave metabolic profiles that permitted discrimination not only between the three species but also between different plant parts of the same species. The discrimination was based on the presence of alkaloids and fatty acids (Frédérich 2004)

Metabolomics is applied in the study of extracts of crude drugs used in different types of traditional medicine, aiming at elucidation of the scientific basis for their alleged medicinal activity. For reasons descibed below, the hitherto used technique of bioassay-guided isolation (page 83) to find one or a few pharmacologically active compounds is not always successful. Plant metabolomics allows study of the relation between the composition of the extracts and their biological effects. Metabolitic fingerprints of the extracts can be obtained by the above-described metabolomic methods without isolation of the metabolites. The results are stored in a plant metabolomic database. The bioactivity of the extracts is studied in various ways (clinical testing, animal studies, effects on enzymes, receptors, etc.) and the results linked to the metabolite profile and analysed by means of multivariate data analysis. In this way it is possible to calculate which compounds or groups of compounds are associated with the highest bioactivity (Wang 2005a).

Metabolomics techniques, particularly NMR, can also be used for mapping body fluids (urine, blood, saliva, lymph, synovial fluid, etc.) and faeces in order to study the effects of various drug treatments on these. An example is a study of the influence of ingestion of chamomile tea on the metabolite profile of urine. Compared with the profile obtained before ingestion of chamomile, urine excretion of hippurate, glycine and an unknown metabolite were increased and the excretion of creatinine decreased. The increased excretion of hippurate could be related to the content of flavonols in chamomile and also the unknown metabolite probably originated from chamomile. The decreased excretion of creatinine was assumed to be due to the antioxidative activity of chamomile (Wang 2005b).

Synergy   As mentioned above, one approach to find new drugs or leads for new drugs involves study of plant products used in various types of traditional medicine (page 22). This research has hitherto been based on the observation of a pharmacological effect of an extract of the plant part used, followed by bioassay-guided fractionation (page 83), leading to isolation of one or several compounds which exert the observed effect. This has been successful in many cases, e.g. the discovery of the antimalarial drug artemisinin (page 408). This approach has, however, been criticized because it does not take into account the possibility of synergetic effects between the different constituents in a plant extract.

Synergy is the combined effect of substances that exceeds the sum of their individual effects and can apply to either increased therapeutic effect, a reduced profile of side effects or, preferably (and logically), to both (Williamson 2001). One example is the antibiotic synercid (page 554), which is a combination of the streptogramins dalfopristin and quinipristin in the ratio 7:3. Individually the two compounds have only bacteriostatic effect but when combined they become bactericidal and are used for treatment of antibiotic-resistant *Streptococcus* strains and vancomycin-resistant *Enterococcus faecium*. Proving synergy is difficult since the concept has a precise mathematical definition according to the method used to prove it. The best method is the isobole method. An *isobole* is an "iso-effect" curve in which a combination of compounds (x,y)

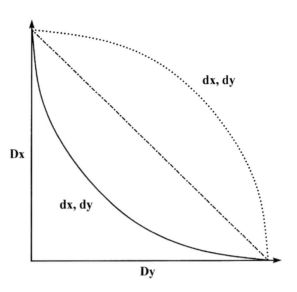

Fig. 2. The isobole method of identifying synergy and antagonism. Dx and Dy are the individual doses of x and y, dx and dy are the doses of x and y in the mixture. The dashed line shows zero interaction, i.e. all combinations of doses do produce this effect if no interaction occurs. The "concave isobole", shown by a solid line, shows synergy. The "convex isobole"(dotted line) shows antagonism. Adopted from Williamson (2001), modified by the author

is represented on a graph (Fig. 2), the axes of which are the dose-axes of the individual agents Dx and Dy.

The isobole is defined as the line joining the points representing the combination to those on the dose axes representing the individual doses with the same effect as the combination. If no interaction occurs between the agents, the isobole is a straight line; if synergy occurs, the isobole is concave; and if antagonism is present, a convex curve is obtained. If synergy is present at one dose combination and antagonism at another, a complicated wave-like or even elliptical isobole will be obtained.

There are few examples where synergy between different constituents of a crude drug has been proven. This is understandable since this necessitates testing of individual constituents and comparing the activity with an equivalent dose in the mixture. This is prohibitively expensive in terms of time and money and has therefore rarely been done. One case where an isobole curve has been determined for two constituents of a crude drug is a study of the interaction between ginkgolide A and B from an extract of *Ginkgo biloba* L. (page 423). The ginkgolides are antagonists of PAF (platelet-activating factor) and a concave isobole was found for 50 % inhibition of PAF by mixtures of ginkgolide A and B in a platelet aggregation test. Clinical indications of synergy has been published for willow bark (*Salix alba* L.) and a few other crude drugs (For references, see Williamson 2001).

Phytomedicine versus the "silver bullet" approach

Treatment of diseases with extracts of crude herbal drugs, single or mixed, has been described as the "herbal shotgun" approach as opposed to the "silver bullet" method of conventional medicine where one single drug is assumed to act on one specific enzyme or receptor (Duke and Bogenschutz-Godwin 1999, cited in Williamson 2001). Besides the possibility of synergy between constituents in the same crude drug or between constituents of different crude drugs, the active ingredients of the extract(s) may also act on different receptors or enzyme systems and the combined effects contribute favourably to the healing process. If the pharmacological activity of the extract(s) depends on synergy between two or several of its constituents, the "silver bullet" approach might not be successful as the isolated fractions and pure compounds might have no or only weak activity. Although such failures are not often published, some examples can be found in the literature (for references see Williamson 2001). Obviously when such a result is obtained, synergy should be suspected and demonstrated by testing mixtures of the isolated fractions and compounds in the bioassay used.

Traditional Chinese Medicine (TCM) and the Indian Ayurveda medicine often uses mixtures of many different crude drugs for the treatment of a certain medical condition and the investigation of these to find active principles or pharmacological effects to substantiate the claimed activity is almost impossible when using conventional methods.

The systems biology approach opens up new possibilities for the study of both Western phytomedicine, based on the use of extracts of single crude drugs and of the more complicated TCM and Ayurveda medicine. Such studies require the cooperation between experts in the various "-omics" and bioinformatics experts (Cho 2006, Ulrich-Merzenich 2007, Verpoorte 2005, Wang 2005a). Molecular pharmacognosy as a multidisciplinary science is particularly suitable for this kind of research.

Pharmacognosy as a research subject

During the 20th century research in pharmacognosy has focused on very different aspects of the subject. Before the Second World War the botanical aspect dominated. Taxonomy and morphology of plants were important and the main object was botanical identification of crude plant-derived drugs. Then came a period when organic chemistry was dominant and pharmacognosists devoted their time to isolation and structure determination of different compounds, mainly from medicinal plants but also to some extent from animals, fungi and bacteria. Also, development of methods for the quantitative analysis of compounds contained in the crude drugs was pursued in some departments of pharmacognosy. So far, biological and pharmacological aspects of natural products attracted very little interest among pharmacognosists. This changed gradually and in the 1960s and 1970s some pharmacognosists developed an interest in using pharmacological methods to screen organisms, mainly plants, for pharmacologically active constituents and to monitor the isolation of these. Others became interested in the biosynthesis of medically used natural products, and subsequently biotechnology entered the stage of pharmacognosy and some departments introduced tissue and cell culture of medicinal plants as research areas. Consistently, the main objects of pharmacognostic research have been plants and comparatively little research has been done on animals, fungi or bacteria. The important discoveries in these classes of organisms have been made by biochemists, microbiologists and molecular biologists.

In 1997 Jan G. Bruhn and Lars Bohlin at the Division of Pharmacognosy, Uppsala University, Sweden, suggested that pharmacognosy as a research subject should be termed "**molecular pharmacognosy**" and proposed the following definition: "*Pharmacognosy is a molecular science that explores naturally occurring structure–activity relationships with a drug potential*". The term in the sense of Bruhn and Bohlin has a broader meaning than in molecular biology and other sciences which have included molecular as a part of their name to imply a focus on enzymes and genes. Thus the definition of molecular pharmacognosy is to be interpreted as a science that focuses on molecules and involves isolation and determination of the structure of pharmacologically active molecules as well as the study of their biosynthesis, including the enzymes and genes which are involved. Also the term structure–activity relationships is used here in a broader sense than is commonly the case in medicinal chemistry. The definition also restricts the research activity to molecules which could be used as drugs, thereby stressing that

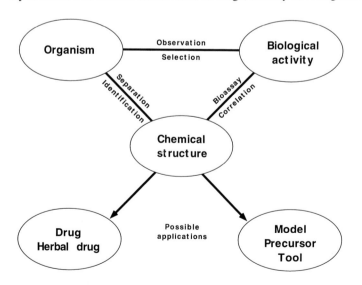

Fig. 3. Explanatory model for molecular pharmacognosy. (After Bruhn 1997, see Further reading). Redrawn by A. Backlund

molecular pharmacognosy is a pharmaceutical research subject. The authors also suggested the explanatory model in Fig. 3.

The starting point is an organism (upper left corner) which displays some kind of biological activity (upper right corner). Bioassay-directed separation from the complex biomass will lead to the identification of a chemical structure (central) which can be correlated to the observed biological activity. This chemical structure can then be used for development of a drug or as a model, precursor or tool in drug research. The model also illustrates the place of herbal drugs and functional foods in pharmacognostic research. The authors also discuss four cases which can result from application of their model. This is illustrated in Fig. 4.

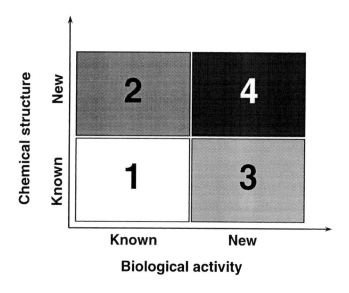

Fig.4. Relationship between chemical structure and biological activity. (After Bruhn 1997, see Further reading). *Redrawn by A. Backlund*

Field 1 illustrates the isolation of compounds with known structures and known activities. Here the only new discovery is that the compounds have been found in an organism where their existence was not previously known. Field 2 illustrates a result which is of greater interest in the search for drugs. Compounds of new structures have been isolated but their biological activity is the same as that exerted by compounds with other structures. Field 3 illustrates the isolation of compounds with known structures found to have a hitherto not known biological activity. Field 4 is clearly of greatest interest as the work has resulted in isolation of compounds with both new structures and new biological activities. Obviously one should develop the methods used so that isolation of compounds from field 3 is avoided or minimized and one should aim at isolating compounds from field 4, although results in fields 2 and 3 are not without interest in the search for drug candidates.

By focusing on the upper triangle of the model presented in Fig. 3, modern pharmacognostic research is in the position to contribute research results of importance to our knowledge of biologically active compounds and their use in the drug discovery process.

The implied prime outcome of application of this model was the finding of lead structures for development of new drugs, influenced by the dogma from *Big Pharma* drug discovery, with bioassay-guided fractionation developing into HTS. The selection of the organisms was often based on the successful discoveries from traditional medicine. The strategy was successful, as during the years 1981–2006, 28 % of new chemical entities, registered as drugs, were directly based on the discovery of natural products, compared with 34 % which were totally synthetic. Another 12 % were synthetic but with natural product pharmacophores or as natural product mimetics. (See Newman 2007 in *Further reading*, page 42).

However, as described above (page 28) HTS of randomly chosen extracts has not given the expected outcome and the number of natural products in the pipeline of the pharmaceutical companies has decreased in later years. The previously (pages 28–37) described fields of chemoinformatics, phylogenetics and systems biology, have been developed since Bruhn and Bohlin presented their model and are likely to change and widen the scope and goals of pharmacognostic research. The model has therefore been amended as illustrated in Fig. 5.

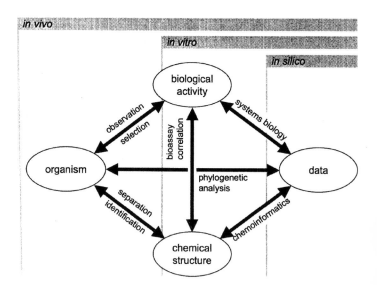

Fig. 5. A revised explanatory model of pharmacognostic research. (From Larsson 2008, see Further reading). *Reprinted with permission from the publisher*

The authors (Larsson *et al.*) have given the following explanation of the model. Four important cornerstones are included: organism, chemical structure, biological activity and data. The arrows repre-

represent the flow information between these. *Organism* and *biological activity* are connected by an observation of the activity and a selection of the organism. In a similar manner the *chemical structure* and *data* are connected by the application of *chemoinformatics* – interpreting as well as guiding. In the wide concept of chemoinformatics all aspects from HTS data to chemography and tertiary protein structures are included. The connection of *biological activity* to *data* by *systems biology* implicates the immense possibilities of implementing wide approaches such as metabolomics and genomics in the field of pharmacognosy.

At present there is a worldwide decline in the number of students who select chemistry and other classical science subjects for higher studies. There is therefore a need to reform the education in these subjects, both on the undergraduate and graduate level. As outlined above, modern pharmacognosy is a truly multidisciplinary subject, which should be attractive to the new generation of students as it bridges the gap between chemistry and biology. Pharmacognosy can thus serve as a template for the revitalization of the educational programmes in science by blending different disciplines, allowing teaching about complex questions in a multidisciplinary way (see Bohlin 2007 in *Further reading* page 44).

Further reading

Pharmacognosy textbooks

Bruneton, J. *Pharmacognosy, Phytochemistry, Medicinal Plants.* 2nd Edn Lavoisier, Paris (1999).

Dewick, P.M. *Medicinal Natural Products. A Biosynthetic Approach.* John Wiley & Sons, Chichester, England (2001).

Evans, W.C. *Trease and Evans' Pharmacognosy.* WB Saunders, London (2001).

Robbers, J.E., Speedie, M.K. and Tyler, V.E. *Pharmacognosy and Pharmacobiotechnology.* Williams and Wilkins, Baltimore (1996).

Natural products in the drug discovery process

Baker, D.D., Chu, M., Oza, U. and Rajgarhia, V. The value of natural products to future pharmaceutical discovery. *Natural Product Reports*, **24**, 1225–1244 (2007).

Cordell, G.A. Biodiversity and drug discovery – a symbiotic relationship. *Phytochemistry*, **55**, 463–480 (2000).

Cox, P.A. and Balick, M.J. The ethnobotanical approach to drug discovery. *Scientific American*, **270**, (6), 82–87 (1994).

Fowler, M.W. Plants, medicines and man. *Journal of the Science of Food and Agriculture*, **86**, 1797–1804 (2006).

Ganesan, A. The impact of natural products upon modern drug discovery. *Current Opinion in Chemical Biology*, **12**, 306–317 (2008).

Intelligent drug design. A *Nature* supplement. *Nature*, **384**, (Suppl. Nov. 7.), 1–26 (1996).

Lam, K.S. New aspects of natural products in drug discovery. *Trends in Microbiology*, **15** (6), 279–289 (2007).

McChesney, J.D., Venkataraman, S.K. and Henri, J.T. Plant natural products: Back to the future or into extinction? *Phytochemistry*, **68**, 2015–2022 (2007).

Molinski, T.F., Dalisay, S.D. Lievens, S.L. and Saludes, J.P. Drug development from marine natural products. *Nature reviews/ Drug discovery*, **8**, 69–85 (2009).

Newman, D.J. and Cragg, G.M. Natural products as sources of new drugs over the last 25 years. *Journal of Natural Products*, **70**, 461–477 (2007)

Newman, D.J., Cragg, G.M. and Snader, K.M. The influence of natural products upon drug discovery. *Natural Products Reports*, **17**, 215–234 (2000).

Schmidt, B., Ribnicky, D.M., Poulev, A., Logendra, S., Cefalu, W.T., Raskin, I. A natural history of botanical therapeutics. *Metabolism Clinical and Experimental*, **57** (Suppl. 1), S3–S9 (2008).

Sneader, W. *Drug Prototypes and their Exploitation*. John Wiley & Sons, Chichester, England (1996). [A historical approach.]

Verpoorte, R. Pharmacognosy in the New Millennium: leadfinding and biotechnology. *Journal of Pharmacy and Pharmacology*, **52**, 253–262 (2000).

Natural products from marine organisms

Brusca, R.C. and Brusca, R.J. *Invertebrates*. Sinauer associates, Sunderland, 2002.

Cragg, G.M., Kingston, D.G.I. and Newman, D.J. (Eds.). *Anticancer agents from Natural Products*. Taylor and Francis, Boca Raton, USA, 2005.

Fusetani, N. (Ed.). *Drugs from the Sea*. Karger AG., Basel, Switzerland, 2000.

Molinski, T.F., Dalisay, D.S., Lievens, S.L. and Saludes, J.P. Drug development from marine natural products. *Nature reviews/ Drug discovery*, **8**, 69–85 (2009).

Chemoinformatics and phylogeny

Larsson, J., Gottfries, J., Bohlin, L. and Backlund, A. Expanding the ChemGPS chemical space with natural products. *Journal of Natural Products*, **68** (7), 985–991 (2005).

Larsson, J., Gottfries, J., Muresan, S. and Backlund, A. ChemGPS-NP: Tuned for navigation in biologically relevant chemical space. *Journal of Natural Products*, **70** (5), 789–794 (2007a).

Larsson, S. The "new" chemosystematics: Phylogeny and phyto-chemistry. *Phytochemistry*, **68**, 2903–2907 (2007b).

Oprea, T.I. and Gottfries, J. Chemography: The art of navigating in chemical space. *Journal of Combinatorial Chemistry*, **3**, 157–166 (2001).

Systems biology

Alterovitz, G., Liu, J., Chow, J. and Ramoni, M.F. Automation, par-allelism, and robotics for proteomics. *Proteomics*, **6**, 4016–4022 (2006).

Bradshaw, R.A. and Burlingame, A.L. From proteins to proteomics *IUBMB Life*, **57** (4/5), 267–272 (2005).

Chavan, P., Joshi, K. and Patwardhan, B. DNA microarrays in herbal drug research. *eCAM*, **3** (4), 447–457 (2006).

Chiou, S.-J., Yen, J.-H., Fang, C.-L., Chen, H.-L. and Lin, T.-Y. Authentication of medicinal herbs using PCR-amplified ITS2 with specific primers. *Planta Medica*, **73**, 1421–1426 (2007).

Cho, C.R., Labow, M., Reinhardt, M., van Oostrum, J. and Peitsch, M.C. The application of systems biology to drug discovery, *Current Opinion in Chemical Biology*, **10**, 294–302 (2006).

Frédérich, M., Choi, Y.H., Angenot, L., Harnischfeger, G., Lefeber, A.W.M. and Verpoorte, R. Metabolomic analysis of *Strychnos nux-vomica*, *Strychnos icaja* and *Strychnos ignatii* extracts by ^{1}H-nuclear magnetic resonance spectrometry and multivariate analysis techniques. *Phytochemistry*, **65**, 1993–2001 (2004).

Roy, S., Khanna, S., Shah, H., Rink, C., Phillips, C., Preuss, H., Subbaraju, G.V., Trimurtulu, G., Krishnaraju, A.V., Bagchi, M., Bagchi, D. and Sen, C.K. Human genome screen to iden-tify the genetic basis of the anti-inflammatory effects of *Boswellia* in microvascular endothelial cells. *DNA and Cell Biology*, **24** (4), 244–255 (2005).

Shyur, L.-F. and Yang, N.-S. Metabolomics for phytomedicine research and drug development. *Current Opinion in Chemical Biology*, **12**, 66–71 (2008).

Sumner, L.W., Mendes, P. and Dixon, R.A. Plant metabolomics: large-scale phytochemistry in the functional genomics era. *Phytochemistry* **62**, 817–836 (2003).

Ulrich-Merzenich, G., Zeitler, H., Jobst, D., Panek, D., Vetter, H. and Wagner, H. Application of the "-Omic-" technologies in phytomedicine. *Phytomedicine*, **14**, 70–82 (2007).

Verpoorte, R., Choi, Y.H. and Kim, H.K. Ethnopharmacology and systems biology: A perfect holistic match. *Journal of Ethno-pharmacology*, **100**, 53–56 (2005).

Wagner H. and Ulrich-Merzenich G. Synergy research: Appro-aching a new generation of phytopharmaceuticals. *Phyto-medicine*, **16**, 97–110 (2009).

Wang, M., Lamers, R.-J.A.N., Korthout, H.A.A.J., van Nesselrooij, J.H.J., Witkamp, R.F., van der Heiden, R., Voshol, P.J.,

Havekes, L.M., Verpoorte, R. and van der Greef, J. Metabolomics in the context of systems biology: Bridging traditional Chinese medicine and molecular pharmacology. *Phytotherapy Research*, **19**, 173–182 (2005a).

Wang, Y., Tang, H., Nicholson, J.K., Hylands, P.J., Sampson, J. and Holmes, E. A metabonomic strategy for the detection of the metabolic effects of chamomile (*Matricaria recutita* L.) ingestion. *Journal of Agricultural and Food Chemistry*, **53**, 191–196 (2005b).

Williamson, E.M. Synergy and other interactions in phytomedicines. *Phytomedicine*, **8** (5), 401–409 (2001).

Yang, S.Y., Kim, H.K., Lefeber, A.W.M., Erkelens, C., Angelova, N., Choi, Y.H. and Verpoorte, R. Application of two-dimensional nuclear magnetic resonance spectroscopy to quality control of Ginseng commercial products. *Planta Medica*, **72**, 364–369 (2006).

Zhang, Y.-B., Shaw, P.-C., Sze, C.-W., Wang, Z.-T. and Tong, Y. Molecular authentication of Chinese herbal materials. *Journal of Food and Drug Analysis*, **15** (1), 1–9 (2007).

Pharmacognosy as a research subject

Bohlin, L., Göransson, U. and Backlund, A. Modern pharmacognosy: Connecting biology and chemistry. *Pure and Applied Chemistry*, **79** (4), 763–744 (2007).

Bruhn, J.G. and Bohlin, L. Molecular pharmacognosy: an exploratory model. *Drug Discovery Today*, **2**, 243–246 (1997).

Larsson, S., Backlund, A. and Bohlin, L. Reappraising a decade old explanatory model for pharmacognosy. *Phytochemistry Letters,* **1**, 131–134 (2008).

2. Plant-derived Crude Drugs and Herbal Remedies

As outlined on (page 20) crude drugs are used as raw materials for isolation of the pure compounds which are ingredients of modern conventional drugs. Plant-derived crude drugs are seldom contained in conventional drugs but are constituents of herbal remedies (page 78).

Nomenclature

The names of the plant drugs are often in pharmaceutical Latin. In English-speaking countries, however, the names are usually in English. The later editions of *The European Pharmacopoeia* (English version) primarily use the English names of the crude drugs and give the pharmaceutical names as an alternative. This praxis will also be followed in this book.

The pharmaceutical names generally consist of two words. One of these is related to the scientific names of the plant from which the drug derives while the second indicates the plant part (bark, leaf, etc.) used. The following terms are used to indicate the parts of plants:

Radix = root. The term does not completely coincide with the botanical concept. A drug termed a radix may sometimes also contain rhizomes.

Rhizoma = rhizome, a subterranean stem, generally carrying lateral roots.

Tuber = a nutritious subterranean organ, which, in a botanical sense, is a rhizome. A tuber is a thick organ, mainly consisting of parenchymatous storage tissue (generally containing starch) and a small proportion of lignified elements.

Bulbus = onion. Botanically, an onion is a stem, surrounded by thick nutritious leaves that are usually low in chlorophyll content.

Lignum = wood. Drugs for which this term is used are obtained from plants with secondary thickening and consist of the woody parts of the xylem. The term is not always botanically quite correct; for example, quassiae lignum also contains the thin bark, albeit only a small amount.

Cortex = bark. Barks are obtained from plants with secondary thickening and, unlike the botanical definition of the term, they consist of all the tissues outside the cambium. Such drugs can be collected from roots, stems and branches.

Folium   = leaf, consists of the middle leaves of the plant.

Flos   = flower. The crude drug may consist of single flowers and/or entire inflorescences.

Fructus   = fruit. The pharmacognostical term is not always synonymous with the botanical one. Thus, the crude drug Cynosbati fructus cum semen (rose hips) is, botanically speaking, a swollen receptacle carrying the true fruits (nuts). Also the second part of the term – semen – is thus not correct from a botanical standpoint as semen is the pharmaceutical term for seed (see below). There is also another crude drug consisting only of the receptacle without fruits. The pharmaceutical name for this crude drug is Cynosbati fructus sine semen.

Pericarpium   = fruit peel or pericarp which is the common botanical term.

Semen   = seed. The drug can consist either of the seed, as removed from the fruit, or of a part of the seed, as in Colae semen, which does not contain the testa or seed coat.

Herba   = herb. The crude drug consists of the aerial parts of the plant; thus, stems as well as leaves, flowers and fruits, if any, are included.

Aetheroleum   = essential or volatile oil is a product obtained from plant material. It usually possesses a distinctive odour and consists of a complex mixture of comparatively volatile components.

Oleum   = oil, is a fixed oil prepared from plant material by pressing.

Pyroleum   = tar, is prepared by dry distillation of plant material.

Resina   = resin, is obtained either from secretory structures in certain plants or by distillation of a balsam (see below). In the latter case, it is the residue after distillation.

Balsamum   = balsam, is a solution of resin in a volatile oil and is generally produced by special cells in the plant.

A scientific plant name consists of three parts: a genus name, a species epithet and an author's name (usually abbreviated).

The word in the pharmaceutical name of the crude drug which is related to the scientific name of the mother plant can be derived either from the genus name or from the species epithet. The two words can appear in different order. An example is the name of

Valerian root. In the *European Pharmacopoeia* this crude drug is denoted as Valerianae radix, but the reverse order, Radix valerianae, is also sometimes used. In this name the word radix (in the nominative) denotes a root drug and valerianae (in the genitive) indicates that the root comes from a plant of the genus *Valeriana*. An example of a name where the botanically related part is the species epithet is Cocae folium, Coca leaf, which is obtained from the shrub *Erythroxylum coca* Lam. There are also cases where two crude drugs can derive from different species of the same genus and the pharmaceutical names of these drugs can then be constructed in different ways. The genus *Digitalis* has more than 100 species of which two – *Digitalis purpurea* L. and *Digitalis lanata* Ehrh. – are of pharmaceutical interest. The leaves of both species are used as crude drugs which are called Digitalis purpureae folium (Digitalis leaf) and Digitalis lanatae folium (Digitalis lanata leaf) respectively. Another example is the bark of the species *Rhamnus frangula* L. and *Rhamnus purshiana* DC. These are called Frangulae cortex (Frangula bark) and Rhamni purshianae cortex (Cascara). Some drugs are not named in accordance with these models but have special names, such as Opium, Curare, Gallae, etc.

As seen above there are no distinct rules for the pharmaceutical names of crude plant drugs. In contrast to plants which are exactly defined by the scientific plant names, crude drugs are not unequivocally defined by the pharmaceutical names. These must therefore be connected to a definition which can be found in monographs of pharmacopoeias and handbooks (see list of references). In these the complete scientific name of the mother plant is given, which, together with a statement of the plant part used, constitutes an unmistakable definition of the crude drug. Thus Belladonnae folium (Belladonna leaf) is the leaf of *Atropa belladonna* L. and Strychni semen (Nux vomica) is the seed of *Strychnos nux-vomica* L. Pharmaceutical names are found mostly for crude drugs which are or have been used in Europe. A universal nomenclature for plant-derived crude drugs, applicable all over the world, could consist of the scientific plant name together with the Latin or English name of the plant part. Such a system has been adopted by WHO for ATC (Anatomical Therapeutic Chemical) classification of herbal remedies. (Guidelines for Herbal ATC classification and the Herbal ATC index. The Uppsala Monitoring Centre. WHO Collaborating Centre for International Drug Monitoring. Uppsala 2004).

Production of crude drugs from medicinal plants

Crude drugs may be prepared from wild or cultivated plants. These are generally referred to as *medicinal plants*. The method of production used in each individual case is largely determined by economic factors. It may be advisable to collect the drug from wild plants, if these are abundant and wages are comparatively low, whereas high wages and a shortage of wild material make cultivation a cheaper alternative. Thus, in Mexico, *Dioscorea* roots (raw material for steroids) are collected almost exclusively from wild plants, whereas in Europe, the leaves of the foxglove (*Digitalis purpurea*) are obtained almost entirely from cultivated plants. Besides economic factors the choice between collection from wild plants and cultivation may also be influenced by environmental considerations. A high demand for a certain wild-growing species can cause its extinction. A recent example is the discovery of the cancer remedy paclitaxel in the bark of the Western yew (*Taxus brevifolia*), a small tree native to Western North America. When the marketing of this drug started, the demand for the bark became so great that the species became threatened and other, renewable, sources had to be found (see page 432). In the future many medicinal plants, which at present are obtained from wild-growing species, may have to be cultivated both for environmental reasons and because of increased demands for a uniform quality of the crude drugs produced from them.

Crude drugs are collected or cultivated all over the world.

Cultivation of medicinal plants

In principle there is not much difference between growing medicinal plants and growing other plants as in agriculture and horticulture. There are several advantages in growing medicinal plants rather than collecting the drug from the wild. Soil conditions, shade, moisture and plant diseases can be more readily controlled, so as to ensure the optimum development of the plants. Harvesting is facilitated, as the individual plants are in approximately the same stage of development and grow closely together in smaller areas. It is easier to deal with the material after harvesting. Drying can be done quickly and efficiently so that the pharmacologically active constituents remain unchanged. All these factors facilitate production of crude drugs with a high and uniform quality.

Another advantage of cultivation of medicinal plants is that the extraction of the desired constituents can be performed directly in association with the cultivation, e.g. when producing essential oils. Finally, cultivation can be combined with plant breeding, which may lead to higher yields of desired constituents.

Many factors affect the development of the plants as well as their content of pharmacologically active compounds. Detailed knowledge concerning their importance in each individual case is required in order for cultivation to give the best possible result. These factors can be divided into two groups: extrinsic ones, such as climate and soil, and intrinsic ones, such as genotypes.

Climate

Temperature, rainfall, hours of daylight and altitude are climatic factors of great importance for the development of plants. In general, plants are not able to withstand too drastic changes in climatic conditions. There are exceptions, however. The opium poppy, (*Papaver somniferum* L.) normally grows in a temperate or subtropical climate (the Mediterranean countries, the Balkans, Turkey) but can easily be grown in Nordic countries, producing the same quantities and kinds of alkaloids there. There are also examples of unusual sensitivity to climatic factors. *Cinchona succirubra* Pav. ex Klotzsch is normally encountered at a relatively high altitude (1 000–3 000 m). It also grows well at lower altitudes, but a much lower yield of alkaloids is then produced.

Soil

The chemical and physical properties of soils show great variation. Soil is a mixture of mineral particles, formed through weathering of rock, and organic components, *humus*, derived from decaying organisms. Rich soils contain 1.5–5 % humus, poor ones less than 0.5 %.

The water-binding capacity of soil, which is very important for plants, depends on the particle size of the soil components. A soil consisting mainly of fine particles (2–20 μm) is termed *clay. Sand* consists of larger particles (20 μm–2 mm), and *gravel* of coarse grains (2–20 mm). Mixtures are also found, such as so-called sandy clay. Clays have a large capacity for binding water (up to 40 % by volume) and a low permeability to air and water, whereas sandy soil dries quickly and is highly permeable.

The pH value of the soil, which is also important for the development of many plants, is largely dependent on the lime content. Soils that are rich in humus and poor in lime are acidic, whereas a high content of lime gives a high pH value. The various soil properties are analogous to climatic factors in that different plants are adapted to, and thrive on, different kinds of soil. However, most plants grow well on neutral soils that are relatively rich in humus and consist of a mixture of fine and coarser particles, thus combining water-binding capacity with permeability to air.

As already mentioned, the lime content has a great influence on the pH of soil. Moreover, calcium is a necessary element for the development of most plants and an ample supply of lime is favourable for many plants. Among medicinal plants, *Digitalis purpurea* L. is an exception in this respect. It is a calcifuge plant, growing poorly in calcareous ground.

Plants absorb nutritive salts and water from the soil through the fine hairs of their roots. The soil must not be too compact but allow the roots to penetrate and spread sufficiently. A clay soil offers more resistance in this respect than does a soft sandy soil. It is necessary to work and break up the soil efficiently by ploughing and harrowing before sowing, to facilitate the development of the root system. Even during the vegetation period some plants may require the soil to be broken up, e.g. by hoeing. A soft soil also facilitates gaseous exchange between the roots and surrounding soil.

The nutritive salts that have been absorbed by plants are, of course, removed from the locality on harvesting. The supply of these salts in soil is not unlimited, and what is removed has to be replaced by fertilizing. N, P, and K are the most important elements and have to be supplied in relatively large quantities. A large number of other elements are required in so-called trace amounts (trace elements). The humus content of the soil ought also to be maintained by ploughing in plant material that is not going to be used for other purposes. Finally, the microflora of the soil is very important for plants. Manure is an excellent means of improving the soil, because it contains nutritive salts as well as humus-forming material and microorganisms. However, natural manure is normally not accessible in sufficient quantity, but must often be supplemented with inorganic salts. Rational fertilizing should be based on soil analyses showing the current content of nutrients.

Irrigation. Control of weeds, pests and diseases

All plants need water to develop properly. If rainfall is not sufficient, the plantation must be irrigated. This can be done either by directing water in furrows close to the plants or by sprinklers. A good, cheap supply of water is therefore a prerequisite for the successful cultivation of medicinal plants.

Weeds are a constant threat to the cultivated plants. In particular during the first stages of development weeds tend to grow faster than the plants in cultivation and might easily take over completely if not removed. Weeding has to be done manually for most medicinal plants as suitable herbicides are not available. Thinning must also be done for some crops, e.g. the opium poppy, *Papaver somniferum* L. Both weeding and thinning are expensive operations.

Attacks by grazing insects such as aphids, thrips, cutworms and mites cause damage to aerial parts of the plants and underground parts may be attacked by worms and nematodes. These are just a few of the many organisms that may attack and threaten plants in cultivation. Many plant diseases such as browning of the leaves and root rot are caused by various fungi. A great many substances are available to combat such attacks but must be used with care. It is essential that no residues of insecticides and other chemicals remain in the plant parts harvested for production of crude drugs. Biological combat is a relatively new trend in insect control. This is performed by application of an enemy to the pest which is not harmful for the plant.

Propagation of plants

Propagation by seeds

Plants can multiply through seeds or vegetatively. Many seeds fall to the ground at the end of the vegetation period and germinate only at the beginning of the next vegetation season. Many seeds have a built-in latency mechanism that allows them to germinate only after a "period of rest", corresponding to the normal interval between ripening and germination. Special external conditions are often required, e.g. the seed may have to be exposed to a low temperature for a certain period of time (corresponding to a winter period) before it can germinate. This fact may be of importance for growing medicinal plants, especially when attempts are made to grow them in climates that differ from that of their natural habitat.

Seeds may be sown directly in the field. But many seeds germinate slowly, and it may then be better first to raise seedlings in a greenhouse and to ground these only when they have reached a sufficient degree of development. It is advisable before sowing to determine the viability of the seed by planting a definite number of seeds and counting the number of seedlings that come up. The ability to germinate may decrease very considerably during storage of seeds. In some cases, it is necessary to sow the seed directly after ripening, in order to obtain satisfactory germination rates. An example is *Colchicum autumnale* L. (meadow saffron): up to 30 % of the seeds will germinate if sown immediately after ripening, but only up to 5 % if left uncovered for a few days before sowing. In such cases vegetative production (see below) is preferable.

Vegetative reproduction

Vegetative reproduction can take place in several ways. Many plants breed naturally both through seeds and by developing vegetative organs such as bulbs or subterranean stems (stolons, rhizomes). When cultivating such plants, it is often advantageous to propagate them by means of the vegetative organs, partly because one can then be completely sure that they show the same genetic makeup. This is, of course, a necessary condition for the propagation of hybrids (see next chapter). Sexual breeding through seeds resulting from outcrossing with one or several other individuals produces a large number of variants, corresponding to the genomes of the original parents as well as to different combinations of these. With plants that do not naturally produce vegetative organs, vegetative multiplication can often be induced artificially. The simplest method is by *division*. A well developed plant with several stems and attached root systems is divided and each part is replanted. The new plants will grow and develop new roots and stems, which may in turn be divided, and so on. *Rheum palmatum* L. and *Gentiana lutea* L. are examples of medicinal plants that are multiplied in this way. Branches of certain plants can develop new root systems if bent down to the ground, kept there, and covered with soil. After some time, the branch can be cut off from the original plant and replanted, resulting in a normal plant. This method is known as layering and is used for vegetative propagation of *Rhamnus purshiana* DC.

Another very common method is the production of *cuttings*. Some plants are able to develop root systems from detached shoots if these are placed in a suitable environment, with good access to water, and preferably, a somewhat increased temperature. *Mentha piperita* L. and *Erythroxylum coca* Lam. are examples of medicinal plants that can be propagated in this way. When this type of propagation is used, the development of roots can sometimes be stimulated by addition of natural or synthetic growth regulators. Thus, *Cinchona* species are multiplied vegetatively by treatment of the shoots with indole-3-butyric acid in very low concentrations. The method saves considerable time in the development of the plant as compared with its development from seed and, moreover, avoids the period as seedling when plants are most sensitive to attacks by pests as well as guarantees that the genetic material is unchanged.

Plant breeding

When cultivating medicinal plants the extrinsic conditions should be as favourable as possible, so that the plants develop optimally and produce the largest possible amount of pharmacologically active constituents. However, the limits set by the plant genome can not be exceeded, and to optimize the yield it is vital to ensure that plants with the most beneficial properties are selected for cultivation. This process is called *plant breeding*. Plant breeding is practised on a large scale with economically important cereals and with other food-producing plants, and the technique developed in this field can also be used for medicinal plants. Plant breeding may also aim at the improvement of other properties such as resistance to pests and diseases, uniform ripening, diminished seed spreading and absence of thorns.

Improvement of the properties of cultivated plants can be achieved by three methods: *selection, cross-breeding* and *induced mutations*.

Selection

The content in a plant species of a certain substance of interest, e.g. an alkaloid or a glycoside, depends on genes coding for this compound as well as on promotors and regulation system for transcription and translation. If the species has not been subjected to any genetic manipulation, it is generally composed of individuals with variations in the set up of these factors. Thus some plants may have a high content, others a medium and some a very low content. The first part of a programme to improve the yield of the desired compound is *selection* of the most productive individuals in the population. Therefore a great number of plants are analysed, and the ones with the highest content are propagated and their offspring analysed. This is continued for several generations until a population has been created which has a high and constant content of the desired compound.

In breeding for increased yield of active constituents it is necessary to have access to sensitive and rapid methods of analysis. During selection each individual plant must be analysed separately and the amount of plant material taken for analysis must be small enough not to harm the plant, allowing offspring to be produced from plants with high contents of the desired active constituents. Experiments on the opium poppy, *Papaver somniferum* L., serve as an example. In one population of this plant, the average morphine content of the seed capsules was 0.39 %. Six individuals, however, contained 0.7 %. Seeds from these specimens were sown separately, and in the offspring all individuals carrying less than 0.7 % morphine were discarded. After some years a line was obtained whose capsules had an average morphine content of 0.77 %, or twice the original average.

Selection of intact plants is time-consuming and costly. The use of tissue cultures (see page 101) can speed up the process and thus reduce the costs.

Cross-breeding

Prior to cell division, the DNA of the cell nucleus, the chemical carrier of genes, is concentrated in the *chromosomes*, structures that have long been recognized because they can be observed with an ordinary light microscope after suitable staining. Every organism has a constant, usually even, number of chromosomes in all its somatic cells. These chromosomes often have different appearance, but pairs that look similar can usually be distinguished. Such cells are known as having a *diploid* set of chromosomes, with the chromosome number $2n$, where n is an integer. When gametes are formed, the two chromosomes of each pair go to different gametes, each of which thus acquires a set of chromosomes half as large as that of the somatic cell (chromosome number n, *haploid* set of chromosomes). When the zygote is formed by fusion of two gametes, the initial chromosome number ($2n$) is restored, so that this number is kept constant from generation to generation. If the two gametes forming the new plant originate from the same plant, or from two plants the genes of which are identical, so called *genets*, the new plant will have exactly the same genomic properties as the parents. If, however, the gametes originate from two different individuals, races, variants or species of plants, the properties of the offspring will differ from those of both of the parents. Due to the presence of recessive hereditary factors, the offspring might also have properties which are not manifested in either of the parents. The result

53

depends on how the chromosomes from the two plants are combined. It is possible that such a crossing results in properties which, from the breeder's point of view, are better than those of either of the two parent plants. One problem is, however, that when the gametes from two different genets are combined, the next generation will not retain the properties because new genetic combinations will result. Propagation of hybrids must therefore be performed by vegetative reproduction. Only in obligate self-pollinating plants are the results from propagation by seeds expected to be stable. Most higher plants do, however, show some degree of out-crossing, i.e. pollination by other individuals, which will result in a zygote with a new combination of hereditary traits. In cases where the plant population is genetically homogeneous this will not necessarily pose a problem. From out-crossing of hybridization, i.e. pollination by another species, deliberate or by chance, occasionally highly interesting combinations are created. One example of this is *Triticum aestivum* L., the common wheat, the origin of which has been traced by genetic studies from a mixed origin of *Triticum monococcum* L. ($2n=14$) and *Triticum turgidum* L. ($2n=28$), the latter which in turn appears to be a hybrid involving *Aegilops speltoides* Tausch ($2n=14$), resulting in *Triticum aestivum* L. ($2n=42$) which is generally considered to be an allotetraploid.

Induced mutations

Modification or changes in genes, which can be transferred to new generations of cells are known as *mutations*. If mutations occur in reproductional organs, these can even be transferred to new generations of plants, and by this process, which is called fixation, become a new genetic trait of a species, in rare cases even resulting in the formation of a new species. Mutations can involve either the nucleotide sequence of DNA itself or changes in the number of chromosomes in the plant.

The total number – $2n$ – of chromosomes differs widely from one organism to another and may be between $2n = 2$ in the nematode *Caenorhabditis elegans* and $2n = $ ca. 1260 in *Ophioglossum reticulatum* L., a fern. Common numbers in higher plants are 12–32. If cell division is disturbed, offspring organisms with higher chromosome numbers ($3n$, $4n$), may arise. This phenomenon is called *polyploidy*. Such mutations may arise spontaneously and it can be shown that if two plants are very similar but exhibit certain constant differences, one of them may be, for example, a tetraploid ($4n$) form of the other. Thus, *Valeriana officinalis* L. occurs in more than one form, with chromosome numbers $2n$, $4n$ and $8n$. The above mentioned *Ophioglossum reticulatum* L. is presumed to be an 84-ploid, i.e. with $84n$. Polyploidy in plants can also be established artificially, e.g. by treatment with the alkaloid colchicine (page 617).

Polyploidy changes the properties of the organisms. In most cases, all or some of the organs are enlarged. The content of active constituents may also be affected, in both positive and negative directions, and qualitative changes may occur. The changes cannot be predicted, but they must be determined by trial and error in each case. It has been demonstrated that some plants of the family *Solanaceae* (*Atropa, Datura*) with the chromosome number 4*n* produce larger amounts of alkaloids than the original 2*n* forms, but that the relative proportions of the main alkaloids are not affected. With *Digitalis lanata* Erh., the content of lanatosides A and B in relation to other lanatosides is higher in the 4*n* form than in the natural 2*n* form. A higher content of active constituents need not necessarily mean that such a mutation is an economic success, because other factors, e.g. reduced foliage, may work in the opposite direction.

Mutations caused by a direct effect on the DNA nucleotide sequence structure can be induced by radioactive radiation, X-rays or various chemicals. Such mutations are often lethal, but may in exceptional cases be favourable in that some undesirable property of the organism may be eliminated. Following the treatment, an extensive selection programme must be carried out to find the favourable mutations. Numerous studies on mutations of this kind in cereals have resulted in the production of many new varieties with better properties than the starting material. No spectacular results have thus far been achieved with medicinal plants, perhaps mainly because considerably less work has been done than with the commercially and vitally important cereals. The attempts made include studies on the opium poppy, *Papaver somniferum* L., in which irradiation with ^{60}Co has given rise to a variety with a morphine content of 0.52 %, as compared with 0.32 % in the starting material.

An increase of the content of desired constituents can be achieved either by increasing the total mass of the plant parts at a constant percentage of the component or by increasing the percentage at constant mass. The latter case is generally preferred by the industry as it facilitates processing.

The above described methods for production of mutants are classical and have been used for a long time in plant breeding. The developments within molecular biology and gene technique have provided much more advanced methods for manipulation of plant genes. These will be described in the next chapter under the heading "Transgenic plants" (page 102).

The new variety of plant which results from plant breeding is called a *cultivar*. A new cultivar must be officially registered before it can be put on the market. States monitor different kinds of institutions where comparative trials of a new medicinal plant variety can take place over several years. If the testing is positive, state agricultural offices accept the cultivar and issue the registration certificate. Breeders are also obliged to maintain the varieties by continuous selection after cultivar approval. This work is also controlled and supervised by the controlling institutes.

Chemical races

As mentioned above individuals in a population of plants which has not been subjected to genetic manipulation may differ considerably in the genes which code for a certain constituent, causing the plant to have a high or a low content of that constituent. As the plant with a low content does not differ from the plant with a high content in any of the morphological characters on which botanical classification is based, the two plants can be regarded as belonging to two different chemical races, which can be distinguished only through studies of the chemical constituents of the plants. The phenomenon of *chemical races* is not restricted to quantitative differences but can also manifest itself qualitatively so that the chemical constituents of one race are completely different from those of the other. The occurrence of chemical races can cause problems in medicinal plant research. Thus there are examples where an interesting compound was found in a certain batch of plant material, but when another batch was examined the compound could not be found again. This has for example been the case if the two batches have been collected at places located at a considerable distance from each other. The concept chemical race is controversial among systematic botanists as it has not been shown that the phenomenon depends on differences in the genes coding for the substance. Various hypothesis have been proposed in attempts to find an explanation.

Collecting and harvesting medicinal plants

Suitable time for collection

The amount of a constituent is usually not constant throughout the life of a plant. The stage at which a plant is collected or harvested is, therefore, very important for maximizing the yield of the desired constituent. Examples of seasonal variation are found in *Ephedra* species. Their content of the alkaloid ephedrine varies considerably with the season and reaches its maximum in the autumn. Seasonal variations have also been noticed for *Rhamnus purshiana* DC, whose content of anthraquinone glycosides varies greatly during the year. In perennial plants, the age of the plant may be important. The differences are sometimes not only quantitative but also qualitative. The Australian species *Duboisia myoporoides* R. Br. is an example. The leaves of a specimen of this tree was found in October to contain 3 % hyoscyamine, but in April practically all of this compound had been converted into scopolamine.

Rules for collection

In the ideal case, thorough studies will have shown exactly at what time the plant contains the largest amount of the constituents desired, so that the drug material can be collected or harvested accordingly. In many cases, however, such exact information is not available. It is then generally assumed that the material is best collected when the organ in question has reached its optimal state of development. The following general rules are based on such assumptions.

Roots and rhizomes are collected at the end of the vegetation period, i.e. usually in the autumn. In most cases they must be washed free of adhering soil and sand.

Bark is collected in the spring, mainly for the practical reason that the cambium shows its maximum activity in the spring, producing an abundance of parenchymatous cells that are not yet differentiated. These cells are soft and it is therefore easy to strip the bark which is located outside the cambium.

Leaves and herbs are collected at the flowering stage.

Flowers are usually gathered when fully developed. In certain cases, as with cloves (from *Eugenia caryophyllata Thunb.*), for example, the unopened flower buds are picked.

Fruits and *seeds* are collected when fully ripe.

Methods of collection

Medicinal plants must largely be collected by hand. This is in particular true in the case of wild plants. With cultivation on a large scale, it may be possible to use modern agricultural harvesters, but in some cases, e.g. with barks, manual collection is unavoidable. Thus, the cost of drug production is largely the cost of the labour involved.

Preservation of plant material

It is sometimes possible to treat harvested material immediately, so as to obtain pure compounds or concentrates of the constituents. Mostly, however, this cannot be done on the spot, and the plant material must first be preserved so that the active compounds will remain unchanged during transport and storage.

The cells of living plants contain not only low molecular-weight compounds and enzymes, but they also have many kinds of barriers that keep these constituents apart. When the plant dies, the barriers are quickly broken down and the enzymes then get the opportunity to promote various chemical changes in the other cell constituents, e.g. by oxidation or hydrolysis. Preservation aims at limiting these processes as far as possible.

Drying

The most common method for preserving plant material is drying. Enzymic processes take place in aqueous solution. Rapid removal of the water from the cell will, therefore, largely prevent degradation of the cell constituents. Drying also decreases the risk of external attack, e.g. by fungi. Living plant material has a high water content: leaves may contain 60–90 % water, roots and rhizomes 70–85 %, and wood 40–50 %. The lowest percentage, often no more than 5–10 %, is found in seeds. Hence, the constituents of seeds do not run the same risk of degradation as those of other organs, which is one of the reasons for the extreme endurance of some seeds, e.g. those of henbane (*Hyoscyamus niger* L.) which has been brought to germination after several hundred years in seed dormancy. To stop the enzymic processes, the water content must be brought down to about 10 %. Moreover, this must be done quickly, in other words at raised temperatures and with rapid and efficient removal of the water vapour.

Drying is, of course, promoted by high temperatures, but on the other hand the constituents desired are often sensitive to heat. For this reason, the choice of temperature must be a balancing act between the necessity for quick drying and the sensitivity to heat of the constituents.

The most efficient drying is achieved in large driers of the tunnel type. The plant material is spread out on shallow trays, which are placed on mobile racks and passed into a tunnel where they meet a stream of warm air. Since the racks move against the warm air, the plant material is exposed to dry air all the time. The moist air is removed through vents. The air temperature is kept at 20–40 °C for thin materials such as leaves, but is often raised to 60–70 °C for plant parts that are more difficult to dry, e.g. roots and barks. In addition, thick roots are often sliced or split for easier drying. There are also other techniques, all based on the principle that the plant material meets dry, warm air.

When the crude drug has been collected under primitive conditions, without access to a drier, it must be dried in the open. Even then, the material should be spread out in shallow layers with good ventilation to facilitate the drying. The choice of sunshine or shade is determined by the sensitivity to light of the constituents.

In a dried drug the enzymes are not destroyed but only rendered inactive due to the low water content. As soon as water is added, they become active again. Hence, dried drugs must be protected from moisture during storage.

Freeze-drying

Freeze-drying (lyophilization) is a very mild method. Frozen material is placed in an evacuated apparatus which has a cold surface maintained at –60 to –80 °C. Water vapour from the frozen material then passes rapidly to the cold surface. The energy of evaporation comes primarily from the material being dried, which remains frozen even though the container is kept at room temperature or higher. The end product reaches the temperature of the surroundings only when it is completely dry. The method requires a relatively complicated apparatus and is much more expensive than hot-air drying. For this reason, it is not used as a routine method, but it is very important for drying heat-sensitive substances, e.g. antibiotics and proteins.

Stabilization

As mentioned above, drying only implies delaying enzymic reactions – the enzymes are left intact and will start functioning again as soon as sufficient water is present. Moreover, considerable enzymic degradation of the constituents may take place during the drying process. On long storage, enzymic reactions will slowly destroy the constituents, because the last traces of water can never be removed. In order to avoid this degradation, the enzymes should be destroyed before drying, a process usually called *stabilization*. This is in effect denaturation of the enzymes by treatment with heat, the most common method being brief exposure (a few minutes only) of the plant material to ethanol vapour under pressure (0.5 atm.). The operation is carried out in a preheated autoclave, so that the ethanol does not condense on the plant material. The merits of drug stabilization are often disputed and the method is nowadays not much used. It may be of value for the isolation of compounds that are very susceptible to enzymic degradation, but the increase in cost as compared with conventional drying is seldom covered by the increased yield of desired compounds.

Fermentation

Enzymic transformation of the original plant constituents is on the other hand desirable in some cases. The fresh material is then placed in thick layers, sometimes covered and often exposed to raised temperatures (30–40 °C) and humidity, so as to accelerate the enzymic processes. This treatment is usually called *fermentation*. The fermented product must, of course, be dried afterwards to prevent attack by microorganisms, e.g. fungi. Fermentation is mostly used to remove bitter or unpleasant-tasting substances or to promote the formation of aromatic compounds with a pleasant smell or taste. It is mainly applied to drugs used as spices or stimulants, e.g. vanilla, (*Vanilla planifolia* Andr.), tea (*Camellia thea* Link.) and cacao (*Theobroma cacao* L.). An example of its importance in medicine is the preparation of the sapogenin hecogenin (starting material for the synthesis of steroid hormones) by fermentation of the pressjuice from the leaves of *Agave rigida* var. *sisalana* (Perrine) Engelm., which is obtained as a by-product of the preparation of fibre for ropes and sacks (sisal hemp).

Storage of crude drugs

As is evident from earlier sections, the stability of a crude drug is usually limited, because of slow enzymic changes in the constituents. However, there are great differences in the stability of crude drugs. Drugs containing glycosides and esters are usually less stable than those contaning alkaloids. Drugs with essential oils deteriorate rather quickly through evaporation, oxidation and polymerization of the substances constituting the essential oil; tannins, on the other hand, have an almost unlimited durability. In order to keep crude drugs as long as possible, it is essential to store them in a dry condition in carefully closed containers. It is also advisable to exclude light, because – even if it does not affect the active constituents – it almost always causes changes in the appearance of the drug, especially loss of colour. Often, it is also necessary to protect the drug against attack by insects.

Quality control of crude drugs

Quality specifications for crude drugs are given in pharmacopoeias and handbooks (see list of references). The specifications are usually presented as a *monograph* of the crude drug. A monograph usually comprises the following items:

1. The name and origin of the crude drug.
2. Characters.
3. Identity, comprising macro- and microscopic morphological characters and chemical tests.
4. Purity tests.
5. Quantitative determinations.
6. Instructions for storage.

Name and origin

The name of the crude drug is given in Latin as previously described (page 45). The vernacular name in the national language is also given. The use of the Latin names is decreasing. Thus the *European Pharmacopeia* has the English name first and gives the Latin name as an alternative. Then follows information on the plant part which constitutes the drug and the scientific name of the plant from which the drug derives including the author's name. Information on requirements for the content of certain constituents can also be given here.

Characters

Under this heading organoleptic properties are described, such as colour, smell and taste.

Identity

The identity of a crude drug is determined through visual inspection (macromorphology) and study of micromorphological details. The monograph gives a detailed description of the macromorphological characters such as shape, size, colour, surface characteristics, texture, fracture and appearance of the cut surface. The observation of these sometimes requires the use of a pocket lens. The monograph also gives a detailed description of the micromorphological characters. To study these a microscope must be used and a suitable preparation of the crude drug must be made. The crude drug is usually dry and brittle and must first be rendered soft by soaking in water for several hours or overnight. Preparations of *leaves* and *flowers* can then be made as follows: a small piece of the material is placed on a glass slide, a few drops of a clarifying agent such as chloral hydrate solution (50 g in 20 ml of water) is added and after application of a cover-glass the preparation is heated to boiling in a small flame. For preparations of hard plant parts such as *wood, stems, rhizomes, roots, barks, fruits* and *seeds* these must be sectioned by cutting. Transverse and longitudinal cuttings should in each case be made. For simple preparations the cutting can be made with a razor blade. For more sophisticated work one must use a special machine, a *microtome*. Cutting in a microtome requires stabilization of weak tissues (such as parenchyma) by freezing of the object or embedding in paraffin wax or other suitable agents. Following cutting, the section is placed on a glass slide and boiled with the clarifying agent as described above. Specimens prepared with chloral hydrate are suitable for observation of parenchyma cells, lignified structures, trichomes and crystals. Starch is destroyed by this reagent and must be studied in preparations made with water and without boiling. Starch can be made visible by staining with iodine, which produces a blue colour. A number of other reagents are also available for staining of cell walls and different cell contents. *Powdered crude drugs* do not require any special handling. A small amount of the powder is placed on a glass slide and after mixing with a few drops of the reagent to be used and application of a cover-glass, the mount is ready to be examined under the microscope.

For definitive identification, comparison with authentic reference material is needed. A laboratory carrying out the quality assurance of crude drugs requires a well-documented collection of reference materials.

Detailed descriptions of microscopic technique can be found in several text- and handbooks as well as illustrations of macro- and microscopic details of a large number of crude drugs (see list of references).

In addition to investigation of the macro- and microscopic characters modern pharmacopoeias also require identification by thin-layer chromatography (TLC) of an extract of the crude drug. Special colour reactions are sometimes also employed, e.g. for the presence of tropane alkaloids in Belladonna leaf and Stramonium leaf.

Purity tests

Tests for purity comprise: determination of foreign matter, ash, loss on drying, extractable matter, heavy metals, residues of herbicides and pesticides, contamination by aflatoxins and microbial contamination. For plants collected or cultivated in areas which are subject to radioactive downfall, e.g. after a nuclear power plant failure, it may also be necessary to determine the amount of radioactive contamination. Tests for heavy metals, residues of herbicides and pesticides, aflatoxins and radioactive contamination are generally not included in the pharmacopoeia monographs but should nevertheless be performed to ensure good quality of the crude drug.

Foreign matter is any material which does not normally belong to the plant part constituting the crude drug, such as soil, sand, stones, moulds, insects, and other animal contamination including animal excreta. The sample should also be free of any plant parts not belonging to the drug, regardless of whether they originate from the same plant as the drug or from another species. Such parts are an indication of adulteration, which might be dangerous if they come from a poisonous plant. To detect foreign matter it is necessary to investigate a sample that is big enough. For uncut drugs a suitable size is 500 g of roots, rhizomes and bark and 250 g of leaves, flowers, seeds and fruits. Testing for foreign matter in powdered crude drugs must be done under the microscope. It is very important to look out for foreign tissue fragments, cell types and cell contents. It is, however, almost impossible to establish admixture of plant materials that lack characteristic cell types and cell contents. The more finely powdered the drug, the more difficult the problem. Most monographs define a limit for the amount of foreign matter which is allowed to be present in the crude drug. The *European Pharmacopoeia* requires that a crude drug contains no more than 2 % (m/m) foreign matter and that it be free from fungi, insects or other animal contamination.

Ash is the residue obtained after combustion of the plant material. This ash is composed of carbonates, phosphates, chlorides and sulphates mainly of Na, K, Mg, Ca and Fe. Silica and silicates are also important constituents. Ash is determined as *total ash, acid-insoluble ash* and *sulphated ash*. Total ash is that obtained upon incineration of the plant material. Acid-insoluble ash is the residue obtained when the total ash is treated with hydrochloric acid and the insoluble matter ignited. The acid-insoluble ash consists mainly of silica and silicates and the percentage is a measure of the amount of contamination by soil and sand. Sulphated ash is obtained by incinerating a mixture of the crude drug and sulphuric acid.

Determination of *loss on drying* gives an indication of the content of moisture in the crude drug. The test can be carried out by heating to 100–105 °C. By this procedure any volatile oil present in the plant material will be included in the result. For such drugs azeotropic distillation is a better method. The plant material is distilled with a solvent such as toluene or xylene which does not mix with water but which dissolves volatile oil. Upon cooling, water separates from the solvent and its volume can be determined.

Determination of *water-soluble extractive* is important for crude drugs which are to be used by the consumer for preparation of infusions – "teas" – for direct consumption. A too low figure can also be an indication of adulteration with previously extracted plant material. For determination of this parameter, the plant material is extracted with water under well defined conditions, and, after evaporation of the solvent, the weight of the residue is determined and the percentage calculated. For some crude drugs determination of the content of extractable matter can also be prescribed for solvents other than water.

Tests for contamination by *heavy metals* are usually not required by the pharmacopoeias, but might be needed for crude drugs which are used directly by the consumer for preparation of "teas". The test can be performed by combustion of the plant material in a mixture of nitric and perchloric acid followed by atomic absorption spectrophotometry of the residue.

Tests for contamination with *herbicides* and *insecticides* (=*pesticides*) are important, in particular for crude drugs originating from cultivations. Drugs collected from wild-growing plants can also be contaminated if they grow close to an agricultural area. The same methods used for determination of pesticides and herbicides in food can usually be employed and the limits allowed are usually of the same order of magnitude. The *European Pharmacopoeia* describes methods for determination of pesticide residues and lists limits for common pesticides.

Aflatoxins are produced by fungi of the genus *Aspergillus* and are extremely poisonous (see page 292). As mentioned above, crude drugs are not allowed to be contaminated by any visible fungi but should nevertheless be tested for the presence of aflatoxins because small amounts of fungi are difficult to detect, in particular if they are hidden inside the plant tissues. The same methods which are used for testing foods can be used for crude drugs and the same limits should be applied.

Tests for *microbial contamination* are described in the pharmacopoeias. All crude drugs are contaminated to a greater or lesser extent by microorganisms of various kinds. Investigation has shown contents varying from 700 in Quillaja bark to 2.5 million bacteria per gram in Digitalis leaf. Especially subterranean organs, such as roots and rhizomes, and leaves covered with trichomes may become heavily contaminated. When the crude drugs are taken directly to a facility for extraction of pure constituents, this microbial contamination causes little harm. However, when the drugs are used in other processes in the pharmaceutical industry or in pharmacies, where a high level of general hygiene is required, heavy contamination is undesirable. It is, of course, essential that crude drugs which are going to be used directly as medicines, either whole or powdered, meet the requirements laid down. The *European Pharmacopoeia* has set the following limits for the microbial contamination of crude drugs:

Total viable aerobic count: Not more than 10^5 bacteria and not more than 10^4 fungi per gram or per millilitre.
Not more than 10^3 enterobacteria and certain other Gram-negative bacteria per gram or per millilitre.
Absence of *Escherichia coli* when tested in 1.0 gram or 1.0 millilitre.
Absence of *Salmonella* when tested in 10.0 grams or 10.0 millilitres.

For crude drugs to which boiling water is added before use (see preparation of "teas" below) the following limits should be met:
Total viable aerobic count: Not more than 10^7 bacteria and not more than 10^5 fungi per gram or per millilitre.
Not more than 10^2 *Escherichia coli* per gram or per millilitre.

Quantitative determinations

In principle, the determination of the content of active constituents does not differ from the analysis of other pharmaceuticals. Many pharmacopoeias describe only rather crude methods for quantitative determinations such as gravimetry or colorimetric determinations following treatment of a partially purified extract with a suitable reagent. The biggest problem in quantitative analysis of constituents in a crude drug is to separate these from the large amount of non-active constituents present. Modern HPLC-methods (High-Performance Liquid Chromatography) give excellent results even with comparatively crude extracts of the plant material. For volatile constituents, such as these of volatile oils, GLC (Gas-Liquid Chromatography) is often a good method.

For many crude drugs which are used for preparation of so-called herbal remedies (see page 78) the compound(s) responsible for the alleged medicinal effect(s) are not known. For such drugs, determination of the content of some other known compound may be used to ascertain the quality of the material. For this determination to be of any value the compound chosen for analysis should be as specific for the drug as possible.

Instructions for storage

Storage in well closed containers, protected from light is usually prescribed (see also Storage of crude drugs page 59).

Sterilization of crude drugs

Different methods for sterilizing drugs have been tested. The two most important ones are treatment with ethylene oxide and exposure to γ-rays. However, neither method is universally applicable.

Treatment of plant materials with ethylene oxide may result in a considerable increase in weight (up to 15 %). This is due partly to physical absorption of the gas and partly to chemical reactions between it and various compounds present in the material. As a result, there may be changes in the content of the active constituents. Thus, while ethylene oxide causes a decrease in the alkaloid content of Belladonna leaf, the content of morphine in Opium and of the glycosides in Digitalis leaf is unaffected. On the other hand, the viscocity of mucilages is decreased.

There is also a risk that after treatment with ethylene oxide toxic and/or mutagenic or carcinogenic compounds may be formed. 2-Chloroethanol and chloroethyl esters of linoleic and gallic acids are examples of such compounds which have been found in treated plant materials.

γ-Irradiation reduces the morphine content of Opium, and, as with ethylene oxide, the viscosity of mucilages is lowered and the glycoside content of Digitalis leaf is not affected. Little is known about the effects of γ-irradiation on the possible formation of toxic compounds in plant materials, and thorough investigation is required before the method can be recommended.

Preparations of crude drugs

As mentioned previously (page 20), the most important use of crude drugs today is for extraction of pure, pharmacologically active compounds to be incorporated into tablets and other ready-made drugs. There is, however, still a market for simple preparations of crude drugs like teas and extracts which are sold as herbal remedies, mostly for self-medication (page 67).

Grinding of crude drugs

Regardless of whether the crude drug is to be used for isolation of a pure compound or for manufacture of a simple preparation, the first operation that must be performed is grinding of the plant material to a powder of suitable particle size. It is important that the particles are of as uniform a size as possible. Excessive dust can clog percolators and result in a turbid extract which is hard to clarify. Large particles take a longer time for complete extraction than small ones and large differences in particle size thus slow down the extraction process. Several types of machines are available for grinding crude drugs. The first operation consists of reducing the size of the plant parts so that they can be fed into the mill proper. This is done in *cutters* where the material is placed on a conveyor belt and forced through a slit on the other side of which a knife is moving up and down thus cutting the material into pieces of a suitable size. A common type of mill for grinding crude drugs is the *hammer mill*. This mill consists of a rotor to which pendulum beaters are attached. The

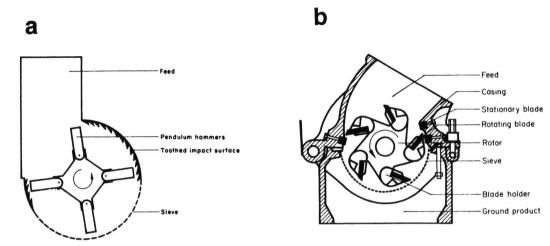

a

Feed

Pendulum hammers
Toothed impact surface

Sieve

b

Feed
Casing
Stationary blade
Rotating blade
Rotor
Sieve

Blade holder
Ground product

Fig. 6. a): Principle of the hammer mill. b): Section through a knife mill. (From P.H. List and P.C. Smith "Phytopharmaceutical Technology", Blackwell Science, Oxford, 1989. Reproduced with permission from the publishers)

hammers, following the rotation of the shaft to which they are attached, are arranged radially and break up the material that is introduced into the rotor chamber. On the walls of the chamber is a grid, which determines the size of the particles passing through it. Another type of mill which is suitable for grinding crude drugs is the *knife mill* in which the material is cut between fixed and rotating knives to particles small enough to pass the grid in the wall of the rotor house. Knife mills are useful for production of low-dust powders of leaves, barks and roots for subsequent percolation or maceration.

Very fine powders can be obtained by grinding in a *tooth mill* consisting of two discs, one of which is rotating and the other stationary. Teeth are arrayed on the discs in concentric rows which brush against each other during movement. The drug is ground as it passes between the discs. The particle size is determined by the distance between the two discs and by the reciprocal velocity between them.

Grinding produces a certain amount of heat which must be observed when grinding crude drugs containing heat-sensitive compounds. Mills cooled with liquid nitrogen are available for such purposes. Seeds and fruits can be made brittle by cooling in liquid nitrogen and are then easily ground in a cooled mill. Cold grinding is also preferable for crude drugs containing volatile oils. For details considering the construction of the mills and which mill should be used to solve a particular problem the reader is referred to the pamphlets and advice of the various manufacturers.

Following grinding, the material must be sifted to ensure the proper particle size. Although the mills are equipped with sieves, the output usually contains some coarse particles which must be

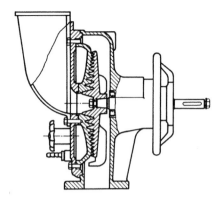

Fig. 7. Cross-sectional view of a toothed disc mill with horizontal axle. (From P.H. List and P.C. Smith "Phytopharmaceutical Technology", Blackwell Science, Oxford, 1989. Reproduced with permission from the publishers)

reground. Sifting can be performed according to two different principles: *sieving* and *blast sifting*. In *sieving* the material is passed through a sieve of suitable mesh size giving two fractions. The fraction passing the sieve consists of particles with a size smaller than or corresponding to the mesh size. The remaining fraction consists of coarser particles which are returned to the mill for continued grinding. Mesh sizes are defined by the pharmacopoeias, which also give directions for construction of the sieves. The distribution of particle sizes in the powder can be determined by *analytical sieving*, using a battery of sieves placed on top of each other and ordered by decreasing mesh size. A weighed quantity of the powder is placed on the coarsest sieve and the assembly of sieves is shaken for a certain period of time. The remains on each sieve are then weighed and the percentage of the total amount calculated. In *blast sifting* the material to be classified is blown with compressed air into an apparatus which allows the particles to sediment according to their weight. Coarse, heavy particles settle fast whereas small, light particles stay for a long time in the air stream.

Herbal "Teas"

These preparations consist of coarse powders (particle size 2–4 mm). A herbal tea may contain only one crude drug but it may also be a mixture. The consumer usually prepares his remedy as an ***infusion***, i.e. by pouring boiling water over the plant material, stirring and allowing the mixture to steep for a short time, whereupon the plant parts are removed by decantation or filtration and the aqueous extract drunk while it is still warm. The recipes for preparation of herbal teas are usually rather imprecise with respect to the dosage. The quantity to be extracted is usually stated in terms of level or heaped teaspoonfuls. On an average a heaped teaspoonful of a crude drug weighs about 2.5 g but there are considerable differences depending on which plant parts are used. Thus this volume of chamomile flowers weighs only 1 g, of a leaf drug 1.5 g and of a root or a bark about 4.5 g. The amount of water to be taken is usually stated as a cup which means a volume of 150–250 ml. Herbal teas are sometimes prepared as *decoctions* which means that the plant parts are boiled with the prescribed quantity of water. In some cases, in particular for crude drugs containing mucilage, an extract is prepared with cold water.

Herbal teas are often marketed packed in teabags containing the right dose for preparation of one cup of tea. "Instant herbal teas" are dried aqueous extracts of crude drugs. They are administered in dose-bags and the remedy is prepared by dissolving the content of one bag in a prescribed quantity of water.

As water is the solvent for preparation of remedies from herbal teas one might expect that these remedies contain only very polar substances. However, investigations, aiming at isolation of pharmacologically active compounds from the preparations, have shown that an aqueous extract contains compounds which, when obtained in a pure state, turn out to be almost insoluble in water. The reason for this is that an aqueous extract of a plant material is very complicated and contains compounds which act as solubilizers for less polar compounds.

It is very important that crude drugs which are used as herbal teas meet the quality specifications described above. As they are used for preparation of aqueous extracts, the specifications for water-soluble extractive are particularly significant.

Extracts

Extracts can be defined as preparations of crude drugs which contain all the constituents which are soluble in the solvent used in making the extract. In *dry* extracts (*extracta sicca*) all solvent has been removed. *Soft* extracts (*extracta spissa*) and *fluid* extracts (*extracta fluida*) are prepared with mixtures of water and ethanol as solvent. A soft extract contains 15–25 % residual water. A fluid extract is concentrated to such an extent that the soluble constituents of one part of the crude drug are contained in one or two parts of the extract. *Tinctures* are prepared by extraction of the crude drug with five to ten parts of ethanol of varying concentration, without concentration of the final product. For both extracts and tinctures the weight-ratio drug/extract should always be stated.

Thus if 100 g of a crude drug yields 20 g of dry extract the ratio is 5:1. Consequently, if the same amount of crude drug is used to prepare 1000 g of tincture, the ratio is 1:10. By definition the crude drug/extract ratio for a fluid extract is 1:1 or 1:2.

Several factors influence the extraction process. Plant constituents are usually contained inside the cells. The solvent used for extraction must therefore diffuse into the cell to dissolve the desired compounds, whereupon the solution must pass the cell wall in the opposite direction and mix with the surrounding liquid. An equilibrium is established between the solute inside the cells and the solvent surrounding the fragmented plant tissues. The speed with which this equilibrium is established depends on *temperature, pH, particle size and the movement of the solvent*. The pH is of great importance for charged constituents. The temperature and particle size have a direct influence on the speed with which equilibrium is attained. Fine particles and high temperature result in a fast establishment of equilibrium. Movement of the solvent relative to the particles facilitates diffusion of the solvent and the solution through the cell walls.

Choice of solvent

The ideal solvent for a certain pharmacologically active constituent should:

1. Be highly selective for the compound to be extracted.
2. Have a high capacity for extraction in terms of coefficient of saturation of the compound in the medium.
3. Not react with the extracted compound or with other compounds in the plant material.
4. Have a low price.
5. Be harmless to man and to the environment.
6. Be completely volatile.

High selectivity is often difficult to obtain and is not always desired, especially for extraction of plants for which the pharmacologically active compounds are not well known. Even for well-known constituents a solvent with high capacity but low selectivity is often preferred to one with high selectivity but low capacity. Aliphatic alcohols with up to three carbon atoms, or mixtures of the alcohols with water, are the solvents with the greatest extractive power for almost all natural substances of low molecular weight like alkaloids, saponins and flavonoids. They also comply with requisites 3–6. According to the pharmacopoeias, ethyl alcohol is the solvent of choice for obtaining classic extracts such as tinctures and fluid, soft and dry extracts. The ethanol is usually mixed with water to induce swelling of the plant particles and to increase the porosity of the cell walls which facilitates the diffusion of extracted substances from inside the cells to the surrounding solvent. For extraction of barks, roots, woody parts and seeds the ideal alcohol/water ratio is about 7:3 or 8:2. For leaves or aerial green parts the ratio 1:1 is usually preferred in order to avoid extraction of chlorophyll.

Extraction procedures

(See list of references for more detailed descriptions of the equipment required for the various extraction procedures.)

Maceration This is the simplest procedure for obtaining an extract and is suitable both for small quantities of drug and for industrial production. Simple maceration is performed at room temperature by mixing the ground drug with the solvent (drug/solvent ratio: 1:5 or 1:10) and leaving the mixture for several days with occasional shaking or stirring. The extract is then separated from the plant particles by straining. The procedure is repeated once or twice with fresh solvent. Finally the last residue of extract is pressed out of the plant particles using a mechanical press or a centrifuge. Kinetic maceration is also performed at room temperature and differs from simple maceration in that the mixture of crude drug and solvent is stirred continuously, usually overnight. With this exception the procedure is performed as described for simple maceration. In industry kinetic maceration is dominant and is performed in closed extractors, equipped with stirrers. Movement of solvent and drug can also be achieved by rotating of the extractor along its axis. Such extractors can have a volume of up to 4 000 litres. For extraction at elevated temperatures the extractors can be equipped with heating jackets or internal heating coils.

Maceration does not result in complete extraction of the desired compounds. Despite careful pressing of the residual plant material a certain amount of solute cannot be recovered.

Percolation   *Simple percolation* is a procedure in which the plant material is packed in a tube-like percolator which is fitted with a filter sieve at the bottom. Fresh solvent is fed from the top until the extract recovered at the bottom of the tube does not contain any solute. This is a slow and costly process requiring large quantities of fresh solvent. The economy can be improved by *repercolation* where a battery of percolators are connected in series. When the first percolator in the series is exhausted, it is emptied and loaded with a new batch of crude drug whereupon it is placed last in the series. Fresh solvent then enters the formerly second percolator until that one is exhausted and so on. In this way the extract leaving the battery is always saturated with the solutes.

A technical problem in percolation is to ensure an equal flow of solvent through the mass of crude drug powder. The drug should not be too finely ground to allow a reasonably fast passage of the solvent. A particle size of 1–3 mm is usually sufficient. Before the material is loaded into the percolator it should be moistened with the solvent and allowed to swell. It is then carefully packed into the percolator in such a way that the layer formed is as uniform as possible. Solvent is administered at the top and passes through the drug. The extract is collected at the bottom or is passed on to the next percolator if a battery is used. Transport of solvent can be achieved by gravity or by pumping.

Percolation on an industrial scale requires a substantial amount of labor for filling and emptying the percolators. Also, recovery of solvent from the exhausted plant material can pose a problem.

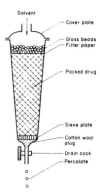

Fig. 8. Percolator. (From P.H. List and P.C. Smith "Phytopharmaceutical Technology", Blackwell Science, Oxford, 1989. Reproduced with permission from the publishers)

Countercurrent extraction   This is a continuous process in which the plant material moves against the solvent. Several types of extractors are available. In the *screw extractor* the plant material is transported by a screw through a tube and meets the solvent which is pumped in the opposite direction. Other extractors are the *carousel extractor*, the *U-extractor* and the *radial pressure extractor*. For details regarding the construction and operation of these extractors the reader is referred to technical handbooks (see list of references). Countercurrent extraction is a suitable procedure for production of large amounts of extracts on an industrial scale.

Extraction with supercritical fluids   At a sufficiently low temperature a gas may be made to liquefy by applying pressure to reduce the volume. However, there is a temperature above which it is impossible to liquefy the gas no matter how great a pressure is applied. This temperature is called the *critical temperature*. The minimum pressure necessary to bring about liquefaction at the critical temperature is called the *critical pressure*. The combination of critical pressure and critical temperature is characteristic of the particular substance and is called the *critical point*. Gases at temperatures and pressures above the critical point

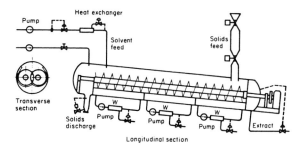

Fig. 9. Countercurrent screw extractor. (From P.H. List and P.C. Smith "Phytopharmaceutical Technology", Blackwell Science, Oxford, 1989. Reproduced with permission from the publishers)

are called *supercritical gases or supercritical fluids*. As the names imply, a substance in the supercritical phase is neither a true liquid nor a true gas but has some of the properties of each. These properties can be varied within wide limits by a suitable choice of pressure and temperature. At high temperature and low pressure the density is low and the supercritical fluid behaves more like a gas, but at low temperature and high pressure the density is increased and it increasingly assumes the properties of a liquid. Other properties, e.g. the dielectric constant, are also altered as well as the density. The properties of the supercritical fluids make them potentially useful for extraction and chromatography. The solvation strengths of many supercritical fluids approach those of liquid solvents, hence many substances will dissolve in them. Supercritical fluids have diffusion coefficients which are close to those of gases and thus they are able to transport dissolved solutes through materials very quickly. The low viscosity of supercritical fluids also allows them to flow easily through small openings. Perhaps the most dramatic property of the supercritical fluid state is the absence of a surface meniscus. The supercritical fluid exhibits no surface tension properties. Only gases which can be converted into the supercritical state at attainable pressures and temperatures can be considered for extraction use. Table 1 gives the critical points for a selection of fluids:

Table 1. Critical temperatures and pressures

Fluid	Critical temp.(°C)	Critical pressure (bar)
Ethylene	9.3	50.4
Carbon dioxide	31.1	73.8
Ethane	32.3	48.8
Nitrous oxide	36.5	72.7
Propylene	91.9	46.2
Propane	96.7	42.5
Ammonia	132.5	112.8
Hexane	234.2	30.3
Water	374.2	220.5

Obviously in their supercritical states water and hexane are unsuitable for extraction of crude drugs because of the high critical temperatures. Also propane and propylene have high critical temperatures like ammonia which also is unsuitable because of its corrosivity. Ethylene and ethane have low critical parameters but are hazardous because of their flammability. Nitrous oxide is quite polar and has reasonably accessive values for the critical temperature and pressure. However, violent explosions have been reported when using this supercritical fluid. Only supercritical carbon dioxide combines easily attainable critical parameters with safety and has been used extensively for extraction of crude drugs and other plant material.

The equipment for extraction with supercritical fluids consists of a tank for supply of the gas, a pump and heat exhanger to control the pressure and temperature of the gas, an extraction chamber containing the plant material to be extracted and a separator in which the pressure and temperature of the extract-containing gas are lowered to achieve separation of extract and gas. The gas is then returned to the pump and circulated again. There are many possibilities to change the pressure and temperature both in the extractor and in the separator and it is thus possible to limit the extraction to comprise only certain desired substances or classes of substances. Other advantages of extraction with supercritical carbon dioxide are freedom of the resultant extracts from solvent, complete physiological harmlessness and low price of the extracting agent. The method is quite widespread in the food industry for extraction of fats and oils from seeds and other materials and is also used in the flavour and perfume industry for extraction of flavours and fragrances from natural products. The method has been applied to decaffeination of coffee. Examples from the pharmaceutical industry are so far quite few but the use of the method for extraction of crude drugs can be expected to increase.

Treatment of the residual plant material

The exhausted plant material is usually pressed to recover as much of the extract as possible. Even after pressing it still contains an appreciable amount of solvent. When operating on a large scale this solvent has to be recovered, which can be done by heating the plant material and condensation of the resulting vapours. This improves the economy of the process and is also of importance from an environmental point of view.

Purification and concentration of extracts

The extract obtained by one of the above described procedures often contains solid particles which have been carried over from the crude drug. These must be removed before further treatment of the extract. The methods used are *decantation, centrifugation* and *filtration.* Decantation is the oldest known process for clarifying liquids containing solids in suspension. The cloudy liquid is left to stand, protected from light and evaporation. Under the influence of gravity the particles sediment and the clear liquid can be decanted or siphoned off. Sedimentation is a very slow process which can take several days or even weeks. Sedimentation can be accelerated by application of a centrifugal force. Several types of centrifuges are available and for industrial applications continuously operating centrifuges are suitable. These are constructed to allow continuous removal of the sediment. Filtration is usually carried out in filter presses in which the extract is forced through the filter by pumping. These presses operate discontinuously as the process has to be stopped and the filters cleaned or exchanged when they become clogged by the solid particles. Addition of *filter aids* such as kiesel-guhr can prevent too rapid clogging by very small particles. For details concerning construction and operation of centrifuges and filter presses the reader is referred to manufacturers' pamphlets and relevant literature (see list of references).

For the manufacture of fluid and soft extracts the clarified extract must be concentrated. Preparation of a dry extract requires complete removal of the solvent. Concentration is a tricky stage in the process in which many chemically labile compounds may undergo degradation, mainly due to the temperature. Concentration *in vacuo* is therefore the preferred method by which the extract can be kept at 25–30°C during the whole procedure. Several types of concentrators are available (for detailed descriptions see the relevant literature in the list of references). *The Kestner concentrator* consists of a number of comparatively narrow, long tubes which are evacuated and heated from the outside. The extract is introduced at the lower ends of the tubes and climbs rapidly upwards under quick evaporation. The upper part of the concentrator contains a chamber for separation of vapour and the concentrated extract, which returns to the inlet where it is mixed with a new portion of unconcentrated extract and climbs the heated tubes a second time. The procedure is continued until the desired degree of concentration is achieved. Other types of concentrators spread the extract in a thin film from which the solvent rapidly evaporates. The film moves over the heated surface at such a speed that the concentrated extract is prevented from sticking and does not reach excessive temperatures. One common type of concentrator operating according to this principle is the *rotary evaporator* (Fig. 10), which is extensively used in laboratories but which is also available at larger sizes for industrial use. The rotary evaporator consists of an evacuated round flask connected to a condenser. The flask rotates in a hot-water bath. The extract to be

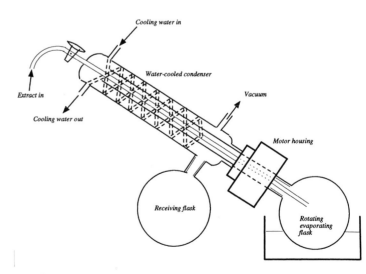

Fig. 10. Rotary evaporator

concentrated is continuously introduced into the flask at a speed corresponding to the speed of evaporation and the volume of the concentrated extract is thus kept constant during the operation. For further details concerning construction and operation of concentrators the reader is referred to special technical handbooks (see list of references) and to manufacturer's specifications.

Drying of extracts

The concentrators described above can be used for production of fluid and soft extracts but are not suitable for complete drying of an extract. Most extracts on the market today are dry extracts for reasons of easier storage and transport but also because of their increased stability. The classical apparatus for drying is the cabinet drier in which the extract is placed on shelves in a cabinet. Hot air (60–80 °C) is blown over the shelves. This method can only be used for extracts containing heat-stable compounds. For thermolabile extracts the cabinet must operate under a vacuum. Heat is in this case supplied by electrical heaters under the shelves. The drawback of cabinet driers is that drying is often not uniform. The material must therefore be taken out of the drier for grinding and mixing whereupon it is returned for further drying. Drying in cabinet driers is therefore labour intensive and expensive.

Drying in atomizers (spray drying) is a method more suitable for industrial production. An atomizer consists of a tower-shaped drying chamber into which a stream of filtered and heated air is introduced from above. The extract (dry residue 20–30 %) is pumped into the chamber through a dispersor which forms small droplets from which the solvent instantaneously evaporates on contact with the hot air. The dried product precipitates to the bottom of the tower where it can be collected or it can be discharged together with the

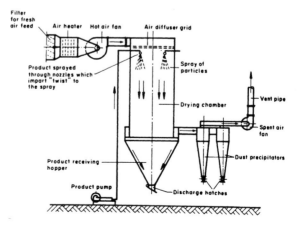

Fig. 11. Flow scheme of a unidirectional spraying tower. (From P.H. List and P.C. Smith "Phytopharmaceutical Technology", Blackwell Science, Oxford, 1989. Reproduced with permission from the publishers)

air to one or more cyclones for separation. Because of the rapid evaporation the extract does not reach a particularly high temperature. Depending on the properties of the extract it may be necessary to mix it with additives such as glucose to facilitate the process.

Quality control

The extract should always be prepared from a plant material of well known quality according to a pharmacopoeia or other realiable monograph. It is usually required that the weight ratio between crude drug and extract be stated: e.g. "dry extract of Senna leaf (8:1)". This means that 1 g of the dry extract corresponds to 8 g of crude drug. The quality control of *a dry* extract should comprise the following parameters:

Identity: Besides organoleptic qualities (colour, smell, taste) a chromatographic fingerprint (thin layer chromatogram, HPLC chromatogram) should be compared with that of a corresponding standard extract. If the pharmacologically active constituents are known, the test should indicate their presence.

Moisture: Max. 5 % determined by drying at 105 °C to constant weight or by Karl Fischer titration.

Solvent residues: Ethanol max. 0.5 %. For noxious solvents such as methanol and methylethyl ketone 0.05 % and 0.002 % are the limits. GLC is the preferred method for determination of solvent residues in dry extracts.

Heavy metals: Contamination with heavy metals should preferably be determined by atomic absorption spectrophotometry following combustion of the material. Commonly accepted limits are: Cd <0.02 ppm, Pb <0.05 ppm and Hg 0.005 ppm (parts per million). Many pharmacopoeias require only a colorimetric test for the total content of heavy metals, usually 20–60 ppm.

Content of pharmacologically active compound(s): To ensure a reliable dosage it is very important that the content of the pharmacologically active constituent(s) is determined precisely. Determination should be done by a sensitive and specific method. Modern HPLC-methods are suitable for determination of specific substances in the complex mixture of compounds which constitute a plant extract. For volatile compounds GLC is a suitable method. Manufacturers of extracts often standardize their products to ascertain that the content of the active compound(s) always lies within certain limits. Standardization can be performed by mixing with lactose or other inert substances.

For extracts of crude drugs, the active constituent(s) of which are not known, determination of the content of a compound which is characteristic of the drug might suffice to ensure the quality of the preparation.

Microbial contamination: The *European Pharmacopoeia* prescribes that preparations for oral and rectal administration should meet the following requirements (Category 3 A):

Total viable aerobic count: 10^3 aerobic bacteria and 10^2 fungi per gram or millilitre.

Absence of *Escherichia coli* when tested on 1.0 gram or 1.0 millilitre of the product.

For extracts for which antimicrobial pretreatment is not feasible and for which the relevant authority accepts a microbial contamination of the raw material exceeding 10^3 viable microorganisms per gram or per millilitre the following limits should be met (Category 3 B):

Total viable aerobic count: not more than 10^4 aerobic bacteria and not more than 10^2 fungi per gram or per millilitre.

Not more than 10^2 enterobacteria and certain other Gram-negative bacteria per gram or per millilitre.

Absence of *Salmonella* when tested on 10.0 gram or 10.0 millilitre.

Absence of *Escherichia coli* when tested on 1.0 gram or 1.0 millilitre.

Absence of *Staphylococcus aureus* when tested on 1.0 gram or 1.0 millilitre.

Tests for aflatoxins, pesticides and radioactive contamination: These tests should be performed if the crude drug used for preparation of the extract has not been subjected to them.

The quality control of *fluid* and *soft extracts* should comprise the same parameters as described above except determination of moisture for which determination of dry matter should be substituted.

Biopharmaceutical aspects on extracts of crude drugs as remedies

Crude extracts are very complicated mixtures, which in addition to the pharmacologically active substance(s) also contain many compounds that are not directly responsible for the activity of the extract. These compounds are usually referred to as *inactive substances*. Inactive substances can, however, influence the activity of the pharmacologically active compounds in several ways. They can solubilize the active compounds thus facilitating their absorption from the intestine. It is often found when isolating an active substance from an aqueous extract of a crude drug that the active compound is not, or is only very slightly, soluble in water although the extract itself has pronounced pharmacological activity. This is explained by the presence in the extract of inactive compounds, such as surface-active agents (e.g. triterpenoid saponins, page 441), which act as solubilizers for the active substance. An example is the content of the flavonoid rutin (page 335) in aqueous extracts of *Fagopyrum esculentum* Moench, which is much higher than can be expected from the solubility of pure rutin in water. Besides solubilization, inactive compounds can also influence the solubility of active compounds by formation of associates. Another mechanism by which inactive compounds can influence the pharmacological activity of an active substance is by interference with membrane permeability. These influences of inactive compounds are of importance when comparing extracts of a crude drug prepared by different extraction procedures. It is not certain that they are pharmacologically equivalent although the content of the active compound(s) is the same.

Herbal remedies

Definitions and trade regulations

The legislation concerning drugs is complicated and varies from country to country. Several categories of drugs are on the market. A *conventional drug* is a drug which complies with legal regulations concerning quality and proofs of safety and efficacy. The demands on safety and efficacy are severe and usually require extensive pre-clinical and clinical studies. In addition to conventional drugs of natural origin, such as penicillin and digitoxin, there are also other natural products available.

Natural remedies are products which contain crude drugs (see page 17) or partly purified extracts of these, with comparatively mild effect, which are suitable for self-medication. A natural remedy can also contain minerals, salt solutions and bacterial cultures of natural origin. The main part of the natural remedies are *herbal remedies* which can consist of a herbal tea (page 67) or be a preparation (solution, syrup, tablet, capsule, etc.) containing an extract of a plant-derived crude drug.

Herbal remedies are used by the layman for self-medication of various – usually mild – diseases, but are not always recognized in professional medicine. The use of herbal remedies has a long tradition in Germany, France and Italy where many products are sold both as conventional drugs and as over-the-counter preparations in pharmacies as well as in healthfood shops. In the Scandinavian countries – on the other hand – herbal remedies were for a long time looked upon with suspicion by the medical profession and strong efforts were made to prevent their marketing. Herbal remedies could not be registered as conventional drugs but were still on the market, particularly in healthfood shops. Their labels could not contain any information on medical use, but this was available as leaflets provided by the dealer. The demand from the public, to have access to herbal and other natural remedies for self-medication, accelerated in the 1970s as a reaction to the side-effects of several conventional drugs. There was a common belief that natural remedies were safer than the synthetic substances contained in many conventional drugs. This is, however, not necessarily true. In many countries the need for regulation of the trade with herbal remedies and other natural, non-conventional remedies became evident. There was no control of the quality of the preparations and the marketing was often not accurate, with exaggerated claims of efficacy. The problem was that these products could not be registered as conventional drugs, due to lack of sufficiently strict documentation of safety and efficacy. As natural remedies could not be patented, the manufacturers could not afford the extensive preclinical and clinical testing needed for the registration of a conventional drug. The non-conventional drugs had, however, been used for a long time, based on traditional experience and this was taken as a base for registration of them as a special category of drugs, termed *natural remedies*. In Sweden the new legislation for natural remedies became effective from January 1, 1978 and was applicable to preparations of crude drugs of herbal or animal origin, minerals, salt solutions and bacterial cultures of natural origin. Similar legislation has been adopted by other countries and in 1999 the European Community passed a directive permitting greater flexibility in the use of bibliographic data in proving the safety and efficacy of so-called well established medicines. This was followed by a directive in 2004 which created a simplified registration procedure for all traditional herbal medicines that, despite long use, did not fulfil the requirements for classification as well-established medicines.

The European Community requires a herbal remedy to be of high quality, i.e. the crude drug and an extract of it must comply with the requirements described previously (pages 60–64 and 76–77). The remedy must be manufactured in accordance with GMP (Good Manufactoring Practice) and is required to have a specified strength and dosage. It shall be intended for self-medication, i.e. the proposed indications for use must not require the diagnostic or therapeutic supervision of a medical practitioner. The documentation of safety and efficacy need not be as rigorous as is required for a conventional drug. Two levels are recognized – *traditional herbal medicinal products* and *herbal medicinal products with well established medicinal use.*

For approval as a traditional herbal medicinal product it is sufficient to provide documentation showing medical use throughout a period of at least 30 years, including at least 15 years within the EU, and with data that are sufficient to prove safety and plausability of the pharmacological effects or efficacy based on longstanding use and experience. This legislation is applicable to herbal remedies containing crude drugs which contain compounds with comparatively mild pharmacological effect and which are thus suitable for self-medication. A crude plant drug cannot be accepted as a constituent of a traditional herbal medicinal product if it contains poisonous compounds or substances with strong pharmacological activity, e.g. Digitalis leaf (page 467), Colchicum seed (page 619) or Belladonna leaf (page 636).

Products that have been used for shorter periods can be approved as herbal medicinal products with well established medicinal use if they have been used in the EU for at least 10 years and if a detailed scientific bibliography addressing non-clinical and clinical characteristics is provided. The documentation must cover all aspects of the safety and/or efficacy assessment and must include or refer to a review of the relevant literature, including pre- and post-marketing studies and experience in the form of epidemiological studies and in particular, comparative epidemiological studies. For this group of herbal remedies the indications allowed are not restricted to suitability for self-medication. The only difference between these products and the conventional remedies is that the documentation concerning safety and efficacy must not be based entirely on product-specific preclinical and clinical tests but can consist of a scientific bibliography.

Outside the European Community the regulation of the trade in non-conventional remedies is very variable. In the USA herbal remedies are generally regarded as food supplements and thus subject to less strict requirements than in Europe, with respect to quality and efficacy. In 2004 a new regulatory category – botanical drugs – was released. Botanical drugs are complex extracts from plants to be used for treatment of disease. They are clinically evaluated for safety and efficacy just as conventional drugs, but the process for botanical drugs can be expedited because of the history of safe human use. Botanical drugs are highly but not completely characterized and are produced under the same strictly regulated conditions as conventional pharmaceuticals. Botanical drugs, such as senna and psyllium, can be sold under the FDA's over-the-counter drug monograph system (See the reference Schmidt 2008 in chapter 1, *Further reading*, page 42).

In Canada the trade in herbal remedies is regulated by the Natural Products Regulations, effective from January 1, 2004. These regulations resemble the directives of the European Community, but traditional use must encompass a time period of 50 years. Also here products for which traditional use is not documented can be approved if a detailed scientific bibliography, covering safety and efficacy, is provided.

Side effects

Herbal remedies are regarded by the public as safer and with less side effects than synthetic drugs. This is probably true for properly controlled preparations, but for preparations which have not been controlled there are big risks involved, due to mistaken identification of the crude drug, adulteration with other crude drugs or contamination with heavy metals, pesticides or mycotoxins. Addition of strong-acting conventional drugs has also been reported. However, in most clinical studies of controlled herbal remedies some side effects are almost always reported. As with other drugs, the prescribed dose should not be exceeded. Examples of herbal remedies which should be used with particular caution are anthraquinone-containing drugs (page 316) which are used as laxatives. They can cause pains and cramps of the abdomen. Excessive use results in severe diarrhoea with loss of electrolytes and water and long-term use can cause albuminuria and haematuria.

Interactions

Herbal remedies can interact with conventional drugs. A patient taking a herbal remedy should therefore always inform his (her) physician about this. Interactions between herbal remedies and conventional drugs are insufficiently investigated. One of the best studied herbals is hypericum (St. John's Wort, page 325), which has been found to impair the pharmacological activity of a number of drugs. Some of these interactions are very serious and can be life-threatening, e.g. with the immunosuppressant cyclosporin and with the HIV-1 suppressor indinavir. These interactions are due to activation of a receptor that regulates expression of cytochrome P450-3A4 monooxygenase, an enzyme which is involved in the oxidative metabolism of more than 50 % of all drugs (page 327). Also other herbal remedies have been found to interact with cytochrome P450 iso-enzymes, e.g. liquorice root (page 442).

Interactions are particularly important for drugs with a narrow therapeutic index such as the anticoagulant warfarin. Hypericum has been found to reduce the anticoagulant activity of warfarin and there are case reports of potentiation of the activity by ginkgo (page 423) and ginseng (page 447).

Information sources

A list of plant-derived crude drugs, described in this book, which are used for preparation of herbal remedies is presented as Appendix I (page 748). More detailed information, including dosage, contraindications, side effects and interactions, can be obtained from the handbooks listed under *Further reading* (page 87).

Isolation of pure compounds from extracts of crude drugs and other organisms

Isolation of compounds with known properties

An extract of an organism is a very complicated mixture of which the desired compound often constitutes only a small part. When the properties of the compound to be isolated are known, the isolation procedure can be optimized. Polarity, charge, molecular size and stability are parameters of importance for designation of the isolation procedure. The polarity determines the choice of solvent for extraction (page 69) and is of importance in chromatographic separations. A basic or acidic compound is easier to isolate than a neutral substance. High-molecular-weight compounds can be separated from those of low molecular weight at an early stage of the isolation procedure, but they are often difficult to separate from each other. Substances which are heat-labile or easily hydrolysed require special precautions during isolation.

There are very few cases where a pure compound can be obtained from an extract in a simple way only involving a few separation stages. An example is an old method for isolation of caffeine from tea leaves which requires only the following six steps:

1. Extraction of the tea leaves with hot water.
2. Precipitation of undesired compounds (proteins, glycosides, tannins etc.) with lead acetate.
3. Precipitation of excess lead with H_2SO_4.
4. Decolourizing of the filtrate with charcoal.
5. Extraction with chloroform of the filtrate from step 4.
6. Evaporation of the solvent and recrystallization of caffeine.

This simple isolation method illustrates several general principles. The solvent used for extraction (water) is not the best solvent for caffeine but the solubility is sufficient and the choice of a cheap solvent contributes to the economy of the isolation procedure which is important for an industrial process. In steps 2 and 3 it is desirable to get rid of the bulk of undesired compounds at an early stage of the separation, thereby facilitating the extraction of the desired compounds. Precipitation with lead acetate is an old method which is avoided today because of the toxicity of lead compounds and the difficulty of disposing of the waste in an environment-friendly way. Precipitation of proteins and other high-molecular-weight impurities by addition of 10 volumes of acetone is often a good method which can be used as a first step in the purification of aqueous extracts. It should be observed, however, that acetone is not a completely inert solvent and can react with some compounds resulting in the formation of artefacts. The decolourization step (4) is often needed to get rid of coloured compounds which have a strong tendency to adhere to the substances one wants to isolate. Step 5 illustrates the use of an efficient and selective solvent for the final extraction from an aqueous filtrate containing the desired compound. Chloroform is, however, dangerous to human health and is comparatively expensive. Step 5 also illustrates separation by partitioning between two immiscible liquids. This principle is often used for the preliminary steps in a separation procedure.

As mentioned above, charged compounds are more easy to isolate than neutral ones. Ion exchangers can often be used. Thus basic substances in a solution can be isolated by passing the solution through a column packed with a cation exchanger. Acidic and neutral compounds are not absorbed and can be washed out of the column. The adsorbed substances are then eluted with a suitable buffer solution. The same procedure can be used for isolation of acids by employing an anion exchanger.

The solubility of charged compounds is very dependent on the pH of the solvent. Alkaloids, for example, are easily soluble in water at low pH, where they form salts, but at high pH they are usually precipitated as free bases. However, these are soluble in organic solvents which do not dissolve alkaloid salts. The isolation of alkaloids is described in more detail on page 607.

Following preliminary purification, chromatographic methods must generally be used to obtain a pure substance. A great number of chromatographic methods are available and the reader is referred to handbooks and textbooks for detailed information on the principles and techniques (see reference list). On the laboratory scale, TLC and HPLC can often be used both for analysis and for isolation of small amounts of substance. Also preparative HPLC-systems are available. GLC is useful for volatile compounds, also for preparative purposes. Droplet counter-current chromatography (DCCC) and centrifugal partition chromatography (CPC, also abbreviated CCCC– Centrifugal Counter-Current Chromatography) are partition systems which operate without a carrier phase and thus minimize the risk of loss of compounds by adsorption to the carrier. For separation according to molecular size, various gel materials, e.g. Sephadex, are available. The principle for the separation is that large molecules cannot enter the gel matrix whereas small molecules are partitioned between the matrix and the surrounding solvent. The gel is packed into a column and a solution of the material to be separated is applied on top. During elution the large molecules pass the column quickly as they stay outside the matrix and follow the solvent front. The small molecules are delayed because of the partitioning process.

Charged molecules can also be separated by electrophoresis in an electric field where positively charged compounds travel towards the cathode and those with a negative charge go towards the anode. The process can be performed in a suitable buffer solution contained in a gel. The separated compounds are localized by various methods such as UV, blotting and staining. Electrophoresis is mainly an analytical method which is used extensively for peptides, proteins and nucleic acids. It can, however, also be used for isolation of small amounts of substances.

Bioassay-guided isolation

In a programme aiming at the discovery of leads for new drugs from natural sources, extracts are prepared from various organisms. These extracts are tested for pharmacological activity in various test systems, the design of which depends on the aim of the project. The next step is isolation of the compound(s) responsible for the discovered activity. This process is hampered by the lack of knowledge of the chemical properties of the compounds. The isolation process must therefore be monitored by testing all isolated fractions for the activity. Only the fractions which give a positive result in the test are subjected to further separation steps and eventually the pure active compound(s) is (are) obtained. This procedure is termed *bioassay-guided isolation*. As a great number of fractions have to be tested, the test method used must be easy to perform and the amount of substance required for a test must be small. A comparatively cumbersome pharmacological test system might have been used to discover a particular effect, e.g. antidiarrhoeal activity, involving use of several groups of animals and statistical evaluation of the results. Such a system cannot be used for monitoring an isolation procedure. More suitable are systems which use isolated organs, cells, enzymes or isolated receptors as test objects. The system can either be broad, i.e. involve several different reaction mechanisms, or it can be specific for only one pharmacological effect. The unspecific systems are useful in a search for pharmacologically active compounds in general, while a more specific system must be used when searching for a compound with a particular effect, e.g. in an extract of a plant with proved clinical efficacy against a certain disease.

An example of an unspecific test system is the isolated guinea pig ileum. This organ contains at least 10 different nerve systems. A piece of the ileum is placed in an organ bath, containing a suitable buffer solution, and is connected to a transducer which is hooked up to an amplifier and a recorder by which the movements of the ileum can be monitored. A pharmacologically active compound which is added to the bath can affect any one of the nerves, causing either a contraction of the ileum or inhibition of artificially induced contractions. For evaluation of a substance, extract, fraction or pure compound, one must thus perform two different experiments. The first experiment is an observation of the ability of the added material to cause contractions of the ileum. In the second experiment the ileum is caused to contract by electric impulses and one observes if these contractions are inhibited by the substance added. It is possible to use the same piece of ileum for several experiments as long as the baseline is recovered following rinsing of the organ bath. The dimensions of the organ bath can be so chosen that only small amounts of the material to be tested are needed for a response.

Because of the complex innervation of the guinea pig ileum (some pharmacologists have described it as a "mini brain"), this test object is particularly useful for detection of compounds with new structures and new pharmacological activities.

A bioassay suitable for discovery of anti-inflammatory drugs is an example of a specific test system. This bioassay detects compounds that inhibit prostaglandin H synthase (cyclooxygenase, COX), an enzyme that converts arachidonic acid into prostaglandins and thromboxanes which play an important role as inflammatory mediators

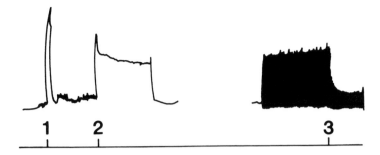

Fig. 12. Responses of the isolated guinea pig ileum to two different phar-macologically active compounds (A and B). 1: Contraction caused by his-tamine (control). 2: Contraction caused by compound A. 3: Inhibition of electrically induced contractions caused by compound B

(page 259). In this assay a solution of the enzyme in a buffer solution, containing suitable cofactors, is incubated for 15 minutes at 37 °C with $1-^{14}C$-arachidonic acid and the compound to be tested. The reaction is terminated by addition of HCl and a solution of unlabelled prostaglandin E_2 + prostaglandin $F_{2\alpha}$ is added as carrier. The unmetabolized arachidonic acid is separated from the prostaglandins by column chromatography, and following elution of the prostaglandin fraction, the radioactivity of this is determined with a scintillation spectrometer. The result is compared with that of a reference experiment without the test compound. If the test material contains COX inhibitor(s), the radioactivity of the prostaglandin fraction from the test should be lower than that of the reference prostaglandin fraction.

Ecological observations in the marine environment and the discovery of organisms with specific properties have led to the development of useful and unique biological test models *in vitro* and *in vivo*.

A common *in vitro* model is based on eggs from Brine shrimp (*Artemia salina*), which is used to evaluate the toxicity of extracts and compounds. It is a simple test but useful in the field to get a preliminary answer on biological activity of an extract from a collected organism. However, the bioassay is very unspecific and other more sophisticated tests are needed for further evaluation of biological activity.

The search for compounds that may act by a new mechanism or on a new biological target (biological novelty) requires multi-target functional assays, e.g. *in vivo* assays using whole organisms. The barnacle settlement assay is an *in vivo* model using cypris larvae from different *Balanus* species to study the effect of test compounds on settlement and mortality of the larvae. The cypris larvae are small, but well organized for finding a suitable substrate for settlement in their normal environment. The larvae have a nervous system for central coordination. The model has been used in studies of biofouling and to guide the isolation of antifouling agents from different marine organisms. Furthermore, recent studies have shown the involvment of serotonin in attachment and metamorphosis of barnacle larvae, which has led to further studies on receptor level and the involvement of biogenic amines in this process. Further understanding of the role of biogenic amines in the cypris larvae and the correlation to the human cell could lead to a useful *in vivo* model for pharmacological evaluation of drug lead candidates.

In the early 1970s the zebrafish (*Danio rerio*) was used as the first vertebrate assay for genetic screening. Further research, focused on organ development, led to characterization of genes in vertebrate pathways and established the zebrafish as a model for studies of human disease. It is today also a relevant and validated model organism for *in vivo* drug discovery. The model has several advantages such as the small size of embryos and larvae, high number of offspring in a short time, transparent and fast development of embryos and larvae. These are attractive features for a model suitable for medium-to-high-throughput screening using microtitre plates. Historically the zebrafish has been used for toxicity studies of agrochemicals and detection of environmental contaminants. More recently it has been used, early in the drug discovery process, to aid in selection of non-toxic and safe lead candidates.

Recent studies have shown that the zebrafish model, in combination with advanced separation and detection techniques, can be used as an *in vivo* assay for bioactivity-guided purification of very small amounts of complex natural extracts for discovery of compounds with specific biological activity. A combination of experiments in zebrafish and confirmation in mammalian cells is used to explore the mode of action of natural products and other small molecules, which can be useful in the area of chemical genetics.

When starting a bioassay-guided isolation of pharmacologically active compounds from an extract, nothing is known of the chemical properties of the compound(s) causing the pharmacological effect displayed in the test system used. It is therefore important to get some preliminary information about polarity, charge, molecular size and stability of the active compounds in order to design a rational isolation procedure. Above we discussed how to apply such knowledge to an isolation procedure. The methods described can also be used to acquire the desired knowledge by testing all the obtained fractions for activity in the chosen pharmacological test system.

Industrial high-throughput screening of extracts

As mentioned previously (page 28) a technique called **high-throughput screening (HTS)** is used by the pharmaceutical industry for detection of leads to new drugs. This technique can be applied to detect pharmacological activities of extracts, fermentation broths or cell cultures. It can also be used to monitor the isolation of active compounds from such material. HTS is defined as *the process by which large numbers of compounds can be tested, in an* *automated fashion, for activity as inhibitors (antagonists) or activators (agonists) of a particular biological target, such as a cell-surface receptor or a metabolic enzyme.* The primary role of HTS is to detect lead compounds and to supply directions for their optimization. Usually the targets used have originally been employed in benchtop experiments and HTS has been developed from these by automation and design of robotics workstations, thereby permitting performance of a large number of tests in a short time. The development and improvement of HTS has progressed extremely fast. At the beginning of the 1990s a staff of 20 people were able to screen around 1.5 million samples in a year (averaging about 75 000 samples/researcher). Today four researchers using a fully automated robotics technology can screen 50 000 samples a day or around 2.5 million samples a year. The present trend is miniaturization and increasing the number of wells in the microtitre plates used, from the previously common 96-well plates via 384-well and 864-well formats up to 1 536-wells on each plate, systems completely dependent on automated pipetting and handling. Besides enhancing throughput, miniaturization will lower the cost of screening by allowing smaller sample volumes. The great amount of data generated also requires a computerized system for handling. Fully automated workstations – from sample dispensing to data collection – are now available and are designed for round-the-clock operation. Many types of *in vitro* assays can readily be converted to HTS. Even binding activity, such as ligand–receptor interaction or protein–protein interaction can be assessed using proximity-dependent transfer methods. Cell-based assays are also employed.

Further reading

Pharmacopoeias and handbooks

Council of Europe. *The European Pharmacopoeia,* 5th edn, 2004.
Hager's Handbuch der Pharmazeutischen Praxis. 5. Auflage. Bd. 1–10. Springer Verlag, Heidelberg (1990–1995).
Jackson B.P. and Snowdon, D.W. *Powdered Vegetable Drugs. An Atlas of Microscopy for Use in the Identification and Authentication of some Plant Materials Employed as Medicinal Agents.* J. & A. Churchill Ltd, London (1968).

Preparations of crude drugs

Bevan, C.D. and Marshall, P.S. The use of supercritical fluids in the isolation of natural products. *Natural Products Reports,* **11** (5), 451–466 (1994).
Eder, M. and Mehnert, W. Bedeutung pflanzlicher Begleitstoffe in Extrakten. *Pharmazie* **53** (5), 285–293 (1998).

List P.H. and Schmidt, P.C. *Phytopharmaceutical Technology.* Blackwell Science, Oxford (1989).

Wijesekera, R.O.B. *The Medicinal Plant Industry.* CRC Press Inc., Boca Raton, Florida (1991).

Herbal remedies

Barnes, J., Anderson, L.A. and Phillipson, J.D. *Herbal Medicines – A Guide for Health-care Professionals.* 3rd Edn. The Pharmaceutical Press, London (2007).

Blumenthal, M. (Ed.). *Herbal Medicine: The Expanded Commission E Monographs.* American Botanical Council, Austin, Texas (2000).

ESCOP Monographs. The scientific foundation for herbal medicinal products. 2nd Edn. The European Scientific Cooperative on Phytotherapy, London (2003).

Ioannides, C. Topics in xenobiochemistry. Pharmacokinetic interactions between herbal remedies and medicinal drugs. *Xenobiotica*, **32** (6), 451–478 (2002).

WHO Monographs on Selected Medicinal Plants. Vol 1 and 2. World Health Organization, Geneva (1999, 2002).

Wichtl, M. (Ed.) *Herbal Drugs and Phytochemicals*, 3rd Edn. Medpharm GmbH, Scientific publishers, Stuttgart (2004).

Screening for pharmacological activity and isolation of pure compounds

Barros, T.P., Alderton, W.K., Reynolds, H.M., Roach, A.G. and Berghmans, S. Zebrafish: an emerging technology for *in vivo* pharmacological assessment to identify potential safety liabilities in early drug discovery. *British Journal of Pharmacology,* **154**, 1400–1413 (2008).

Broach, J.R. and Thorner, J. High-throughput screening for drug discovery. *Nature,* **384** (Suppl. Nov. 7), 14–16 (1996).

Claeson, P. and Bohlin, L. Some aspects of bioassay methods in natural-product research aimed at drug lead discovery. *Trends in Biotechnology,* **15**, 245–248 (1997).

Hostettmann, K. (Ed.). *Methods in Plant Biochemistry,* **6**, Academic Press (1985).

Hostettmann, K., Marston, A. and Hostettmann, M. *Preparative Chromatography Techniques.* (2nd Edn.), Springer Verlag, Berlin (1998).

Meyer, B.N., Ferrigni, N.R., Putman, J.G., Jacobson, L.B., Nichols, D.G. and Mc Laughlin, J.L. Brine shrimp: a convenient general bioassay for active plant constituents. *Planta Medica,* **45**, 3–34 (1982).

Qiu, J.-W., Hung, O.S. and Qian, P.-Y. An improved barnacle attachment inhibition assay. *Biofouling,* **24** (4), 259–266 (2008).

3. Biotechnological Drug Production

Production of Antibiotics

Antibiotics are drugs used to treat infections caused by bacteria or fungi and are characterized by their ability to kill or inhibit the growth of these without causing harm to the patient. Most antibiotics are natural products, produced by fungi or bacteria. In 1929, *Alexander Fleming* discovered that fungi can produce compounds which are harmful to bacteria. A culture of a *Staphylococcus* species in a forgotten petri dish had become contaminated with a fungus and Fleming noted that around each colony of the fungus there was a zone void of bacteria. The fungus thus appeared to produce and excrete a compound which inhibited the growth of the bacteria. The fungus was identified as a rare strain of a *Penicillium* species, at that time known as *Penicillium notatum* Westling but for which the current scientific name is *Penicillium chrysogenum*. This species was first discovered at The Royal Pharmaceutical Institute in Stockholm, Sweden by a pharmacognosist – Richard Westling – who, however, did not make the observations that Fleming did. Fleming prepared subcultures of *Penicillium chrysogenum* and found that the antibacterial activity was present in a filtrate of the broth in which the fungus had been grown for one to two weeks at room temperature. This filtrate was given the name *penicillin* and the first technical use of the preparation was for killing bacteria interfering with the production of a vaccine against influenza. Fleming's discovery was not pursued further until the Second World War, when there was a pronounced shortage of antibacterial drugs for treatment of infected wounds. Large-scale production of penicillin was hampered by the fact that the original strain of *Penicillium chrysogenum* could only be grown in surface cultures and purification of the crude penicillin also proved to be extremely difficult. The situation improved when it was discovered that a strain *Penicillium chrysogenum*, growing on a cantaloupe melon in the fruit market in Peoria (Illinois, USA), gave a high yield of penicillin and could be grown in deep tanks. The first clinically used preparations of penicillin were far from pure. Intensive efforts at purification and crystallization resulted in the isolation of pure benzylpenicillin (also called penicillin G, page 561) the chemical structure of which was determined by X-ray crystallography in 1945. Benzylpenicillin turned out to be an excellent drug for the treatment of infections. This success initiated a search for other compounds with an antibiotic effect. At present, more than 6 000 such compounds have been isolated and about 90–100 of these are produced commercially for clinical use. One problem in therapy with antibiotics is that prolonged use can cause the infecting bacteria to develop resistance to the drug, which then becomes ineffective. There is thus a constant need for new antibiotics to replace those which cannot be used due to the formation of resistant bacterial strains. Much research is still going on with the aim of finding new antibiotics, not only from fungi and bacteria, but also from a number of marine organisms and higher plants.

Structurally, antibiotics are a very heterogeneous group of natural products. They may be carbohydrates, macrocyclic lactones of acetate origin, quinones, amino acids, or peptides, as well as steroids, etc. In this book we shall treat the various antibiotics in the chapters which deal with their dominant biosynthetic pathway, but as their structures are often very complicated and derive from a combination of several pathways, their classification is sometimes a matter of opinion.

Most antibiotics are produced by cultivation of the producing organisms under controlled conditions in closed vessels called *fermenters or bioreactors*. These are often of considerable size and bioreactors with a volume of 200 000 litres are used in the pharmaceutical industry for penicillin production.

The growth curve of microorganisms

A culture of bacteria which are transferred to a new medium and allowed to grow under stationary conditions develops as illustrated by the graph in Fig. 13.

Four different phases, A–D, are marked on the graph. Phase A is the *lag phase*, during which the bacteria adapt themselves to the new medium. There is no change in the number of individuals in this phase. The duration of the lag phase depends on the conditions in the culture used for inoculation. If the bacteria are in their growth phase (phase B) when transferred to the new medium and it does not differ too much from the old one, phase A may be very short or totally absent so that the culture starts with phase B. If, on the other hand, the bacteria in the mother culture had stopped growing due to lack of substrate, the lag phase may be fairly long.

Phase B is called the *log phase* in which the microorganisms grow logarithmically, causing a doubling per unit of time of their biomass. As long as there is an excess of substrate, the growth rate is maximal and independent of the concentration of the substrate.

In phase C – *the stationary phase* – the substrate has been used up, the growth rate is zero, but the number of organisms remains constant.

Phase D is the *death phase* in which the organisms die due to lack of energy and the curve falls.

There can sometimes be more than one log phase on the graph. This is caused by the presence in the medium of more than one substrate which can be used by the organism. The substrate which is

Log. number of organisms

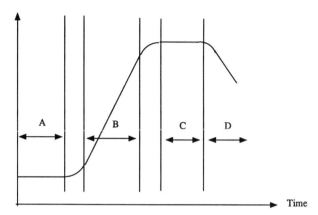

Fig. 13. The growth curve of bacteria. Fungi do not always follow this type of curve

the easiest to use is consumed first, bringing about the first log phase. When this substrate is finished, the curve falls, but after a while, when the organism has adapted itself to the second substrate, the curve rises again to form a second log phase.

The slope of the curve during the log phase indicates the specific growth rate (μ), which is a function of three parameters – the concentration of the limiting substrate, S, the *maximum specific growth rate*, μ_m, and a substrate-specific constant, K_s, in accordance with the Monod equation:

$$\mu = \mu_m \frac{S}{K_s + S}$$

named after *Jacques Monod* who derived it. If there is an excess of substrate, μ becomes equal to μ_m. The constant K_s is the substrate concentration at which half the maximum specific growth rate is obtained.

The maximum specific growth rate is an important factor in the productivity of an industrial process. The higher the value of μ_m, the higher the productivity. The value of μ_m depends both on the properties of the organism and on the conditions of the fermentation process, including the temperature or the substrate. The substrate is also of importance. A simple substrate, e.g. glucose, gives a higher μ_m than a complicated substrate such as a polysaccharide. In the latter case, the microorganism must first degrade the polysaccharide to glucose or another simple sugar, before it can be used as the substrate. An example is given in Table 2 below, which shows values of μ_m for the fungus *Fusarium gramineum* grown with three different substrates.[1]

---

[1] Andersson, C., Longton J., Maddix C., Scammel G.W. and Solomons G.L.. The growth of microfungi on carbohydrates, pp. 314-329. In Tannenbaum, S.R. and D.I.C. Wang (eds.), *Single cell protein*, Vol. II. MIT Press, Cambridge, PA, 1975.

Table 2. Maximum specific growth rates for three substrates

Substrate	Number of glucose units	μ_m/h	Time (h) for doubling the number of microorganisms
Glucose	1	0.28	2.48
Maltose	2	0.22	3.15
Maltotriose	3	0.18	3.85

Energy metabolism and production of metabolites in microorganisms

Metabolites, e.g. antibiotics, can be produced in three different types of process that can be distinguished as follows:

Type 1 (Fig. 14A) The desired product is formed simultaneously with primary metabolites and from the same substrate. Growth, carbohydrate consumption, and product formation follow parallel curves. The desired product is formed either directly from the substrate (A → product) or via intermediates (A → B → C → product). An example of a type 1 process is the fermentation of sugar to ethanol.

Type 2 (Fig. 14B) The same substrate is used for formation of the product desired and for primary metabolism, but the product is formed via a different metabolic pathway:
In this type of process, the graph shows two maxima. At first there is consumption of carbohydrate, accompanied by

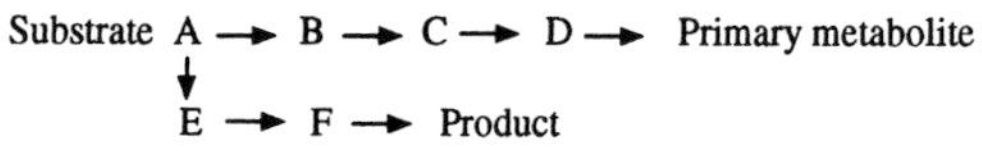

Substrate A → B → C → D → Primary metabolite
↓
E → F → Product

growth, but no product is formed. Product formation starts later, when growth has decreased, and is accompanied by an increase in carbohydrate consumption. The production of citric acid and of some amino acids by fermentation are examples of type 2 processes.

Type 3 (Fig. 14C). Antibiotics and vitamins are usually formed in fermentation processes of this type. Here, the process starts with growth of the organism and formation of primary metabolites, using mainly carbohydrate as the substrate. Only when most of the carbohydrate has been consumed, and the growth rate has decreased, does the formation of the desired product start. This is not formed from carbohydrate, but from a different substrate.

To obtain the maximum yield in a fermentation process, it is thus very important to know to which type the fermentation process belongs and to monitor both the consumption of carbohydrate and the formation of the desired product.

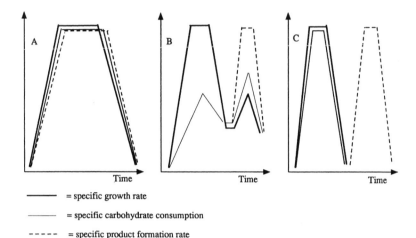

———— = specific growth rate

———— = specific carbohydrate consumption

- - - - = specific product formation rate

Fig. 14. Primary metabolism and production of desired secondary metabolites in different types of fermentation processes

Technical aspects of the production of antibiotics

The production of antibiotics involves the following four steps:
1. Development and storage of the inoculum.
2. Preparation of the preculture.
3. Production fermentation.
4. Isolation of the antibiotic produced.

The inoculum

Following the discovery of a clinically useful antibiotic much research is needed to optimize the formation of the compound by the organism before an industrial process for production of the antibiotic can be developed. This is comparable to the breeding of medicinal plants for the highest yield of a substance to be used as a drug (see page 52). The methods are comparable, e.g. selection and formation of mutants by irradiation or treatment with chemicals. The increase in yield obtained by such development programmes can be very high. An example is the formation of penicillin from *Penicillium notatum*. The original strain of this fungus produced 1.2 mg of penicillin/litre of fermentation medium, while the strains of *Penicillium chrysogenum* used industrially today yield 50 g/litre of medium.

A high-yielding strain of an organism is a very valuable asset for

93

an industrial company involved in the production of antibiotics. It is necessary to store it so that it does not become mutated or modified and thus changed into a low-producing variety. A number of sub-cultures are therefore produced, which are stored under different conditions depending on the stage in the production at which they are to be used. Three different methods for storage are available:

1. *Freeze-drying* (lyophilization). Protective substances like skimmed milk or sucrose are added to the culture before freezing, to prevent formation of ice crystals in the cells. Freeze-drying is the safest method for storing bacteria or fungi. A freeze-dried culture, stored at low temperature, can be kept almost indefinitely.

2. *Deep-freezing* (–18 °C to –80 °C) or storage in liquid nitrogen (–196 °C) is the most common method. A deep-frozen culture keeps for several years. Some 95 % of the organisms die, but the remaining 5 % are sufficient to start a new culture.

3. *Storage in a refrigerator* (+2 °C to +6 °C) is not as good. The culture has to be re-inoculated every 8–16 weeks and the risk of contamination with other organisms is considerable.

Preparation of the preculture

Preparation of precultures starts on a small scale. The stored inoculum is transferred to a suitable medium in Erlenmeyer flasks, which are shaken continuously. With a freeze-dried inoculum, growth starts 4–10 days after the transfer. Deep-frozen bacteria need 4–48 hours to start growth, but deep-frozen fungi need 1–7 days. For cultures stored in refrigerators growth commences within minutes.

The culture is then transferred from the Erlenmeyer flasks to a small fermenter with a volume between 200 and 6 000 litres, in order to prepare an inoculum for the production fermentation which proceeds in bioreactors with volumes of up to 200 m³.

Production fermentation

This is a tricky process which needs strict control of many parameters. Medium and apparatus must be sterile before the process can be started. Sterilization is usually performed with heat. Heat-sensitive components of the medium are dissolved separately, sterile-filtered and added to the sterilized medium. Heat sterilization is expensive in energy. A continuous two-step process is usually employed, in which the cold non-sterile solution is preheated by the hot, sterile solution in a heat-exchanger. In this way, about 90 % of the energy required for batch-wise sterilization can be saved. The demand for sterile conditions is also reflected in the construction of the fermenter. It should have as few openings as possible. All parts must be accessible for sterilization by steam and gaskets for stirrers need to be so constructed that contaminants cannot enter from the outside.

It is also necessary to provide for an effective exchange of gases during fermentation. Oxygen must be supplied and the carbon dioxide formed during the fermentation must be able to escape. This is effected by stirrers, which also help to dissipate the heat formed during the fermentation process. The oxygen is supplied as sterile air and large quantities are needed. A fermenter with a volume of 50 m³ needs 3 000 m³ of air per hour. Filtering the air through glass wool, packed in big, replaceable cartridges, is the most common method for sterilizing air.

The composition of the medium must be adjusted to the requirements of the organism used. A source of carbohydrate, e.g. glucose, sucrose, starch or dextrin, is needed. For the production of penicillin, lactose is used, because glucose acts as an inhibitor of the process. Another general requirement is a source of nitrogen. Soyameal, yeast extract or corn-steep liquor are frequently used. Buffer salts are added to keep the pH at a suitable level.

The composition of the medium is crucial to a successful fermentation and is therefore a well-kept secret of each company.

Isolation of the antibiotic produced

The fermentation is terminated when an optimal quantity of antibiotic has been produced (see page 92). The desired substance is present either inside the producing organism or in the medium. The first step in the isolation process is therefore separation of the medium and the organism. This is done by filtration or centrifugation. Filtration is performed continuously in big drum filters, if necessary with the addition of filter aids such as kieselguhr.

If the antibiotic is present in the medium, the next step is concentration. For an electrically charged compound, an ion-exchanger can be used. Other methods involve extraction with organic solvents, selective adsorption on solid media, etc. The problems are of the same kind as those encountered in the isolation of drugs from plants. Chromatographic separations are often required for the final stages. As in all industrial processes, the aim is to find a process which gives the highest yield at the lowest cost.

If the antibiotic is located inside the organisms, these must first be disintegrated. This can be done with physical, chemical or biological methods. Physical methods are grinding, pressing or treatment with ultrasonic waves, an efficient but expensive method. Chemical methods involve osmosis or treatment with acid, acetone or toluene. The most important biological method is enzymic degradation of the cell wall. Following disintegration of the organisms, the antibiotic can be extracted as described above for antibiotics present in the medium.

Plant Tissue and Cell Culture (Plant Biotechnology)

Cultivation of a plant tissue on an artificial medium was first described in 1939 and major work in the field started in the 1950s. Today, plant tissue and cell culture is an important technique in several areas, such as the commercial production of ornamental plants, plant breeding and studies in the biosynthesis of secondary metabolites. It is also of interest for the industrial production of plant-derived natural substances, including drugs.

The basis for plant tissue and cell culture is that each plant cell, irrespective of the organ from which it derives, contains all the genetic information relating to the whole plant. It is thus possible to start with a single cell, which is allowed to multiply by division and form a tissue of loosely attached cells called a *callus*. These cells can be kept growing on a semi-solid substrate or they can be suspended in a liquid medium, forming a *suspension culture*. By addition of suitable hormones and growth substances, the callus cells can be forced to differentiate into organs such as roots, stems, leaves, etc. and eventually into whole plants.

Callus cultures

As mentioned above, a callus culture can be started from a single cell. In practice, however, a callus culture is usually started with a small piece of plant tissue derived from a part of the plant where cell division is in progress. Suitable parts are: tips of stems, roots or leaves; cotyledons from very young plants; tissue from the ovary or stamens; and parts of the cambial zone of roots or stems. The plant part selected to establish a culture is termed the *explant*. It is necessary to prevent infection of the callus tissue by microorganisms. The surface of the explant must therefore be sterilized, but the process must be mild enough not to harm the plant cells. A common technique is to wash the explant with water and a mild soap solution, after which it is transferred to a solution of sodium or calcium hypochlorite where it is left for between 5 and 30 minutes. Addition of a wetting agent facilitates penetration of the sterilizing liquid into hairy areas or other difficult-to-reach parts. From then on, all operations must be performed in a sterile environment, e.g. in a laminar-flow transfer cabinet. After sterilization, the plant part is rinsed with sterile water and aseptically transferred to the sterile growth medium. This is usually an agar gel in a petri dish or a flask. Plant cells in a callus culture are seldom autotrophic. The medium must therefore contain not only salts, trace elements and nitrogen, but must also have an organic carbon source, as well as vitamins, which cannot be synthesized by the callus tissue. In 1962, T. Murashige and F. Skoog published the composition of a growth medium (Table 3) which has since become very widely used as a quasi-universal medium in most laboratories. It is usually referred to as *the Murashige–Skoog medium* (the MS medium). The compositions of several other media have also been published and they may in special cases give better results.

The salt concentration in this medium is comparatively high. In special cases, the concentration of some of the components (particularly NH_4^+) may have to be reduced. Sucrose is the most commonly used carbon source. Besides ammonium nitrate, other sources of nitrogen are usually added, such as a protein hydrolysate, e.g. enzymic casein hydrolysate or pure amino acids such as L-arginine, L-aspartic acid, L-asparagine, glycine, L-glutamic acid or L-glutamine. Of the vitamins added, only thiamine seems to be a consistent requirement for the *in vitro* growth of plant tissues.

Table 3. The composition of the Murashige–Skoog medium[1]

A. Mineral salts	Concentration (mg/l)
NH_4NO_3	1 650
KNO_3	1 900
$CaCl_2 \cdot 2\ H_2O$	440
$MgSO_4 \cdot 7\ H_2O$	370
KH_2PO_4	170
Na_2EDTA	37.3
$FeSO_4 \cdot 7\ H_2O$	27.8
H_3BO_3	6.2
$MnSO_4 \cdot 4\ H_2O$	22.3
$ZnSO_4 \cdot 4\ H_2O$	8.6
KI	0.83
$Na_2MoO_4 \cdot 2\ H_2O$	0.25
$CuSO_4 \cdot 5\ H_2O$	0.025
$CoCl_2 \cdot 6\ H_2O$	0.025

B. Organic constituents	Concentration (mg/l)
Saccharose	30 000
Edamin[2]	1 000
Glycine	2
Indole-3-acetic acid	1–30
Kinetin	0.04–10
Agar	10 000
myo-Inositol	100
Nicotinic acid	0.5
Pyridoxine.HCl	0.5
Thiamine.HCl	0.1

The pH of the medium is adjusted to 5.7–5.8 by addition of HCl or NaOH.

---

[1]T. Murashige and F. Skoog. (1962) A revised medium for rapid growth and bio assays with tobacco tissue cultures. *Physiologia Plantarum* **15**, 473–497.
[2]A pancreatic digest of lactalbumin (Sheffield Chemical Co., Norwich, N.Y., USA).

Several classes of compounds regulate growth and development in intact plants. In tissue culture work, the two most important groups of growth regulators are auxins and cytokinins. Auxins are organic compounds which stimulate the lengthwise growth of stems and have the opposite effect on roots. Cytokinins promote cell division.

IAA

The natural auxin is indole-3-acetic acid (IAA), but several synthetic compounds have been found to be more active and two of them are frequently used in tissue culture work: 2,4-dichlorophenoxyacetic acid (2,4-D) and α-naphthaleneacetic acid (NAA).The archetype cytokinin is N^6-furfuryladenine (kinetin). Other naturally occurring cytokinins are N^6-(Δ^2– isopentenyl)-adenine (2iP) and N^6-trans-Δ^2-hydroxymethyl-γ-methylallyladenine (zeatin).

2,4-D

The concentrations of auxins and cytokinins in the medium are very important for the growth and development of the cells. By varying these concentrations, unorganized growth, growth plus formation of roots, or growth plus formation of shoots can be achieved.

Some time after the transfer of the explant to the medium, an undifferentiated tissue begins to form round the starting material, the time required varying from some weeks to six months. The tissue is transferred to fresh agar gel or to a liquid medium if a suspension culture is desired. 2,4-D is added if the cells are to remain as a callus. To keep a callus culture alive, it is necessary to transfer the cells to fresh medium at regular intervals.

NAA

The callus is grown on a semi-solid substrate if it is to be stored for future use or if it is to be used for the regeneration of organs or new plants. Callus cultures are usually stored in the dark, as exposure to light for prolonged time can induce photosynthesis and subsequent differentiation. For the production of secondary metabolites, a suspension culture is generally used.

Suspension cultures

In a callus culture, where the cells are in contact with one another, they form an unorganized tissue. In a suspension culture the cells are free in the medium or form small aggregates. To grow plant cells as a suspension culture, the same medium can be used as for a callus culture, the only difference being that the agar is omitted. Suspension cultures are usually started from callus cultures, but they can also be established directly from sterilized explants. The friability of the callus culture determines how easy it is to start a suspension culture from it. If it is very friable, it is enough to shake the callus in the culture medium. Increasing the concentration of auxin in the medium can result in increased friability. Sometimes, the callus tissue needs a period of time to adapt to the liquid medium before it becomes friable enough.

Kinetin

Organ cultures and regeneration of plants

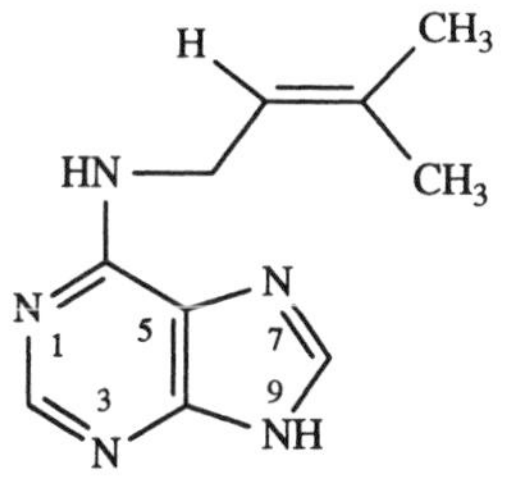

Plant organs and even whole plants can be regenerated from callus cultures. Which organ is developed depends on the balance between the concentration of auxin and cytokinin. The salt concentration in the medium and the light conditions are also of importance.

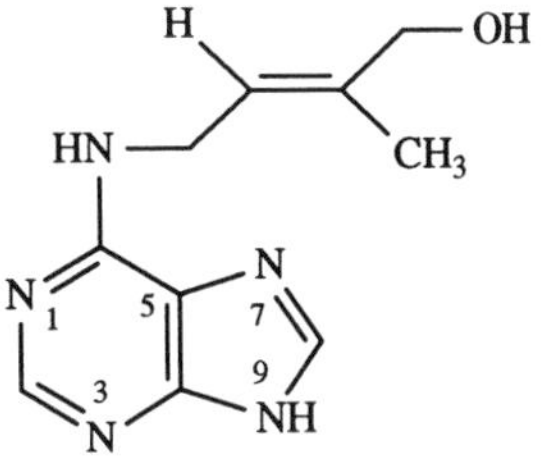

Shoots are regenerated in the presence of light and in a medium of high osmolarity containing chelating agents. An exogenous supply of carbohydrates is usually necessary. The auxin level should be lower than the cytokinin level.

Roots are formed instead of shoots if the auxin concentration is increased relative to the cytokinin concentration.

New plants can be formed from a shoot culture by reduction of the concentration of sucrose and salt. This causes induction of *adventitious roots*. Addition of auxin, as well as increasing the light intensity, is also beneficial for rooting. When the roots are well developed the new plant can be transferred to sterile soil and allowed to continue its development in the normal way.

Another method of regenerating plants from a callus culture is *somatic embryogenesis*. The formation of somatic embryos is accomplished by transferring callus from a medium containing an auxin to a medium lacking auxin. Addition of anti-auxin may increase the number of embryos formed. Somatic embryos are morphologically indistinguishable from sexual embryos.

2iP

Zeatin

Environmental factors of importance for plant tissue and cell culture

Temperature

The optimum temperature for *in vitro* growth of plant cells is in the range 25–30° C, but there can be considerable differences between plant species.

Aeration

Plant cells in a culture are aerobic and must therefore have access to oxygen in order to grow properly. A callus culture is usually grown on a semi-solid medium in a petri dish or a flask. The container must be closed to prevent infection by microorganisms, but it must permit exchange of gases with the outside environment. This is no problem for cultures in petri dishes. In flask cultures, the flask is usually closed with a plug of cotton wool or some other material, e.g. parafilm, which permits exchange of gases but prevents microorganisms from entering the flask.

For suspension cultures, aeration presents a more difficult problem. On a laboratory scale, the culture is usually contained in Erlenmeyer flasks which are placed on a shaker, which keeps the contents moving and facilitates the uptake of oxygen. Aeration can also be performed by blowing air into the suspension. For large-scale cultures, specially designed bioreactors are used. The volume of such bioreactors can be in the range 5l to 75 m^3. These bioreactors are basically constructed in the same way as fermenters for the production of antibiotics from microorganisms (page 94), but aeration presents a more difficult problem. In fermenters for the production of antibiotics, aeration is usually accomplished by rotating turbines with blades of varying design, operating at a comparatively high speed. These stirrers produce a high shear, which the small microorganisms can withstand, but which disrupts the bigger and softer plant cells. Stirrers for plant cell bioreactors must therefore be designed so that they produce as low a shear as possible. Turbine stirrers must therefore be run at low speed and the paddles must be designed to give low shear.

In *bubble bioreactors* stirring and aeration are combined by passing a stream of sterile air through the medium. In *air-lift bioreactors* the air rises in a special draft tube in the bioreactor, thereby causing a lifting action which results in thorough mixing under conditions of low shear.

Light

Visible light is defined as electromagnetic radiation with wavelengths between 390 and 760 nm. Although plant cells in tissue culture normally do not carry out photosynthesis, they are dependent on light for their growth and development. It is therefore necessary to control the light conditions with respect to intensity, wavelength and exposure time. Intensity can be defined as *irradiance,* i.e. W/m^2 or *illuminance,* i.e. lux. Measurement of irradiance is not spectrally defined but illuminance is a photometric term indicating the level of visible light as the human eye would see it. Plant cells in tissue culture are not sensitive to light in the same manner as the human eye and light should therefore be measured as irradiance at defined wavelengths rather than as illuminance. Two light sources which give the same illuminance can differ very much in irradiance measured as $W/m^2/(nm)$. The irradiance should be measured inside the container in which the culture is situated, as much of the light reaching the container will be lost by absorption or reflection. In order to enable experiments to be reproduced in other laboratories, if irradiance is not measured, it is necessary to give an exact description of the light source and its geometric position in relation to the plant material, as well as a description of the culture and the bioreactor.

The irradiance needed to allow plant tissue or cell cultures to develop is much lower than that needed by a photosynthesizing plant. Too high levels can be detrimental to the culture. The best light source for tissue cultures is the fluorescent lamp, of which there are many types with different spectral emission characteristics. Controlling the light conditions is easy for callus cultures in petri dishes, but for suspension cultures in shaking flasks or in bioreactors this presents severe technical problems. As the culture grows, the cell mass absorbs so much of the light that only the layer closest to the wall of the container receives enough light.

The influence of light on plant cultures has been studied at the cellular, tissue and organ levels. Thus, callus growth and shoot initiation can be either enhanced or inhibited depending on the wavelength and irradiance. High near-UV and blue light have been found to inhibit the growth of callus cultures. Red light can cause enhancement or inhibition of callus growth depending on the species. Similar effects have also been found with suspension cultures. Light also influences organogenesis in organ cultures. Both enhancement and inhibition have been observed depending on wavelength and species.

Light is also important for the production of metabolites by plant tissue and cell cultures, and so are irradiance, wavelength and exposure time. Thus, comparison of a *Digitalis lanata* Erh. culture, grown in complete darkness, with another which has been exposed to light for 2–3 periods of 15 minutes each, showed that the cardenolide content of the exposed culture was about 10 times higher than that of the culture which had been kept in darkness all the time.

Tissue and cell culture in plant breeding

Somaclonal variation

The walls of the cells growing in a callus or suspension culture can be dissolved by treatment with enzymes such as cellulases and pectinases. The protoplasts formed are then dispersed on an agar plate, where each protoplast regenerates the cell wall and divides to form a colony of cells. These colonies can then be transferred to other media on which the cells continue to grow and on which they can be developed into organs or new plants as described above. The cell lines deriving from the different protoplasts vary considerably in their biological properties. This phenomenon is called *somaclonal variation* and is mainly a result of the mutations continuously occurring in all living cells. Somaclonal variation is used in plant breeding for selecting desired properties in plants, such as resistance to certain diseases or herbicides, or increased production of various secondary metabolites. It is also possible to fuse protoplasts to form new cells, a technique which increases the chances of new genetic combinations. New cells can also be grown from parts of protoplasts, e.g. fusion of the nucleus of one protoplast with cytoplasm and organelles from another.

Micropropagation of plants

Plants with special properties resulting from various techniques of plant breeding must often not be allowed to multiply sexually, as this may cause loss of the desired properties in a part of the offspring. Such plants must be propagated by vegetative techniques in order to preserve the genetic characteristics in the progeny (see page 51). Micropropagation is a technique for vegetative propagation involving the production of callus cultures from explants taken from meristematic zones of the parent plant. New plants are then formed from the callus cultures by the techniques described above. This method is much used for the production of ornamental and other commercially important plants. A new method is the production of so-called synthetic seeds, formed by enclosing somatic embryos in a coat of alginate, containing nutrients, pesticides, hormones and other factors which will enable the new plant to develop rapidly when the propagule is planted.

Transgenic plants

Organisms containing genes not belonging to their normal genome are known as transgenic organisms. Foreign genes can be introduced by several vectors, e.g. into dicotyledonous plant cells with the aid of soil bacteria of the genus *Agrobacterium*, plant pathogens which cause plant diseases commonly known as crown gall tumour and hairy roots. *Plasmids* are circular, double-stranded DNA molecules which are separated from the chromosomal DNA. In these bacteria the plasmids are divided into three regions called the *Vir* region, the T_L region and the T_R region. Plant cells with a low resistance (due to wounding, drought, etc.) are infected by the bacteria. Following the infection, the DNA contained in one or both of the two *T* regions of the plasmid is transferred from the bacterium and incorporated into the nuclear DNA of the host plant. The incorporation is a very complex process mediated by a number of *Vir* proteins resulting from expression of genetic information contained in the *Vir* region of the plasmid. Expression of the genetic information in the incorporated *T* region causes morphological changes in the plant on the site of the infection. Plasmids from *Agrobacterium tumefaciens* cause development of crown gall tumours and are called *Ti-plasmids* (Tumour-inducing plasmids). Plants infected with *Agrobacterium rhizogenes* develop a large number of lateral roots (a phenomenon known as hairy roots) as a result of expression of the genetic information in *Ri-plasmids* (Root-inducing plasmids). Beside morphological changes, synthesis of novel metabolites – *opines* – is also induced in the infected plant cells. The opines are essential for the bacteria as nutrients but they are not known to have any importance for the plant cells. There is, however, some evidence that they might affect the production of ordinary metabolites of the plant. The opines are derivatives of amino acids and are of importance for classification of the plasmids contained in various strains of *Agrobacterium*. Thus Ri-plasmids in *Agrobacterium rhizogenes* can be grouped into two main classes: agropine- and mannopine-type strains, depending on which type of opine they cause their host plant to produce.

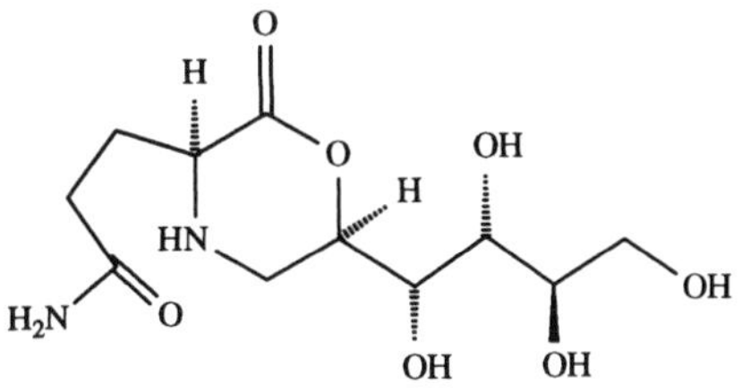

Agropine

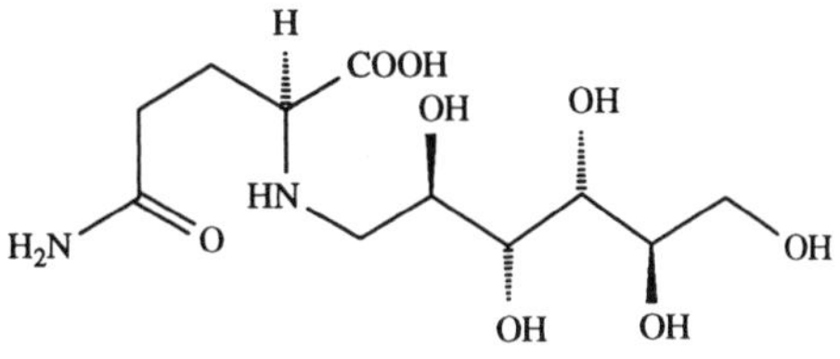

Mannopine

Several other opines have been found in the two classes. Octopine and nopaline are representatives of corresponding opine families produced by expression of the genetic information in plasmids from *Agrobacterium tumefaciens*:

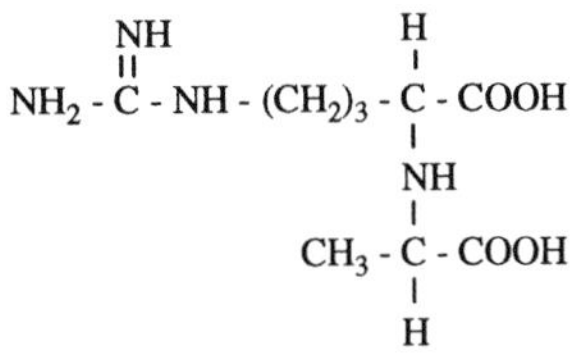

Octopine

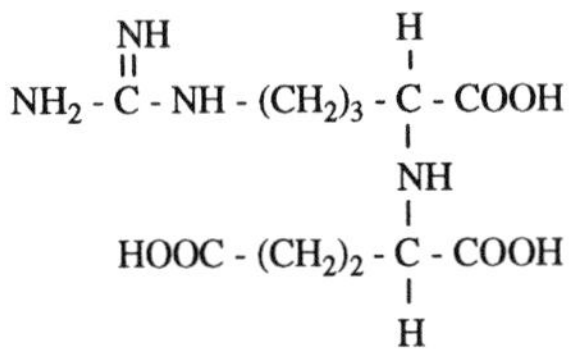

Nopaline

Ti-plasmids from *Agrobacterium tumefaciens* were the first tool to be used for introduction of foreign genes in plants. These plasmids can be isolated, and by recombinant DNA techniques a foreign gene can be introduced within the *T* region of the plasmid. The manipulated plasmid is then re-introduced into *Agrobacterium tumefaciens* and a culture of the transformed bacteria is produced which is used to infect a plant resulting in development of crown galls. Plants can be regenerated from the tumour tissue and can now produce the metabolite corresponding to the genetic information introduced via the plasmid. Regeneration of plants from the crown gall tissue is, however, difficult and hairy roots have to a great extent superseded crown galls for development of transgenic plants.

The hairy roots, which, as described above, develop in dicotyledonous plants at the site of infection with *Agrobacterium rhizogenes,* have proved to be very useful tools for genetic experiments. They can be grown in artificial media in organ cultures and can be used for production of secondary metabolites on a large scale (page 108). The Ri-plasmids can be transformed with foreign genes in the same way as the Ti-plasmids and bacteria containing the transformed plasmids can be used to infect plants which then develop hairy roots. Regeneration of plants from these hairy roots is much easier than regeneration from crown galls.

Using the techniques of cell and tissue culture, tissues of plants (e.g. small leaf discs) can be grown in the presence of *Agro-*

bacterium spp. containing manipulated plasmids, which introduce the desired genetic information into the cells of the tissues. Transgenic plants can then be regenerated from the tissues with conventional techniques.

The above described techniques for introduction of foreign genes in plants are suitable for dicotyledons. For monocotyledons other techniques have to be used, such as bombardment with DNA-coated tungsten microprojectiles (0.6 μm diameter) which are shot into plant cells with a special particle gun. The technique permits introduction of genes into a wide range of tissues including suspension and callus cultures, tissues isolated directly from plants or even tissues of whole seedlings.

The development of methods to transfer foreign genes into the plant genome has attracted great interest for the improvement of commercially important crop species. Thus many attempts have been made to improve the resistance to microorganisms, pests or herbicides, to increase the biomass and grain size or to generally improve the quality of the plants.

The potential for the use of transgenic plants for production of drugs is also very large. Not only can the yield of metabolites normally produced by a medicinal plant be increased, but the plant can also be made to produce completely new metabolites.

Virtually any natural compound can be produced by a suitable plant provided that full knowledge of the genetic machinery involved in its biosynthesis is available. Thus methods have been developed for production of antibodies and vaccines by plants, e.g. in *Nicotiana tabacum.* An important advantage of plants over other recombinant expression systems for production of antibodies is the ability to assemble full-length heavy chains with light chains to form full-length antibodies efficiently. Full-length antibodies are not readily assembled in bacterial expression systems, and bivalent antibody molecules can only be produced in *Escherichia coli* by complex molecular engineering.

Other advantages of production of antibodies by plants are the ease of storage and distribution of transgenic lines and the fact that the process of growing the recombinant protein does not require skilled labour. Consequently, the potential is high for protein production on an agricultural scale at extremely modest cost. Economic production of kilogram quantities would open many new areas for use of antibodies in industry and for medical purposes. Hitherto, monoclonal antibodies have been used against human diseases with varying degrees of success and a consistent problem arises from the need to administer sufficient amounts of the antibody to overcome the rapid rate of clearance from the body. This can be overcome by production of antibodies in plants. It has been estimated that by expressing antibody in soybean (*Glycine max* Merrill.), at a level of 1% of total plant protein, 1 kg of antibody could be produced at an approximate cost of US$ 100, a cost which certainly can be further reduced with development of improved vectors and purification procedures. The expression of therapeutic antibodies in edible plants or of targeting antibody expression to seeds, tubers or fruit for oral delivery is another promising concept. Antibody delivery through foodstuffs has already been successful in the oral cavity and it has also been demonstrated that recombinant antigens could be delivered to the gastrointestinal tract via plant material, inducing an immune response. The production of antibodies in plants could also be of great importance for the Third World where existing agricultural infrastructure could be used rather than building new pharmaceutical factories

Transgenic plants can also be used for the production of vaccines. This is of particular interest to developing countries as the vaccines can be produced in edible plants. In developing countries the use of ordinary vaccines involves several problems, such as high cost and the need for refrigeration installations. Access to vaccines contained in edible plant parts, such as potatoes (*Solanum tuberosum* L.) or bananas (*Musa paradisica* L.) would overcome many of these problems. Enterotoxic *Escherichia coli* causes acute watery diarrhoea by colonizing the small intestine and producing one or more heat-labile multimeric enterotoxins. Heat labile enterotoxin B (LT-B) has been produced in tobacco (*Nicotiana tabacum* L.) and in potato tubers. In clinical experiments, feeding transgenic potato expressing the LT-B subunit to humans resulted in development of mucosal and systemic immune responses against LT-B. Another important target for current vaccine efforts is hepatitis B, a virus that can cause chronic liver disease. Recombinant vaccines produced in yeast are commercially available and potatoes that express hepatitis B surface antigen have been developed. Current efforts are directed against expressing this antigen in bananas, which have the advantage that they can be eaten raw. Banana vaccines, delivered as a purée, would cost only a small fraction of the price of traditional vaccines. In the future banana vaccines could be produced against a range of diseases, including measles, polio, diphtheria, yellow fever and certain types of viral diarrhoea.

Industrial production of natural products by plant tissue and cell cultures

Plant-derived, pharmacologically active, natural products are in most cases obtained by extraction from crude drugs. From an industrial point of view, the production of natural products by cultivating plant cells in fermenters (see the production of antibiotics from microorganisms, page 90), instead of from crude drugs, would offer several advantages such as: independence of climatic factors and political circumstances, continuous supply of material for extraction, and the possibility of regulating the procedure so that the desired compound is obtained as the principal metabolite. Although the first reports on growing plant cells on an artificial medium were published as early as 1939, only a few substances are currently produced industrially from plant cell and tissue cultures. Examples are: ergot alkaloids from the fungus *Claviceps purpurea* (see page 679), the alkaloid berberine from *Coptis orientalis* (page 645), and the red pigment shikonin from *Lithospermum erythrorhizon* (*Boraginaceae*). Berberine is used for the treatment of diarrhoea and shikonin is a remedy for skin diseases and is also an ingredient in lipsticks. Both compounds are produced and used mainly in Japan. Sanguinarine, rosmarinic acid and *Ginseng* saponines are other compounds which can be produced by plant cell and tissue cultures. Procedures for the commercial production of paclitaxel (Taxol®) and scopolamine by this technique are in advanced stages of development and may soon be employed for the manufacture of these important drugs.

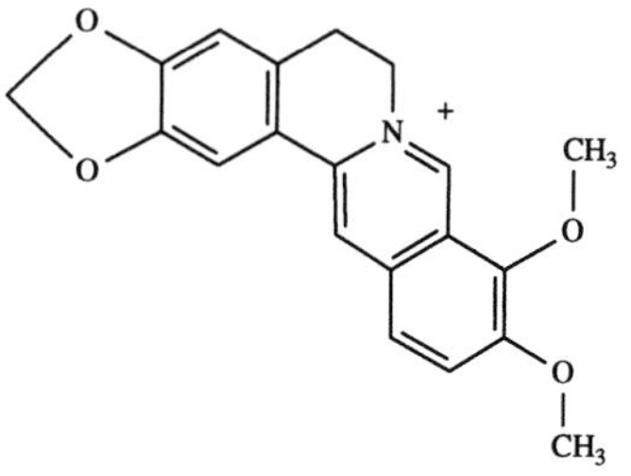

Berberine

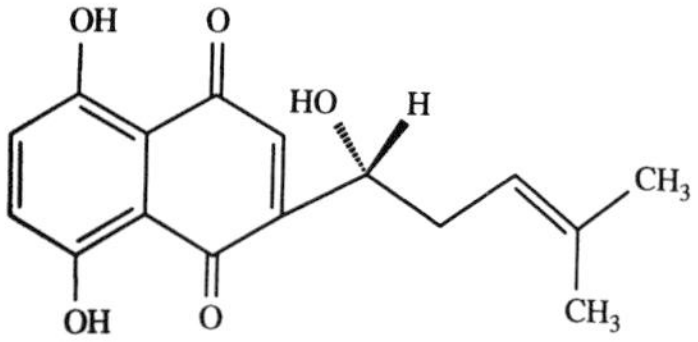

Shikonin

That plant cell and tissue cultures have not yet become of importance for the production of pharmacologically active natural products is mainly due to the low yields of the desired substances. The reason for the low concentrations of secondary metabolites in cultures, compared with the concentrations of the same compounds in fully developed plants, is not completely understood. Production of secondary metabolites seems to be associated with the development of organs (root, stem, leaf, etc.), but for an industrial process the development of organs is not convenient. It is technically much easier to handle an undifferentiated mass of cells than to grow organs or whole plants on an artificial medium. A breakthrough in this field will depend on basic research in molecular biology to clarify how the plants regulate the formation of secondary metabolites and how this is connected with the development of organs. It might then be possible to separate these two functions and to find the conditions necessary for the production of high concentrations of secondary metabolites by undifferentiated masses of plant cells. Much work in this field is being done in many laboratories throughout the world and a solution of the problem can be expected in the not too distant future.

Methods for increasing the production of secondary metabolites

The methods that are used today for increasing the production of secondary metabolites are: optimization of media and environmental factors, selection of high-yielding cell lines, elicitation and cultivation of organs, mainly hairy roots.

Optimization of media and environmental factors

The composition of the medium, with respect to both salts and growth factors, has a great influence on the production of secondary metabolites. Thus, successive additions of glucose to the substrate can increase accumulation of secondary metabolites as compared with using a fixed concentration. Varying the amounts of trace elements, such as Mn^{2+}, can also cause considerable changes in the production of secondary metabolites. Because so many different factors are involved, optimization of the medium is a difficult and time-consuming task.

Light conditions are also of great importance. The difference in cardenolide production by *Digitalis lanata* Erh. cells which were kept in the dark compared with that of cells exposed to short intervals of light, has been mentioned above.

Selection of high-yielding cell lines

This method is essentially the same technique as is used in traditional plant breeding, where a great number of plants are analysed for the desired metabolite and those plants which have the highest content are selected for propagation (see page 52). A tissue culture should therefore be started from an explant of a high-yielding variety of the plant. The yield can then be further improved by selection of cell lines originating from individual protoplasts as described above, taking advantage of the somaclonal variation. Selection requires the availability of a rapid and sensitive analytical method for the desired metabolite. RIA (radioimmunoassay) and ELISA (enzyme-linked immunosorbent assay) are often suitable, but can sometimes give erroneous results due to cross-reactivity for compounds produced by the cell culture but not normally present in the whole plant. HPLC (high-performance liquid chromatography) is more specific, but is also more time-consuming, and it usually requires more material for the analysis.

One difficulty which is often encountered in experiments to increase the production of secondary metabolites by selection, is that the high-yielding cell line may not be stable. After several transfers to fresh media, the production of metabolites decreases until it is often no higher than average.

Elicitation

Infection of plants by microorganisms often induces production of metabolites which are not normally present. These substances – called phytoalexins – are harmful to the infecting organism and are part of the plant defence system. When the microorganism attacks the plant, certain compounds are released from the cell walls of the microorganism or of the plant. These compounds, which have been termed *elicitors,* activate the genes in the plant which code for the enzymes needed for the synthesis of the phytoalexins. Elicitors have been isolated and characterized from several bacteria and fungi. Glucans, glucomannan and glycoproteins are common elicitors, but some derived from fatty acids have also been identified. Elicitors are released by interaction between the plant cell and the infecting organism. When the plant cell is attacked, it releases extracellular hydrolytic enzymes which attack the cell wall of the infecting organism and release a polysaccharide which acts as an elicitor. It has been found that in many plants microbial attack results in an increase in the activity of chitinase and 1,3-β-glucanase. As chitin is not a component of plant cell walls, but a common component in the cell walls of microorganisms, in particular fungi, as well as the exoskeleton of insects, it appears evident that the plant cell chitinase is a factor in the plant defence system. Elicitors can, however, also be released from the plant cell wall by enzymes from the infecting organism, in these cases known as endogenous elicitors.

Besides the natural elicitors described above, it has also been found that heat, cold, detergents, salts of heavy metals and ultraviolet light can function as elicitors. These factors probably act by causing the stressed plant cell to activate hydrolases which liberate endogenous elicitors.

Addition of elicitors to plant cell cultures can increase production of the desired secondary metabolites. Yeast, glucans, chitin, chitosan and homogenized mycelia from different fungi have been used for this purpose. Jasmonic acid and salicylic acid are signal compounds which are produced in plants as a response to a stress situation (e.g. an infection) and which induce the production of various metabolites aimed at defence against the stress inducer. The methyl esters of these compounds are also used as elicitors. One, often-encountered advantage of the use of elicitors is that more of the desired metabolite is excreted into the medium, which facilitates the extraction of the product.

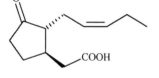

Jasmonic acid

Salicylic acid

Organ culture

As mentioned above, changing the proportions of auxins and cytokinins in the medium can induce organ formation. Production of secondary metabolites is generally greater in organ cultures than in callus or suspension cultures, but handling organ cultures on an industrial scale is difficult. As mentioned above (page 103) the

Agrobacterium-introduced hairy roots are often good producers of secondary metabolites and, on a dry weight basis, their content can sometimes be higher than in the normal plant. Hairy roots can be cultivated in hormone-free media because genes in the Ri T-DNA regulate the balance of endogenous hormones. The cultures are genetically and biochemically stable in contrast to suspension cultures. Hairy roots can, however, only be used for production of compounds which in the intact plant are biosynthesized by roots.

Special fermenters have been designed for hairy-root cultures, in which the roots are distributed on a network of stainless steel rods and the medium is sprayed continuously over the roots.

Future prospects for plant biotechnology in the production of drugs

Isolation of enzymes in biosynthetic pathways

As mentioned above (page 106) one reason for the limited use of cell and tissue culture in the production of drugs is our incomplete knowledge of the enzymes which catalyse the various biosynthetic stages and how these are regulated. Cell suspension cultures have proved useful for the isolation of enzymes, as these are generally present in larger amounts in the culture than in the intact plant. One fascinating example of the potential of plant biotechnology in this area is the elucidation at the enzyme level of the complete biosynthetic pathway leading to berberine as described in chapter 10, page 645.

Isolation and characterization of the enzymes involved in biosynthetic processes is important not only for elucidation of biosynthetic pathways but could also be of great importance for drug development. The enzymes could be used in combination with biomimetic organic synthesis for production of known compounds which are difficult to synthesize with conventional chemical methods. One can also envisage biomimetic synthesis of new compounds with a potential as drugs. Knowledge of the biosynthetic enzymes will permit identification of the genes that code for them and thus open the field for gene technology involving cloning of the genes and their expression in bacteria to permit production of large amounts of the enzymes at a much lower cost than by isolation from cells or organisms.

Discovery of new pharmacologically active compounds

Undifferentiated plant cells often produce compounds which are not found in the organs of the differentiated plant. These might have pharmacological activity and their isolation and investigation could open up new roads to drugs. An attractive aspect is that such products could be patented and thus the problem of patent rights could be overcome, which often deters industry from exploring plant-derived products. Another use of plant cell cultures in the search for new drugs is direct screening of the cultures with highly sensitive receptor binding assays. An example is screening for compounds that react with the benzodiazepine receptor (from rat brain) which led to the isolation of a highly active compound – 2-hydroxy-1-methoxy-dibenzoquinoline-4,5-dione – from a cell culture of *Aristolochia macrophylla* Duch. (CIBA-GEIGY Ltd, Basel, Switzerland, unpublished). It was not previously known that this type of compound can react with the benzodiazepine receptor but this discovery might lead to synthesis of a new class of substances to be used as anxiolytic drugs.

Transgenic plants as factories for drugs

As described above (page 105) transgenic plants have been developed for production of antibodies and vaccines. This approach is virtually without limits and might very well lead to production of a number of difficult-to-synthesize drugs which are now produced with less effective methods. A couple of examples of development in this field have recently been published: *Human a-1-antitrypsin* is a protein of therapeutic potential in cystic fibrosis, liver disease and haemorrhages. Rice plants (*Oryza sativa* L.) that produce this protein have been developed and clinical trials with protein extracted from the malted grain of this variety of rice have been started by the company Applied Phytologics (Sacramento CA, USA) in 1998. The company hopes to have regulatory approval for transgenic medical products by 2004. Gaucher's disease is a recessively inherited lysosomal storage disorder, resulting from deficiences of *lysosomal hydrolase glucocerebrosidase enzyme*. A drug developed from enzyme purified from human placentas is highly effective at reducing clinical symptoms. The enzyme, however, is one of the world's most expensive drugs as 10–12 tonnes/year of placentas are needed for production of enough glucocerebrosidase for the treatment of one patient. A patent for production of the enzyme in transgenic tobacco was filed in 1999, which brings hope for substantial reductions in the costs for treatment of Gaucher's disease.

Biopharmaceuticals derived from transgenic plants will have to meet the same standards of performance and safety as biopharmaceuticals originating from other systems. As plants can often produce human equivalent proteins, there seems little reason why this should not be the case. In some cases purification procedures will need to be developed to ensure that pharmaceuticals produced in non-edible plants, such as tobacco, are not co-purified with other potentially toxic or antigenic plant metabolites. Many of the companies developing transgenic plant expression systems have therefore chosen edible plants such as maize (*Zea mays* L.) after surveying various crops for potential protein recovery. In general, one would perhaps expect biopharmaceuticals from transgenic plants to be safer than their counterparts derived from animal-based sources, which have the potential for contamination with human pathogens.

Another concern in the development of "plant factories" for biopharmaceuticals is that such compounds are pharmacologically active at low concentrations and might be toxic at higher ones. Many have physiochemical properties that might cause them to persist in the environment or bioaccumulate in living organisms, possibly damaging non-target organisms. If pharmaceutical plants are to be grown in the open environment, methods of preventing environmental "contamination" must be developed. Such strategies may involve isolation from neighbouring crops, the use of self-pollinating plants and the use of male sterile lines.

The use of transgenic plants is subject to an intense political debate (particularly concerning plants for food production) in which there are many overtones that are not always founded on scientific principles. The debate has resulted in the introduction of many laws and statutes which hinder development in this field. It is therefore difficult to foresee the chances for the establishment of drug production based on transgenic plants.

Metabolic engineering of secondary metabolites

Metabolic engineering of secondary metabolites has been achieved for antibiotics, e.g. derivatives of erythromycin (page 277). For plant metabolites this approach has hitherto met with very limited success. The main reason for this is insufficient knowledge of the biochemical pathway that leads to the desired compound as well as of important factors for regulation of the pathway. As described on page 142 the biosynthesis of a pharmacologically active compound generally proceeds in several steps, each catalysed by a specific enzyme which is encoded by a particular gene. If the complete pathway is known, it is possible to manipulate these genes in various ways, e.g. by enhancement of the expression of a rate-limiting enzyme. The discovery of transcription factors that regulate the whole pathway opens up the possibility to enhance the expression or activity of all the genes involved in the pathway by manipulation of these factors. Pharmacologically active secondary metabolites are often stored intracellularly, e.g. in vacuoles, to which they are located by transporter proteins (page 145). Overexpression of such proteins can be of importance to hinder damage of cells by increased concentrations of metabolites, obtained by genetic engineering.

Examples of metabolic engineering of secondary metabolites are modifications of anthocyanin and flavonoid pathways, leading to changes in flower colour or production of increased levels of antioxidative flavonols in tomato. Introduction of a gene encoding hyoscyamine-6S-dioxygenase (page 633) from *Hyoscyamus niger*

L. into *Atropa belladonna* L. resulted in a transgenic plant where practically all hyoscyamine was converted to the more valuable scopolamine. Genetic engineering has also shown that it is possible to obtain high amounts of scopolamine in hairy-root cultures of *Hyoscyamus niger*, which might become profitable for industrial production of this alkaloid.

In the future, application of the techniques of systems biology (page 31), particularly genomics and metabolomics, might overcome the difficulties hitherto experienced in metabolic engineering of plant metabolites and lead to industrial production by cell and organ culture of known, as well as new, medicinally useful metabolites from plants.

Further reading

Antibiotics

Clark, R.W. *The Life of Ernst Chain: Penicillin and Beyond.* Weidenfeld and Nicholson, London (1985).

Hare, R. *The Birth of Penicillin.* Allen and Unwin, London (1970).

Lancini, G.C. and Lorenzetti, R. *Biotechnology of Antibiotics and other Microbial Metabolites.* Plenum, New York (1994).

Lancini, G., Parenti, F. and Gallo, G.G. *Antibiotics – a Multidisciplinary Approach.* Plenum, New York (1995).

Macfarlane, R.G. *Alexander Fleming: the Man and the Myth.* Chatto and Windus, London (1984).

Patnaik, P.R. Penicillin fermentation: mechanisms and models for industrial-scale bioreactors. *Critical Reviews in Biotechnology,* **20** (1), 1–15 (2000).

Sneader, W. *Drug Prototypes and their exploitation. Chapter 15, Antibiotics.* John Wiley & Sons Ltd, Chichester (1996).

Plant biotechnology, general review

Banthorpe, D.V. Secondary metabolism in plant tissue culture: scope and limitations. *Natural Products Reports,* **11** (3), 303–328 (1994).

Techniques of plant and tissue culture

Biotechnology in Agriculture and Forestry. **Vols. 1–62**, Springer Verlag, Berlin (1986–2008).

Evans, D.E., Coleman, J.O.D. and Kearns, A. *Plant Cell Culture.* Taylor & Francis, London (2003).

Smith, R.H. *Plant Tissue Culture. Techniques and Experiments.* Academic Press, London (2000).

Pharmaceutical applications

Kutney, J.P. Plant cell culture combined with chemistry: a powerful route to complex natural products. *Accounts of Chemical Research*, **26**, 559–566 (1993).

Verpoorte, R., van der Heijden, R. and Schripsema, J. Plant cell biotechnology for the production of alkaloids: present status and prospects. *Journal of Natural Products,* **56**, 186–207 (1993).

Oksman-Caldentey, K.-M. and Hiltunen, R. Transgenic crops for improved pharmaceutical products. *Field Crops Research,* **45**, 57–69 (1996).

Transgenic plants

Giddings, G., Allison, G., Brooks, D. and Carter, A. Transgenic plants as factories for biopharmaceuticals. *Nature Biotechnology*, **18**, 1151–1155 (2000).

Haq, T.A., Mason, H.S., Clements, J.D. and Arntzen, C.J. Oral immunization with a recombinant bacterial antigen produced in transgenic plants. *Science,* **268**, 714–719 (1995).

Hiatt, A. Antibodies produced in plants. *Nature,* **344**, 469–470 (1990).

Ma, J.K.-C., Lehner, T., Stabila, P., Fux, C.I. and Hiatt, A. Assembly of monoclonal antibodies with IgG1 and IgA heavy chain domains in transgenic tobacco plants. *European Journal of Immunology*, **24**, 131–138 (1994).

Ma, J. K.-C. and Hein, M.B. Antibody production and engineering in plants. *Annals of the New York Academy of Sciences,* **79** (2), 72–81 (1996).

Moffat, A.S. Exploring transgenic plants as a new vaccine source. *Science*, **268**, 658–660 (1995).

Tinland, B. The integration of T-DNA into plant genomes. *Trends in Plant Science*, **1** (6), 178–184 (1996).

Metabolic engineering of secondary metabolites

Gómez-Galera, S., Pelacho, A.M., Gené, A., Capell, T. and Christou, P. The genetic manipulation of medicinal and aromatic plants. *Plant Cell Reports*, **26**, 1689–1715 (2007).

Oksman-Caldentey, K.-M. and Inzé, D. Plant cell factories in the post-genomic era: new ways to produce designer secondary metabolites. *Trends in Plant Science*, **9**, 433–440 (2004).

4. Formation of Pharmacologically Active Compounds in Plants – Biosynthesis

Photosynthesis

Photosynthesis is the process which transforms solar energy into chemical energy that can be used by living cells. Without photosynthesis, life on earth would not be possible. The enormous amount of chemical energy that is trapped in fossil fuels – coal, oil, gas, etc. – is also the result of photosynthesis, since the fossil fuels originate from plants and animals that once lived on earth. When we exploit this energy source today, we are utilizing energy that was emitted from the sun in ancient times and captured by the process of photosynthesis.

Photosynthesis also provides the primary building material for all living organisms as it involves the fixation of carbon in the form of organic compounds. All plant constituents, including the ones which are pharmacologically active and utilized as drugs, are originally derived from the primary products of photosynthesis.

Photosynthesis takes place in all higher plants. It also occurs in other eukaryotes, such as green, brown and red algae and in some unicellular organisms, e.g. euglenoids, dinoflagellates and diatoms. The process of photosynthesis is believed to have been invented by a group of eubacteria commonly referred to as purple bacteria more than two billion years ago. The same enzymatic machinery and principles were consequently refined among green bacteria and cyanobacteria. At some point in history, by processes yet obscure, a cyanobacteria became domesticated by a eukaryote cell, forming the first chloroplast. During evolution the chloroplast lost parts of its genome, but retained the sections necessary for e.g. photosynthesis and protein replication, and the chloroplast was inherited in the same way as the mitochondria along the evolutionary lineages. Among many primitive unicellular organisms (sometimes erroneously referred to as "protists") such as the euglenoids, dinoflagellates and apicomplexa (including the malaria parasites *Plasmodium*) the presence of chloroplasts have been demonstrated from molecular and microscopical data. Among more complex organisms such as Phaeophyta, the brown algae, Rhodophyta, the red algae, and Chlorophyta, all green plants, the chloroplasts are ubiquitous, eventually to be lost in the evolutionary lineage leading to fungi and animals.

The general photosynthetic reaction

All photosynthetic activity can be described as a reaction between an electron donor and an electron acceptor under the influence of light, the energy of which is absorbed, giving a product with high chemical energy.

$$H_2Don + A \xrightarrow{\text{light radiation}} H_2A + Don$$

donor   acceptor   Energy-rich by-product
                                          product

With all photosynthesizing organisms except a few bacteria, the donor is water (H_2O) and the by-product oxygen (O_2). Photosynthesizing bacteria use other donors, e.g. hydrogen sulphide, thiosulphate, hydrogen, lactic acid or isopropyl alcohol, giving other by-products than oxygen. Carbon dioxide is the most common acceptor, and is used as the carbon source for the synthesis of all organic constituents. However, other acceptors are also employed.

Photosynthetic organs in higher plants

In higher plants the photosynthetic apparatus consists of *chloroplasts,* a special type of plastid occurring in the cells of those organs in which photosynthesis takes place. Chloroplasts are 3–10 μm in diameter and usually spherical or ellipsoidal. They can be isolated by centrifugation and have a structure similar to that of mitochondria, with the exception that the chloroplast has three membranes. Studies with the electron microscope have led to the recognition of four major structural features or compartments in the chloroplast: (*1*) A pair of outer limiting membranes, called the *envelope*; (*2*) a background matrix or *stroma*; (*3*) a highly organized internal system of membranes, called *thylakoids* and (*4*) the *lumen*, which is the interior space of the thylakoid.

The two membranes forming the envelope are 5.0 to 7.5 nm thick and the distance between them – the *intermembrane space* – is 10 nm wide. The outer envelope membrane does not serve as a permeability barrier and thus the intermembrane space is freely accessible to metabolites from the cytosol of the leaf cell. The inner envelope membrane, on the other hand, is semipermeable and regulates the exchange of metabolites between the chloroplast and the cytosol. The amorphous *stroma* is predominantly composed of protein and contains all the enzymes required for photosynthetic carbon reduction – the dark reactions of photosynthesis (page 128). Embedded in the stroma is a complex system of individual pairs of parallel membranes that are fused, forming an enclosed structure resembling flattened sacs, called *thylakoids*. In some regions adjacent thylakoids are stacked, forming *grana*. The thylakoids contain chlorophyll, the light-harvesting system and the reaction centres for the primary, light-dependent reactions (see below). The *lumen*, which is the interior space of the thylakoid, is in direct contact with the site of water oxidation which causes formation of oxygen (page 125). It also functions as a reservoir for protons that are pumped across the thylakoid membrane during electron transport.

The chloroplasts also have a genome of their own, which is inherited as duplicate copies when the chloroplasts divide. The chloroplast genome (cpDNA) of higher plants is a circular double-stranded ordinary DNA molecule of approximately 150 kbp, as compared with the 3 mbp of free-living cyanobacteria. The reduction and dense packing of the cpDNA as well as the important role of the chloroplasts in plant growth and production have made it a suitable object for study of genomic evolution. Today (2002) there are more than 20 complete chloroplast genomes available on the public databases, including those of a number of aberrant organisms. A very large proportion of all nucleotide sequencing performed for evolutionary studies of plants have been made on the chloroplast genomes, mainly due to the appealing property of the genes occuring only in single copies in the cpDNA, with the exception for the base-by-base identical inverted repeat regions which exhibit one copy each of the included genes.

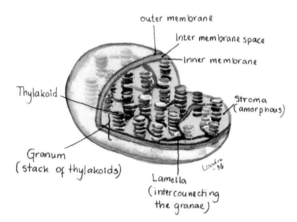

Fig. 15. Schematic drawing of the three-dimensional structure of a chloroplast. Figure drawn by Louise Karlsson

Light reactions and dark reactions

Two discrete steps can be recognized in the photosynthetic process. The first step involves the absorption of light energy and ends with the production of ATP and NADPH. As these processes take place only in the presence of light, they have been termed *light reactions*. The second step consists of a series of reactions in which ATP and NADPH are used for the reduction of CO_2 and the formation of glyceraldehyde-3-phosphate. These processes can occur in the dark and have therefore been termed *dark reactions*.

Light reactions; photosystems I and II

In higher plants the light reactions are associated with a set of enzymatic complexes and enzymes which are located in the thylakoid membranes. These are: photosystem I (PSI), photosystem II (PSII), the oxygen-evolving complex, the light-harvesting complex, the cytochrome b_6/f complex, ATP-synthase, and ferredoxin-NADP-reductase. In association with photosynthetic pigments and other cofactors, these components participate in the oxidation of water according to the equation

$$2\ H_2O \rightarrow 4\ H^+ + 4\ e^- + O_2$$

and subsequent employment of the liberated electrons and protons in reactions – driven by the energy of light – which eventually result in the production of ATP and NADPH. The arrangement of these complexes in the thylakoid membrane is schematically presented in Fig. 16.

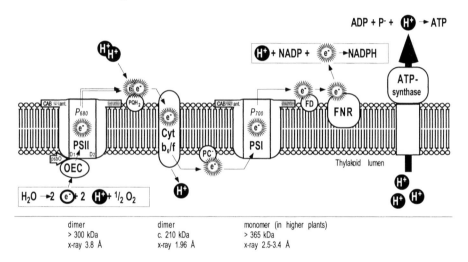

dimer	dimer	monomer (in higher plants)
> 300 kDa	c. 210 kDa	> 365 kDa
x-ray 3.8 Å	x-ray 1.96 Å	x-ray 2.5-3.4 Å

Fig. 16. Organization of the photosynthetic enzyme complexes in the thylakoid membrane. **OEC** = *the oxygen-evolving complex,* **PsbO** = *polypeptide stabilizing the Mn cluster,* **PSII** = *photosystem II,* D_1+D_2 = *protein heterodimer (see text),* **PSI** = *photosystem I,* **CAB-ant** = *external antennas of the photosystems,* **Cytb$_6$/f** = *the cytochrome b_6/f complex,* **FNR** = ferredoxin-NADP reductase, **PC** = *plastocyanin,* PQH_2 = *hydroquinone equivalent of plastocyanin,* **FD** = *ferredoxin.* P_{680} *and* P_{700} = *specialized chlorophyll dimers.* e^- = *electron. The electrontransport from OEC to NADP is schematically indicated. A more detailed scheme is presented in Fig. 17 (page 123). See text for further details. Figure drawn by Anders Backlund*

Photosynthetic pigments

The most important photosynthetic pigment, present in all photosynthesizing organisms, is *chlorophyll*. Higher plants contain two

117

kinds of chlorophyll: *a* and *b*. The formula below shows their structures. Four substituted pyrrole rings are arranged round a magnesium atom, which is coordinately bound to the nitrogen atoms of the pyrrole rings to form a stable, planar complex. Chlorophyll also contains a long hydrophobic side-chain, consisting of the alcohol phytol, which esterifies a propionic acid residue attached to ring IV. Chlorophylls absorb light energy efficiently as they contain many conjugated double bonds.

The pyrrole ring system is biosynthesized in the chloroplasts and glutamic acid is the ultimate precursor, which via 5-aminolaevulinic acid (ALA) forms uro'gen III (= uroporphyrinogen III or uroporphobilinogen III). Uro'gen III is the precursor not only of chlorophyll but also of haems, sirohaem, factor F_{430} and vitamin B_{12}. The pathway from glutamic acid to uro'gen III is treated in the description of the biosynthesis of vitamin B_{12} (page 580). The rest of the extremely complicated pathway has to a great extent been elucidated both at the enzyme and at the gene levels but a description falls outside the scope of this book. For reviews see Reinbothe (1996) and Bolivar (2006) in *Further reading* page 146.

The side-chain – phytol – is a diterpene derivative. Geranylgeranyl diphosphate (page 413) is involved in formation of the ester linkage and reduction takes place after attachment to the chlorophyll molecule.

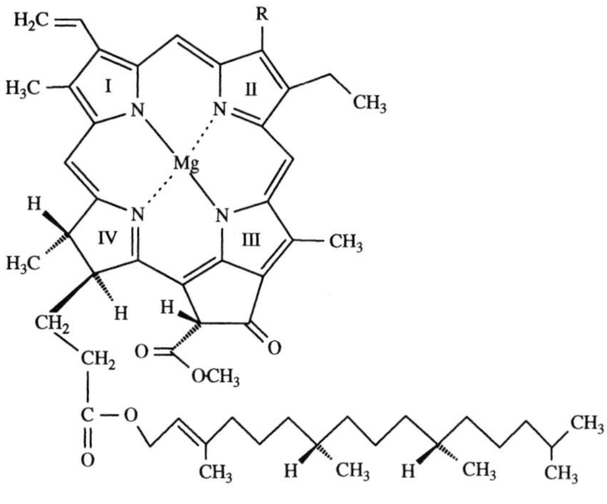

Chlorophyll a: R = Me. Chlorophyll b: R = CHO

Both photosystems have one specialized chlorophyll dimer, called P_{700} in PSI and P_{680} in PSII, which can transfer the energy of absorbed light to an electron, thus enabling it to be used in the chemical processes described on page 122. The names P_{680} and P_{700} are derived from the observation of a bleaching maximum at 680 nm and 700 nm respectively, associated with the photo-oxidation of the pigments.

Besides chlorophyll, the photosystems also contain a number of other pigments that can act as receptors of light energy in their light-harvesting antennas, e.g. *carotenoids*, which are long polyiso-prenoid molecules with numerous conjugated double bonds. Each end of the chain carries a substituted cyclohexene ring. An example of a carotenoid is ß-carotene, the structure of which is shown below.

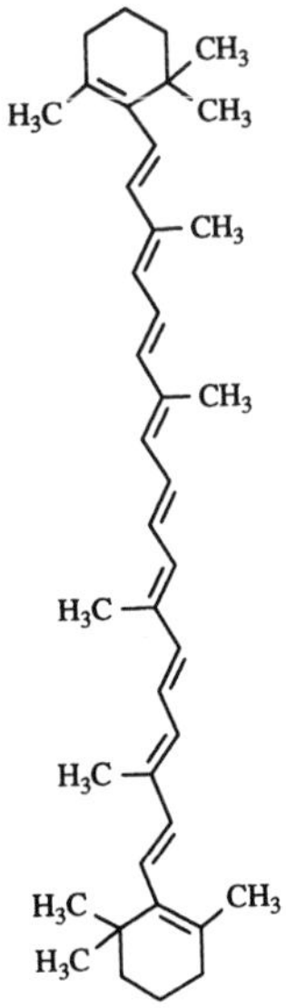

β-Carotene

The structures of photosystems I and II

Photosystems I and II (PSI and PSII) are the key-factors in the photosynthetic process. Their structures are very complicated and have been subject to intense studies, where all available methods for structure analysis of big protein complexes have been used, including electron cryo-microscopy and X-ray crystallography. Also studies in molecular genetics and biochemistry have contributed to the present understanding of the architecture of the photosystems which, however, is not yet complete. The following is a very brief description of the essential parts of the systems. For more details and figures see references in *Further reading* on page 146.

PSII   The genes for more than 20 protein subunits have been identified in PSIIs from various photosynthesizing organisms. The crystal structure has been determined for PSII from the thermophilic cyanobacterium *Synechococcus elongatus* at 3.8 Å resolution. PSII is a homodimer, each monomer of which contains at least 17 subunits, 14 of which are located within the thylakoid membrane while three protrude on the luminal side of the membrane. Two proteins, D_1 (=PsbA) and D_2 (=PsbD), form a heterodimer which binds most of the electron transfer cofactors, including the specialized chlorophyll dimer P_{680} and is regarded as the reaction centre core of PSII. D_1 binds the Mn cluster and other components associated with the oxygen-evolving complex (see page 125). Each of the D_1 and D_2 proteins has a molecular size of 39 kDa and contains 5 transmembrane helices. Two other proteins – CP43 (=PsbC) and CP47 (=PsbB) – with molecular sizes 52 and 56 kDa respectively, contain 6 helices each. CP43 and CP47 bind 12 and14 molecules, respectively, of chlorophyll *a* and form the inner light-harvesting complex of PSII. Other subunits, also located within the thylakoid membrane, are PsbE and PsbF (= the α- and β- subunits of cytochrome *b*-559), PsbH, PsbI, PsbJ, PsbK, PsbL PsbM, PsbN and PsbX. Located on the luminal side are cytochrome *c*-550 (=PsbV), PsbU and the polypeptide PsbO which stabilizes the Mn cluster.

PSI   The native form of photosystem I in cyanobacteria is a trimer, whereas it is generally agreed that this photosystem is monomeric in higher plants. The genes for 15 protein subunits of PSI monomers have been identified and denoted alphabetically as *PsaA–PsaX*. A cyanobacterial monomer consists of 11–12 subunits and plant and algal complexes contain 3 additional proteins. The crystal structure has been determined at 2.5 Å resolution for PSI from *Synechococcus elongatus*. Each monomer contains 12 protein subunits, 9 of which (PsaA, PsaB, PsaF, PsaI, PsaJ, PsaK, PsaL, PsaM and PsaX) are located within the thylakoid membrane, whereas 3 (PsaC, PsaD and PsaE) protrude on the stromal side. PsaA and PsaB are high-molecular-mass proteins (82–83 kDa) while the other proteins are small (<20 kDa). PsaA and PsaB contain 11 transmembrane helices each. These are divided into an amino-terminal domain (six α–helices) and a carboxy-terminal domain (5 α–helices). Coordinated to the C-terminal domains of PsaA and PsaB are the factors of the electron transport chain of PSI (see below), including the specialized chlorophyll dimer P_{700} which, in analogy with P_{680} of PSII, transfers the energy of absorbed light to an electron. PsaA and PsaB also bind 79 chlorophyll *a* molecules which form the inner antenna of PSI. Most of these chlorophylls are located in the N-terminal domains of the two subunits. In addition to the chlorophylls, the internal antenna contains 22 β-carotene molecules which might serve two functions in photosynthesis: light harvesting in the 450–570 nm range, where chlorophyll *a* has a poor absorption, and photoprotection by quenching excited chlorophyll *a* triplet states, preventing formation of toxic singlet oxygen. Another 11 chlorophyll *a* molecules are bound by the small subunits PsaJ, PsaK, PsaL, PsaM and PsaX. PSI differs from PSII with respect to the arrangement of the inner antenna, which in PSII is located in separate subunits (CP43 and CP47) and consists of a smaller number of chlorophyll molecules. However, structurally there is a resemblance in that the six N-terminal transmembrane helices of PSI are arranged in a manner similar to the arrangement of the transmembrane helices of CP43 and CP47 in PSII. The proteins PsaC, PsaD and PsaE form the stromal domain.

In addition to the protein subunits, PSI also contains 4 lipids, associated with PsaA/PsaB. Three of these are phospholipids (I, II, IV) and one (III) is a galactolipid. The location of lipids I and II close to the core of PSI and the binding of a chlorophyll *a* molecule in the antenna by lipid III indicate that the lipids are integral to and functionally important in PSI and that they are not artefacts formed during preparation.

The light-harvesting complex

As mentioned above, the photosystems contain an inner light-harvesting complex containing about 90 molecules of chlorophyll *a* in photosystem I and 25–50 molecules in photosystem II. However, most of the light energy which drives the photosynthetic process is taken up by an outer light-harvesting complex, consisting of a number of chlorophyll-binding proteins – called Chl *a/b* proteins or CAB proteins – which surround the photosystems, forming their external "antennas". They contain not only chlorophyll *a* and *b* but also carotenoids and about 70 % of the pigments which are involved in photosynthesis are found in these complexes. The genes that encode the proteins in higher plants are known and have been matched to the corresponding proteins by peptide sequencing. Ten proteins designated Lhca1–Lhca4 and Lhcb1–Lhcb6, having similar molecular masses ranging from 22 to 28 kDa, are the building blocks of six main separable pigment–protein complexes. The most important of these is LHCII which contains Lhcb1, Lhcb2 and Lhcb3 and most probably exists as a trimer *in vivo*. LHCII can transfer exitation energy both to PSI and PSII. It is associated with PSII but under certain conditions it can dissociate from it and migrate independently between stacked and unstacked regions of the thylakoid membrane. The structure of pea LHCII has been determined by electron crystal-

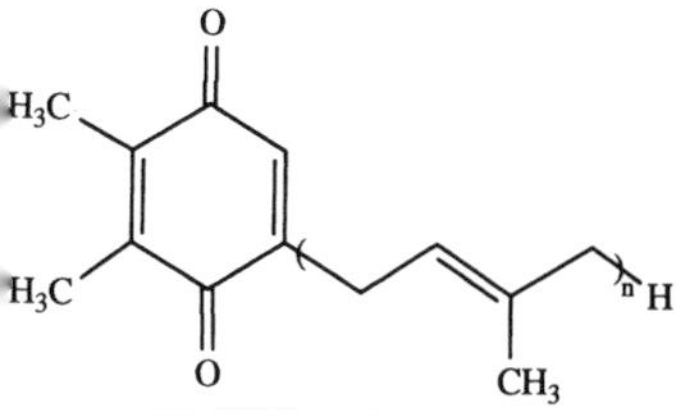

n = 1, 2, 3, 4, 8, 9, 10
Plastoquinone A: n = 9

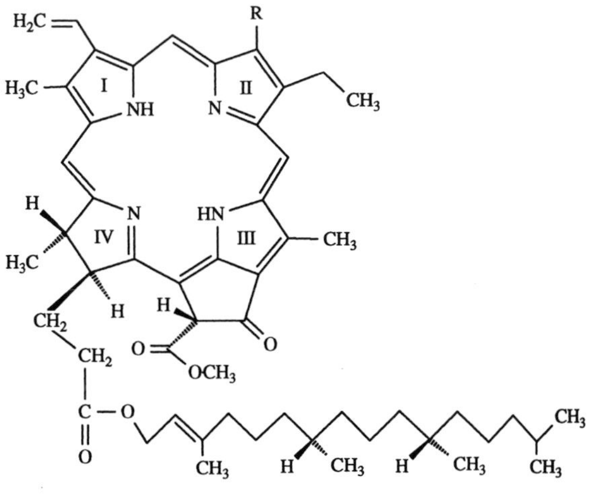

Phaeophytin a: R = Me. Phaeophytin b: R = CHO

lography to 3.4 Å resolution. For details see the review by Green and Durnford in *Further reading* on page 146. Lhcb4 (CP29), Lhcb5 (CP26) and Lhcb6 (CP24) are also associated with PSII. Associated with PSI are the proteins Lhca1 – Lhca4.

The cytochrome b_6/f complex

Like the previously described complexes PSI, PSII and the light-harvesting complex, the cytochrome b_6/f complex is a multimeric, large integral membrane protein. It functions as an intermediate in the transfer of electrons from PSII to PSI and the translocation of protons across the membrane. The complex contains four proteins with molecular masses ranging from 18 to 31 kDa. Three of these – cytochrome f, cytochrome b_6 and the Rieske Fe_2S_2 protein – have redox prosthetic groups which are active in the electron transport (see below). In addition to these four proteins there are at least three additional small (5 kDa) hydrophobic polypeptides present, the function of which is unknown. The functional form of the complex is a dimer with molecular mass 210 kDa, believed to contain 22–24 transmembrane helices. A large amount of primary sequence data has been obtained for the proteins of the complex. Best known is cytochrome f, for which the crystal structure has been determined at a resolution of 1.96 Å. For details see a review by Cramer *et al.* in *Further reading*, page 146.

Electron transport in photosynthesis

The energy absorbed by the antennae in the photosystems is used to transport electrons from water (oxidation-reduction potential of the water/oxygen complex = +0.8 V) via a large number of intermediate compounds to $NADP^+$ (oxidation-reduction potential –0.3 V) with a net gain in electrochemical potential of 1.1 V (at pH 7). The electrons are obtained from water by the activity of the oxygen-evolving complex (see below). The easiest way to understand the process is to start at the antenna of photosystem II, and to imagine that we follow one electron on its path from the reaction centre of photosystem II through all the intermediate compounds to $NADP^+$. This is, of course, not what is happening in reality, as each reacting compound is surrounded by a cloud of electrons, any of which can be passed on to the next stage in the reaction sequence. It is thus very unlikely that the electron which leaves the cloud in a particular reaction step is the same electron which entered the cloud in the preceding step. The energy of two light quanta, absorbed by the pigments in the antenna, is transferred to the special chlorophyll molecule P_{680}, in the protein complex D_1/D_2 of photosystem II. This converts P_{680} to a strong reductant – P_{680}^* – which has the unique power to release one electron. In less than 10 picoseconds (ps) this electron is transferred to the first acceptor in the electron-transport chain – *phaeophytin* (Phaeo in Fig. 17a), a derivative of chlorophyll

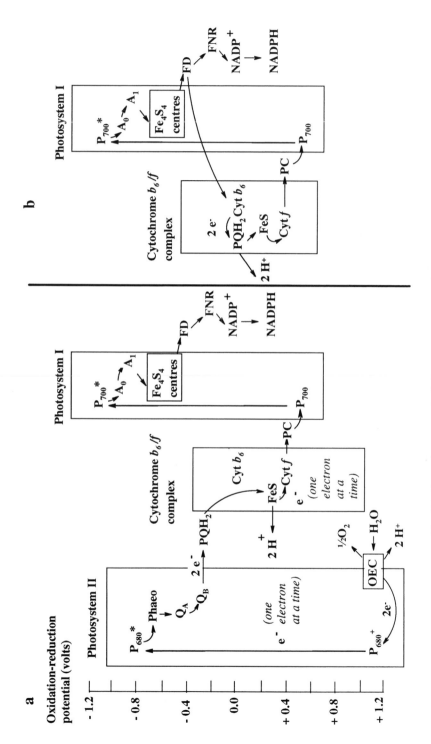

Fig. 17. Energy transport in light reactions. a. Linear flow of electrons.
b. Cyclic electron transport

lacking the Mg atom. Within about 100 ps, the electron leaves phaeophytin and is transferred to the electron acceptor Q_A (a quinone) to form the semiquinone anion Q_A^-. Q_A^- rapidly donates its electron to the next acceptor Q_B, which is plastoquinone, transiently bound to protein D_1 in PSII. The binding stabilizes the formed anion Q_B^-. There is evidence that one Fe_2^+ ion is indirectly involved in this transfer, probably stabilizing the semiquinone anion Q_A^- in the thylakoid membrane.

Q_B^- can accept a second electron (originating from a new reaction sequence P_{680} − phaeophytin − Q_A) to form Q_B^{2-}. At this level, the light quantum-activated one-electron process is converted into a two-electron process. Q_B is the "two-electron gate" of the process. At the same time, protonation occurs, forming the hydroquinone equivalent (PQH_2) of plastoquinone (PQ). The protons used in the reaction originate from the stroma side of the thylakoid membrane. The hydroquinone leaves the binding site on D_1/D_2 and becomes part of the membrane pool. In the next step of the electron-transport chain, PQH_2 enters the cytochrome b_6/f complex, where it transfers its electrons to the Rieske Fe_2S_2 protein, thereby liberating its two protons. The electrons travel on, one at a time, first to cytochrome f, and finally to the soluble copper protein *plastocyanin* (PC). The two liberated protons leave the complex on the lumen side of the thylakoid membrane and, as they originally came from the stroma side, an electrochemical proton gradient is generated across the thylakoid membrane. This gradient is a driving force in the synthesis of ATP by *ATP synthase* which is coupled to the electron-transport chain in a similar way as in mitochondria (see below). Plastoquinone returns first to the membrane pool and then to the protein from which PQH_2 was previously liberated, thus re-forming Q_B. PC is a protein, consisting of a single protein chain of 99 amino-acid residues, which contains a coordinately bound copper atom and occurs in the membrane lumen. Formation of reduced plastocyanin (PC^-) completes the activity of the cytochrome b_6/f complex. PC^- enters PSI where its electron is used to reduce the oxidized form of the special chlorophyll molecule P_{700} of photosystem I. P_{700} has the same function as the P_{680} of PS II, i.e. it is excited by light quanta from the antenna of PS I and is raised to a high energy level − P_{700}^* − from which it releases an electron that is then transferred to a primary acceptor (Fig. 17a). The oxidized P_{700} returns to its original energy level and, after receiving an electron from a new reduced plastocyanin, is ready for a new round.

The primary electron acceptor of PSI − A_0 − is a specialized chlorophyll *a* molecule. The second acceptor − A_1 − is a phylloquinone (vitamin K_1) molecule. The first two electron transfers − from P_{700}^* to A_0 and from A_0^- to A_1^- are extremely fast. The subsequent electron acceptors are three bound *iron-sulphur centres* each of which contains 4 iron and 4 loosely bound sulphur atoms. From these iron-sulphur centres, the electron is transferred to ferredoxin (FD), a soluble Fe-S protein of molecular mass 10 kDa. The reduced ferredoxin transfers its electron to *ferredoxin-NADP reductase* (FNR), a thylakoid-bound FAD-flavoprotein, which, in a two-step process with addition of H^+ reduces NADP to NADPH. This last molecule is released into the medium and used in the dark reactions.

The electron transport described above is a straightforward process in which electrons are transported from water to $NADP^+$ reducing this compound to NADPH to be used for the subsequent reduction of CO_2 in the dark reactions. The process also generates a proton gradient across the thylakoid membrane which drives ATP synthesis. There is, however, also a cyclic electron transport described in the literature. This process (Fig. 17b), which involves only PSI and the cytochrome b_6/f complex, generates only the proton gradient for ATP formation and is not involved in the production of NADPH. In these reactions ferredoxin does not transfer its electron to FNR but to cytochrome b_6 in the cytochrome b_6/f complex. For the continuation of the process two electrons are needed which are transferred from cytochrome b_6 to plastoquinone with formation of PQH_2. The continuation of the process is the same as in the linear electron transport, resulting in formation of the proton gradient needed for ATP synthesis and formation of PC^- which returns to PSI. The cyclic electron transport can utilize light of wavelengths which can not be used in linear electron transport. It is suggested to be a process that enables the chloroplast to direct all the incoming energy to ATP synthesis at occcasions when there is a higher demand for ATP than for carbohydrates.

The oxygen-evolving complex of photosystem II and the photolysis of water

The protein D_1 in the core complex of photosystem II contains a redox-active tyrosine residue (Y161, denoted as *YZ*), which together with a Mn_4 cluster, Ca^{2+} and Cl^- comprises the oxygen-evolving complex (*OEC*). The exact arrangement of the manganese ions in the Mn_4 cluster is unknown, but electron paramagnetic resonance and X-ray absorption spectroscopy data suggest that it is a tetramer of μ-oxo-bridged units with Mn–Mn separations of 2.7 and 3.3 Å.

In the sections above we followed the flow of electrons from P_{680} along the electron-transport chain of photosystem II to P_{700} of photosystem I and then along the electron-transport chain of that system to the final production of NADPH. The original source of these electrons is water, and we shall now look at the reactivity of the OEC to see how the release of electrons from water is effected according to the overall formula

$$2\ H_2O \longrightarrow 4\ H^+ + 4\ e^- + O_2$$

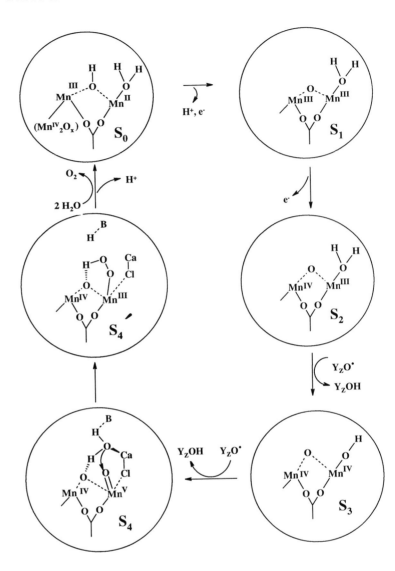

Fig. 18. Proposed S-state cycle. Solid arrows indicate light-driven steps and dashed arrows denote spontaneous steps. Steps that involve H-atom abstraction by $Y_z^\bullet$ are emphasized by including the reduction of $Y_z O^\bullet$ in the figure. The role of a protein residue acting as a base, B, and $Ca\text{-}OH_2$ in the O-O bond-forming step is shown in the S4 - S4'step but the proposed structure of $Ca\text{-}OH_2$ and Cl would be present in all of the earlier S-states as well. Redrawn from Vrettos, J.S. et al. Biochimica et Biophysica Acta, 1503, 229–245 (2001)

The electrons released in the process are used for replacement of the electrons which have been transported from P_{680} as described above and the protons are used for generation of the electrochemical proton gradient across the thylakoid membrane, needed for the synthesis of ATP. With respect to energy formation, oxygen is to be regarded as a by-product, but its formation by the activity of the OEC associated to photosystem II is a prerequisite for all organisms with aerobic metabolism, as is the subsequently formed ozone-layer an absolute demand for all DNA-encoded life exposed to sunlight.

When the electron of the excited P_{680}^* is transferred to phaeophytin, oxidized P_{680} ($= P_{680}^+$) is formed. P_{680}^+ is a strong oxidant which quickly removes an electron from Y_Z forming the neutral radical $Y_Z^\cdot$ which in turn oxidizes the tetranuclear manganese cluster (Mn_4) through a cycle of five "S_n states" ($n = 0$–4). Fig. 18 illustrates the current concept of this procedure. Only two of the four manganese atoms participate in the process.

Proton-coupled electron transfers in the $S_2 - S_3$ and $S_3 - S_4$ transitions form a terminal Mn(V)=O moiety. Nucleophilic attack on this electron-deficient Mn(V)=O by a calcium-bound water molecule results in a Mn(III)-OOH species, which effects the release of O_2 resulting in a return of the Mn-cluster to the S_0 state.

Elucidation of the structure and function of the OEC is a very complicated problem which has taken many years of work to even begin to understand. The authors, cited above (Fig. 18), have proposed a structural and functional model of the OEC that incorporates elements of the protein environment and is in good agreement with both the spectroscopic and the biochemical data. Following the publication of the X-ray crystal structure of PSII at 3.5 and 3.0 Å in 2004–2005, the authors have revised their model to fit in with the new information. The crystallographic model suggests a well-defined proton exit channel of hydrophobic residues leading to the lumenal exterior of PSII. This channel is on the opposite side of the Mn_4 cluster from Y_z (D_1-Y161) and leads directly *away* from Y_z. This casts doubt on Yz's function as the catalytic base in water splitting, and an arginine residue (CP43-Arg357), which is very close to the OEC, has been suggested to play the role of the thermodynamically indispensable redox-coupled base (see McEvoy, 2006 in *Further reading* page 147).

Photosynthetic phosphorylation. Formation of ATP

In the above description of the transport of electrons through photosystem II, the passage of the electrons through the cytochrome b_6/f complex is accompanied by protonation and deprotonation. This process gives rise to an electrochemical proton-gradient across the membrane of the thylakoid, which drives the formation of ATP. There is evidence that these protons and the protons originating from the decomposition of water do not join the internal water phase but are confined within the thylakoid membrane and together act to drive the formation of ATP by flowing through the ATP synthase complex. This complex is partly embedded in the thylakoid membrane and consists of two parts, denoted as CF_0 and CF_1. CF_1 is located above the thylakoid membrane on the stroma side. CF_0 is embedded in the membrane and the two parts are interconnected by a short stalk. Both proteins contain a number of subunits. CF_1 has 5 different types denoted by the Greek letters α, β, γ, δ and ε. There are altogether 9 subunits comprising 3 α, 3 β and one of each of the γ, δ and ε subunits. In CF_0 there are 4 types of subunits denoted as a, b, b´ and c (in cyanobacteria) or correspondingly IV, I, II and III in chloroplasts. The total number is uncertain but the minimal stoichiometry for the CF_0 of chloroplasts has been suggested to be 12 units of type III and one unit of each of types I, II and IV. The estimated molecular mass of CF_1 is ca. 390 kDa and that of CF_0 ca. 160 kDa.

Protons do not readily pass through the thylakoid membrane, but, as described above, protons are pumped across the thylakoid membrane from the stroma into the lumen in combination with the electron transport from Q_b in PSII to plastocyanin via the cytochrome b_6/f complex. This results in a difference in proton concentration (ΔpH) across the membrane which may be quite large, amounting to three or four orders of magnitude. As the protons carry a positive charge ΔpH causes an electrical potential gradient across the membrane with a higher positive charge on the lumen side than on the stroma side. ATP synthase utilizes the combined membrane potential and the proton gradient – called the *proton motive force* – as a driving force for synthesis of ATP. The active site for ATP synthesis is CF_1 while CF_0 acts as a proton channel across the membrane, transferring the energy of the gradient to CF_1 where it is utilized to phosphorylate ADP to ATP. The combination of the electron process and the activity of ATP synthase causes a cyclic transport of protons from the stroma into the lumen and from the lumen back to the stroma.

Dark reactions

In the dark reactions, the products of the light reactions – ATP and NADPH – are used for the reduction of CO_2 to form carbohydrates. The dark reactions have been mapped in great detail by Melvin Calvin (Nobel Prize in 1961) and his collaborators and have been termed *the reductive pentose phosphate cycle*. Other names are *the Calvin-Benson cycle or the C3 cycle*. The method used employed green algae which were irradiated for brief periods (a few seconds) in the presence of $^{14}CO_2$ and then quickly killed and extracted. The radioactive products formed were isolated chromatographically and identified. By comparing the products formed on very brief irradiation with those appearing after slightly longer irradiation, it was possible to determine the course of the biosynthetic reactions. Thus it was found that the first stable product of photosynthesis – formed after as little as two seconds – was 3-phospho-D-glycerate. This product contains three carbon atoms and the name the *C3 cycle* is derived from this fact. 3-Phospho-D-glycerate is formed by condensation of atmospheric CO_2 with one molecule of D-ribulose-1,5-bisphosphate followed by immediate cleavage of the condensation product into two molecules of 3-phospho-D-glycerate. This reaction

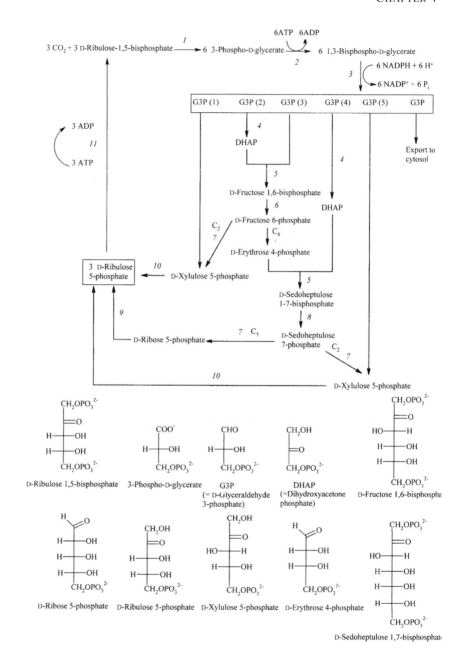

Fig. 19. The reductive pentose phosphate cycle. Enzymes indicated by the italic numbers are: **1**: ribulose-bisphosphate carboxylase (Rubisco) (EC 4.1.1.39), **2**: phosphoglycerate kinase (EC 2.7.2.3), **3**: glyceraldehyde-3-phosphate dehydrogenase (NADP+, phosphorylating) (EC 1.2.1.13), **4**: triose-phosphate isomerase (EC 5.3.1.1), **5**: fructose-bisphosphate aldolase (EC 4.1.2.13), **6**: fructose-bisphosphatase (EC 3.1.3.11), **7**: transketolase (EC 2.2.1.1), **8**: sedoheptulose-bisphosphatase (EC 3.1.3.37), **9**: ribose-5-phosphate isomerase (EC 5.3.1.6), **10**: ribulose-phosphate 3-epimerase (EC 5.1.3.1), **11**: phosphoribulokinase (EC 2.7.1.19).
The structures are drawn as Fischer projections

129

is catalysed by the enzyme *ribulose-bisphosphate carboxylase* (*Rubisco*, also spelled *RuBisCO: EC 4.1.1.39*) and is the key-reaction for photosynthesis whereby atmospheric CO_2 is fixed into organic matter. Rubisco is the most abundant protein in the world, accounting for approximately 50 % of the chloroplast protein in most leaves. The enzyme requires Mg^{2+} and CO_2 is not only a substrate but also an activator for Rubisco. The name "dark reactions" for the reactions in the C3 cycle is not completely adequate. It was created in the early days of photosynthetic studies to distinguish CO_2-binding reactions, which might take place also in the dark, from the reactions causing formation of ATP and NADPH, which are *directly* dependent on *light energy* and were thus called "light reactions".

The continuation of the C3 cycle is outlined in Fig. 19. Three molecules of CO_2 and three molecules of D-ribulose 1,5-bisphosphate yield six molecules of D-glyceraldehyde 3-phosphate (G3P). Five molecules of G3P (numbered 1–5 in Fig. 19) are used in a complicated series of reactions involving eight enzymes and the formation of 4-, 5-, 6- and 7-carbon sugars. The net effect of these reactions is the regeneration of D-ribulose 1,5-bisphosphate which is used for the next turn of the cycle.

Two G3P molecules (numbers 2 and 3 in Fig. 19) are in the centre of the reactions. G3P (2) forms dihydroxyacetone phosphate DHAP, which reacts with G3P (3) forming D-fructose 1,6-bisphosphate, which is hydrolysed to D-fructose 6-phosphate and inorganic phosphate. A *transketolase* (EC 2.2.1.1) catalyses a reaction between G3P (1) and D-fructose 6-phosphate resulting in the formation of D-xylulose 5-phosphate and D-erythrose 4-phosphate. D-Xylulose 5-phosphate is converted to D-ribulose 5-phosphate and D-erythrose 4-phosphate reacts with DHAP (formed from G3P (4)), yielding D-sedoheptulose 1,7-bisphosphate, which is hydrolysed to D-sedoheptulose 7-phosphate. The transketolase (EC 2.2.1.1) catalyses a reaction between C3P (5) and D-sedoheptulose 7-phosphate causing the formation of D-ribose 5-phosphate and D-xylulose 5-phosphate, which both are transformed to D-ribulose 5-phosphate. The described reactions have thus resulted in formation of three molecules of D-ribulose 5-phosphate, which are phosphorylated to three molecules of D-ribulose 1,5-bisphosphate and a new turn of the cycle is ready to start.

As outlined above the enzyme *transketolase* (EC 2.2.1.1) catalyses reactions between the aldose G3P and the ketoses D-fructose 6-phosphate and D-sedoheptulose 7-phosphate with formation of the ketose D-xylulose 5-phosphate and the aldoses D-erythrose 4-phosphate and D-ribose 5-phosphate. Fig. 20 illustrates these reactions.

The ultimate goal of the photosynthetic process is to provide energy from the light reactions and organic carbon from the dark reactions for production of all primary and secondary metabolites which the plant needs to grow and sustain life. The sixth molecule of G3P is the primary product in this process. It can be exported to the cytosol where it is used for synthesis of sucrose (see page 150). Sucrose contains 12 carbon atoms and the corresponding number of CO_2 molecules is needed for its formation. Thus four turns of the C3 cycle are required for production of one molecule of sucrose. Sucrose is transported from the cytosol of the leaf cells to other parts of the plant where it can be used as starting material for all biosynthetic processes. G3P can also be used within the chloroplast for synthesis of starch, which is a storage form for carbon and energy needed for respiration and other metabolic processes during periods when photosynthesis is not sufficient. Sucrose may also be temporarily stored in the vacuole of the cell for the same purpose.

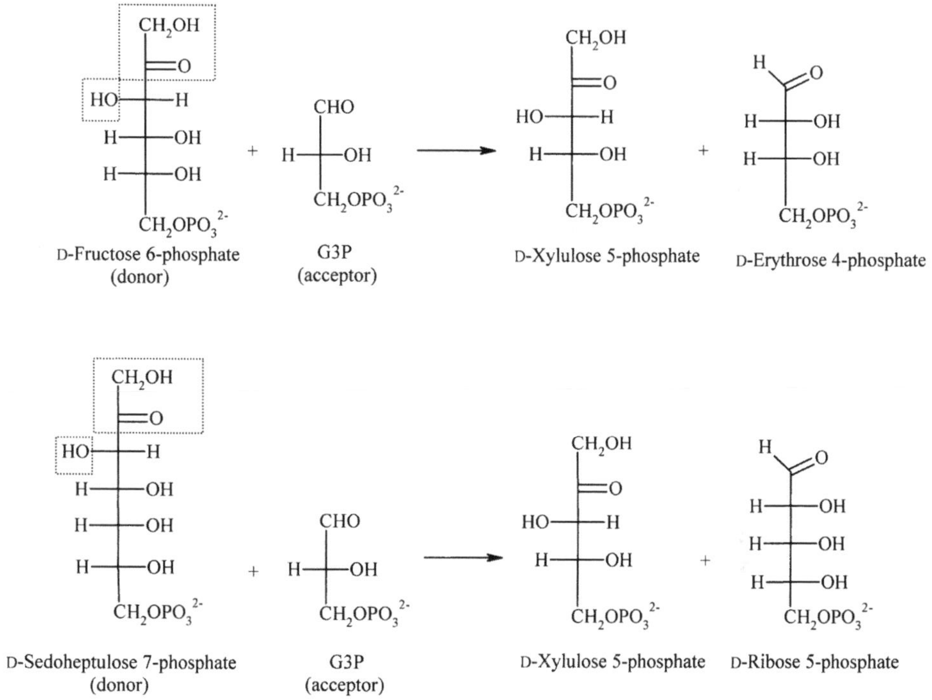

D-Fructose 6-phosphate (donor) + G3P (acceptor) → D-Xylulose 5-phosphate + D-Erythrose 4-phosphate

D-Sedoheptulose 7-phosphate (donor) + G3P (acceptor) → D-Xylulose 5-phosphate + D-Ribose 5-phosphate

Fig. 20. Transketolase reactions (the structures are drawn as Fischer projections)

Photorespiration

C₃ Plants

The first reaction in the reductive pentose phosphate pathway, as described above, is the formation of the three-carbon product D-glyceraldehyde 3-phosphate. Most plants fix carbon by this pathway and are therefore called *C₃ plants*. C₃ plants, however, are hampered in their fixation of CO_2 by a process called *photorespiration*. This process depends on the dualistic nature and ability of Rubisco also to act as an oxygenase. In this reaction, molecular oxygen is bound to the enzyme and reacts with D-ribulose 1,5-bisphosphate to form 2-phosphoglycolate and 3-phospho-D-glycerate.

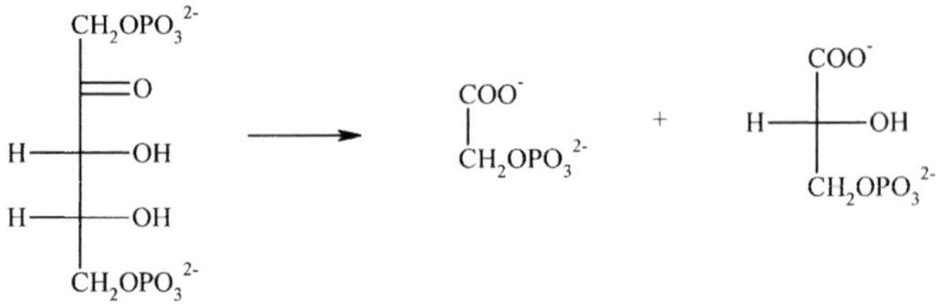

D-Ribulose 1,5-bisphosphate   2-Phosphoglycolate   3-Phospho-D-glycerate

2-Phosphoglycolate cannot be used in the reductive phosphate pathway, but two molecules of the substance react further along a cyclic pathway involving glycine, serine and glycerate to give one molecule of 3-phospho-D-glycerate, one molecule of CO_2, and one molecule of inorganic phosphate:

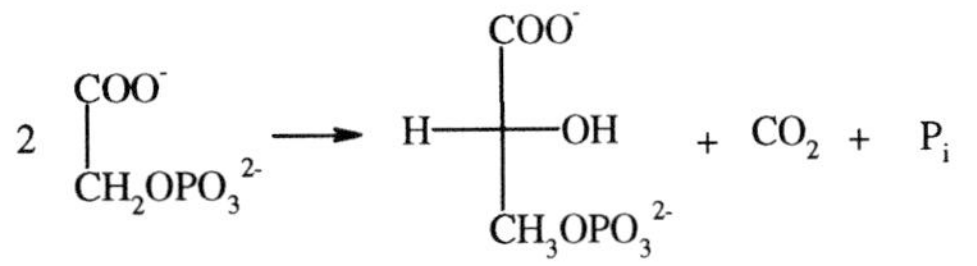

Photorespiration thus competes with carbon dioxide assimilation and makes photosynthesis less efficient. More than 30 % of the carbon fixed by photosynthesis is lost by photorespiration in normal C_3 plants. The process is stimulated by light and raised temperatures and is thus especially unfavourable for plants growing in a hot climate.

CAM and C_4 plants

Plants living under hot and dry conditions are in an especially unfavourable situation with respect to photosynthesis. Not only is photorespiration favoured by the high temperature, but access to water is limited, which forces the plant to close its stomata in order to retain water. Many plants in such areas have adapted themselves to the scarcity of water by decreasing their leaf surface and by storing water in fleshy tissues. Such plants are called *succulents*. Typical succulents are *cacti* (family *Cactaceae*) and stonecrops, *Sedum* species (*Crassulaceae*). In addition to saving water, some succulents have also adjusted themselves to optimizing photosynthesis by introducing a preliminary carboxylation step before the reductive pentose phosphate pathway. This procedure has been called *Crassulacean Acid Metabolism* (CAM) and the plants having this function are called *CAM plants*. CAM is a procedure for fixing CO_2 at night, when the temperature is low and the stomata can be kept open, allowing atmospheric CO_2 into the mesophyll, but when the normal CO_2-fixing procedure cannot work because of the absence of light. The mechanism for CAM is the following:

CO_2 reacts with phosphoenolpyruvate (PEP) to form oxaloacetate. The reaction is catalysed by the enzyme *phosphoenolpyruvate carboxylase* (EC 4.1.1.31), which is present in the cytoplasm of the leaf cells. Oxaloacetate is reduced by another enzyme – *malate dehydrogenase (NADP+)* (EC 1.1.1.82) – to malic acid, which is stored in the vacuoles of the leaf cells.

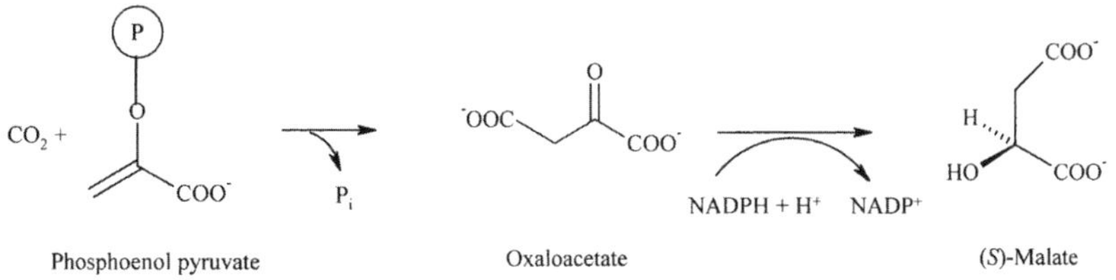

Phosphoenol pyruvate   Oxaloacetate   (*S*)-Malate

In daytime, when light is available but the stomata have to be closed, malate is transported back to the cytoplasm, where it is decarboxylated by: *malate dehydrogenase (oxaloacetate-decarboxylating)* (EC 1.1.1.40) to yield CO_2 and pyruvate. The CO_2 liberated is then used in the normal photosynthetic reactions in the chloroplasts.

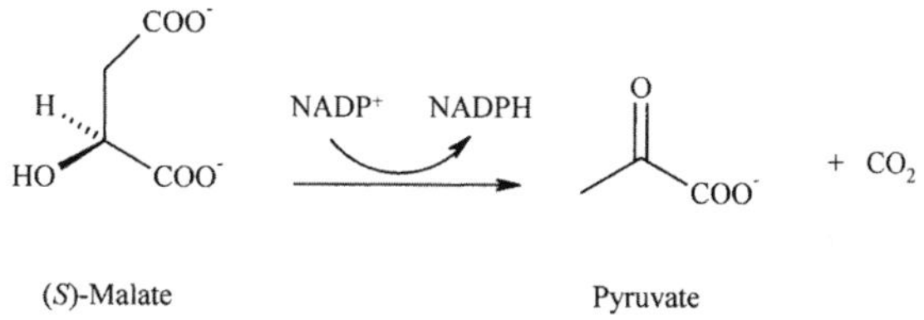

(*S*)-Malate   Pyruvate

As mentioned above, CAM is a mechanism for CO_2 fixation which is typical of succulent plants. However, there is a variation of this process which has been adopted by other plants living in hot climates, e.g. sugar cane (*Saccharum officinarum* L.) and maize (*Zea mays* L.). This alternative has been called the C_4 *mechanism* and the plants in which it works are called C_4 *plants*. The C_4 process differs from CAM in that it functions during daytime. The series of reactions starts in the same way as CAM, i.e. by fixing CO_2 to PEP with the formation of oxaloacetate, which is subsequently transformed into malate. This process takes place in the mesophyll, but, instead of storing the malate in the vacuoles as CAM plants do, the C_4 plants immediately transport it to specialized cells surrounding the vascular bundles in the leaf. These cells contain chloroplasts and form a tissue around the vascular bundles which is easily detected microscopically. This structure was first described by German botanists, who termed it *Kranz* anatomy, a reference to the appearance of the tissue which in a transverse section resembles a garland

(Kranz = garland). In the Kranz cells, malate is broken down to form pyruvate and CO_2, which is then fixed by the normal pentose phosphate pathway. Obviously, photorespiration must also be operating in C_4 plants, but its effects are not observable. Neither high leaf temperature nor oxygen depresses the assimilation of carbon in these plants. The reason for this is probably that the pentose phosphate pathway operates in a well insulated area: any CO_2 formed within the cells of the vascular bundle as a consequence of glycolate degradation is likely to be trapped by another PEP molecule on its way through the cytoplasm. C_4 plants can utilize very high light intensities and can maintain a positive carbon balance even when the concentration of CO_2 is very low. C_4 plants can grow very rapidly and thus economically important C_4 plants, like sugar cane and maize, can give a much higher yield per unit of area than do C_3 plants.

Biosynthetic pathways

Photosynthesis transforms CO_2 and H_2O into carbohydrates, which fulfil many important functions in the plant. An example is cellulose which functions as a building material. The controlled metabolism of carbohydrates in glycolysis and in the citric acid cycle releases the energy needed for all biological reactions, and carbohydrates are the source of carbon for the synthesis of organic compounds in plants. Biosynthetic reactions in plants have been studied most intensely since the Second World War and it is now known that all plant constituents are formed via a few general pathways, as illustrated in Fig. 21. As seen from this figure, carbohydrates are degraded to pyruvic acid, which is oxidized to acetate that can condense to form fatty acids and polyketides (also aromatic ones). A different biosynthetic pathway leads from acetate via the condensation product mevalonic acid to sesquiterpenes, triterpenes and steroids. Monoterpenes, diterpenes and tetraterpenes are formed from glyceraldehyde-3-phosphate – which is a primary product of photosynthesis – and pyruvic acid, along a newly discovered pathway – the non-mevalonate pathway. A third route ends in the formation of certain amino acids. Other amino acids are formed directly from pyruvic acid. Yet another route, the so-called shikimic acid pathway, leads from carbohydrate to aromatic amino acids. It is named after the key product, shikimic acid, which is formed from the tetrose erythrose and pyruvic acid. Shikimic acid is the starting material for the biosynthesis of tannins and can also, via chorismic acid (formed by reaction with a second molecule of pyruvic acid), give rise to aromatic amino acids. The nitrogen required for amino-acid biosynthesis comes originally from air. Atmospheric nitrogen is reduced to ammonia by nitrogen-fixing bacteria, thus making it accessible for the biosynthetic processes taking place in higher plants. Bacteria are estimated to fix about 100 million tons of nitrogen per annum;

94 % of the nitrogen contained in human food is probably originally fixed by bacteria, and the remaining 6 % is derived from synthetic fertilizers.

An important biosynthetic pathway leads from amino acids to proteins, alkaloids and purines, and the last, in combination with carbohydrates and inorganic phosphate, form the vital nucleic acids.

In addition to the groups of compounds formed along the main routes, schematically outlined here, many important natural compounds are formed by the combination of substances derived from different main pathways. It is also worth mentioning that the main biosynthetic pathways are common to all studied organisms and that the difference in this respect between autotrophic plants and animals is the ability to photosynthesize which has been lost by evo-

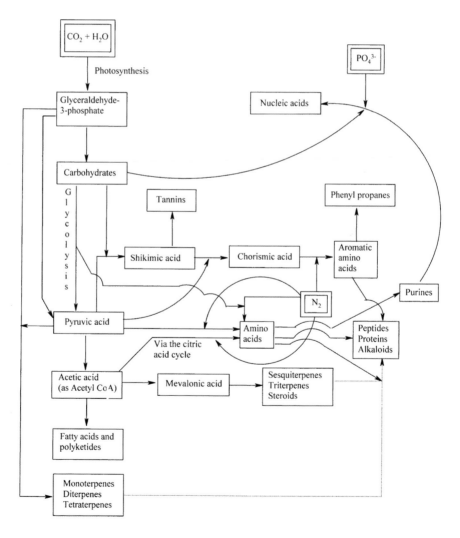

Fig. 21. Overview of general biosynthetic pathways

lutionary processes in animals. It follows that photosynthesizing organisms constitute a necessary prerequisite for all organic life, photosynthesis being the only process by which CO_2 is transformed into vitally important organic compounds.

Glycolysis and the citric acid cycle

The pathways illustrated in Fig. 21 will be dealt with in detail in chapters 5–11. However, as pyruvic acid is a key-compound in these pathways, a description of its formation is merited here. Also its further fate in *the citric acid cycle*, which generates several compounds used in various biosynthetic reactions, will be described. Pyruvic acid is ionized under physiological conditions in the plant and will therefore be referred to as pyruvate in the following text.

Pyruvate is formed from storage carbohydrates (sucrose and

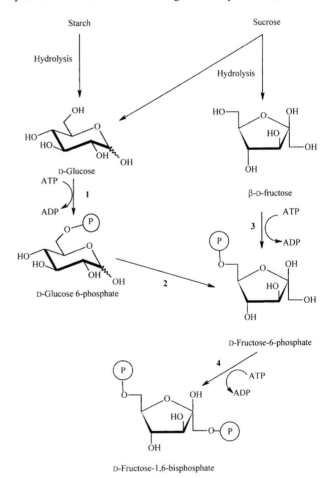

Fig. 22. Formation of D-fructose-1,6-bisphosphate from starch and sucrose. Participating enzymes **1**: hexokinase *(EC 2.7.1.1)*, **2**: glucose-6-phosphate isomerase *(EC 5.3.1.9)*, **3**: fructokinase *(EC 2.7.1.4)*, **4**: 6-phosphofructokinase *(EC2.7.1.11)*

starch, (pages 150 and 153), which are hydrolysed to yield glucose and fructose. Glucose and fructose are phosphorylated yielding D-glucose 6-phosphate and D-fructose 6-phosphate, respectively. D-Glucose 6-phosphate is isomerized to D-fructose 6-phosphate. The fructose 6-phosphate formed in these reactions is finally phosphorylated to form D-fructose 1,6-bisphosphate (FBP, Fig. 22). These reactions require energy in the form of ATP (two molecules for each molecule of FBP formed).

Glycolysis   The energy-rich FBP is now ready to enter glycolysis, a series of reactions which result in formation of pyruvic acid and recovery of the two molecules of ATP previously used, as illustrated in Fig. 23. The first step is cleavage of FBP to form dihydroxyacetone phosphate and D-glyceraldehyde 3-phosphate, a reaction catalysed by the enzyme *fructose-bisphosphate aldolase* (EC 4.1.2.13). Dihydroxyacetone phosphate is converted to D-glyceraldehyde 3-phosphate (G3P) by *triose-phosphate isomerase* (EC 5.3.1.1). One molecule of FBP thus generates two molecules of G3P, and eventually two molecules of pyruvate. G3P is further phosphorylated to 1,3-bisphospho-D-glycerate in a reaction catalysed by *glyceraldehyde-3-phosphate dehydrogenase (phosphorylating)* (EC 1.2.1.12). In this reaction the phosphate group originates from inorganic phosphate and NAD^+ is reduced to NADH. *Phosphoglycerate kinase* (EC 2.7.2.3) catalyses removal of one phosphate group (which is transferred to ADP with formation of ATP) forming 3-phospho-D-glycerate, which is converted to 2-phospho-D-glycerate by *phosphoglycerate mutase* (EC 5.4.2.1). This compound is then reduced to phosphoenol pyruvate by *phosphopyruvate hydratase* (EC 4.2.1. 11). Finally the phosphate group is transferred to ADP by *pyruvate kinase* (EC 2.7.1.40) with formation of pyruvate and a second molecule of ATP.

Pyruvate can of course also be formed from de novo synthetized G3P in direct connection with photosynthesis (see Fig 19).

The netresult of glycolysis of one molecule of glucose or fructose is formation of two molecules of pyruvate + 2ATP + 2 NADH. Glycolysis is thus not only a method for the plant to generate pyruvate as a starting material for a number of biosynthetic pathways, but also to recover a part of the energy (in the form of ATP) that is stored in the carbohydrates formed during photosynthesis. The reducing NADH formed, is used in other biosynthetic procedures or is – in the presence of oxygen – metabolized in the mitochondria to produce ATP. Phosphoenolpyruvate is a starting material for the biosynthesis of aromatic amino acids (page 202) and can also be formed from pyruvate by the reversed reaction.

The citric acid cycle   As seen from Fig. 21 several biosynthetic pathways start with pyruvic acid. One of these involves decarboxylation and oxidation to acetic acid and transfer of the acetyl group to coenzyme A with formation of acetyl coenzyme A (acetyl CoA). The reaction is catalysed by the *pyruvate dehydrogenase complex*, which is very large and consists of the three enzymes *pyruvate dehydrogenase (acetyl-*

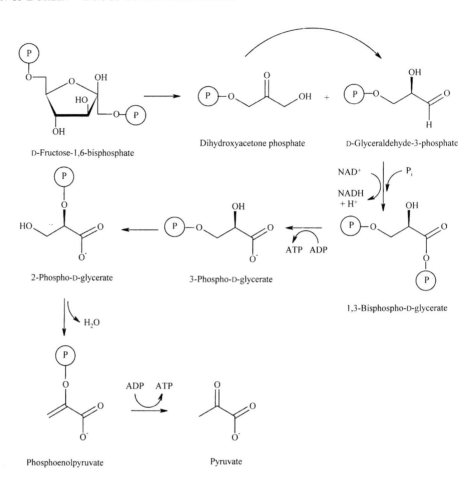

Fig. 23. Formation of pyruvate from D-fructose 1,6-bisphosphate

transferring) (EC 1.2.4.1), *dihydrolipoyl dehydrogenase* (EC 1.8.1.4) and *dihydrolipoyllysine-residue acetyltransferase* (EC 2.3.1.12). Each of these is composed of several polypeptide chains and the complex requires thiamine diphosphate, lipoic acid and FAD as cofactors. The overall progress of this complicated reaction is illustrated in Fig. 24.

AcetylCoA is the reactive form of acetic acid that is the starting material for the acylpolymevalonate pathway to fatty acids and polyketides (page 237) and the mevalonate branch of the isopentenyl diphosphate pathway (page 367) leading to sesquiterpenes,

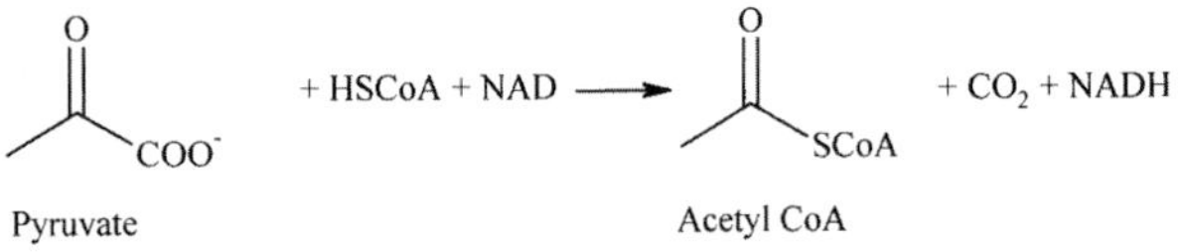

Fig. 24. Formation of acetyl CoA from pyruvate

triterpenes and steroids. Acetyl CoA is also further metabolized in the citric acid cycle, which generates energy as GTP (guanosine triphosphate) and reducing potential as reduced flavin adenine dinucleotide (FADH2) and reduced nicotinamide adenine dinucleotide (NADH). In addition several substances are formed which are utilized in other biosynthetic reactions such as 2-oxoglutarate, oxaloacetate and succinyl CoA which participate in formation of amino acids (chapter 8, page 486).

The first reaction in the citric acid cycle is formation of citrate from acetyl CoA and oxaloacetate, catalysed by *citrate* (Si)-*synthase* (EC 2.3.3.1). Citrate is isomerized to *iso*-citrate via *cis*-aconitate by the enzyme *aconitate hydratase* (EC 4.2.1.3). In the next reaction *isocitrate dehydrogenase (NAD+)* (EC 1.1.1.41) oxidizes *iso*-citrate to 2-oxoglutarate with formation of CO_2 and NADH. In a complicated reaction, catalysed by a complex of the three enzymes *oxoglutarate dehydrogenase (succinyl-transferring)* (EC 1.2.4.2), *dihydrolipoyllysine-residue succinyltransferase* (EC 2.3.1.61) and *dihydrolipoamide dehydrogenase* (EC 1.8.1.4), 2-oxoglutarate is oxidized and reacts with CoA, forming succinyl CoA, CO_2 and NADPH. The following reaction releases CoA from succinyl CoA with formation of GTP from GDP and inorganic phosphate and is catalysed by *succinate-CoA ligase (GDP-forming)* (EC 6.2.1.4). The succinate formed is oxidized to fumarate by *succinate dehydrogenase (ubiquinone)* (EC 1.3.5.1). In the next reaction *fumarate hydratase* (EC 4.2.1.2) transforms fumarate to (*S*)-malate. In the final reaction of the cycle, oxaloacetate is regenerated from malate, enabling a new turn of the cycle to start. The catalysing enzyme in this reaction is *malate dehydrogenase* (EC 1.1.1.37).

Investigation of biosynthetic pathways

On page 128 and the following pages the dark reactions of photosynthesis were described, and it was already mentioned that these were elucidated by allowing green algae to photosynthesize for shorter or longer periods in the presence of $^{14}CO_2$, followed by isolation of the compounds formed, and determination of their radioactivity. This permitted conclusions about which compounds were formed first (after only very brief illumination) and which substances were derived from these and appeared later in the biosynthetic sequence. The use of radioactive isotopes for labelling supposed precursors in a biosynthetic pathway has for a long time been the dominating method for investigation of how different organisms synthesize various secondary metabolites. However, before this can be done, it is necessary to isolate and determine the structure of all compounds in the organism which can be assumed to be involved in the biosynthesis.

Suppose that one wants to find out how a plant makes a particular alkaloid. The work will then start by isolation of a fraction containing all alkaloids present in the plant, a task which is compara-

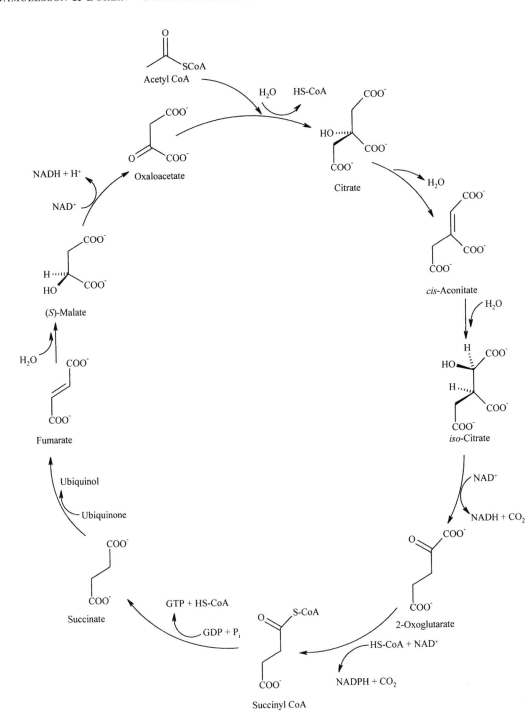

Fig. 25. The citric acid cycle

tively easy to perform (see page 607). The individual alkaloids contained in the fraction are then separated and their structures determined. From these structures it is often possible to guess a plausi-

ble biosynthetic sequence for the formation of the particular alkaloid. Let us assume that the fraction contains five alkaloids A, B, C, D and E and that E is the alkaloid we are interested in. Let us also suppose that the structures of the four substances indicate that B, C, D and E are formed from A. This hypothesis can then be investigated by synthesis of a variety of compound A where one suitable carbon atom consists of radioactive ^{14}C instead of the normal ^{12}C. This variety of compound A is then fed to the plant which produces the alkaloids and after a suitable time the alkaloid fraction is isolated and separated to yield B, C, D and E. If our hypothesis was right these should now be radioactive.

The SI unit which is used for radioactive disintegration rate is the *becquerel* (Bq) which is defined as one disintegration per second (d.p.s.). This is a very small unit and one often uses the megabecquerel (MBq) which is 10^6 d.p.s. or the gigabecquerel (GBq) = 10^9 d.p.s. Formerly the unit *curie* = 3.7×10^{10} d.p.s. or its subunits millicurie (3.7×10^7 d.p.s) and microcurie (3.7×10^4 d.p.s.) were used. The *specific activity* of a radioactive compound is defined as the radioactive disintegration rate per millimole of the compound which thus should be expressed as Bq/mM, MBq/mM or GBq/mM. In the literature one often sees the specific activity expressed in the older units mCi/mM (millicurie per millimole) or μCi/mM (microcurie per millimole). Also the expression dpm/mM (disintegrations per minute per millimole) is encountered.

In the above example determination of the specific radioactivity of each compound permits calculation of how much of the labelled A has been incorporated into each of the three other compounds and from these figures the sequence of the biosynthesis can be deduced. The incorporation rate in % is calculated as ([specific activity of the product] $\times$ 100)/(specific activity of the precursor). Suppose the following figures for the incorporation rates were obtained: B: 0.25 %, C: 0.75 %, D: 0.10 % and E 0.05 %. Such low incorporation rates are generally encountered when working with living plants because of difficulties in introducing the precursor into the biosynthetically active compartments of the plant. When working with bacteria, fungi or cell cultures of plants much higher incorporation rates are usually obtained. Based on the incorporation rates the biosynthetic sequence should be: A→C→B→D→E. This is of course a very simple example and such unambiguous results are seldom obtained in practice. For such a comparatively long sequence the incorporation in the last compound (E) is often so low that it cannot be determined. One must then synthesize labelled compounds that are closer to this substance in the chain (e.g. B and D in the example) and show that these are incorporated with good yields. It might also be necessary to degrade the compounds chemically to see that the atom or atoms that were labelled in the precursor end up in the right place in the structure of the compound under study. Labelling at more than one place in the structure, either by the same isotope or by different isotopes, might also become necessary. The deduced sequence must also be chemically reasonable, i.e. the reactions which are supposed to take place must be in accordance with known organic chemical reaction mechanisms.

^{14}C and ^{3}H (tritium) are the most commonly used radioactive labels in biosynthetic experiments. Both isotopes are β-emitters of low or medium toxicity and are comparatively easy to handle. These isotopes are detected and assayed by liquid scintillation counters, which are instruments that utilize the conversion of the kinetic energy of a particle into a fleeting pulse of light as the result of its penetrating a suitable luminescent substance. Use of liquid scintillation media, consisting of a solvent in which the excitation occurs and a fluorescent solute which emits light to actuate a photomultiplier, permits the sample to be incorporated in the same solute, and, hence, attain optimum geometry between sample and scintillator. Modern instruments are fully automatic and it is possible also to measure mixed radiations such as ^{14}C and ^{3}H. The instrument is connected to a suitable ratemeter which records the counts over a given period of time.

The development of advanced NMR techniques has made it possible to use the non-radioactive carbon isotope ^{13}C for labelling purposes. Also other stable isotopes such as ^{2}H, ^{15}N and ^{18}O can be used. These can be detected by mass spectroscopy.

Biosynthetically active enzymes

The reactions which lead from the precursor A to the final product E in the example above are catalysed by enzymes which often have a very well defined substrate specificity. Each enzyme is encoded by a unique gene. Taking this fact into consideration the exemplified pathway can be rewritten as follows:

$$A \xrightarrow[\textit{Enzyme 1}]{\text{Gene 1}} C \xrightarrow[\textit{Enzyme 2}]{\text{Gene 2}} B \xrightarrow[\textit{Enzyme 3}]{\text{Gene 3}} D \xrightarrow[\textit{Enzyme 4}]{\text{Gene 4}} E$$

Much useful information has been obtained by the mapping of biosynthetic pathways using tracer techniques as described above. However, deductions made by this technique are sometimes ambiguous and the final establishment of a pathway requires identification of the enzymes which catalyse each individual reaction and of the genes which encode these enzymes. Modern techniques for isolation and characterization of proteins, and the development of molecular biology and genetics have now made it possible to study biosynthetic events also at the level of enzymes and genes. Progress has been particularly prominent for natural products produced by bacteria, as isolation of enzymes from a fermentation liquid is comparatively easy and as a number of bacteria genome have now been completely sequenced, allowing for comparisons. In tissues of plants the enzymes are generally present in rather small amounts and it is rather difficult to obtain quantities great enough to allow the characterization of a biosynthetic pathway at the enzyme level. Some biosynthetic enzymes are membrane-bound and their isolation requires the use of detergents and other techniques to liberate them from the membranes. The use of cell cultures has facilitated the isolation of biosynthetic enzymes from plants (page 109). The isolation and characterization of all the enzymes involved in the biosynthesis of the alkaloids berberine and morphine is worth mentioning (pages 645, 652). However, the development of modern gene technology has opened up new possibilities in this field. Identification of the individual genes encoding the biosynthetic enzymes is now possible, and it is possible to clone them or their open reading frames. These clones can then be expressed in various organisms, e.g. *Escherichia coli*, which can be grown on a large scale permitting isolation of large amounts of the enzyme (up to 1 g of enzyme per litre of cells is possible). Also, characterization of the enzymes with respect to amino acid sequence and molecular size is facilitated by gene technology. It is much easier to determine the sequences of nucleotides than the amino acid sequence of proteins and from the nucleotide sequence of the open reading frame encoding a biosynthetic enzyme, its amino acid sequence is easily deduced and its molecular weight can be calculated.

The availability of biosynthetic enzymes in large enough quantities permits synthesis of natural compounds in cell-free systems and it is sometimes possible to perform a "one pot" synthesis by mixing all the enzymes which are needed for a particular biosynthesis with the precursor of the endproduct. For instance, this has been done for hydrogenobyrinic acid which is a close precursor for vitamin B_{12} (page 587). Identification of the genes encoding the enzymes also opens up new fields in the search for new drugs from natural products, as it is possible to manipulate the genes, thereby obtaining enzymes which catalyse the biosynthesis of variants of the original compound which might have better properties. This is discussed further on pages 277 and 538.

Through these developments the study of biosynthetic pathways has become a subject which is not only of interest from a theoretical point of view, but which is also of great practical importance, particularly for the discovery of new drugs of natural origin. However, much work in this field remains to be done. For many plant-derived natural products, in particular, our knowledge is still restricted to what has been learnt from tracer experiments.

Enzyme nomenclature

The nomenclature for enzymes is complex and confusing The names usually indicate which type of reaction the enzymes catalyse, but in the literature several names for an enzyme catalysing a specific reaction are found. It also happens that the same name has been given to enzymes that catalyse different reactions. To bring order in this mess, the Nomenclature Committee of the International Union

of Biochemistry and Molecular Biology (NC-IUBMB) has laid down rules for the classification and nomenclature of enzymes. These rules are available at the URL:
http://www.chem.qmul.ac.uk/iubmb/enzyme/

The classification of enzymes is based on the chemical reactions that they catalyse. Each enzyme is given a codenumber, consisting of the letters "EC" followed by four numbers, separated by full stops. Those numbers represent a progressively finer classification of the enzyme. Each enzyme is also given a recommended name as well as a systematic name that uniquely defines the reaction catalysed.

All enzymes can be classified in one of the following classes, which are represented by the first number in the EC code:

Class	Name	Reaction catalysed
1	Oxidoreductases	Oxidation/reduction, transfer of H and O atoms or electrons from one substance to another
2	Transferases	Transfer of a functional group from one substance to another. The group may be a methyl-, acyl-, amino- or phosphate group
3	Hydrolases	Formation of two products from a substrate by hydrolysis
4	Lyases	Non-hydrolytic addition or removal of groups from substrates. C-C, C-N, C-O or C-S bonds may be cleaved
5	Isomerases	Intramolecular rearrangement, i.e. isomerization changes within a single molecule
6	Ligases	Joining together of two molecules by synthesis of new C-O, C-S, C-N or C-C bonds with simultaneous hydrolysis of a diphosphate bond in ATP or a similar triphosphate

The next two numbers in the EC code represent subclasses and sub-subclasses for finer classification. The fourth number is the serial number of the enzyme in its sub-subclass.

Example: EC 4.1.3.27
Accepted name: anthranilate synthase
Systematic name: chorismate pyruvate-lyase (amino accepting; anthranilate-forming)
Catalysed reaction: chorismate + L-glutamine = anthranilate + pyruvate + L-glutamate (the first step in the biosynthesis of tryptophan (page 206)
EC 4 Lyases
EC 4.1 Carbon-carbon lyases
EC 4.1.3 Oxo-acid lyases
EC 4.1.3.27 Individual number for anthranilate synthase

The above-mentioned site of the Nomenclature Committee of the International Union of Biochemistry and Molecular Biology also provides a database from which information on individual enzymes can be obtained. The base can be searched by the enzyme code or an enzyme name. One can also search for enzymes involved in biosynthetic pathways, for enzymes catalysing formation of a specific compound or for enzymes isolated from specific organisms. This database also provides a number of links to other databases containing information on enzymes.

In this book the accepted enzyme names are used and the corresponding EC codenumber is indicated. For enzymes which have not yet been classified in the EC system the names are used which were given by the author of the publication from which the presented information was obtained.

Transporters of secondary metabolites

In plants, pharmacologically active secondary metabolites are often transported from the organ in which they are produced to another organ where they exert their action or are stored. An example is *Atropa belladonna* L. which makes tropane alkaloids in the roots and then transports them to the aerial parts (page 605). Another example is nicotine, which is produced in the roots and transported to the aerial parts and accumulated in the leaves. Vacuoles, which often occupy 40-90 % of the inner volume of plant cells, are important storage organs. Storage in vacuoles serves two purposes: protection of the metabolite from catabolism and protection of the cell from damage, caused by a too high concentration of the metabolite. Transport of secondary metabolites across cell membranes involves employment of particular membrane proteins – transporter proteins – and can be achieved by two different mechanisms: H^+-gradient dependent secondary transport via H^+ antiport (= counter-transport) or directly energized primary transport by ATP-binding cassette (ABC transporters). Antiport or counter-transport is a term to describe transport in opposite directions of two different ions or molecules. ABC transporters are located on the outer membrane of the cell or on intracellular organelles. They bind ATP and use the energy released by hydrolysis of ATP to transport molecules across cell membranes. ABC transporters are found in a range of organisms from bacteria to humans. Most of the transported substrates are hydrophobic compounds although some can transport metal ions, peptides and sugars. ABC transporters have been extensively studied in connection with their contribution to the resistance of tumour cells to chemotherapy drugs.

The membrane transport of plant secondary metabolites is a newly developing research area and it has been found that ABC transporters are involved in some plant systems. As for the examples cited above, no specific transporters have been identified so far (2007). In these examples the situation is particularly complicated as the alkaloid should be loaded into xylem tissue and unloaded at mesophyll tissue where it is finally accumulated in the vacuoles. This implies transport across at least three different membranes, plasma membranes in the root and in the leaf and the vacuolar membranes of mesophyll cells. *Coptis japonica* (*Ranunculaceae*) is a plant which produces the alkaloid berberine (page 645) in the roots and translocates it to the rhizome, where it is trapped by an ABC transporter in the plasma membrane and transported into the cytosol of the rhizome cell. This is an unique property as ABC importers have hitherto been found only in prokaryotes. The alkaloid is further transported from the cytosol into a vacuole, probably by an H+ antiporter.

Classification of natural products

Natural products have previously been classified on the basis of their chemical structures (carbohydrates, steroids, etc.), by their physiological effects (vitamins, antibiotics, etc.), or by their occurrence (flower pigments, heartwood constituents, lichen acids, etc.). In this textbook, pharmacologically active natural products have been grouped according to biosynthetic principles. This places them in a natural context and it also emphasizes the importance of knowledge of the biosynthetic pathways for natural products in the process of drug discovery as outlined above.

Further reading

Photosynthesis

Allen, J.F. Photosynthesis of ATP – electrons, proton pumps, rotors, and poise. *Cell*, **110**, 273–276 (2002).

Armstrong, F.A. Why did Nature choose manganese to make oxygen? *Philosophical Transactions of The Royal Society B*, **363**, 1263–1270 (2008).

Barber, J. The structure of photosystem I. *Nature Structural Biology*, **8**, 577–579 (2001).

Bolivar, D.W. Recent advances in chlorophyll biosynthesis. *Photosynthesis Research*, **90**, 173–194 (2006).

Chitnis, P.R. Photosystem I: Function and physiology. *Annual Review of Plant Physiology and Plant Molecular Biology*, **52**, 593–626 (2001).

Cramer, W.A., Soriano, G.M., Ponomarev, M,. Huang, D., Zhang, H., Martinez, S.E. and Smith, J.L. Some new structural aspects and old controversies concerning the cytochrome $b_6 f$ complex of oxygenic photosynthesis. *Annual Review of Plant Physiology and Molecular Biology*, **47**, 477–508 (1996).

Diner, B.A. and Rappaport, F. Structure, dynamics, and energetics of the primary photochemistry of photosystem II of oxygenic photosynthesis. *Annual Review of Plant Biology*, **53**, 551–580 (2002).

Fromme, P., Jordan, P. and Krauss, N. Structure of photosystem I. *Biochimica et Biophysica Acta (BBA) – Bioenergetics*, **1507**, 5–31 (2001).

Green, B.R. and Durnford, D.G. The chlorophyll-carotenoid proteins of oxygenic photosynthesis. *Annual Review of Plant Physiology and Molecular Biology*, **47**, 685–714 (1996)

Hopkins, G.W. *Introduction to Plant Physiology*. John Wiley & Sons, Inc., New York (1995).

Jordan, P., Fromme, P., Witt, H.T., Klukas, O., Saenger, W. and Krauss, N. Three-dimensional structure of cyanobacterial photosystem I at 2.5 Å resolution. *Nature*, **411**, 909–917 (2001).

McEvoy, J.P. and Brudvig, G.W. Water-splitting chemistry of photosystem II. *Chemical Reviews*, **106** (11), 4455–4483 (2006).

Photosynthetic water oxidation. *Biochimica et Biophysica Acta*, Special issue, **1503** (1–2), 1–259 (2001).

Rawsthorne, S. Towards an understanding of C3-C4 photosynthesis. *Essays in Biochemistry*, **27** (46), 135–146 (1992).

Reinbothe, S. and Reinbothe, C. The regulation of enzymes involved in chlorophyll biosynthesis. *European Journal of Biochemistry*, **237** (2), 323–343 (1996).

Rhee, K.-H. Photosystem II: The solid structural Era. *Annual Review of Biophysics and Molecular Structure*, **30**, 307–328 (2001).

Van Walraven, H.S. and Bakels, R.H. Function and regulation of cyanobacterial and chloroplast synthase. *Physiologia Plantarum*, **96**, 526–532 (1996).

Zouni, A., Witt, H.-T., Kern, J., Fromme, P., Krauss, N., Saenger, W. and Orth, P. Crystal structure of photosystem II from *Synechococcus elongatus* at 3.8 Å resolution. *Nature*, **409**, 739–743 (2001).

Biosynthetic pathways

Thomas, R. Biogenetic speculation and biosynthetic advances. *Natural Product Reports*, **21**, 224–248 (2004).

Transporter mechanisms

Hollenstein, K., Dawson, R.JP. and Locher, K.P. Structure and mechanism of ABC transporter proteins. *Current Opinion in Structural Biology*, **17**, 412–418 (2007).

Yazaki, K. Transporters of secondary metabolites. *Current Opinion in Plant Biology*, **8**, 301–307 (2005).

Yazaki, K. ABC transporters involved in the transport of plant secondary metabolites. *FEBS Letters* **580**, 1183–1191 (2006).

5. Carbohydrates

General carbohydrate chemistry is dealt with in organic chemistry. Here, we shall concentrate on the production of medicinally and pharmaceutically important carbohydrates, i.e. their occurrence in plants and their isolation from plant materials. Some biosynthetic aspects will also be discussed, and so will the sugar alcohols such as sorbitol and mannitol, which are biosynthetically related to carbohydrates. Certain structurally complicated carbohydrate derivatives, produced by microorganisms and having antibiotic properties, will also be considered, in connection with a general discussion of antibiotics and their production.

As shown in the previous chapter, carbohydrates are products of plant photosynthesis formed from water and carbon dioxide. They can be grouped into *sugars* and *polysaccharides*. Sugars, which are water-soluble and more or less sweet in taste, are either *monosaccharides*, such as glucose, fructose and fucose, or *oligosaccharides*, containing up to five or six monosaccharide units, which may be of one or different types. Thus, *disaccharides* consist of two monosaccharide residues bound to each other, *trisaccharides* contain three such residues, and so on. Monosaccharides can be classified according to the number of carbon atoms they contain. Thus, *trioses, tetroses, pentoses, hexoses* and *heptoses* are C_3 to C_7 compounds. Polysaccharides are macromolecules, containing a large number of monosaccharide residues. Their solubility in water is very low and they have no taste. Economically important polysaccharides are starch and cellulose, both of which are built up from the hexose glucose. More complex groups of polysaccharides are gums and mucilages, which on hydrolysis give not only pentoses and hexoses, but also uronic acids which are oxidation products of polysaccharides.

Gums and mucilages form a more complex group of polysaccharides. On hydrolysis they give both hexoses and pentoses, and also uronic acids, which are oxidation products of carbohydrates.

Glucose (Glucosum, D-glucose, dextrose, grape sugar)

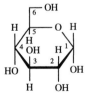

α-D-Glucopyranose

Monosaccharides

Considerable amounts of free glucose occur in many fruits, for example grapes. Technically, glucose is produced by acid hydrolysis of corn or potato starch under pressure. The mixture obtained is neutralized, passed through a filter press, decolourized with active carbon, and concentrated. The glucose crystallizes and is further purified by recrystallization.

Pure crystalline glucose is an important ingredient in intravenous infusions, where it serves as a source of energy. It is also used in large quantities in the food and confectionery industries, often in the form of a thick syrup, Glucosum liquidum.

**Fructose (Fructosum,
D-fructose, laevulose,
fruit sugar)**

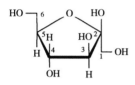

β-D-Fructofuranose

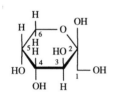

β-D-Fructopyranose

Fructose is a ketohexose occurring in the free form in many fruits and in honey. It is also a component of higher saccharides such as sucrose and inulin (see below). In Nature, fructose occurs only in the D-form and like glucose, it shows mutarotation. In the case of glucose this is due to interconversion between the α- and β-forms, but in fructose it is caused by interconversion between the pyranose and furanose structures. Crystalline fructose is the pyranose form and a freshly prepared aqueous solution has $[\alpha]_D^{20} = -132°$, a value that quickly changes to -92.4°, owing to partial conversion to the furanose form. At 25 °C, a fructose solution at equilibrium contains 22 % fructofuranose. Fructose residues built into oligosaccharides and polysaccharides have the furanose structure.

In industry, fructose is prepared by hydrolysis of the disaccharide sucrose or the polysaccharide inulin and is isolated from the hydrolysis mixture by precipitation with lime. The lime–fructose complex is decomposed by precipitation of the lime as an insoluble salt (carbonate, phosphate, oxalate or sulphate). When inulin is used as the source, it does not need to be isolated from the plant material, and press juice or extracts containing inulin can be hydrolysed direct. In order to obtain a good yield of fructose, it is necessary to work at low temperatures and to avoid excess acid and alkali.

Fructose is more soluble in water and has a sweeter taste than glucose. It is used as a sweetener for diabetics and, in infusions, for parenteral nutrition.

Crude drug

Honey (Mel). Honey is a sweet viscid fluid, produced as a reserve food by the honeybee, *Apis mellifica*. The honeybee is a social insect, and a colony consists of 10 000–50 000 individuals and sometimes even more. Each colony contains one fertile female – the queen bee; a few males – the drones – and a large number of sterile females – workers. From time immemorial man has known the art of keeping bees for their production of honey and wax. Special housing is provided for the bees: hives, which are constructed in such a way that it is easy to collect the products desired. In the hive, workers build honeycombs with hexagonal cells in which larvae hatch and develop. Wax for the honeycombs is secreted from cells under the abdomen of the bee. Some of the honeycomb cells are used for storing honey. To produce honey, workers visit flowering plants and suck up the nectar from the flowers by means of a specially shaped proboscis. At the same time, using special structures on their hind legs, they collect pollen, which is also used for food. Bees play an important role as pollination vectors for many plants. The nectar, taken from flowers by bees, consists largely of the disaccharide sucrose, which is then hydrolysed by enzymes in the honey sac of the bees. The construction of modern beehives facilitates the harvesting, because the bees are forced to store the honey in the cells of a special part of the honeycomb. At harvesting these parts are removed and the honey recovered by centrifugation. The wax is also of economic importance and is taken care of (see the chapter on waxes, page 255).

The mixture of glucose and fructose formed in the honey sac of the bee constitutes 70–80 % of the honey. The content of fructose is sometimes higher than that of glucose, due to presence of free fructose in the nectar. Other constituents of honey are water, sucrose (up to 10 %), pigments and volatile oils, giving the honey its special aroma. The colour, aroma and consistency of honey depend on the plant species from which the nectar has been collected.

Some honey is used pharmaceutically for flavouring drug mixtures and in folk medicine honey is used as an expectorant and cough remedy. Its main use, however, is as a sweetener and flavouring in the food and confectionery industries.

Lincomycin

Lincomycin is an example of an *antibiotic* (see page 89), which structurally can be regarded as an octose related to galactose. Carbon atom 6 carries an amino group, which forms an amide with 1-methyl-4-propyl-2-pyrrolidine-carboxylic acid. The anomeric carbon bears an -S-CH₃ group and the compound is thus also an example of an *S*-glycoside (see page 180).

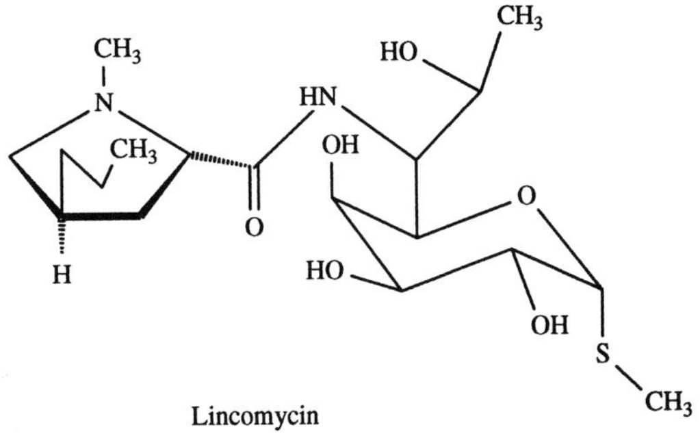

Lincomycin

Lincomycin is obtained from cultures of the bacterium *Streptomyces lincolnensis*. It is a remedy for infections caused by staphylococci.

Disaccharides

Sucrose (cane sugar, saccharose, Saccharum)

Sucrose is a sugar that occurs in many plants. The main sources for its industrial preparation are the sugar cane *Saccharum officinarum* L. (*Poaceae*) and the sugar beet *Beta vulgaris* L. (*Chenopodiaceae*).

Biosynthesis

Sucrose is the first disaccharide to appear in free form as a reaction product after photosynthesis, and carbohydrate transport within the plant is largely the transport of sucrose. The sugar is probably the

α-D-Glucopyranosido-β-D-fructofuranose

most common precursor of polysaccharides. The starting material for the biosynthesis of sucrose (Fig. 26) is D-glyceraldehyde 3-phosphate (G3P), which is produced in the chloroplast by the photosynthetic C3 cycle (page 128). G3P is exported through the membranes of the chloroplast envelope by an antiport protein which delivers G3P to the cytosol in exchange for phosphate on a one-for-one basis. Four molecules of G3P are required for the synthesis of one molecule of sucrose. Two molecules of G3P are converted to dihydroxyacetone phosphate (a reaction that possibly may take place in the chloroplast before export) and react with the two remaining molecules of G3P to form two molecules of D-fructose 1.6-bisphosphate, which are converted to D-fructose 6-phosphate in the same way as in the C3 cycle as illustrated in Fig. 19, page 129. One of the molecules of D-fructose 6-phosphate is converted to D-glucose 6-phosphate by *glucose-6-phosphate isomerase* (EC 5.3.1.9) and further to D-glucose 1-phosphate by *phosphoglucomutase* (EC 5.4.2.2). D-Glucose 1-phosphate is activated by UTP (page 738) to form UDP-glucose in a reaction catalysed by *UTP-glucose-1-phosphate uridylyltransferase* (EC 2.7.7.9). Catalysed by *sucrose-phosphate synthase* (EC 2.4.1.14) the second molecule of D-fructose 6-phosphate then reacts with UDP-glucose to form

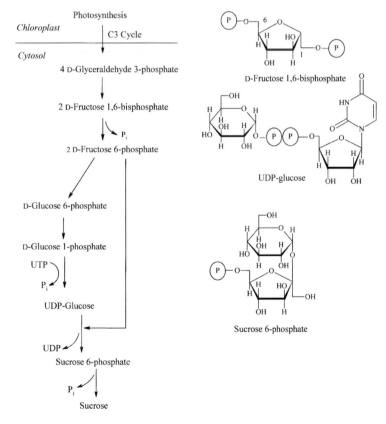

Fig. 26. Biosynthesis of sucrose

sucrose 6-phosphate which is split into sucrose and inorganic phosphate by *sucrose-phosphatase* (EC 3.1.3.24).

Production of sugar from sugar cane

Sugar cane is cultivated on a large scale in many tropical and subtropical countries such as Cuba, Puerto Rico, the Philippines, Hawaii, Indonesia, India and the southern United States. Sugar beet is grown in colder climates and used for sugar production in Europe. The preparation of sucrose from sugar cane or sugar beet is a unique case of the large-scale production of a pure natural compound from a plant material.

Sugar cane is usually raised from cuttings and is ready for harvesting 1–2 years after planting. The plants are then 5–6 m high and the sugary stems are 2.5–5 cm in diameter. The cane is cut by hand or machine, the leaves are removed, and the cane is cut into pieces, from which the juice is pressed out between rollers. The waste material, known as bagasse, is often used as fuel. The press juice is treated with lime and heated, so that the plant acids are neutralized and the proteins precipitated. The juice is clarified, filtered, concentrated to a thick syrup, and put into a vacuum evaporator, where most of the water is removed and the sugar crystallizes. When the water content is reduced to about 10 %, the mixture is centrifuged to separate the sugar crystals from the mother liquor, *molasses*. The molasses so obtained contains a good deal of sugar, which is recovered in two more evaporation stages. The molasses finally remaining after the third evaporation is used for cattle fodder or alcohol fermentation.

The crude sugar obtained from this process contains approximately 95 % sucrose and is brown in colour. For further purification (refining), the crystals are washed with small amounts of water in a centrifuge. They are then dissolved and the weakly acidic solution is neutralized with lime, mixed with kieselguhr, and filtered. The filtrate is decolourized by passing it through a layer of animal charcoal (active carbon). The expended carbon is recovered, washed, heated to glowing, and re-used. The decolourized sugar solution is concentrated and on crystallization gives pure colourless sugar containing at least 99.9 % sucrose. This is one of the purest organic compounds produced industrially on a large scale.

Production of sugar from sugar beet

Sugar beet can be grown in colder climates than sugar cane. It is the dominant source of sugar in Europe and is also used in the United States. Sugar beet seeds are sown early in spring and the beets are ready for harvesting just before the ground freezes in late autumn. They then contain approximately 15 % sucrose. In the refinery, the sugar beet is cut into strips and extracted with warm water in a countercurrent process. The extracted plant material is used for cattle feed. The extract – which contains ca. 15 % sucrose – is heated with lime. Carbon dioxide is then added to precipitate the calcium, and the solution is filtered, all in a continuous process. The treatment with lime brings about a substantial degree of purification, and the solution is then further purified by means of active carbon, concentrated and crystallized, as described above. The molasses can be used for cattle fodder or for yeast production. Some sugar works apply the so-called *Steffen process* by which residual sugar in molasses is recovered as an insoluble addition product with lime, which is then used for the lime-treatment of the crude extract. The saccharated lime is cleaved in the subsequent treatment with carbon dioxide and the sugar is liberated.

In pharmacy, sucrose is used in the form of a concentrated aqueous solution (syrup) as a vehicle for many medicines. The syrup acts as a preservative (fungi and bacteria cannot grow in a concentrated sugar solution) and the sugar masks the often bitter or pungent taste of the drug. Sugar is also used as a constituent of tablets and pills, or as a starting material for the preparation of invert sugar, i.e. the mixture of glucose and fructose obtained on hydrolysis. Invert sugar is used in intravenous infusions for parenteral nutrition as the human body uses fructose more fully than glucose. Sucrose to be used for this purpose is required to be free from pyrogens.

Lactose
(Lactosum, milk sugar)

4-O-β-D-Galactopyranosido-α-D-glucopyranose

Lactose constitutes 2–8 % of the milk of mammals. Cow's milk contains 3–5 % and human milk 4–8 % of lactose. This sugar is obtained as a by-product of cheese production. Fresh milk is treated with rennet, an extract from calves' stomachs containing the enzyme rennin, which makes the milk proteins coagulate into curd which is then processed to cheese. The mother liquor – the whey – is concentrated, giving crude lactose, which is decolourized with active carbon and recrystallized. Lactose has a sweetish taste and is used in pharmaceutical technology as a constituent of tablets and for diluting highly active compounds, such as atropine, when small amounts of these are to be present in a medicine.

Polysaccharides

Starch (Amylum)

Starch is a plant product formed by fusion of a large number of glucose units arising from photosynthesis. It can be found in most parts of plants: in stems, roots, fruits and seeds. Starch consists of grains of various sizes (2–170 μm) and often of characteristic shape for different species. Under the microscope the grains often show distinct concentric striations around a dark spot, the *hilum*. These striations are the result of the deposition of successive layers of starch, which differ in their diffraction, around the hilum. The presence and shape of starch grains are often important criteria in the microscopic identification of plant materials. Many plants store starch as a reserve food. It can be stored in different parts of the plant, but most commonly in subterranean organs or in fruits and seeds. Such plants with large starch deposits are the most important sources of human food, e.g. cereals and potatoes.

Pure starch is prepared by mechanical disintegration of the starchy parts of plants, followed by suspension in water to remove tissue fragments. The starch is then separated from the water by centrifugation and dried to a water content of about 10 %. Flour pre-pared from the grains of wheat, rye, etc., by grinding and sifting, is not pure starch but contains fragments of the seed coat, embryo and endosperm.

Starch is a mixture of two structurally somewhat different poly-saccharides, *amylose* and *amylopectin*. Amylose consists of a straight chain of 250–300 D-glucose residues, linked together through α-1,4-bonds. The molecular mass is about 10^6 Da.

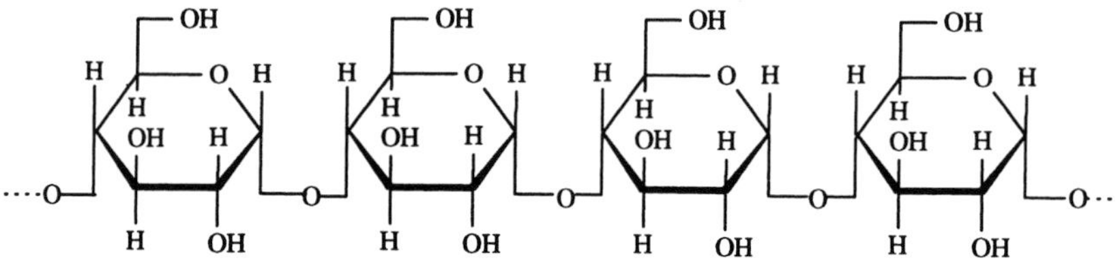

Amylose

In amylopectin, most of the glucose units are linked by α-1,4-bonds, but the structure is complicated by branching via α-1,6-link-ages. Amylopectin is a much larger molecule with a molecular mass of about 10^9 Da.

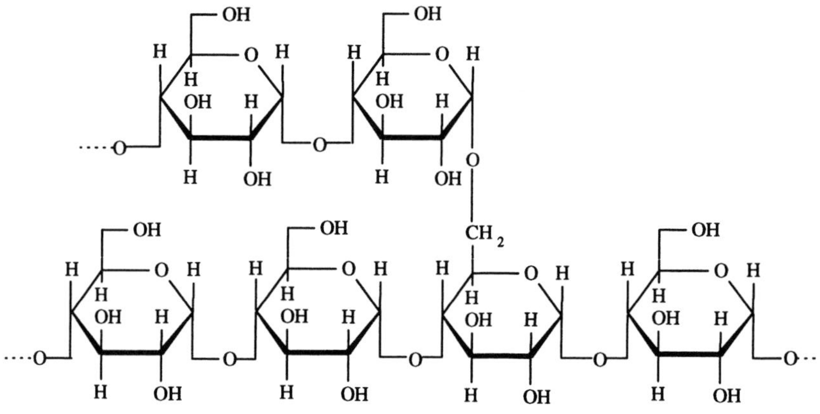

Amylopectin

About 4 % of the linkages between the glucose units in amylopectin are of the 1,6-type. Also present are varying amounts of ester-bound phosphoric acid, which is important for the technical properties of starch, e.g. paste formation.

The proportions of amylose and amylopectin in different starches vary, but a ratio of 1:3 is common.

Biosynthesis of starch

As mentioned previously (page 131), starch is formed in the chloroplasts where it serves as a reserve of carbon and energy, needed during times when photosynthesis is not active. The starch which is found in storage organs, such as seeds, roots, rhizomes and tubers, is formed in non-photosynthetic amyloplasts at the site of accumulation. The source of this starch is sucrose, originally produced in the cytosol of leaf cells and transported to the storage organs.

Starch biosynthesis in storage organs of dicotyledonous plants

When the sucrose enters the cytosol of storage cells in dicotyledonous plants, it is hydrolysed by the enzyme *sucrose synthase* (EC 2.4.1.13) to yield fructose and glucose. This enzyme also catalyses the reversible reaction between glucose and UDP with formation of UDP-glucose (Fig. 27). UDP-glucose is then converted to D-glucose 1-phosphate by *UTP-glucose-1-phosphate uridylyltransferase* (EC 2.7.7.9, also called UDP-glucose-pyrophosphorylase or UGPase). The enzyme *phosphoglucomutase* (EC 5.4.2.2) transforms D-glucose 1-phosphate to D-glucose 6-phosphate, which is translocated into the amyloplast via a phosphate translocator. In the amyloplast the same enzyme (EC 5.4.2.2) converts D-glucose 6-phosphate into D-glucose 1-phosphate, which serves as a substrate for *glucose 1-phosphate adenylyltransferase* (EC 2.7.7.27, also called ADP glucose pyrophosphorylase or AGP-ase) which catalyses attachment of an ADP-residue from ATP yielding ADP-glucose. The amyloplast cannot synthesize ATP, which has to be imported from the cytosol.

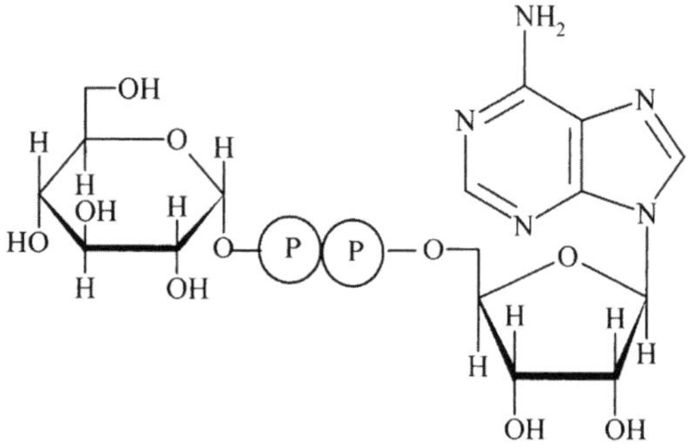

ADP-glucose

ADP-glucose is the starting material for formation both of amylose and of amylopectin. The enzyme *starch synthase* (EC 2.4.1.21) transfers the glucosyl unit from ADP-glucose first to a glucose molecule and then to the reducing end of a growing glucose chain via α, 1 → 4 linkages. Amylopectin is formed as a result of the combined action of starch synthase and *1,4-α-glucan branching enzyme* (EC 2.4.1.18, also called starch branching enzyme). The starch synthase first catalyses formation of an 1,4-α-glucan of suitable length, whereupon the starch branching enzyme catalyses hydrolysis of an α-1-4 bond and transfer of the released reducing end to a C_6-glucosyl unit, resulting in an α, 1 → 6 branch point.

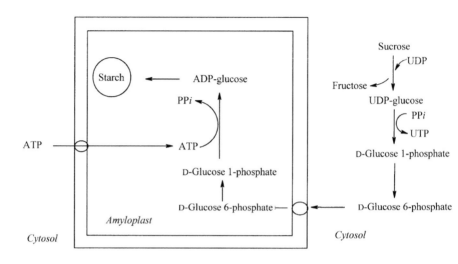

Fig. 27. Biosynthesis of starch in storage organs of dicotyledonous plants

Starch biosynthesis in cereal endosperms

In contrast to the conditions in the storage organs of dicotyledonous plants most of the AGP-ase in cereal endosperms is cytosolic. In the plastid, the AGP-ase activity is only 10 % of the total activity and is not sufficient to catalyse a normal rate of starch synthesis. Most of the ADP-glucose is produced in the cytosol and transported into the

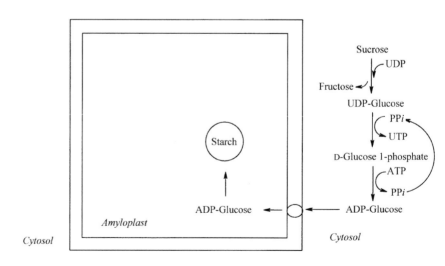

Fig. 28. Biosynthesis of starch in cereal endosperms

amyloplast where starch synthase and starch branching enzyme catalyse the formation of amylose and amylopectin (Fig. 28).

An alternative model for *starch biosynthesis* The above-described classical models for biosynthesis of starch in storage organs suffers from some inconsistencies. Thus a dramatic reduction of UGP-ase activity in transgenic potato tubers does not affect the starch content, implying that UGP-ase is not involved in starch biosynthesis. The function of UGP-ase in heterotrophic tissues of maize has been found to synthesize UDP-glucose rather than to degrade it. Other experiments have yielded results indicating that hexose phosphates are not directly involved in the starch biosynthetic process. These and other inconsistencies, together with the detection that both chloroplasts and amyloplasts are able to import ADP-glucose from the cytosol, that sucrose synthase can accept adenosine diphosphate effectively with production of ADP-glucose, and that the function of UGP-ase is to synthesize UDP-glucose rather than to degrade it, have led to the proposal of an alternative model for starch biosynthesis in storage organs as illustrated in Fig. 29.

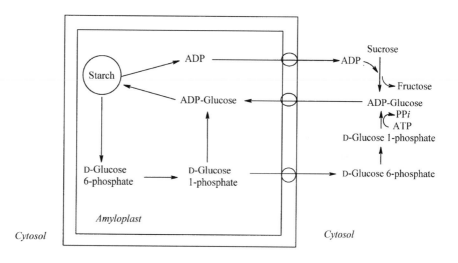

Fig. 29. An alternative model of starch biosynthesis in storage organs

According to this model sucrose synthase degrades sucrose and converts it to ADP-glucose which is transported into the amyloplast where it – as in the classical model – acts as a substrate for synthesis of amylose and amylopectin. The model also takes into account available evidence of concurrently occurring breakdown of starch in amyloplasts, which are equipped with enzymes capable of degrading starch. These enzymes degrade starch to glucose 6-phosphate. Glucose 6-phosphate is converted to glucose 1-phosphate which is used for synthesis of ADP-glucose by plastidic AGP-ase. Glucose

6-phosphate can also be transported out of the amyloplast into the cytosol, where it is converted to glucose 1-phosphate and then to ADP-glucose by cytosolic AGP-ase. Degradation of starch in the amyloplast also yields ADP, which is transported to the cyto-sol and used for the synthesis of ADP-glucose by sucrose synthase.

The evidence for an alternative model of starch biosynthesis in storage organs also prompts a change of the classic model for starch synthesis in leaves, according to which the triose phosphates produced in the Calvin–Benson cycle are used both within the chloroplast for synthesis of transient starch and exported to the cytosol for synthesis of sucrose. The alternative model predicts that all triose phosphate synthesized in the chhloroplast is exported to the cytosol where it is used for synthesis of sucrose. The transient starch in the chloroplast is formed in the same way as in the storage amyloplasts from cytosolic sucrose

In both the classical models and the alternative model the fructose generated by the breakdown of sucrose by sucrose synthase can be diverted towards glycolysis or recycled in sucrose biosynthesis.

The helical structure of starch

Starch granules show birefringence under polarized light which indicates a long-range internal order. X-ray studies show distinctive diffraction patterns arising from a crystalline structure. Crystallographic studies show that starch occurs in three allomorphic forms known as A, B and V. The A and B forms both have left-handed double helices with six glucose units per turn. The A and B forms appear to differ only in the packing of the starch helices. The V form is actually a whole family of allomorphs adopted in the presence of internally absorbed small molecules such as iodine, DMSO or *n*-butanol. All of the V forms have a single helix structure, also left-handed, with approximately an 8 Å drop per six glucose units that constitutes one turn of the helix. The double helix structure of starch is shown in Fig. 30.

The double helix arises directly from the preferred orientation of two adjacent glucose molecules connected by an α-(1-4) glycosidic linkage. Figure 30(a) depicts the structure of maltose, showing the torsion angles (curved arrows) across the glycosidic bond holding the two glucose units together. These angles were obtained by experimentation with molecular mechanics modelling programs which showed a strong energy minimum at the given angles. If a series of glucose units is connected maintaining these torsion angles, a left-handed helix is obtained. Figure 30(b) shows two helices entwined to give the double helix of A-type or B-type starch. Each diamond represents a glucose unit. A portion of the A-type or B-type starch double helix, generated using molecular mechanics modelling is shown in Fig. 30(c) Each strand is 8 units long. One strand has been coloured a uniform dark grey to distinguish it from the other. The black atoms in the light strand are oxygen atoms, the light grey atoms are carbons and the small near-white atoms are hydrogens. In the A and B forms of starch the helix

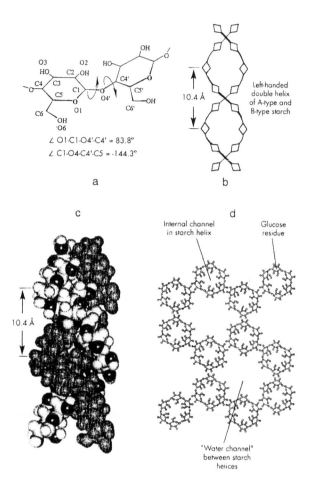

$\angle$ O1-C1-O4'-C4' = 83.8°
$\angle$ C1-O4-C4'-C5 = -144.3°

Left-handed
double helix
of A-type and
B-type starch

10.4 Å

a   b

c   d

Internal channel   Glucose
in starch helix   residue

10.4 Å

"Water channel"
between starch
helices

Fig. 30. The double helix structure of starch. From Journal of Chemical Education 77 (8) Robert D. Hancock and Bryan J. Tarbet: "The other Double Helix – the fascinating chemistry of starch", pages 988–992, (2000). Used with permission from the Journal of Chemical Education and the Division of Chemical Education Inc. © 2000

has the hydroxymethyl groups of C-6 oriented somewhat into the central cavity and the planes of the glucose units are oriented at a considerable angle to the long axis of the helix. The orientation of the glucose units relative to the long axis of the double helix is shown in Fig. 30(d), which depicts the packing of starch helices in crystalline type B starch, viewed along the long axes of the helices. The central cavity of each helix is too small to accommodate water molecules. The large "water channels", each surrounded by six starch helices, are filled to a variable extent with water molecules (not shown).

The orientation of the C-6 hydroxymethyl groups into the central cavity of the helix means that the central cavity of the A and B forms of starch is too small to accommodate molecules. In contrast the V form has the planes of the rings of the glucose units aligned more closely in the wall of the helix and has a larger central cavity that is largely hydrophobic, enabling starch to absorb small linear hydrophobic molecules or hydrophobic parts of molecules.

The layered structure of starch grains that can be observed microscopically reflects the laying down of alternate layers of crystalline and non-crystalline starch during the growth of the starch grain. The amylopectin molecules extend all the way from the hilum to the surface of the starch granule, accounting for the large size of these molecules. The packing of molecules from a central point to produce a roughly spherical grain will be most efficient if these have an approximate wedge shape in cross-section. To achieve this shape the growing amylopectin molecules undergo alternating linear and branching growth to give crystalline and non-crystalline growth rings. This occurs in the chloroplasts where the granules are formed by the action of starch synthases or starch branching enzymes. Thus, at any one time, all of the amylopectin molecules are undergoing *either* branched growth *or* linear growth. It is the layers of branching growth in the amylopectin that periodically increase the width of the molecule and produce the wedge shape.

The role of amylose may simply be that of a packing material, filling in gaps in the structure of the growing granule. Amylose is responsible for the blue colour given with iodine, a well-known identification reaction for starch. The blue colour is connected to the helical structure of the amylose molecule, within which iodine can be enclosed in the form of chain molecules, each containing 15 iodine atoms at a distance of 0.31 nm from one another. The branching of amylopectin approximately every 20 glucose units interrupts the linear array of iodine molecules which are held together by van der Waals' forces. Thus iodine is much more weakly absorbed by amylopectin than by amylose and gives a much less intense pink colour.

Swelling and hydrolysis of starch   At room temperature starch normally contains about 20 % by weight of water, which occupies the water channels. Starch is insoluble in cold water, but when a mixture of starch and water is warmed below the gelatinization temperature of about 60 °C more water reversibly enters the starch structure. The non-crystalline layers absorb water more readily than the crystalline layers. Above the gelatinizing temperature the helix unwinds and becomes solvated forming a gel.

If heated with strong acids, starch is hydrolysed to the end product glucose. Hydrolysis under milder conditions yields *dextrins*, relatively high-molecular-weight fragments of starch molecules. The course of the hydrolysis can be followed by means of the iodine reaction: the colour changes from blue to red or brown, when so-called erythrodextrins are formed. The products arising on continued hydrolysis give no colour. So-called *soluble starch* is a product obtained on mild hydrolysis of starch under strictly controlled conditions. With cold water, it forms an opaque solution that gives the blue colour reaction with

160

iodine. Soluble starch is used as a reagent in iodometric titrations. Starch can also be hydrolysed enzymically, by amylases.

α-Amylase: (EC 3.2.1.1.) occurs in saliva and pancreatic juice. It breaks the α-1,4-linkages in starch. Amylose is broken down to glucose (13 %) and maltose (87 %). The latter is a disaccharide built up from two glucose residues. In contrast, the enzyme cannot split the 1,6-linkages in amylopectin, which therefore gives a mixture of glucose, maltose and larger fragments (amylodextrin).

β-Amylase: (EC 3.2.1.2) occurs in plants and can be isolated from wheat, barley or sweet potatoes. It breaks α-1,4-linkages with the formation of maltose.

Oligo-1,6-glucosidase (EC 3.2.1.10) which can be isolated from yeast, severs α-1,6-linkages and can be used to supplement the other amylases.

Starch is used as a filler and disintegrant in tablets. It is also a constituent of medicinal powders and face powders.

Crude drugs **Potato starch** (Solani amylum) is prepared from tubers of the potato, *Solanum tuberosum* L. (*Solanaceae*).

Wheat starch (Tritici amylum) is obtained from the ears of *Triticum aestivum* L. (*Poaceae*).

Maize starch (Maydis amylum) is produced from fruits of *Zea mays* L. (*Poaceae*).

Rice starch (Oryzae amylum) is prepared from rice, *Oryza sativa* L. (*Poaceae*). It consists of very small grains (4–6 μm) and is used in particular in powders.

Dextran Dextrans are a group of polysaccharides consisting of D-glucose units that are held together by α-1,6-linkages, with a few ramifications via 1,2-, 1,3- or 1,4- linkages.

Dextran is formed by the action of the enzyme *dextran sucrase* (EC 2.4.1.5) on sucrose and consists of a gummy mass which can be drawn out into thick threads. Many different bacteria contain dextran-sucrase. One such, *Leuconostoc mesenteroides,* is employed in the industrial preparation of dextrans. The pharmaceutical company Pharmacia Corporation (now part of Amersham Biosciences) has developed a special strain – B-512, which yields a product with a low degree of branching, a desirable feature for medicinal use.

Crude dextran is a mixture of polysaccharides with molecular weights varying from a few kDa up to several MDa. Only about 1 % of the crude product ranges in molecular weight between 10 000 and 100 000, which is the clinically useful region. Therefore, the crude material is subjected to partial acid hydrolysis, followed by fractionation, to give products with the molecular size desired. The pharmacological properties of dextran are largely dependent on the

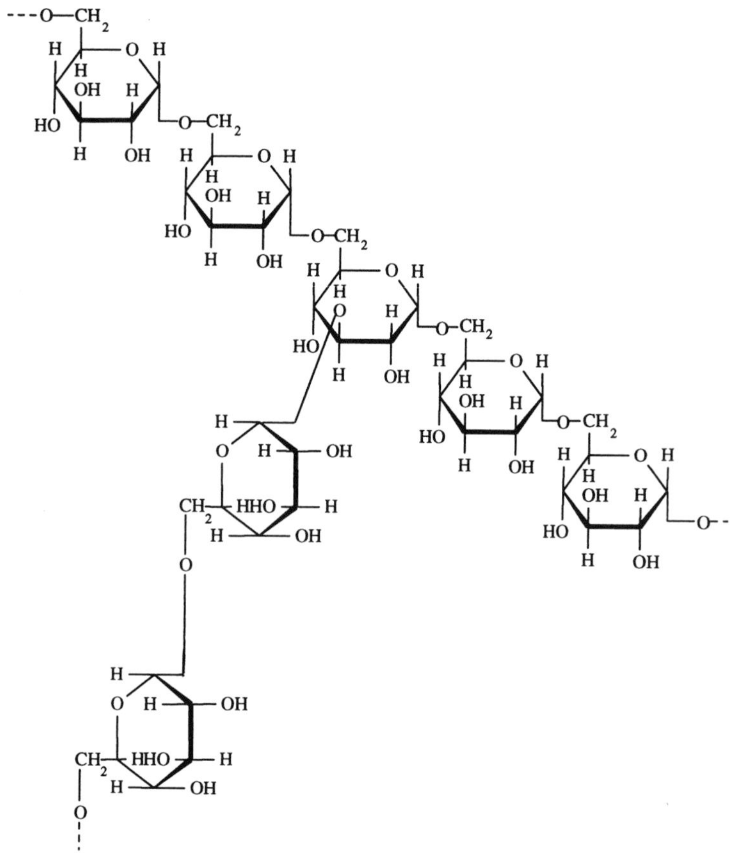

Dextran

molecular weight. Thus, dextran with a molecular weight of 100 000 causes aggregation of red blood corpuscles, whereas dextran with a molecular weight of about 40 000 counteracts this or breaks the aggregates. The product is sold under the name Macrodex and is of two kinds: Macrodex® with an average molecular weight ($\overline{M}_w$) of 70 000, and Rheomacrodex® with $\overline{M}_w$ 40 000. The molecular-weight distribution in both preparations has to be kept constant and is carefully checked.

Dextran is used clinically as a plasma expander in the treatment or prevention of shock caused by bleeding or serious burns. It can replace albumin in maintaining the colloid-osmotic pressure and can therefore be used as a substitute for blood in transfusions. It also has an anti-thrombogenic effect, i.e. prevents blood clotting, and improves capillary circulation. The latter property is especially pronounced in dextran with a low average molecular weight (Rheomacrodex).

Inulin Inulin is a D-fructofuranose polymer, in which ca. 35 fructose residues are linked together by β-2,1-linkages. The chain also contains a terminal glucose residue.

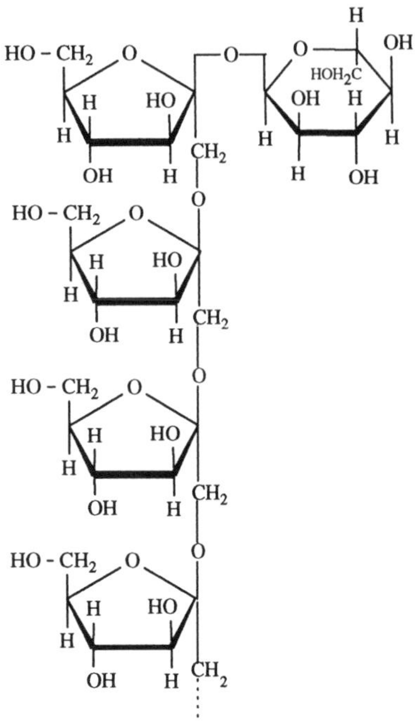

Inulin

Many plants belonging to the families *Asteraceae* and *Campanulaceae* contain inulin rather than starch as a reserve food. It is obtained from the roots of plants of the *Asteraceae*, e.g. from the genera *Inula* (whence the name) (Plate 1, page 211), *Dahlia*, *Helianthus* (artichoke), *Cichorium* (chicory), *Taraxacum* (dandelion) and *Arctium* (burdock). The method of isolating inulin is based on its solubility in hot water and precipitation on cooling. Inulin is used as a raw material for fructose production, and also – in 10 % sterile aqueous solution – as a diagnostic aid for checking kidney function. The latter use is based on the fact that inulin, when injected into the bloodstream, is neither absorbed nor decomposed but secreted unchanged through the kidney. As no reabsorption takes place in the kidney tubuli, the secretion rate (as determined by urine analysis) is a measure of the efficiency of filtration by the kidney glomeruli. This gives important information about kidney function.

Cellulose   Cellulose is the main constituent of the cell walls of most plants. It is present in young meristematic cells as well as in the thickened walls of fibres. Many other polysaccharides (hemicelluloses) are also normal constituents of cell walls. Practically pure cellulose is found in flax fibres and in cotton-seed hairs, both of which have been used for spinning into thread for thousands of years.

Cellulose molecules consist of long chains of D-glucose residues joined by β-1,4-linkages. The molecular weight is of the order of 1 million. Cellulose molecules occur in a tight parallel arrangement in

163

the cell wall and are kept together by hydrogen bonds and van der Waals' forces to form a unit, a *micelle*, which is about 6 nm in diameter and 75 nm long. The micelles form *microfibrils*, which are about 10–30 nm in diameter. These are not entirely solid, but have spaces between the individual micelles. The microfibrils, in turn, can be grouped in bundles, called *macrofibrils* with a diameter of about 0.5 µm and about 4 µm in length.

Crude drug   **Cotton** (Gossypium) is the seed hair of *Gossypium* species (*Malvaceae*). Cotton is an ancient cultivated plant. It is grown in tropical and subtropical countries, especially in Egypt and elsewhere in Africa, India, South America, the West Indies and the southern United States. Three species dominate: *G. arboreum* L., cultivated especially in the Old World, *G. barbardense* L. in the New World and *G. herbaceum* L. worldwide. *Gossypium* species may ecologically be either annual or perennial, but are commercially always re-cultivated annually, in order to prevent attack by noxious insects. The plants grow to a height of 1–1.5 m and have large yellow, pink or red flowers, from which the fruit capsules develop. On ripening, the capsule opens up lengthwise and cotton emerges. It consists of long white hairs, arising from the seed coat, which aid dispersal of the seeds. The capsule contains a large number of seeds, but these are relatively small and the most prominent feature is the large tuft of seed hairs. Hairs and seeds are collected by machines and separated from one another. The seeds are pressed to give cotton-seed oil. While most of the hair goes to the textile industry for spinning and weaving, a minor part is worked up into cotton wool for use as a surgical dressing, **Absorbent cotton** (Lanugo gossypii absorbens). For this, fragments of the fruit wall, broken hairs, etc., are first removed in special machines. The fat is then eliminated by washing with a weakly alkaline soap solution, hereby ensuring good water-absorbing properties. Finally, the cotton is bleached, washed with acid and water, dried and carded.

Derivates of cellulose   Most of the numerous hydroxyl groups in cellulose are reactive and can easily be etherified or esterified. Cellulose ethers swell considerably in water, giving clear or turbid, viscous colloidal solutions. Cellulose ethers are used as bulk laxatives. They are not absorbed, but swell and exert pressure on the intestinal walls, which stimulates defaecation. Bulk laxatives have a mild effect and should be the first choice for treatment of constipation. They should, however, not be used in cases of pathological constriction of the gastrointestinal tract, suspected presence of ileus or severely variable Diabetes mellitus. They should always be taken together with a sufficient quantity of water to allow swelling. No side effects are generally encountered. Bulk laxatives can delay the absorption of other medication, taken at the same time. There may be a need for reduction of the insulin dosage in diabetics. Cellulose compounds are also used in pharmaceutical technology, e.g. as tablet disintegrants.

Methylcellulose, containing 27.5–31.5 % methoxyl groups, is used as a laxative and as a slimming agent. It has no nutritive value, but swells in the stomach and gives a feeling of fullness.

Ethylcellulose, with 44–51 % ethoxy groups, is used in tablet production.

Cellulose acetate phthalate is a reaction product obtained from partially acetylated cellulose by treatment with phthalic anhydride. It is used as a coating for tablets designed to resist gastric juice.

Carboxymethylcellulose is a polycarboxymethyl ether of cellulose in which some of the hydroxyl groups have been replaced by HOOC-CH$_2$- functions. Carboxymethylcellulose is acidic and has ion-exchange properties which make it useful for separating proteins. The sodium salt of this cellulose derivative is used as a thickening agent and as a laxative.

Hyaluronan Hyaluronan (hyaluronic acid, hyaluronate) was first isolated from the vitreous body of the eye and the name is derived from the Greek word *hyalos* which means glassy.

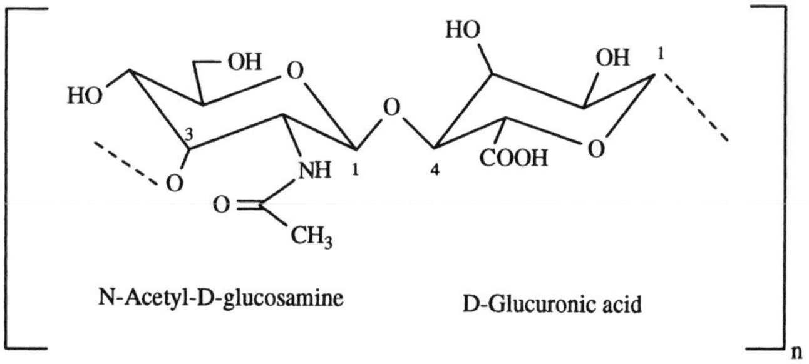

N-Acetyl-D-glucosamine   D-Glucuronic acid

Hyaluronan

The substance is a linear polysaccharide containing alternate N-acetylglucosamine and glucuronic acid residues. The glucosamine linkages are β-1,4 and the glucuronide linkages are β-1,3. Hyaluronan has been found to be present in every mammalian tissue or tissue fluid so far analysed, but the concentration is subject to great variation. Loose connective tissues contain the highest concentration. In the human body, the umbilical cord and the synovial fluid contain 4.1 g/l and 1.4–3.6 g/l, respectively, whereas blood serum contains only 0.01–0.1 mg/l. The comb of the cock contains 1–2 % hyaluronan and is used as a material from which to isolate the compound.

Hyaluronan is biosynthesized in the cell membrane from UDP-*N*-acetylglucosamine and UDP-glucuronic acid. A UDP residue is attached to the reducing end of the chain. When the UDP is liberated the hyaluronan chain is transferred to a UDP-sugar precursor. In this way, the chain grows out from the membrane into the extracellular space.

165

Electron microscopy has shown that hyaluronan has a linear chain structure. The molecular weight depends on the source from which the compound is extracted. Hyaluronan from soft connective tissue has a molecular weight of 1–10 million. Hyaluronan is stabilized by hydrogen bonds, and X-ray diffraction studies suggest a helical ribbon-like conformation incorporating four hydrogen bonds per disaccharide, parallel to the axis of the helix. In solution, the molecule has the shape of an extended, highly solvated coil. At a concentration of 1 g/l the molecules start to entangle and at higher concentrations they form a continuous network with unique rheological properties. The material is both elastic and viscous, which allows a solution to be drawn into a syringe and injected through the hypodermic needle into the tissues, where it regains its viscosity.

Hyaluronan is used for viscosurgery of the eye in intraocular lens implantations, extra- and intracapsular lens extractions, penetrating keratoplasty, retinal detachment and glaucoma. The use of hyaluronan prevents mechanical damage to the sensitive tissues and its unique properties permit manipulation of a detached retina without touching it with instruments. Another advantage is that hyaluronan is a natural component of the eye tissues and is thus degraded by the enzymes which are normally present in them.

Hyaluronan is also used in veterinary practice for the treatment of diseases of the joints in racehorses and has also been tried in human medicine.

For medicinal purposes hyaluronan must be extensively purified to remove an inflammatory fraction. A non-inflammatory fraction of sodium hyaluronate is obtained from the combs of cocks by a patented procedure in which hyaluronan is extracted with a 20:1 mixture of water and chloroform. The inflammatory fraction is removed by a series of extractions with chloroform at different pHs and ion concentrations. Pure hyaluronan is finally obtained by precipitation with ethanol, and is washed with acetone, and vacuum dried.

The sodium salt of hyaluronan is marketed by Pfizer as a 1 % sterile buffered solution, under the trade name *Healon*®. The molecular weight of this fraction is 2–4 million and the viscosity of the solution is 100 000–300 000 cSt.

Heparin Heparin is a mixture of polysaccharides consisting of uronic acid residues alternating with D-glucosamine. The amino group of D-glucosamine can be acetylated or sulphated and some of the hydroxyl groups both in D-glucosamine and in the uronic acids are sulphated in a variable manner. The molecular mass of the polysaccharides is in the range 5–25 kDa. The sulphate groups and the carboxyl groups make heparin strongly acidic. 60–90 % of the sequence of heparin is made up of repeating, trisulphated, disaccharide units, composed of l-iduronic acid and D-glucosamine (A in Fig. 31). The remaining 10–40 % of the polymer is composed of other disaccharide variants, e.g. B in Fig. 31.

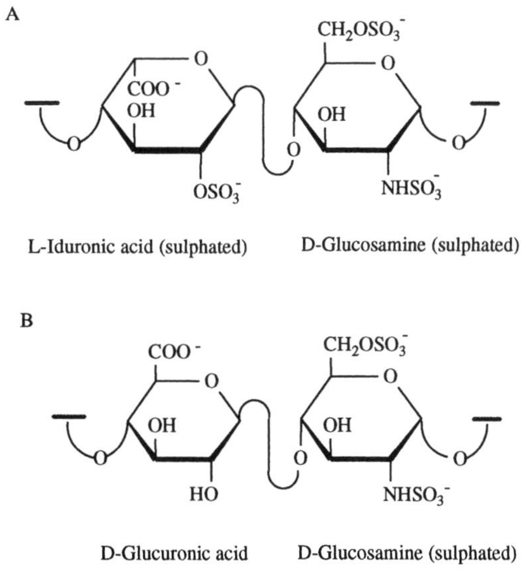

Fig. 31. Disaccharide units contained in polysaccharide chains of heparin. Unit A is most frequent (see text)

Heparin is covalently bound to glycoproteins in the mast cells of both mammals and humans and is obtained from bovine lung or porcine intestines. The tissue is ground and subjected to autolysis for 24 hours or to alkaline hydrolysis that liberates heparin which is extracted with an alkaline solution of ammonium sulphate. Following heat treatment and filtration for removal of proteins, the filtrate is acidified, precipitating heparin. The crude heparin is purified by washing with dilute sulphuric acid, suspension in ethanol for removal of fats, and treatment with trypsin to remove any remaining proteins. Further purification is obtained by dissolving the heparin in water and precipitating it by addition of ethanol. Heparin can then be crystallized as the barium salt from which it is transformed into the sodium salt, which is used for medical purposes.

Biosynthesis   Heparin is biosynthesized as a proteoglycan (molecular size 750–1 000 kDa) consisting of a central core protein, from which long polysaccharide chains (molecular size 60–100 kDa) extend. The synthesis of the core protein takes place in the rough endoplasmic reticulum and the attachment of sugars is located to the Golgi apparatus. Biosynthesis of the polysaccharide chains involves three phases: chain initiation, polymerization and polymer modification.

During chain initiation a D-xylose (Xyl) residue, two D-galactose (Gal) residues and a D-glucuronic acid (GlcA) residue are added in a stepwise manner to specific serine residues of the core protein, with formation of the tetrasaccharide β-GlcA-(1,3)- β-Gal-(1,3)- β-Gal-(1,4)- β-Xyl-. Polymerization then takes place by alternating addition of 1→4 linked D-glucosamine-N-acetate (GlcNAc) and GlcA monosaccharide units to the linker tetrasaccharide with for-

167

mation of [GlcA-(1,4)-GlcNac]n polymer chains. These reach a length of approximately 300 sugar units before the synthesis stops. Transformation of the chains starts with replacement of the Nac groups of the GlcNAc residues with sulphate groups. These reactions start at random sites on the polymer chain and continue from those sites along the chain until isolated GlcNac residues are encountered, i.e. the GlcNac residue in the sequence GlcNS-(1,4)-GlcA-(1,4)- GlcNac-(1,4)-GlcA-(1,4)-GlcNS where the last GlcNS residue is the starting site for a previous N-deacetylation/N-sulphation reaction. The presence of isolated GlcNac residues indicates that they cannot be a substrate for N-*acetylglucosamine-deacetylase* (EC 3.5.1.33). Heparin does not contain any unsubstituted D-glucosamine residues which indicates that N-deacetylation and N-sulphation are joined reactions *in vivo*. All following reactions depend on N-sulphate groups for substrate recognition. N-deacetylation and N-sulphation are thus key reactions that determinine the final fine structure of heparin and are followed by conversion of GlcA into iduronic acid (IdoA), catalysed by *heparosan*-N-*sulphate-glucuronate 5-epimerase* (EC 5.1.3.17). A strict requirement for C-5 epimerization is that the GlcA substrate be attached to the reducing end of a *GlcNS* residue. Thus in the sequences GlcNS-(1,4)-GlcA-(1,4)-GlcNS and GlcNS-(1,4)-GlcA-(1,4)-GlcNAc, GlcA can be epimerized with formation of the sequences GlcNS-(1,4)-IdoA-(1,4)-GlcNS and GlcNS-(1,4)-IdoA-(1,4)-GlcNAc. Epimerization, however, is not complete and the corresponding sequences with GlcA instead of IdoA can also be found in heparin. A sequence such as GlcNAc-(1,4)-GlcA-(1,4)-GlcNS can not be epimerized due to the specificity of the enzyme. Following epimerization, the IdoA residues are 2-O-sulphated and GlcNS residues are 6-O-sulphated to an extent which depends on the substrate specificity of the 2-O- and 6-O-*sulphotransferases*. Other O-sulphation reactions also occur but to a much lesser extent. 3-O-Sulphation of GlcNS and D-glucosamine-N-sulphate-6-O-sulphate residues is particularly important as this modification is required for anticoagulant activity (see Fig. 32).

When modification of the polysaccharide chains is complete, these are cleaved at some of the GlcA residues by *heparin lyase* (EC 4.2.2.7) resulting in shorter, free heparin chains. Because of the uneven distribution of the GlcA residues and because cleavage does not occur at all GlcA residues, the mixture of cleaved heparin chains is polydisperse with molecular sizes ranging from 5–25 kDa. This

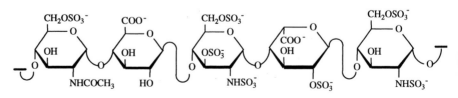

Fig. 32. The antithrombin binding site within heparin

polydispersity, in combination with the variations in modification of the polysaccharide chains during biosynthesis, accounts for the complicated composition of natural heparin.

Heparin is a blood anticoagulant which acts by binding to thrombin and antithrombin thereby accelerating the rate of thrombin inactivation. A pentasaccharide with an unusual sequence has been defined as the antithrombin binding site (Fig. 32).

The sodium salt of heparin is used clinically to dissolve thrombi and to prevent thrombosis in connection with surgery. It is active *in vitro* and is used to prevent blood coagulation in research preparations. Low-molecular mass (4–5 kDa) heparin preparations are obtained by controlled, partial, chemical or enzymatic hydrolysis of heparin. They have better bioavailability and longer duration of action than the natural heparin.

Gums and mucilages

Gums and mucilages are less well-defined polysaccharide derivatives. They may contain both pentose and hexose residues, as well as their oxidation products, uronic acid units. The carboxyl groups are usually neutralized by calcium or magnesium ions. Attempts have been made to distinguish between gums and mucilages on the basis of their behaviour in contact with water: gums are normally fairly soluble, whereas mucilages do not dissolve but only swell to form a viscous mass.

Gums have also been considered as pathological products; on the other hand, mucilages are normal plant constituents. However, it has been difficult to apply this classification in a consistent way. Many products in the group are intermediate between the two classes.

Gums and mucilages can be produced in various ways in the plant:
1. By formation from the middle lamella of cells. This is common in algae, e.g. agar.
2. By formation from the entire cell wall, e.g. the seed-coat epidermis of flax.
3. In special mucilage-secreting cells, as in squill.
4. As a product of cell-wall decomposition, e.g. tragacanth, gum arabic and sterculia gum.

Gums and mucilages are of great technical use in pharmacy as tablet binders and disintegrants, as emulsifiers, and as thickeners. Some that are insoluble in water, but undergo considerable swelling, are used as bulk laxatives, in the same way as cellulose derivatives (page 164).

Crude drugs **Agar** (Agar) is obtained by extraction of red algae (*Rhodophyta*), in particular the two genera *Gelidium* and *Gracilaria*, seaweeds which grow along the coasts of both the Pacific and Atlantic Oceans. Before the Second World War, Japan was the main producer of agar, but nowadays it comes from many other countries: New Zealand, Australia, South Africa, Argentina, Chile, Spain and the United States. The second largest producer and exporter, after Japan, is Spain.

The algae used for agar production in Japan are largely cultivated. Poles are driven into the bottom of the sea. Algae attach themselves to these poles, grow, and are harvested at intervals by removing the poles and scraping them clean. The algae can also be collected by divers, or at low tide. Seaweed that has floated ashore during heavy gales can also be used; in Spain, 75 % of the raw material is obtained in this way. The algae are dried, washed with fresh water, and extracted by boiling with slightly acidified water. The viscous extract is filtered and cooled. The gel formed is cut into bars, which are then pressed through a wire net, giving a product in the form of strips. Water is removed by repeated freezing and thawing and final drying at 35 °C. This production method is practised mainly in winter, taking advantage of the low temperatures. A more modern technique is to cut the gel into flakes and dry them with warm air in tall towers.

Agar is not a uniform product; 10 % comprises water, ash, nitrogenous compounds, and traces of fat, and 90 % is polysaccharide, which can be separated into two fractions: *agarose* and *agaropectin*.

Agarose is a polysaccharide consisting of a chain of alternating D-galactose and 3,6-anhydro-L-galactose residues.

Agaropectin has a more complicated structure and consists of uronic acids and D-galactose residues, partially esterified with sulphuric and pyruvic acids. It is acidic in character, though occurring in salt form in the agar.

Agar is used as an emulsifying agent, as a tablet disintegrant, and as a culture medium in bacteriology. It is also used as a thickener in the food industry. Pure agarose and agarose derivatives are used for chromatographic, electrophoretic, and affinity-chromatographic separations in biochemistry.

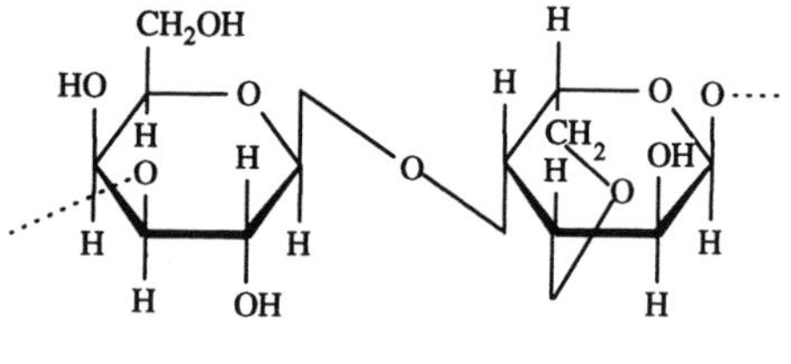

Agarose

Alginic acid is a cell-wall constituent in brown algae. This polysaccharide consists of D-mannuronic acid units which are linked together by β-1,4-linkages, and varying amounts of L-guluronic acid residues, also 1,4-linked. The polysaccharide chain is believed to consist partly of homogeneous sections containing either mannuronic or guluronic acid residues and partly of sections in which the two uronic acids alternate.

Several species of brown algae, e.g. *Macrocystis pyrifera* (L.)

D-Mannuronic acid

L-Guluronic acid

Agardh. and *Laminaria* species, are used in the preparation of alginic acid, which is carried out in many countries, e.g. Japan, the United States, Canada and Scotland. The algae are collected by hand or with special machines, dried, ground, and extracted with dilute sodium carbonate solution. Alginic acid is precipitated from the extract by acidification or as a calcium salt which is then decomposed by addition of hydrochloric acid. The alginic acid is neutralized with sodium carbonate and sodium alginate is precipitated with ethanol.

The substance is a colourless and tasteless powder. With water it forms a viscous colloidal solution, and large amounts are used as a thickening agent in the food industry. Pharmaceutically, alginates are used in ointments, pastes and tablets.

Calcium alginate, which is insoluble in water, is employed in the preparation of styptic gauze and pads. Calcium alginate threads are prepared by forcing sodium alginate solution through a fine spinneret into a bath of acidified calcium chloride, thereby precipitating calcium alginate in the form of long threads that can be spun and woven. The fabric or wool is then dipped in a sodium chloride solution. Sodium ions replace calcium ions on the surface of the fibres and the final product is a calcium alginate material with a covering of sodium alginate. The latter reacts with the calcium ions in blood, and the calcium alginate precipitated forms a paste that absorbs blood and blood corpuscles and facilitates coagulation. Haemostatic dressings of calcium alginate are non-irritant and non-poisonous, can be sterilized in autoclaves or with hot air, and are absorbed by the body, so that they may be left in the wound after operations.

Tragacanth (Tragacantha) is a mucilage obtained from *Astragalus* species (*Fabaceae*) growing in Asia Minor, mainly Iran. The plants are small thorny bushes found in mountainous areas. On wounding the stem, the cell walls of the pith and medullary rays are transformed into a mucilage that oozes out and dries to a solid mass, the shape of the pieces being determined by the shape of the incisions. The best quality has the form of ribbons, showing striations due to the way in which it gradually exudes from the incision and dries.

Tragacanth consists of a water-soluble part, *tragacanthin*, and a larger amount of water-insoluble material, *bassorin*, which swells considerably in water. Like tragacanthin, bassorin consists of polysaccharides containing both sugar and uronic-acid residues, which, in the case of bassorin, are partly methylated. Tragacanth is used as a lubricant for rubber gloves worn during the examination of body cavities, as an emulsifier, and in suspensions. It is also widely applied in the textile and cosmetics industries. Bassorin finds employment as a laxative.

Acacia (Acaciae gummi, Gum arabic) is a gum obtained from *Acacia senegal* (L.) Willd. (*Fabaceae*) and closely related species.

Acacia senegal is a spiny tree, up to 6 m high, growing in the

171

Sudan and Senegal. A large proportion of the acacia gum available on the market comes from plantations in Kordofan (Sudan). Harvesting starts with an incision through the bark to the cambium. The bark above and below the cut is then removed so that an area approximately 7×90 cm of the cambium is laid bare. This operation stimulates the formation of gum, which forms as droplets in the cut. These droplets dry to opaque balls with a crackled surface. The gum is produced during the dry period in February and March, because it is formed only in dry weather and by trees growing in dry locations. The gum is collected 20–30 days after making the incision.

A major constituent of acacia gum comprises the calcium, potassium and magnesium salts of arabic acid. This acid is a branched-chain polysaccharide, the building blocks of which are D-galactose, L-arabinose, L-rhamnose and D-glucuronic acid. Acacia gum also contains oxidizing enzymes, such as peroxidases. For this reason, easily oxidized compounds should not be mixed with this drug.

Acacia gum is employed in cough mixtures, to cover bitter and acrid tastes. In pharmaceutical technology, it is also used as an emulsifier, as a constituent of suspensions, and in tablet production. Technically, acacia is utilized as an adhesive, in the production of Indian ink and water-colours, and in making chocolates and sweets.

The **carob bean** (Ceratoniae fructus) is the fruit of *Ceratonia siliqua* L. (*Fabaceae*), a tree growing in the Mediterranean area. The fruits are pods, up to 20 cm long and 3 cm wide, containing several seeds embedded in a pulp that is soft when fresh but becomes hard on drying. The fruit contains invert sugar (13 %), sucrose (20 %), other sugars (4 %), proteins (1–2 %), pectin (2–3 %) and mucilage (3 %). It is used as a food because of its high sugar content and pleasant taste. Powdered and dried carob beans, suspended in water, are a remedy against diarrhoea, which is used especially for infants.

The endosperm of the seed contains a water-soluble mucilage, which constitutes about 40 % of the weight of the seed. To isolate the mucilage, the seeds are boiled with a 4 % sodium carbonate solution, washed, and passed through rollers to crush the seed-coat. The very hard endosperm is not crushed but can then be separated from the other parts of the seed. Up to 90 % of the endosperm consists of the polysaccharide *carubin* (Ceratoniae gummi), which can be extracted with water and isolated by subsequent evaporation. It is a galactomannan with a molecular weight of approximately 310 000. On hydrolysis, it gives 16–20 % D-galactose and 80–84 % D-mannose. Carubin is used technically in the textile and paper industries. Medicinally, it is given to prevent vomiting in infants, the effect of the mucilage being to thicken the stomach contents.

Guar (Cyamopsidis seminis pulvis) is obtained from the endosperm of the seed of *Cyamopsis psoraloides* DC (*Fabaceae*), a plant long cultivated in India and Pakistan and nowadays also grown in the United States. This herb is about 60 cm in height and its fruits are 5–12.5 cm long pods, containing 5–6 round, light brown seeds. In production of guar, the seed is quickly passed through a flame to loosen the seed coat, which is then crushed, to free the endosperm. After removal of the embryo, the endosperm is milled to give guar which is then autoclaved to destroy enzymes present in any residual embryo fragments and to kill contaminating microorganisms. Guar contains about 86 % water-soluble mucilage, which can be extracted and fractionated by precipitation with ethanol. The main constituent is a galactomannan (see below).

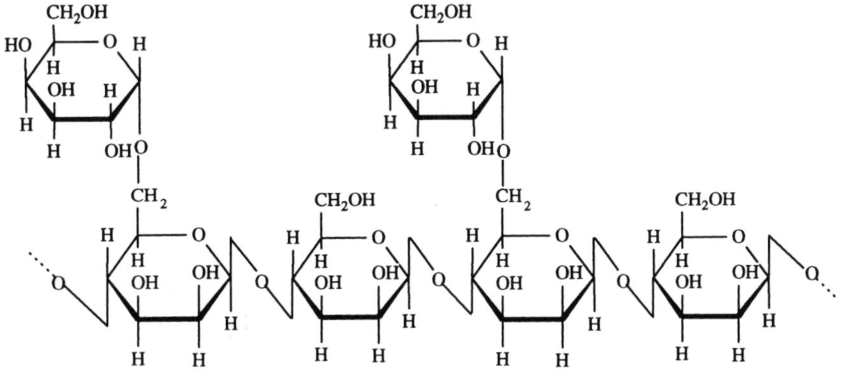

Guar gum (partial structure)

Guar finds use in several very different industries: mining, paper, textiles and food. In pharmacy, it is used as a disintegrant in tablets, as an emulsifier, and as a thickener. It is also employed as a laxative. Administration of guar before meals reduces the glucose content in the blood of diabetics. Guar is therefore of value in the treatment of this disease. It also reduces the cholesterol level in blood serum, both in healthy persons and in diabetics: the higher the original concentration, the more pronounced the reduction.

Guar galactomannan (Guar galactomannanum). This product is obtained by partial hydrolysis of Guar. It consists mainly of polysaccharides of D-galactose and D-mannose at molecular ratios of 1:1.4 to 1:2. Structurally, the molecules are built of a linear main chain of β-(1→4)-glycosidically linked mannopyranoses and single α-(1→6)-glycosidically linked galactopyranoses.

Sterculia gum (Sterculiae gummi, Karaya gum, Indian tragacanth). The source of this gum is *Sterculia urens* Roxb. and related species (*Sterculiaceae*). *Sterculia* species are trees, native to India. The gum is exuded through natural cracks in the trunk or after incisions which reach into the heartwood. The gum can be collected all the year round, but collection is normally done during the period March–June. The gum exuded solidifies to semi-transparent, colourless or yellowish to light reddish brown, drop-shaped lumps with a faint odour of acetic acid. Sterculia gum does not dissolve in water but undergoes considerable swelling. It is used as a laxative and technically as a substitute for the more expensive tragacanth.

Psyllium seed (Psyllii semen, Plantago seed). The crude drug consists of the ripe seed of *Plantago psyllium* L. (previously known as *P. afra* L.) or *P. arenaria* Waldst. et Kit (= *P. indica* L). (*Plantaginaceae*). *Plantago psyllium* and *P. arenaria* are annual herbs growing in the Mediterranean area and cultivated mainly in southern France. The seed is small (2–3 mm long) and dark brown and looks like a flea – hence the name (psyllium = flea). The epidermis of the seed coat contains a mucilage that swells when the seed is soaked in water and this ability to swell makes the drug useful as a bulk laxative. The entire seed is used. Ground seeds should not be employed, because the pigment of the seed coat may be absorbed and accumulated in the kidney tubules.

In India, a different species, *Plantago ovata* Forsskal, is cultivated which yields **Indian psyllium** or **ispaghula seed** (Plantaginis ovatae semen), lighter in colour than ordinary psyllium. The separated seed coat of ispaghula, **Ispaghula husk** (Plantaginis ovatae seminis tegumentum), is on the market as a constituent of various laxatives. Foreign material and fat are removed from the seed coat by sifting and extraction with isopropyl alcohol.

Linseed (Lini semen) is obtained from flax, *Linum usitatissimum* L. (*Linaceae*) and is one of the oldest of cultivated plants, as it has been grown for its fibre throughout much of the world for thousands of years. Flax is an annual herb, about 60 cm high, with blue flowers. The epidermis of the seed coat contains mucilage and linseed is used as a laxative, just like psyllium. Linseed also contains a fixed oil; see (page 252).

Marshmallow root (Althaeae radix) comes from marshmallow, *Althaea officinalis* L. (*Malvaceae*), a perennial 1–2 m tall herb with an erect woody stem, which grows in central Europe and which has been introduced into the United States. The drug is obtained from the roots of both wild and cultivated plants that are at least two years old. Late in the autumn the roots are harvested, and the cork and outer parts of the bark are removed. The peeled roots are dried quickly at a raised temperature (ca. 40 °C). Too slow drying gives a yellow product. The crude drug is often cut into small cubes to facilitate production of a starch-free mucilage. The root contains a fair amount of starch, which escapes from the cells on powdering or chopping, and gives a turbid mucilage when the drug is macerated with cold water. In the form of small cubes, however, the starch present on the surface of the crude drug can easily be washed away and the washed pieces will give a clear mucilage.

Marshmallow root is used as a vehicle in cough mixtures and

to cover the bitter or pungent taste of other drugs (cf. Acacia gum). A cold-water extract is used as a herbal remedy for treatment of respiratory catarrh with irritating cough. It is, however, not an expectorant and can also be used for treatment of gastritis. No contraindications or side effects are reported. *Interactions*: The absorption of other drugs, taken simultaneously, may be delayed.

Reduction products of carbohydrates: sugar alcohols

Several polyalcohols occurring in Nature can chemically be regarded as reduction products of hexoses. Among higher plants, D-sorbitol is found in the families *Rosaceae* and *Plantaginaceae* whereas D-mannitol is present in species of several families e.g. *Oleaceae*, *Apiaceae* and *Scrophulariaceae*. D-Sorbitol can be looked upon as a reduction product of D-glucose and D-mannitol as the corresponding derivative of D-mannose. The following structures are drawn as Fischer projections.

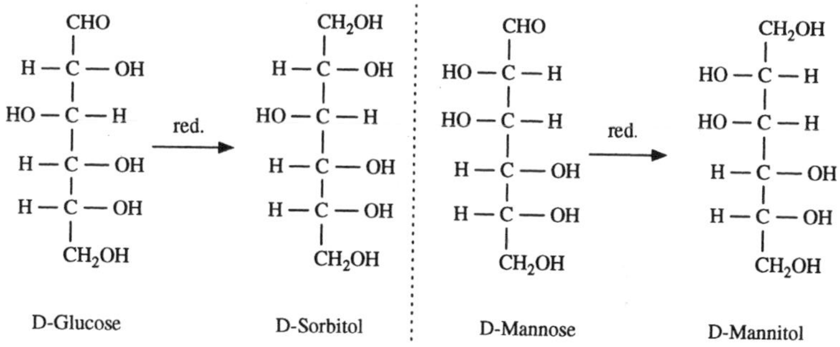

	red.		:		red.	
CHO		CH$_2$OH		CHO		CH$_2$OH
H—C—OH		H—C—OH		HO—C—H		HO—C—H
HO—C—H		HO—C—H		HO—C—H		HO—C—H
H—C—OH		H—C—OH		H—C—OH		H—C—OH
H—C—OH		H—C—OH		H—C—OH		H—C—OH
CH$_2$OH		CH$_2$OH		CH$_2$OH		CH$_2$OH
D-Glucose		D-Sorbitol		D-Mannose		D-Mannitol

In sorbitol-producing plants this compound is used instead of sucrose for transportation of the primary products of photosynthesis from the leaf to other parts of the plant, where carbon is needed for synthetic purposes. As described previously (page 151) D-glyceraldehyde-3-phosphate (G3P) is exported from the chloroplast to the cytosol of the leaf where it is converted to D-fructose 6-phosphate and D-glucose 6-phosphate, which ultimately form sucrose. In those plants which use sorbitol as the main transporter of carbon, D-glucose 6-phosphate is reduced to sorbitol-6-phosphate by *aldose-6-phosphate reductase (NADPH)* (EC 1.1.1.200). Finally *sorbitol-6-phosphatase* (EC 3.1.3.50) removes the phosphate group to yield sorbitol.

D-Mannitol is also used as a transport carbohydrate and is formed in connection with photosynthesis in a way similar to that of sorbitol. Here too D-fructose 6-phosphate is the starting material which is isomerized to D-mannose 6-phosphate by *mannose-6-phosphate*

175

isomerase (EC 5.3.1.8). This compound is reduced to D-mannitol-1-phosphate by a NADPH-dependent *mannose-6-phosphate 6-reductase* (EC 1.1.1.224). Finally the phosphate group is removed by *mannitol-1-phosphatase* (EC 3.1.3.22) and D-mannitol is formed.

Sorbitol   Sorbitol is named after *Sorbus aucuparia* L. (*Rosaceae*), rowan tree or mountain ash, the fruits of which contain a large amount of the compound. Sorbitol is used as a sweetener for diabetics, as a mild laxative, and as an ingredient of irrigations used in urology. It is also a starting material for the *semi*-synthesis of ascorbic acid (vitamin C) and of wetting and emulsifying agents (Tween). Sorbitol is prepared by the catalytic reduction of glucose.

Mannitol   Mannitol can be obtained either by reduction of mannose or by isolation from the crude drug **Manna**, which is obtained from the manna ash, *Fraxinus ornus* L. (*Oleaceae*). This is a small tree of southern Europe, cultivated particularly in Sicily. Manna is exuded following cuts made in the bark of ca. eight-year-old trees and dries to a brownish white crystalline product. It contains 50–60 % mannitol and is used as a mild laxative. Like bulk laxatives, mannitol is not absorbed and the effect is due to the swelling brought about by the absorption of water.

Pure mannitol is used in the same way as inulin (page 163) to test kidney function. An infusion containing 150 g/l of mannitol is employed as a diuretic to treat anuria occurring as a side effect of surgery. It is also applied in treating oedema and increased intracranial pressure, effects which are due to osmosis. Mannitol is also an ingredient of irrigating fluids used in urology.

Natural products related to carbohydrates

Ascorbic acid (vitamin C)   Ascorbic acid occurs in fruits and vegetables. It is a vitamin that humans like monkeys and guinea-pigs are unable to biosynthesize and instead must take with their food. Lack of ascorbic acid gives rise to scurvy, a deficiency disease which used to affect sailors, living on salted food for long periods. The symptoms are bleeding,

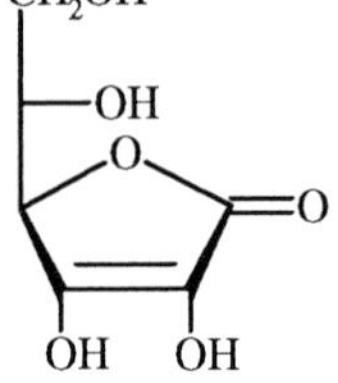

L-Ascorbic acid

changes in connective tissues and in the skeleton, as well as loosening of the teeth.

Ascorbic acid is produced synthetically from glucose via sorbitol (page 176).

Biosynthesis in plants The biosynthesis of L-ascorbic acid in plants has been a puzzle for a long time. The pathway outlined in Fig 33 was published in 1998 and was regarded as a breakthrough, accommodating all hitherto unsettled issues. The starting material is D-glucose 6-phosphate which is isomerized to D-fructose 6-phosphate and then further transformed to D-mannose 6-phosphate. The phosphate group is moved to carbon 1 of mannose, yielding D-mannose 1-phosphate. This sugar is activated by reaction with GTP forming GDP-D-mannose, which is epimerized at carbons 3 and 5 yielding GDP-L-galactose which loses GMP and forms L-galactose 1-phosphate

D-Glucose 6-phosphate   D-Fructose 6-phosphate   D-Mannose 6-phosphate   D-Mannose 1-phosphate

L-Galactose 1-phosphate   GDP-L-Galactose   GDP-D-Mannose

L-Galactose   L-Galactonolactone   L-Ascorbic acid

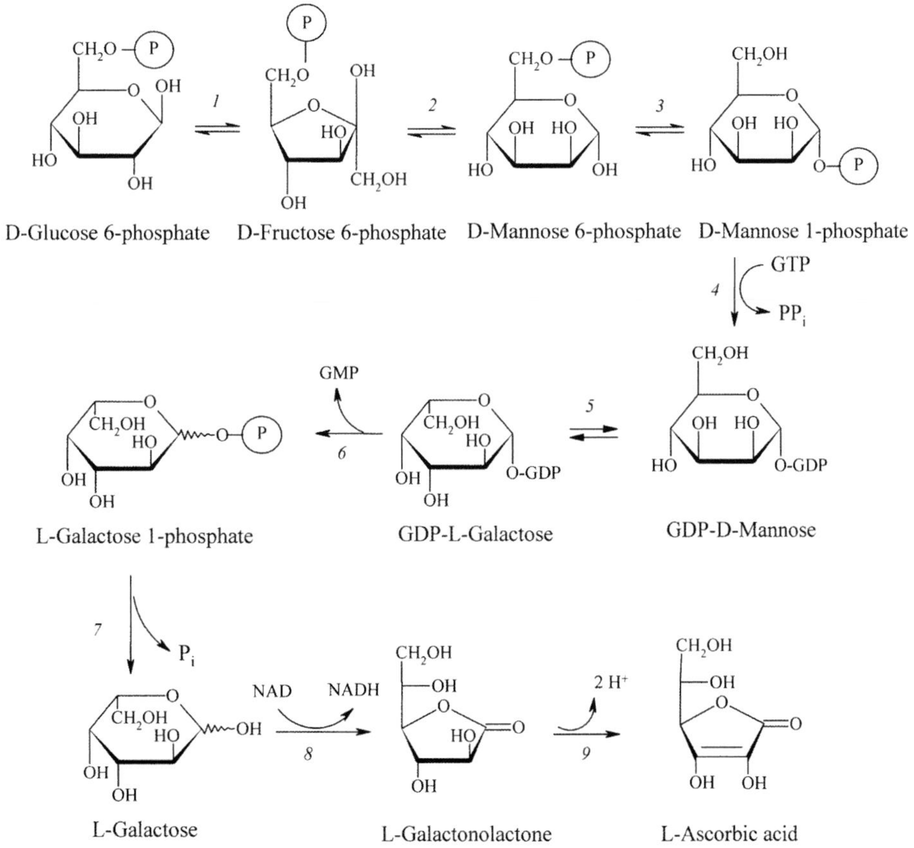

Fig. 33. Biosynthesis of ascorbic acid in plants. The participating enzymes are: **1**: glucose-6-phosphate isomerase (EC 5.3.1.9); **2**: mannose-6-phosphate isomerase (EC 5.3.1.8); **3**: phosphomannomutase (EC 5.4.2.8); **4**: mannose-1-phosphate guanylyltransferase (EC 2.7.7.13); **5**: GDP-mannose 3,5-epimerase (EC 5.1.3.18); **6**: GDP-L-galactose phosphodiesterase **7**: L-galactose 1-P phosphatase; **8**: L-galactose dehydrogenase **9**: L-galactono-1,4-lactone dehydrogenase

Guanosine triphosphate (GTP)

which is dephosphorylated to yield L-galactose. L-Galactose is oxidized to L-galactono-1,4-lactone, which is the immediate precursor to L-ascorbic acid. The L-galactono-1,4-lactone is probably formed via the unstable L-galactono-1,5-lactone.

This pathway for the biosynthesis of L-ascorbic acid has been challenged on account of the following unexpected experimental facts.

1. The enzyme, mannose-6-phosphate isomerase (EC 5.3.1.8), which links D-mannose metabolism with common D-glucose/fructose intermediates of general metabolism is often not expressed in plants. The origin of the D-mannose 6-phosphate in the route to ascorbic acid is therefore unclear.

2. The enzyme GDP-mannose 3,5-epimerase (EC 5.1.3.18), which catalyses reaction 5, has been found to form a previously unsuspected novel product: GDP-L-gulose in addition to the well-known GDP-L-galactose.

3. The highly specific L-galactono-1,4-lactone dehydrogenase (reaction 9) is not the only enzyme for the last step of ascorbic acid synthesis in plants.

The discovery of the ability of GDP-mannose 3,5-epimerase to produce GDP-L-gulose led to the proposal of a variation of the pathway – the L-gulose pathway – where GDP-mannose is transformed to GDP-L-gulose, which forms L-gulono-1,4-lactone that is finally converted to L-ascorbic acid.

An enzyme called VTC2 has recently been discovered in *Arabidopsis thaliana*, which acts as a unique GDP-L-galactose/GDP-D-glucose:hexose 1-phosphate guanylyltransferase of broad specificity and catalyses both guanylyltransferase and phosphorylase reactions. It is possible that the main flow to L-ascorbic acid in plants does not proceed via D-mannose phosphates and D-mannose 6-phosphate isomerase (as illustrated in Fig 33). The main route might instead be that VTC2 in combination with a GDP-D-mannose 2″-epimerase utilizes D-glucose phosphates that are converted to L-hexoses for the synthesis of L-ascorbic acid. A detailed description of the so-called VTC cycle for the linkage of photosynthesis with the biosynthesis of L-ascorbic acid and the cell-wall metabolism in

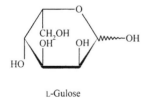

L-Gulose

L-Gulono-1,4-lactone

plants is presented by Wolucka (2007). See *Further reading* page 192.

Crude drug **Rose hips** (Cynosbati fructus sine semini). The drug rose hips consists of the dried receptacle (thalamus) of the dog rose, *Rosa canina* L. (*Rosaceae*), a common plant in Europe. In the drug, the fruits ("seeds") and most of the hairs covering the inside of the receptacle have been removed. Fresh rose hips contain up to 5 % ascorbic acid, but much of this is destroyed during drying, so that the final content in the drug is about 1 %. Carotenoids and flavonoid substances with vitamin P effect are also present.

Rose hips are used as a source of vitamin C in vitamin preparations and as a herbal remedy for treatment of osteoarthritis. The effect on the joint pain and stiffness of this disease has been demostrated in several clinical studies. A galactolipid named GOPO® has been identified as an active compound. GOPO consists of glycerol, esterified with two tri-unsaturated C_{18} fatty acids and connected to galactose with a glycoside bond at C-3. The full chemical name is (2S)-1,2-di-O[(9Z, 12Z, 15Z)-octadeca-9,12,15-trienoyl]-3-O-β-D-galactopyranosyl glycerol.

Rosehip extract and GOPO inhibit chemotaxis and chemiluminescence of human PMNs (polymorphonuclear leukocytes) *in vitro*, an effect that was used for bioassay-guided isolation of GOPO. In clinical studies daily intake of a standardized powder of rose hips resulted in decreased chemotaxis of PMNs and a reduced level of the inflammatory marker acute phase C-reactive protein (CRP) in the patients. Rosehip powder also alleviated pain and improved general well-being, sleep quality and mood. To ensure activity against osteoarthritis rose hips should be standardized for content of GOPO.

Glycosides

Glycosides consist of a sugar residue covalently bound to a different structure called the *aglycone*. There are many different types of aglycones. The sugar residue is in its cyclic form and the point of attachment is the hydroxyl group of the hemiacetal function. The

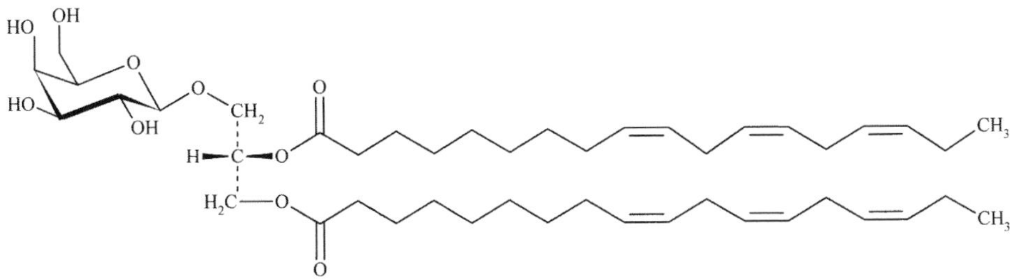

GOPO[R] = 2(*S*)-1,2-di-*O*-[(9*Z*,12*Z*,15*Z*)-octadeca-9,12,15-trienoyl-3-*O*-β-D-galactopyranosyl glycerol

sugar moiety can be joined to the aglycone in various ways. The most common bridging atom is oxygen (O-*glycoside*), but it can also be sulphur (S-*glycoside*), nitrogen (N-*glycoside*) or carbon (C-*glycoside*). α-*Glycosides* and β-*glycosides* are distinguished by the configuration of the hemiacetal hydroxyl group. The majority of naturally-occurring glycosides are β-glycosides.

O-Glycosides can easily be cleaved into sugar and aglycone by hydrolysis with acids or enzymes. Almost all plants that contain glycosides also contain enzymes that bring about their hydrolysis (glycosidases). This fact has to be borne in mind when glycosides are to be isolated from plant materials. In order to obtain the glycosides intact (primary glycosides = genuine glycosides), the plant material must be treated so that contact between enzymes and glycosides is avoided as far as possible. Living cells contain barriers between enzymes and their substrates, but these barriers are broken when the cells die. Enzymes require water in order to function. Rapid drying of the plant material may, therefore, decrease the risk of glycoside hydrolysis. Another method is to grind the fresh plant material with solid ammonium sulphate. The salt precipitates the enzymes, so that they cannot act upon the glycosides, which can then be extracted with organic solvents.

Glycosides are usually soluble in water and in polar organic solvents, whereas aglycones are normally insoluble or only slightly soluble in water. It is often very difficult to isolate intact glycosides because of their polar character and the fact that aglycones are often neutral compounds lacking acidic or basic properties.

Biosynthesis Glycoside biosynthesis includes the aglycone as well as the linkage between sugar and aglycone. The biosynthesis of an aglycone depends, of course, on its structure. The coupling of the sugar and the aglycone, however, takes place in the same way, whatever the nature of the aglycone. The sugar is phosphorylated to give a sugar-1-phosphate, which reacts with uridine triphosphate (UTP) to form uridine diphosphate-sugar and inorganic phosphate. The UDP-sugar then reacts with the aglycone, giving the glycoside and free UDP.

UTP + sugar 1-phosphate ——> UDP-sugar + PPi
UDP-sugar + aglycone ——> glycoside + UDP

The sugar part can then be extended by attaching further sugar residues to the first one.

Many important drugs are glycosides. Their pharmacological effects are largely determined by the structure of the aglycone, but are modified by the sugar part. For this reason, and because this book classifies pharmacologically-active natural products according to biosynthetic principles, glycosides will be treated together with their aglycones and not as a group by itself, as in most current textbooks of pharmacognosy and phytochemistry.

Aminoglycoside antibiotics

These antibiotics, which are also called aminocyclitol antibiotics, are produced by soil actinomycetes of the genera *Streptomyces, Micromonospora* and *Bacillus*. Their structures are composed of an aminocyclitol moiety, to which various amino sugars are attached. The most common aminocyclitol derivatives are streptamine and 2-deoxystreptamine, which occur as two enantiomers, D and L as illustrated in Fig. 34.

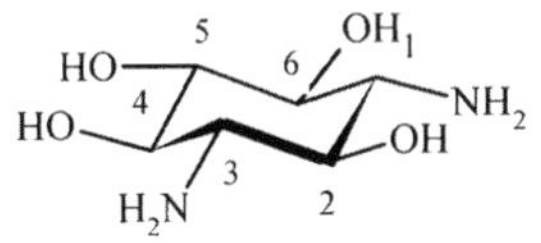

D-Streptamine

Fig. 34. Structure of D-streptamine. For the L-enantiomer the numbering of the carbon atoms in the ring is counter-clockwise

The aminoglycosides are highly water-soluble and have a basic character. Following oral administration they are poorly absorbed. Only about 1 % of an oral dose can pass into the bloodstream by way of the gastrointestinal tract. This makes them useful for treatment of intestinal infections. For treatment of systemic infections they are parenterally administered. The blood–brain barrier excludes aminoglycosides from the central nervous system. The aminoglycoside antibiotics are bactericidal and are particularly effective against Gram-negative bacteria. They act by binding to the A-site decoding region of the bacterial 16S ribosomal RNA, thereby causing mistranslation of mRNA or premature termination of protein synthesis, leading to cell death.

Aminoglycosides are generally used for specialized purposes, not as drugs of first resort. They are almost always used in combination with a β-lactam antibiotic to take advantage of the synergism between these two classes of antibiotics. A drawback is their associated nephro- and ototoxicity and to a lesser extent neuromuscular blockade. Ototoxicity and nephrotoxicity are more likely to occur when therapy is continued for more than 5 days, at higher doses, in the elderly and with renal insufficiency. Nephrotoxicity is dose-dependent and generally reversible. The ototoxicity might lead to irreversible vestibular and/or cochlear damage.

Resistance to aminoglycoside antibiotics is caused by bacterial enzymes that modify the structure of the antibiotics, such as aminoglycoside phosphotransferases, adenylyltransferases and acetyltransferases. Resistance can also be caused by decreased uptake and/or accumulation of the drugs in the bacteria and by alteration of the ribosomal binding sites.

Streptomycin   Clinically important aminoglycoside antibiotics are streptomycin,

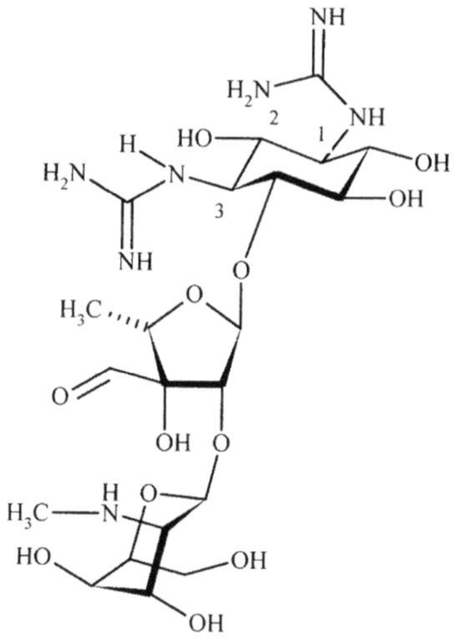

Streptomycin

gentamicin, neomycin, tobramycin and the semisynthetic compound amikacin.

In streptomycin the amino-groups of streptamine are bound as guanidino substituents, forming streptidine. At C-4 streptidine is glycosidically bound to L-streptose (which is a derivative of L-lyxose) which in turn forms a glycoside bond to *N*-methyl-L-glucosamine.

Streptomycin is produced by *Streptomyces griseus* and was the first clinically useful aminoglycoside antibiotic. It was introduced in 1944 as the first remedy against tuberculosis for which it remained as a first-line drug for many years. Nowadays it is used as a second-line agent for this disease because of its toxicity and because of the availability of synthetic drugs such as isoniazid and ethambutol and the semisynthetic antibiotic rifampicin (page 342).

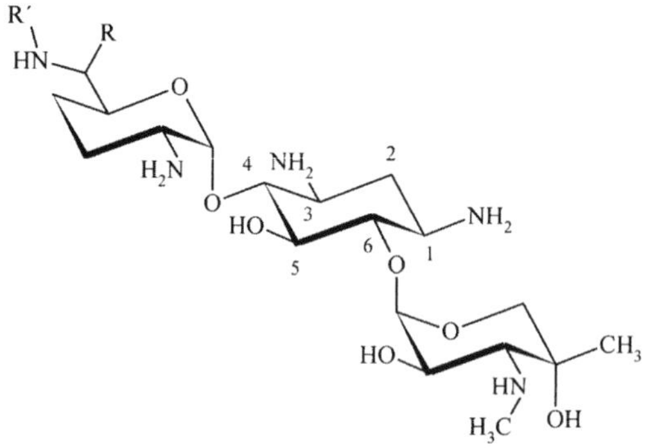

Gentamicin C1: R = R′ = CH$_3$
       C1a: R = R′ = H
       C2: R = CH$_3$, R′ = H

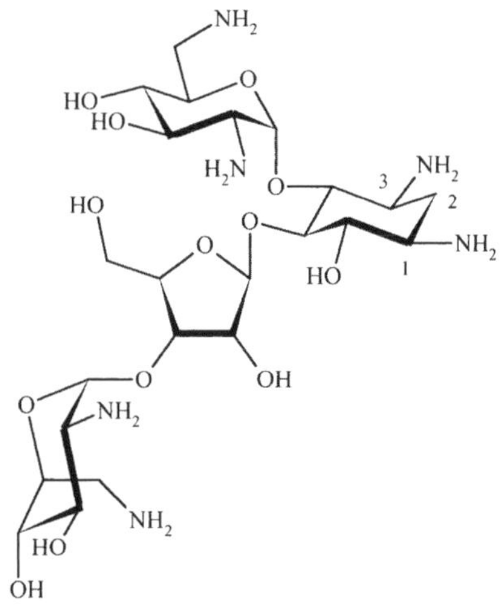

Neomycin B

Gentamicin Gentamicin is a mixture of aminoglycosides which is produced by *Micromonospora purpurea*. The main components are gentamicin C1, C1a and C2. In the gentamicins the aminocyclitol derivative 2-deoxy-streptamine forms a glycoside bond with L-garosamine (3-deoxy-4-C-methyl-3-methylamino-β-L-arabinose) at C-6 and with a derivative of D-purpurosamine at C -4.

Gentamicin is active against Gram-negative bacteria and against some strains of staphylococci. It is used to treat severe systemic infections, often in combination with other antibacterials such as penicillin for enterococcal and streptococcal infections and metronidazole or clindamycin for mixed aerobic–anaerobic infections.

Neomycin Neomycin is produced by *Streptomyces fradiae*, and like gentamicin it is a mixture of several aminoglycosides with neomycin B as the main compoent. Other components are neomycin A and neomycin C (a stereoisomer of neomycin B). In neomycin B, C-5 of 2-deoxy-streptamine forms a glycosidic bond with D-ribose, which in turn is bound to L-neosamine B. The sugar connected at C-4 is neosamine C.

Neomycin is used for topical treatment of infections of the skin, ear and eye caused by staphylococci and other organisms. It is not indicated for the treatment of systemic infections because it can cause irreversible ototoxicity.

Kanamycin and amikacin Kanamycin A is obtained from cultures of *Streptomyces kanamyceticus*. In this compound 2-deoxy-streptamine forms a glycoside bond with 6-amino-6-deoxy-α-D-glucose at C-6 and with 3-amino-3-deoxy-α-D-glucose at C-4. In the semisynthetic compound amikacin the amino-group at C3 in kanamycin A is substituted with a 4-amino-L-2-hydroxybutyryl group.

Amikacin is less toxic than the parent molecule and is resistant to many enzymes that inactivate gentamicin and tobramycin. It is used in combinations with other antibiotics against *Staphylococcus aureus* and certain species of *Streptococcus*. Strains of multidrug-resistant *Mycobacterium tuberculosis*, including streptomycin-resistant strains are usually susceptible to amikacin.

Kanamycin is not used clinically any more.

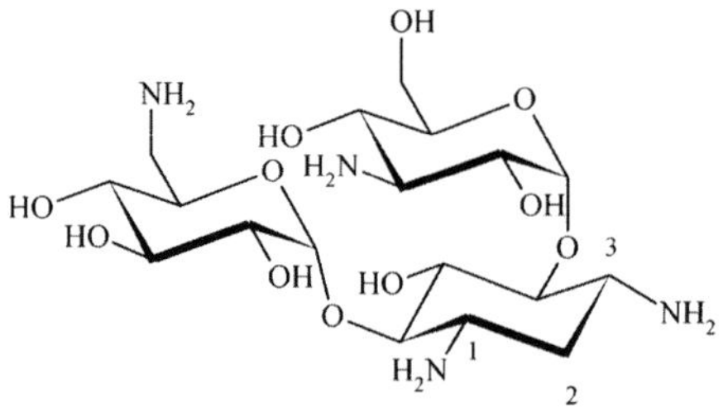

Kanamycin A

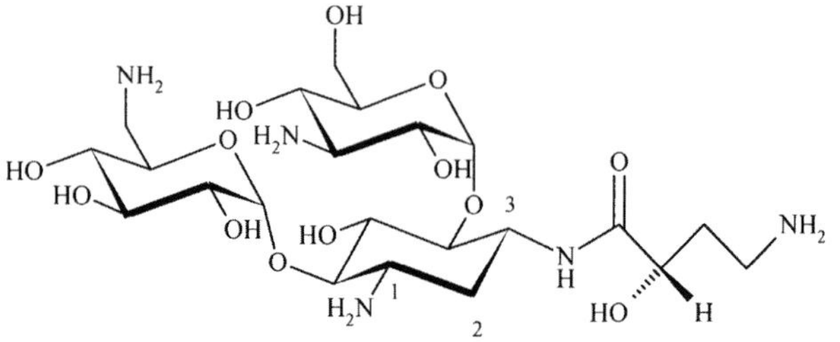

Amikacin

Tobramycin   Tobramycin is produced by *Streptomyces tenebrarius* and consists of 2-deoxy-streptamine, glycosidically bound to 3-deoxy,3-amino-D-glucose at C-6 and to nebrosamine (2,6-diamino-2,3,6-trideoxy-D-*ribo*-hexose) at C-4.

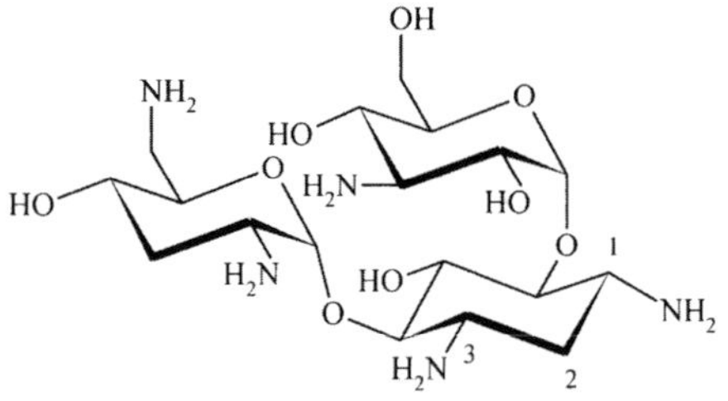

Tobramycin

It is used for treatment of infections with *Staphylococcus aureus* and certain species of *Streptococcus*. Tobramycin is also active against some *Mycobacterium* species but not against anaerobic bacteria.

Biosynthesis   Two pathways are recognized for the biosynthesis of the clinically used aminoglycoside antibiotics.

Streptomycin is synthesized via the *myo*-inositol pathway and gentamycin, neomycin, tobramycin and kanamycin are formed via the 2-deoxy-*scyllo*-inosose pathway.

The myo-*inositol pathway*   This pathway has been elucidated by exploration of the over 29 genes involved in the biosynthesis. The organization of these genes and the mechanism for control of the production of streptomycin are known (see reference Flatt (2007) in *Further reading* page 193). Some of the enzymes have been biochemically characterized. The biosynthesis involves formation of the streptamine-derived cyclitol streptidine 6-phosphate and the two sugar-units L-dihydrostreptose and *N*-methyl-L-glucosamine from glucose 6-phosphate, followed by connection of these units to form streptomycin.

Formation of streptidine 6 phosphate (Fig. 35).

The enzyme myo-*inositol-1-phosphate synthase* (MIP) (EC 5.5.1.4) converts D-glucose 6-phosphate to *myo*-inositol 1-phosphate. This enzyme is not clustered with the streptomycin biosynthetic genes, indicating that this reaction step is mediated by an MIP which is also involved in primary metabolism. *myo*-Inositol 1-phosphate loses the phosphate group to form *myo*-inositol, which is oxidized to *scyllo*-inosose followed by transamination to form *scyllo*-inosamine. This compound is phosphorylated to yield *scyllo*-inosamine 4-phosphate followed by amidino transfer to form amidino-*scyllo*-inosamine 4-phosphate. Conversion of this compound to streptidine 6-phosphate is accomplished by a second round of oxidation, transamination and amidination.

Formation of deoxythymidine 5´-diphosphate-L-dihydrostreptose (dTDP-L-dihydrostreptose) (Fig. 36).

Also here the starting material is glucose 6-phosphate which is converted to glucose 1-phosphate and activated by attachment of deoxythymidine 5´-diphosphate to C-1-OH with formation of dTDP-D-glucose. This is oxidized to 4-keto-6-deoxy-dTDP-D-glucose, which is epimerized to dTDP-4-keto-L-rhamnose, which is reduced to dTDP-L-rhamnose and finally converted to dTDP-L-dihydrostreptose.

Formation of cytidine 5´-diphosphate-N-methyl-L-glucosamine (CDP-N-methyl-L-glucosamine) (Fig. 37).

As in the formation of the two previous compounds this biosynthesis also starts with glucose 6-phosphate. The starting material is converted to glucose 1-phosphate, which is activated by cytidine 5´-diphosphate to form CDP-D-glucose. This is oxidized at C-2, transaminated and *N*-methylated, whereupon several not yet elucidated steps yield CDP-*N*-methyl-L-glucosamine.

Connecting the three ring systems to form streptomycin (Fig. 38).

Streptidin 6-phosphate reacts with dTDP-L-dihydrostreptose with formation of *O*-1,4-α-L-dihydrostreptosyl-streptidine 6-phosphate, which in turn reacts with CDP-*N*-methyl-L-glucosamine, yielding dihydrostreptomycin 6-phosphate. Dehydrogenation yields streptomycin 6-phosphate which is dephosphorylated by an extracellularly-localized streptomycin phosphatase to yield strepto-mycin.

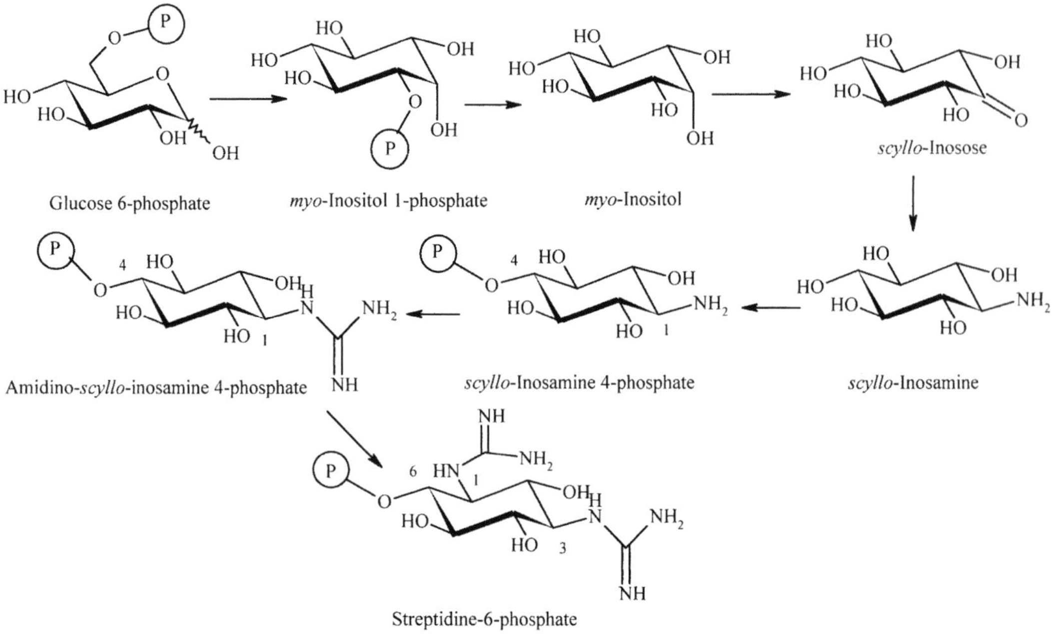

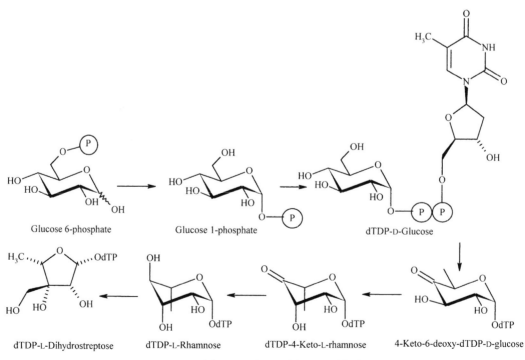

Fig. 35. Biosynthesis of streptidine 6-phosphate

Fig. 36. Biosynthesis of dTDP-L-dihydrostreptose

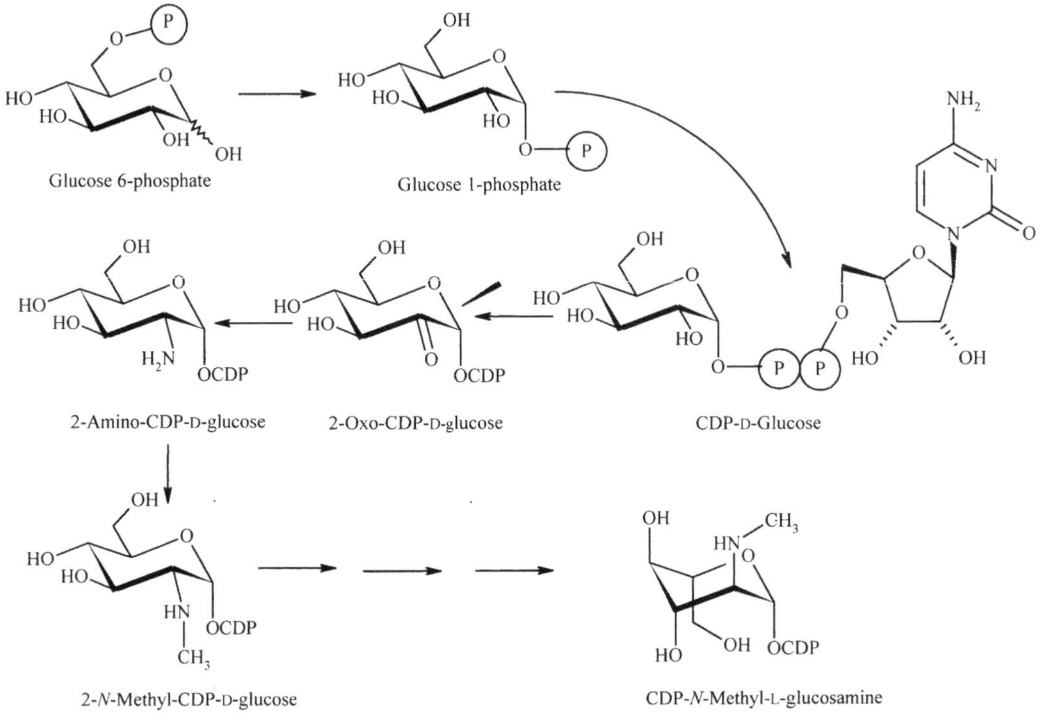

Fig. 37. Biosynthesis of CDP-N-methyl-L-glucosamine

The 2-deoxy-scyllo-inosose pathway

This pathway (Fig. 39) leads to formation of the aminoglycoside antibiotics which have 2-deoxystreptamine as their aminocyclitol moiety. Also for this pathway gene clusters for the biosynthesis of the various antibiotics have been detected and studied. Many of the participating enzymes have been biochemically characterized.

2-Deoxystreptamine (2-DOS) is formed from D-glucose 6-phosphate, which is converted to 2-deoxy-scyllo-*inosose* by the enzyme 2-*deoxy*-scyllo-*inosose synthase* (2-DOI synthase). Transamination yields 2-deoxy*scyllo*-inosamine, which is oxidized to 2-deoxy-3-keto-*scyllo*-inosamine from which 2-deoxystreptamine is formed by a final transamina tion step.

The enzyme 2-DOI synthase was first purified from the bacterium *Bacillus circulans*, which produces the aminoglycoside antibiotic butirosin (not used clinically) and its gene in this organism –*btrC* – was identified. Homologues of *btrC* have been identified in all gene clusters of organisms producing aminoglycosides containing 2-DOS as the aminocyclitol.

The continuation of the biosynthesis of 2-DOS antibiotics involves biosynthesis of the various amino sugars and their connection to C-4 + C-5 or C-4 + C-6 of 2-DOS. Biosynthetic gene clusters have been characterized for the 2-DOS-derived aminoglycosides, genes coding for enzymes involved in the pathways have been expressed and biosynthetic routes to different antibiotics have been proposed. For more information, see the reference Flatt (2007) in *Further reading* on page 193.

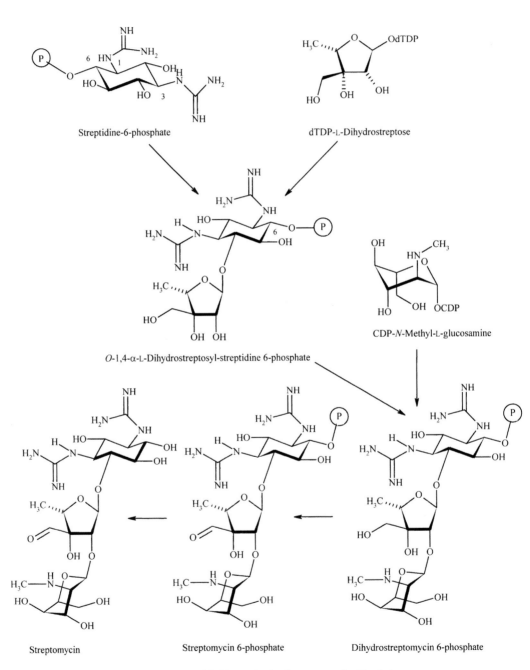

Fig. 38. Biosynthesis of streptomycin. Final steps

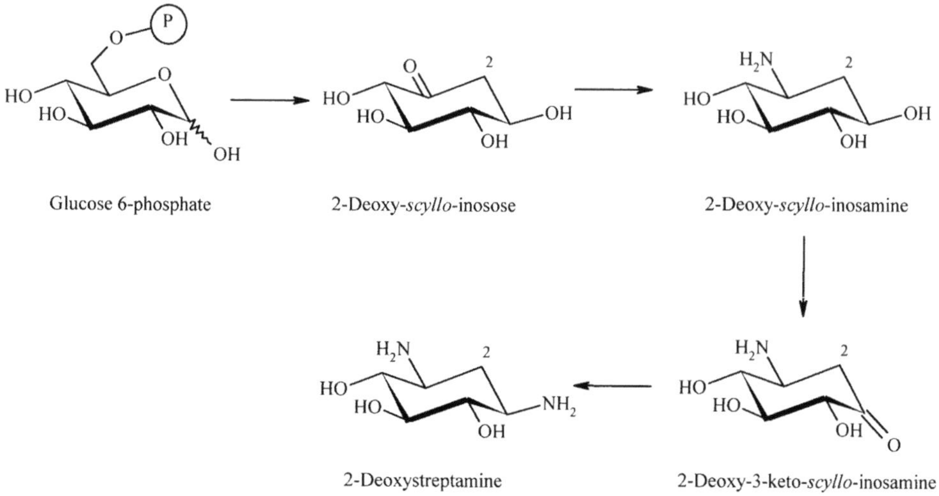

Fig. 39. Biosynthesis of 2-deoxystreptamine (2-DOS)

Acarbose

Acarbose (Glucobay®, Precose®, *O*-{4,6-dideoxy-4[1s-(1,4,6/5)-4,5,6-trihydroxy-3-hydroxymethyl-2-cyclohexen-1-yl]-amino-α-D-glucopyranosyl}-(1→4)-*O*-α-D-glucopyranosyl-(1→4)-D-gluco-pyranose) is structurally related to the aminocyclitol glycosides described above and is composed of the C7-aminocyclitol valien-amine, which forms an N-glycoside with a trisaccharide formed from 4,6-dideoxyglucose and maltose.

Acarbose is produced by the actinomycete *Actinoplanes* sp. SE50 and is an inhibitor of α-glucosidases, especially sucrase. It is used for treatment of type 2 diabetes mellitus and its effect is due to retardation of the digestion and absorption of carbohydrates in the small intestine, whereby the increase in blood-glucose concentration after a carbohydrate load is reduced. As the carbohydrates remain in the intestine, bacteria in the colon will digest the complex carbohydrates, causing flatulence and diarrhoea, which are common side effects of treatment with acarbose.

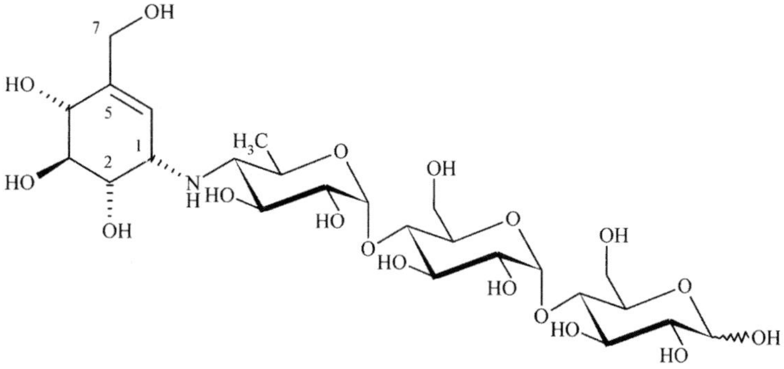

Acarbose

189

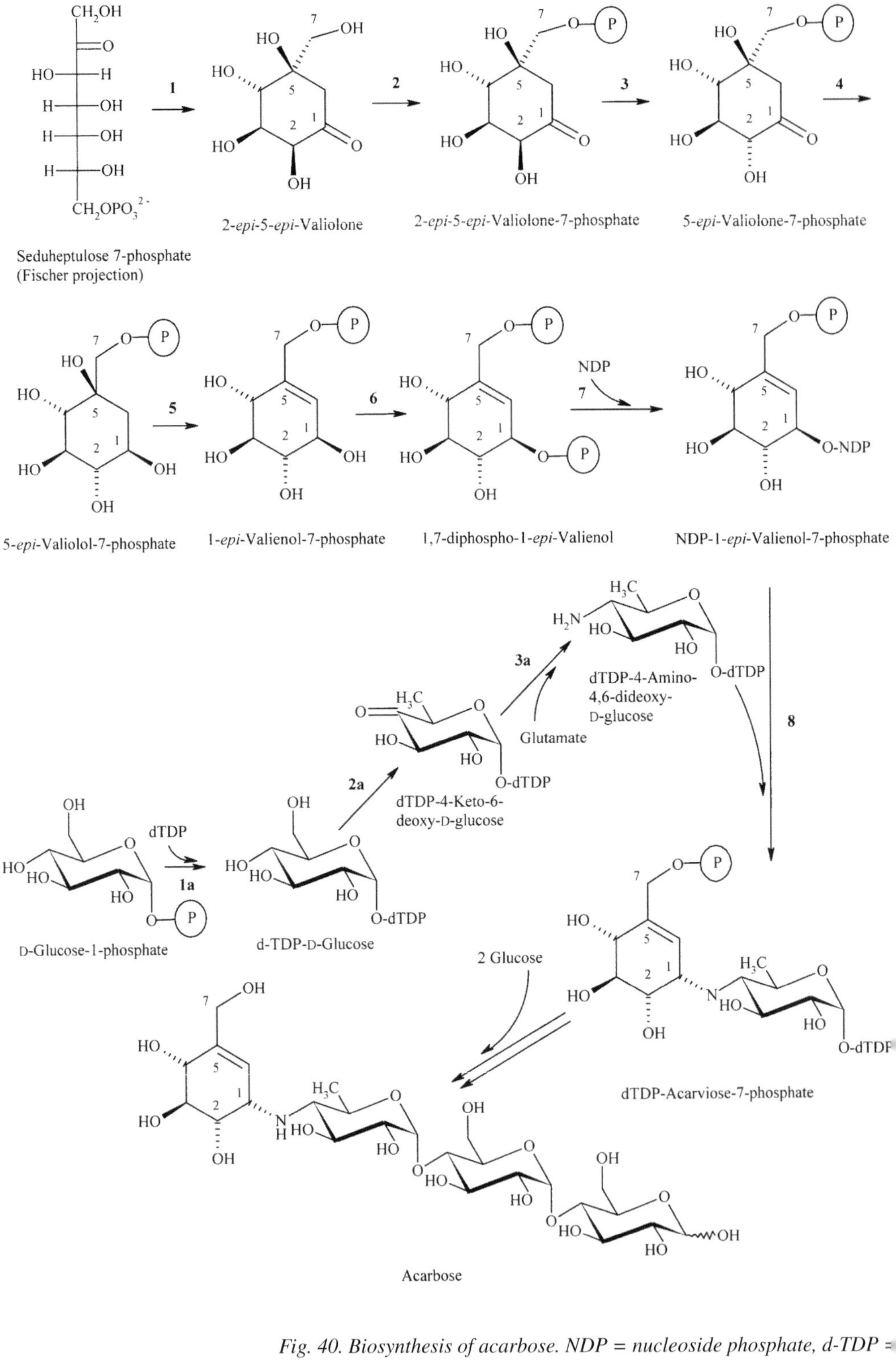

*Fig. 40. Biosynthesis of acarbose. NDP = nucleoside phosphate, d-TDP =
deoxy-thymidine diphosphate*

Biosynthesis The C_7N aminocyclitol moiety of acarbose is derived from a C7 sugar phosphate (2-*epi*-5-*epi*-valiolone), which is formed from sedoheptulose, emanating from the pentose phosphate pathway (page 128). The first part of the biosynthesis involves 7 different reactions (**1–7** in Fig. 40) and results in formation of NDP-1-*epi*-valienol-7-phosphate. The nitrogen originates from glutamate and is introduced at formation of an *N*-glucoside between NDP-1-*epi*-valienol-7-phosphate and the sugar dTP-4-amino-4,6-dideoxy-D-glucose, yielding dTDP-acarviose-7-phosphate (reaction **8**). The amino sugar is formed from D-glucose-1-phosphate. Finally acarbose is formed by the addition of two more D-glucose units. This biosynthetic pathway is only a proposal since many of the enzymes and the genes that code for them have not yet been completely characterized. From the gene cluster of *Actinoplanes* sp. SE50, genes coding for the enzymes which perform reactions 1–4 in Fig. 40 have been identified and expressed. The deduced amino-acid sequence of the epimerase, which catalyses reaction 3, shows no similarities with any known epimerases and is presumably the first representative of a new class of epimerases. The following steps (5–8) are speculative. Enzymes which catalyse the formation of the deoxy-sugar moiety of acarbose (reactions **1a–3a**) have been identified from the corresponding genes and expressed.

Ethyl alcohol, C_2H_5OH

Ethyl alcohol (ethanol) is not a carbohydrate but is produced industrially from various carbohydrates by fermentation. Commonly used raw materials are sugar cane, molasses, grain and potatoes. Ethanol can also be obtained by fermentation of glucose and mannose contained in waste products from production of paper pulp by the sulphite process. Sugar cane and molasses contain sucrose which is directly fermented by yeast. Grain and potatoes contain starch which must be broken down to glucose by enzymes such as *diastase* from barley malt, which converts starch to maltose and *maltase* from yeast, which hydrolyses maltose to glucose.

For industrial production of ethanol from starch-containing products the material is boiled to cause the starch to form a paste. After cooling, enzymes are added and the glucose formed is extracted and fermented with yeast for 3 days. The ethanol concentration in the mixture is now 8–9 %. Further concentration is obtained by distillation, yielding a crude product containing 85–90 % ethanol but also methanol, propanol, butanol, amyl alcohols and small amounts of aldehydes. These are removed by repeated distillation yielding a pure product containing 95.5 % ethanol and 4.5 % water.

The general reaction for fermentation of glucose is:

$$C_6H_{12}O_6 \quad \xrightarrow{\text{enzyme}} \quad 2\ C_2H_5OH + 2\ CO_2$$

The process is anaerobic.

In pharmacy ethanol is used as a solvent. Mixtures of ethanol and water are used for production of plant extracts (page 69).

Ethanol can also be produced synthetically from ethylene by addition of water:

$$C_2H_4 + H_2O \rightarrow C_2H_5OH.$$

Synthetic ethanol is not used in pharmacy and medicine.

Further reading

Starch

Hancock, R.D. and Tarbert, B.J. The other double helix – the fascinating chemistry of starch. *Journal of Chemical Education*, **77** (8), 988–992 (2000).

Smith, A.M., Denyer, K. and Martin, C. The synthesis of the starch granule. *Annual Review of Plant Physiology and Plant Molecular Biology*, **48**, 67–87 (1997).

Heparin

Linhardt, R.J. Heparin: An important drug enters its seventh decade. *Chemistry & Industry*, 21 Jan., 45–50 (1991).

Rabenstein, D.L. Heparin and heparan sulfate: structure and function. *Natural Products Reports*, **19**, 312–331 (2002).

Mannitol

Gao, Z. and Loescher, W.H. NADPH supply and mannitol biosynthesis. Characterization, cloning and regulation of the non-reversible glyceraldehyde-3-phosphate dehydrogenase in celery leaves. *Plant Physiology,* **124**, 321–330 (2000).

Gutierrez, A.J.E. and Gaudillere J.P. Distribution, métabolisme et rôle du sorbitol chez les plantes supérieures. Synthèse. *Agronomie*, **16**, 281–298 (1996).

Ascorbic acid

Wolucka, B.A., Van Montagu, M. TheVT2 cycle and the *de novo* biosynthesis pathways for vitamin C in plants: An opinion. *Phytochemistry*, **68**, 2602–2613 (2007)

Loewus, F.A. Biosynthesis and metabolism of ascorbic acid in plants and of analogs of ascorbic acid in fungi. *Phytochemistry*, **52**, 193–210 (1999).

Smirnoff, N. Biosynthesis of ascorbic acid in plants: A renaissance. *Annual Review of Plant Physiology and Plant Molecular Biology*, **52**, 437–467 (2001).

Rose hips

Kharazmi, A. Laboratory and preclinical studies on the anti-inflammatory and anti-oxidant properties of rosehip powder – Identification and characterization of the active component GOPO®. *Osteoarthritis and Cartilage*, **16** Suppl. 1, S5–S7 (2008).

Aminoglycoside antibiotics

Busscher, G.F., Rutjes, F.P.J.T and van Delft, F.L. 2-Deoxystreptamine: central scaffold of aminoglycoside antibiotics. *Chemical Reviews*, **105** (3), 775–791 (2005).

Flatt, P.M. and Mahmud, T. Biosynthesis of aminocyclitol-aminoglycoside antibiotics and related compounds. *Natural Product Reports*, **24**, 358–292 (2007).

Forge, A. and Schacht, J. Aminoglycoside antibiotics. *Audiology and Neurotology*, **5**, 3–22 (2000).

Llewellyn, N.M. and Spencer, J.B. Biosynthesis of 2-deoxystreptamine-containing aminoglycoside antibiotics. *Natural Product Reports*, **23**, 864–874 (2006).

Acarbose

Wehmeier, U.F. and Piepersberg, W. Biotechnology and molecular biology of the α-glucosidase inhibitor acarbose. *Applied Microbiology and Biotechnology*, **63**, 613–625 (2004).

6. Natural Products Derived Biosynthetically from Shikimic Acid

Shikimic acid

Shikimic acid was first isolated in 1885, and named after *shikimi-no-ki*, the Japanese name of the tree, *Illicium anisatum* L. (*Illiciaceae*), from which it was first prepared.

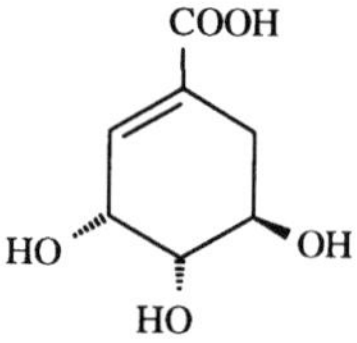

Shikimic acid

As shown in the scheme on page 135, several biosynthetic pathways lead from the primary products of photosynthesis – carbohydrates – to other natural products. Shikimic acid is a key substance in one of these routes which consequently has been called ***the shikimic acid pathway***. This pathway is one of the two biosynthetic reaction series in which aromatic compounds are formed in Nature. The other one starts with acetate and will be dealt with in the next chapter. Complicated aromatic substances may contain partial structures derived from either of these two routes.

The principal part played by shikimic acid in the biosynthesis of aromatic amino acids was discovered during work with mutants of the bacterium *Escherichia coli*. Normal strains of this bacterium grow well on media containing glucose as the only source of carbon. Exposure to γ-radiation, however, gave rise to strains that could not grow on glucose alone but needed one or more aromatic amino acids for their development. One of these mutants required not only the three common aromatic amino acids, tyrosine, phenylalanine and tryptophan, but also *o*-aminobenzoic acid and *p*-hydroxybenzoic acid. This fact indicated that the five compounds mentioned might be derived from a common "starting compound" – a *precursor*. This precursor was found to be shikimic acid, which alone could replace the other five acids in the culture medium. Continued studies showed that shikimic acid is a universal precursor in aromatic amino acid biosynthesis, not only in bacteria but also in fungi and higher plants. The shikimic acid pathway is not operative in animals, which thus are dependent on an external supply of the aromatic amino acids. (For human beings phenylalanine and tryptophan are essential amino acids while tyrosine can be synthesized from phenylalanine in the liver).

The aromatic amino acids are used as building blocks in proteins and peptides, but in plants they are also precursors of a great number of secondary products such as alkaloids, phenols, hydroxycinnamic acids, phenylpropanes, coumarins, chromones, xanthones, stilbenes, flavonoids, lignans, neolignans, lignins and others. Shikimic acid is also the starting point of a reaction series that leads via gallic acid to tannins, which are found in varying amounts in practically all plants.

Localization of the shikimic acid pathway

In plants, proteins are synthesized in three different compartments: in the cytoplasm, in the chloroplast and in the mitochondria. Therefore, the aromatic amino acids must either be synthesized *in situ* in the respective compartments or they are synthesized outside the compartments and have to be imported. There is no evidence that the aromatic amino acids are synthesized in the mitochondria. It has been well established that all three aromatic amino acids can be synthesized in the chloroplasts and this has been confirmed by isolation from chloroplasts of cDNAs and genes which code for enzymes of the shikimic acid pathway. It is a matter of debate, however, if there is, in addition, a complete pathway in the cytoplasm. The identification of isozymes of cytosolic origin has mainly been based on biochemical fractionation studies and the activities have never been purified to homogeneity. Nor have their respective cDNAs or genes been determined. Until this has been done the question of the existence of two pathways remains open. Phenylpropanoids, on the other hand, seem to be synthesized solely in the cytoplasm. If there is no complete pathway in the cytoplasm, the phenylalanine required for the formation of the phenylpropanoids must be transported from the chloroplasts to the cytoplasm.

The enzymes of the shikimic acid pathway

The enzymes which catalyse the reactions in the pathway have all been isolated and characterized. As mentioned above, it has also been possible to isolate cDNAs and genes coding for several of these enzymes. In plants most of the steps in the pathway are catalysed by separable, monofunctional, enzymes. The situation is different in bacteria and fungi. In bacteria the enzymes of the pathway are separable up to chorismate, whereas the steps from chorismate to the aromatic amino acids are catalysed by multifunctional enzyme complexes. In some fungi the activities of five consecutive steps of the prechorismatic pathway reside on a single polypeptide called the *arom* complex and multifunctional complexes are involved also in the postchorismatic steps.

The enzymes are large protein molecules with apparent molecular masses in the range of 50–220 kDa.

Biosynthesis of shikimic acid

The biosynthesis of shikimic acid starts with an aldol condensation of phosphoenol pyruvate and the carbohydrate 4-phospho-D-erythrose, both of which are glucose metabolites. At the pH prevalent in the plant, the acids involved in this and the following reactions are ionized. The catalyst is *DAHP synthase (= 3-deoxy-7-phosphoheptulonate synthase* (EC 2.5.1.54)) and the product is a heptose derivative – 7-phospho-3-deoxy-D-*arabino*-heptulosonate – (also called 3-deoxy-D-*arabino*-heptulosonate-7-phosphate (DAHP). *DAHP synthase* enzymes fall into two distinct classes according to whether they are plant derived or microbial in origin. In microorganisms isozymes of DAHP synthase are characterized by different

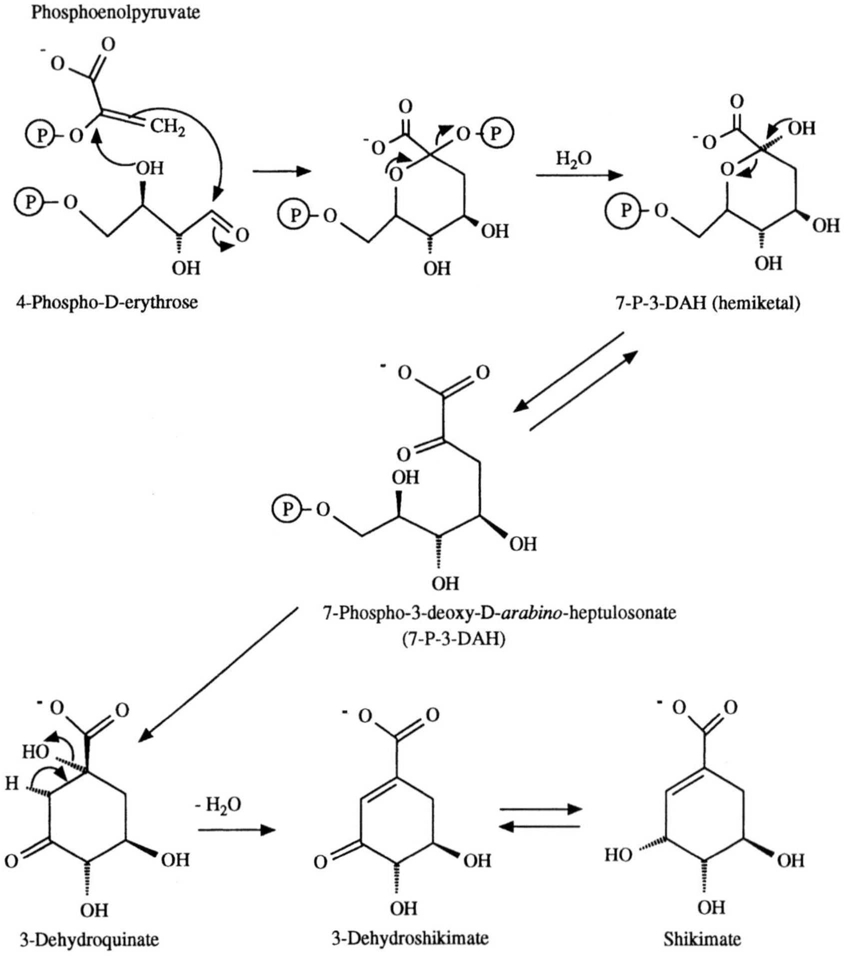

Fig. 41. Biosynthesis of shikimic acid

sensitivities towards feedback inhibition by the aromatic amino acids L-phenylalanine, L-tyrosine and L-tryptophan. DAHP synthase from microorganisms has been extensively studied both as a protein and on the gene level. Determination of the crystal structure of a phenylalanine regulated DAHP synthase from *Escherichia coli* has allowed conclusions concerning the structure of the active site of the enzyme.

In plants DAHP synthase occurs in several different forms (isozymes), which are differentiated by their requirement for the divalent cations Mn^{2+} or Co^{2+}. For *tobacco (Nicotiana* sp.), spinach (*Spinacia oleracea* L.) and parsley (*Petroselinum crispum* (Miller) A.W. Hill) the Mn^{2+} stimulated DAHP synthase isozyme has been assigned to the chloroplast whereas the Co^{2+}requiring isozyme seems to be located in the cytoplasm. The plastide-localized isozyme seems to be specific for 4-phospho-D-erythrose but the cytosolic enzyme also uses a wide range of other aldehydes as substrates and its involvment in DAHP synthesis is questionable. cDNA clones encoding chloroplast-specific DAHP synthases have been isolated from several plant species (*Solanum tuberosum, Nicotiana tabacum, Arabidopsis thaliana, Lycopersicon esculentum*).

A sequence of reactions (oxidation, β-elimination, reduction and an intramolecular aldol condensation) transform DAHP into 3-dehydroquinate. The catalysing enzyme is *3-dehydroquinate synthase* (EC 4.2.3.4). Some of the reactions appear to occur non-enzymatically and the enzyme is probably responsible only for the oxidation and reduction steps. 3-Dehydroquinate synthase requires Co^{2+} and bound NAD^+ as cofactor. The enzyme from bacteria has been characterized both as a protein, including crystallographic studies, and with respect to encoding genes. From tomato *Lycopersicon esculentum* Miller a cDNA clone coding for 3-dehydroquinate synthase has been isolated and sequenced.

In a reaction catalysed by *3-dehydroquinate dehydratase* (EC 4.2.1.10) (DHQase), 3-dehydroquinate loses water to form 3-dehydroshikimate. From studies, which have been performed mainly in bacteria and fungi, the existence of two distinct classes of dehydroquinate dehydratase enzymes has been proved. These are denoted as types I and II and have quite different physical and biochemical properties. Type I enzymes are typically dimers with a subunit mass of 27 kDa while type II enzymes are dodecamers with the subunit mass 16 kDa. DHQases of type I fulfil a biosynthetic role but those of type II can function as either biosynthetic or catabolic enzymes. In bacteria DHQase is monofunctional, whereas in plants it is part of a bifunctional enzyme with SORase (see below). Also the mechanisms for the elimination of water are different for the two types of DHQase. Type I enzymes use a Shiff base mechanism between the substrate and an active-site lysine (Fig. 42, upper part). Also involved are a histidine residue and a glutamic acid residue. Histidine acts as a base to deprotonate lysine which then reacts with the keto group of the substrate forming an imine inermediate. *Cis*-dehydration of this intermediate then yields dehydroshikimate. The liberated proton is accepted by the glutamic acid residue. Type II enzymes catalyse a *trans*-dehydration of 3-dehydroquinate via an enolate intermediate (Fig. 42, lower part).

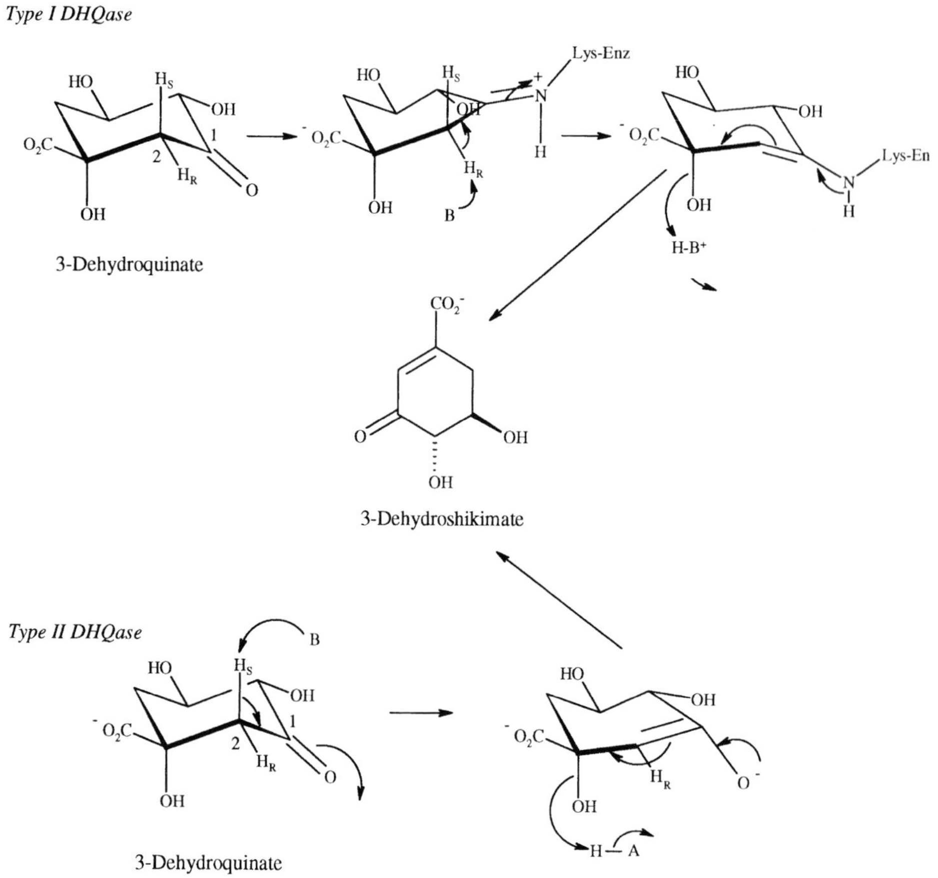

Fig. 42. Elimination of water from 3-dehydroquinate

3-Dehydroshikimate is finally reduced to shikimate with the aid of *shikimate dehydrogenase* (EC 1.1.1.25, also called SORase), a reaction which requires $NADP^+$. The reaction is reversible. SORase has been isolated and characterized from microorganisms. In higher plants DHQase and SORase reside on the same polypeptide, which thus is a bifunctional enzyme, and has been isolated from peas (*Pisum sativum* L.) and purified to homogeneity with an apparent molecular mass of 59 kDa. The dehydroquinase part of this enzyme has been established as a type I enzyme. A complete amino acid sequence has been determined for the DHQase domain and for a part of the SORase domain. A cDNA encoding for this bifunctional enzyme in tobacco (*Nicotiana tabacum* L.) has been cloned and sequenced.

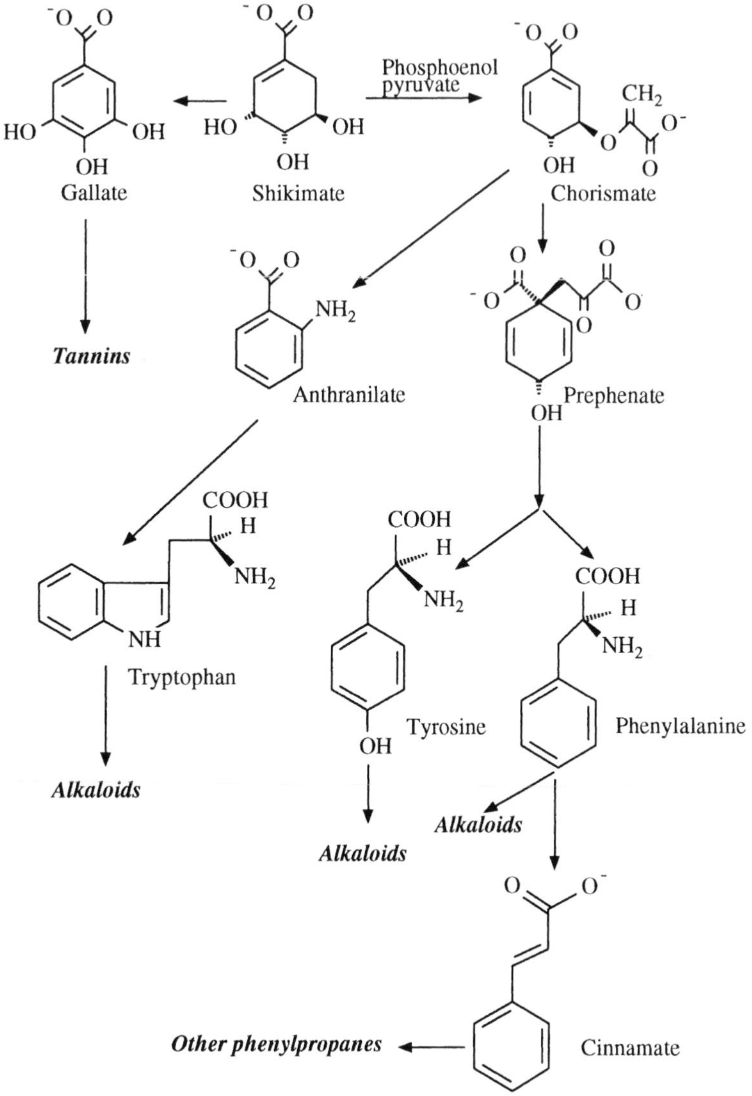

Fig. 43. Outline of biosynthetic routes from shikimic acid

Shikimate is the starting point of various biosynthetic routes leading to aromatic amino acids and gallic acid (Fig. 43).

Gallic acid and tannins

Gallic acid is a building block of many tannins. These are plant constituents that are able to form insoluble and indigestible compounds with proteins. Gallic acid is formed from shikimic acid by way of 3-dehydroshikimate:

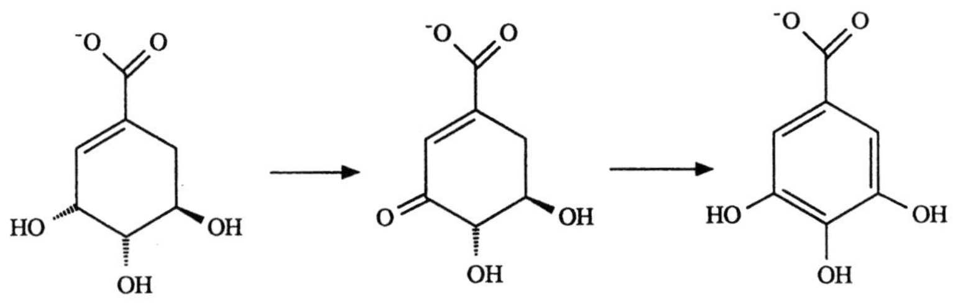

Shikimate   3-Dehydroshikimate   Gallate

This seems to be the main reaction route both in bacteria and in higher plants. Some studies indicate, however, that an alternative pathway, via phenylalanine and other phenylpropanes, also operates in higher plants.

Tannins are a very heterogeneous group of natural products, occurring in larger or smaller amounts in practically all plants. In those plants which contain especially large amounts, they are often confined to certain organs, such as leaves, fruit, bark or wood. Tannins are often present in unripe fruits, but disappear during ripening. The function of tannins in plants is not known with certainty, but it seems possible that they are an integrated part of the plant's chemical defence, protecting against both grazing and invasion by unspecifically destroying enzymes. The tannin-producers protect themselves by extensive compartmentalization of the tannins.

Tannins are amorphous substances, which form colloidal, acidic aqueous solutions with an astringent taste. With iron salts, tannins form dark blue or greenish black, soluble compounds (ink), and they are precipitated by salts of many other metals, such as copper, lead and tin. The most important property of tannins is their ability to form insoluble compounds with proteins. This reaction is the basis for their extensive use in leather production. The same property is the reason for their limited medical use, which includes the treatment of diarrhoea, bleeding gums and, to some extent, skin injuries.

Tannins can be divided into two groups: *hydrolysable tannins*, which are split into simpler molecules on treatment with acids or enzymes, and *condensed tannins* (catechin tannins), which give complex insoluble products on similar treatment. Hydrolysable tannins can be subdivided into *gallotannins*, which yield sugar and gallic acid on hydrolysis, and *ellagitannins*, which furnish not only sugar and gallic acid, but also ellagic acid. Condensed tannins are complex polymers, whose chemistry is only partly known. Their building blocks include catechins and flavonoids, which are often esterified with gallic acid.

Biosynthetically, flavonoids and catechins are derived both from acetate (page 329) and from shikimic acid.

Crude drugs   **Galls** (Gallae). Galls are vegetative growths produced on plants by the action of insects or, in some instances, fungi. Their shape is usually that of a round or oval ball, but tufts and bud-like shapes also occur, and they are found on leaves as well as on twigs, shoots, apices and inflorescenses.

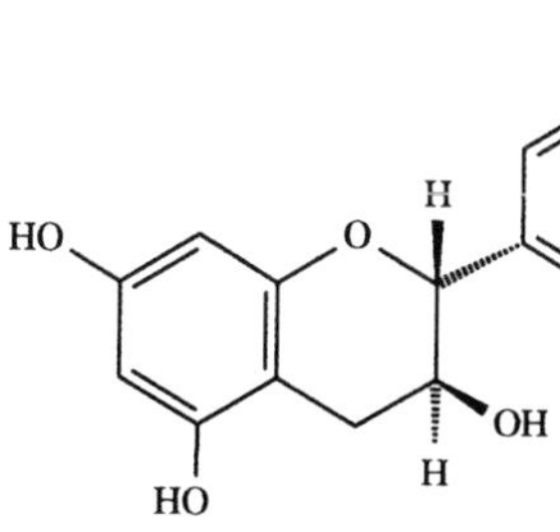

Catechin 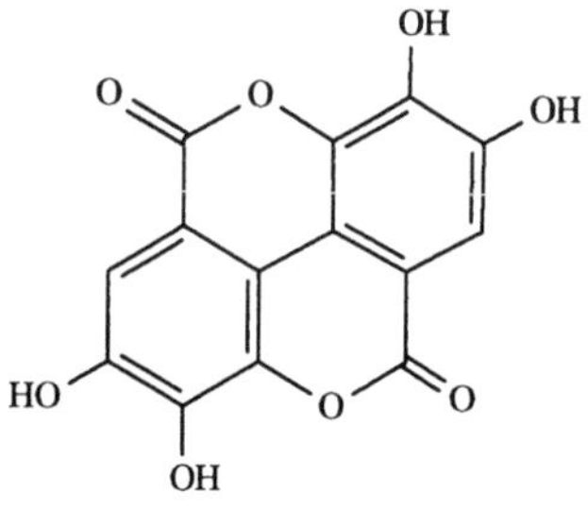   Ellagic acid

Turkish galls arise on young branches of *Quercus lusitanica* Lam. (*Fagaceae*) as a result of attack by the gall-wasp *Adleria gallae-tinctoria*. *Quercus lusitanica* is a tree or shrub, about 3 m high, naturally occurring in Greece, Turkey, Syria and Iran. The gall-wasp bores a hole in a branch of the tree and deposits an egg. As the insect develops, it produces an enzyme that hydrolyses the starch in the surrounding cells to sugar, which it then uses as a food. The tissue attacked grows abnormally and forms a gall, the cells of which produce gallic acid and tannins, and thus protects the developing gall-wasp larvae. When the insect is mature and exits from the gall, the latter contains 50–70 % *tannin*.

Tannin is prepared from powdered galls by extraction with a mixture of ether, ethanol and water. Two layers are formed. The

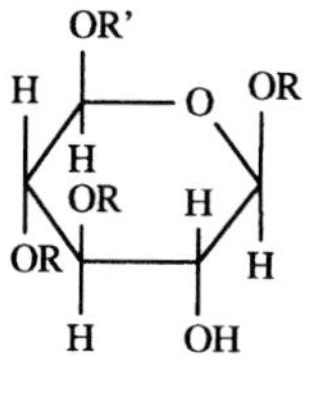

Tannin (R = gallic acid, R' = *m*-trigallic acid)

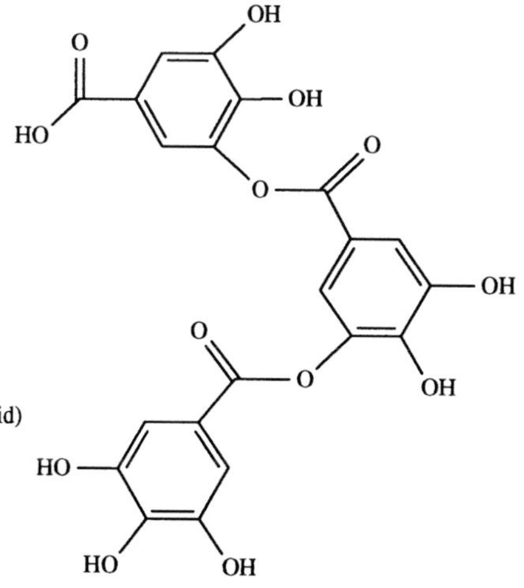

m-Trigallic acid

201

ether layer contains free gallic acid, whereas the tannin is present in the aqueous layer, from which it can be isolated. Purified tannin is a mixture of gallates of glucose, the composition varying with the origin. One example is 1,3,4-trigalloylglucose with a *m*-trigalloyl ester group at C-6 of the glucose unit. This type of ester bond between the carboxyl group in one molecule and a hydroxyl group in another molecule of the same compound is called a *depside* bond.

Chinese tannin is obtained from galls on *Rhus semialata* Murr (*Anacardiaceae*), resulting from attack by the aphid (*Melaphis sinensis*). Also Chinese tannin consists of a mixture of gallates of glucose, e.g. 1,3,4,6-tetragalloylglucose with the hydroxyl group at C-2 esterified with *m*-trigallic acid.

Tannin is used for the local treatment of catarrh and infections, as well as to stop local bleeding. It can also be used as a remedy against alkaloid poisoning, as most alkaloids give insoluble products with tannin. The technical uses include the preparation of inks, dyeing of textiles, and tanning to make special types of leather.

Hamamelis leaf (Hamamelidis folium) consists of the dried leaves of *Hamamelis virginiana* L. (*Hamamelidaceae*), an up to 7 m high tree occurring in eastern North America. Hamamelis leaf contains gallotannins and is used in the form of infusions or extracts for treatment of diarrhoea, in ointments and suppositories against haemorrhoids and as a mild astringent for the topical treatment of skin inflammation. Both the leaves and the bark of *Hamamelis virginiana* are also ingredients in face lotions.

The aromatic amino acids phenyl-alanine, tyrosine and tryptophan

The aromatic amino acids are building blocks in proteins and peptides, but they are also important intermediates in the biosynthesis of many other natural products such as alkaloids, phenols, phenolic acids, acetophenols, phenylacetic and hydroxycinnamic acids, phenylpropanes, (*iso*)coumarins, chromones, xanthones, stilbenes, flavonoids, lignans, neolignans, lignins and others. The aromatic amino acids are formed from shikimate which is first transformed to a common precursor – *chorismate* – as outlined in Fig. 44.

The first step is phosphorylation of shikimate with ATP to yield 3-phosphoshikimate. The reaction is catalysed by the enzyme *shikimate kinase* (EC 2.7.1.71). The enzyme has been isolated from bacteria and fungi and the amino acid sequences of the enzymes from these sources have been determined. The genes *aroK* and *aroL* which encode two isozymes (shikimate kinase I and II) have been isolated from *Escherichia coli* and sequenced. The type I kinase has a much lower affinity for substrate than the type II kinase and there is growing evidence that the true function of type I is not the phosphorylation of shikimate but the catalysis of some other biological process. A gene – *AroI* – encoding the enzyme has also been cloned and sequenced from *Bacillus subtilis*. The enzyme has been isolated from chloroplasts of spinach and was observed to be very unstable. A cDNA clone coding for shikimate kinase has been isolated from tomato and expressed *in vitro*.

The 5-hydroxyl group of 3-phosphoshikimate then reacts with one molecule of phosphoenol pyruvate forming 3-phospho-enolpyruvylshikimate (= 5-enolpyruvyl-shikimate 3-phosphate, EPSP). This reaction, which proceeds with the help of *3-phospho-shikimate 1-carboxyvinyltransferase* (EC 2.5.1.19, also called EPSP synthase), involves an unstable, transient, enzyme-bound

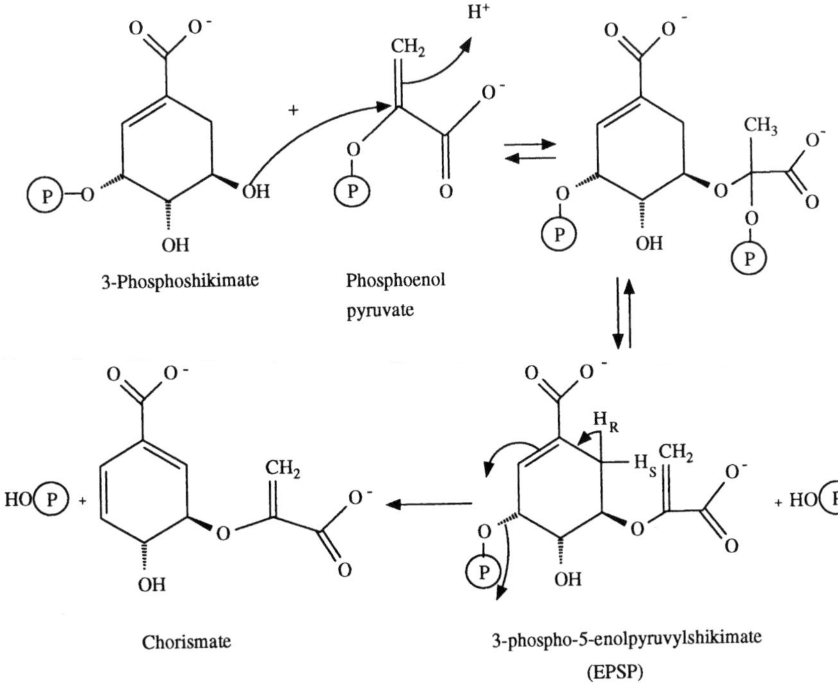

3-Phosphoshikimate   Phosphoenol pyruvate

Chorismate   3-phospho-5-enolpyruvylshikimate (EPSP)

Fig. 44. Biosynthesis of chorismic acid

intermediate. The three-dimensional structure of the enzyme from *Escherichia coli* has been solved by crystallographic techniques. Several EPSP synthase genes and cDNA clones have been isolated from higher plants. All EPSP synthase specific genes and cDNAs isolated so far encode enzymes occurring in the chloroplast. The broad-spectrum, non-selective herbicide *glyphosate* [*N*-(phospho-nomethyl)-glycine] is an inhibitor for EPSP synthase which acts by binding to the PEP binding site of the enzyme. Plant cells can adapt to the presence of glyphosate by overproducing EPSP synthase or by developing a herbicide-resistant form of the enzyme.

Chorismate synthase (EC 4.2.3.5) catalyses the transformation of EPSP to chorismate by 1,4-*trans*-elimination of hydrogen and phosphate. Chorismate synthase requires a reduced flavin for activity although the reaction does not involve an overall change in oxidation state. Some enzymes, e.g. those from the bread mould *Neurospora crassa* and from *Bacillus subtilis,* are bifunctional and have an associated flavin reductase that utilizes NADPH to produce the reduced flavin. Others, e.g. enzymes from *Escherichia coli* and from plants, are monofunctional and must be supplied with a reduced flavin. Chorismate synthase has been characterized from the fungus *Neurospora crassa*, as well as from the bacteria *Escherichia coli* and *Salmonella typhi,* where also the nucleotide sequence of the *aroC* gene encoding the enzyme has been determined from *E. coli* and *S. typhi.* Chorismate synthase has also been studied in higher plants. Thus it has been purified to homogeneity from *i.a.* a *Corydalis glauca* Pursh cell culture. This enzyme is a dimer and the molecular weight of the subunit has been estimated by gel electrophoresis to be 42 kDa. A corresponding cDNA clone has been isolated from the same plant and from tomato.

Biosynthesis of tyrosine and phenylalanine

The pathway for the formation of these amino acids is different in bacteria and fungi compared with that in plants. The processes are outlined in Fig. 45. The first reaction is the conversion of chorismate into prephenate, which is a Claisen-like rearrangement that transfers the phosphoenolpyruvate-derived side chain so that it becomes directly bonded to the carbocycle. The reaction is catalysed by the enzyme *chorismate mutase* (CM, EC 5.4.99.5). In microorganisms this enzyme can occur as a monofunctional enzyme (e.g. in *Bacillus subtilis*), but it can also occur as part of a bifunctional enzyme as in *Escherichia coli*, where it is combined with the next enzyme in the pathway – *prephenate dehydratase* (EC 4.2.1.51) (for formation of phenylalanine) or *prephenate dehydrogenase* (EC 1.3.1.12) (for formation of tyrosine, see below). Both enzyme complexes have been cloned and sequenced and so has the monofunctional enzyme from *Bacillus subtilis*. The crystal structure of these enzymes has also been determined and their active sites located and characterized.

Chorismate mutase has been obtained from baker's yeast (*Saccharomyces cerevisiae*) and its crystal structure has been determined. The active site of this enzyme has been located and characterized.

In higher plants CM-activity has been found in a variety of plants often existing as two or even three separable isozymes. A chorismate mutase of chloroplast origin has been produced by cloning the corresponding gene from *Arabidopsis thaliana* and has been expressed in yeast. From the same plant and with a similar technique, what appears to be a cytosolic isozyme to CM has been identified.

Biosynthesis of tyrosine and phenylalanine in bacteria and fungi

Several pathways may exist depending on the organism and often more than one route may operate in a particular species as a result of the enzyme activities available. The presently accepted concepts are the following.

For formation of phenylalanine the enzyme prephenate dehydratase converts prephenate into phenylpyruvate which is transformed to phenylalanine by transamination, involving the amino group of glutamate, in a reaction catalysed by *aromatic-amino-acid transaminase* (EC 2.6.1.57).

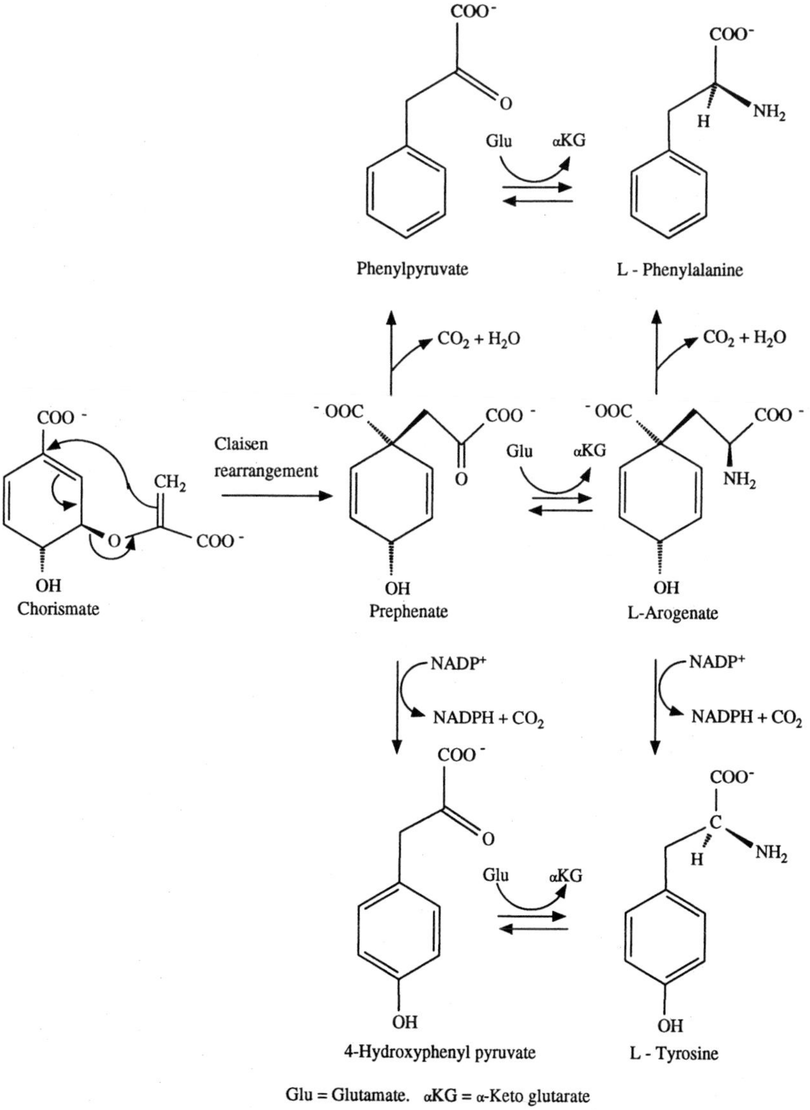

Fig. 45. Biosynthesis of phenylalanine and tyrosine from chorismate

For formation of tyrosine the enzyme prephenate dehydrogenase catalyses the formation of 4-hydroxyphenylpyruvate from prephenate. Transamination (glutamate) of this compound in a reaction catalysed by the enzyme *tyrosine transaminase* (EC 2.6.1.5) yields tyrosine. The two aromatic aminotransferases may in reality be one single enzyme which can accept both phenylpyruvate and 4-hydroxyphenylpyruvate as a substrate. Such an enzyme (EC 2.6.1.57) has been obtained from *Escherichia coli*.

Biosynthesis of tyrosine and phenylalanine in plants

In plants transamination reaction catalysed by *glutamate-prephenate transaminase* (EC 2.6.1.79) transforms prephenate into L-arogenate. The enzyme has been isolated and purified to homogeneity from cell cultures of Anchusa officinalis L. and has an apparent mass of 220 kDa, with subunits of molecular masses 44 kDa and 57 kDa, respectively, indicating a possible $\alpha_2\beta_2$ subunit structure.

Phenylalanine and tyrosine are formed from L-arogenate in reactions catalysed by *arogenate dehydratase* (EC 4.2.1.91) and *arogenate dehydrogenase* (EC 1.3.1.78), respectively. All transamination reactions in these pathways are reversible.

Biosynthesis of tryptophan

The first step in the biosynthesis of tryptophan (Fig. 46) is transformation of chorismate into anthranilate. This is a complicated reaction, catalysed by *anthranilate synthase* (EC 4.1.3.27), in which the enolpyruvyl side-chain of chorismate is eliminated accompanied by donation of an amino group from glutamine. The reaction requires the presence of Mg^{2+}. Anthranilate synthase is composed of three subunit activities. One, called *TrpG* is a glutamine amidotransferase which catalyses the release of ammonia from glutamine. The next subunit – *2-amino-2-deoxyisochorismate synthase* – catalyses the incorporation of ammonia into chorismate at the 2-position. Anthranilate is formed from the product by removal of the side chain, catalysed by the subunit *2-amino-2-deoxyisochorismate lyase*. The combined action of the last two subunits has also been termed subunit *TrpE*. Anthranilate synthase has been studied extensively both as a protein and at the cDNA and gene levels in bacteria and fungi as well as in higher plants.

The anthranilate formed reacts with 5-phosphoribosyl 1-diphosphate to yield 5-phosphoribosyl-anthranilate which is then transformed to 5-phospho-1-(o-carboxyphenylamino)-1-deoxyribulose. The enzymes catalysing these reactions are *anthranilate phosphoribosyltransferase* (APRT, EC 2.4.2.18), and *phosphoribosylanthranilate isomerase* (PRAI, EC 5.3.1.24), respectively

The next reaction is a decarboxylation followed by ring closure of the side chain yielding 1-indolyl-3-phosphoglycerol. The catalysing enzyme is *indol-3-glycerol-phosphate-synthase* (InGPS, EC 4. 1. 1. 48). In bacteria the genes for the enzymes APRT, PRAI and InGPS have been characterized and found to be organized in apparently un-linked clusters or together with the other genes required for trypto-phan biosynthesis. In plants the knowledge of these three enzymes is more limited. The best studied plant is *Arabidopsis thaliana* from which an APRT-specific cDNA clone has been isolated and the cor-responding gene has been cloned and sequenced. Three genes

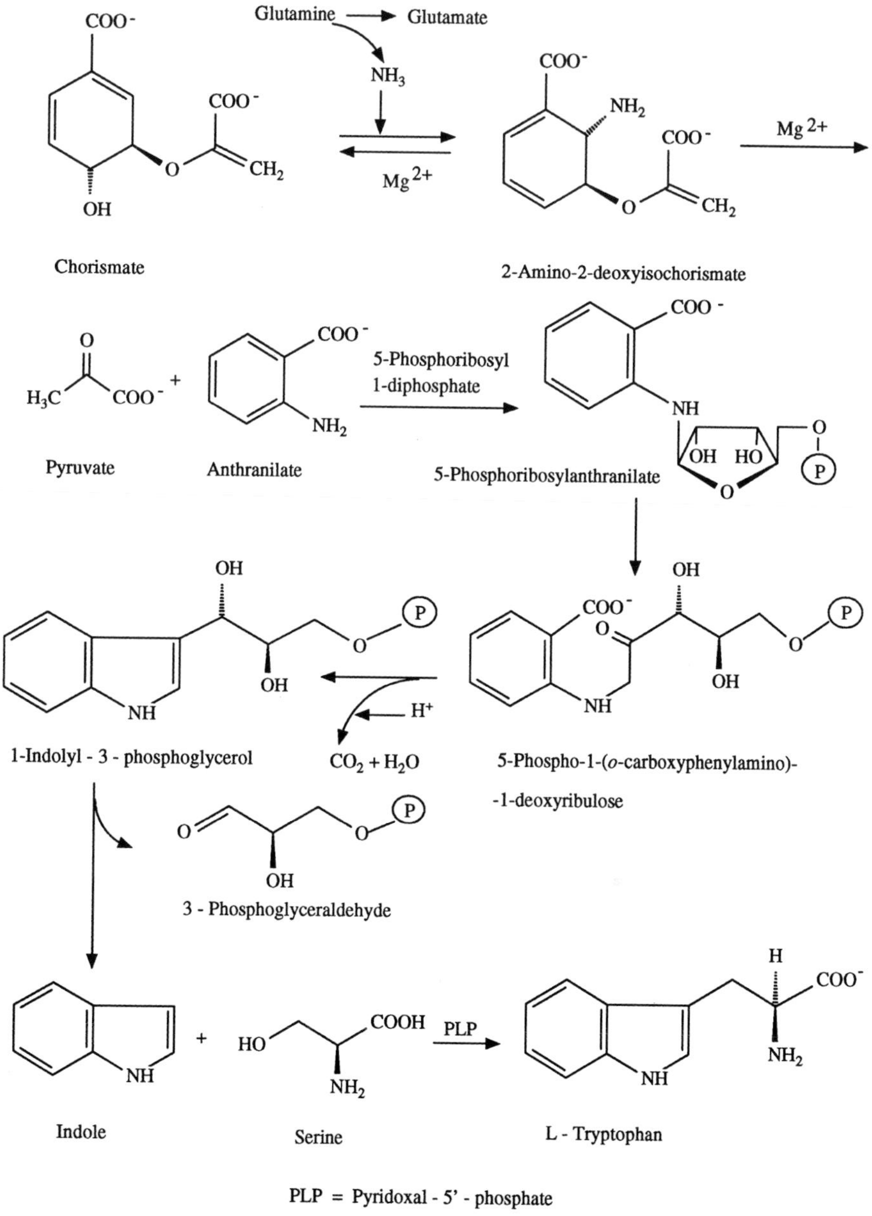

Fig. 46. Biosynthesis of tryptophan from chorismate

207

encoding PRAI have been characterized and the amino acid sequence for the encoded protein has been deduced. A cDNA encoding InGPS has also been isolated.

In a reaction catalysed by *tryptophan synthase* (EC 4.2.1.20) the phosphoglycerol side-chain is replaced by serine, giving tryptophan. Tryptophan synthase is a multienzyme which is composed of two α-subunits and one β_2 dimeric subunit. The α-units catalyse the reaction from indolyl-3-phosphoglycerol to indole and 3-phosphoglyceraldehyde. The β-subunits catalyse the conversion of indole and serine to tryptophan in a reaction requiring pyridoxal-5'-phosphate. Indole is not normally released as an intermediate but is channelled between the active sites. X-ray analysis of the enzyme has shown that there is a hydrophobic tunnel connecting the two catalytic sites.

Tryptophan synthase has been extensively studied in bacteria both on the protein and gene level. As regards plants, cDNAs encoding the α-protein of tryptophan synthase have been isolated from *Arabidopsis thaliana* and from *Zea mays*. The genes coding for the two β-subunits have also been characterized from these two plants. A gene encoding the β-subunit of tryptophan synthase has been cloned and characterized from the camptothechin-producer *Camptotheca acuminata* (see page 696). The predicted gene product shows a high degree of similarity to corresponding proteins from maize and *Arabidopsis thaliana*. Correlations between the expression of this gene and the time and location of camptothecin production suggest that tryptophan and indole alkaloid synthesis are regulated in a coordinated fashion.

As already mentioned, tryptophan is a precursor of alkaloids. However, alkaloid biosynthesis often includes precursors from other biosynthetic pathways, e.g. acetate and mevalonate. We shall therefore return to this topic in a later chapter (page 604).

Phenylpropanes

Phenylpropanes are aromatic compounds with a propyl side-chain attached to the aromatic ring, which can be derived directly from phenylalanine. The ring often carries oxygenated substituents (hydroxyl, methoxyl and methylenedioxy groups) in the *para*-position. Natural products in which the side-chain has been shortened or removed can also be derived from typical phenylpropanes.

Typical phenylpropanes

Cinnamic acid is formed from phenylalanine by deamination, catalysed by the enzyme *phenylalanine ammonia lyase* (EC 4.3.1.24) (PAL). The (*E*)-cinnamic acid formed is a precursor of other phenylpropanoids, such as lignin, flavonoids and coumarins. PAL has been purified and characterized from a number of plants. It is a

Cinnamic acid

$$CH= CH- C\overset{OH}{\underset{O}{}}$$

Cinnamic acid

large protein – 240 to 330 KDa – composed of four subunits. PAL genes have been sequenced and organization and structure established from several plants including *Arabidopsis thaliana* (three PAL-genes), *Petroselinum crispum* (four PAL-genes), *Phaseolus* spp., *Citrus limon, Nicotiana tabacum, Oryza sativa* and *Triticum æstivum* (two PAL-genes). The active site of PAL contains a dehydroalanine residue which participates in the elimination of ammonia. This residue is formed post-translationally by modification of a serine residue.

Crude drugs   **Peru balsam** (Balsamum peruvianum) is obtained from *Myroxylon pereirae* Klotzsch, also known as *M. balsamum* var. *pereire* (Royle) Harms (*Fabaceae*), a 25 m high tree growing in Central America, with a centre of distribution along the coast of El Salvador. The balsam is a pathological product, formed when the tree is injured. The bark is loosened on four parts of the circumference of the trunk by beating and burning with torches. To allow the tree to survive, the bark is left intact between the segments of loosened bark.

After about a week, the damaged parts of the bark fall off and the balsam starts to ooze out and is soaked up in rags wound round the trunk. The balsam is recovered by boiling the rags in water, from which it separates on cooling and can be collected.

Peru balsam is produced in El Salvador and Honduras and is a viscous, dark brown fluid with a vanilla smell. The main constituents (55–66 %) are esters of cinnamic and benzoic acids, especially benzyl cinnamate (cinnamein), cinnamyl cinnamate (styracin) and benzyl benzoate. Resins and small amounts of vanillin and free cinnamic acid are also present. Peru balsam is an antiseptic component of ointments and is used against skin diseases caused by parasites.

Tolu balsam (Balsamum tolutanum) is a similar product, derived from *Myroxylon toluiferum* H.B. & K. (*Fabaceae*). This tree occurs in Colombia and Venezuela. The product is obtained after making incisions in the bark and is a thick paste which solidifies and gradually becomes hard and brittle. Tolu balsam contains up to 80 % of esters of cinnamic acid with resin alcohols (complex, high-molecular-weight alcohols of incompletely known structure). A small quantity of free vanillin gives the product its characteristic smell. Tolu balsam is used in cough mixtures, for flavouring medicines, and also in making sweets and perfumes.

Cinnamaldehyde

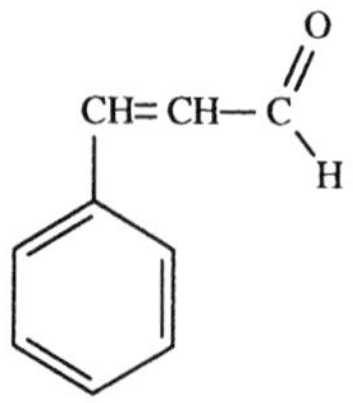

Cinnamaldehyde

Crude drugs **Cinnamon** (Cinnamomi cortex) is the peeled bark of *Cinnamomum zeylanicum* Blume (*Lauraceae*) (also known as *C. verum* J. Presl), a small tree with yellow flowers, cultivated in Sri Lanka, southern India, the Seychelles, West Indies and Brazil. In plantations, it is kept in the form of a bush by coppicing and the shoots that are allowed to grow out are harvested when about 18 months old. The bark is separated, and, after removal of the outer layer, carefully dried. Some fermentation takes place during the drying procedure, which gives the drug a brown colour. The bark contains a volatile oil (0.5–1.5 %) in special oil cells which is recovered by steam distillation (see below). Its main constituent is cinnamaldehyde (65–75 %). Besides the volatile oil from the bark (Cinnamomi zeylanici corticis aetheroleum), the *European Pharmacopoeia* also lists volatile oil from the leaf (Cinnamomi zeylanici folii aetheroleum). This oil contains mainly phenols (eugenol) which constitute 70–95 % of the weight.

Cinnamon and its oil are used as spices and for flavouring drugs. In traditional medicine, cinnamon is used as a decoction for treatment of stomach complaints. Due to the content in the bark of small amounts of the liver and kidney toxin coumarin, this use is not recommended.

Chinese cinnamon (Cinnamomi chinensis cortex, cassiae cortex, cassia) from *Cinnamomum cassia* Blume (previously known as *Cinnamomum aromaticum* Nees), indigenous to China and cultivated in Japan, Central and South America, and Indonesia, is also used as a spice. Chinese cinnamon is not peeled and it contains mucilage as well as volatile oil which is considered to be inferior to that of cinnamon.

Volatile oils (essential oils, aetherolea)

Cinnamon oil is an example of a large group of natural products, for which the collective terms volatile oils and *essential oils* are often used. The corresponding Latin term is *aetheroleum*. Volatile oils are very complex, aromatic-smelling, volatile mixtures containing many different compounds. Most of the essential oils have a high refractive index and they are often optically active. These properties are used for their identification and quality control. Essential oils have low water solubility but are readily soluble in organic solvents.

Plate 1.
Inula helenium L. Inulin can be prepared from the root of this and other *Asteraceae* (page 163). Photograph by Gunnar Samuelsson.

Plate 2.
Aesculus hippocastanum L. The bark and leaves contain aesculin which is used in sun-tan preparations (page 230). The seed cotyledons contain aescin, a triterpenoid saponin which is a remedy for chronic venous insufficiency (page 445). Stockholm, Sweden. Photograph by Gunnar Samuelsson.

211

Chemically and biosynthetically, they are not a homogeneous group. Phenylpropanes (from shikimic acid) and terpenes (from acetate) are very common constituents, but substances of other types, e.g. sulphur-containing compounds (mustard oils), may also be present.

Plants that are rich in essential oils (0.01–10 % dry weight) are found in about 30 % of plant families. High contents of essential oils are especially common in members of the *Apiaceae, Laminaceae, Lauraceae, Myrtaceae* and *Rutaceae.*

Essential oils occur in plants in specially developed organs of various kinds, e.g. glandular hairs on leaves, stems and flowers. Secretory ducts or cavities or oil cells (schizogenous or lysigenous) may also be present in the plant tissue.

The oils are mostly isolated from the plant material by steam distillation, which can be carried out in various ways. In *distillation with water*, the plant material is mixed with water, which is then boiled; the steam and accompanying essential-oil vapour are condensed and the oil separated from the water. This procedure cannot be applied to oils that are decomposed on prolonged boiling. With such oils, *hydrosteam distillation* is used – the plant material is mixed with water, but the mixture is not boiled; instead, steam is passed into it from an external source. During its passage through the plant material, the steam carries away the volatile oil. The distillate is condensed and separated as in the previous method. Fresh plant material can be distilled with steam alone – *direct steam distillation*. In this case, the material is spread out on a grating in the distillation vessel and steam enters from below. On passing through the plant material, the steam takes the volatile oil along with it. Steam distillation is a simple method that can be used on a large scale. However, it is not applicable if the volatile oil contains readily hydrolysable substances, e.g. esters, or substances that are oxidized or undergo other changes at raised temperatures. Other methods must then be used, such as expression or extraction with organic solvents.

Volatile oils and plant materials containing such oils are natural products of great economic importance. Their main use is in perfumery and cosmetics and in food processing (spices). In pharmaceutics, volatile oils are used to give drugs an agreeable smell or taste. A few essential oils or pure compounds isolated from such oils are used directly as medicines.

Anethole

Anethole is a phenylpropane derivative carrying a methoxyl group in the p-position to the side-chain. Biosynthetically, the compound is derived from cinnamic acid, which is first oxidized to 4-hydroxycinnamic acid, a reaction catalysed by the enzyme trans-cinnamate 4-monooxygenase (EC 1.14.13.11). The hydroxyl group acquires the methyl group from the amino acid methionine, a common methyl donor in biosynthetic reactions. 4-Methoxycinnamic acid is then reduced in two steps to anethole (Fig. 47).

$CH=CH-CH_3$

OCH_3

Anethole

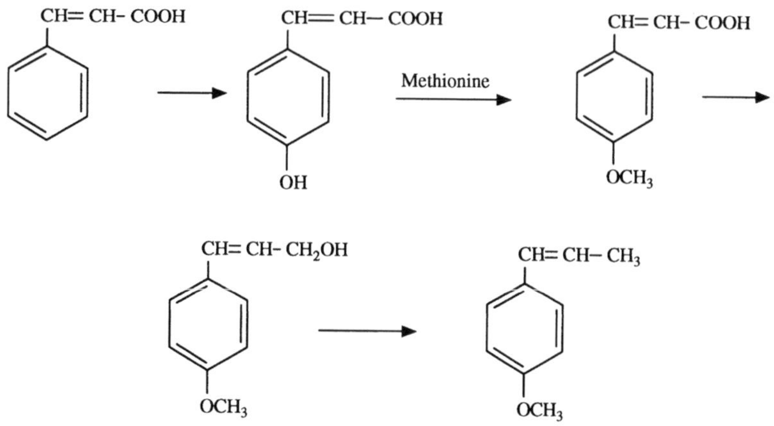

Fig. 47. Biosynthesis of anethole

Crude drugs   **Anise oil** (Anisi aetheroleum) is a volatile oil obtained from **aniseed** (Anisi fructus) or **star-anise** (Anisi stellati fructus). Aniseed is the schizocarp fruit of *Pimpinella anisum* L. (*Apiaceae*), a herb originating from the east Mediterranean but since long extensively cultivated in Europe (Spain, Italy, Germany and southern Russia) and in South America. Aniseed contains 1–3 % volatile oil.

Star-anise is the fruit of *Illicium verum* Hook. *f.* (*Illiciaceae*), a 4–5 m high *tree grown* in southern China, southern Vietnam, Japan, the Philippines and recently also in Jamaica. The fruit of star-anise contains 2.5–5 % volatile oil. Both fruits are used as spices and as raw materials for obtaining anise oil by steam distillation. The composition of the oil is approximately the same, whichever of the two plants is used. The main constituent is anethole (80–90 %). Other constituents are methylchavicol, an isomer of anethole, and *p*-methoxyphenylacetone. Anise oil from star-anise also contains small quantities of terpenes. Due to the high content of anethole, oil of anise solidifies at temperatures below 15 °C. The tendency of the oil to solidify decreases on ageing, due to oxidation and the occurrence of polymerization processes, reactions which are promoted by air and light. One of the polymerization products is dianethole, a stilbene derivative with oestrogenic properties.

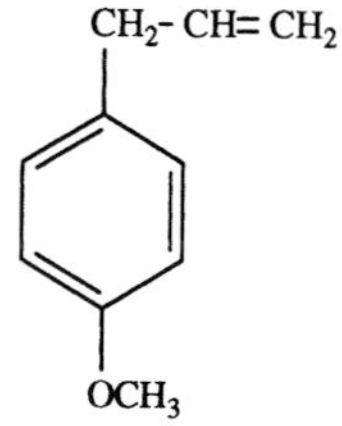

Methylchavicol

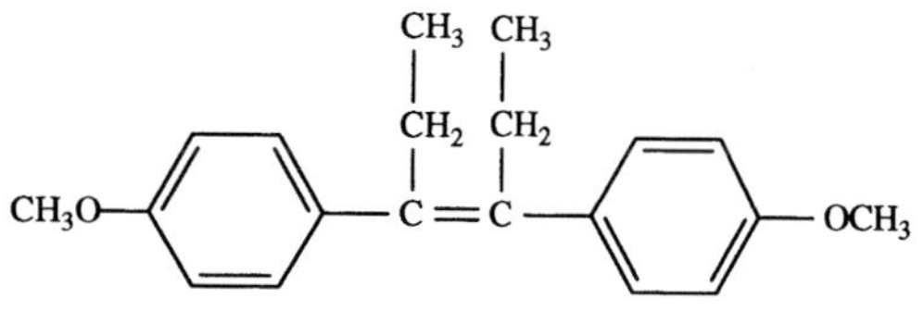

Dianethole

213

Anise oil is used in pharmacy for taste improvement and as an expectorant in cough mixtures. Both aniseed and anise oil are used in herbal remedies for relief of flatulence and for treatment of catarrhs of the respiratory tract.

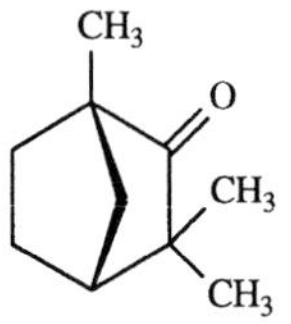

CH₃

O

CH₃

CH₃

Fenchone

Sweet fennel (Foeniculi dulcis fructus) is the schizocarpic fruit of *Foeniculum vulgare* Miller ssp. *vulgare* var. *dulce* (Miller) Thellung (*Apiaceae*), a plant which grows wild in the Mediterranean area is widely cultivated. Sweet fennel contains not less than 2 % volatile oil which can be obtained by steam distillation. The oil contains 80 % or more of anethol and less than 5 % of fenchone, a bicyclic monoterpene (see page 372).

Bitter fennel (Foeniculi amari fructus) is the fruit of *Foeniculum vulgare* Miller ssp. *vulgare* var. *vulgare* (*Apiaceae*). Also this plant originates from the Mediterranean area and is cultivated. The fruits contain at least 4 % of volatile oil which, in comparison with the oil from Sweet fennel, has a lower content of anethole (60 %) and a higher content (at least 15 %) of fenchone. The higher content of fenchone results in a bitter taste of the fruit and the oil.

Fennels are used as a spice and their oil as taste-improving agents. They are also used in herbal remedies for relief of flatulence and as expectorants.

Eugenol
4-Allyl-2-methoxyphenol

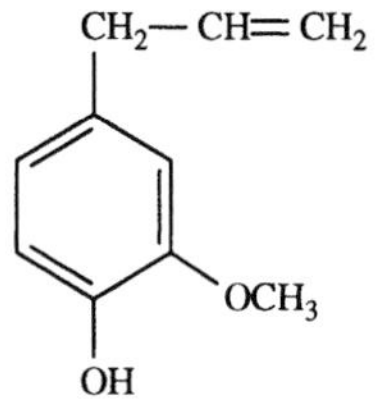

$CH_2-CH=CH_2$

OCH₃

OH

Eugenol

Crude drugs **Clove oil** (Caryophylli floris aetheroleum) is obtained from **clove** (Caryophylli flos), which is the dried flower-buds of *Eugenia caryophyllata* Thunb. (*Myrtaceae*), also known under the names *Syzygium aromaticum* (L.) Merril. et Perry and *Caryophyllus aromaticus* L. This is a tree, 10–15 m high, originating from the Moluccas. Nowadays, the main producing countries are Madagascar, Indonesia and Brazil. The flower-buds are collected when they start changing colour from green to pink, and they are carefully dried in the sun, which takes 4–5 days. Cloves are used as a spice. More than half the world production goes to the tobacco industry for flavouring cigarettes. The volatile oil is prepared from the crude drug by steam distillation. Its main constituent is eugenol (84–95 %). Also present are sesquiterpenes (α- and β-caryophyllenes) and small amounts of esters, ketones and alcohols. Being a phenol, eugenol can readily be isolated from clove oil by treatment with alkali. It forms a water-soluble salt, which can easily be separated from the other oil components.

Clove oil and eugenol are used in dental practice as disinfectants with a certain analgesic effect.

Myristicin

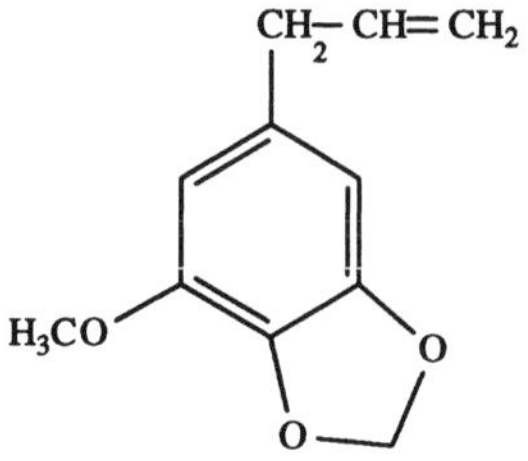

Myristicin

Crude drug   **Nutmeg** (Myristicae semen) is the kernel of the dried, ripe seed of *Myristica fragrans* Houtt. (*Myristicaceae*). The nutmeg tree is 10–20 m high, with a natural origin in the Moluccas but cultivated in Indonesia, Malaysia, Sri Lanka and the West Indies. It is dioecious and in plantations the number of male trees is reduced to about 10 % of the total. The fruit is yellow and pear-shaped. On ripening, the pericarp splits, disclosing the black seed surrounded by a red aril which is separated and dried to give the crude drug **mace** (Macis). The seed is usually dried in an oven until the kernel shrinks and can be heard to rattle in the testa (seed shell). The testa is then crushed and separated from the kernel, which constitutes the nutmeg drug. Nutmeg is often dipped into a slurry of lime as a precaution against insect attack.

Nutmeg contains 30–40 % fat and 5–15 % volatile oil. Terpenes constitute 60–80 % of this oil and myristicin about 8 %. Mace oil has a similar composition, but its terpene content is even higher. Both drugs are used as spices.

If taken in large quantity nutmeg has a hallucinogenic effect, which is followed by unpleasant side-effects such as reddening of the skin, palpitations and decreased salivation. These reactions are thought to be due to the myristicin contained in the drug.

Nutmeg oil (Myristicae fragrantis aetheroleum) is the volatile oil of nutmeg.

Rosavin

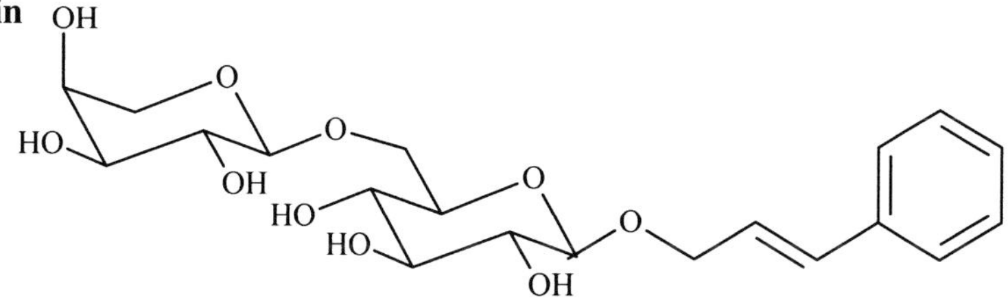

Rosavin = cinnamyl-(6′-*O*-α-L-arabinopyranosyl)-*O*-β-D-glucopyranoside

Crude drug   **Rhodiola** (Rose root, Golden root, Arctic root) is the dried rhizome and root of *Rhodiola rosea* L. (*Crassulaceae*). The plant is a perennial herb, up to 30 cm in height, with fleshy, grey-green leaves and yellow blossoms, growing in tight tufts. *Rhodiola rosea* grows at high altitudes in the Arctic areas of Europe and Asia. It is also found in mountainous parts of Europe, such as the Alps, the Pyrenees and Carpathian Mountains. Although its natural habitat is at high altitudes it can also be grown on low land. The crude drug smells of roses.

Rhodiola contains the phenylpropane derivatives rosavin (see structure above), rosin and rosarin. Also present is the phenylethanoid tyrosol and its glucoside salidroside.

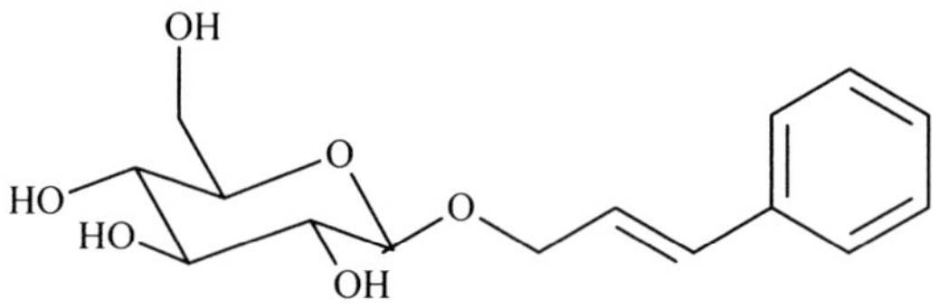

Rosin = cinnamyl-*O*-β-D-glucopyranoside

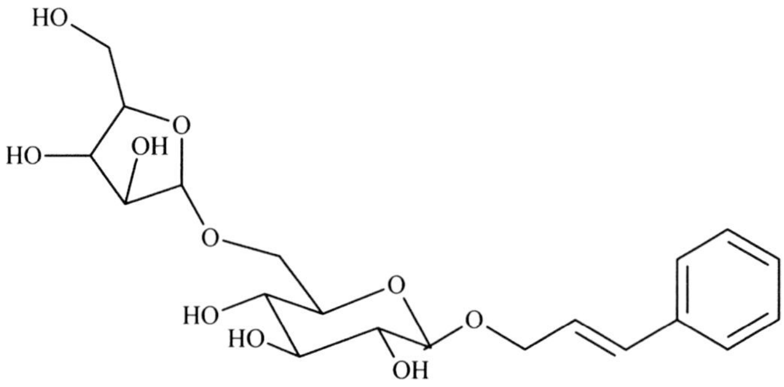

Rosarin = cinnamyl-(6′-*O*-α-L-arabinofuranosyl)-*O*-β-D-glucopyranoside

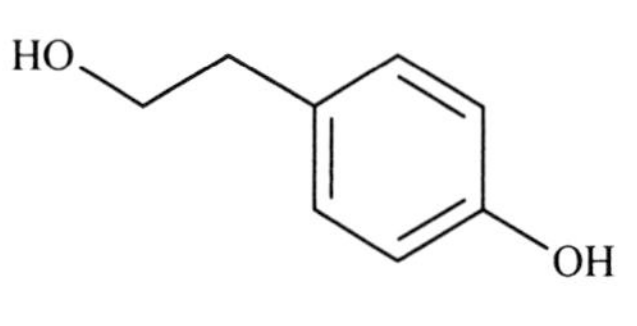

p-Tyrosol

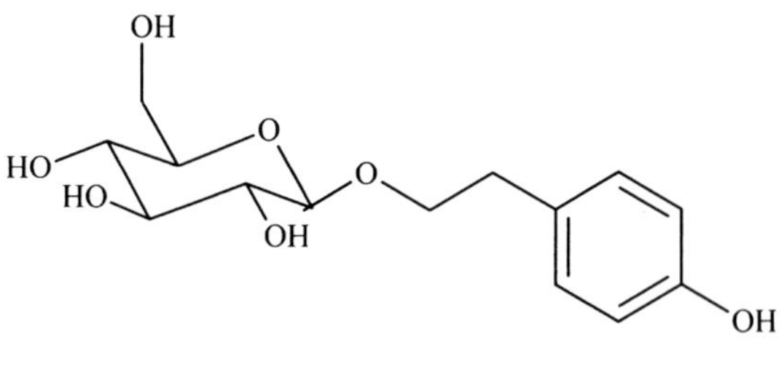

Salidroside = *p*-hydroxyphenylethyl-*O*-β-D-glucopyranoside

Rhodiola is a herbal remedy for treatment of fatigue and to improve work capacity. It is classified as an adaptogen. For a description of this concept see page 226. In controlled clinical studies single-dose administration of aqueous ethanolic extracts of rhodiola have been found to effectively increase mental performance and physical working capacity within 30 minutes of administration. The stimulating effect continued for at least 4–6 hours. The phenylpropanoids and –ethanoids are the active compounds in the extract. Rhodiola extracts used to be standardized on their content of salidroside, but since this compund is present in a number of *Rhodiola* species it is now recomended to standardize the extracts also for rosavin, rosin and rosarin which are all specific to *Rhodiola rosea*.

Contraindications: Should not be given to individuals with bipolar disorder who are vunerable to becoming maniac when given antidepressants or stimulants. *Side effects*: No reports are available. *Interactions*: None documented but additive effects with other stimulants may be encountered.

Podophyllotoxin and peltatins

Podophyllotoxin and the peltatins are *lignans*, a group of natural products which are essentially cinnamyl alcohol dimers, though further cyclization and other modifications create a wide range of structural types. Lignans were previously thought to be formed by oxidative coupling of two cinnamic-acid residues involving nonspecific H_2O_2-requiring peroxidases. However, recent studies have demonstrated that an enzymic stereospecific coupling mechanism must operate, leading to enantiomerically pure products.

Biosynthesis

The phenylpropane derivative coniferylalcohol is probably the ultimate precursor for podophyllotoxin and the peltatins. An intermediate in the biosynthesis of these compounds is matairesinol, the biosynthesis of which has been elucidated using *Forsythia* x *intermedia* Zabel (*Oleaceae*) as a model system. In the presence of a one-electron oxidant such as laccase, and a "*dirigent protein*", two molecules of *E*-coniferyl alcohol form (+)-pinoresinol via regio- and stereoselective intermolecular 8.8'-coupling of two putative enzyme-bound intermediate radicals (A, Fig. 48). The NADH-dependent, bifunctional enzyme *pinorecinol/lariciresinol reductase* then performs sequential stereoselective reduction of (+)-pinoresinol, resulting in the formation of (+)-lariciresinol which is consecutively transformed to (-)-secoisolariciresinol. The stereoselectivity of this process results in inversion of the configuration at C-2 and C-5 of pinoresinol, a process which is envisaged to occur either by a concerted S_N2 mechanism or via reduction of an intermediate quinomethane. Stereoselective dehydrogenation of (-)-secoisolariciresinol, catalysed by NAD-dependent *secoisolariciresinol dehydrogenase*, then yields (-)-matairesinol.

All the proteins and enzymes participating in this pathway have been purified from *Forsythia* x *intermedia* and the corresponding genes cloned and fully functional recombinant proteins expressed. It has also been shown that the same pathway operates in *Linum flavum* (*Linaceae*), a plant which ultimately produces 5-methoxy-podophyllotoxin. From *Podophyllum peltatum* (*Berberidaceae*), one of the two commercially important podophyllotoxin-producing

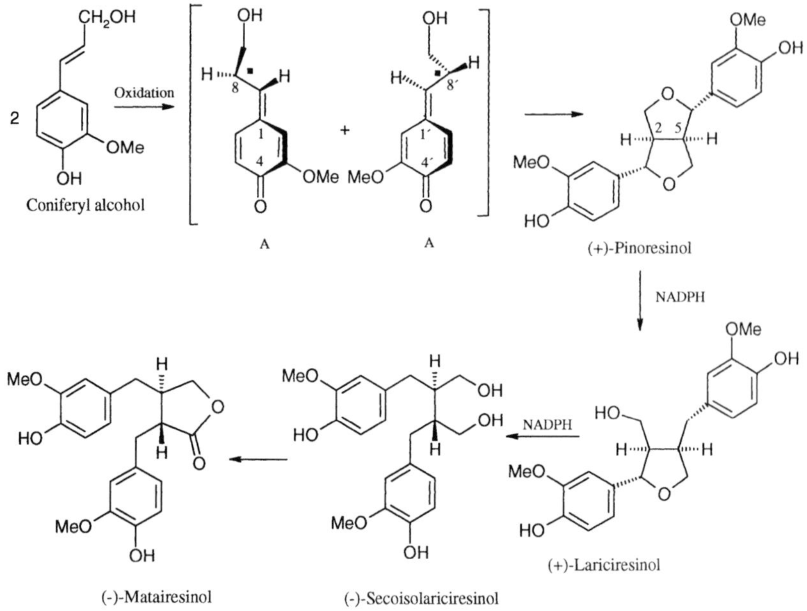

Fig. 48. Formation of (-)-matairesinol from E-coniferylalcohol

plants (see page 221), the gene encoding the dirigent protein has been isolated and expressed in an insect cell expression system, yielding a protein with a molecular mass of approximately 26 kDa which cross-reacted with *Forsythia* x *intermedia* dirigent polyclonal antibodies, thus strengthening the hypothesis that the biosynthetic pathway to podophyllotoxin also operates via (-)-matairesinol.

(-)-[14]C Matairesinol is efficiently incorporated into (-)-podophyllotoxin, β-peltatin, 4'-demethylpodophyllotoxin and α-peltatin by *Podophyllum* species. Matairesinol is converted either to yatein or 4'-demethylyatein. Yatein then forms podophyllotoxin and β-peltatin while 4'-demethylyatein is the precursor for 4'-demethylpodophyllotoxin and α-peltatin (Fig. 49).

Preparations of podophyllotoxin (0.5 % alcoholic solution or 0.15–0.30 % podophyllotoxin cream) are used against *Condyloma-ta acuminata*, a sexually transmitted disease, caused by a papilloma virus, manifesting itself as warty excrescences on the sexual organs. The mechanism for the activity against the warts is that podophyllotoxin stops nucleoside transport across the cell membrane which leads to death of the infected cell and, consequently, of the papilloma virus.

Both the peltatins and podophyllotoxin have an inhibitory effect on tumours and have therefore been extensively studied. The presence of a *trans*-γ-lactone ring has been found to be essential for the activity of these compounds. Treatment with weak alkali causes epimerization to the stable *cis*-form, which is physiologically inactive. The mechanism for the tumour inhibiting effect of podophyllotoxin is its activity as a microtubule inhibitor.

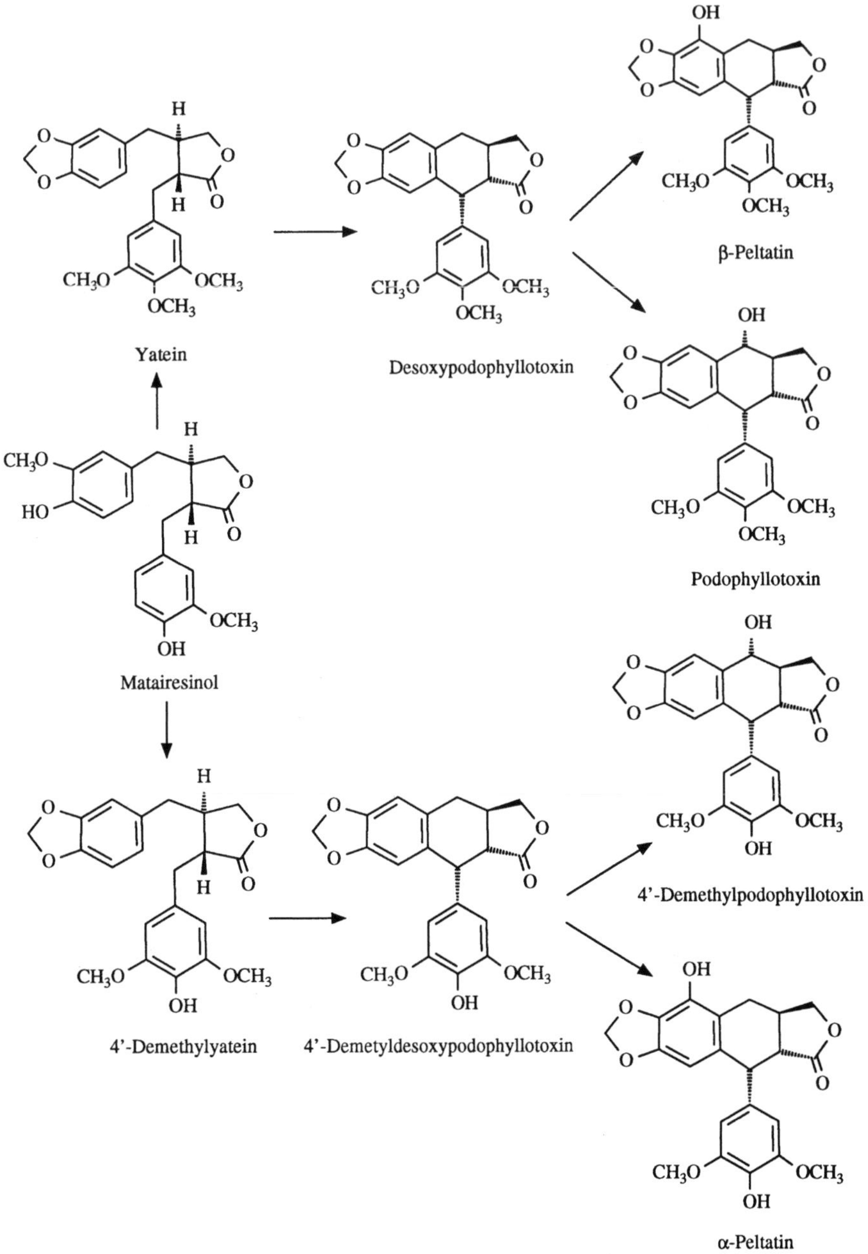

Fig. 49. Biosynthesis of podyphyllotoxin and peltatins from matairesinol

Microtubules and microtubule inhibitors
Microtubules are organized structures that are found in all nucleated cells and participate in a number of activities such as cell division, cell movement, maintenance of cell shape, intracellular transport, signal transduction and initiation of DNA synthesis. Microtubules are long tubular structures with an outside diameter of about 24 nanometres and a central lumen about 15 nm in diameter. Their length is indeterminate but is always much greater than their width,

219

often many micrometres. Examined with the electron microscope, cross-sections of microtubules look like circles. At very high magnifications each circle is found to be composed of usually 13 circular subunits. These subunits are strands, called protofilaments, each one consisting of a linear assembly of dimers of two very similar proteins: the a- and b-tubulins. Both tubulins are globular molecules, each with a molecular mass of about 55 kDa. Microtubules are labile structures; the equilibrium between the microtubule and the pool of free tubulin results from continuous assembly/disassembly at both ends of the microtubule. Microtubule inhibitors – such as podophyllotoxin, colchicine (page 617), vincristine and vinblastine (page 681) act by binding to tubulin and inhibiting formation of protofilaments and microtubules. They also promote disassembly of microtubules. Podophyllotoxin causes complete inhibition of tubuline polymerization at concentrations as low as 5 mM. Microtubule inhibitors cause a range of effects, including arrest of the mitotic spindle, arrest of malignant cell invasion and inhibition or acceleration of the release of enzymes and hormones.

Paclitaxel (page 426) is a microtubule inhibitor with a different mechanism of action. It binds specifically to the β-subunit of tubulin, preferentially in the formed microtubule rather than to the free β-tubulin or to tubulin dimers. It thereby stabilizes the mictotubules and interferes in the equilibrium between tubulin and microtubules.

Teniposide, etoposide and etopophos

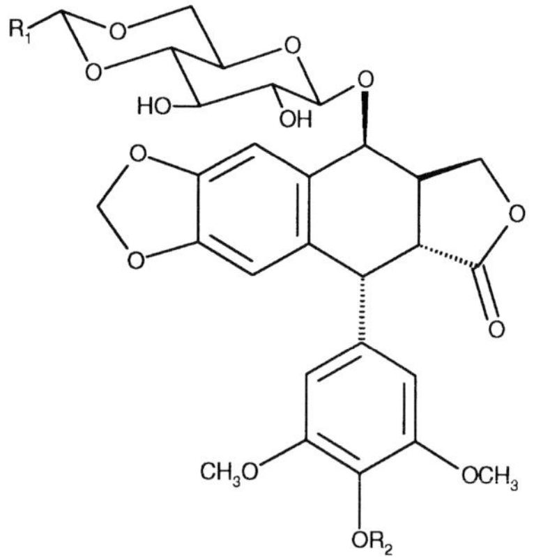

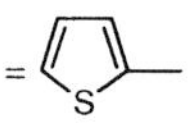

Teniposide: $R_1 =$; $R_2 = H$
Etoposide: $R_1 = CH_3$; $R_2 = H$
Etopophos: $R_1 = CH_3$; $R_2 = PO(OH)_2$

Podophyllotoxin is not used as a tumour inhibiting drug, but the research on the compound has resulted in the introduction of three other cancer remedies: teniposide and etopophos, which are glycosidic derivatives of epi-*podophyllotoxin*, the C-9 β-hydroxy isomer of podophyllotoxin. Teniposide is used as a remedy against malignant neoplasms and cancer of the bladder and etoposide is used against small-cell lung cancers, some types of leukaemia, and Hodgkin's disease. Etopophos is a water-soluble phosphate ester prodrug of etoposide. The prodrug can be administered in higher doses than etoposide as a short intravenous injection, whereafter it is rapidly converted to the parent compound by plasma phosphatases and thus constitutes an improved formulation of etoposide. These compounds have no affinity for tubulin and thus have no effect on microtubule assembly at clinically relevant levels. In contrast to podophyllotoxin they cause an irreversible premitotic block in the late S and early G_2 phases. This results from binding to topoisomerase II, an enzyme required for the unwinding of DNA during replication. Topoisomerase II forms a transient, covalent DNA protein link, the cleavable complex, which allows one double strand of DNA to pass through a temporary break in another double strand, Etoposide (and teniposide) binds and stabilizes the cleavable complex preventing repair of the double-strand breaks.

These compounds are manufactured from podophyllotoxin, which is obtained mainly from the Indian species *Podyphyllum emodi* Wall. (see Indian podophyllin below). Increasing demand constitutes a threat to this species, the natural populations of which have been drastically reduced. No economic synthesis of podophyllotoxin has been developed but attempts are in progress to produce the compound in cell cultures, for example from *Linum* species (particularly *L. album*) which accumulate podophyllotoxin and 6-methoxypodophyllotoxin. Also cultivation of the American species *Podophyllum peltatum* L. is under consideration.

Crude drugs **Podophyllin** (Podophyllinum) is obtained by ethanol extraction of the dried root and rhizome of *Podophyllum peltatum* L. (*Berberidaceae*). The ethanol extract is concentrated and dilute hydrochloric acid is then added. The precipitate formed is washed, dried and ground to a powder, the crude drug podophyllin. *Podophyllum peltatum* is a low herb, growing in hardwood forests in the eastern United States and Canada. The plant has a 1 m long rhizome and two large peltate leaves. Podophyllin contains approximately 20 % podophyllotoxin, 10 % β-peltatin and 5 % α-peltatin. Podophyllin was formerly used as a drastic purgative. The purgative effect is mainly due to the peltatins present. Podophyllin is a very toxic substance and internally a cumulative dose of only 0.5 mg/kg of bodyweight can be lethal. Oral or dermatological poisonings of different podophyllin preparations are highly insidious. On the fourth or fifth day at the earliest, difficulty in breathing and paralysis of the extremities occur. If the patient survives, the paralysis can persist for 18 months.

A 25 % solution in ethanol or benzoic tincture of the drug is employed as a remedy against *Condylomata acuminata* (see above). This solution is applied to the warts and must be washed off with water in 2 hours. The solution is not stable and the remedy must therefore be prepared *ex tempore* for each occasion. The low stability of the preparation makes the dosage uncertain and another disadvantage of the treatment is the high toxicity at the treated site. These risks require that applications be made only by an experienced physician or nurse. A safer treatment is the application of remedies containing pure podophyllotoxin (see above).

Indian podophyllin is obtained from an Indian species – *Podophyllum emodi* Wall. (also known as *P. hexandrum* Royle) – in the same way as from the American plant. Indian podophyllin contains up to 40 % podophyllotoxin but no peltatins. As a consequence, its purging effect is weaker than that of American podophyllin. Indian podophyllin is a raw material for isolation of podophyllotoxin.

Syringaresinol diglucoside (Eleutheroside E)

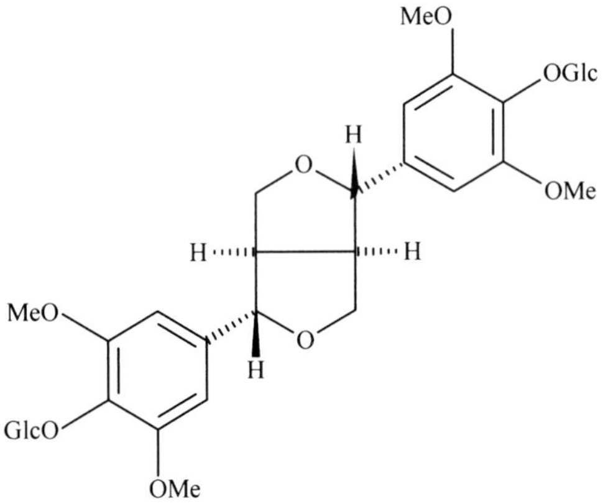

Syringaresinol diglucoside (Eleutheroside E; Glc = glucose)

Crude drugs **Eleutherococcus** (Russian ginseng, Siberian ginseng) is the dried root of *Eleutherococcus senticosus* (Rupr. & Maxim.) Maxim. (*Araliaceae*). The plant is a thorny bush, growing in western Siberia, in Korea and in the Shanxi Province in China. The chemistry of this crude drug is complicated. Besides the syringaresinol diglucoside other lignans, phenylpropanoids, coumarins, triterpenoids and polysaccharides have been isolated. Many of these have been denoted as eleutherosides, separated from each other by a letter suffix, e.g. eleutheroside A, B, etc. This is unfortunate since it indicates a uniform class of compounds which it is not. Thus eleutheroside A and K are triterpenoids (daucosterol and β-hederin, respectively), eleutheroside B is the phenylpropanoid syringin and eleutheroside B1 is the coumarin isofraxidin-7-glucoside.

Eleutherococcus is a herbal remedy for treatment of asthenia and fatigue, based on its property as an adaptogen. For a more detailed description of the concept adaptogen see page 226. A large number of uncontrolled, as well as placebo-controlled, randomized, double-blind clinical studies have shown that extracts of Eleutherococcus are efficient in increasing mental and physical work capacity in situations of fatigue and stress. Very little is known of the contribu-

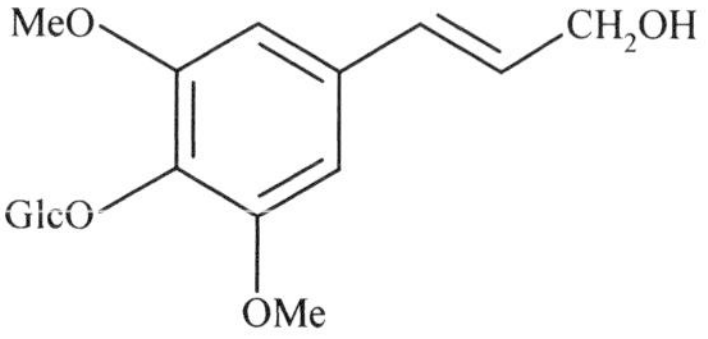

MeO

CH$_2$OH

GlcO

OMe

Syringin (Eleutheroside B; Glc = glucose)

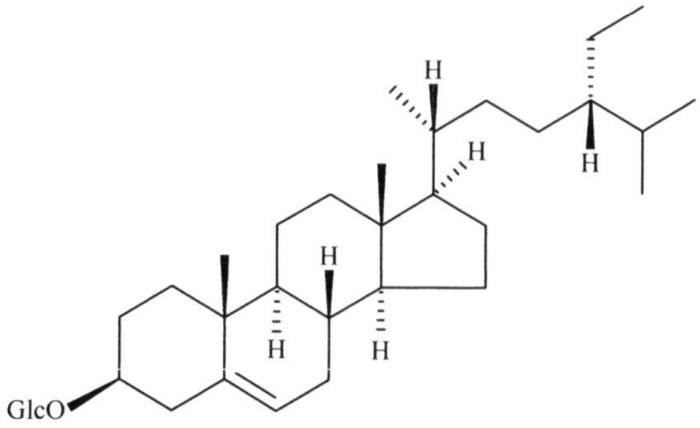

H

H

H

H

H

H

H

H

GlcO

Daucosterol (Eleutheroside A; Glc = glucose)

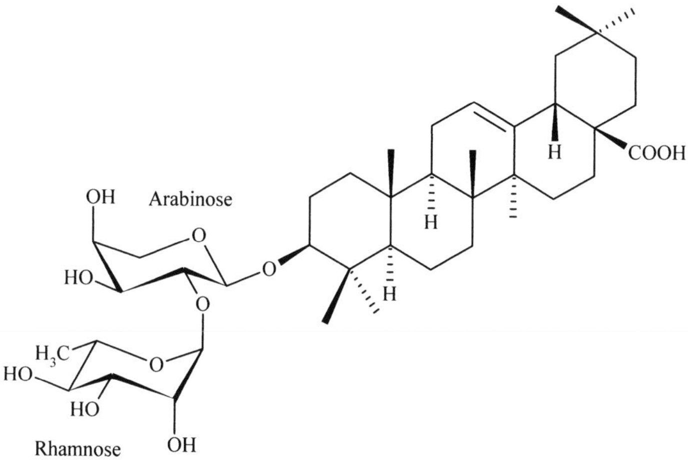

OH   Arabinose

O

HO

O

H$_3$C

O

HO

HO   OH

Rhamnose

O

H

COOH

H

H

MeO

GlcO

OMe

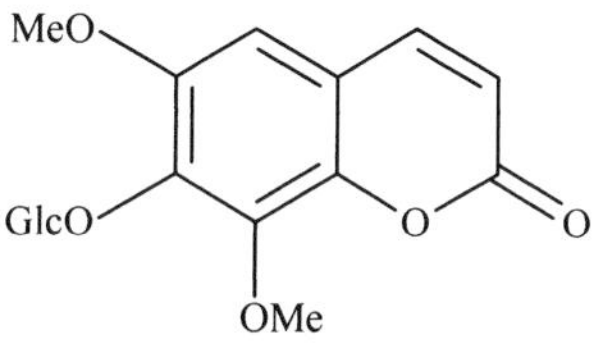

Isofraxidin-7-glucoside
(Eleutheroside B$_1$;Glc = glucose)

β-Hederin (Eleutheroside K)

tions of the above mentioned compounds to the adaptogenic effect. The lignans and the phenylpropanoids have a structural resemblance to cathecolamines which could suggest an effect on the sympathoadrenal system and possibly imply an effect in the early stages of the stress response.

Contraindications: none reported. *Side effects*: No serious adverse effects are reported. *Interactions*: Not investigated.

Schisandrin B   The berries of *Schisandra chinensis* (Turcz.) Baill. contain a number of lignans with adaptogenic and hepatoprotective activities. Schizandrin B is one of the major components of this mixture. The methylenedioxy group is important for the hepatoprotective activity of the compound.

Crude drug   **Schisandra** is the dried, ripe fruit of *Schisandra chinensis* (Turcz.) Baill. The plant is a woody climber which can be up to 8 m long. It

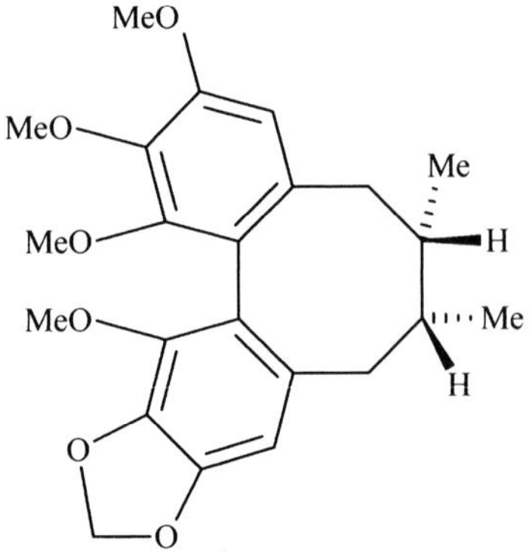

Schisandrin B (Me = CH$_3$)

is native to East Asia, Korea and Japan and grows in deciduous and mixed forests on slopes with mountainous thickets and along river banks up to 2 500 m above sea level. A detailed macroscopic and microscopic description is obtainable in *The American Herbal Pharmacopoeia* (see *Further reading* below). The ripe fruits are generally picked in the autumn. They adhere rightly to the stalks, and are removed after drying in the sun.

The primary pharmacological active compounds of schisandra are about 40 dibenzo[a,c]cyclooctadiene lignans. Their nomenclature is inconsistent. Russian scientists name them schisandrins while in Japan they are called gomisins and Chinese investigators refer to them as wuweizisus and/or wuweizi esters. Besides the above-mentioned schisandrin B the main compounds are: schisandrol A and B, schisandrin A, schisantherin A and B and gomisin N. Their structures differ among the substituents in the two benzene rings and in the octane ring as indicated in Fig. 50.

Extracts of schisandra are used as herbal remedies for treatment of fatigue and for enhancement of physical performance and

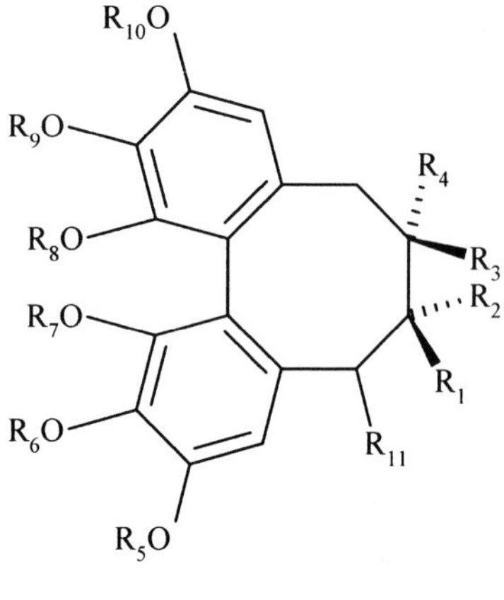

Compound (synonyms)	R_1	R_2	R_3	R_4	R_5	R_6	R_7	R_8	R_9	R_{10}	R_{11}
Schisandrol A (schisand-rin wuweizichun A)	OH	Me	H	Me	Me	Me	Me	Me	Me	Me	—
Schisandrol B (wuwei-zichun B, gomisin A)	OH	Me	H	Me	Me	Me	Me	Me	-CH$_2$-		—
Schisandrin A (deoxy-schisandrin, wuweizisu A)	H	Me	H	Me	Me	Me	Me	Me	Me	Me	—
Schisandrin B (wuweizisu B, γ-schisandrin)	H	Me	H	Me	-CH$_2$-		Me	Me	Me	Me	—
Schisantherin A (wuwei-ziester A, gomisin C, schisandrer A	OH	Me	Me	H	Me	Me	Me	Me	-CH$_2$-		β-O-benzoyl
Schisantherin B (wuweizi-ester B, gomisin B, schi-sandrer B	OH	Me	Me	H	Me	Me	Me	Me	-CH$_2$-		β-O-angeloyl
Gomisin N (*pseudo-γ-schisandrin*)	H	Me	H	Me	-CH$_2$-		Me	Me	Me	Me	—

Fig. 50. Structures of lignans contained in schisandra. Me = CH$_3$. –CH$_2$- indicates that two substituents are involved in formation of a methylene dioxy ring

endurance owing to their adaptogenic properties (for a description of the concept adaptogen see page 226). Clinical studies, mainly performed in the former Soviet Union, have indicated efficacy for this indication but they are of poor quality.

Schisandra has been used in China since antiquity for various purposes. Recent Chinese pharmacological research has focused on its hepatoprotective activity, which is attributed to the lignan fraction. The hepatoprotective effect of ethanol extracts as well as of isolated lignans has been well documented in animal studies and shown to be due to an apparent strong antioxidant activity, specifically enhancing the hepatic gluthathione antioxidant system. The lignans are also able to induce liver microsomal cytochrome P-450 and to stimulate biosynthesis of protein and liver glycogen. The positive effects were confirmed in clinical studies and have resulted in development of an anti-hepatotoxic lignan derivative, which is widely used in China for the treatment of chronic viral- and drug-induced hepatitis.

Contraindications: High gastric acidity or peptic ulcers. High intracranial pressure or epilepsy. Use of schisandra is not recommended for pregnant women. *Side effects*: Heartburn, stomach pain, allergic skin rashes. *Interactions*: No clinical effects reported, but studies in animals indicate that schisandra can enhance CNS-inhibitory effects of chlorpromazine, reserpine and pentobarbital and antagonize the CNS-stimulatory effect of caffeine and amphetamine.

Adaptogens

The concept adaptogen was originally created in the Former Soviet Union in the 1950s to describe remedies that increase the resistance of organisms to stress in experimental studies. In the original definition adaptogens are non-specific remedies that increase resistance to a very broad spectrum of harmful factors – "stressors"– of different physical, chemical and biological natures. An adaptogen must have a normalizing effect, i.e. counteracting or preventing disturbances brought about by stressors, and it must be innocuous with a broad range of therapeutical effects without causing any disturbance (other than very marginally) to the normal functioning of the organism. This definition has been updated and today adaptogens are defined as *a new class of metabolic regulators which increase the ability of an organism to adapt to environmental factors and to avoid damage from such factors.*

The concept was not generally recognized in Western countries as it seemed to be in clear contrast to some of the key-concepts of modern pharmacology: potency, selectivity and with efficacy balanced by an accepted level of toxicity. In 1998, however, the term adaptogen was allowed as a functional claim for certain products by the US Food and Drug Administration and it is now a generally accepted concept. Crude drugs that meet the criteria of being adaptogens are Eleutherococcus (page 222), Rhodiola (page 216), Schisandra (page 223) and Ginseng (page 447).

The mechanisms of action of adaptogens are difficult to define and to rationalize. The immune system and the stress system are the two regulatory systems responsible for stimulus–response coupling and adaptogens are intricately involved in the action of these. Stress is a defence response of an organism to external factors resulting in the stimulation of formation of endogenous activating messengers such as cathecholamines, prostaglandins, cytokines, nitric oxid (NO), PAF, etc., that in turn activate the energetic and other resources of the organism. This is the so-called "switch-on" system, which involves the efferent sympatho-adrenal system (SAS) and the hypothalamus–pituitary–adrenal (HPA) axis, as well as various mediators at the cellular, organ and system level. The switch-on system is balanced by a "switch-off" system that protects cells and the whole organism from over-reacting to the activating messengers. This system includes some important enzymes and mediators of intra- and extra-cellular communications at the cellular, organ and system levels. At the cellular level active components are the antioxidant system – superoxide dismutase, catalase, glutathioneperoxidase, eicosanoids and NO. At the organ level eicosanoids and NO are active and at the system level important factors are corticotropin-releasing factor (CRF), corticosteroids, prostaglandin E_2 and NO. When the stress system is in the normal state (*homeostasis*) the activities of the switch-on and switch-off system balance each other at a certain level of equilibrium which reflects the "reactivity" of the stress system, i.e. its sensitivity to a stressor and the degree of protection of the organism against damaging effects. Adaption to a stressor results in decreased sensitivity to the stressor and the basal levels of the mediators of switch-on and switch-off increase but their ratio does not change and *heterostasis* is achieved.

Plant adaptogens can be defined as agents which reduce the damaging effects of various stressors by virtue of reduction of the reactivity of the host defence system: they adapt an organism to the stress and have a curative effect in stress-induced disorders. They increase the production of both deactivating (cortisol) and activating (NO) messengers of the stress system and increase the capacity of this system to respond to external signals at the higher level of the equilibrium (heterostasis). It thus seems unlikely to expect that adaptogens have one or a few targets as their active principles are directed towards the regulatory systems that are common for all the tissues involved in the regulation of homeostasis.

Repeated administration of adaptogens vs single dose administration

The repeated administration of adaptogens gives an effect analogous to that produced by repeated physical exercise, conveying the organism from homeostasis to heterostasis. This effect is mainly associated with the HPA axis which is believed to play a primary role in the reactions of the body to repeated stress and adaption. Repeat dose administration of adaptogens has been shown to be of value in sports medicine and can lead to increased endurance, e.g. of long-distance runners, or to a more rapid recovery from a strenous event. It is also of value for patients suffering from chronic disease or a disturbed state. It should be noted that the stress-protective effect obtained by repeat dose administration is not the result of inhibition of the stress response but of adaptive changes in the organism as a response to the repeated stress-agonistic effect of the drug. Adaptogens are stress-agonists and not stress-antagonists.

Administration of adaptogens in a single dose is of value in situations that require a rapid response to strain or a stressful situation. The effects are here associated with the sympatho-adrenal (SAS) system, which provides a rapid response mechanism that controls the acute response of the organism to the stressor, resulting in increased levels of catecholamines, neuropeptides, ATP, NO and eicosanoids. Suitable crude drugs for this purpose are rhodiola, eleutherococcus and schisandra. They can also be used for repeated administration. Ginseng (page 447), on the other hand, gives an adaptive effect only after repeated administration for periods of 1–4 weeks. These differences in effect can be correlated to the content of active compounds in the crude drugs. Rhodiola, eleutherococcus and schisandra contain phenolic compounds, particularly phenylpropane and phenylethane derivatives, which are structurally related to catecholamines and presumably play important roles in the SAS and CNS systems. Ginseng – on the other hand – contains tetracyclic terpenes (ginsenosides) which are structurally similar to corticosteroids and might play key roles in the HPA axis- mediated regulation of the immune and neuroendocrine systems

Adaptogens as stimulants

There are very important differences between the stimulating effects of adaptogens and those of other stimulants of the CNS as summarized in the following table:

Effect	Stimulants	Adaptogens
Recovery process after exhaustive physical load	Low	High
Energy depletion	Yes	No
Performance in stress	Decrease	Increase
Survival in stress	Decrease	Increase
Quality of arousal	Bad	Good
Insomnia	Yes	No
Side effects	Yes	No
DNA/RNA and protein synthesis	Decrease	Increase

Stimulants, defined as drugs that increase the activity of the sympathetic nervous system, produce a sense of euphoria and can be used to increase alertness and the ability to concentrate on mental tasks. Stimulants such as caffeine, nicotine, amphetamines and cocaine are also used, and sometimes abused, to boost endurance and productivity. However, long-term stimulant abuse can impair mental function and lead to psychotic symptoms. Furthermore, traditional stimulants that possess addiction, tolerance and abuse potential produce a negative effect on sleep structure, and cause rebound hypersomnolence or "come down" effects. By definition, plant adaptogens do not exhibit such negative effects: in fact one plant adaptogen, that derived from *Rhodiola rosea*, has been shown significantly to regulate high-altitude sleep disorders and to improve sleep quality. Plant adaptogens stimulate the nervous system by mechanisms that are totally different from those of traditional stimulants, being associated rather with metabolic regulation of various elements of the stress-system and modulation of stimulus–response coupling). Depending on the mediators of the stress-system involved in the adaptogen-induced stress-response, an immediate (single dose effect) or a long term (after multiple administration) stimulating effect may be observed.

Coumarins and furanocoumarins

Coumarins and furanocoumarins are derivatives of 5,6-benzo-2-pyrone (α-chromone). The various substances differ mainly in the substituents attached to the benzene ring (OH, OCH$_3$, CH$_3$). The biosynthesis of this group of natural products starts with *trans*-cinnamic acid, which is oxidized to *o*-coumaric acid, and this is followed by formation of the glucoside. This compound (*o*-coumaric acid glucoside) is isomerized to the corresponding *cis*-compound, *o*-coumarinic acid glucoside, which gives coumarin by ring closure.

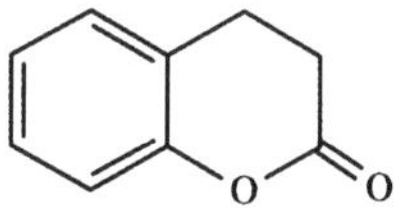

Furanocoumarins contain an additional ring, a furan ring, the carbon atoms of which derive from isopentenyl diphosphate (page 366).

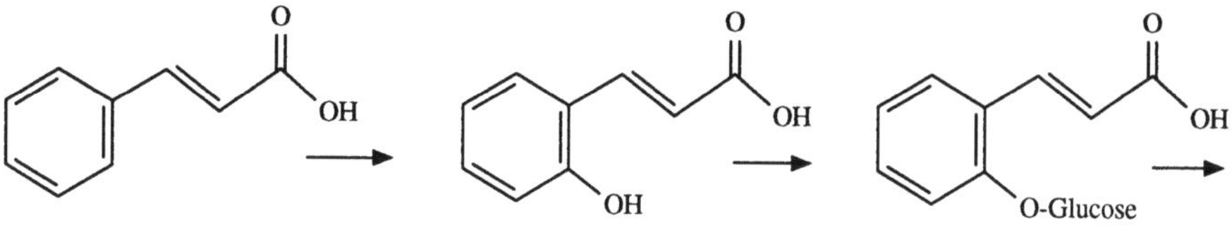

5,6-Benzo-2-pyrone

Trans-Cinnamic acid · *o*-Coumaric acid · *o*-Coumaric acid glucoside

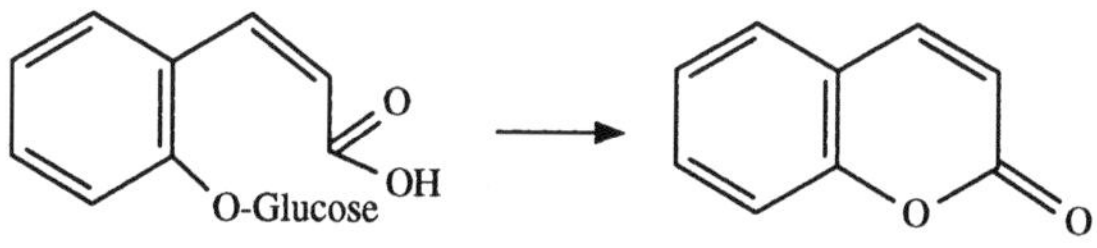

o-Coumarinic acid glucoside · Coumarin

Psoralen and angelicin are examples of the two types of furanocoumarins, linear and angular; in the former the furan ring is attached at the 6,7-position and in the latter at the 7,8-position.

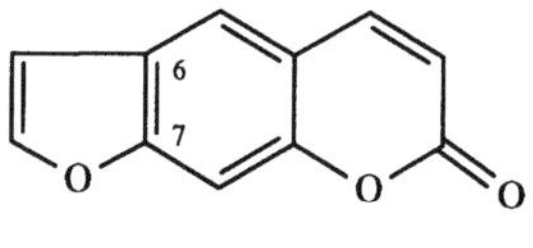

Psoralen 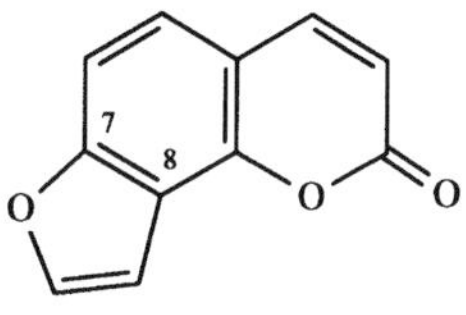 · Angelicin

Coumarin

Coumarin derivatives occur in roots, fruits and seeds of plants belonging to the families Apiaceae, Asteraceae, Fabaceae, Lamiaceae, Moraceae, Poaceae, Rutaceae and Solanaceae. They are especially common in the Apiaceae. Furanocoumarins are encountered mainly in the Apiaceae, Fabaceae, Moraceae and Rutaceae.

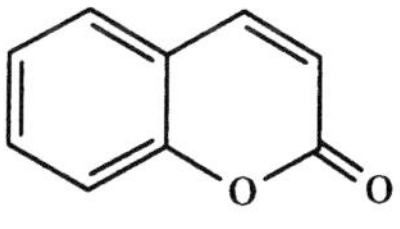

Coumarin was first isolated from the Tonka-bean, i.e. the seed of *Dipteryx odorata* Willd. (*Fabaceae*). This tree grows in Guyana and its local name is *coumarouna*, from which the substance got its name. Tonka-beans were formerly used as a spice, particularly in rum. When the plants wither, coumarin is formed through hydrolysis of the

Coumarin

229

corresponding glycoside. It is coumarin that gives new-mown hay its characteristic smell. Coumarin is no longer used as a flavouring and taste-improving agent in pharmacy and the food industry, since it has been found to have liver-damaging and carcinogenic properties.

Umbelliferone, aesculetin

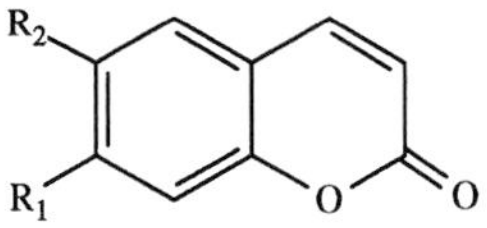

Umbelliferone: R_1 = OH, R_2 = H

Aesculetin: R_1 = R_2 = OH

These coumarin derivatives are employed for protection against the sun, because they absorb short-wave UV radiation (280–315 nm) that is harmful to the skin, but transmit the long-wave UV radiation (315–400 nm), that gives the brown suntan.

Umbelliferone is found in many plants and can be prepared through distillation of resins from certain *Apiaceae*.

Aesculetin is the aglycone of aesculin (6β-D-glucosyloxy-7-hydroxycoumarin). This glycoside can be extracted from the bark and leaves of the horse chestnut, *Aesculus hippocastanum* L. (*Sapindaceae*) (Plate 2, page 211). Both the glycoside and the aglycone are used in sun-tan preparations.

Dicoumarol

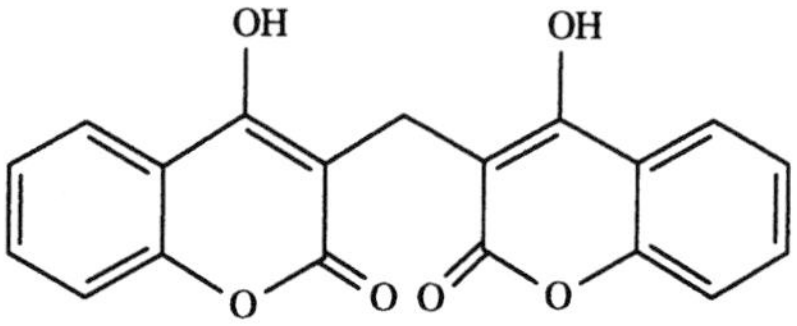

Dicoumarol

Dicoumarol occurs for example in mouldy sweet clover, *Melilotus officinalis* (Lam.) (*Fabaceae*), in which it is formed by the influence of microorganisms on the coumarin derivatives present. In cattle that have been eating mouldy sweet clover, the coagulating power of the blood is drastically decreased, which leads to fatal haemorrhage (sweet clover disease). This discovery led to the introduction of dicoumarol into medicine as an anti-bloodclotting agent for the prevention of thrombosis. Dicoumarol acts by replacing vitamin K (page 418) as the apoenzyme in an active enzyme complex. Nowadays, dicoumarol is produced synthetically.

Psoralens

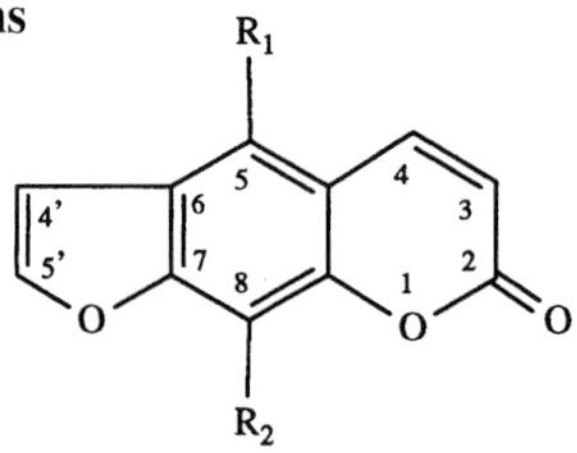

Psoralen: R_1 = R_2 = H

Bergapten: R_1 = OCH_3, R_2 = H

Xanthotoxin (8-MOP): R_1 = H, R_2 = OCH_3

Imperatorin: R_1 = H, R_2 = $CH_2-CH=C\begin{smallmatrix}CH_3\\CH_3\end{smallmatrix}$

Psoralen and its derivatives occur in plants belonging to the families *Rutaceae*, e.g. bergamot (*Citrus aurantium* L.) and lemon (*Citrus limon* (L.) Burm. f.), *Apiaceae*, e.g. celery (*Apium graveolens* L.) and parsnip [*Petroselinum crispum* (Miller) A.W. Hill], *Fabaceae*, e.g. *Psoralea corylifolia* L., and *Moraceae*, e.g. fig (*Ficus carica*

L.). The treatment of vitiligo, a disease characterized by the presence of white spots on the skin, with an extract of *Ammi majus* L. (*Apiaceae*) or the fruits of *Psoralea corylifolia* has long been known. An extract of the fruit of *Ammi majus* is either given orally or painted on the unpigmented spot. The patient is then exposed to sunshine for 1–2 hours. After 7–12 hours, reddening, and sometimes blisters, appear on the irradiated parts of the skin, and are followed two days later by the development of a strong brown colouring. The "enhanced" tanning cannot be obtained by using the plant extract alone without subsequent irradiation, and exposure to the sun alone has only a slight effect. This form of therapy, based on the joint action of medicine and irradiation, is called photochemotherapy.

The active principles in these plant extracts are psoralen, in *Psoralea corylifolia*, and xanthotoxin, in *Ammi majus*, and the effect is shown by other plants containing 6,7-furanocoumarins. Dermatoses may appear after the consumption of or direct contact of the skin with such plants, if immediately followed by exposure to the sun. Only 6,7-furanocoumarins produce this effect, whereas the isomeric 7,8-furanocoumarins are inactive.

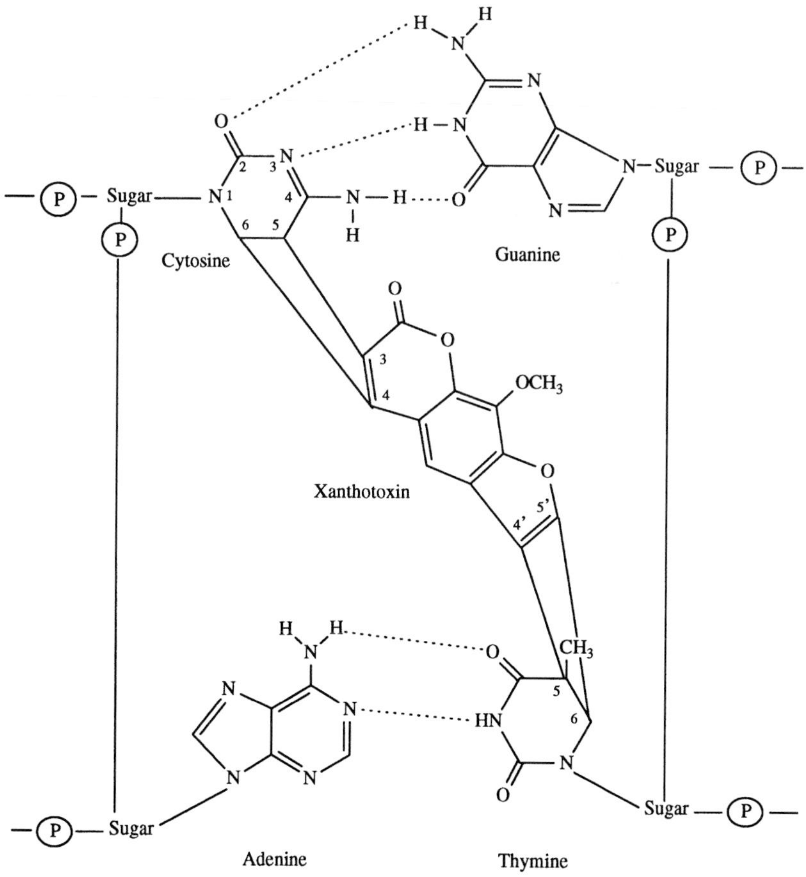

The mechanism of photosensitization by furanocoumarins is not fully known. However, the first step seems to be that the substances interfere with DNA by forming bridges between its base pairs. The double bonds of psoralen are first activated by the UV irradiation. The 3,4-double bond can then react with the 5,6-double bond of the pyrimidine base in a DNA base pair, while corresponding reactions take place between the 4',5'-double bond of the psoralen molecule and the pyrimidine base of another base pair.

This bridge-building inhibits the replication and transcription of DNA and, consequently, the synthesis of RNA and proteins and the occurrence of cell division.

The inhibition of DNA synthesis by the psoralens is believed to be the most important factor in their beneficial effect on psoriasis. The cause of this non-infectious skin disease is unknown. It is characterized by an abnormal production of the outermost layer of the skin, which forms scales and peels off, often in large amounts. The most widely used substances are xanthotoxin (8-MOP), taken orally, and the synthetically prepared 4,5',8-trimethylpsoralen (TMP, trioxalen), for topical application or in baths. UV light with wavelengths of 320–380 nm (UV A) is used for the irradiation.

Crude drug **Ammi fruit** (Ammi majoris fructus) is the schizocarp fruit of *Ammi majus* L. (*Apiaceae*). This plant is naturally occurring in the Mediterranean region and is also cultivated in Argentina and Australia. The mericarps are 2–2.5 mm long and 0.5–0.75 mm broad. They contain xanthotoxin (0.5 %), imperatorin (0.3 %) and bergapten (0.04 %). Extracts of the fruit can be used for the treatment of vitiligo and for the preparation of xanthotoxin.

Substances formed from phenylpropanes by shortening of the side-chain

The side-chain of phenylpropanes can be shortened by β-oxidation. An example is the formation of *p*-hydroxy-benzoic acid from 4-hydroxycinnamic acid (*p*-coumaric acid):

The reaction involves initial activation of 4-hydroxycinnamic acid by formation of a coenzyme A ester which is then subjected to

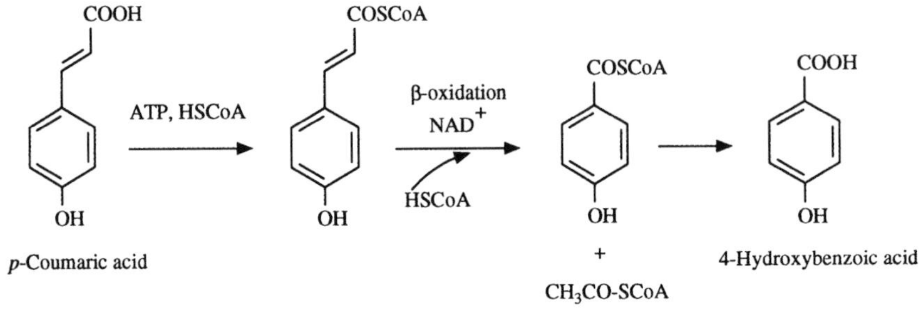

p-Coumaric acid   ATP, HSCoA   β-oxidation NAD$^+$ HSCoA   + CH$_3$CO-SCoA   4-Hydroxybenzoic acid

β-oxidation, yielding 4-hydroxybenzoyl CoA and acetyl CoA which are rapidly hydrolysed to the corresponding acids probably via thio-esterases. The oxidation may proceed until the entire side-chain has disappeared. This side-chain elimination can also occur at an early stage in the biosynthetic process that normally leads to phenylalanine and other phenylpropanes. Thus *p*-hydroxybenzoic acid can also be formed directly from prephenate (see page 204).

Some natural products that are formed by these two routes will be discussed here.

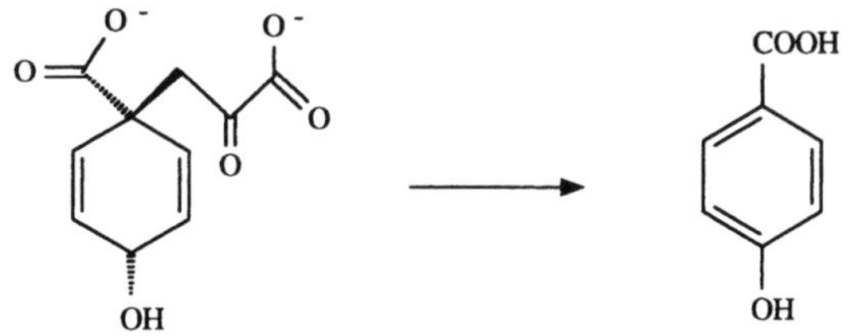

Prephenate   *p*-Hydroxybenzoic acid

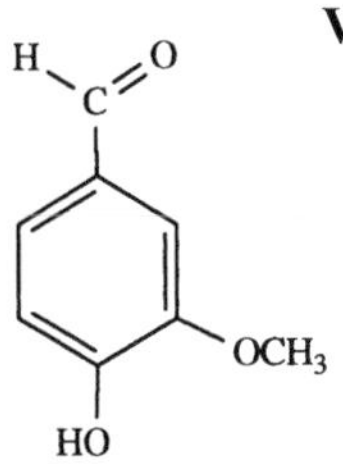

Vanillin

Vanillin

Vanillin occurs in several crude drugs. Its presence in Peru balsam and in Tolu balsam has already been mentioned. The most important source of natural vanillin is the spice vanilla. Vanillin is easily synthesized from eugenol or lignin as starting material. Lignin is a complex polymer, formed by oxidative coupling of phenylpropane derivatives. It occurs in wood, where it surrounds and strengthens the cellulose fibres. Degradation products of lignin, formed as by-products during the manufacture of paper from wood, are nowadays used for the large-scale preparation of vanillin.

Crude drug

Vanilla (Vanillae fructus) is the fully grown but unripe fruit of *Vanilla planifolia* Andr. (*Orchidaceae*), which has been subjected to a fermentation process. *Vanilla planifolia* is a climber from Central America, Mexico, and the northern part of South America, which is grown in most tropical areas. When cultivated outside its home region, the plant requires artificial fertilization, because the shape of its flowers is adapted to pollination by humming-birds. The fruits are harvested when they begin to turn yellow and are then subjected to a long fermentation process (curing). The process is initiated by dipping the fruits in hot water for some minutes (scalding), followed by sunning during which they are spread out in the sun until they are hot. They are then wrapped in sheets and packed in well-insulated cases overnight (sweating). The process of sunning and sweating is repeated for 1 to 2 weeks. Then follows drying indoors for 2 to 4 weeks. The drying prevents the beans from being spoiled by fungi and allows other aroma-producing reactions to take place. Finally the vanilla-pods are kept in a conditioned room for several months until they are ready for shipping. The complete process can last up to 6 months from the harvest.

The aroma develops during the fermentation. The fresh fruit contains vanillin in the form of the odourless glucoside vanilloside (glucovanillin). This compound is split during the fermentation, and vanillin with its characteristic smell is set free.

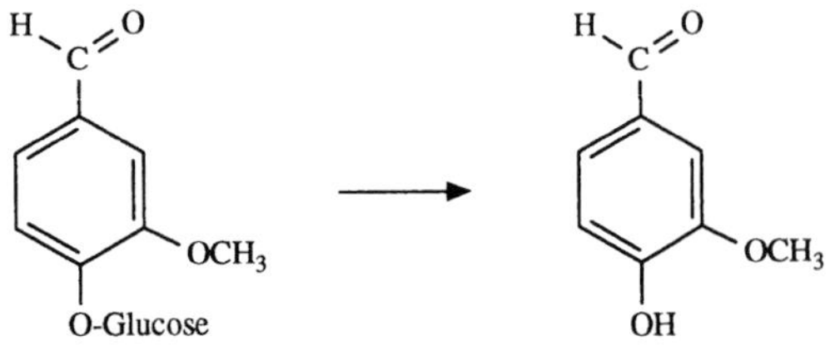

Vanilloside   Vanillin

The fresh fruit also contains a glucoside of vanillyl alcohol, vanilloloside. The alcohol is liberated during the fermentation and oxidized to vanillin.

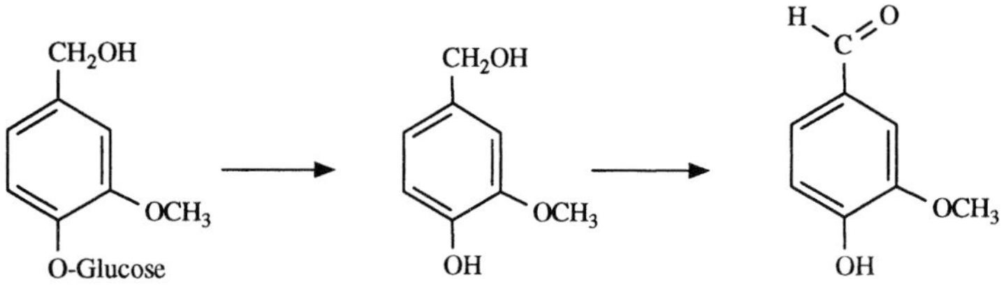

Vanilloloside   Vanillyl alcohol   Vanillin

Vanilla contains 1–4 % vanillin, plus smaller amounts of a large number of other compounds, which modify the fragrance of the vanilla, giving a finer aroma and taste than pure vanillin.

Coniferyl benzoate

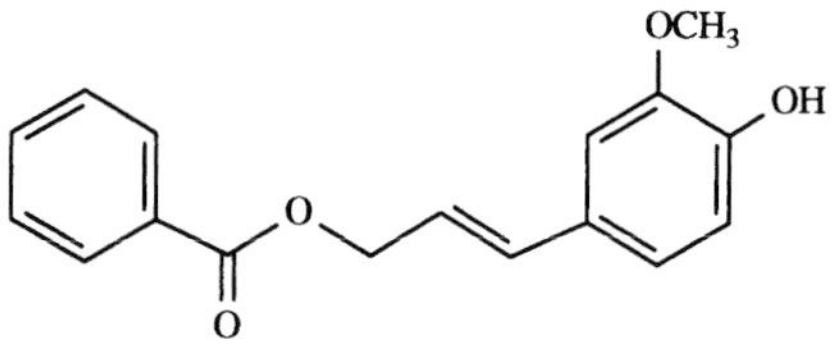

Coniferyl benzoate

Benzoin (Siam benzoin, Benzoe tonkinensis, Gum benzoin) is a resin from *Styrax tonkinensis* Craib ex Hartwich (*Styracaceae*), an East Asian tree, cultivated in Thailand, Vietnam, Laos and Sumatra. The drug is obtained from 6–10-year-old trees by making incisions through the bark to the wood. The tissue formed when the wound heals contains secretory ducts which produce a yellowish white balsam that drips out of the wound and solidifies. The product first emerging is discarded and only the resin formed later is used as a drug. It consists of brown, rounded or flattened pieces with a milky white, porcelain-like fracture and a smell of vanilla. 60–70 % of the drug is coniferyl benzoate. Other constituents are free benzoic acid (10–20 %), cinnamyl benzoate (2 %) and vanillin (1 %). Benzoin is used as a preservative for fat, e.g. lard (page 255), and as a disinfectant in mouthwashes, etc. It is also used in the perfumery industry.

Further reading

Reviews

Dewick, P.M. The biosynthesis of shikimate metabolites. *Natural Products Reports*, **18**, 334–355 (2001). Previous reports: *ibid.* **17**, 269–292 (2000); **16**, 525–560 (1999); **15**, 17–58 (1998). **12**, 579–607 (1995); **12**, 101–133 (1995); **11**, 173–203 (1994); and older.

Knaggs, A.R. The biosynthesis of shikimate metabolites. *Natural Products Reports*, **20**, 119–136 (2003).

Schmid, J. and Amrhein, N. Molecular organization of the shikimate pathway in higher plants. *Phytochemistry*, **39** (4), 737–749 (1995).

Rhodiola

Brown, R.P., Gerbarg, P.L. and Ramazanov, Z. *Rhodiola rosea*: A phytomedicinal overview. *HerbalGram (The Journal of the American Botanical Council)*, **56**, 40–52 (2002).

Tolonen, A., Pakonen, M., Hohtola, A. and Jalonen, J. Phenylpropanoid glycosides from *Rhodiola rosea*. *Chemical and Pharmaceutical Bulletin*, **51** (4), 467–470 (2003).

Podophyllotoxin

Bohlin, L. and Rosén, B. Podophyllotoxin derivatives: drug discovery and development. *Drug Discovery Today*, **1**, 343–351 (1996).

Broomhead, A.J., Rahman, M.M.A., Dewick, P.M., Jackson, D.E. and Lucas, J.A. Matairesinol as precursor of podophyllum lignans. *Phytochemistry*, **30** (5), 1489–1492 (1991).

Canel, C., Moraes, R.M., Dayan, F.E. and Ferreira, D. Molecules of interest. Podophyllotoxin. *Phytochemistry*, **54**, 115–120 (2000).

Xia, Z.-Q., Costa, M.A., Proctor, J., Davin, L.B. and Lewis, N.G. Dirigent-mediated podophyllotoxin biosynthesis in *Linum flavum and Podophyllum peltatum*. *Phytochemistry*, **55**, 537–549 (2000).

Schisandra

Schisandra berry. Monograph, October 1999. *American Herbal Pharmacopoeia*. P.O. Box 5159. Santa Cruz, CA 95063, USA

Adaptogens

Panossian, A., Wikman, G. and Wagner, H. Plant adaptogens III Earlier and more recent aspects and concepts on their mode of action. *Phytomedicine*, **6** (4), 287–300 (1999).

Panossian, A. and Wagner, H. Stimulating effect of adaptogens: an overview with particular reference to their efficacy following single dose administration. *Phytotherapy Research*, **19**, 819–838 (2005).

Wagner, H., Nörr, H. and Winterhoff, H. Plant adaptogens. *Phytomedicine*, **1**, 63–76 (1994).

7. Natural Products Derived Biosynthetically from Acetate

Acetate, formed from carbohydrate via pyruvic acid, is, as shown in the scheme on page 135, the starting material for the biosynthesis of many important natural products. Two main routes originate with acetate: the *acylpolymalonate pathway*, leading to fatty acids and polyketides, and the *isopentenyl diphosphate pathway*, which gives terpenes and steroids.

7a. The acylpolymalonate pathway

This reaction complex is catalysed by enzymes containing coenzyme A (page 720).

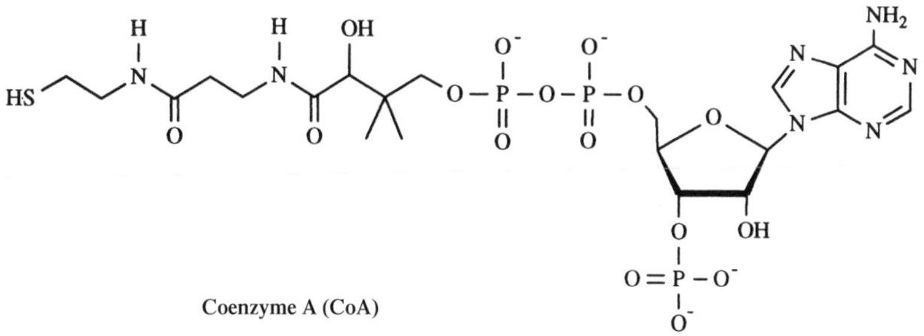

Coenzyme A (CoA)

Acetylcoenzyme A, CH_3CO-SCoA, formed on oxidative degradation of sugar (page 138), constitutes the basic unit for the biosynthesis of natural products along this pathway. Condensation of acetate units gives long chains or ring structures. This was first shown in studies on the biosynthesis of fatty acids, where it was found that labelled acetate was a good precursor for fatty acids and that every carbon atom in a fatty acid came from acetate. In the middle of the 1950s A.J. Birch and R. Robinson independently published the hypothesis that acetate units, when linked to each other without reduction, give a β-polyketomethylenic acid, which can be condensed further to aromatic substances with an oxygen atom attached to every second carbon atom (Fig. 51).

This hypothesis was based on the presence of such structures in many natural products. It was shown that a large number of natural products could be derived via the same biosynthetic route, where different acid residues, RCO-, constitute the first unit in the β-polyketone chain RCO-$(CH_2CO)_n OH$ and aldol condensations, alkylations, reductions, dehydrations and oxidations take place.

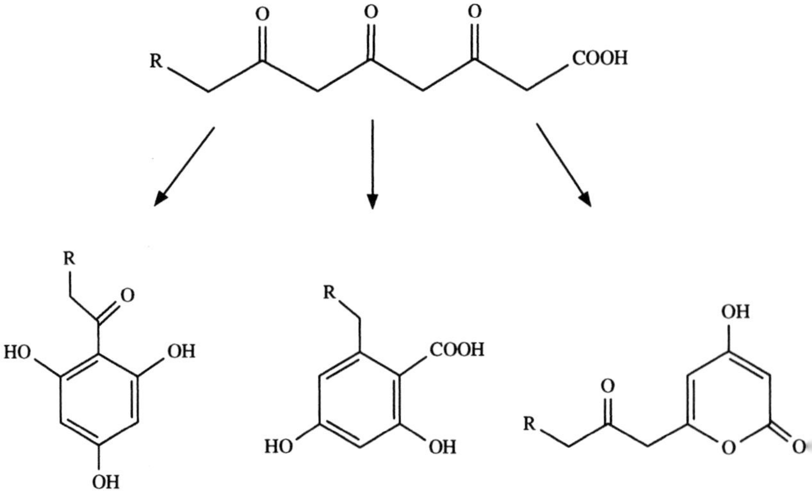

Fig. 51. Cyclic compounds formed by condensation of a β-polyke-tomethylenic acid.

Birch suggested the name *polyketides* for the substances concerned. This term is sometimes used for all compounds formed by the coupling of acetate or similar units, even if most of the oxygen functions have been reduced, as in fatty acids. The hypothesis was later verified when it was shown that labelled acetate was incorporated into several aromatic compounds in accordance with the hypothesis, and also that those steps in the condensation process which utilize one acyl-CoA and several malonyl-CoA units (see below) were the same in fatty-acid and in polyketide biosynthesis.

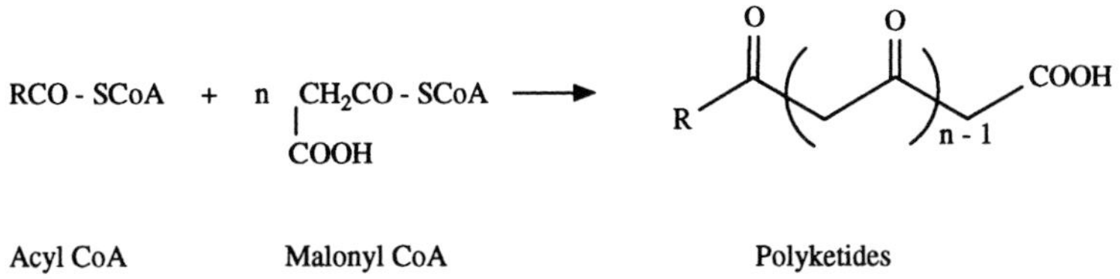

This fact is the reason why this biosynthetic route is named the *acylpolymalonate pathway*.

Fatty acids

Saturated fatty acids

Fatty acids can be either *saturated* or *unsaturated* with a varying number of carbon atoms in the chain. The most common saturated fatty acids are:

Butyric acid	$CH_3(CH_2)_2COOH$
Caproic acid	$CH_3(CH_2)_4COOH$
Caprylic acid	$CH_3(CH_2)_6COOH$
Capric acid	$CH_3(CH_2)_8COOH$
Lauric acid	$CH_3(CH_2)_{10}COOH$
Myristic acid	$CH_3(CH_2)_{12}COOH$
Palmitic acid	$CH_3(CH_2)_{14}COOH$
Stearic acid	$CH_3(CH_2)_{16}COOH$
Arachidic acid	$CH_3(CH_2)_{18}COOH$

In organisms saturated fatty acids very seldom occur in the free state. They are mainly found as esters with glycerol (fats and oils, page 249) or other alcohols of higher molecular weight such as waxes (page 255).

Biosynthesis of saturated fatty acids

In the biosynthesis of saturated fatty acids acetyl-CoA and malonyl-CoA are used as building blocks. Malonyl-CoA is formed by carboxylation of acetyl-CoA. Both CO_2 and HCO_3^- can function as carboxylating agents. The reaction is catalysed by the enzyme *acetyl- CoA-carboxylase* (EC 6.4.1.2) which uses biotin (page 577) as a coenzyme. CO_2 is first attached to biotin and then transferred to acetyl-CoA. The reaction also requires the energy formed on degradation of ATP to ADP.

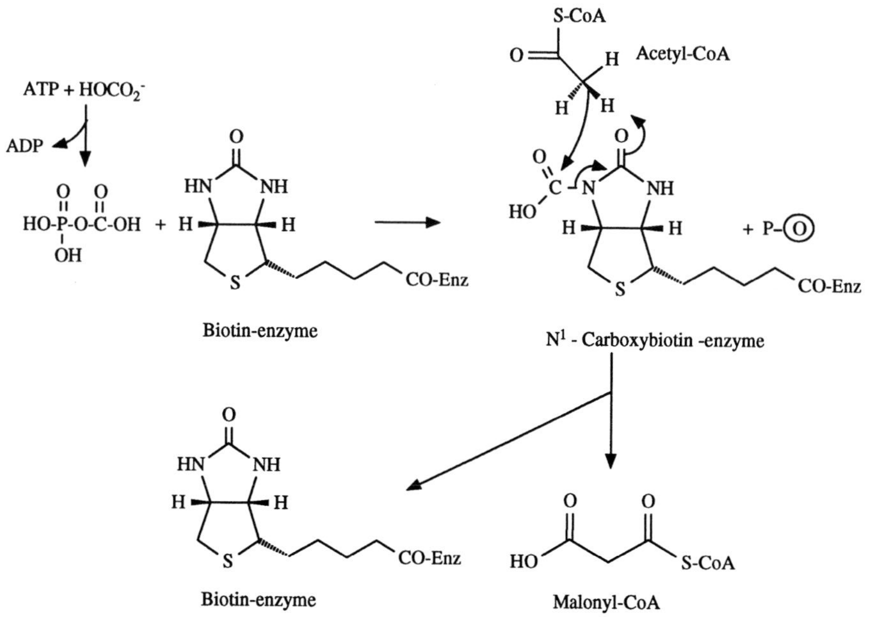

Starting with the acetyl group from acetyl-CoA, the fatty acids are constructed by stepwise addition of C_2-units derived from malonyl-CoA. The process also involves successive reductions of the car-

239

bonyl groups of the condensation products. The overall reaction for formation of palmitic acid is presented as an example:

$$CH_3 - \overset{\overset{\displaystyle O}{\|}}{C} - S - CoA + 7\ HO - \overset{\overset{\displaystyle O}{\|}}{C} - CH_2 - \overset{\overset{\displaystyle O}{\|}}{C} - S - CoA + 14\ NADPH + 14\ H^+ \longrightarrow$$

$$CH_3 - CH_2 - (CH_2 - CH_2)_6 - CH_2 - COOH + 14\ NADP^+ + 6\ H_2O + 8\ CoA - SH + 7\ CO_2$$

The reactions are catalysed by enzymes which are parts of a complex called *Fatty Acid Synthase*. This complex can either be integrated into a single multifunctional polypeptide (*Fatty Acid Synthase I = FAS I*) (EC 2.3.1.85) or the enzymes can be dissociable (*Fatty Acid Synthase II = FAS II*). The biosynthesis of fatty acids is catalysed by FAS I in fungi, birds and mammals and by FAS II in bacteria and plants.

Fatty acid synthase II (FAS II)

FAS II is a dissociable complex which is found in bacteria, plants and parasites. Each component is encoded by a separate gene that produces a unique, soluble enzyme that catalyses a single step in the pathway. The complex contains *Acyl Carrier Protein (ACP)*, a small protein (MW 8 847) with 4´-phosphopantetheine as the prosthetic group, and whose function is to carry growing fatty acid chains along the biosynthetic pathway.

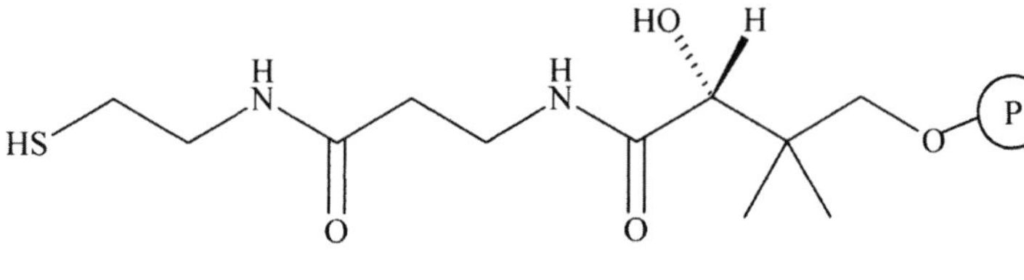

4´-Phosphopantetheine

The model system for the development of FAS II biochemistry is *Escherichia coli*. All genes encoding the different enzymes of the pathway, have been identified and the corresponding proteins characterized.

The gene *acpP* encodes the apo-protein of ACP, which receives its 4´-phosphopantetheine prosthetic group from CoA in a reaction catalysed by *holo-[acyl-carrier-protein] synthase* (EC 2.7.8.7). The first reaction in the biosynthesis of a fatty acid is a reaction (Fig. 52, reaction 1) in which the malonyl group of malonyl-CoA is transferred to ACP by the enzyme *[acyl-carrier-protein] S-malonyltransferase* (EC 2.3.1.39, also known as FabD or malonyl-CoA:ACP transacylase), forming malonyl-ACP. Following these initiation reactions, the chain elongation steps, involving condensation and β-carbon processing are ready to start. The first condensation step involves formation of acetoacetyl-ACP from acetyl CoA and malonyl-ACP in a reaction catalysed by *β-ketoacyl-acyl-carrier-protein synthase III* (EC 2.3.1.180, also known as FabH or β-ketoacyl-ACP synthase III) and proceeding with loss of CO_2 (Fig. 52, reaction 2). The acetoacetyl-ACP formed then enters three β-carbon processing reactions, which remove the oxygen atom at C_3. The first of these is catalysed by *3-oxoacyl-[acyl-carrier-protein] reductase* (EC 1.1.1.100, also known as β-ketoacyl-ACP reductase

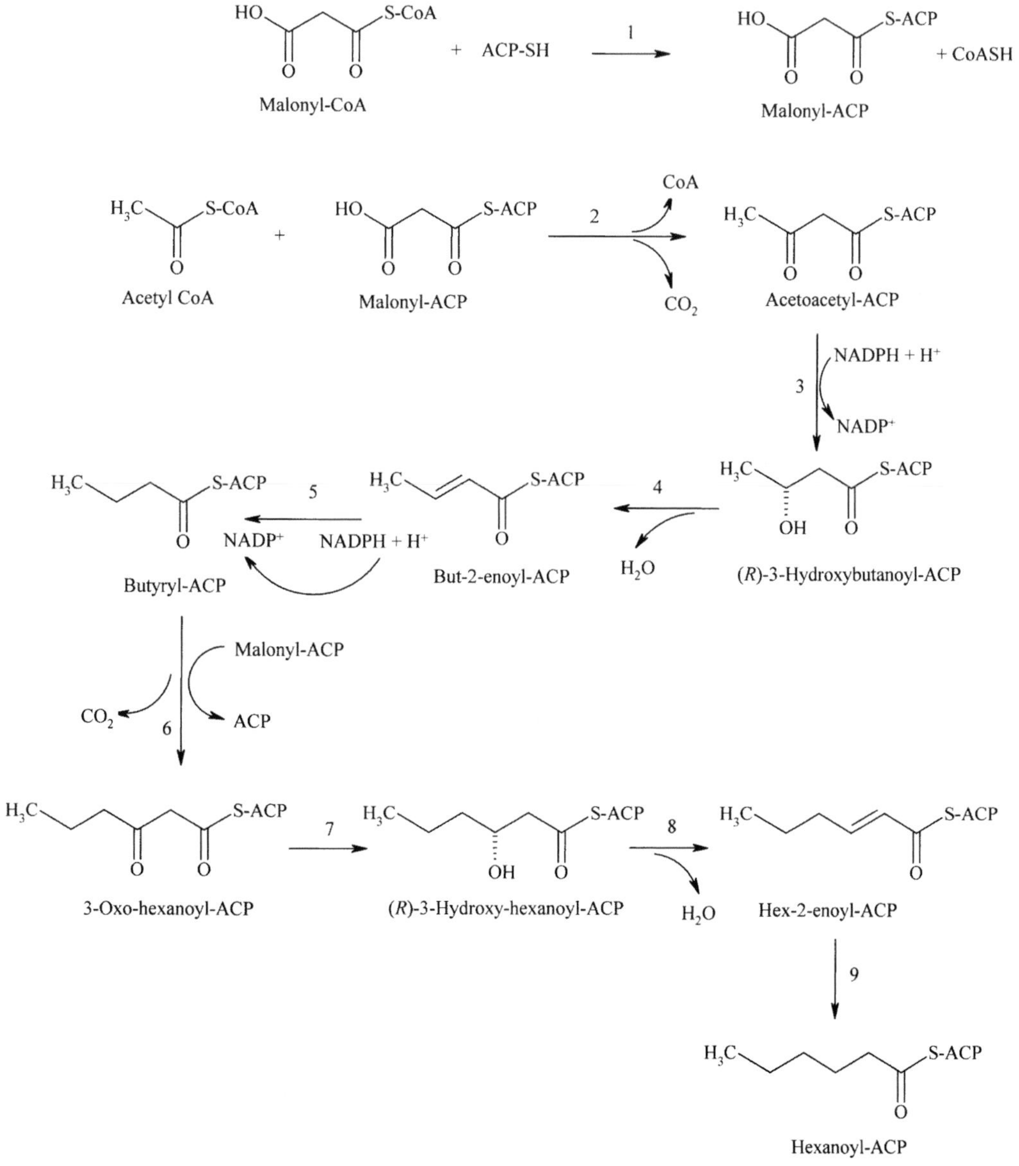

Fig. 52. *The first steps in fatty acid biosynthesis, catalysed by FAS II enzymes*

241

or FabG) resulting in formation of (*R*)-3-hydroxybutanoyl-ACP (Fig. 52, reaction 3), which, in the next reaction (Fig. 52, reaction 4), is dehydrated to but-2-enoyl-ACP by *crotonoyl-[acyl-carrier-protein] hydratase* (EC 4.2.1.58). Finally *enoyl-[acyl-carrier-protein] reductase (NADH)* (EC 1.3.1.9), also known as NADH-dependent enoyl-ACP reductase (or FabI) reduces but-2-enoyl-ACP to yield butyryl-ACP (Fig. 52, reaction 5).

Elongation of the fatty acid chain by two more carbons is effected by condensation of butyryl-ACP with another molecule of malonyl-ACP (Fig. 52, reaction 6), a reaction analogous to reaction 2 in Fig. 52 and which is catalysed by *β-ketoacyl-acyl-carrier-protein synthase I* (EC 2.3.1.41), also known as 3-oxoacyl:ACP synthase I or FabB. This enzyme is also active in all further chain elongations. β-Carbon processing reactions (Fig. 52, reactions 7, 8 and 9), form hexanoyl-ACP, which enters the next elongation step. The biosynthesis proceeds in the same way until the final length of the fatty acid chain has been reached, when ACP is released from the product by an *acyloyl-[acyl-carrier-protein] hydrolase*.

FAS II can produce fatty acids of different chain lengths. The enzymes for the condensation and β-carbon processing reactions are the same all the way with exeption of the dehydration reactions (reactions 4 and 8 in Fig. 52). *Crotonoyl-[acyl-carrier-protein] hydratase* (EC 4.2.1.58) is specific for short chain length 3-hydroxy-acyl-[acyl-carrier-protein] derivatives (C_4 to C_8). The derivative with 10 carbon atoms is dehydrated by *3-hydroxydecanoyl-[acyl-carrier protein] dehydratase* (EC 4.2.1.60) and derivatives with chain lengths of 12 to 16 carbons are dehydrated by *3-hydroxy-palmitoyl-[acyl-carrier-protein]dehydratase* (EC 4.2.1.61).

Besides fatty acids of different chain lengths FAS II can produce unsaturated fatty acids, branched-chain fatty acids and hydroxy fatty acids.

Fatty acid synthase I (FAS I)

In fungi the functional components are distributed between two nonidentical polypeptides of 205 kDa and 207 kDa, which are encoded in two genes without evident linkage. In mammals and birds FAS I is encoded by a single gene which may have evolved by fusion of component genes. Here the enzymes are integrated into a single multifunctional polypeptide which in the total complex forms a dimer where the two monomer units are arranged head to tail. In rats the size of the monomer is 272 kDa. In avian FAS I, as represented by the complex from chicken liver, the size of the subunit is 267 kDa.

The complex involves seven enzyme activities which perform reactions that are similar to those described for fatty acid synthase II above. The biosynthesis of the fatty acid is initiated by a reaction catalysed by the domains *acetyl transacylase* in which the acetyl group from acetyl-CoA becomes attached to the enzyme (for mechanism and attachment points see page 246):

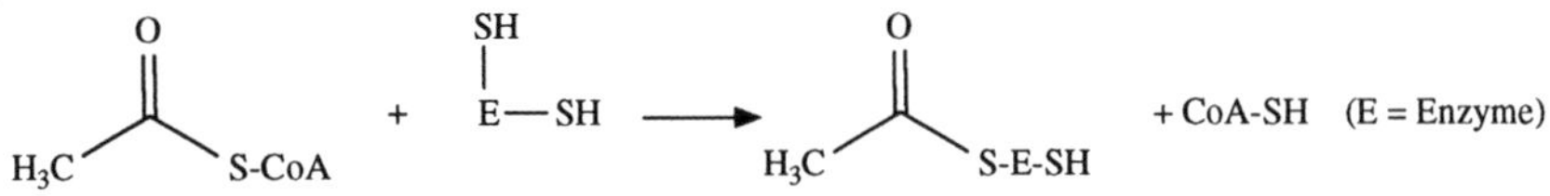

(reaction 1: acetyl-S-CoA + E—SH → acetyl-S-E-SH + CoA-SH (E = Enzyme)) (1)

In the next reaction, mediated by *malonyl transacylase*, malonyl-CoA is added to the enzyme complex:

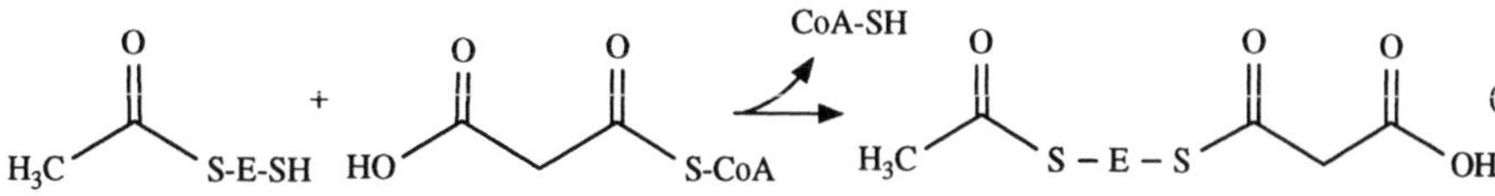

(reaction 2) (2)

The activities of acetyl transferase and malonyl transferase can sometimes (e.g. in FAS I from chicken liver) be combined in one protein.

Then follows a condensation reaction in which CO_2 is released. This reaction is catalysed by the domain *β-ketoacyl synthase*.

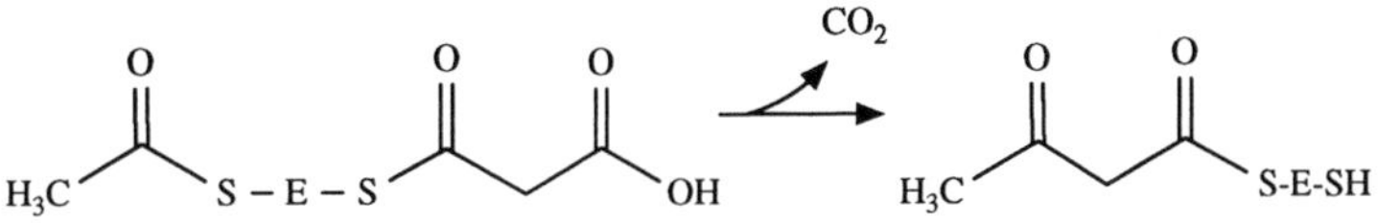

(reaction 3) (3)

Three consecutive steps, analogous to reactions 3–5 in Fig. 52, result in formation of a saturated acyl derivative. These reactions are catalysed successively by the domains *β-ketoacyl reductase*, *β-hydroxyacyl dehydratase* and *enoyl reductase*.

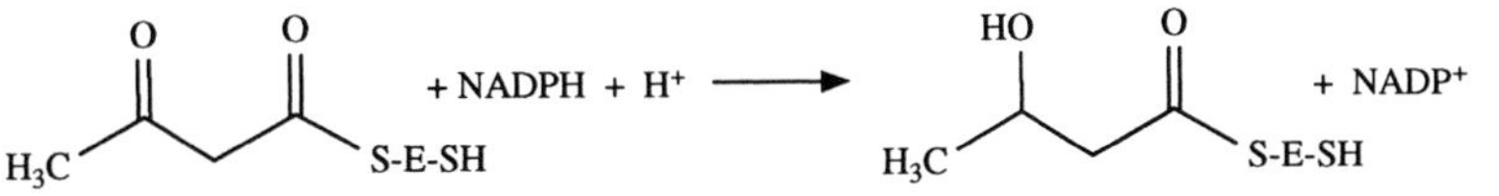

(reaction 4: + NADPH + H⁺ → ... + NADP⁺) (4)

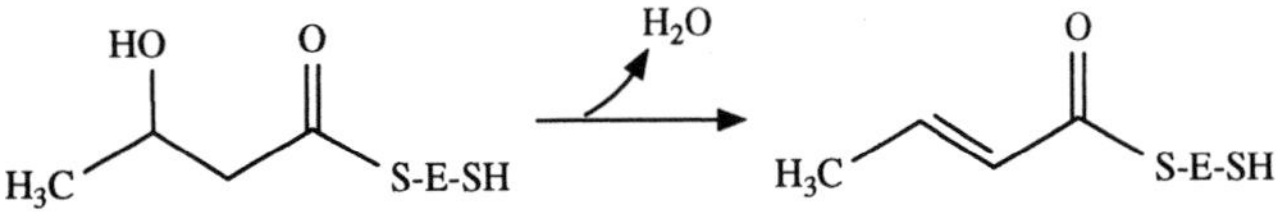

(reaction 5: + H_2O → ...) (5)

(reaction 6: + NADPH + H⁺ → ... + NADP⁺) (6)

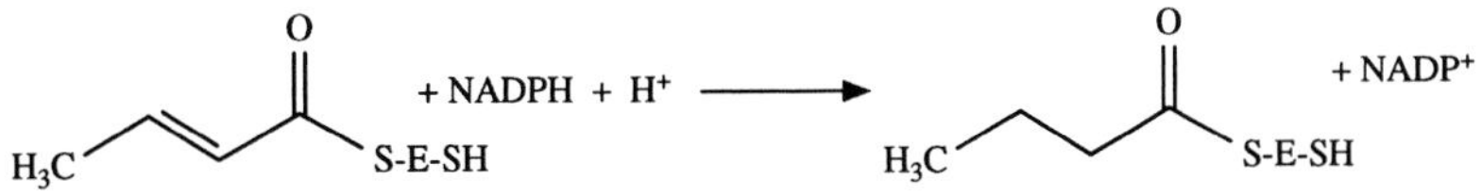

The acyl derivative formed then acts as a primer for further elongation and reduction cycles to yield hexadecanoyl-ACP, which is hydrolysed to the free palmitic acid by a *thioesterase domain* (see reaction on the next page).

Usually FAS I produces only palmitic acid, in contrast to FAS II which can produce fatty acids of varying chain length as well as unsaturated fatty acids, branched-chain fatty acids and hydroxy fatty acids.

243

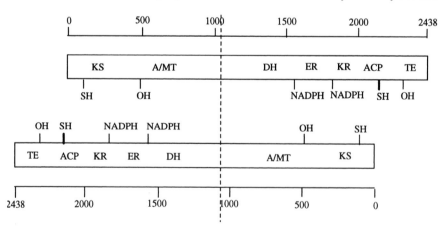

Palmitic acid

Structure of FAS I

The cDNA for the FAS I enzyme from liver of chicken (*Gallus gallus*) has been sequenced and the corresponding amino acid sequence determined. Each of the two identical chains contains 2 438 amino acid residues. The regions containing the enzyme's active sites have been identified. A functional map for FAS I from chicken is presented in Fig. 53. FAS I from rat has also been studied and the amino acid sequence shows 78 % homology with the enzyme from chicken liver.

Starting from the amino-terminal end of each chain the catalysing domains are arranged in the following order (abbreviations in parentheses refer to Fig. 53): β-ketoacyl synthase (KS), acetyl- and malonyl transacylase (A/MT), β-hydroxyacyl dehydratase (DH), enoyl reductase (ER), β-ketoacyl reductase (KR) and thioesterase (TE). In chicken liver FAS I the acetyl transacylase and

KS = β-Ketoacyl synthase. A/MT = Acetyl/Malonyl transacylase. DH = β-Hydroxyacyl dehydratase

ER = Enoyl reductase. KR = β-Ketoacyl reductase. ACP = "Acyl carrier protein". TE = Thioesterase

SH = Cysteine .   OH = Serine.   SH = 4'-phosphopantetheine

The figures indicate the numbers of the amino acid residues

Fig. 53. Functional map of FAS I from chicken liver

malonyl transacylase activities are combined in a single enzyme. Between β-ketoacyl reductase and thioesterase is a region (ACP) in which the cofactor *4'-phosphopantetheine* (page 240) is bound to a serine-OH in the protein through phosphate.

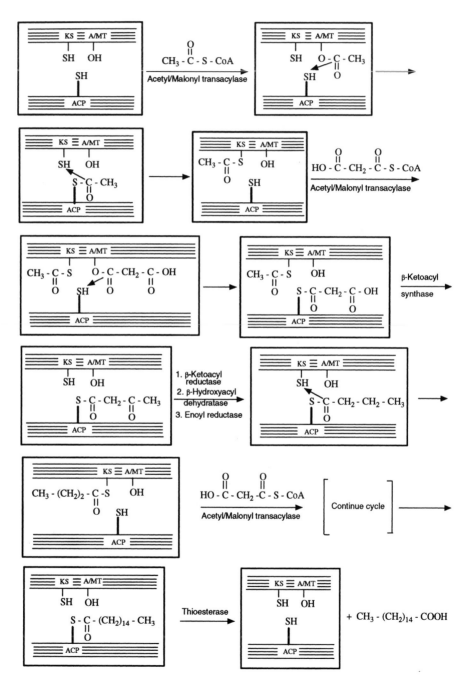

Fig. 54. Schematic representation of the mechanism of palmitic acid biosynthesis by chicken liver FAS I (see Fig. 53)

This region plays a very important role in fatty acid synthesis as it is a site where the growing fatty acid chain is attached and provided with a long flexible arm which enables it to reach the active site of each enzyme in the complex. The different chemical reactions can thus be performed without releasing intermediates from the enzyme. This region corresponds to the acyl carrier protein (ACP), which is the protein carrying out this function in bacteria and plants.

Mechanisms of action of FAS I

Because of the head to tail arrangement of the two protein chains, the enzyme complex has two independent catalytic centres in which the domains containing β-ketoacyl synthase, acetyl transacylase and malonyl transacylase in one chain cooperate with the domains forthe remaining four enzymes in the other chain. This is indicated by the dashed vertical line in Fig. 53. In each of the two centres three active sites can be recognized which covalently bind the reaction intermediates. Fig. 54 is a schematic representation of the mechanism of the biosynthesis of palmitic acid by chicken liver FAS I. In the first step the primer – acetyl-CoA – is loaded as an oxyester on to a serine OH in the combined acetyl/malonyl transacylase. It is then transferred to 4'-phosphopantetheine in the "ACP" region of the other chain, where a thioester linkage is formed. From there it moves to a cysteine sulphhydryl group on the β-ketoacyl synthase and malonyl-CoA is loaded on the serine OH in acetyl/malonyl transacylase and then transferred to 4'-phosphopantetheine. The acetyl group leaves the cystein-SH and is subjected to decarboxylative condensation with the malonyl group. Following reduction the elongated chain is then attached to the SH-group on β-ketoacyl synthase and a new malonyl-CoA molecule enters into the process. Condensation and reduction reactions thus take place with the growing chain attached to 4'-phosphopantetheine and the cystein-SH of β-ketoacyl synthase serves for storing the chain while a new malonyl unit is attached as a thioester to 4'-phosphopantetheine. When the fatty acid chain has reached its final length it is released from 4'-phosphopantetheine by the enzyme *thioesterase*.

The functional map of FAS I, depicted in Fig. 53, has been questioned as a result of experiments involving mutant complementation studies, which indicated that structural and functional contacts can be made between the ACP domain and the KS and A/MT domains both across the subunit interface and within individual subunits. Contact within individual subunits is not possible if the two subunits are located head to-tail and fully extended, as indicated in Fig. 53. Another model has therefore been proposed that retains the original concept of head-to tail oriented subunits, but places the KS and A/MT domains near the centre of the dimer, where they can freely access the ACP domains of both subunits. The domains responsible for the β-carbon processing reactions are located at opposite ends of

Substrate loading and condensation

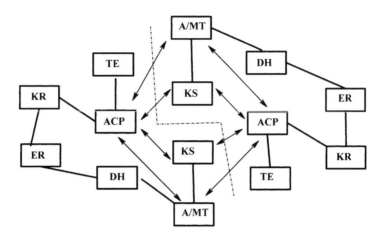

β-Carbon processing and chain-termination

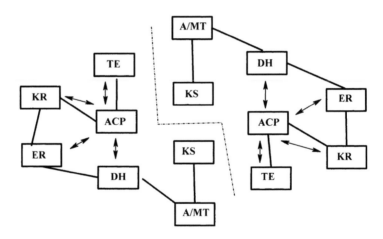

Fig. 55. Proposed alternative functional map for FAS I (From Smith (2003)
See Further reading *page 358*

the dimer and therefore have access to only one of the ACP domains. Since each KS domain is able to access both ACP domains it is possible for the growing acyl chains to switch from one phosphopantetheine to the other during the elongation process.

Unsaturated fatty acids

Unsaturated fatty acids can have one or more double bonds. Fatty acids with two or more double bonds are commonly called *polyunsaturated fatty acids*. Examples of unsaturated fatty acids are:

Name	Structure	Chain length, position of double bonds. Stereochemistry (c = cis)
Oleic acid	$CH_3(CH_2)_7\text{-}CH=CH\text{-}(CH_2)_7COOH$	18:1 (9c)
Linoleic acid	$CH_3(CH_2)_4\text{-}CH=CH\text{-}CH_2\text{-}CH=CH\text{-}(CH_2)_7COOH$	18:2 (9c, 12c)
α-Linolenic acid	$CH_3\text{-}(CH_2\text{-}CH=CH)_3\text{-}(CH_2)_7COOH$	18:3 (9c,12c,15c)
γ-Linolenic acid	$CH_3\text{-}(CH_2)_3\text{-}(CH_2\text{-}CH=CH)_3\text{-}(CH_2)_4\text{-}COOH$	18:3 (6c,9c,12c)
Arachidonic acid	$CH_3\text{-}(CH_2)_3\text{-}(CH_2\text{-}CH=CH)_4\text{-}(CH_2)_3COOH$	20:4(5c,8c,11c,14c)
Eicosapentaenoic acid (EPA)	$CH_3\text{-}(CH_2\text{-}CH=CH)_5\text{-}(CH_2)_3\text{-}COOH$	20:5(5c,8c,11c,14c,17c)

A convenient way to characterize the unsaturated fatty acids is indicated to the right of the structures above. The first figure indicates the chain length, the second the number of double bonds and the following figures denote the position of the double bonds, counted from the carboxyl terminus. It is also possible to indicate the position of the double bonds by counting from the methyl terminus (the ω end). In that system only the position of the first double bond is indicated in addition to the number of carbon atoms, and the number of double bonds. An example is linoleic acid which is denoted 18:2,ω-6. The position of the other double bonds is inferred from the knowledge that following the first double bond the sequence - CH2-CH=CH- is repeated for as many times as there are additional double bonds. In virtually all cases the stereochemistry of the double bond is *cis* (Z).

Unsaturated fatty acids are ususally biosynthesized by dehydrogenation of saturated ones, e.g. oleic acid from stearic acid. A Δ^9-*desaturase* which introduces a *cis* double bond into a saturated fatty acid is present in most eukaryotic organisms. This enzyme requires NADPH and molecular O_2 as cofactors. Oleic acid and linoleic acid are directly derived from stearic acid. Desaturation proceeds in sequence: 18:0 → 18:1(9) → 18:2(9,12). α-Linolenic acid is biosynthesized from lauric acid into which three double bonds are introduced followed by extension of the chain: 12:0 → → → 12:3(3,6,9) → 14:3 (5,8,11) → 16:3(7,10,13) → 18:3(9,12,15). Mammals cannot synthesize linoleic acid and α-linolenic acid which are very important for the construction of membranes and as precursors for prostaglandins (see page 259). These fatty acids are therefore called **essential fatty acids** and must be provided by the diet.

Like the saturated fatty acids, unsaturated fatty acids are seldom found in their free form in Nature. They are usually encountered as esters with glycerol in fats and oils. Glycerides of saturated fatty acids are solid while those of unsaturated fatty acids are liquids (oils).

Other derivatives of fatty acids

Dehydrogenation and/or oxidation of fatty acids may give rise to other long-chain natural products. The biosynthesis of long-chain alcohols, such as cetyl alcohol ($C_{16}H_{33}OH$, hexadecanol) and myricyl alcohol ($C_{30}H_{61}OH$, tricosanol), is closely connected with fatty acid biosynthesis. Long-chain alkanes and those acetylene derivatives which occur in plants of the families *Apiaceae* and *Asteraceae,* for example, are also derived biosynthetically from unsaturated fatty acids.

Fatty acids can also be transformed into heterocyclic compounds. An example is the alkaloid coniine, which is biosynthesized from acetate via 5-oxo-octanoate (see page 622).

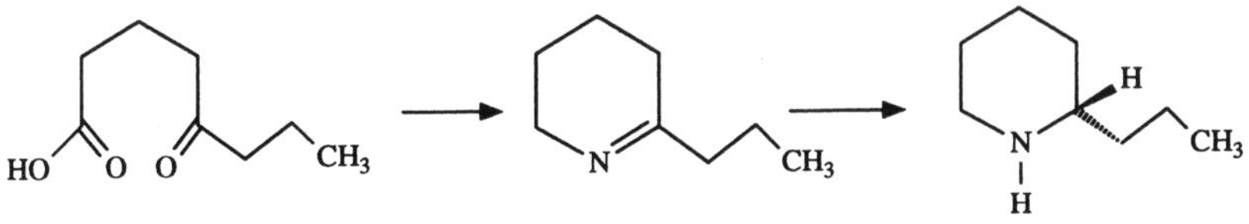

(+)-Coniine

Fats and waxes

Fats and waxes are alcohol esters of long-chain fatty acids. The two groups differ mainly as regards the alcohol. In fats this is always glycerol, in waxes an alcohol of higher molecular weight, e.g. cetyl alcohol. The biosynthesis of high-molecular-weight aliphatic alcohols is the same as that of fatty acids, described above.

Fats

The alcohol component of fats – glycerol – derives from the glycolysis product dihydroxyacetone phosphate, which is reduced to glycerol-1-phosphate by an NADH-dependent dehydrogenase. Triglycerides are then formed from glycerol-1-phosphate which first reacts with two fatty acyl-CoA molecules to give a fatty acid diglyceride phosphate, which then loses its phosphate residue and reacts with one more acyl-CoA molecule to form a triglyceride:

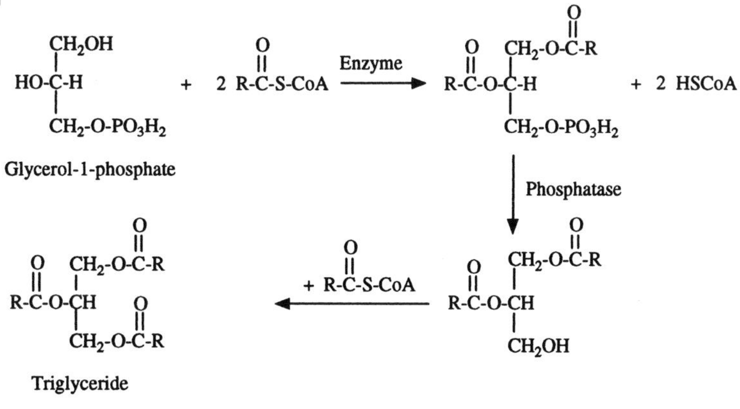

Fig. 56. Biosynthesis of triglycerides

Triglycerides can be either homogeneous, when all three hydroxyl groups of the glycerol are esterified with the same fatty acid, or mixed, when there are different fatty acid residues in the triglyceride. The properties of the fats are determined mainly by the fatty acid residues occurring in them. Solid fats contain essentially saturated fatty acids, whereas oils have a high content of unsaturated ones.

Fats occur in both plants and animals, normally as reserve nutrients. In plants, fat can be found in all organs, but large amounts occur only in fruits and seeds. In animals, fat is usually stored in a special tissue beneath the skin.

In general, fats are obtained by pressing. Pressing at room temperature yields the best quality. If heat is applied, the yield is usually higher, but the quality is lowered. Fat can also be isolated by boiling in water or extraction with organic solvents.

Most of the fats prepared are used for food. Fats have no pronounced pharmacological effects and are, therefore, not drugs as such. In medicine they are mainly used as vehicles for drugs intended for external application (ointments and liniments).

Crude drugs **Olive oil** (Olivae oleum) is obtained from the ripe fruit of the olive, *Olea europaea* L. (*Oleaceae*). The olive is a small evergreen tree, which rarely becomes taller than 10 metres but can live for a long time. Its origin is found in the east Mediterranean and it has been cultivated for thousands of years, nowadays also in south-western United States and many other subtropical countries. A large number of varieties exist and the fruits show great variation in size, colour and oil content. The fruit is a drupe and most of the oil is located in the fleshy mesocarp. The oil is obtained by pressing, the quality depending on the pressure and temperature applied. The finest oil is produced at low pressure and room temperature. Olive oil consists chiefly of glycerides of oleic acid (65–75 %), palmitic acid (10–15 %) and linoleic acid (9–15 %). Glycerides of stearic, myristic and arachidic acids are present in smaller amounts.

Arachis oil (Peanut oil, Arachidis oleum) is prepared from the peeled seed of the peanut, *Arachis hypogaea* L. (*Fabaceae*) (Plate 3, page 253). This is an annual herb, native to northern South America but now widely cultivated in the southern United States, China, India and West Africa. After fertilization, the plant buries the developing fruits – pods – in the soil, where they ripen. Each pod contains 1–3 reddish brown seeds with a thin papery testa. The seed contains 40–50 % oil, mainly localized in the two cotyledons which are transformed into storage organs. To isolate the oil the seeds are first shelled mechanically, followed by pressing. The press cake is rich in proteins and is used as cattle-feed. Arachis oil is prepared in large quantities to meet the demands of the food industry. It consists of glycerides of oleic acid (50–65 %), linoleic acid (18–30 %), palmitic acid (8–10 %) and C_{20}, C_{22} and C_{24} fatty acids (6–7 %).

Sesame oil (Sesami oleum) is obtained by expression from the seeds of *Sesamum indicum* L. (*Pedaliaceae*), an annual herb grown in India, China, Japan, Africa, the West Indies and the United States. The sesame seed is small (weight 2–4 mg), ovoid or egg-shaped, white, yellow or reddish brown in colour. It contains about 50 % oil, 22 % protein and 4 % mucilage, and constitutes an important food-source in India. The oil is a mixture of glycerides of oleic (43 %), linoleic (43 %), palmitic (9 %) and stearic acids (4 %). It has good keeping qualities due to the presence of a phenolic substance, *sesamol*, which is a hydrolysis product of the lignan *sesamolin*, occurring in the unsaponifiable fraction of the oil. Sesamolin is used as a synergist for pyrethrum insecticides (see page 391).

Almond oil (Amygdalae oleum) is obtained from the seeds of *Prunus dulcis* (Mill.) D.A. Webb (*Rosaceae*); see also page 593. Almond trees are among our oldest known cultivated plants, and have been domesticated for several thousand years. Both bitter and sweet almonds can be used for the production of almond oil, which contains glycerides of oleic (77 %), linoleic (17 %), palmitic (5 %) and myristic acids (1 %).

Cottonseed oil (Gossypii oleum) comes from the seeds of cotton, *Gossypium herbaceum* L. and other *Gossypium* species (*Malvaceae*); see page 164. The oil has glycerides of linoleic (45 %), oleic (30 %), palmitic (20 %), myristic (3 %), stearic (1 %) and arachidic acids (1 %).

Soyabean oil (Soiae oleum). The source of this oil is *Glycine max* Merrill. (*Fabaceae*), a herb native to East Asia and nowadays cultivated on a large scale in the United States as well as in South America and India. It requires a relatively warm climate with fairly wet summers, so that cultivation in Europe has not been very successful. Soyabean production in the United States has increased rapidly and is now financially as important as the production of wheat and maize. Brazil and India are growing increasing amounts and are likely to be important sources in the future.

The plant is a bushy, hairy, annual herb, 0.5–1 m high, bearing small, blue violet flowers. The fruit is a hairy pod, about 10 cm long, containing 2–5 round, pea-like seeds, the colour of which can vary from off-white to yellowish green or brownish black.

The seed contains 35 % carbohydrates, up to 50 % protein and 20 % oil which can be expressed. The oil consists of glycerides of linoleic acid (50 %), oleic acid (30 %), linolenic acid (7 %) and saturated fatty acids (14 %), mainly palmitic and stearic acids. The linolenic acid content makes soyabean oil a drying oil, i.e it is oxidized by atmospheric oxygen and becomes semi-solid.

Soyabeans are – in spite of their relatively low oil content – one of the most important raw materials for the margarine industry, in particular in the United States, where it is the starting material for

75 % of the margarine produced. It is the high content of unsaturated fatty acids that makes the oil particularly suitable for margarine production. Soyabean oil has many uses in the food industry and forms half the cooking fat produced in the United States and 15 % of the total nutritional fat used throughout the world.

The unsaponifiable fraction of soyabean oil contains the sterols stigmasterol and sitosterol, which are important starting materials for the synthesis of steroid hormones (see page 456).

The press cake is rich in high-grade protein and is used for cattle-feed. In the Far East, it is an important traditional foodstuff, and in recent years much work has been devoted to the preparation of protein-rich products for human consumption from the residue left after the oil has been expressed. Several soya protein products ("soya bacon", "hamburger protein", etc.) are nowadays available on the market, particularly in the United States. The cost of protein from soyabeans is much lower than that of animal protein. The production of bacon requires 10 times, and beef 15–20 times the agricultural area needed to produce the same amount of "soya bacon" and "soya beef".

Maize oil (Maydis oleum) is obtained from the embryo of maize, *Zea mays* L. (*Poaceae*). The embryo is recovered in combination with production of maize starch by wet milling, a process in which the maize is first treated with water under controlled conditions (steeping) for up to 40 hours. Following gentle milling to break up the kernel, the embryos are separated and the remaining slurry of starch, proteins and fibers is further processed to yield maize starch (page 161). Maize oil is obtained from the oil-bearing embryos by pressing and contains glycerides of linoleic (50 %), oleic (37 %), palmitic (10 %) and stearic (3 %) acids . The high content of linoleic acid makes maize oil a valuable foodstuff, which is used in the margarine industry to increase the content of polyunsaturated fatty acids in margarine.

Linseed oil (Lini oleum) is derived from the seed of flax, *Linum usitatissimum* L. (*Linaceae*). Linseed has been described above (page 174) as a source of mucilage, which is present in the testa. The seed also contains 30–40 % of a fixed oil, obtained by crushing and pressing the seed, which is used in the preparation of liniments.

Linseed oil contains glycerides of linolenic (50 %), oleic (22 %) and linoleic (18 %) acids. The oil is thus very rich in polyunsaturated fatty acids and is easily oxidized by atmospheric oxygen to a solid material, a property used in the formulation of paints. The oil is boiled with metal oxides, e.g. iron oxide or manganese dioxide ("boiled linseed oil") and the salts formed facilitate the oxidation, making the paint dry to form an elastic layer.

Castor oil (Ricini oleum) is the fixed oil from ripe seeds of *Ricinus communis* L. (*Euphorbiaceae*, Plate 4, page 253). *Ricinus commu-*

Plate 3.
Arachis hypogaea L. Peanut, from which peanut (arachis) oil is pressed (page 250). From: Köhler's Medizinal-Pflanzen (Gera Unternhaus 1887).

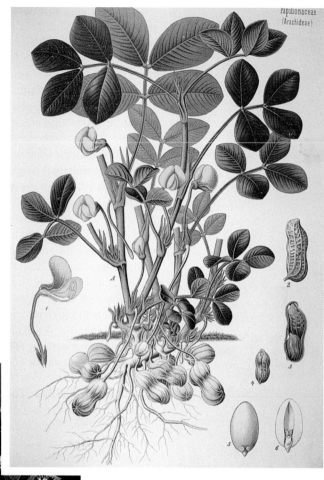

Plate 4.
Ricinus communis L. Plant with fruits. The fruits contain seeds from which castor oil is pressed (page 252). The seeds also contain ricin, a type II RIP toxin (page 513). Photographed by Gunnar Samuelsson outside the city of Monterrey, Mexico.

nis is a perennial plant which under good conditions can reach a height of up to 15 m. Its origin is found in South Asia but nowadays it is common and cultivated in all tropical and subtropical countries. The fruit of *Ricinus communis* is a capsule with three carpels which holds one seed in each carpel. The seed is somewhat compressed, ellipsoidal, 8–15 mm long and 4–7 mm broad. The testa is hard and beautifully marbled with a colour that can vary from grey to brown.

For the production of castor oil, the hard testa is first crushed by pressing between rollers and then separated from the endosperm and embryo, which yield the oil. The oil expressed at room temperature is for medicinal purposes, while that obtained at higher temperatures is a lower-grade product for technical use.

The main component (75 %) of castor oil is the triglyceride of ricinoleic acid, an unsaturated, hydroxylated fatty acid, CH_3-$(CH_2)_5$-$CHOH$-CH_2-$CH=CH$-$(CH_2)_7$-$COOH$. The other components of castor oil are mixed glycerides containing two ricinoleic acid residues, but the third fatty acid may be oleic acid, linoleic acid or a saturated fatty acid, such as stearic or palmitic acid.

In doses of about 15 ml, castor oil is used as a purgative. The effect is due to the ricinoleic acid liberated by lipases in the duodenum. Ricinoleic acid has multiple effects on the intestinal mucosa. It can increase mucosal permeability and cause cytotoxicity, associated with release of eicosanoids, platelet activating factor, other autacoids and nitric oxide. In addition, ricinoleic acid disrupts normal intestinal motility. The combination of these effects on the mucosa and smooth muscle of the gut account for its laxative action.

Castor oil also has a wide range of technical uses. It is relatively soluble in ethanol and is therefore used in beauty products (hair lotions, etc.). It is also a useful lubricant for high-speed rotary combustion engines running at widely varying temperatures.

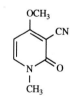

Ricinine

A toxic glycoprotein – *ricin* – is also present in the seed. It remains in the marc after expression of the oil, but as a safety precaution castor oil is always treated with steam to denature any ricin that might be present. The properties of ricin are discussed on page 513.

Castor seeds also contain ricinine, an alkaloid with only weak pharmacological effects.

Cocoa butter (Cacao oleum) is obtained from the seeds of *Theobroma cacao* L. (*Sterculiaceae*), as a by-product from the production of cocoa powder. Since the main interest in cocoa is as a purine drug, a more detailed description of the plant and of the procedure for making cocoa is to be found in a later chapter (page 736).

Cocoa butter consists of glycerides of oleic (37 %), stearic (34 %), palmitic (26 %) and linoleic acids (2 %). The glycerides usually contain one unsaturated and two saturated fatty acid residues. About 57 % of the total glycerides is oleopalmitostearin, 22 % is oleodistearin and 4 % is oleodipalmitin. Cocoa butter is a solid at room temperature, but melts at 30–35 °C. It is used as a suppository base.

Lard (Adeps suillus) is an animal fat, obtained from the peritoneum and kidneys of the pig, *Sus scrofa,* var. *domesticus*, by warming to 57 °C. The molten lard is washed with water and dried at a low temperature.

Lard is a mixture of glycerides of oleic (48 %), palmitic (28 %), linoleic (11 %), stearic (9 %) and myristic (3 %) acids, which is solid at room temperature. It readily becomes rancid and needs to be preserved by the addition of anti-oxidants. Lard is used as an ointment base.

Waxes

Like fats, waxes are esters of fatty acids. The alcohol, however, is not glycerol but usually a long-chain, high-molecular-weight alcohol. In plants, waxes are generally found covering the external parts, like the epidermis of leaves and fruits, where their main function is to prevent the loss of water. Wax is also produced by insects, e.g. the honeycombs of bees (*Apies* spp.) and wasps (*Vespa* spp.).

Wax is used in pharmacy to make soft ointments harder, and to prepare lip salves. The technical uses of waxes are substantial, e.g. in shoe polishes and car waxes.

Crude drugs **Beeswax** is available in two qualities: *yellow beeswax* (Cera flava) and *white beeswax* (Cera alba). Beeswax consists of honeycomb wax from beehives (see Honey, page 149), which has been melted in water and cast in moulds. This produces the unrefined yellow wax. The white variety is obtained by subsequent bleaching thin, cast strips of yellow wax in the sun until the yellow colour fades. The main component of beeswax is myricyl palmitate, i.e. the ester of palmitic acid with myricyl alcohol, $C_{30}H_{61}OH$. Free cerotic acid ($C_{25}H_{51}COOH$) constitutes about 15 % of the wax.

Spermaceti (Cetaceum) is a wax obtained from the sperm whale, *Physeter catodon,* which lives in tropical and subtropical seas, especially the Indian and Pacific Oceans. This whale reaches a length of 20 m and has an enormous head that makes up one-third of the body. In the front part of the head there is a large cavity containing an oil-like fluid. When the whale has been killed, the cavity is opened and the fluid is drawn off. On cooling, it forms a solid and a liquid layer. The solid material represents about 11 % of the total amount of oil, which may fill 10 barrels from a single whale. The solid portion is pressed, melted, filtered and treated with boiling alkali to saponify any remaining oil. The end product is a white, somewhat glossy wax, the crude drug spermaceti, comprising a mixture of the cetyl esters of mainly lauric, myristic, palmitic and stearic acids.

Phospholipids

Phospholipids are derivatives of glycerol-1-phosphate (page 249) where the two hydroxyl groups are esterified with fatty acids and the phosphate group is esterified with an alcohol, usually choline, ethanolamine, serine or inositol (Fig. 57). Phospholipids are important components of cell membranes and as there are both polar and non-polar regions in their structure they can act like detergents. One important phospholipid is *Platelet Activating Factor* (PAF, page 426) which has a number of biological effects and is of importance in inflammation and allergic reactions.

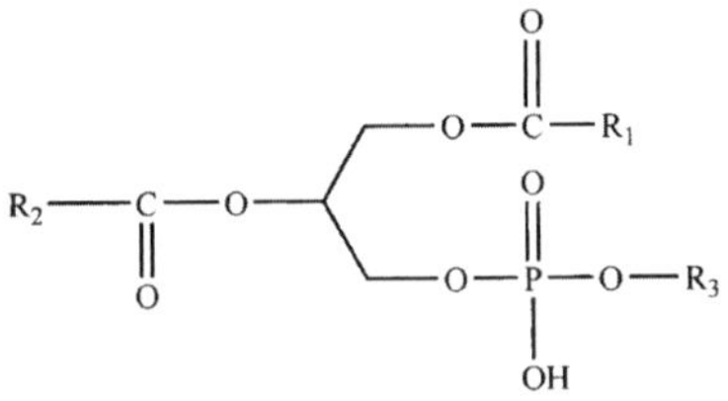

R_1 and R_2 = residues of fatty acids (saturated or unsaturated)

Phosphatidylcholine: $R_3 = CH_2CH_2N(CH_3)_3^+$

Phosphatidylethanolamine: $R_3 = CH_2CH_2NH_3^+$

Phosphatidylserine: $R_3 = CH_2CHCOOH$, NH_2

Phosphatidylinositol: $R_3 =$

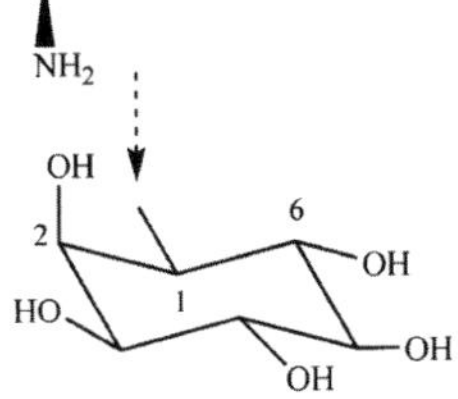

Fig. 57. Structures of phospholipids

Eicosanoids

The eicosanoids are derivatives of C_{20} polyunsaturated fatty acids which comprise three groups of biologically very active compounds: the *prostaglandins*, the *thromboxanes* and the *leukotrienes*.

Prostaglandins

Prostaglandins owe their name to the fact that they were first discovered in seminal fluid. It was soon shown, however, that they are produced not only in the prostate gland but in practically all mammal organs although in very small amounts. They are active at very low concentrations, almost in the same range as hormones, and they have influence on a wide range of organs such as the cardiovascular system, the intestinal tract, the uterus and the bronchi. The concentration of prostaglandins can increase rapidly as a result of various stimuli which can be either physical (e.g. inflammation, tissue damage, allergic manifestations, cold, poisons) or physiological such as hormones (histamine, bradykinin, arginine, vasopressin, platelet activating factor, angiotensin II, etc.) and proteases, e.g. thrombin.

Structurally the prostaglandins can be considered as derivatives of a hypothetic C_{20} acid – **prostanoic acid** – which consists of a cyclopentane ring with two side-chains comprising 7 and 8 carbon atoms, respectively. The carboxyl group is attached to the C_7 chain. About 15 different prostaglandins are known. They can be arranged into six different series (denoted D, E, F, G, H and I) depending on the oxidation of the cyclopentane ring:

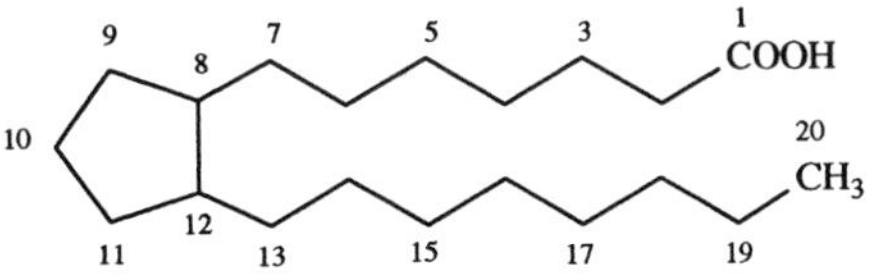

Prostanoic acid

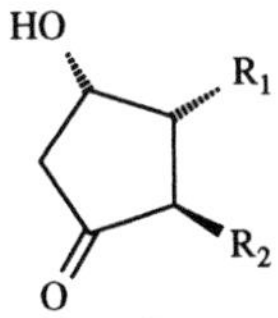

PGD

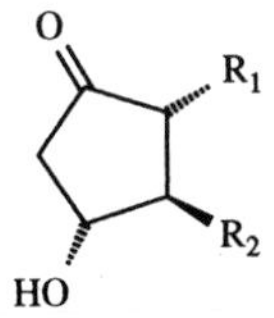

PGE

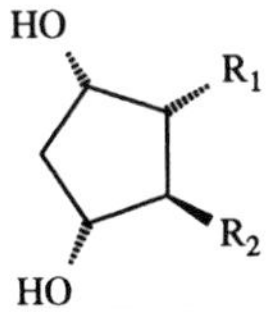

PGF

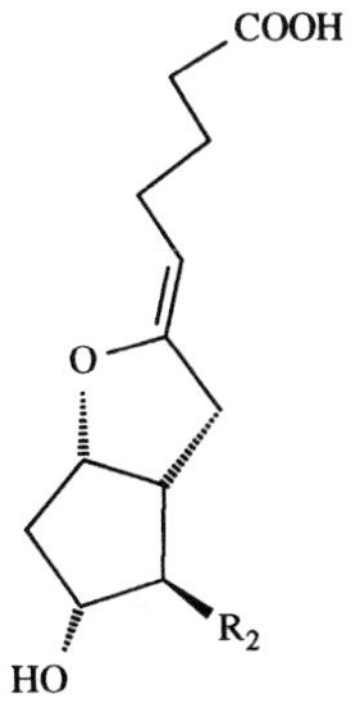

PGI

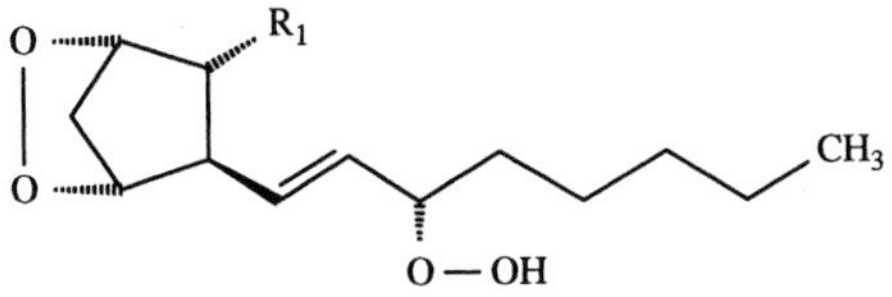

PGG

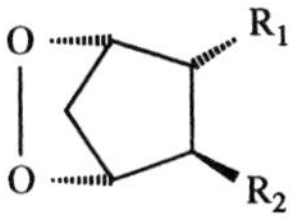

PGH

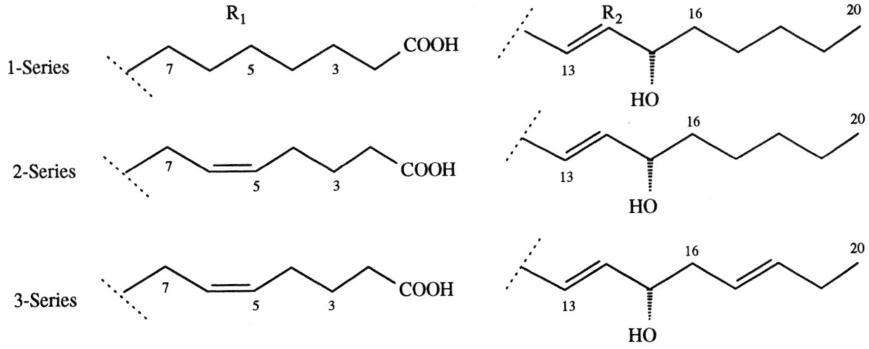

Within these series the arrangement is based on the number of double bonds in the chains. This number is written as an index to the serial letter. The term prostaglandin is often abbreviated as PG.

PG E$_2$, for example, denotes a prostaglandin of the E-series with two double bonds. In the 1-series there is no double bond in the C$_7$ chain and one (C-13 – C-14) in the C$_8$ chain. In the 2-series there is one double bond (C-5 – C-6) in the C$_7$ chain and one in the C$_8$ chain (C-13 – C-14). In the 3-series the C$_7$ chain has one double bond (C-5 – C-6) and the C$_8$ chain two double bonds (C-13 – C-14 and C-17 – C-18). The C$_8$-side chain has an α-OH group at C-15, with the exception of the G-series where the substituent at that position is –O–OH. The substituent at C-11 is always in the α-position. The substituent at C-9 is α in natural prostaglandins but is β in some synthetic derivatives. For a complete denomination the letter α or β is therefore written after the index-figure indicating the number of double bonds. Thus PGF$_{2α}$ denotes a prostaglandin of the F-series with two double bonds and the substituent at C-9 in the α-position. In the I-series the oxygen substituent at C-9 forms an ether bridge with carbon 6. The most widely distributed prostaglandins are: PGE$_1$, PGF$_{1a}$, PGD$_2$, PGE$_2$ and PGF$_2$.

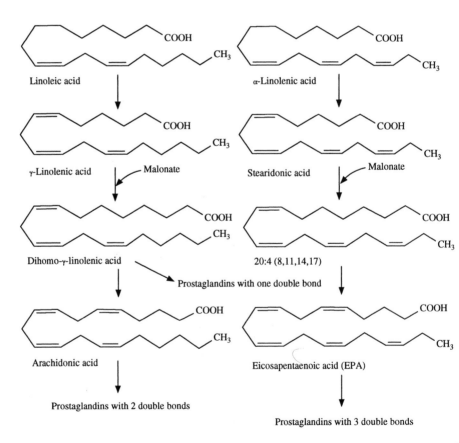

Fig. 58. Biosynthesis of dihomo-γ-linolenic acid, arachidonic acid and eicosapentaenoic acid

Biosynthesis of the prostaglandins

The ultimate precursors for the prostaglandins are the essential fatty acids linoleic acid and α-linolenic acid which cannot be synthesized by mammals and must thus be provided by the food. For biosynthesis of prostaglandins with one and two double bonds linoleic acid is first converted to γ-linolenic acid, which, after chain extension, yields **dihomo-γ-linoleic acid** ($\Delta^{8,11,14}$-eicosatrienoic acid). This acid is the precursor of prostaglandins with one double bond, but for biosynthesis of prostaglandins with two double bonds it must first be converted to **arachidonic acid**. Prostaglandins with three double bonds derive from α-linolenic acid which is dehydrogenated to stearidonic acid [18:4(6,9,12,15)] which by chain elongation yields the polyunsaturated fatty acid [20:4(8,11,14,17)]. This is further dehydrogenated to **eicosapentaenoic acid** [EPA, 20:5(5,8,11,14,17)], which is the precursor of prostaglandins with three double bonds.

Oxidation and radical reaction processes convert the three polyunsaturated amino acids into the prostaglandins. The best studied of these pathways is the one leading from arachidonic acid to prostaglandins with two double bonds. In the body arachidonic acid is stored in phospholipids (page 256), e.g. phosphatidyl inositol or phosphatidyl choline. *Phosphoinositide phospholipase C* (EC 3.1.4.11) and *phospholipase A$_2$* (EC3.1.1.4) are two enzymes which mobilize arachidonic acid from these phospholipids. Phosphoinositide phospholipase C specifically releases phosphoinositol from the glycerol ester and arachidonic acid is liberated through the concerted action of *lipoprotein lipase* (EC 3.1.1.34) and *acylglycerol lipase* (EC 3.1.1.23). Phospholipase A$_2$ directly liberates arachidonic acid from phosphatidylcholine.

The enzyme *prostaglandin-endoperoxide synthase* (EC 1.14. 99.1, also known as PGH synthase, prostaglandin H synthase, cyclooxygenase or COX) converts arachidonic acid via PGG$_2$ to PGH$_2$. PGH synthase is an integral membrane glycoprotein which has an associated haem group. The enzyme has been purified from vesicular glands of sheep (*Ovis aries*) or cattle (*Bos taurus*) following solubilization with detergents. In detergent solution the protein exists as a homodimer with a subunit molecular mass of 72 kDa. cDNAs encoding PGH synthase have been isolated from sheep, mouse and human and the amino acid sequences for the encoded enzymes have been deduced. The molecular masses calculated from the amino acid sequences are about 65.5 kDa. The difference between the empirically determined molecular mass and that calculated from the amino acid sequences is attributed to the presence of two to three mannose carbohydrate side-chains per subunit.

PGH synthase exhibits both a *bis*-oxygenase (cyclooxygenase) activity, which causes formation of PGG$_2$, and a peroxidase activity, which reduces the hydroperoxyl group at C-15 yielding PGH$_2$ (Fig. 59). The two activities occur at distinct but interacting sites on the enzyme. Thus non-steroidal anti-inflammatory agents like

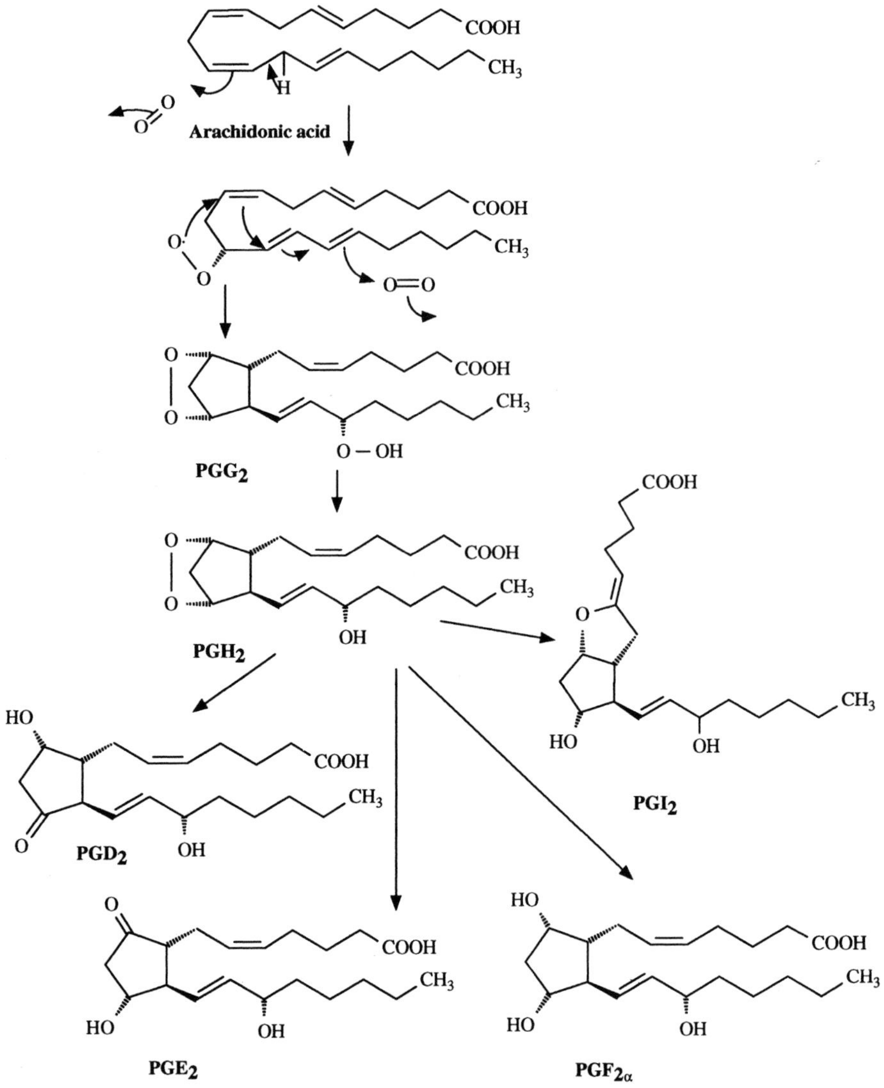

Fig. 59. Biosynthesis of prostaglandins from arachidonic acid

aspirin competitively inhibit the binding of arachidonic acid to the cyclooxygenase site but have no influence on the peroxidase activity. In the formation of PGG_2 a hydrogen atom is removed from C-13 producing a carbon radical which is trapped by molecular oxygen at C-11. Serial cyclization of the 11-hydroperoxyl radical yields a bicyclic peroxide. A second molecule of oxygen then reacts with a carbon radical at C-15 yielding PGG_2 which thus contains both a cyclic peroxide and an acyclic peroxide. The peroxidase activity of PGH synthase finally cleaves the acyclic peroxide yielding PGH_2 which is the precursor for the D, E, F and I series of prostaglandins with two double bonds.

PGH synthase also converts dihomo-γ-linolenic acid and EPA to PGH$_1$ and PGH$_3$, respectively, thus providing precursors also for the prostaglandins with one and three double bonds.

Non-oxidative rearrangement of prostaglandins of the H series yields prostaglandins of the D, E and I series. Enzymes catalysing formation of PGD$_2$ have been partly characterized. PGE derivatives can be formed non-enzymatically from the PGH precursors but there is also evidence for enzyme-mediated PGE formation. Biosynthesis of PGI$_2$ from PGH$_2$ is catalysed by *Prostaglandin-I synthase* (EC 5.3.99.4, also known as PGI synthase). This enzyme has been purified from bovine and porcine aorta and found to have a subunit molecular weight of 50 kDa. PGI synthase also converts PGH$_3$ to PGI$_3$ but PGH$_1$ is not a substrate.

PGF$_2$ formation requires a net two electron reduction of PGH$_2$. There are two epimers of PGF$_2$ which have biological activity and which are found in biological fluids; one is PGF$_{2\alpha}$ and the other is 9α,11β-PGF$_2$. There are three potential mechanisms for the formation of these PGF$_2$ derivatives: (1) reduction of PGH$_2$ to PGF$_2$ by an endoperoxide reductase, (2) reduction of PGD$_2$ to 9α,11β-PGF$_2$ by an 11-ketoreductase and (3) reduction of PGE$_2$ to PGF$_2$ by a 9-ketoreductase. Proteins have been isolated which catalyse each of these reactions. They have, however, rather broad substrate specificities and the K_{cat}/K_m values for prostaglandin substrates are often unfavourable. It is therefore doubtful if any of these proteins represent PGF synthases.

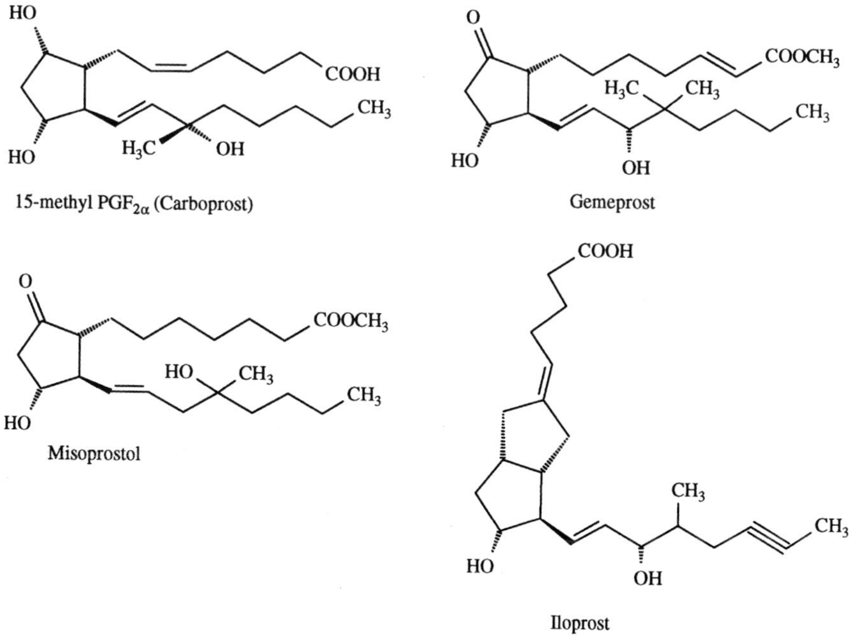

Fig. 60. Structures of "unnatural" prostaglandins

Prostaglandins as drugs

Prostaglandins control many important physiological processes and are thus compounds with a significant drug potential, but because of their activity in many different areas the risk of undesired side effects is also high. Control of the biosynthesis of prostaglandins is important for treatment of inflammations. The anti-inflammatory activity of steroids is due to their inhibitory effect on the release of arachidonic acid from storage phospholipids. Non-steroidal anti-inflammatory drugs (aspirin, indomethacin, ibuprofen) exert their activity by inhibition of the early steps in the biosynthetic pathway from unsaturated fatty acids to prostaglandins.

Prostaglandins to be used as drugs are obtained by total synthesis. The most common use is in obstetrics. **PGE$_2$** (dinoprostone) is applied locally to widen the cervix prior to surgical abortion and also close to parturition when the cervix-status is unfavourable. The use of **PGF$_{2\alpha}$** (dinoprost) is restricted to abortions as it has a half-life of only 10 minutes. **15-methyl PGF$_{2\alpha}$** (carboprost) is a more stable derivative and effective at low dosage. It is used to induce abortions and to treat post-partum haemorrhage within 24 hours following parturition. **Gemeprost** is a non-natural prostaglandin derivative which is administered intravaginally and used to soften and dilatate the cervix in early abortions.

PGE$_1$ (alprostadil) has vasodilator properties and is used for treatment of newborn infants with congenital heart defects prior to corrective surgery. The drug is metabolized very rapidly and must be administered by continuous intravenous infusion. It is also used to treat male impotence by direct injection into the penis before intercourse. The analogue **misoprostol** is active orally and is used for inhibition of gastric secretion in the treatment of gastric and duodenal ulcers, particularly those resulting from treatment with non-steroidal anti-inflammatory agents.

PGI$_2$ (prostacyclin, epoprostenol) inhibits platelet aggregation and is employed to inhibit blood clotting during renal dialysis. It has a very short half-life (about 3 minutes) and must be infused continuously. **Iloprost** is a stable carbocyclic analogue which is used to treat thrombotic diseases.

Thromboxanes

Thromboxane A$_2$ (TXA$_2$) is a potent thrombogenic agent and vasoconstrictor. It was originally identified as a product of arachidonate metabolism by human platelets but has also been found to be synthesized by lung and macrophages. TXA$_2$ is an unstable compound with a half-life of approximately 30 seconds at 37 °C.

TXA$_2$ is biosynthesized from PGH$_2$ in a reaction catalysed by *thromboxane-A synthase* (EC 5.3.99.5), an enzyme which has been purified to apparent homogeneity from porcine lung and from human platelets. The molecular mass of the platelet enzyme has been reported to be 58 kDa.

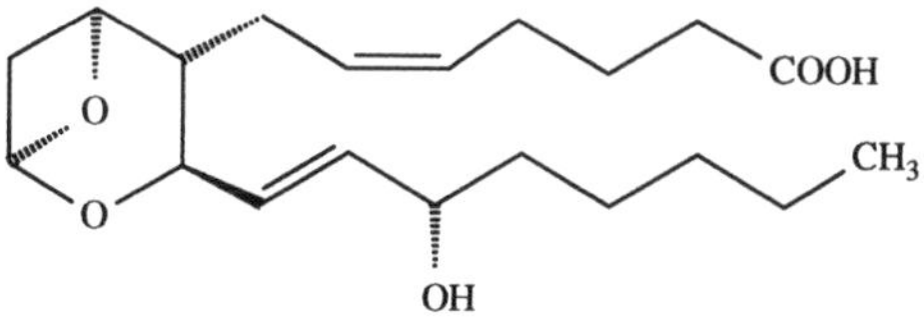

Thromboxane A$_2$

As previously mentioned (page 260) aspirin interacts with the cyclooxygenase activity of PGH synthase by competing with arachidonic acid for the binding site. In addition aspirin causes acetylation of a serine hydroxyl group thereby causing irreversible inhibition of the cyclooxygenase. When this has happened new activity can only occur through synthesis of new enzyme. This is of particular importance for platelets which in contrast to other cells cannot synthesize PGH synthase. Ingestion of about 70 mg of aspirin (about 1/7 of the dose used for treating headache) is sufficient to inactivate more than 95 % of the platelet enzyme and the cyclooxygenase activity is lost for the life of the platelet – about five days – and consequently no TXA2 can be formed. On the other hand the biosynthesis of PGI2 – which is an inhibitor of platelet aggregation – is not influenced very much by aspirin treatment, as endothelial cells can synthesize new PGH synthase quite rapidly. The net pharmacological effect of the aspirin dose is thus selective inhibition of the thrombogenic TXA2 without any particular influence on the anti-thrombogenic agent PGI2. This is the biochemical base for the use of low doses of aspirin to prevent formation of thrombi. This anti-thrombogenic effect of aspirin is unique and is not exhibited by other non-steroidal anti-inflammatory drugs.

Leukotrienes

Leukotrienes are derivatives of arachidonic acid containing – as their name implies – three conjugated double bonds. The suffix leuko- indicates that these compounds were first detected as products liberated following activation of leukocytes. Five naturally occurring leukotrienes have thus far been identified and denoted LTA$_4$ – LTE$_4$. The index figure indicates that the compounds contain four double bonds.

Biosynthesis of the leukotrienes

The enzyme *arachidonate 5-lipoxygenase* (EC 1.13.11.34, also known as 5-lipoxygenase or 5-LO) catalyses stereospecific abstraction of hydrogen at C-7 of arachidonic acid followed by oxygenation with molecular oxygen at C-5. The reaction product is 5-hydroperoxy-6,8,11,14-eicosatetraenoic acid (5-HPETE) which is converted to the unstable leukotriene A$_4$ (LTA$_4$, Fig. 61), in a reaction involving loss of water and formation of an epoxide ring. This reaction is also catalysed by *5-LO*, which thus is an enzyme with dual activities. *5-LO* has been

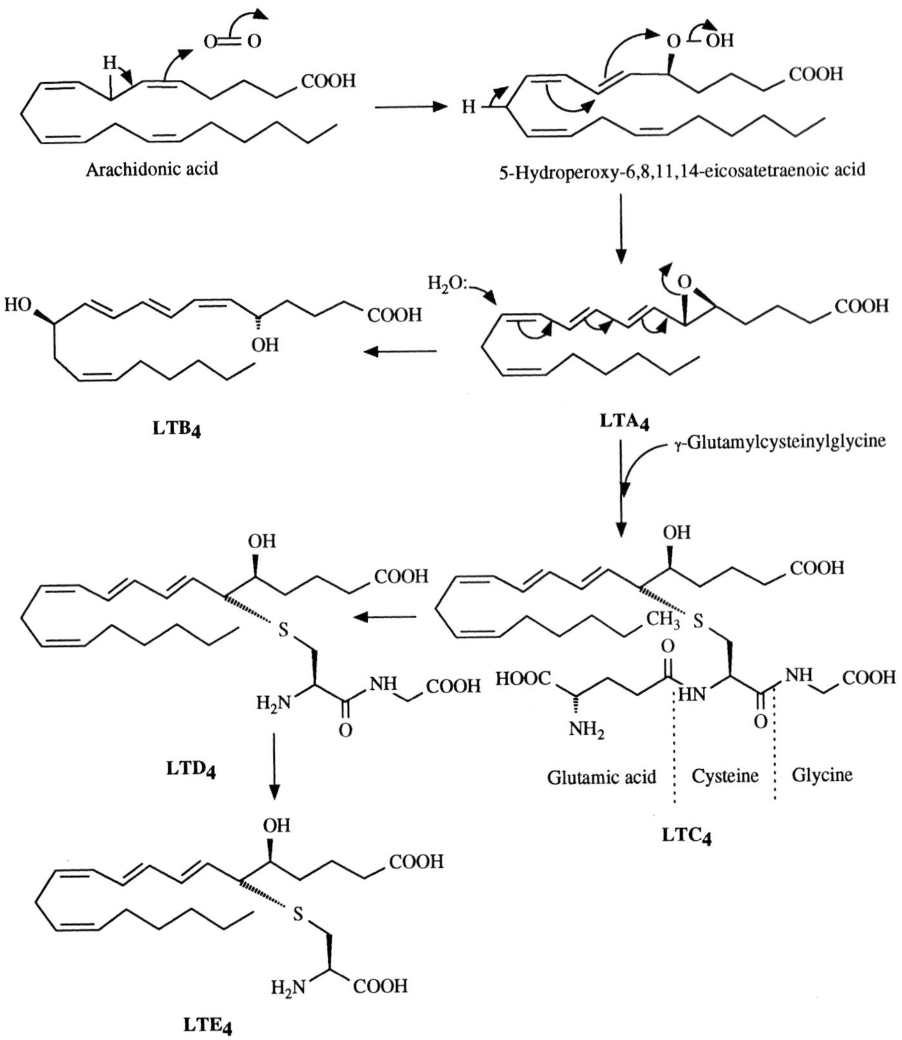

Fig. 61. Biosynthesis of the leukotrienes

purified from many sources including man, rat, pig and potato tubers. The cDNA encoding the human enzyme has been cloned from human placenta and has been expressed in a baculovirus/insect cell system. The expressed enzyme was able to synthesize both 5-HPETE and LTA$_4$ and in immunoblotting analysis it was indistinguishable from *5-LO* isolated from human leukocytes. Calculated from the cDNA sequence the enzyme is composed of 673 amino acid residues with a molecular mass of 78 kDa, a figure that corresponds well with the molecular mass range of 75–80 kDa determined for the enzymes isolated from various sources. *5-LO* is a soluble enzyme which in the resting cell is located in the cytosol. In response to agents which stimulate synthesis of leukotrienes and in the presence of Ca^{2+}, the enzyme translocates from the cytosol to a membrane compartment where it interacts with a membrane protein called *FLAP* (*Five-Lipoxygenase*

Activating Protein). FLAP specifically binds arachidonic acid and transfers this compound to *5-LO* which then performs the synthesis of LTA$_4$. FLAP has been isolated from rat and human leukocytes and a cDNA encoding the protein has been isolated, sequenced and expressed in the same virus/insect system previously used for expression of the *5-LO* cDNA. The nucleotide sequence of the gene of the rat cDNA corresponds to a protein composed of 161 amino acid residues and with a molecular mass of 18 kDa. The gene of the human cDNA is similar to that of the rat cDNA.

Hydrolysis of LTA$_4$, catalysed by *leukotriene-A$_4$ hydrolase* (EC3.3.2.6, also known as LTA$_4$-hydrolase), yields leukotriene B$_4$ (LTB$_4$). LTA$_4$-hydrolase is a cytosolic enzyme which was first isolated from human and rat leukocytes and was determined to have a molecular mass of 68–70 kDa. It has subsequently been found in many other tissues from several species. cDNA clones have been isolated from human lung, placenta and spleen. The nucleotide sequence of the gene of the human lung cDNA corresponds to a sequence of 610 amino acid residues of a protein with a molecular mass of 69 140 Da. The nucleotide sequence of the ORF of the spleen cDNA also corresponds to the amino acid sequence of a protein composed of 610 amino acid residues but the calculated molecular mass was slightly different (69 153 Da). The cDNA clone has been expressed in *Escherichia coli* yielding a protein with full enzyme activity and structural fidelity. LTA$_4$-hydrolase contains one mole of zinc per mol of enzyme and the zinc is essential for the activity of the enzyme.

In another reaction, catalysed by *leukotriene-C$_4$ synthase* (EC 4.4.1.20, also known as LTC$_4$-synthase), the epoxide of LTA$_4$ is attacked by the nucleophilic sulphur atom of glutathione (γ-glutamylcysteinylglycine), forming leukotriene C$_4$ (LTC$_4$). LTC$_4$-synthase is a membrane bound protein consisting of two equal subunits with the molecular mass 18 kDa. It is a very labile enzyme which has been isolated from the human monocytic leukaemia cell line THP-1. A cDNA clone has been isolated from the same cell line, which gene encodes a 150 amino acid protein with the molecular mass 16 568 Da. The deduced amino acid sequence indicates that the protein is composed predominantly of hydrophobic amino acids. The active enzyme has been expressed in bacterial, insect and mammalian cells. LTC$_4$-synthase is distinct from other known glutathione-*S*-transferases by nucleotide and deduced amino acid sequences.

Loss of glutamic acid catalysed by *γ-glutamyltransferase* (EC 2.3.2.2) yields leukotriene D$_4$ (LTD$_4$) which can lose the glycine residue forming leukotriene E$_4$ (LTE$_4$) in a reaction catalysed by *membrane alanyl aminopeptidase* (EC 3.4.11.2, also known as cysteinylglycine dipeptidase).

Medical importance of leukotrienes

The leukotrienes are involved in acute allergic manifestations, particularly asthma. The *Slow Reacting Substance of Anaphylaxis* (SRSA) has been identified as a mixture of LTC_4, LTD_4 and LTE_4. The leukotrienes are also involved in inflammatory processes connected with rheumatism and are of importance in heart infarct. LTB_4 facilitates migration of leukocytes in inflammation and seems also to have a role in the pathology of psoriasis. Because of the important role that these mediators may play in the pathophysiology of asthma and inflammatory processes, compounds that inhibit their production or action (*5-LO*-inhibitors and FLAP antagonists) are of clinical interest.

Lipstatin

Lipstatin is isolated from the Actinobacterium *Streptomyces toxytricini* and is composed of an unsaturated C_{13} straight carbon chain joined to a straight C_6 chain by a 4-membered β-lactone ring. The structure also comprises an *N*-formylleucine ester. Lipstatin is a potent inhibitor of pancreatic lipases and the tetrahydroderivative **Orlistat** (Xenical®, Alli®) is designed as a drug to treat obesity.

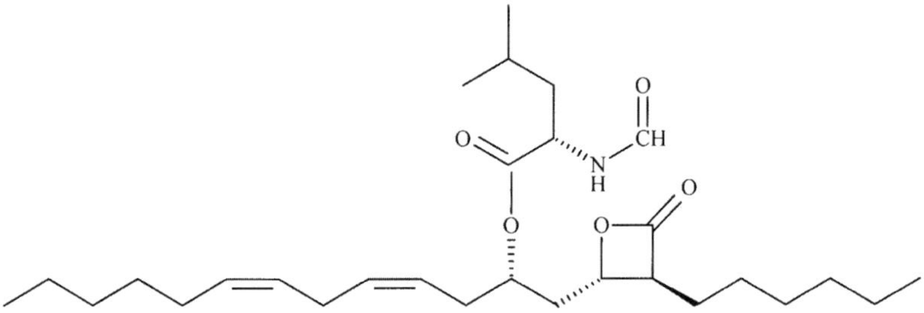

Lipstatin

Lipstatin is proposed to be biosynthesized by a Claisen condensation of the C_{14} carboxylic acid 3-hydroxy-tetradeca-5,8-dienoic acid or its thioester with octanoyl-CoA, followed by esterification with leucine and amidation of the leucine NH_2-group with formic acid. The C_{14} acid is supposed to have been formed by incomplete β-oxidation of the corresponding C_{14} unsaturated fatty acid.

Orlistat inhibits pancreatic lipase whereby triglycerides from the diet are prevented from being hydrolysed into absorbable free fatty acids and are excreted undigested.

Polyketides

The structural variation among this group of substances is very large. Reduction and cyclization of a polyketide chain of moderate length can result in aromatic compounds containing one, two or three rings like 6-methyl-salicylic acid, napthoquinones and anthraquinones. This is a second pathway to aromatic compounds in addi-

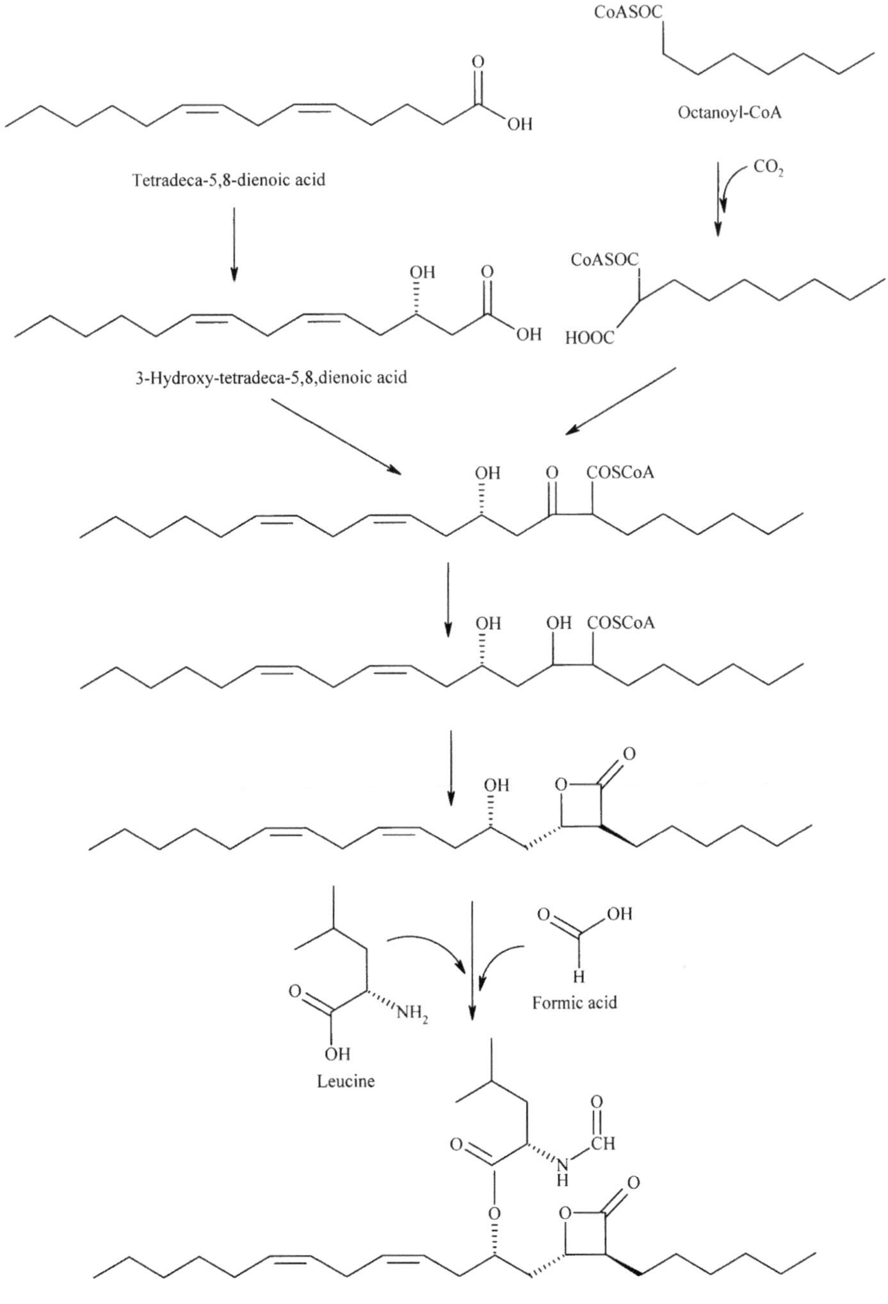

Fig. 62. Proposed biosynthesis of lipstatin

tion to the shikimic acid pathway already discussed (page 194). Ring closure can proceed through aldol- or Claisen reactions and the resulting ring be made aromatic by enolization. The polyketide chain remains attached to the enzyme during these reactions.

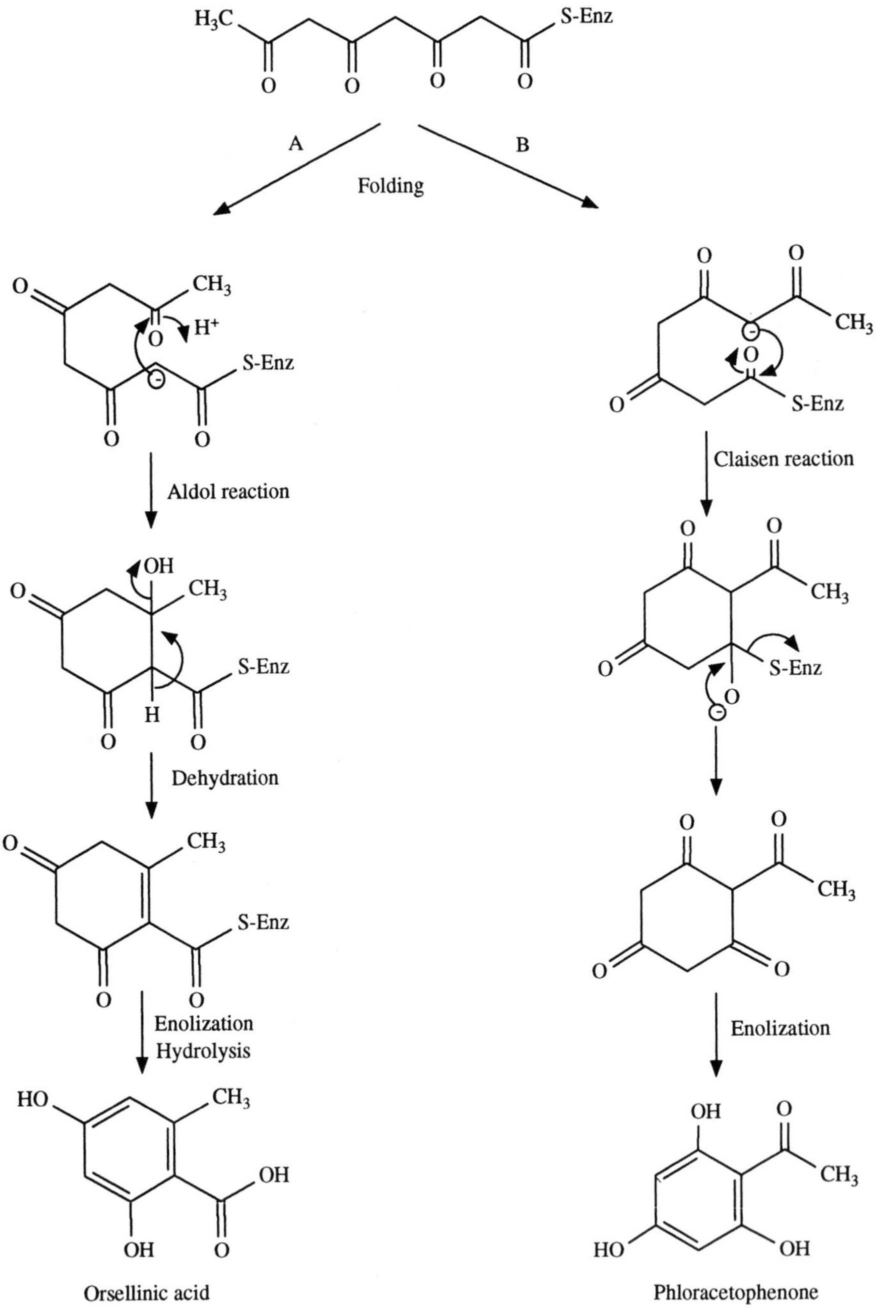

Fig. 63. Formation of aromatic rings from a poly-β-keto chain

As every second carbon in the polyketide is oxygenated, the aromatic product will usually have a pattern of *meta*-hydroxy substitution in contrast to aromatic compounds derived from the shikimic acid pathway which are characterized as having hydroxy groups in *ortho* positions.

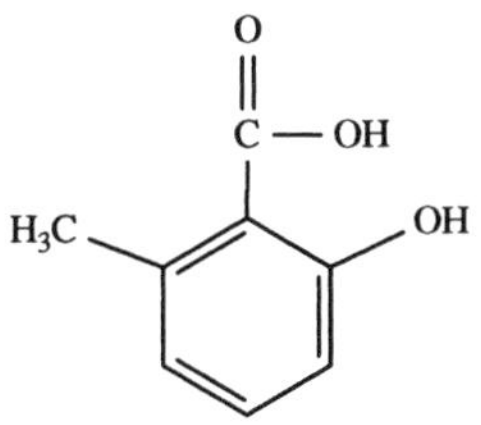

6-Methyl-salicylic acid (a metabolite from *Penicillium patulum*)

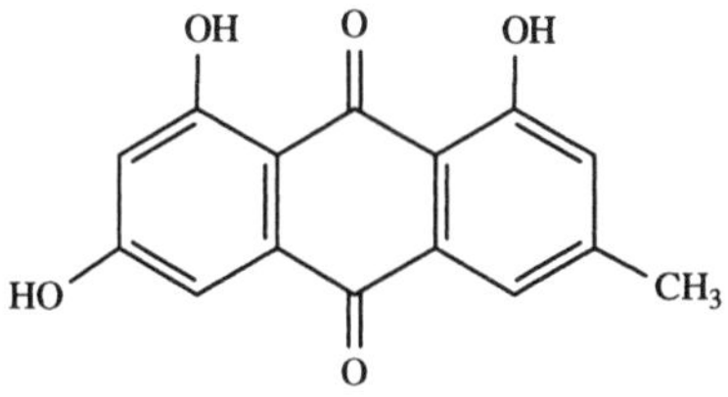

Frangulaemodin (an anthraquinone)

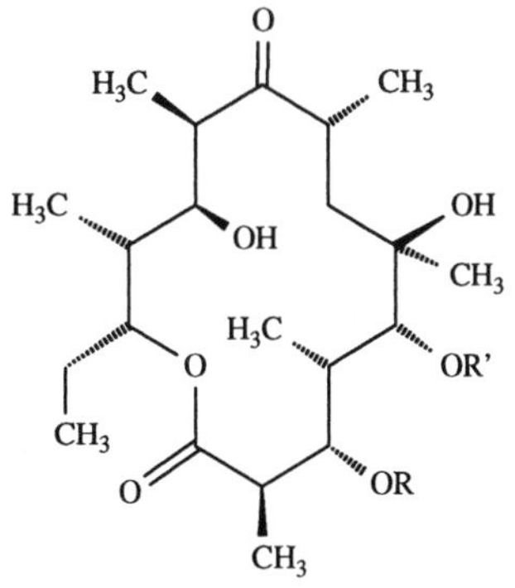

Erythromycin B (a macrolide antibiotic). R = cladinose. R' = desosamine

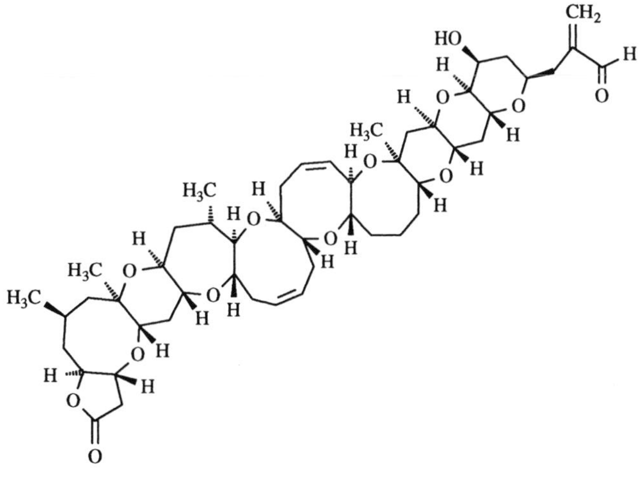

Brevetoxin A (a toxic polyether from the dinoflagellate *Gymnodinium breve*)

Fig. 64. Structures of some polyketides

Longer polyketide chains can give rise to large rings like the macrolides which will be discussed (page 271). Cyclization can also be more extensive resulting in big molecules with many fused 6- or 7-membered rings as in brevetoxin.

Also other acids than acetate can contribute two of their carbons to the growing chain. The remaining carbons will then appear as substituents. Methyl or ethyl substituents usually derive from propionate and butyrate, respectively, but also other rarer groups can occur if a more unusual acid is involved in the condensation process. Methyl groups can also derive from methionine and are introduced by C-alkylation mediated by *S*-adenosyl methionine (page 497). Also the keto groups may be modified. Typically they are reduced to hydroxyls but they may also be completely removed. The side-chain variations, the varied fate of each of the keto groups, the possibility of chirality at one or more carbon atoms and the variations of the chain length are factors that account for the huge diversity in the primary structure of the molecules.

The biosynthesis of polyketides involves stepwise assembly of the carbon chain backbones of polyketides from simple two-, three- and four-carbon building blocks, such as acetyl-CoA, propionyl-CoA and butyryl-CoA and their activated derivatives malonyl-, methylmalonyl- and ethylmalonyl-CoA. This procedure is catalysed by *polyketide synthases (PKSs)* which resemble the fatty acid synthases. In fatty acid biosynthesis each succesive chain elongation step is followed by a fixed sequence of ketoreduction, dehydration and enoyl reduction. In polyketide biosynthesis, however, the individual chain elongation intermediates can undergo all, some, or none of these functional group modifications. The chemical complexity of the products is the result of this, in combination with the use of different starter and chain elongation units. The PKSs need to be more highly programmed than the FASs involving a correct choice of stereochemistry, e.g. *R* or *S* configurations of methyl or hydroxyl groups and *E* or *Z* configurations at double bonds.

There are several types of PKSs. *Type I modular PKSs* are giant megasynthases comprising several, tightly joined, proteins containing a number of *modules*, each which extends the polyketide chain with two carbon atoms. Each module has enzymatically active *domains*, which perform specific operations in the chain-elongating process, for example acyltransferases, acyl carrier proteins and ketoacyl synthases (see erythromycin, page 272). *Type I iterative PKSs* are proteins consisting of a single module of catalytic domains, linked together in essentially the same manner as in a typical module of a type I modular PKS, but where the single set of domains carries out successive chain extensions and processes the β-keto group to different extents in different cycles (see lovastatin, page 298). *Type II PKSs* consist of more loosely bound, dissociable, enzymes. *Type III PKSs* are structurally the simplest PKSs although from a mechanistic point of view they are very complex. These enzymes are comparatively small homodimers with a monomer molecular mass of 42–45 kDa. They function as iterative enzymes that contain two independent active sites, each of which catalyses single or multiple condensation reactions yielding polyketides of different lengths. There are no sequences corresponding to the domains in the other PKSs, but conserved amino residues have been identified and their function in the biosynthetic process elucidated (see chalcone synthase, page 330).

Between 5 000 and 10 000 polyketides are known and about 1 % of these are at present shown to possess biological activity. Pharmaceutically important polyketide drugs include antibiotics, cancer chemotherapeutics, immunosuppressants, cholesterol-lowering agents and fungicides. Their yearly sales worldwide exceed 15 billion US dollars (2000). Thus an enormous value is attached by the pharmaceutical industry to polyketide compounds. More and more new polyketides are created through developing advanced techniques of combinatorial biosynthesis (see page 277) and new directions can be observed, such as attempts to link the biosyntheses of modular polyketides and non-ribosomal peptides (page 531) and the attachment of important deoxy-sugar moieties to various polyketide-derived aglycones.

The following polyketides are of interest in medicine:

A. *Polyketides derived from acetate or propionate*
      Macrolides
      Polyene macrolide antibiotics
      Epothilones
      Griseofulvin
      Aflatoxins
      Mevastatin and lovastatin
      Leptospermone and Nitisinone
      Mupirocin
      Tetracyclines
      Anthracyclines
      Enediyne antibiotics
      Anthraquinones
      Hypericin

B. *Polyketides of mixed biogenetic origin*
      Flavonoids
      Kava pyrones
      Flavonolignans
      Mycophenolic acid
      Ansamycin antibiotics
      Rapamycin
      Tacrolimus (FK 506)
      Rotenonoids
      Khellin

Polyketides derived from acetate or propionate

Macrolides

Macrolides are antibiotic compounds produced by bacteria of the genus *Streptomyces*, the main structural feature of which is a large lactone ring of acetate origin containing 12, 14 or 16 atoms. The structure also includes 1–3 deoxysugar molecules attached to the ring. Macrolides are biosynthesized by type I modular polyketide synthases, i.e. they are organized in a complex of multifunctional proteins (page 270). Depending on the starter unit (acetate, propionate, butyrate, etc.) and the types of extension units (malonate, methylmalonate, ethylmalonate) the lactone ring is supplied with various substituents (methyl, ethyl, etc.). Oxygen heterocycles can also be found fused to the macrolide. These are the result of biosynthesis of a much longer polyketide chain than that actually required for formation of the lactone ring.

Erythromycins

Saccharopolyspora erythraea (formerly *Streptomyces erythreus*) produces a mixture of macrolides of which erythromycin A, B and C are most important. The lactone ring of the erythromycins contains 14 atoms (13 carbon and one oxygen). Erythromycin A, which is used clinically as a broad-spectrum antibiotic, has the following structure:

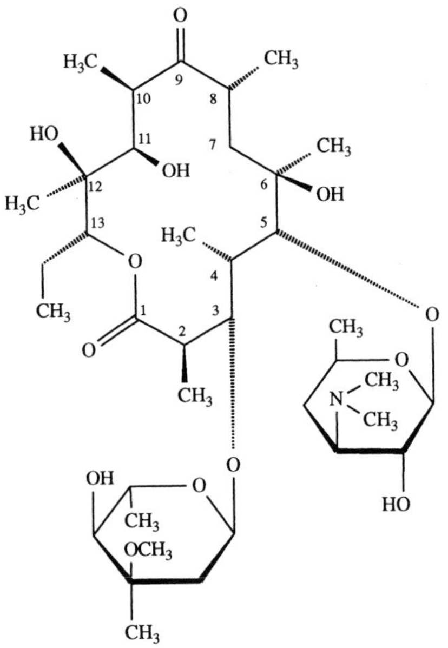

The glycosidically bound sugar residues are L-cladinose (at C-3) and D-desosamine (at C-5).

Erythromycin is used against Gram-positive bacteria and is substituted for penicillin when the patient is allergic to the latter drug. It is also used for the treatment of infections by *Mycoplasma*. The side effects are few, the most common ones being gastrointestinal symptoms. These are dose-dependent and administration of high doses can thus be a problem.

Macrolide antibiotics act by inhibiting protein synthesis, binding to the 50 S subunit of bacterial ribosomes. Erythromycin is known to inhibit the elongation-factor-G-dependent release of deacylated tRNA from the P site of the ribosome.

Derivatives of erythromycin A have been developed with improved acid stability and enhanced activity against bacteria resistant or intermediately sensitive to erythromycin A. **Clarithromycin** is the 6-*O*-methyl derivative with enhanced acid stability and better efficacy than erythromycin A against *Haemophilus influenzae*. In **roxithromycin** the group $CH_3OCH_2CH_2OCH_2O$-N= is substituted for the carbonyl oxygen at C-9. Roxithromycin is acid stable and has excellent bioavailability after oral administration thus compensating for a lower antimicrobial activity than erythromycin A. **Azithromycin** is a semi-synthetic analogue of erythromycin A with an expanded ring system. It is better adsorbed than erythromycin from the gut and has a half-life of around 70 hours, which permits once-daily dosing. Azithromycin is more effective than erythromycin A against a number of Gram-negative organisms, including *Haemophilus influenzae*.

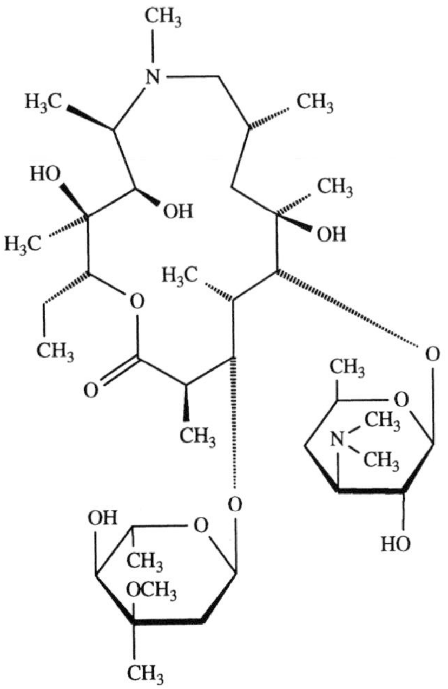

Azithromycin

Telithromycin is a member of a new class of semisynthetic macrolide derivatives – ketolides – which are characterized by a keto group at position 3 of the macrolactone ring. Telithromycin is obtained by removing the 3-L-cladinose from erythromycin A and oxidation of the resulting 3-hydroxyl group. The 3-keto group of the ketolides enables the antibiotics to bind to their target without tripping the inducible resistance to macrolide-lincosamide-strep-togramin$_B$ (MLS$_B$) drugs that many groups of pathogens exhibit. The methoxy group at C-6 confers increased acid stability to the molecule and the C-11,12 carbamate side-chain improves binding of the drug to its target, the 50 S ribosomal subunit. Telithromycin has proved particularly effective for the treatment of infections caused by penicillin- and macrolide-resistant strains of *Streptococcus pneumoniae*.

Azitromycin and telithromycin also have anti-inflammatory activities and have been prescribed for alleviating symptoms of pulmonary pneumonia. The mechanism for this activity is supposed to be the ability of these macrolides to suppress the overabundance of neutrophils in the lungs.

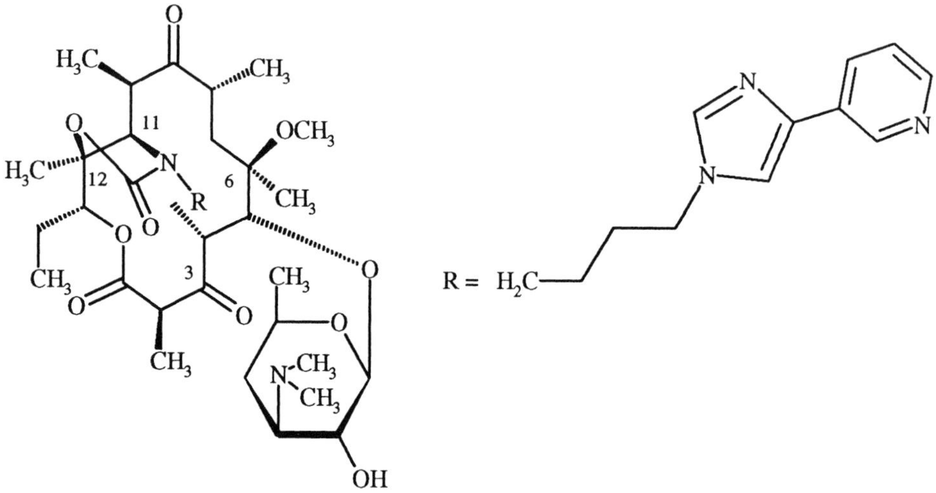

Telithromycin

Biosynthesis of erythromycins

The enzyme *erythronolide synthase* (EC 2.3.1.94, also known as 6-deoxyerythronolide B synthase or DEBS) is a type I modular PKS which catalyses the biosynthesis of 6-deoxyerythronolide B which is the parent aglycone of the erythromycins. The organization of this enzyme complex has been elucidated by application of the techniques of molecular biology. These investigations have established DEBS as the archetypal modular polyketide synthase. It provides the structural and mechanistic case against which other synthases may be compared both *in vitro* and *in vivo*.

DEBS is composed of three large (ca. 3 500 amino acids) multidomain proteins, designated DEBS 1, DEBS 2 and DEBS 3. Each of these three units is coded for by a separate gene denoted *eryAI*, *eryAII* and *eryAIII*, respectively (Fig. 65). Each of the proteins contains two

2(S)-methylmalonyl-CoA

distinct functional units, called *modules*. The modules are arranged in essentially the same order as the actual sequence of condensation and functional group modification reactions. Each module includes a set of domains (AT, KS, ACP, etc. see page 270), containing the enzyme activities which are required for *one* of the six cycles of chain elongation and associated β-ketoreduction reactions that occur during the biosynthesis of 6-deoxyerythronolide B. Each domain is presumed to form a localized globular structure with a catalytic role determined by conserved amino acid residues at its specific active site (*motif*) usually comprising four to six amino acids. In the modules the domains are separated by linker regions comprising up to 100 amino acid residues. These linkers are frequently rich in alanine, proline and charged residues and the particular amino acid composition may allow them to serve either as rigid connectors or to provide flexibility to allow segmental motion of the multienzymes. The KR domains are preceded by long sections of primary amino acid sequence which have not yet been associated with any particular catalytic function. The predicted domain organization of the DEBS proteins is shown in Fig. 66.

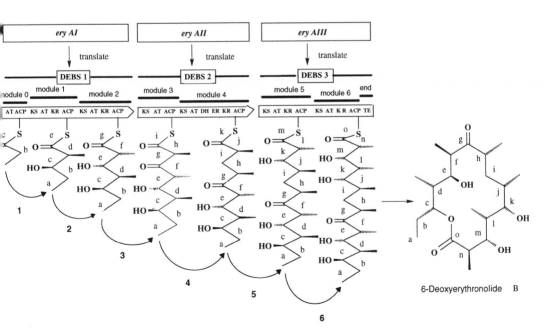

Fig. 65. Genetic model for the 6-deoxyerythronolide B synthase and biosynthesis of 6-deoxyerythronolide B. AT = acyltransferase. ACP = acyl carrier protein. KS = β-ketoacyl synthase. KR = β-ketoacyl reductase. DH = dehydratase. ER = enoyl reductase. TE = thioesterase. The arrows indicate decarboxylative condensation with 2(S)-methylmalonyl-CoA. The carbon atoms of the growing chain are labelled in lower case in alphabetical order. The activity of each module elongates the chain with two carbon atoms. The starter contains carbons a, b and c. Carbons d and e are added by module 1, f and g by module 2 etc.

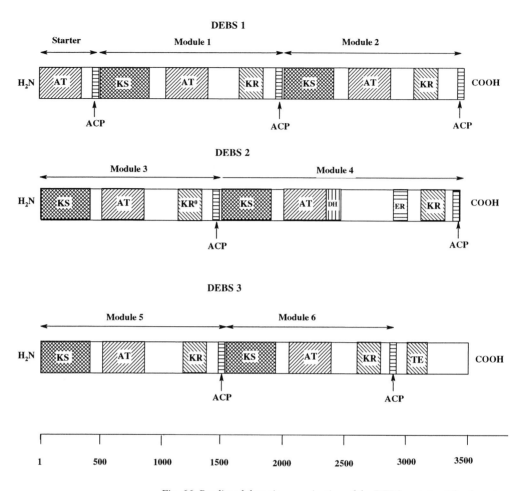

Fig. 66. Predicted domain organization of the DEBS proteins. The domains are represented by boxes whose lengths are in proportion to the sizes of the domains. In module 3 there is a sequence (KR⁰) corresponding to a KR domain which, however, is inactive. The ruler shows the amino acid residue number within the primary structure of the constituents. The linker regions are also given in proportion

The mechanisms of action of DEBS resembles FAS I (page 246) in many ways but there are also important differences. In FAS I the domains for the enzyme activities (AT, KS, ACP, etc.) are used over and over in a cyclic process whereas DEBS functions as an assembly line where each domain has its own special task. The first two domains (AT and ACP) in DEBS 1 (Fig. 65) act as a starter module (module 0). The starter is propionyl-CoA from which the propionyl group is transferred by *propionyl-CoA acyltransferase (AT of module 0)* to the 4'-phosphopantetheine of the *acyl carrier protein (ACP)* in module 0 (compare the mechanism for fatty acid biosynthesis in Fig. 54, page 245). At the same time the *AT* in module 1 transfers a molecule of 2(S)-methylmalonyl-CoA (formed by carboxylation of propionyl-CoA) to the 4'-phosphopantetheine of the

second *ACP*. The propionyl residue (residing on *ACP* in module (0)
is transferred to the active cysteine of β-*ketoacyl synthase* (*KS* in
module 1) and decarboxylative condensation of the propionyl and
methylmalonyl residues (see reaction 3 in the biosynthesis of fatty
acids by FAS I, page 243 and Fig. 54) takes place with formation of
a 2-methyl-3-keto-pentanoyl-ACP thioester. β-*Ketoacyl reductase*
(*KR* in module 1) reduces the 3-carbonyl to a hydroxy group. The
resulting (2*S*,3*R*)-2-methyl-3-hydroxy-pentanoyl group is transferred by the *ACP* on which it resides to *KS* in module 2. A new molecule of methylmalonyl-CoA is transferred by the *AT* of module 2
to the *ACP* of that module, and decarboxylative condensation
between the (2*S*,3*R*)-2-methyl-3-hydroxy-pentanoyl group and the
methylmalonyl residue yields a 7-carbon chain, the carbonyl of
which is reduced by *KR* of module 2 and transferred to the *KS* of
module 3, making it ready to be condensed with a methylmalonyl
residue which in the meantime has been transferred from *AT* of
module 3 to the *ACP* of module 3. The β-ketoacyl reductase in module 3 is inactive and the carbonyl (carbon atom g in Fig. 65) in the
resulting 9-carbon chain is therefore not reduced. Module 4 contains *dehydratase* (*DH*) and *enoyl reductase* (*ER*) and the new carbonyl (carbon atom i, Fig. 65) of the 11-carbon chain, produced by
the normal reactions on that module, is therefore completely
reduced in the same way as in the biosynthesis of fatty acids (page
243). The carbon chain is further extended by modules 5 and 6
resulting in a 15-carbon chain which is released from the *ACP* of
module 6 by a *thioesterase* (*TE*) which also forms the lactone function in 6-deoxy-erythronolide B. (2*S*)-methylmalonyl-CoA is the
exclusive substrate for polyketide chain elongation. Decarboxylative condensation normally occurs with inversion of configuration.
The extender units that show a net retention of configuration (such
as those incorporated by modules 1, 3 and 4) must therefore undergo epimerization either prior to or immediately after the condensation reaction.

The factors responsible for fidelity in the programming of this
polyketide biosynthesis are largely unknown. A hierarchy of biochemical choices must be controlled such as substrate, reaction
sequence, stereochemistry and chain length.

Cloning of the genes coding for DEBS has allowed expression of
the complete erythronolide synthase and establishment of cell-free
systems for synthesis of erythronolide. These systems can accept
other starters than propionyl-CoA and also other intermediates of
chain elongation. This opens up exciting possibilities for the rational production of entirely new polyketide metabolites. Another fascinating possibility to use this enzyme system for the synthesis of
non-natural polyketides is manipulation of the genes encoding the
polyketide synthase. That this is possible was shown in an experiment where the gene *eryAI* (encoding the DEBS 1 protein) was
fused to the part of the *eryAIII* gene encoding the *TE* domain of
DEBS 3. The resulting hybrid gene was introduced into *Strepto-*

myces coelicolor resulting in the production of substantially enhanced amounts of the expected triketide lactone. The protein (DEBS 1-TE) decoded by this synthetic gene could also be expressed, yielding a protein which catalysed formation of the lactone in a cell-free system.

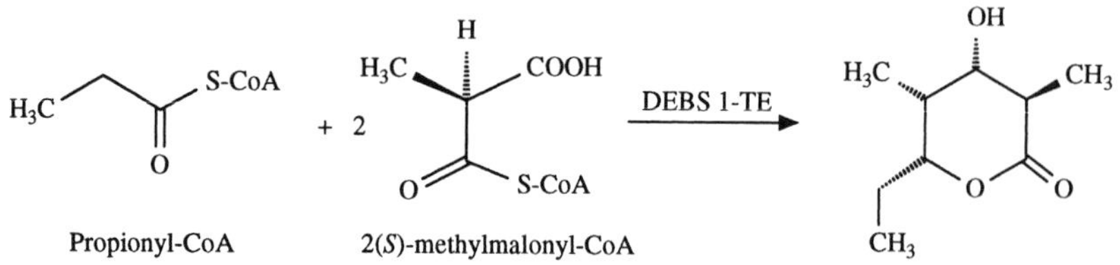

Fig. 67. Biosynthesis of (2R, 4S,5R)-2,4-dimethyl-3,5-dihydroxy-n-heptanoic acid δ-lactone from propionyl-CoA and 2(S)-methylmalonyl-CoA, catalysed by DEBS 1-TE

Experiments have also been performed in which the *TE* domain has been placed at the end of modules 3 and 5. Also here the engineered proteins gave products consistent with truncation of chain extension at the expected stages. Domains and modules from other type 1 PKS systems (e.g. from avermectin) have also been incorporated into DEBS 1-TE resulting in the production of the expected compounds by the hybrid proteins. Intensive research in this area is in progress and new drugs produced by these techniques are expected to be developed within the near future.

The continuation of the biosynthesis of the erythromycins is illustrated in Fig. 68. Oxidation at C-6 converts 6-deoxyerythronolide B to erythronolide B in a reaction catalysed by the enzyme EryF which is a cytochrome P450 hydroxylase. EryF has been overexpressed in *Escherichia coli* and the structure has been determined by X-ray studies. In the next step the sugar L-mycarose is attached to the C-3 hydroxyl group of erythronolide B, yielding 3-*O*-mycarosylerythronolide B, to which D-desosamine is added at the C-5 hydroxyl. The product – erythromycin D – is then oxidized at C-12 yielding erythromycin C, which by *O*-methylation of the 3" hydroxyl of mycarose is converted to erythromycin A. Erythromycin B is produced from erythromycin D by *O*-methylation of mycarose. Erythromycin A can also be synthesized from erythromycin D by hydroxylation at C-12 but this route is considered to be of minor importance.

Avermectins

Avermectins are a series of 16-membered macrocyclic lactones produced by *Streptomyces avermectilis*. These compounds lack antibacterial and antifungal activities but have potent anthelmintic, insecticidal and acaricidal properties. Their toxicity to animals and humans is low. Eight avermectins are known which differ in the

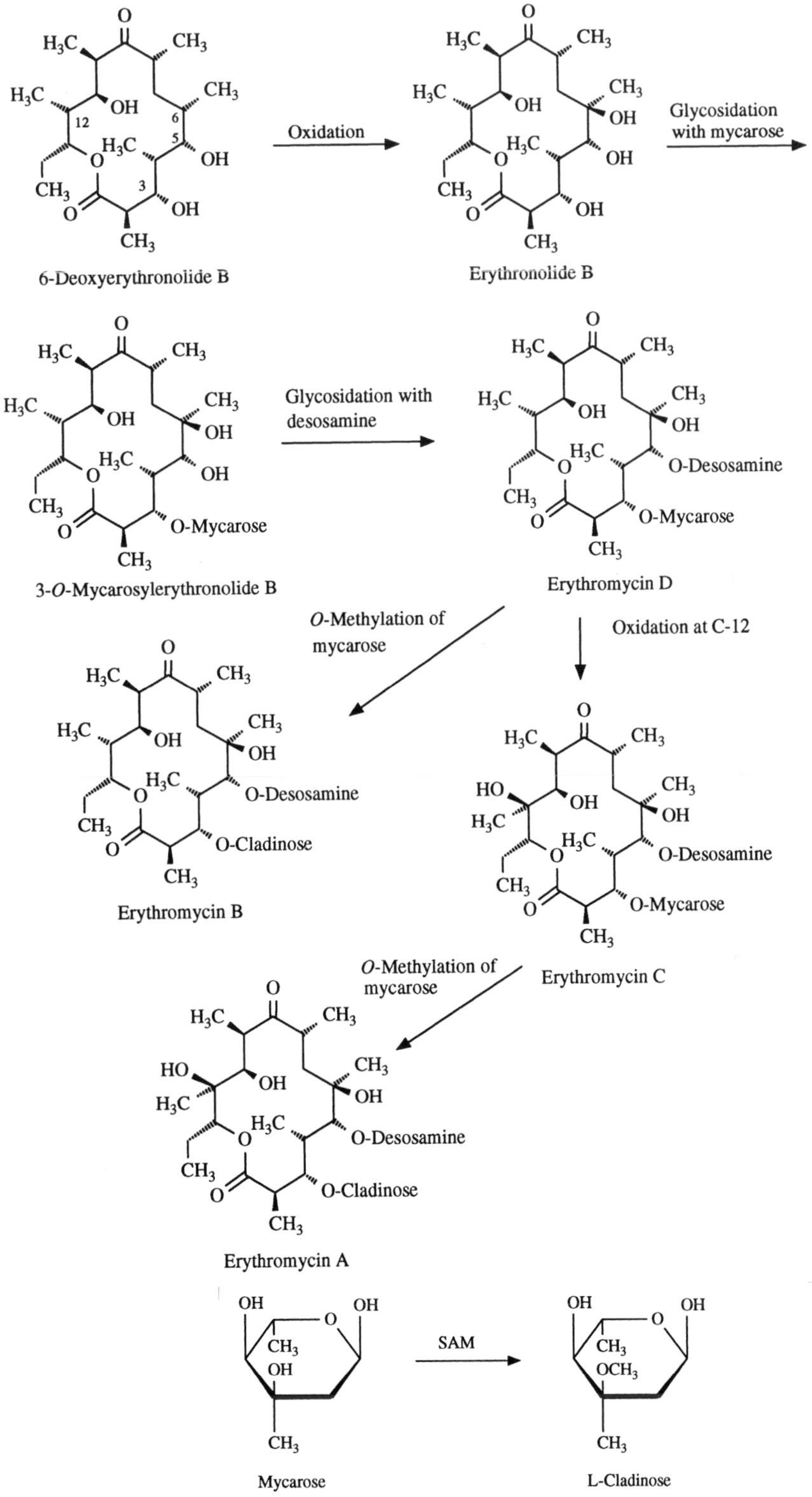

Fig. 68. Biosynthesis of erythromycin A from 6-deoxyerythronolide B

substitution at carbons 5 and 26 as well as bonds and substituents involving carbons 22 and 23 (see structure below). A disaccharide composed of two L-oleandrose units is attached at C-13.

Biosynthesis of avermectins

The biosynthesis of the avermectins is more complicated than that of erythromycin. The starter is 2-methylbutyryl-CoA (avermectins of the *a* series) or 2-methylpropionyl-CoA (avermectins of the *b* series). These starters derive from L-isoleucine and L-valine respectively. In the structures below they are recognized in the branched side-chain at C-25.

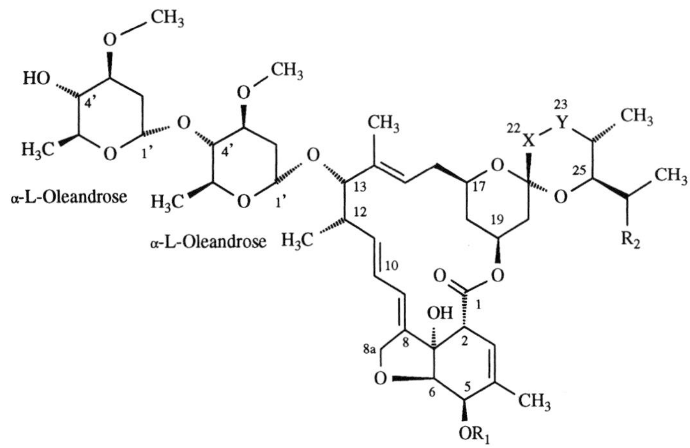

	R$_1$	R$_2$	X - Y
Avermectin A1a	CH$_3$	C$_2$H$_5$	CH = CH
A1b	CH$_3$	CH$_3$	CH = CH
A2a	CH$_3$	C$_2$H$_5$	CH$_2$-CH(OH)
A2b	CH$_3$	CH$_3$	CH$_2$-CH(OH)
B1a	H	C$_2$H$_5$	CH = CH
B1b	H	CH$_3$	CH = CH
B2a	H	C$_2$H$_5$	CH$_2$-CH(OH)
B2b	H	CH$_3$	CH$_2$-CH(OH)

An 82 kbp biosynthetic gene cluster containing 18 ORFs and genes has been analysed. Two pairs of genes – *aveA1/aveA2* and *aveA3/avA4* – are convergently transcribed yielding four proteins denoted as AVES1–4. This complex is also called *6,8a-seco-6,8a-deoxy-5-oxo-avermectin synthase*. AVES1 contains a starter module (0) and two synthetic modules. AVES2 has four modules and AVES3 and AVES4 contain three modules each. These are all large proteins including 3953, 6239, 5532 and 4881 amino acid residues respectively. All domains of the modules have been identified and table 4 gives an overview of their organization.

Table 4. Modules and domains of AVES 1–4

Domain	AVES1 Module No.			AVES2 Module No.				AVES3 Module No.			AVES4 Module No.			
	0	1	2	3	4	5	6	7	8	9	10	11	12	
KS	-	+	+	+	+	+	+	+	+	+	+	+	+	
AT	N	P	A	A	A	A	P	P	A	P	A	P	A	
DH	-	-	½	-	-	-	+	(+)	+	+	-	-	+	
ER	-	,	-	-	-	-	-	-	-	-	-	-	-	
KR	-	+	+	-	+	+	+	+	+	+	(+)	-	+	
ACP	+	+	+	+	+	+	+	+	+	+	+	+	+	TE

Abbreviations: *KS* = β-ketoacyl synthase. *AT* = acyltransferase. *A* = acetate specific acyltransferase. *P* = propionate specific acyltransferase. *N* = 2-methylpropionate or 2-methylbutanoate-specific acyltransferase. *DH* = dehydratase. *ER* = enoyl reductase. *KR* = β-ketoacyl reductase. *ACP* = acyl carrier protein. *TE* = thioesterase. A + sign indicates a functioning domain. A - sign indicates absence of the domain. (+) Indicates an inactive domain with apparently intact sequence. ½ = "partial activity" (see below).

The starter module (0) loads 2-methylpropionate (b-series) or 2-methylbutanoate (a-series) from 2-methylpropionyl-CoA or 2-methylbutanoyl-CoA, respectively, on to the N domain from where it is moved to the ACP domain which transfers it to the KS domain in module 1. Methylmalonate is loaded on to the P domain in module 1 and the synthesis then proceeds through modules 1–6, providing chain elongation in the order P-AAAA-P (P = propionyl, A = acetyl). Then follows formation of acetals involving C-17, C-21 and C-25 (numbers refer to the structure of the final avermectin molecule) as shown in Fig. 69 for the lb series. "Partial activity" of the DH domain in module 2 results in a mixture of compounds with a double bond between C-22 and C-23 (1-series) or the structure $-CH_2-CH(OH)-$ (2-series) at these positions. There is no satisfactory explanation for how a domain can be only partially active. The *DH* of module 2 contains one mismatched amino acid, a serine replacing proline in the motif histidine-$(X)_3$-glycine-$(X)_4$-proline (X = any amino acid). However, introduction of proline at this site did not restore full DH2 activity. The "partial activity" must therefore be attributed to some other regions, or even to that downstream activities act on its substrate prematurely.

Following acetal formation, the synthesis proceeds, by the action of modules 7–12, with addition of P, A, P, A, P and A in the given order. The KR domains in modules 7 and 10 have a sequence which appears fully consistent with activity, but to account for the structure of the polyketide they must be presumed to be inactive. Release of the polyketide formed is accompanied by closure of the macrolide ring and formation of a ring involving C-2 and C-7. The product is 6,8a-seco-6,8a-deoxy-5-oxo-avermectin 1b. Aglucones for avermectins in the series 1a, 2a and 2b are formed in a similar way.

Post-PKS reactions involve C-5 of the aglucone, the carbonyl of which is reduced to a hydroxy group, forming the B-series of avermectins. Further methylation yields the A-series which has a methoxygroup attached to C-5. Oxidation of C-6 and C-8a followed by lactone formation results in a furan ring. Finally two molecules of L-oleandrose are attached to C-13. The genes for these post-PKS tailoring reactions have been identified.

For avermectin the possibility of genetic engineering has been demonstrated and further work is in progress which in the future might yield compounds with useful medical properties.

A mixture of about 85 % avermectin B1a and about 15 % avermectin B1b is termed **abamectin** and is used on agricultural crops

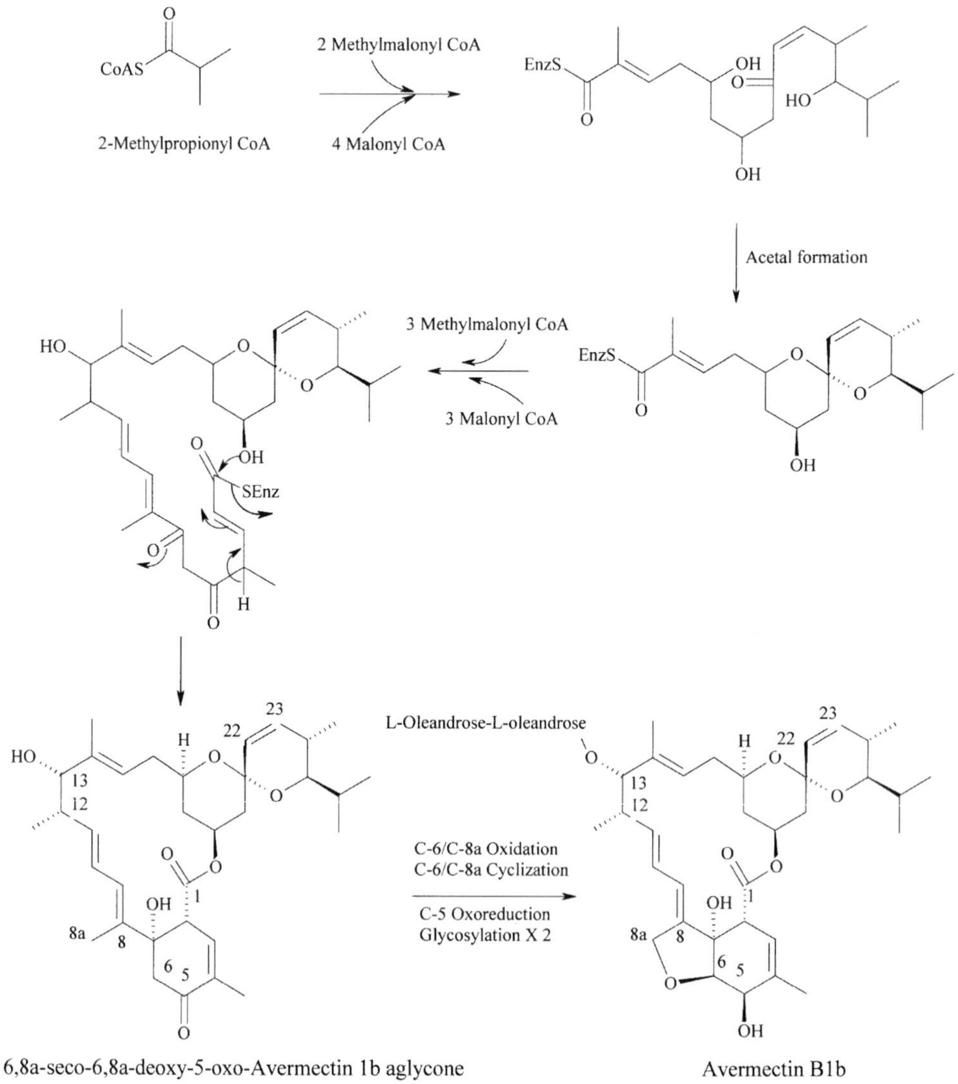

6,8a-seco-6,8a-deoxy-5-oxo-Avermectin 1b aglycone   Avermectin B1b

Fig. 69. Biosynthesis of avermectin B1b

to control mites and insects. **Ivermectin** is a 22,23-dihydro derivative of avermectin B1a which is used as an antiparasitic agent. It is
extremely efficient against nematode and arthropod parasites and is
used in veterinary practice for the health care of livestock and pets.
It is also used for treatment of river blindness which is caused by the
nematode *Onchocera volvulus* and which is the major cause of blindness in Africa.

Spiramycins

The spiramycins are macrolides containing a 16-membered lactone
ring which are produced by *Streptomyces ambofaciens*. One disaccharide composed of D-mycaminose and L-mycarose is attached to
C-5 and another unusual sugar – D-forosamine – forms a glycoside
bond at C-9. Three spiramycins are known, and differ from each
other in the substituents at C-3.

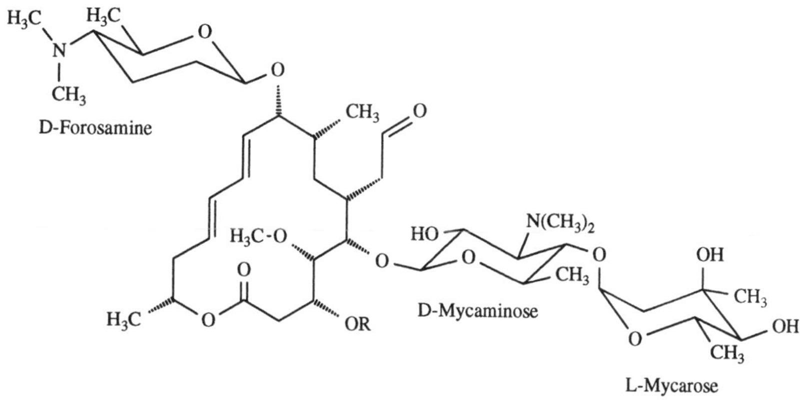

Spiramycin I	R = H
Spiramycin II	R = COCH$_3$
Spiramycin III	R = COCH$_2$CH$_3$

The spiramycins are biosynthesized by a type I PKS using acetate as
the starter and employing five malonyl-CoA, one methylmalonyl-
CoA and one ethylmalonyl-CoA as extension units. The latter forms
the ethyl side-chain which has been oxidized to an aldehyde.

A mixture of the three spiramycins, containing mainly spi-
ramycin I and minor amounts (10–15 %) of II and III, is used to treat
infections by *Staphylococcus* and *Neisseria* spp. in cases when other drugs, e.g. erythromycin, cannot be used.

Polyene macrolide antibiotics

This set of compounds is structurally related to the previous group.
Three to eight conjugated double bonds are present in a 20–44
membered ring. Biosynthesis is performed by large type I modular

polyketide synthases in a manner similar to that of erythromycin (page 274).

Nystatin Nystatin is produced by *Streptomyces noursei* and consists of a mixture of related compounds with nystatin A1 as the principal component. The compound is a ring composed of 38 atoms which contains two + four conjugated double bonds. An unusual type of amino sugar – D-mycosamine – is attached to the ring via a glycoside bond. The commercial product also contains nystatin A2 and A3, which have additional glycoside residues.

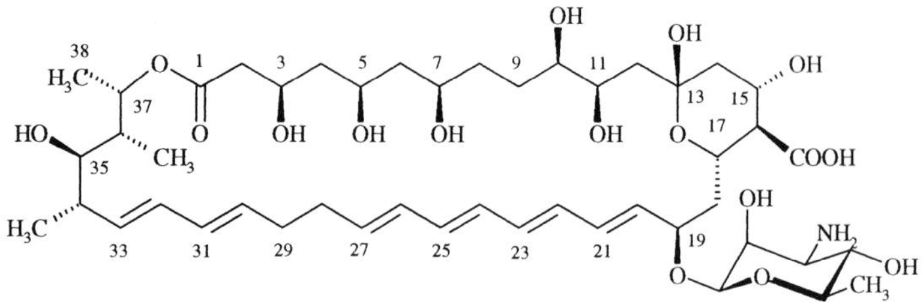

Nystatin A₁

Biosynthesis The elucidation of the biosynthesis of nystatin is a good example of how modern molecular biology techniques can be used to tackle a difficult biosynthetic problem. A DNA region of 123 580 bp from the genome of *Streptomyces noursei* has been amplified, sequenced and shown to be involved in nystatin biosynthesis. As presented in table 5, 22 genes have been putatively identified, encoding a type I modular PKS and enzymes for deoxysugar biosynthesis, modification, transport and regulation.

The polyketide chain which constitutes the backbone of nystatin is assembled by a giant PKS consisting of six closely connected type I modular PKSs designated as proteins NysA, NysB, NysC, NysI, NysJ and NysK. The amino acid sequences of these PKSs have been deduced from the DNA sequences in their parent genes, permitting assignment of modules and domains by homology comparison with amino acid sequences of other known type I PKSs. Domains involved in the biosynthesis of PKSs have been extensively studied and they could therefore be recognized from the presence of conserved amino acid sequence motifs at their active sites, in which certain amino acids are always present at fixed positions. *DH* domains, for example, have the conserved motif Histidine-(X_3)-Glycine-(X_4)-Proline, where X can be any amino acid. The predicted functional features of the PKSs are shown in Fig. 71.

The organization of the genes within the biosynthetic operon is illustrated in Fig. 70.

NysA most probably represents a loading module involved in the initiation of the nystatin aglycone biosynthesis. The presence of a module, specifically used for loading, is unusual. Generally loading

Table 5. Putative genes identified in the nystatin biosynthetic gene cluster of Streptomyces noursei

Gene	Protein	Number of amino acids	Function
nysA	NysA, type I PKS	1 366	Loading module
nysB	NysB, type I PKS	3 192	Modules 1 and 2
nysC	NysC, type I PKS	11 096	Modules 3–8
nysDI	Glycosyltransferase	506	Attachment of mycosamine
nysDII	Aminotransferase	352	Mycosamine biosynthesis
nysDIII	GDP-mannose-4,6-dehydratase	344	Mycosamine biosynthesis
nysE	Thioesterase	251	Release of polyketide chain from PKS
nysF	4´-Phosphopantetheine transferase	245	Post-translational PKS modification
nysG	ABC transporter	605	Efflux of nystatin
nysH	ABC transporter	584	Efflux of nystatin
nysI	NysI, type I PKS	9 477	Modules 9-14
nysJ	NysJ, type I PKS	5 435	Modules 15–17
nysK	NysK, type I PKS	2 066	Module 18 + TE
nysL	P450 monooxygenase	394	Hydroxylation at C-10
nysM	Ferredoxin	64	Electron transfer in P450 system
nysN	P450 monooxygenase	398	Oxidation of methyl group at C-16
nysRI	Transcriptional activator	966	Regulation of nystatin production
nysRII	Transcriptional activator	953	Regulation of nystatin production
nysRIII	Transcriptional activator	927	Regulation of nystatin production
ORF4	Transcriptional activator	210	Regulation
ORF3	Transcriptional repressor	253	Regulation
ORF2	Transcriptional activator	354	Regulation

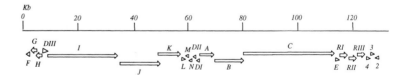

Fig. 70. *Gene organization within the nystatin biosynthetic operon. The* nys *genes are designated with capital letters in italics. The ORFs 2, 3 and 4 are designated only with corresponding numbers. [From Brautaset, T. et al.* Chemistry & Biology, 7 (6), 395–403 (2000)]

modules of type I modular PKSs are fused to the first condensing module. The KSS-domain of NysA contains a serine residue instead of the cysteine residue normally present in KS domains. This KSS-domain might function as a decarboxylase that acts on malonyl ACP

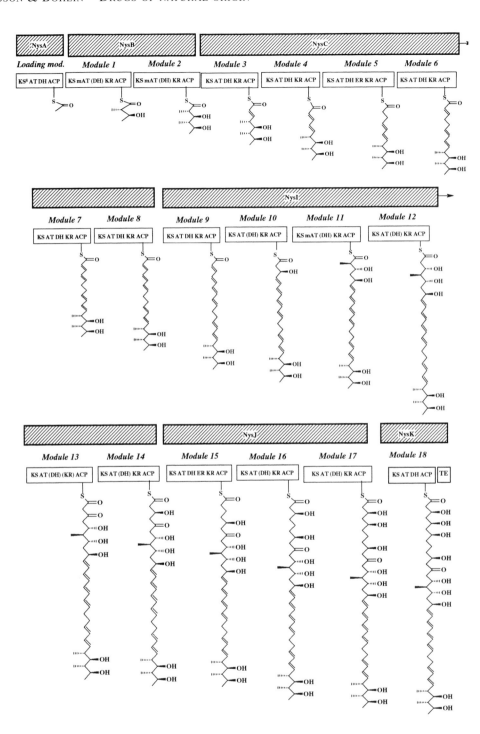

Fig. 71. Biosynthesis of the polyketide chain of nystatin. Each module elongates the chain with two carbon atoms (cf. the biosynthesis of 6-deoxyerythronolide B, Fig. 65). Domain abbreviations: KS^S = ketosynthase with serin substituted for the active site cysteine; KS = ketosynthase; AT = acetate-specific acyltransferase; mAT = propionate-specific acyltransferase; DH = dehydratase; ER = enoyl reductase; KR = β-ketoacylreductase; ACP = acyl carrier protein. Parentheses indicate inactive domains

to generate acetyl starter units in analogy with other known starter modules where the cysteine at the active site is replaced with gluta-mine. NysA thus might act as follows: Malonyl ACP is loaded on to the serine OH in the AT domain (cf. fatty acid biosynthesis Fig. 54, page 245). It is then transferred to the ACP and decarboxylated by KSS yielding an acetate group which can be transferred to KS in module 1 of NysB.

HOOC—S-Enz $\xrightarrow{KS^S}$ —S-Enz + CO_2

Malonate bound to first *ACP* in NysA

The *DH* domain is intact but seems to serve no function and might have been retained in this protein in the process of evolution.

NysB contains modules 1 and 2. The *DH* domains in these modules are inactive because they lack the conserved active-site motif of *DH* domains. The *AT* domains are propionate specific, i.e. they accept methylmalonyl-CoA. NysB is involved in the first two elongation steps of the nystatin polyketide moiety biosyn-thesis.

The NysC protein, with its 11 096 amino acid residues, is one of the largest of the known bacterial polypeptides. It is composed of six modules responsible for elongation steps 3 to 8 in the formation of the nystatin polyketide chain. All *AT* domains are acetate specif-ic. This correlates well with the assumption that NysC incorporates six malonyl-CoA extenders into the nystatin aglycone.

NysI contains modules 9 to 14 and performs elongation steps 9–14 of the nystatin polyketide backbone. Module 11 contains a propionate-specific *AT* domain, which is consistent with incorpora-tion of methylmalonyl-CoA at this elongation step. The *DH* domains in modules 10, 11, 12, 13 and 14 are inactive due to large internal deletions in the active-site motif. The *KR* domain in mod-ule 13 is inactive as it lacks the active-site motif. The latter feature, together with the inactive DH domain in module 11, most probably accounts for the presence of a six-membered ketalic ring (between C-13 and C-17 on the nystatin molecule).

NysJ is responsible for elongation steps 15–17 in the nystatin macrolactone ring assembly. The *DH* domains in modules 16 and 17 are inactive due to large deletions (module 16) or replacement of a conserved histidine residue with tyrosine (module 17). The *ER* domain in module 15 is responsible for the reduction of a double bond between C-8 and C-9.

NysK contains the final 18th module in the nystatin PKS. This module lacks a *KR* domain and the protein has a *TE* domain at the carboxyl terminus, suggesting that it also participates in the release of the completed polyketide chain from the PKS complex. The *DH* domain is apparently intact, although its activity should not be required at the last condensation step.

NysE represents another *TE* domain, the gene for which is located at the end of the nystatin biosynthetic gene cluster. The role for the presence of two *TE* activities remains uncertain. It has been proposed that one of them might be a "proof-reading" enzyme, which clears off certain substrates that would block further extension of the chain by PKS.

Regulation of the biosynthesis process is supposed to be performed by five transcriptional activators encoded by the genes *nysRI, nysRII, nysRIII, ORF4* and *ORF2* plus a transcriptional repressor encoded by *ORF3*.

Biosynthesis of mycosamine is performed by enzymes encoded by genes *nysDII* and *nysDIII*, whereas the gene *nysD1* encodes an enzyme responsible for attachment of mycosamine to the polyketide at C-19. Hydroxylation of C-10 and oxidation of the methyl group at C-16 are performed by P450 monooxygenases encoded by the genes *nysL* and *nysN*, respectively. The *nysM* gene encodes a ferredoxin which presumably constitutes a part of one or both P450 monooxygenases and serves as an electron donor. The genes *nysG* and *nysH* encode polypeptides which are highly similar to transporters of the ABC family and might be involved in ATP-dependent efflux of nystatin. Finally the *nysF* gene encodes an enzyme which is homologous to the 4´-phosphopantetheine transferases which are enzymes that carry out the post-translational modification of the ACP domains on the PKSs that is required for their full functionality.

Nystatin is used as an antifungal agent, the activity of which depends on an interaction with sterols present in the cell membrane of fungi. The formed polyene-sterol complexes are capable of organizing themselves into transmembrane channels, which make the membrane permeable to water and ions causing death of the cell. Nystatin is too toxic to be used intravenously but is used as lozenges for treatment of oral candidiasis and orally for intestinal candidiasis. Nystatin has no effect on bacteria or virus and thus does not affect the normal content of bacteria in the body. Administered as a cream nystatin is used topically for skin infections caused by *Candida albicans*.

Amphotericin B Amphotericin B is produced by *Streptomyces nodosus* from 16 acetate and 3 propionate units. Like nystatin this compound is a ring made up of 38 atoms. It contains seven conjugated double bonds and has a glycoside link to a molecule of *D*-mycosamine.

Biosynthesis The assembly of amphotericin is performed by a type I modular polyketide synthase yielding compound A (Fig. 72) as the first enzyme-free intermediate. Mycosaminylation of the hydroxy group at C-19, insertion of a hydroxy group at C-8, and oxidation of C-41 to a carboxyl group then yields amphotericin B.

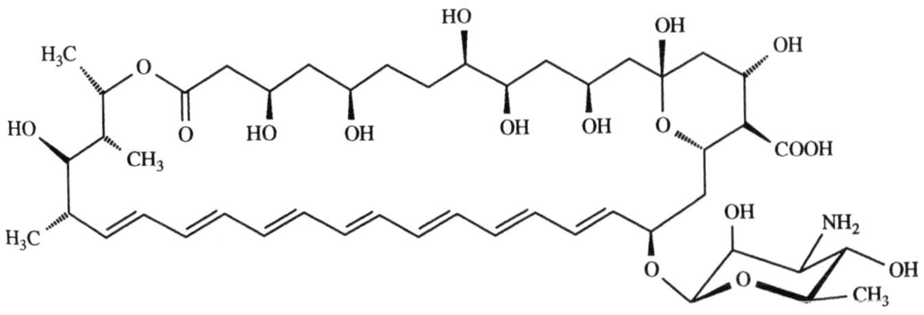

Amphotericin B

Nucleotide sequence analysis of 113 193 bp of a large polyketide synthase operon from DNA of *Streptomyces nodosus* has revealed genes for six large type I modular PKSs as well as genes for two cytochrome P450 enzymes, two transporter proteins and genes

Table 6. Modules and domains of AmphA–AmphK

Domain	AmphA Module No.	AmphB Module No.	AmphC Module No.						
	0	1	2	3	4	5	6	7	8
KS^S	+	-	-	-	-	-	-	-	-
KS	-	+	+	+	+	+	+	+	+
AT	A	P	P	A	A	A	A	A	A
DH	(+)	-	-	+	+	+	+	+	+
ER	-	-	-	-	-	+	-	-	-
KR	-	+	+	+	+	+	+	+	+
ACP	+	+	+	+	+	+	+	+	+

Domain			AmphI Module No.					AmphJ Module No.			AmphK Module No.
			9	10	11	12	13	14	15	16	17	18
KS			+	+	+	+	+	+	+	+	+	+
AT			A	A	P	A	A	A	A	A	A	A
DH			+	-	-	-	-	-	(+)	+	¤	¤
ER			-	-	-	-	-	-	-	+	-	-
KR			+	+	+	+	¤	+	+	+	+	+
ACP			+	+	+	+	+	+	+	+	+	+	TE

Abbreviations: *KS^S* = β-ketoacyl synthase where serine is substituted for cysteine at the active site. *KS* = β-ketoacyl synthase. *AT* = acyltransferase. *A* = acetate-specific acyltransferase. *P* = propionate specific acyltransferase. *DH* = dehydratase. *ER* = enoyl reductase. *KR* = β-ketoacyl reductase. *ACP* = acyl carrier protein. *TE* = thioesterase. A + sign indicates a functioning domain. A - sign indicates absence of the domain. (+) Indicates an inactive domain with apparently intact sequence. ¤ indicates an inactive domain with dysfunctional sequence.

involved in biosynthesis and attachment of mycosamine. The PKS-genes code for six proteins which contain a starter module and 18 synthesizing modules, the domains of which have been identified, allowing deduction of the complete assembly of a polyketide chain corresponding to A in Fig. 72. The biosynthesis is very similar to the biosynthesis of nystatin. Also here the product of one of the PKSgenes (AmphC) is a very large protein, composed of 10 910 amino acids and containing modules 3–8 of the total PKS (see nystatin above). As in nystatin there is also a gene coding for a second *TE*-domain, the function of which is unknown. An overview of the modules and domains of the PKS proteins is presented in table 6.

The results of the studies on the biosynthesis of nystatin and amphotericin B provide the prerequisites for engineered biosynthe-

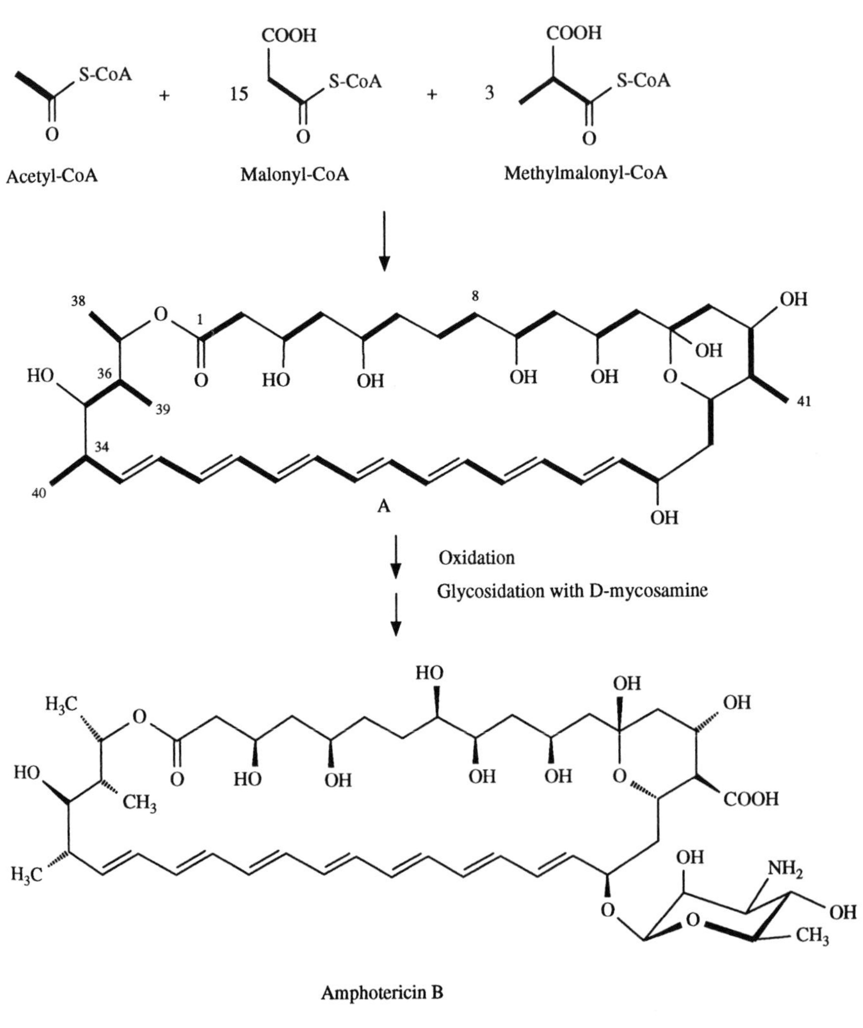

Fig. 72. Biosynthesis of amphotericin B

sis of analogues which might lead to improved drugs with better activity and fewer side effects.

Amphotericin B is used parenterally for the treatment of severe fungal infections. On account of its toxicity, the treatment must be carried out in hospital. Monitoring of hepatic and renal functions is required and this must be combined with routine blood counts and measurement of plasma electrolytes.

Like nystatin, amphotericin enters the fungal cell membrane, where it disrupts membrane integrity by binding to ergosterol. This causes leakage of ions which are critical for the normal functions of the fungal cell.

Epothilones

The main structural feature of the epothilones is a 16-membered lactone ring, but they differ from the macrolides by having no sugar units attached to the ring. These compounds are produced by strains of the soil bacterium *Sorangium cellulosum* and were originally tested for activity against plant pathogenic fungi, but they could not be exploited for this purpose because of phytotoxicity. Later on they were independently found to have taxane-like antitumour activity and one compound – epothilone B – was even more active than paclitaxel (page 426) in the tubulin polymerization assay. It replaced bound paclitaxel from microtubules and its activity against cancer cells was hardly impaired by resistance to taxol and other cytostatics. The structure of the epothilone-tubulin complex has been determined and found to differ from the corresponding paclitaxel-tubulin complex, each ligand exploiting the tubulin-binding pocket in a unique and independent manner. The expectation of finding a common pharmacophore was thus not met.

These results triggered a very extensive research activity resulting in the isolation and structure determination of about 40 epothilones. Many epothilones have been synthesized and analogues have been prepared. Derivatives are also prepared by chemical derivatization of material isolated from natural sources. The natural product epothilone B, five derivatives of natural compounds and one entirely synthetic compound (sagopilone) have entered clinical evaluation. So far (2008) only one of these –ixabepilone (= the lactam of epothilone B, Bristol-Meyers-Squibb) – has been

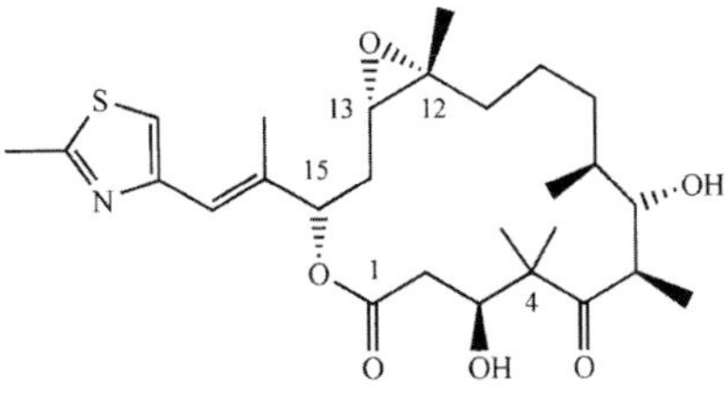

Epothilone B 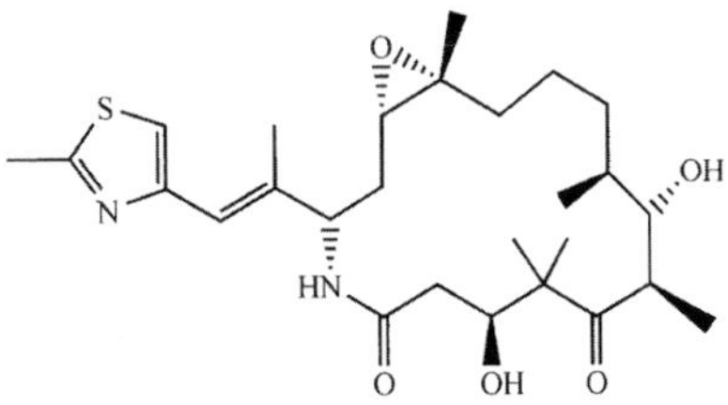   Ixabepilone

approved by the FDA for the treatment of metastatic or advanced breast cancer.

Biosynthesis The epothilones are biosynthesized by an enzyme complex comprising both PKSs and one NRPS (Non-Ribosomal Peptide Synthase, see page 531). The complex is coded for by a cluster of genes spanning over 56 kB. The gene *epoA* codes for a loading domain to which a starter acetate unit is attached. This is transferred to an NRPS, coded for by *epoB*, which catalyses incorporation of cysteine for formation of the 2-methyl-thiazole ring. The rest of the structure is built from acetate and propionate units, assembled by PKSs coded for by the genes *epoC, epoD, epoE* and *epoF*. The epoxy ring in epothilone B is formed by oxidation of a double bond between C-12 and C-13 in the precursor (epothilone C or D), catalysed by a cytochrome P450 enzyme coded for by *epoK*. One of the methyl groups on C-4 is transferred from SAM.

Griseofulvin

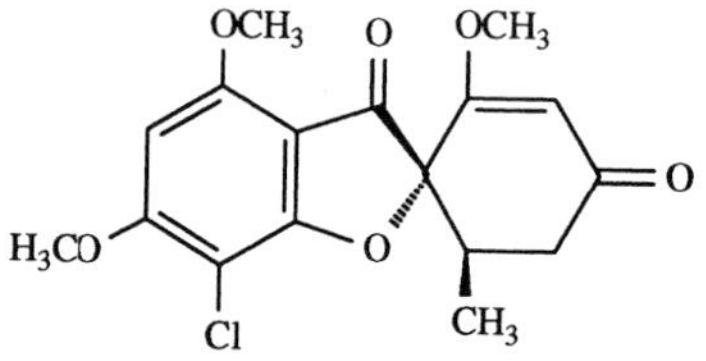

Griseofulvin

Griseofulvin is an antibiotic produced by *Penicillium griseofulvum*, *P. patulum* and other *Penicillium* species. The first step in the
Biosynthesis biosynthesis of griseofulvin is formation of a C_{14}-polyketide by condensation of one acetate and six malonate units. A Claisen reaction yields ring A of a benzophenone intermediate, the second ring of which (C) is formed by an aldol reaction. *O*-methylation yields griseophenone C which is chlorinated to form griseophenone B. Phenolic oxidation and stereospecific radical coupling results in the formation of ring B of a compound, which is converted to griseofulvin by *O*-methylation and stereospecific reduction of a double bond in ring C.

This antibiotic is administered orally against fungal infections of the skin. It is effective against fungi with chitinous cell walls.

Aflatoxins

Aflatoxins are fungal metabolites produced by the fungi *Aspergillus flavus* and *A. parasiticus*, which grow on foodstuffs such as cereals, beans and peas, coconuts, almonds, macaroni and fishmeal. Because of the high toxicity of the aflatoxins, foods which are susceptible to mould infection must be carefully monitored for the presence of these toxins. Also crude drugs should be tested for the absence of aflatoxins.

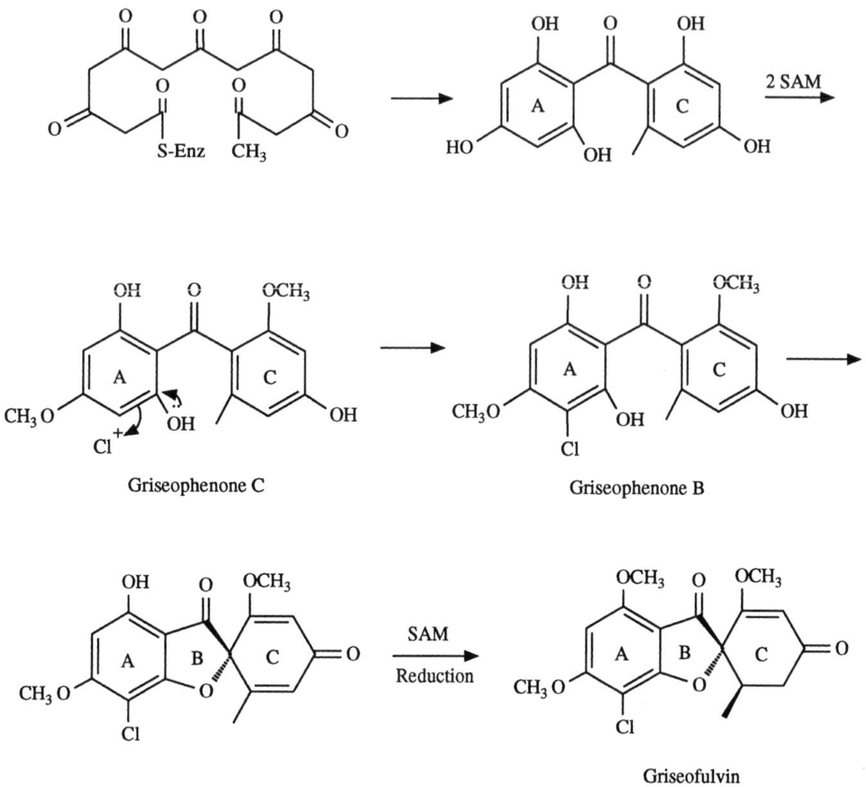

Fig. 73. Biosynthesis of griseofulvin

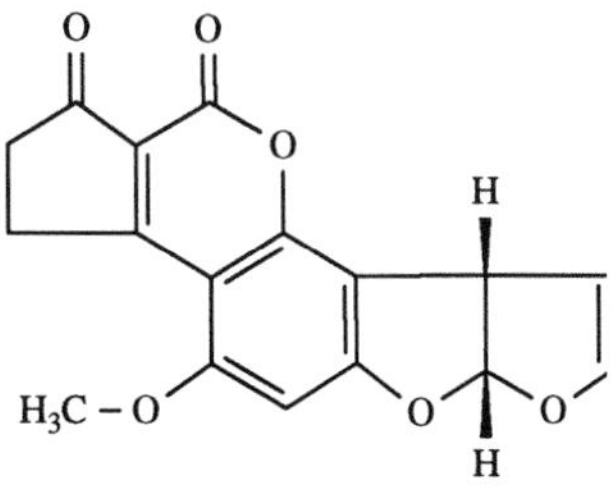

Aflatoxin B₁

The aflatoxin family comprises eight compounds: aflatoxin B_1, aflatoxin B_2, aflatoxin G_1, aflatoxin G_2, aflatoxin M_1, aflatoxin M_2, aflatoxin GM_1 and aflatoxin GM_2. In the G-series the cyclopentenone ring in the B-series is replaced by a lactone ring. Other structural differences between the compounds involve hydroxyl groups and double bonds.

Aflatoxins are highly poisonous substances which damage the liver; they are also carcinogenic, especially aflatoxin B_1. The LD_{50} for most domestic animals is in the range 0.3–2 mg/kg bodyweight. Mice, hamsters and rats are less sensitive (LD_{50} 5–18 mg/kg). The toxicity of aflatoxins was discovered in 1960, when a large number of poultry died after being fed with mouldy peanut flour. The toxic substance in the fodder was found to be aflatoxin. The mutagenic and carcinogenic effects of aflatoxin in humans and other vertebrates have been studied extensively. Cytochromes P450 in the liver and kidney oxidize the dihydrofuran ring system to an *exo*-epoxide, which has an absolute configuration that matches the helical twist of DNA to provide a potent electrophile in the major groove that reacts with N-7 of guanine residues yielding covalent adducts (Fig. 74).

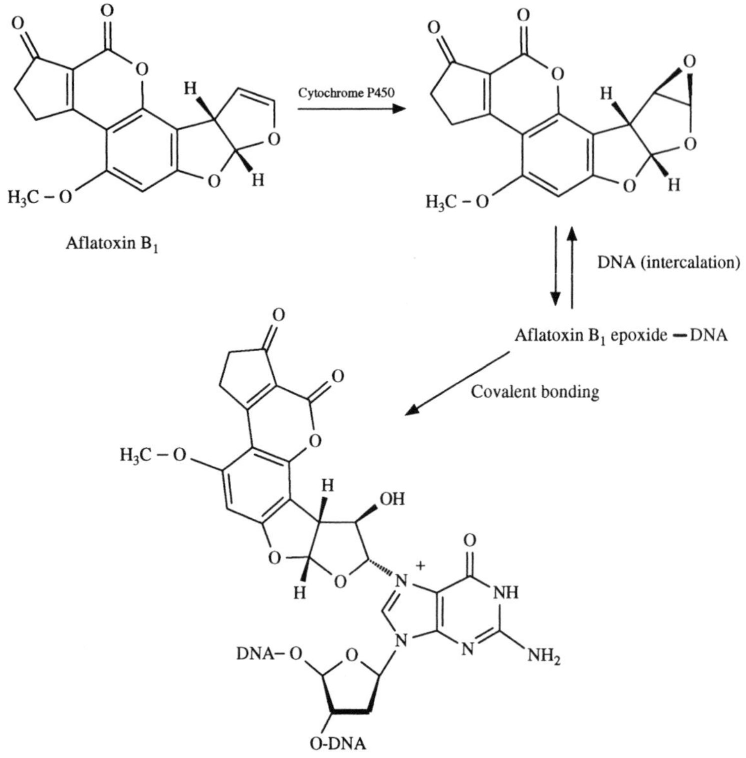

Fig. 74. Oxidation of aflatoxin B1 and formation of a DNA adduct

Biosynthesis   The biosynthesis of the aflatoxins is long and complex and characterized by diverse oxidative rearrangement processes. The current
concept of the aflatoxin pathway is illustrated in Fig. 75. A FAS I-
type fatty acid synthase produces a hexanoyl starter unit (highlighted in Fig. 75) which is subsequently extended by a polyketide synthase yielding a linear polyketide. After formation of the poly-β-
ketoacyl thioester, cyclization of the chain occurs while it is still
connected to the PKS resulting in production of an anthrone having
a hexanoyl and four hydroxyl groups. This compound is further oxidized to form *norsolorinic* acid (NA). The carbonyl in the side-
chain of NA (1´ in Fig. 75) is then reduced to a hydroxyl yielding
(1´ S)-averantin (AVT). A monooxygenase, stereospecific for the 1´
carbon of averantin, hydroxylates the 5´ carbon of (1´S)-averantin
with formation of (1´S, 5R) and (1´S, 5´S) *hydroxyaverantin*. These
compounds are then oxidized to *5´-oxoaverantin* by a cytosol dehydrogenase. 5´-Oxoaverantin is converted to (1´S, 5´S)-*averufin*
(AVF) by a cyclase via intramolecular acetal formation among the
5´-ketone and the two hydroxyl groups 1´-OH and 3-OH.

The continuation of the biosynthesis of aflatoxin is illustrated in
Fig. 76. A monooxygenase, stereospecific for AVF, converts this
compound to *hydroxyversicolorone*, which is oxidized and
rearranged to *versiconal hemiacetal acetate,* which is hydrolysed to

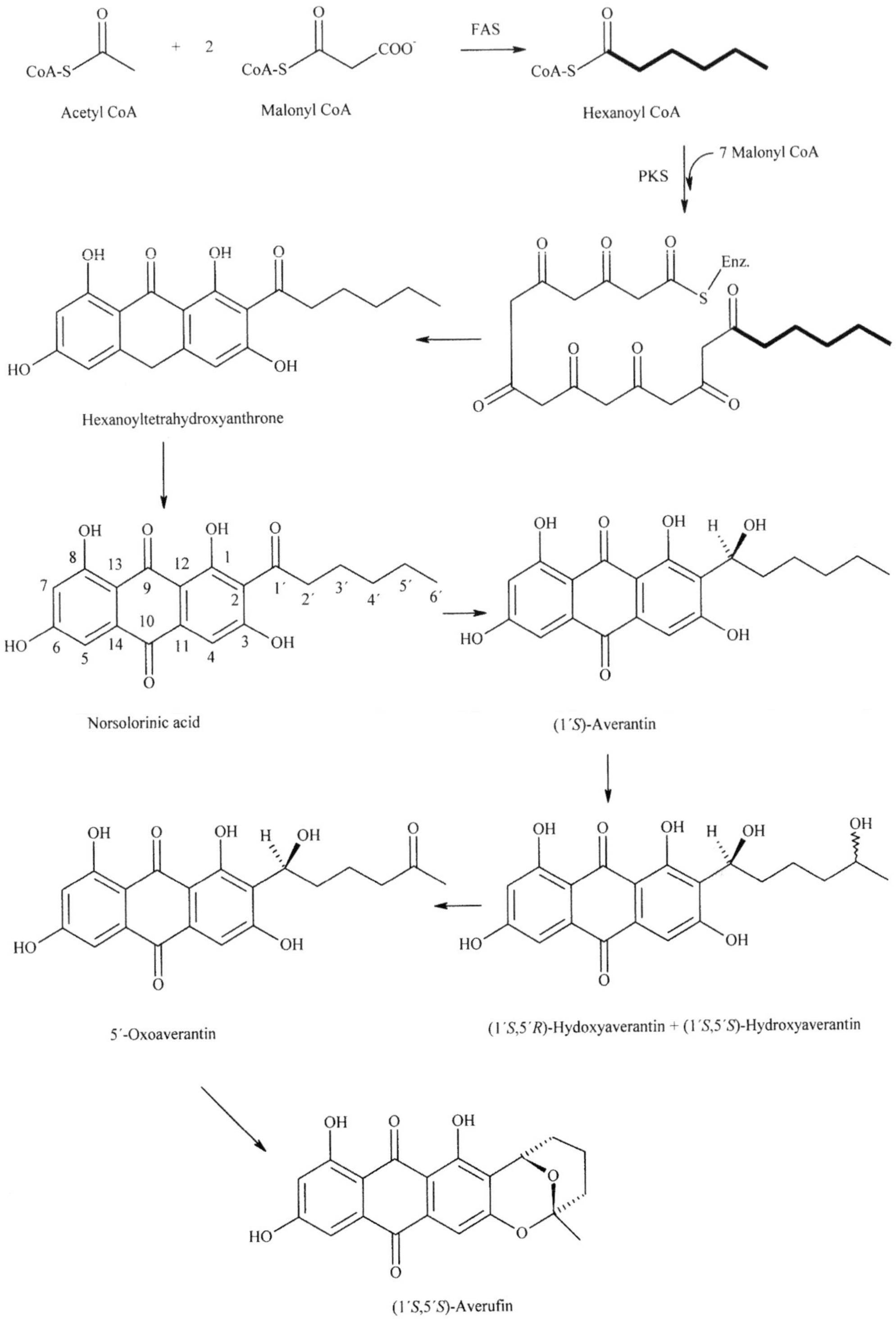

Fig. 75. Biosynthesis of (1´S,5´S)-averufin. FAS = fatty acid synthase. PKS = polyketide synthase

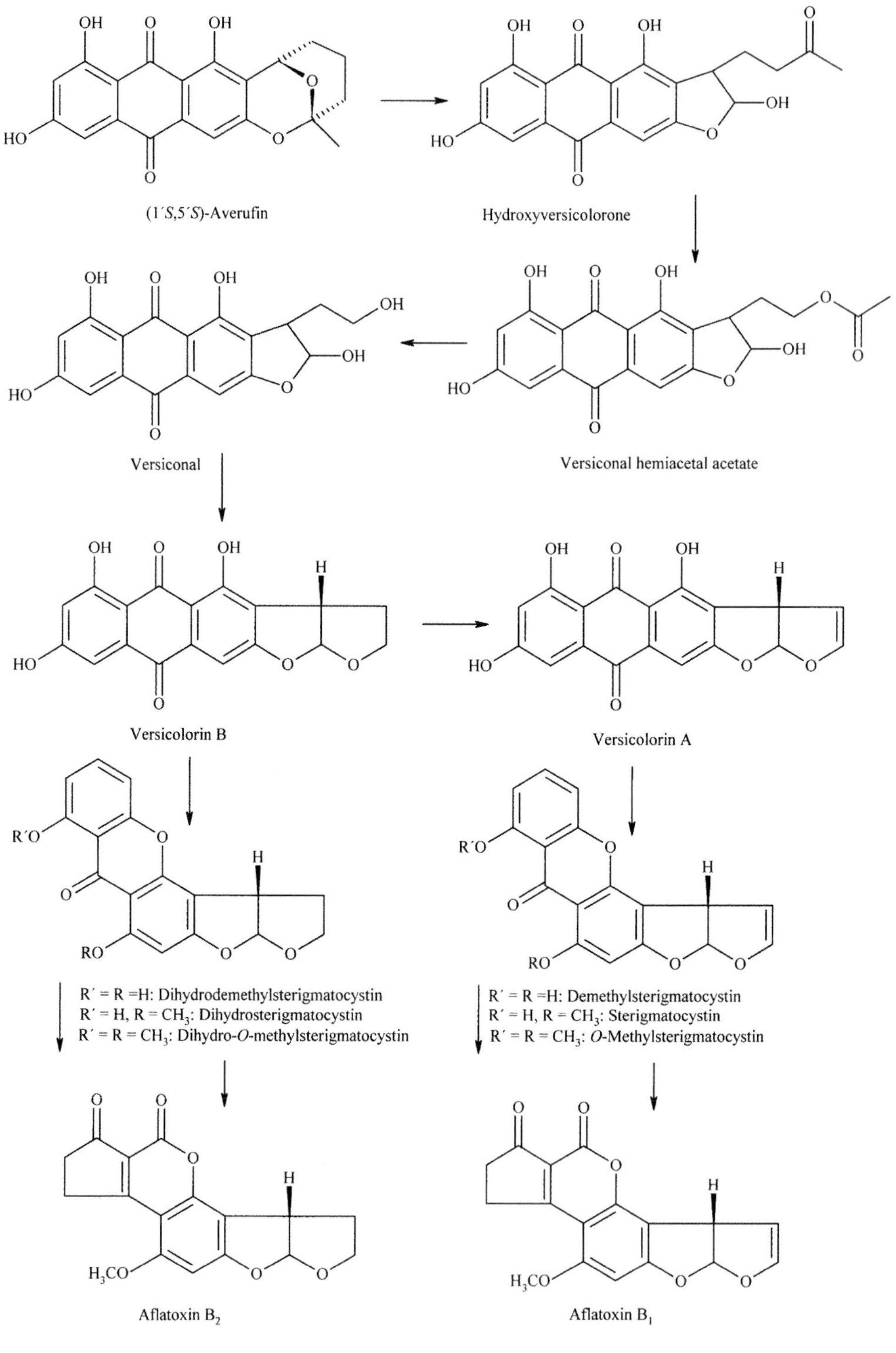

Fig. 76. Biosynthesis of aflatoxins B₁ and B₂ from (1´S,5´S)-averufin

form *versiconal*. Versiconal is converted to *versicolorin* B by a stereospecific cyclase.

This compound represents a juncture in the biosynthetic pathway leading to either the acutely carcinogenic dihydrobisfuran series (e.g. aflatoxin B1) or the relatively non-carcinogenic tetrahydrobisfuran series (e.g. aflatoxin B2) of aflatoxins. Both branches of the pathway involve complicated reactions which proceed via *dihydrosterigmatocystin* and *sterigmatocystin*, respectively, and their methylated derivatives.

The enzymes involved in the biosynthetic pathway to the aflatoxins are only partly isolated and characterized but intensive research, particularly in the molecular biology field, is in progress. The following is a summary of the most important results hitherto achieved.

At least 18 enzyme steps are required for conversion of acetyl CoA to its final products in the B- and G-series of aflatoxins. The genes encoding the enzymes and the transcription factors are located within a huge gene cluster of about 70 kb in the genomes of *Aspergillus flavus* and *A. parasiticus*. Twenty genes have been identified in the cluster and the structures and functions have been experimentally clarified for 16 of these.

Mevastatin and lovastatin

Hypercholesterolaemia is one of the major risk factors for coronary heart disease. Since as much as two-thirds of total body cholesterol is biosynthesized in the liver from acetyl-CoA, an attractive way to lower high levels of plasma cholesterol is to inhibit its biosynthesis, in which hydroxymethylglutaryl-CoA reductase (page 367) is the rate-limiting enzyme. In 1976 a fungal metabolite – *mevastatin* – was isolated from cultures of *Penicillium citrinum* and shown to be a potent inhibitor of HMG-CoA reductase both *in vitro* and *in vivo* and to reduce plasma cholesterol levels in several animal species. Mevastatin was not used clinically because it inhibited biosynthesis of cholesterol not only in the liver but also in the lens and adrenal glands. However, an active metabolite – *pravastatin* – was found to lack this side effect. Pravastatin is the 6-OH derivative of mevastatin

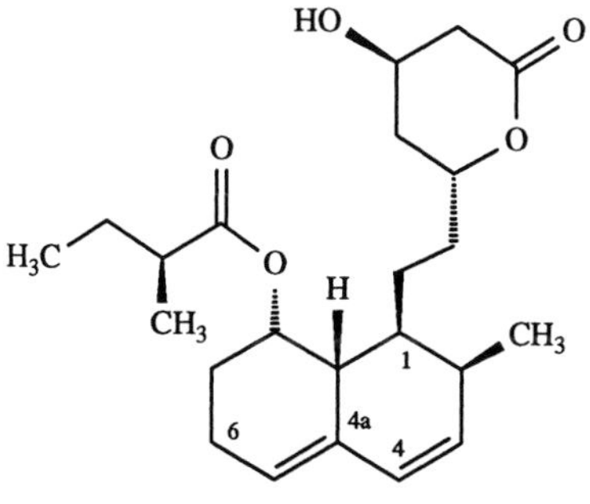

Mevastatin

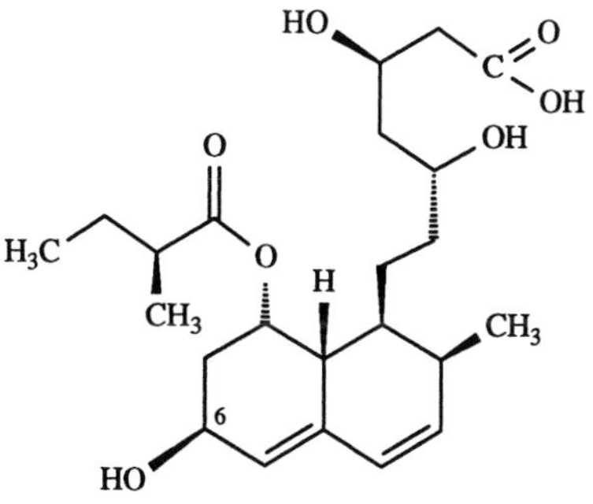

Pravastatin

with an opened lactone ring and is thus much less lipophilic and does not penetrate extrahepatic cells by passive diffusion. The sodium salt of pravastatin is administered orally as a single daily dose for the treatment of severe hyperlipidaemia.

Mevastatin (compactin, ML-236B) is isolated from cultures of *Penicillium citrinum* or *P. brevicompactum* and pravastatin is obtained by fungal hydroxylation of mevastatin. *Mucor hiemalis* SANK 36372 has been found to be one of the most potent fungi for this transformation.

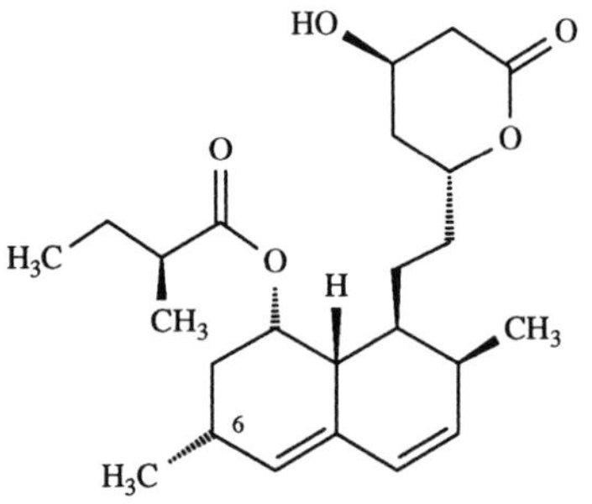

Lovastatin

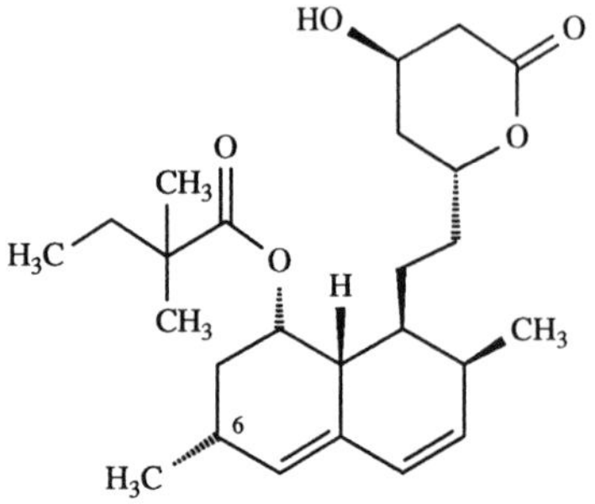

Simvastatin

A related compound called *lovastatin* (monacolin K, mevinolin, MB-530B, MK-803) was isolated a few years later from *Monascus ruber* and *Aspergillus terreus* and shown to have a greater potency as a cholesterol lowering agent than mevastatin. Structurally lovastatin is the 6-methyl derivative of mevastatin. Lovastatin is used clinically but has to a great extent been replaced by *simvastatin*, which is synthesized from lovastatin by methylation of the side-chain to a 2,2-dimethylbutanoyl group. Also simvastatin is administered as one daily oral dose. Lovastatin and simvastatin as such are pharmacologically inactive, but are metabolized in the liver whereby the lactone ring is opened, forming the active hydroxy acids.

Biosynthesis   Lovastatin is composed of two polyketide chains joined through an ester linkage at C-8. One chain is a nonaketide that undergoes cyclization to form an octahydronaphthalene ring system and the other is a diketide, (2R)-2-methylbutyrate. Two different large multifunctional PKSs form the polyketides from an acetate starter unit and malonate extender units. The PKS forming the long polyketide is *lovastatin nonaketide synthase* (EC 2.3.1.161) (LNKS). The enzyme consists of a single module with six domains - KS, AT, DH, ER (thought to be non-functional), KR and ACP (for abbreviations see erythromycin page 275). A methyl transferase (MT) is also present which adds a methyl group from *S*-adenosyl methionine during chain extension. The terminal TE domain, which is often found in FASs and PKSs, is substituted with a putative peptide synthetase

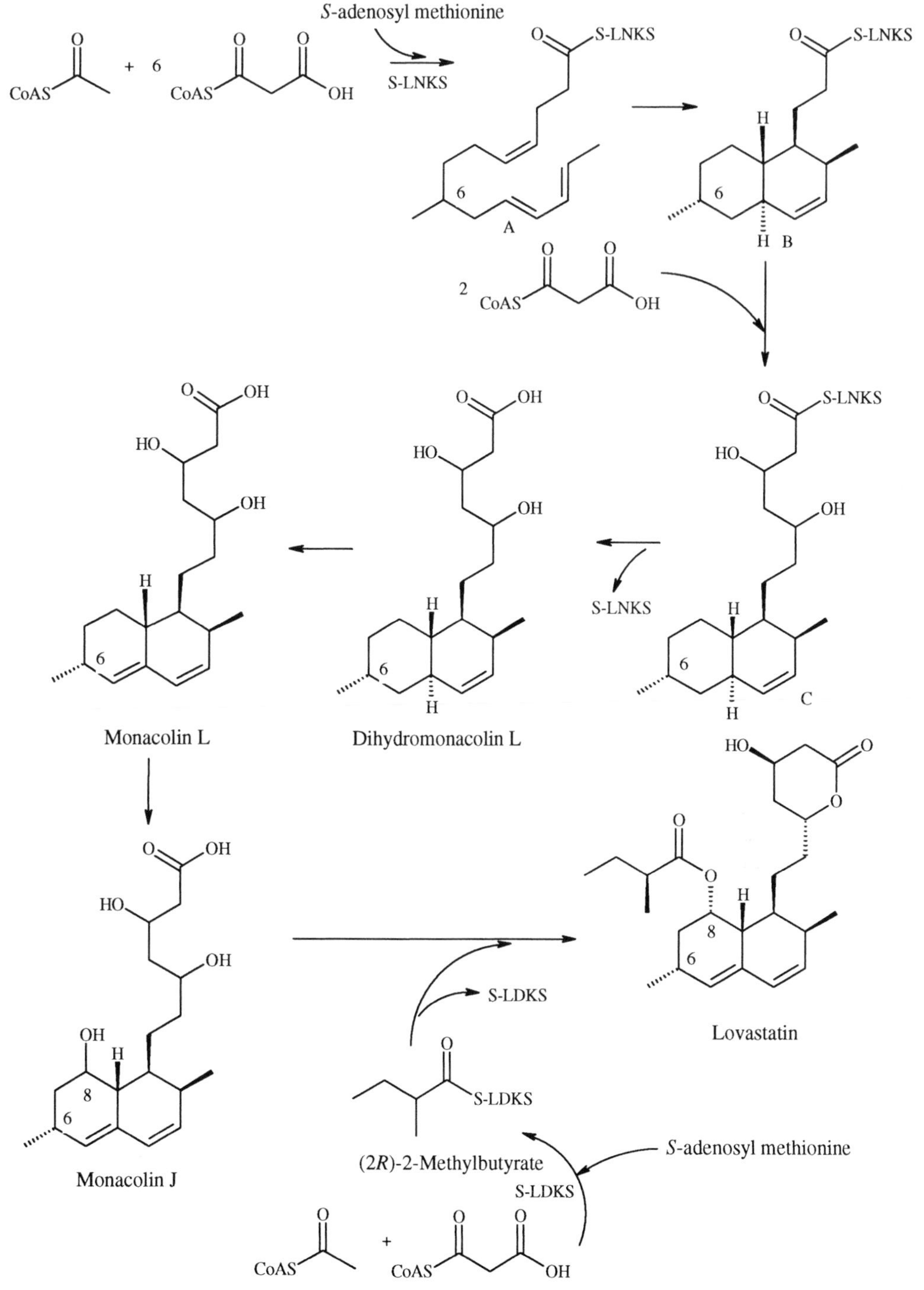

Fig. 77. Biosynthesis of lovastatin. S-LNKS = lovastatin nonaketide synthase. S-LDKS = lovastatin diketide synthase

elongation domain (PSED), similar to the corresponding domains of peptide synthetases (page 534). The function of this domain is unknown as there is no role for amino acids or other nitrogen-containing compounds in lovastatin biosynthesis. It is believed that the LNKS, in cooperation with a discrete type II protein called lovC (A putative ER), synthesizes dihydromonacolin L, the first enzyme-free intermediate in lovastatin biosynthesis. The process is *iterative*, i.e. at least some of the domains act in every successive cycle, in contrast to the function of DEBS (page 275), which contains a distinct module for each round of chain extension. The reactions are illustrated in Fig 77. The methyl at C-6 (lovastatin-numbering) is added by the MT domain at the fourth round of chain extension. An intramolecular Diels-Alder condensation within the heptaketide A forms compound B, which contains the octahydronaphthalene ring system. Two more chain elongation steps give the nonaketide-enzyme-complex C, from which dihydromonacolin L is released and further transformed first to monacolin L and then to monacolin J.

In addition to the methyl group at C-6, dihydromonacolin L contains two hydroxy groups and one double bond. LNKS must therefore be able to use different sets of activities during each chain extension cycle. The enzyme must also have the ability to differentiate between substrates that vary both in their state of reduction and chain length. The basis for this programming remains an intriguing mystery, because it is not obvious how a single set of activities can make the choice of oxidation level at each chain-extending condensation.

The short polyketide chain is assembled by *lovastatin diketide synthase (LDKS)*, which is also a unimodular PKS containing six regular domains (including an active ER domain) and an additional MT domain. This PKS is non-iterative and catalyses a single condensation followed by methylation, ketoreduction, dehydration and enoyl reduction to yield (2R)-2-methylbutyrate, which remains bound to the enzyme until it is attached to monacolin J by a specific transesterase enzyme as illustrated in Fig. 77.

The genes for LNKS and LDKS from *Aspergillus terreus* have been cloned and sequenced. An iterative PKS has also been found to catalyse the biosynthesis of 6-methylsalicylic acid in *Penicillium griseofulvum*.

Leptospermone and Nitisinone

Leptospermone is a natural herbicide which was detected following an observation that relatively few weeds grew under the plant *Callistemon citrinus* Stapf (*Myrtaceae*). Leptospermone, which is a component of the steam-volatile oils from many Australian myrtaceous plants, was excreted by the tree into the surrounding soil, where it acted as a herbicide. Other, synthetic herbicides containing a cyclohexanedione moiety, e.g. mesotrione, were prepared and these, as well as leptospermone, were found to act by inhibition of the enzyme *4-hydroxyphenylpyruvate dioxygenase* (EC

1.13.11. 27), thus establishing the importance for the cyclohexa-nedione structure for the biological activity. 4-Hydroxy-phenylpyruvate dioxygenase is involved in the conversion of tyro-sine to plastoquinone and α-tocopherol in plants. In humans the enzyme functions in tyrosine catabolism, which proceeds along the pathway tyrosine → 4-hydroxyphenylpyruvate → homogen-tisic acid → maleyl acetoacetate → fumaryl acetoacetate → fumarate and acetoacetate. This pathway does not operate nor-mally in the hereditary disease tyrosinaemia type 1, where the enzyme that catalysis the last step in the pathway is non-function-al. This causes accumulation of fumaryl- and maleyl acatoacetate and their saturated derivatives and leads to liver and kidney dam-age. Inhibitors of 4-hydroxyphenylpyruvate dioxygenase stops the catabolism before the formation of homogentisic acid and will thus prevent formation of the toxic compounds. Mesotrione has been tried for treatment of tyrosinaemia type 1 and is active, but further work has resulted in the synthesis of **Nitisinone** (Orfadin®) which has better properties and which is now the first-line treatment for this rare disease. The development of nitisinone is a nice example of how a drug can be developed by combination of research results from different areas.

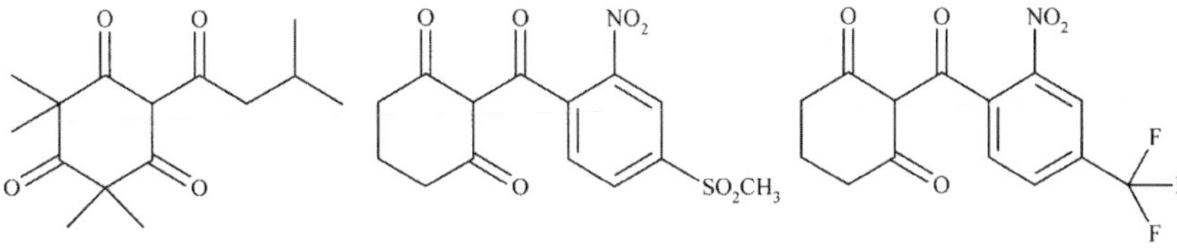

Leptospermone　　　　Mesotrione　　　　Nitisinone

Mupirocin

Mupirocin is a polyketide antibiotic, produced by *Pseudomonas flu-orescens* and is a mixture of several pseudomonic acids, with pseudomonic acid A (PA, mupirocin A) constituting 90 % of the mix-ture. Pseudomonic acid A consists of a C_9 saturated fatty acid *(9-hyd-roxynonanoic acid)*, joined to the C_{17} unsaturated polyketide *monic acid* via an ester linkage.

Beside pseudomonic acid A, mupirocin contains three other pseudomonic acids, denoted PA-B – PA-D. Compared with the structure of PA-A, PA-B has an additional OH-group at C-8, PA-C has a double bond between C-10 and C-11 instead of the epoxide group and PA-D has a double bond at C-4´,5´ in the 9-hydroxy-nonanoic acid part of the molecule.

Mupirocin is bacteriostatic at low concentrations and bacterici-dal at high concentrations. It is mainly active against Gram-positive aerobes, including methicillin-resistant strains of *Staphylococcus*, while Gram-negative organisms (excluding *Haemophilus influen-*

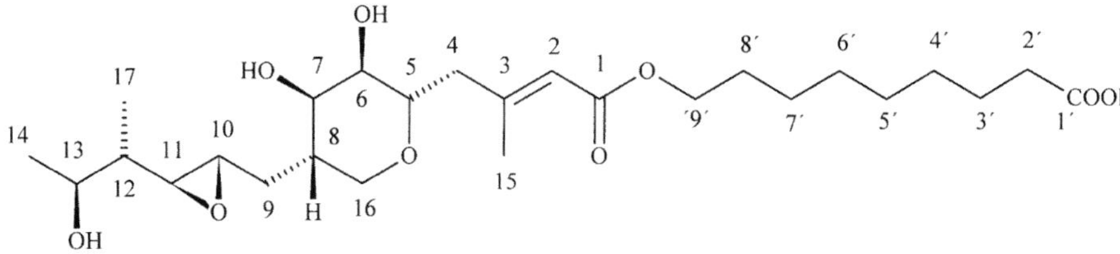

Mupirocin A

zae and *Neisseria* spp.) are insensistive. The antibacterial activity of mupirocin is due to inhibition of bacterial protein synthesis by competitive inhibition of isoleucyl transfer RNA synthetase.

Mupirocinis is used for topical treatment of bacterial skin infections.

Biosynthesis

The biosynthesis of mupirocin is not completely known. A 74 kb biosynthetic gene cluster has been identified in *Pseudomonas fluorescens*. This cluster codes for six multi-domain enzymes and 26 other peptides. Four large type I PKSs have been identified as well as several single function enzymes with sequence similarity to type II PKSs. Mupirocin is therefore believed to be constructed by a mixed type I and type II PKS system. The acyltransferase organization is atypical in that there are only two AT domains and both are found on the same protein denoted as MmpC and coded for by *mmpC*. This gene, as well as three multifunctional PKS genes (*mmpA, mmpB* and *mmpD)* are located in the first half of the cluster. The remaining part of the cluster is referred to as the tailoring region. The genes *mmpA, mmpB* and *mmpD* code for the proteins MmpA, MmpB and MmpD, respectively, which are responsible for the synthesis of the backbone of muporicin. The biosynthesis of monic acids starts on MmpD. One of the AT domains from MmpC may transfer an activated acetyl group from acetyl-CoA to the first ACP domain. The chain is extended by malonyl-CoA, followed by a SAM-dependent methylation at C-12 and reduction of the B-keto group to an alcohol. The dehydration (DH) domain in module 1 is predicted to be non-functional. Module 2 adds another malonyl-CoA unit, followed by ketoreduction and dehydration. In module 3 the chain is extended by two more carbons from malonyl-CoA, followed by SAM-dependent methylation at C-8, ketoreduction and dehydration. Addition of malonyl-CoA in module 4, followed by ketoreduction completes formation of a 12-carbon product, which is transferred to MmpA, where modules 5 and 6 perform two more rounds of extension, resulting in formation of the backbone of a C_{16} heptaketide precursor of monic acid.

The 9-hydroxynonanoic acid is probably generated by MmpB, functioning iteratively with additional ER activity. The nonanoic acid unit could also be derived partially or entirely from outside of the mupirocin cluster. Further modification of the backbone of mupirocin could occur either prior to or after esterification of monic acid and 9-hydroxynonanoic acid.

Tetracyclines

As the name implies, the structure of this group of antibiotics contains four condensed 6-membered rings. The structure of tetracycline can be seen as an example:

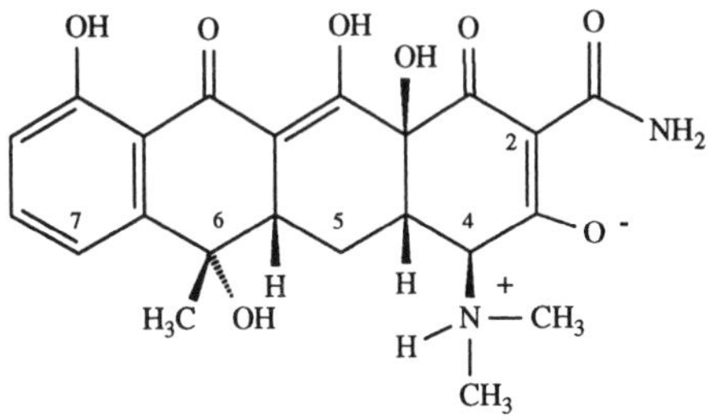

Tetracycline

Other antibiotics of this group have different substituents at positions 5, 6 and 7. The amide group at C-2 and the dimethylamino group at C-4 are structurally interesting features. Tetracyclines are produced by several species of *Streptomyces*. The three most common products are tetracycline, chlortetracycline (aureomycin), which has a chlorine atom at C-7, and oxytetracycline, with a hydroxyl group at C-5. *Streptomyces aureofaciens* produces chlorotetracycline almost exclusively, but in a mutant of this species the chlorination step in the biosynthesis is blocked and only tetracycline is produced. Oxytetracycline is obtained from cultures of *Streptomyces rimosus*.

The antimicrobial spectrum of tetracyclines is wide. Gram-positive and Gram-negative bacteria, aerobic and anaerobic bacteria, as well as *Mycoplasma, Chlamydiae and Rickettsiae*, are all killed as a result of its interaction with protein biosynthesis at the ribosome, where they interfere with the elongation of the peptide chain. However, resistance is easily developed and can be transferred to other antibiotics. The use of tetracycline should be restricted to the treatment of particular infections, like those caused by *Mycoplasma* and urogenital infections caused by *Chlamydiae*. A number of side effects are known, including kidney damage, phototoxic reactions and allergic manifestations. Teratogenic effects have also been observed.

Tigecycline
Tigecycline is a derivative of tetracycline which has activity against a broad range of Gram-positive and Gram-negative bacteria, including those resistant to tetracyclines. It is used for intravenous treatment of complicated skin and skin structure infections or complicated intra-abdominal infections caused by susceptible organisms.

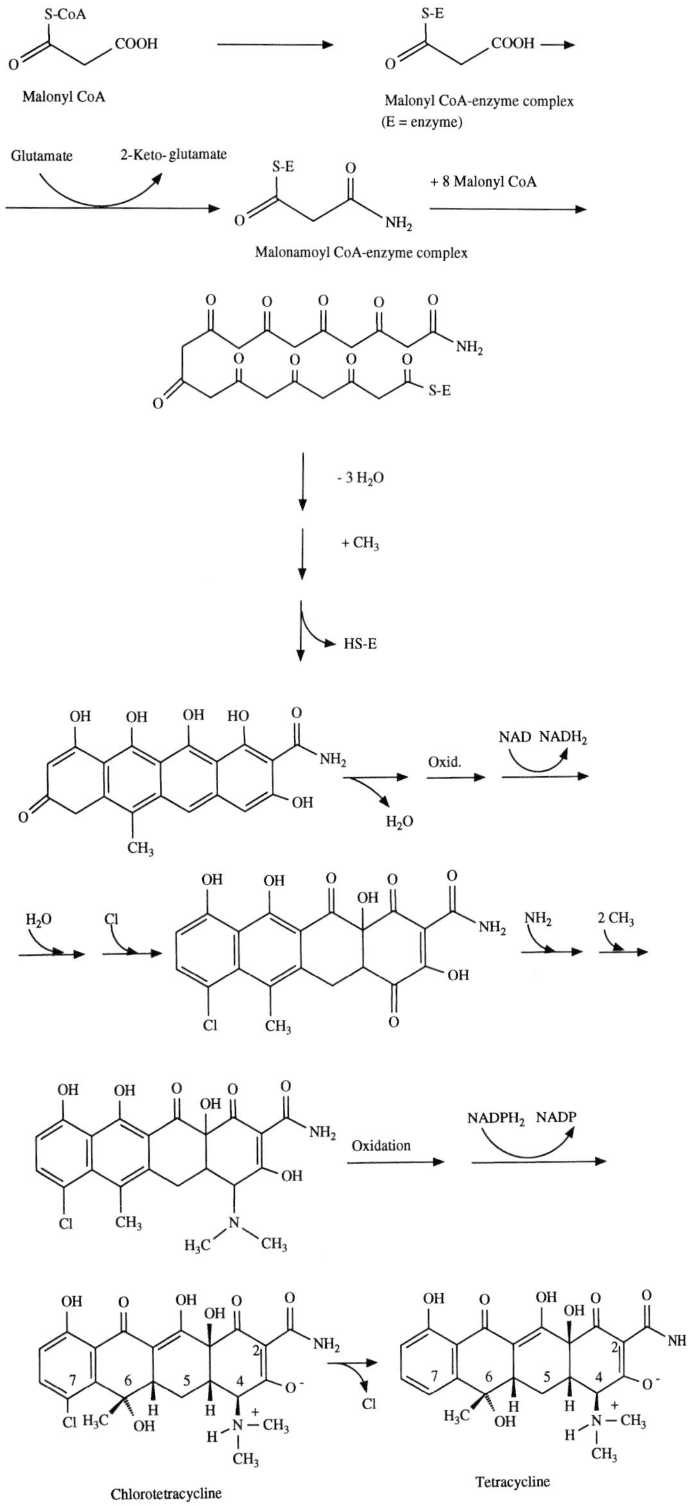

Fig. 78. Biosynthesis of tetracycline

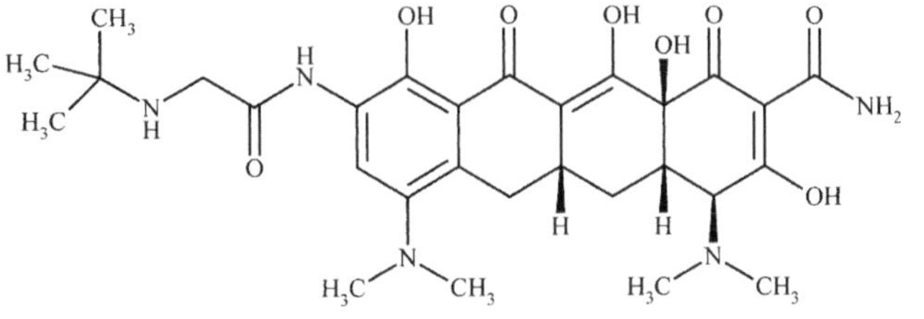

Tigecycline

Biosynthesis of tetracyclines

Tetracyclines are biosynthesized via the acylpolymalonate pathway. The enzyme which catalyses the reactions is a dissociable type II PKS (page 270), where each activity required for the assembly of the polyketide can be attributed to a separate enzyme. Genetic analysis of type II PKSs has revealed that the enzymes are coded for by *discrete* genes arranged adjacent to each other. The complete nucleotide sequences of the genes (*otcY*) which encode enzymes required for the assembly of the carbon chain of oxytetracycline have been determined revealing three genes called *otcY-1*, *otcY-2* and *otcY-3*.

The hitherto accepted pathway for the biosynthesis of tetracycline is depicted in Fig. 78. Malonyl-CoA is thought to react in a transamination process involving glutamate with formation of malonamoyl-CoA. Condensation with 8 molecules of malonyl-CoA yields a polyketide. Reduction and ring closure occur in two steps, and are followed by a series of complicated reactions involving formation of aromatic rings and subsequent de-aromatization of several of these rings. Oxidation, reduction and methylation are also frequent reactions. Altogether, 72 intermediate products have been recognized, but not all of them have been characterized. The amide group of the A-ring is already formed in the first transamination reaction. The dimethylamino group is introduced at a late stage in the chain of biosynthetic reactions. An abbreviated scheme of the biosynthesis is presented on the preceding page.

Results from experiments with the above-described genes for the formation of the carbon chain of oxytetracycline have cast doubt on the role of malonamoyl-CoA as a starter of the chain assembly. They indicate that acetate is the primer and that nitrogen for formation of the amide is introduced subsequently.

Anthracyclines

This group of polyketides is characterized by the presence of a four fused-ring aromatic system which usually forms a glycoside with a deoxy(amino)sugar. Structurally they appear to be related to the tetracyclines. Two natural compounds – **daunorubicin** and **dox-**

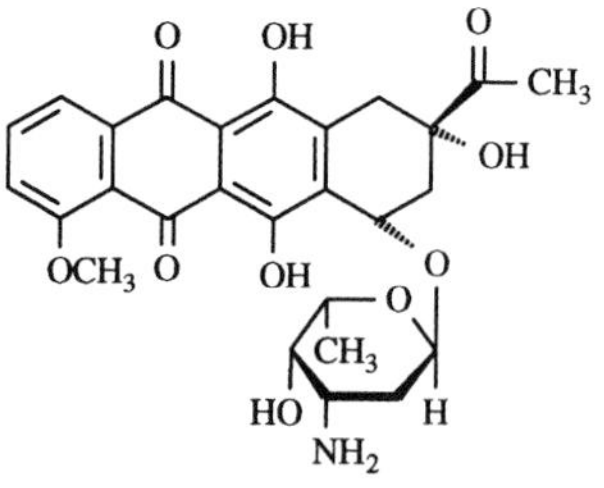

Daunorubricin

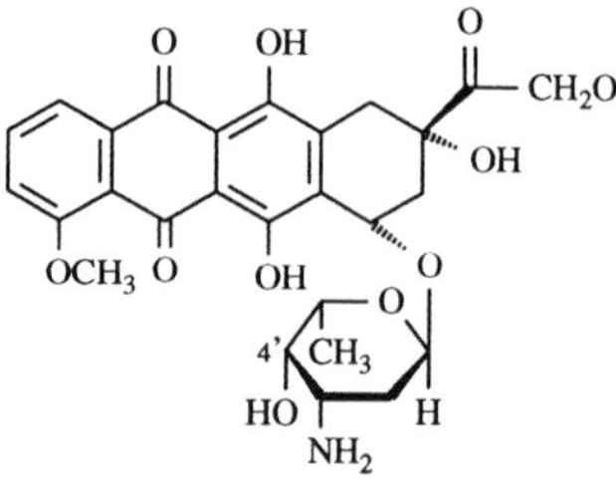

Doxorubicin

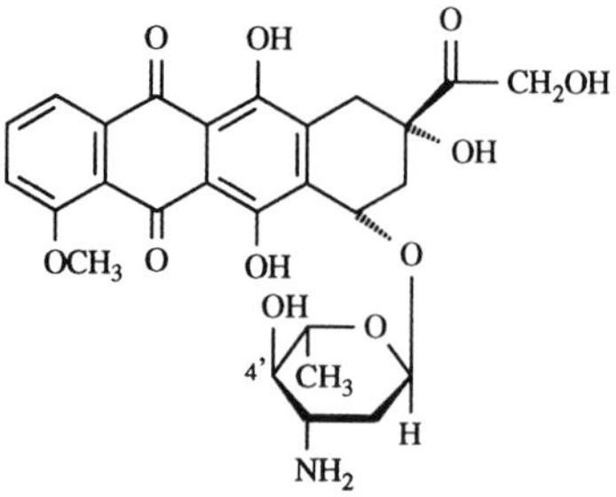

Epirubricin

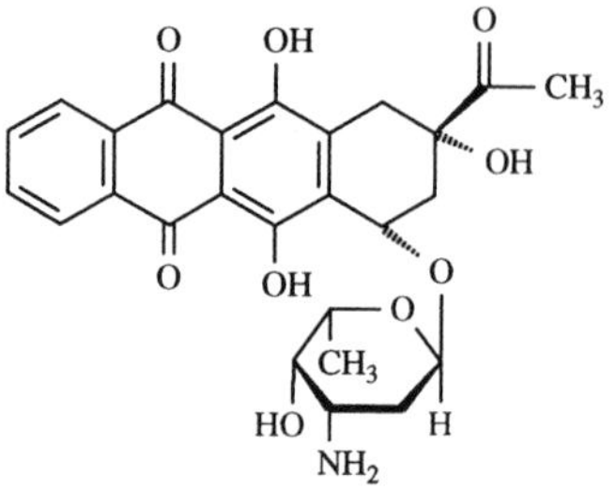

Idarubicin

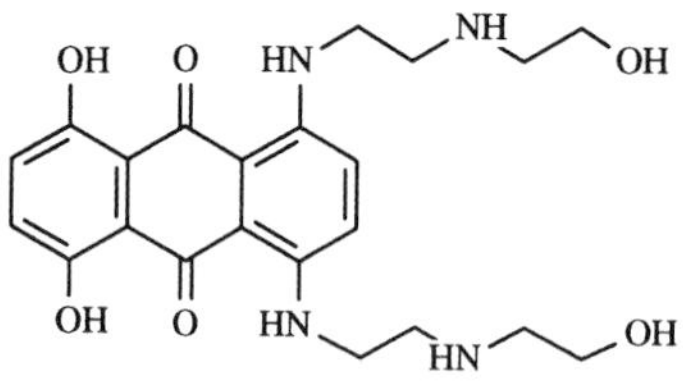

Mitoxantrone

orubicin – are used as antitumour drugs. In addition to these the 4'-epimer of doxorubicin, **epirubicin** and the semisynthetic **idarubricin** as well as the synthetic analogue **mitoxantrone** are used for the same purpose.

Daunorubicin (synonyms: daunomycin, rubidomycin) is produced by cultures of *Streptomyces peucetius* or *S. coeruleorubidus*. Doxorubicin (synonym: adriamycin) is obtained from a mutant of *S. peucetius*.

The anthracyclines are cancer remedies and act by binding between adjacent base-pairs in DNA, thereby inhibiting replication and transcription. **Daunorubicin** is used in the treatment of leukaemia and of the AIDS-related cancer disease Kaposi's sarcoma. **Doxorubicin** (Adriamycin) is a more effective cancer remedy with a broad spectrum of antitumour activity. It is used for treatment of several types of cancer such as breast cancer, various sarcomas, lymphomas and cancers of the uterus, the testis and the bladder. **Epirubicin** is used against breast cancer. All these drugs are administered by injection but **idarubricin** is available in capsules for oral administration against myeloid leukaemia. The synthetic analogue **mitoxantrone** is used for parenteral administration in severe cases of breast cancer, lymphomas and myeloic leukaemia.

All anthracyclines have serious side effects such as nausea and vomiting, bone marrow depression, hair loss and local tissue necrosis. At higher doses cardiotoxicity can be expected.

Biosynthesis of anthracyclines

Like the tetracyclines the anthracyclines are synthesized by a type II PKS system. Isotope-labelling experiments and work with blocked mutants have elucidated the pathway illustrated in Fig. 79 Propionyl-CoA acts as a starter and 9 molecules of malonyl-CoA are used for synthesis of a polyketide. *dpsA*, *dpsB* and *dpsG* have been identified as the genes which encode the enzymes needed for this process, but the corresponding enzymes have not yet been expressed.

The polyketide undergoes cyclizations yielding 12-deoxyaklanoic acid. A ketoreductase, encoded by the gene *dpsE* reduces the carbonyl at C-9 and cyclization is catalysed by an enzyme encoded by *dpsF*. A gene denoted as *dpsH* seems to encode an enzyme that ensures that the proper tricyclic fused-ring system is formed. *dpsC* and *dpsD* are two genes which are assumed to code for enzymes that ensure that propionyl-CoA instead of acetyl-CoA is used by the enzymes encoded by *dpsA*, *dpsB* and *dpsG*. The gene *dnrG* codes for an enzyme which oxidizes C-12 to a carbonyl forming aklanoic acid, the carboxyl of which is O-methylated to yield the methylester. This reaction is catalysed by an enzyme encoded by the gene *dnrC*. An intramolecular Claisen cyclization, catalysed by a *DnrD-cyclase*, yields aklaviketone. DnrD-cyclase, which is encoded by the gene *dnrD*, has been isolated and purified to homogeneity. The C-7 carbonyl of aklaviketone is then reduced by the product of *dnrE*, which is a ketoreductase with a unique, different specificity than that of the ketoreductase encoded by *dpsE* (see above). The product – aklavinone – is then oxidized at C-11 yielding ε-rhodomycinone, a reaction catalysed by the product of the gene *dnrF*. Glycosidation with L-daunosamine yields rhodomycin D. An esterase encoded by the gene *dnrP* hydrolyses the methylester of rhodomycin D and C-13 of the product is oxidized to a carbonyl by DoxA, a cytochrome P450 protein encoded by the gene *doxA*, yielding carminomycin, the C-4 hydroxyl group of which is methylated to yield daunorubicin in a reaction catalysed by the product of the gene *dnrK*. Finally C-14 of daunorubicin is oxidized yielding doxorubicin. This oxidation is also catalysed by DoxA.

Enediynes

Enediynes are a family of natural products with a unique molecular architecture, produced by bacteria of the genera *Streptomyces, Strep-*

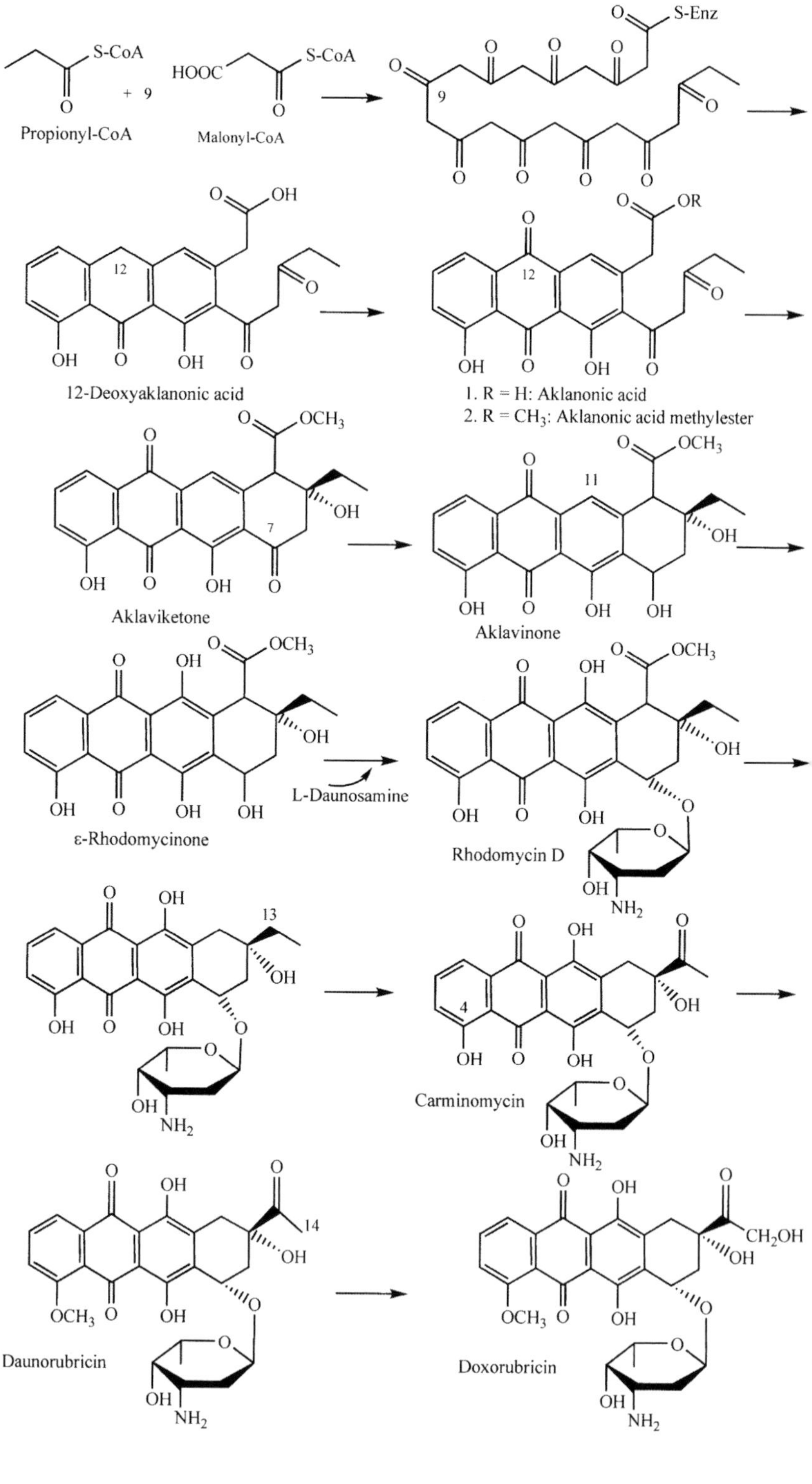

Fig. 79. Biosynthesis of daunorubicin and doxorubicin

tosporangium, Actinomyces, Micromonospora and *Actinomadura* as well as by tunicates of the genera *Polysyncraton* and *Didemnum*. The enediynes have antibiotic and antitumour activity and some are used clinically for treatment of special types of cancer.

Structurally the enediynes are characterized by a core composed of two rings, one of which contains two acetylenic groups, conjugated by (or to) a double bond. To this core a variety of other structures are connected. The endediynes are categorized into two subfamilies, depending on whether the ring that contains the acetylenic groups is built of 9 or 10 carbon atoms (9- and 10-membered enediynes).

9-membered enediynes

The structural core of this family is a bicyclo[7.3.0]dodecdadiynene chromophore which is unstable and requires an associated protein for stabilization. Nine compounds have been isolated and the complete structures determined for five of them. Of medicinal interest is **neocarzinostatin** (NCS) which is produced by *Streptomyces carzinostaticus*. Attached to the chromophore are 2-hydroxy-7-methoxy-5-methyl-1-naphtalenecarboxylic acid, ethylene carbonate and *N*-methylaminfucosamine. The 9-membered ring is oxidized to form an epoxide, which contributes to the instability of the compound.

The chromophore is associated 1:1 with a protein component (NCS apoprotein) which has been characterized as a 113 amino acid polypeptide, based on the genebase sequence and apoprotein crystal structure. The three-dimensional solution structure of intact neocarzinostatin has been determined.

All 9-membered enediynes have strong antitumour activities. Some members are 5 000–8 000 times more active than doxorubicin (page 306). They have, however, delayed toxicity which limits their clinical use. Conjugation of neocarzinostatin with poly(styrene-co-maleic acid/anhydride) (SMA) yields a more stable product with reduced toxicity. The polymer is linked to the apopro-

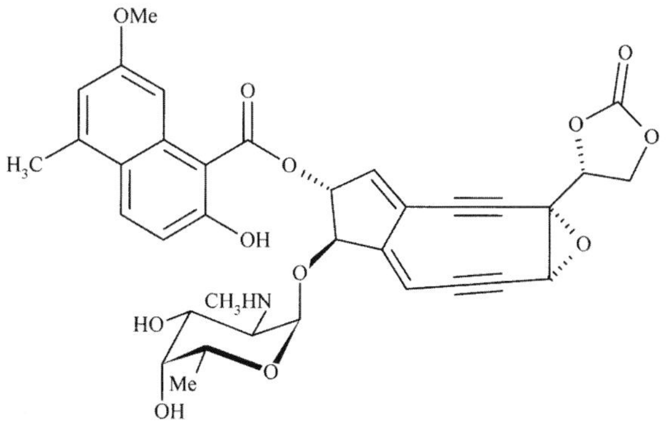

Neocarzinostatin

tein in the conjugate, which is known as SMANCS and is used in Japan to treat hepatoma.

Both the apoprotein and the chromophore are involved in the mechanism of action. The apoptotein serves to stabilize the chromopore and also acts as a transporter whereby the chromophore is delivered to its target tumour, where the aglycone acts as a warhead that binds to the minor groove of DNA, causing single-stranded or double-stranded DNA lesions. The DNA damage causes a significant decrease in DNA replication competency and ultimately leads to cell death (see further under "mechanism of action" page 312).

10-membered enediyenes

In this family the core is a bicyclo[7.3.1]tridecadiynene chromophore which, in contrast to the 9-membered representatives, is stable and does not require attachment to an apoprotein. The complete structures of five compounds are known, and one of them – **calicheamicin γ_1^1** – is of medical interest.

Attached to the core is a complicated carbohydrate structure comprising four glycosidic units (A, B, D and E), a fully substituted iodothiobenzoate moiety (C) and an unusual N-O glycosidic linkage between A and B. Attached to the core is also a methyl trisulphide moiety and an *O*-methyl carbamate group. Calicheamicin γ_1^1 is isolated from *Micronospora echinospora* spp. *calichensis*. It is an antitumour agent which is 1 000 times more potent than doxorubicin, but is too toxic for direct use as a drug and the mode of action is relatively unspecific. A strategy to overcome these drawbacks has been to conjugate calicheamicin to tumour-directed monoclonal antibodies (mAbs). Such a preparation is **gemtuzumab ozogamicin** which under the tradename *Mylotarg* is used clinically for treatment of first-relapse acute myeloid leukaemia (AML) in patients over 60 years of age. Mylotarg is a semisynthetic derivative of calicheamicin γ_1^1 coupled to a humanized mAb (hP67.6), specific for the antigen CD33.

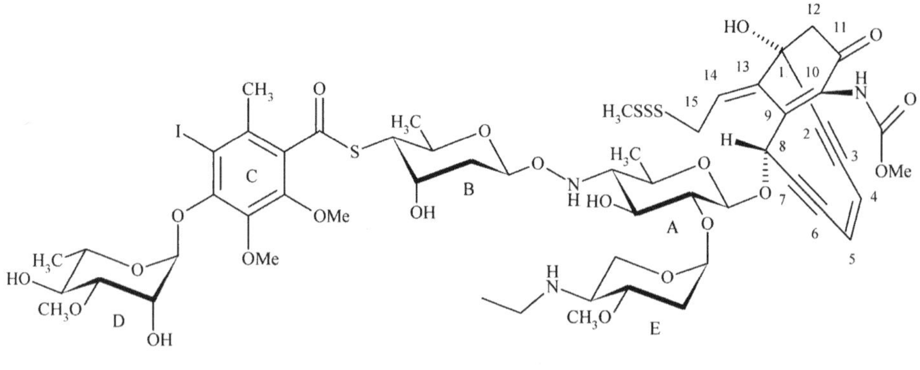

Calicheamicin γ_1^1 (The 10-membered ring comprises carbon atoms 1 - 9 + 13)

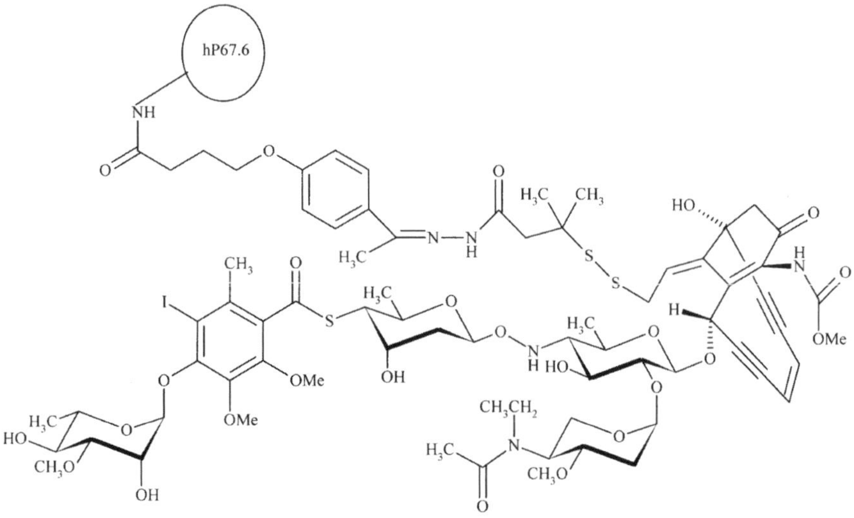

Gemtuzumab ozogamicin

The CD33 antigen is a sialic acid-dependent adhesion protein found on the surface of leukaemic blasts and immature normal cells of myelomonocytic lineage, but not on normal haematopoietic stem cells. The monoclonal antibody hP67.6 is a recombinant humanized Ig4, κ antibody which is produced by mammalian cell suspension culture and is coupled to *N*-acetyl-calicheamicin through an acid-labile linker containing a hydrazone function. Gemtuzumab ozogamicin binds CD 33 with high affinity and rapidly enters lyso-somal vesicles where the acidic pH causes hydrolysis of the hydra-zone structure, releasing calicheamicin that attacks cellular DNA, leading to cell death as further described below (page 312) under the heading "mechanism of action".

Biosynthesis

Early labelling studies showed that the chromophore of the enediynes derives from acetate. The biosynthetic locus coding for calicheamicin has been cloned and characterized and found to con-sist of 74 open reading frames that span more than 90 kb. They include 16 genes of which *O1–O6* are expected to participate in the construction or modification of the orsellinic acid derivative an *E1–E10* in the establishment of the calicheamicin aglycone. Of these genes *O5* and *E8* code for two iterative type I PKSs called CalO5 and CalE8, respectively. From the sequence of *E8* the full-length CalE8 can be deduced to comprise 1 919 amino acids. The five main domains KS, AT, KR, DH and TD have been identified in CalE8 (for explanation of the functions of the domains see text to Fig. 65, page 275). Seven genes (*T1–T7*) are associated with mem-brane transport and 14 (*S1–S14*) with the production of the four unusual sugar units. The locus also contains four glycosyltrans-ferase genes (*G1–G4*), an insertional element (*IS*) and 22 ORFs (*U1–U22*) of unknown function. The complete gene map is presented in the paper by Ahlert *et al.*, see *Further reading*.

The 9-membered enediynes were previously postulated to derive from a fatty acid precursor. However, a gene (*SgcE*) has been identified in the gene locus of the 9-membered enediyne C-1027 which shows a remarkable similarity to *CalE8*, implicating that the biosynthesis of the aglycone of the 9-membered enediynes also proceeds via polyunsaturated polyketide intermediates. An unique iterative type I PKS has been isolated from *Streptomyces carzinostaticus*, which synthesizes the 2-hydroxy-7-methoxy-5-methyl-1-naphtalenecarboxylic acid moiety of neocarzinostatin by condensation of six intact acetate units.

Mechanism of action

In the 9-membered enediynes the apoprotein acts as a transporter and delivers the warhead to DNA by controlled release. In calicheamicin the aryltetrasaccharide delivers the warhead to its target, binding tightly in the minor grove of double-helical DNA. The binding is highly specific for sequences such as 5´-TCCT-3´ and 5-TTTT-3´, and a significant portion of this sequence selectivity is associated with the large and polarizable iodo substituent of the orsellinic acid moiety of calicheamicin. Once the molecule is in the vicinity of DNA, a nucleophile (e.g. glutathione) attacks the central sulphur atom of the trisulphide group, causing the formation of a thiol, which adds intramolecularly to the adjacent α,β-unsaturated ketone embedded within the framework of the aglycone. This reaction causes a significant change in structural geometry, imposing strain on the enediyne ring. This strain is relieved by the following cycloaromatization reaction generating a highly reactive benzenoid diradical. This radical abstracts hydrogen atoms from duplex DNA at the C(5´) position of the cytidine in 5´-TCCT-3´ and the C(4´) position of the nucleotide three base pairs removed on the 3´side of the complementary strand, leading to cleavage of both strands of DNA.

Binding of the chromophore to the minor grove of DNA and electronic rearrangement to form the benzenoid diradical species, followed by abstraction of hydrogen atoms from the deoxyribose of DNA, leading to cleavage of DNA, is also the mechanism of action for the 9-membered enediynes. Also their chromophores require activation. Thiol activation to trigger radical formation is relevant also here, but other activators such as acidic pH and light have been shown to initate electronic rearrangement.

Anthraquinones

Anthraquinone derivatives are the active components in a number of

Fig. 80. The mechanism for cleavage of DNA by calicheamicin

crude drugs with purgative properties. These anthraquinone derivatives are glycosides, often glucosides or rhamnosides. The presence
of the sugar residue is a prerequisite for the pharmacological
effects. Anthraquinones are coloured substances and many of them
are used technically as dyes. The basic structure of these compounds is 9,10-anthraquinone, and their differences lie in the
arrangement of the attached substituents. Examples are: frangula-
emodin, the aglycone of various purgative anthraquinone glycosides, and alizarin, a colouring matter occurring as a glucoside in
madder (*Rubia tinctorum* L., *Rubiaceae*).

Reduced forms of anthraquinones, which exhibit keto-enol tau-
tomerism, are often encountered. The following scheme (Fig. 81)
shows the relationships.

The proportions of the various oxidation stages vary. In fresh
Senna leaves (from *Cassia obovata* Collad or *Cassia acutifolia*
Delile, see below) anthrones dominate, but on drying during preparation of the crude drug dianthrones are formed. In frangula bark the
anthrones are stable. Anthraquinones turn intensely red in alkali, a
reaction that can be used for their quantitative determination (Born-
träger reaction). Anthrones and anthranols do not show this reaction.

Anthraquinones are widespread in the plant kingdom, but have
rarely been reported to occur in mosses, ferns and conifers. With
regard to angiosperms, anthraquinones are found mainly in the *Liliaceae*, *Polygonaceae*, *Rhamnaceae*, *Rubiaceae* and *Fabaceae*.
They have also been encountered in insects, e.g. in the cochineal
louse which produces the anthraquinone dye carmine (see page
322) and in fungi, e.g. *Penicillium* and *Aspergillus* species.

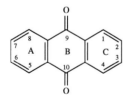

9,10-Anthraquinone

313

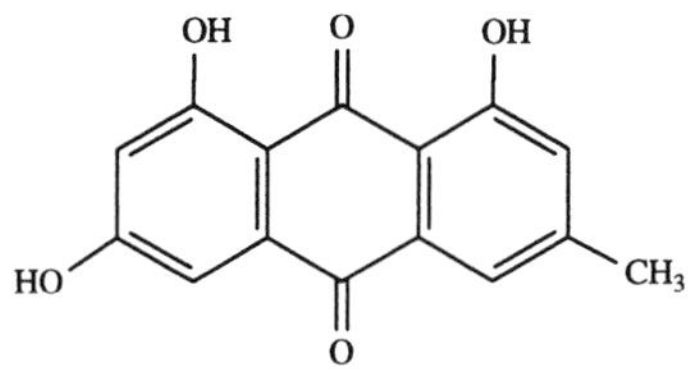

Frangulaemodin

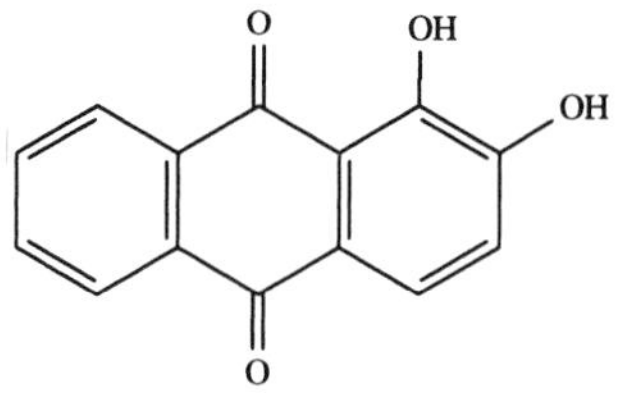

Alizarin

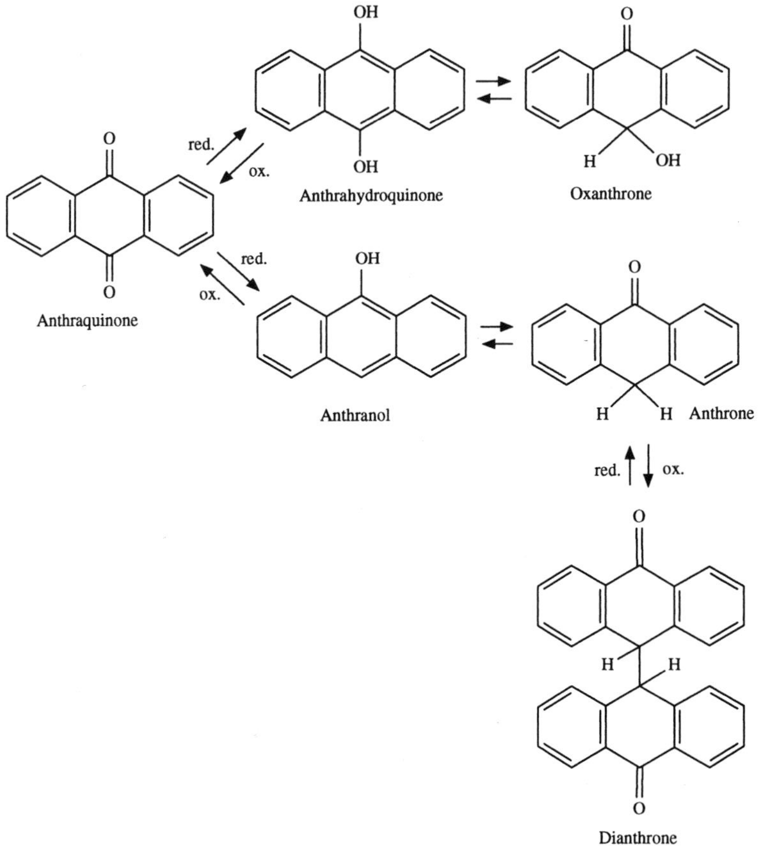

Anthrahydroquinone

Oxanthrone

Anthraquinone

Anthranol

Anthrone

Dianthrone

Fig. 81. Reduction and oxidation processes of anthraquinones

Anthraquinone derivatives can be biosynthesized from acetate via the acylpolymalonate pathway or via shikimic acid. The former is the case in anthraquinone-producing fungi. For higher plants, it has been shown that anthraquinones may be synthesized either way. The studies so far reported show that acetate is the precursor in the *Polygonaceae* and *Rhamnaceae* and that the shikimate route operates in the *Rubiaceae* and *Gesneriaceae*. It remains to be shown which of the routes is the most common among the angiosperms.

The acylpolymalonate pathway of anthraquinone biosynthesis

It was mentioned earlier (page 237) that the biosynthesis of aromatic substances from acetate begins in the same way as fatty acid biosynthesis: acetyl-CoA is carboxylated to give malonyl-CoA and both products are then transferred to an acyl-carrying protein (ACP) and react to form acetoacetyl-ACP and CO_2. In the biosynthesis of fatty acids acetoacetyl-ACP is reduced to butyryl-ACP, which reacts stepwise with further malonyl-ACP units until the fatty acid has reached its final length. In the biosynthesis of anthraquinones and other aromatic substances, on the other hand, the reduction of acetoacetyl-ACP and the corresponding products in the later steps does not take place, but a β-polyketo acid is formed. In anthraquinone biosynthesis, this intermediate reaches a length of 16 carbon atoms and undergoes ring closure, reduction or oxidation to give the final product (Fig. 82). Hitherto no PKSs have been identified in plants which produce medicinally used anthraquinone derivatives.

The shikimic acid pathway of anthraquinone biosynthesis

The existence of this pathway for anthraquinone biosynthesis was first demonstrated in the case of the pigment alizarin and purpurin-carboxylic acid, both of which are found in madder, *Rubia tinctorum*.

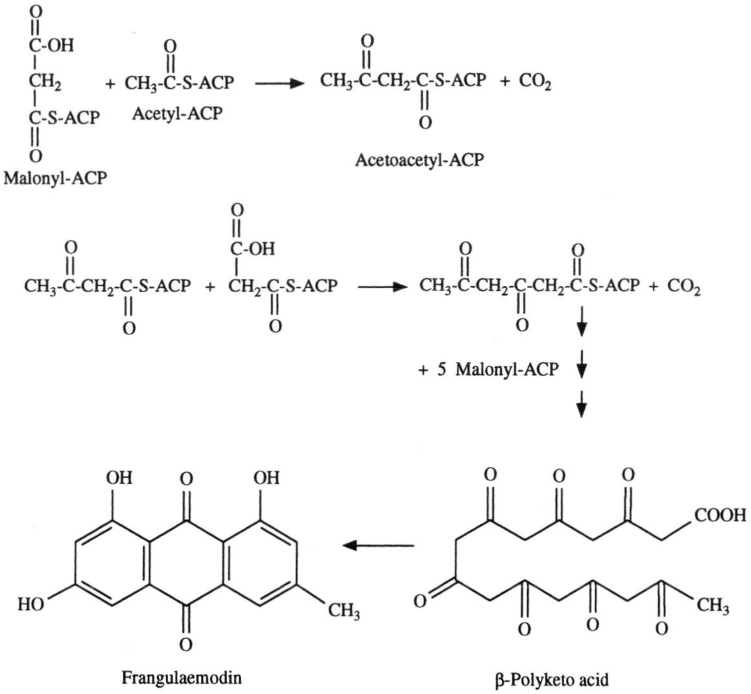

Fig. 82. Biosynthesis of anthraquinones from acetate

In this reaction series (Fig. 83), shikimate reacts with α-ketoglutaric acid, which can be formed either by deamination of glutamic acid or via the citric acid cycle. The product is *o*-succinylbenzoic acid, which, together with mevalonic acid (formed from acetate, see page 367), gives an intermediate that yields alizarin by ring closure. It follows that in anthraquinones formed in this manner, ring A comes from shikimic acid and ring C from acetate. In ring B, C-9 derives from shikimic acid and C-10, C-13 and C-14 from α-ketoglutaric acid.

Medicinal use of anthraquinone drugs

Anthraquinone drugs are used as purgatives. The sugar moiety in the glycosides present increases solubility and facilitates transport to the site of action. The aglycone is the active part of the molecule, but it is mainly the anthrone form that is effective. In the colon, bacteria hydrolyse the glycosides and reduce the liberated aglycones to anthrones which act directly on the large intestine to stimulate peristalsis. Thus, crude drugs containing little or no anthrones nevertheless have laxative activity. However, drugs rich in anthrones, e.g. frangula bark, have too strong an effect and must therefore be stored for a period of time to allow oxidation to the corresponding anthraquinones to take place.

The anthraquinones inhibit stationary and stimulate propulsive contractions of the colon, resulting in an accelerated intestinal passage. The shortened contact time causes a reduction in liquid absorption, which together with stimulation of chloride secretion, increases water and electrolyte content. The result is a soft stool which is easy to defaecate.

Anthraquinone-containing crude drugs are used for preparation of both conventional drugs and herbal remedies, mostly as simple extracts. Sennosides A and B (page 321) are the only pure compounds contained in conventional drugs. Anthraquinone-containing drugs should be used with care and only for short periods of time (1–2 weeks) as they are associated with a number of side effects and interactions. They should not be the first choice for treatment of constipation. Bulk laxatives (page 164) are preferable. The smallest dose which gives a satisfactory effect should be chosen.

Contraindications: intestinal occlusion, acutely inflammatory intestinal diseases and appendicitis.

Side effects: vomiting and spasmodic gastrointestinal complaints. Overdosage and long term use leads to loss of electrolytes, in particular potassium ions, which in turn can cause inhibition of intestinal motility. Long term use can also cause albuminurea and haematuria and, in rare cases, heart arrhythmias, nephropathies, oedemas and accelerated bone deterioration.

Interactions: The loss of K^+-ions in long term use can enhance the effect of cardiac glycosides.

Special precautions: Anthraquinone-containing crude drugs

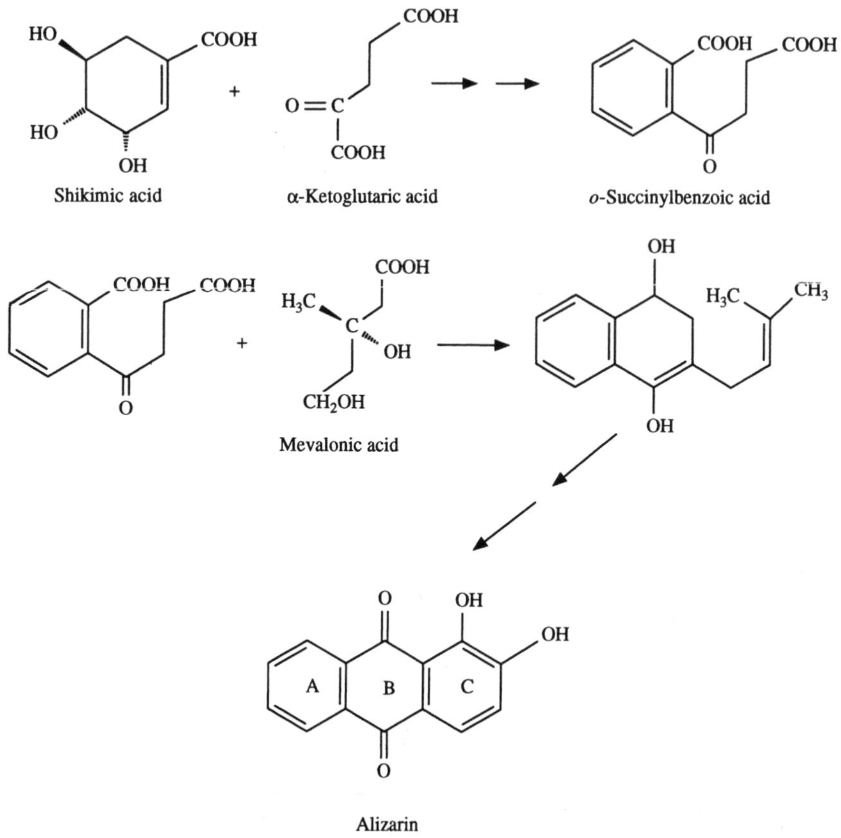

Fig. 83. Biosynthesis of anthraquinones from shikimic acid

should not be used in pregnancy or during nursing. Nor should they be given to children under 12 years of age.

Crude drugs **Frangula bark** (Alder buckthorn bark, Frangulae cortex) is the dried bark of the alder buckthorn, *Rhamnus frangula* L. (*Rhamnaceae*), a shrub native to Europe and western Asia. The plant grows on swampy ground, often around lakes and streams. The fresh bark contains mainly anthrones, which have too strong a purgative action. The drug must therefore be stored for at least a year, to allow the anthrones to become oxidized (see page 314). Stored frangula bark contains at least 7 % anthraquinones, mainly glucofrangulins A and B. Gluco-frangulin A is 6-*O*-α-L-rhamnosylemodin 8-*O*-β-D-glucoside. Glucofrangulin B is an analogue of glucofrangulin A, containing an apiose instead of rhamnose residue.

Other constituents are frangulin A, which is emodin 6-rhamnoside, and frangulin B, which is emodin 6-apioside and frangulin C, which contains xylose instead of rhamnose and apiose. Extracts of frangula bark are used as laxatives.

Cascara (Rhamni purshianae cortex) is obtained from a relative of

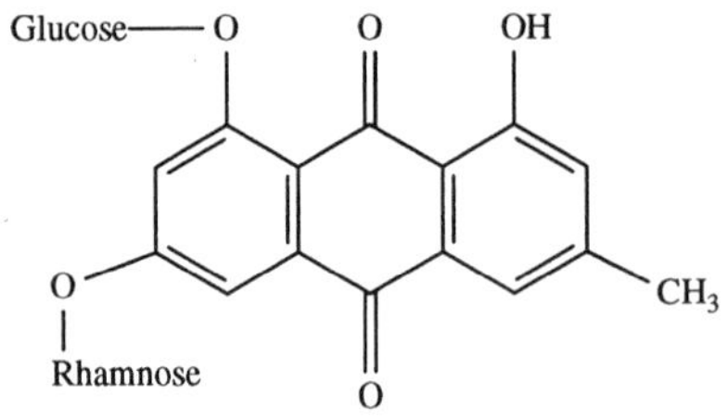

Glucofrangulin A

the alder buckthorn called *Rhamnus purshiana* DC. (*Rhamnaceae*). This is a tree, up to 18 metres high, growing on the west coast of North America. The crude drug is collected in British Columbia, Canada, and in the American States of Washington and Oregon. Most of the drug is obtained from the wild, but attempts are now being made to cultivate the plant because of the risk of extinction. The name cascara bark comes from the Spanish *cascara sagrada,* which means sacred bark. Like alder buckthorn bark, fresh cascara bark contains anthrones and must be stored before use. Stored cascara bark contains 8–10 % anthraquinones, mainly *cascarosides*. These compounds have a glucose residue linked directly to one of the carbon atoms of the anthraquinone skeleton; in other words, they are *C*-glycosides. There are four cascarosides, A–D, forming two stereoisomeric pairs which differ in their aglycones. In cascarosides A and B the aglycone is aloe-emodin anthrone, in C and D chrysophanol anthrone. All four cascarosides have a glucose residue attached to the oxygen at C-8. The *C*-linked glucose residue is located at C-10 in the β-configuration for cascarosides A and C and in the α-configuration for cascarosides B and D. Apart from these main components, cascara bark contains ordinary *O*-glycosides of emodin, aloe-emodin, and chrysophanol, some dianthrones, and a smaller amount of free aglycones. Preparations of cascara are used for treatment of constipation.

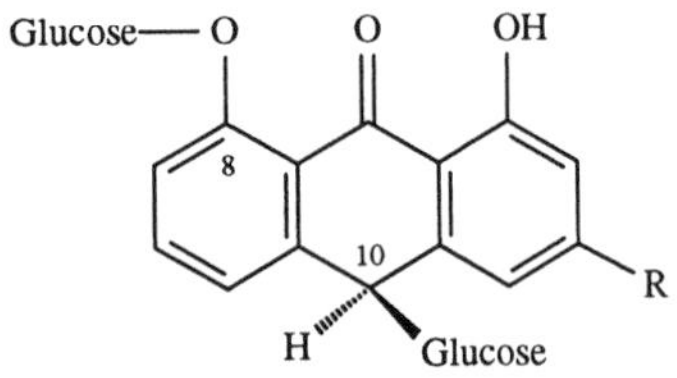

R = CH₂OH: Cascaroside A
R = CH₃: Cascaroside C

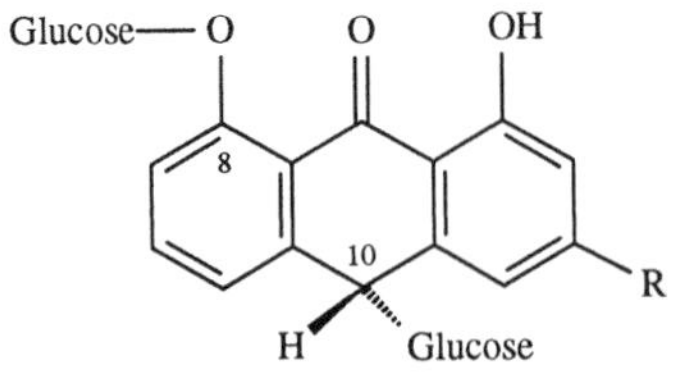

R = CH₂OH: Cascaroside B
R = CH₃: Cascaroside D

Aloes (Aloe) is the solid residue obtained by evaporation of the liquid that comes from cut leaves of plants belonging to the genus *Aloe* (*Liliaceae*). This genus comprises over 350 species, two of which are important for the production of aloes. *Aloe ferox* Mill. grows in

South Africa and affords *Cape aloes* (Aloe capensis), while *Aloe vera* L. (previously known as *Aloe barbadensis* Mill.) of the West Indies gives the product *Barbados aloes* (also known as Curaçao aloes, Aloe barbadensis).

Aloe species are succulents with large, fleshy, water-storing leaves arising from a relatively short stem. The leaves carry large spines at their edges. For the preparation of aloes, the leaves are cut near the base and placed on a watertight bed, on which the juice draining from the leaves is collected. After about 6 hours all the juice has emerged and it is then evaporated on an open fire. Cape aloes is evaporated at a higher temperature and for a longer time than Curaçao aloes. Thus, the end product of the Cape variety is hard and brittle and shows a glassy fracture, whereas Curaçao aloes is softer, with a wax-like fracture. Aloes contains *C*-glycosides of anthraquinones, and resins. The main constituents are *aloin A* and *aloin B* which are stereoisomeric C-10 glucosides of aloe-emodin anthrone. The aloins are difficult to separate and the mixture of them is known as *barbaloin*. Cape aloes also contains aloinosides A and B, which are the corresponding 11-α-L-rhamnosides of the aloins.

Cape aloes contains at least 18 % anthraquinone derivatives, calculated as barbaloin. The corresponding figure for Barbados aloes is 28 %. Aloes has a limited use as a purgative. Aloes is the most effective laxative among the anthraquinone-containing crude drugs and should be used with great care and only for short periods of time.

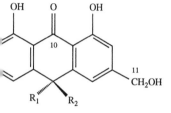

Aloin A: R₁ = H, R₂ = Glucose
Aloin B: R₁ = Glucose, R₂ = H

"Aloe vera" products

A transverse section of an *Aloe* leaf shows an epidermis underneath which is a parenchymatous tissue containing chlorophyll. Next, there follows a layer of large, parenchymatous cells containing mucilage. This layer often makes up three-fifths of the thickness of the leaf. It is followed by two rows of vascular bundles with large pericyclic cells. It is these pericyclic cells that contain the "aloetic"-juice from which aloes is prepared. When the leaf is cut the juice flows out of the cells.

A preparation containing mucilage from the parenchymatous layer of *Aloe vera* leaves is marketed as a herbal remedy for the treatment of sunburn and other skin afflictions. It is also an ingredient in cosmetic preparations. Note the different origins of the crude drug aloes and of the so-called "Aloe vera" preparations. The latter do not contain any anthraquinone derivatives.

Rhubarb (Rhei radix) is the peeled and dried rhizome and root of *Rheum palmatum* L. or of *R. officinale* Baillon (*Polygonaceae*). This plant is a perennial herb, native to the mountainous regions of China. There are about 50 *Rheum* species, which hybridize easily. This makes it difficult to specify exactly the plant source of the crude drug. It is, therefore, common to state that *Rheum palmatum* L. and possibly other species and hybrids of the genus *Rheum*, except *Rheum rhaponticum* L. (rhapontic rhubarb, edible garden rhubarb), supply the drug. Rhubarb is obtained from China and India, as well as from plantations in Europe. In China, rhubarb is collected from wild plants growing in mountainous districts up to 3 000 metres above sea level. The plant is harvested when 6–10 years old. The cortex and part of the wood of the rhizome are removed, so that the crude drug consists mainly of the pith, together with some of the wood. The material, threaded on cords, is often dried over fire.

The rhubarb produced in Europe is often harvested earlier, when 4–5 years old, and consists of smaller pieces. Rhubarb contains a complex mixture of anthraquinone glycosides, the aglycones of which can be grouped as follows:

a.   Anthraquinones not containing a carboxyl group: chrysophanol, aloe-emodin, emodin and physcion.

b.   Anthraquinones containing a carboxyl group: rhein.

c.   Anthrones and dianthrones of the compounds listed above under a and b. Noteworthy glycosides of compounds belonging to this group are the sennosides A and B, which also occur in Sennae folium (page 321). The effect of the drug is largely due to these sennosides.

d.   Heterodianthrones of the above anthrones, e.g. palmidin A, made up of aloe-emodin anthrone and frangula-emodin anthrone, and palmidin B, which is a heterodianthrone of aloe-emodin anthrone and chrysophanol anthrone. The two halves of palmidin C are frangula-emodin anthrone and chrysophanol anthrone, while palmidin D is a heterodianthrone of chrysophanol anthrone and physcion anthrone.

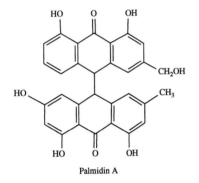

Palmidin A

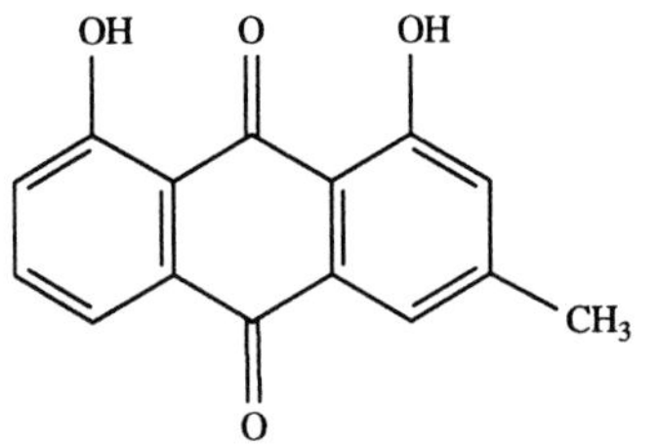

Chrysophanol

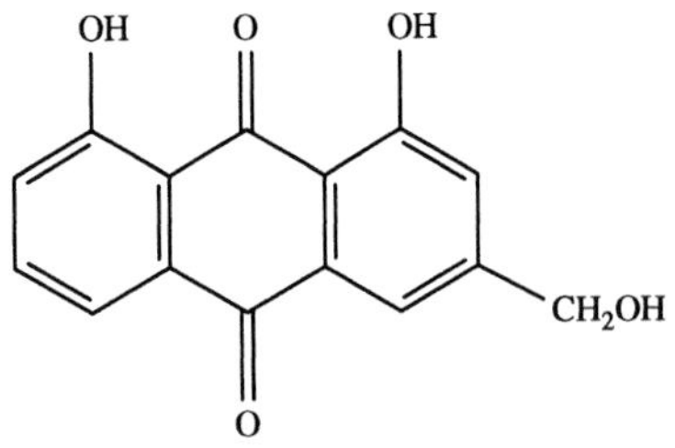

Aloe-emodin

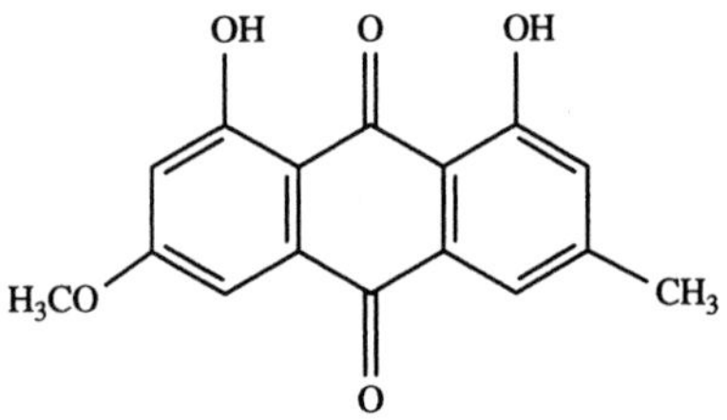

Physcion

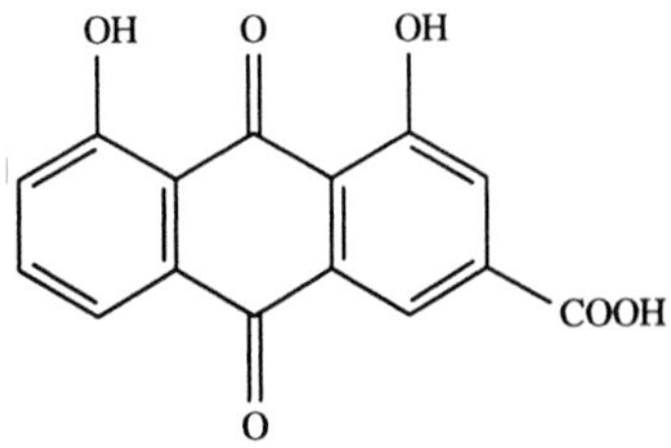

Rhein

The total content of anthraquinone derivatives in rhubarb can amount to 12 %, calculated as rhein, but it is very variable. The *European Pharmacopoeia* requires the content to be not less than 2.2 %. Rhubarb also contains tannins, which counteract the purgative effect of the anthraquinones and make rhubarb a relatively mild laxative.

So-called rhapontic rhubarb comes from *Rheum rhaponticum* L., the rhizome of which is not used as a drug. This rhizome can be distinguished from medicinal rhubarb by its content of the glycoside rhaponticin, the aglycone of which is a stilbene; this compound exhibits a blue fluorescence in UV light.

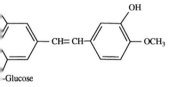

Rhaponticin

Senna leaf (Sennae folium). According to the *European Pharmacopoeia* this drug consists of the leaflets of *Cassia senna* L. (= *C. acutifolia* Delile), known as Alexandrian or Khartoum senna or *Cassia angustifolia* Vahl, known as Tinnevelley senna, (*Fabaceae*) (Plate 5, page 323). For Alexandrian senna, however, this is not a scientifically correct nomenclature. The species *Cassia senna* L. is today divided into the two species *Cassia acutifolia* Delile and *Cassia obovata* Collad. which thus both should be considered as the source of Alexandrian senna. The nomenclature of the plants that yield crude senna drugs is further complicated by the fact that some botanists consider that both Alexandrian and Tinnevelly senna emanate from the same species – *Senna alexandrina* Mill. This name is used by WHO in the *Herbal ATC Index* (page 47). The plants are shrubs with a height of about one metre. *Cassia angustifolia* is cultivated in the south of India, especially in the district of Tinnevelly. Alexandrian senna is cultivated in Sudan from wild or cultivated plants. The collection is performed mainly in September and the branches bearing the pods and leaves are dried in the sun whereupon the leaves and pods are separated by sieving. The leaves, which pass through the sieve together with stalk fragments and sand, are then graded partly by handpicking and partly by means of sieves. Three grades are obtained: (1) whole leaves, (2) a mixture of whole leaves and half-leaves and (3) siftings. The pods are handpicked into various qualities. Tinnevelley senna is obtained from cultivated plants in South India, N.W. Pakistan and Jammu. Pods and leaves are collected very carefully and the leaves are compressed into bales which contain few broken leaflets.

Like rhubarb, senna leaf contains a complex mixture of anthraquinone glycosides, the aglycones being monomeric anthraquinones as well as dianthrones and heterodianthrones. The total content is at least 2.5 %, calculated as sennoside B. Two of the main glycosides are sennosides A and B, the aglycones of which are stereoisomeric rhein dianthrones.

Smaller quantities of sennosides C and D are also found in senna leaf. They are glycosides of a stereoisomeric pair of heterodianthrones based on rhein anthrone and aloe-emodin anthrone.

Pure sennosides A and B are ingredients of conventional drugs.

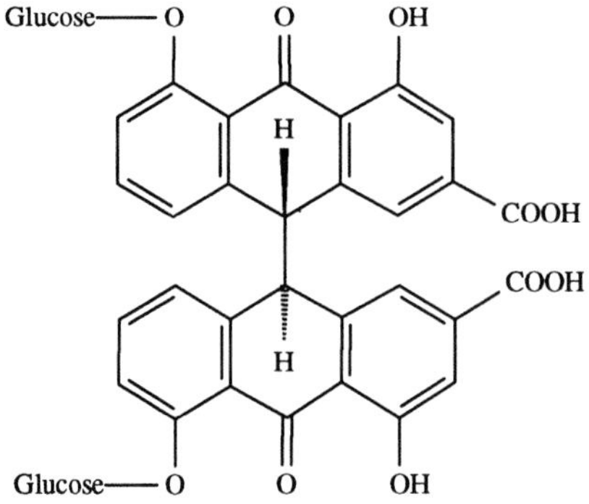

Sennoside A

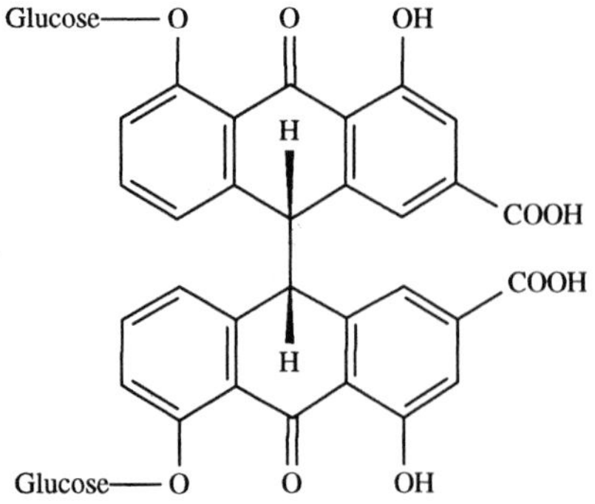

Sennoside B

Senna pods The *European Pharmacopoeia* lists the senna pods as two different crude drugs: Alexandrian senna pods (Sennae fructus acutifoliae) and Tinnevelley senna pods (Sennae fructus angustifoliae) (originating from *Cassia senna* L. and *Cassia angustifolia*, Vahl, respectively). The same remarks concerning the plant names as presented above for Senna leaf are also valid for these crude drugs. The constituents are the same as in senna leaf, with the addition of primary glycosides of rhein dianthrone containing up to 10 sugar residues. Senna fruit has a milder laxative effect than senna leaf.

Senna leaf and senna pods are the most frequently used anthraquinone-containing crude drugs for preparation both of conventional drugs and herbal remedies. They are considered to be safer than the other representatives of this group.

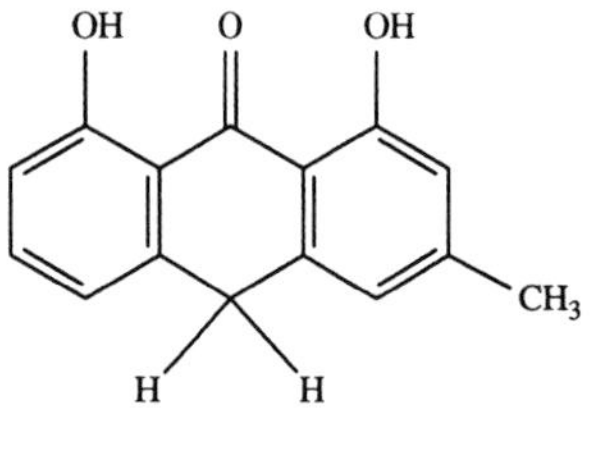

Chrysophanol anthrone

Physcion anthrone

Chrysarobin (Chrysarobinum) is a secretion from cavities in the heartwood of *Andira araroba* Aguiar (*Fabaceae*), a tree which reaches a height of 20 metres, and grows in the forests of Brazil. To obtain the chrysarobin, the tree is cut and the trunk split, so that the secretion can be scraped out. The material obtained (Goa powder) is dissolved in benzene, the solution is filtered, and the benzene is evaporated. The residue constitutes the crude drug chrysarobin, which consists of free anthrones, dianthrones and anthraquinones. The main components are chrysophanol anthrone and physcion anthrone.

Chrysarobin cannot be used as a laxative, because its effect is too violent. It is strongly irritating to the skin and to mucous membranes. It has been used against psoriasis but has nowadays largely been replaced by the synthetic anthranol dithranol (1,8,9-trihydroxy-anthracene).

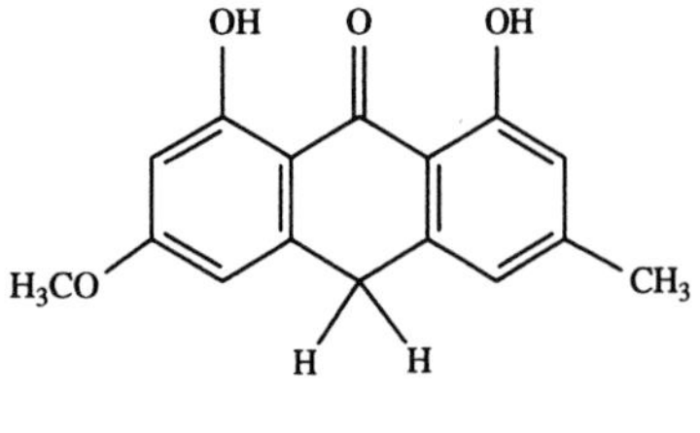

Carmine is a red pigment obtained from the cochineal louse, *Dactylopius coccus*, (Plate 7, page 323) which lives on cactae of the genera *Opuntia* and *Nopalea* (*Cactaceae*), especially *Nopalea*

Legumlnosae.

Cassia acutlfolla Delile.

Plate 5.
Cassia acutifolia Delile. Senna leaf and Senna pods (page 321, 322) are obtained from this plant. From: Köhler's Medizinal-Pflanzen (Gera Unternhaus 1887).

Plate 6.
Hypericum perforatum L. St. Johns Wort. An extract of the aerial parts is used as a herbal remedy against depressions (page 324). Photographed by Gunnar Samuelsson in the Botanical Garden, Uppsala, Sweden.

Plate 7.
A colony of the cochineal insect (*Dactylopius coccus*) feeding on a leaf of *Opuntia* sp. The dried insect is used for preparation of the dye carmine (page 322). Photograph by Gunnar Samuelsson in the island of Fuerteventura (Canary Islands).

Carminic acid

cochenillifera (L.) Salm-Dyck (previously known as *Opuntia cochenillifera* Miller). Its home countries are Mexico and Peru, but it has been brought to the West Indies, the Canary Islands and Spain. The dried female insect – the crude drug **cochineal** – contains about 10 % of an intensely red, water-soluble colouring matter, carminic acid, a *C*-glycoside of an anthraquinone derivative. Cochineal is obtained by removing the insects from the phyllocladia of the cacti with a small brush. They are then killed by plunging them into boiling water, by the heat of a stove, or by exposing them to the fumes of burning sulphur or charcoal. Carmine is a concentrate containing about 50 % carminic acid. Cochineal and carmine are old natural products which have become of interest as colouring matters for lipstick, food, confectionaries, and beverages, e.g. Campari, because they are believed to be less harmful than synthetic pigments.

Hypericin

Hypericin is a red-coloured, dimeric anthraquinone derivative which is present in the leaves and flowers of *Hypericum perforatum* L. (St. John's Wort, Plate 6, page 323).

Biosynthesis

Hypericin is formed from two molecules of emodin anthrone by oxidative phenolic coupling (Fig. 84).

Oxidative phenolic coupling is a widespread phenomenon both in the plant and in the animal kingdoms. Several groups of enzymes catalyse this reaction. They have iron or copper as a prosthetic group and are all able to effect one-electron transfer. Hydrogen peroxide and molecular oxygen, used as oxidants, are ultimately reduced to water, whilst the transition metal catalysts shift between their oxidized and reduced forms. As for the phenol part, the enzyme removes one electron and the phenoxy radical formed can couple in a number of ways.

Hypericin is a photosensitizing agent and causes the so-called "light sickness" in animals feeding on *Hypericum*. The first symptoms are skin reactions which appear when the animals are exposed to light following feeding. The active wavelengths range from 540 to 610 nm with peaks at 550–600 nm. Only animals with white or light-coloured coats are subject to this disease, which manifests itself in symptoms of psychomotor excitement and skin efflorescense in the form of blisters similar to those caused by burns. In grave cases the poisoning results in haemolysis, epileptic fits and the death of the animals.

Hypericin has anti-viral activity against retroviruses such as influenza virus and *Herpes simplex* virus both *in vitro* and *in vivo*. Exposure to fluorescent or visible light enhances the activity. The compound seems to act directly on the virus (possibly on membrane components) and has no activity against the transcription, translation or transport of viral proteins to the cell membrane and it has also no direct effect on the polymerase. The antiviral activity of hypericin is well defined from an experimental point of view, but the potential value for human retroviral-induced diseases remains

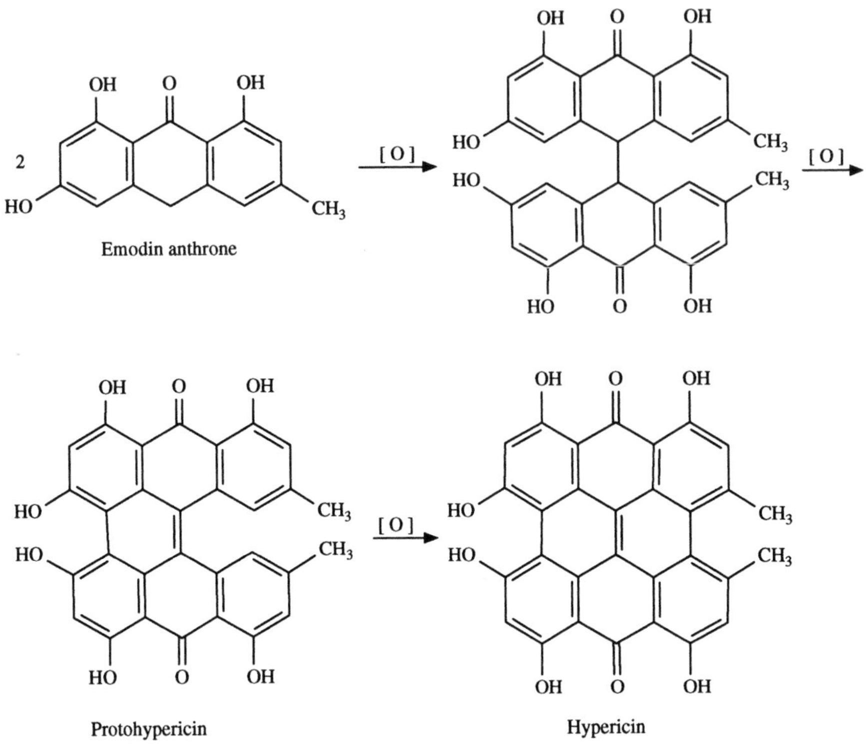

Fig. 84. Biosynthesis of hypericin

unknown and awaits clinical evaluation including pharmacokinetic studies. Also pseudohypericin (see below) has the same antiviral activity as hypericin.

Hypericin has been thought to be responsible for the antidepressant activity of extracts of Hypericum described below.

Crude drug **Hypericum** (Hyperici herba) is the dried, flowering, aerial parts of St. John's Wort, *Hypericum perforatum* L. (*Clusiaceae*). This species is a herbaceous perennial plant which is widely distributed in Europe, Asia and Northern Africa and now also naturalized in the USA. The plant is up to 90 cm tall, upright, branching with opposite, oblong, entire, sessile leaves exhibiting numerous pellucid dots and a few black glandular dots. The numerous, bright yellow flowers have many stamens and form a loose, branching cluster. A great number of black, glandular dots are visible in the petals. These dots (and also those in the leaves) are lysigenous secretory cavities, which contain hypericin. The harvesting period is July–August. The plant must be dried immediately to avoid degradation of the active principles.

Hypericin is the main component among the group of dimeric anthraquinone derivatives present in the plant. Other components are protohypericin, pseudohypericin and cyclopseudohypericin.

The total amount of hypericin and derivatives is 0.05–0.1 %. Emodin anthranol has also been found.

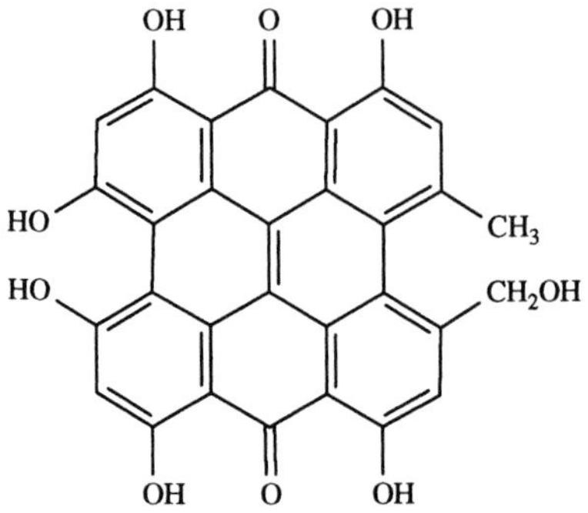

Pseudohypericin

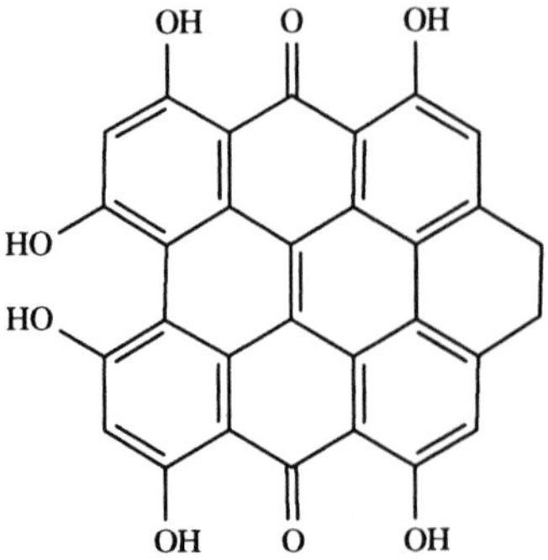

Cyclopseudohypericin

Besides the anthraquinone derivatives the crude drug also contains a mixture of proanthocyanidins (catechin, epicatechin and dimers, trimers, tetramers and high polymers of these compounds), quercetin and quercetin glycosides as well as dimeric flavonoids. Tannins are present in an amount of ca. 10 %. Hyperforin and adhyperforin are phloroglucin derivatives which are present (ca. 3 %) in fresh flowers. Hyperforin is the main component, the amount of which is ten times higher than that of adhyperforin. It is an unstable compound which is to a great extent destroyed during drying of the plant and several oxidation products of hyperforin have been isolated and identified. Hyperforin and its degradation products have antibacterial activity.

Hypericum is a well-known herbal remedy and an extract prepared with vegetable oils has a long reputation as an anti-inflammatory and wound-healing agent. Hypericum is also known for the antidepressant activities of ethanol-water extracts of the crude drug. Controlled clinical trials on a fairly large number of patients has verified that the extract has antidepressive activity in doses of 200–900 mg/day. The clinical indications for an antidepressive acticity of hypericum extract are also supported by results from a number of pharmacological studies. The extract inhibits the synaptosomal uptake of norepinephrine, serotonin and dopamine, induces β-receptor down-regulation when given subchronically to rats and is active in a large variety of behavioural models indicative of antidepressant activity. However, MAO-A and MAO-B inhibiting properties of the extract are probably too weak to contrtibute significantly to its antidepressive activity. The active constituent(s) is not known with certainty. Hypericin might be involved but also the flavonoids have been suggested to be responsible. Most commercial extracts today are standardized on the content of hypericin. There is, however, growing evidence that hyperforin might be the main active component. Thus the pharmacological effects exerted by the hyper-

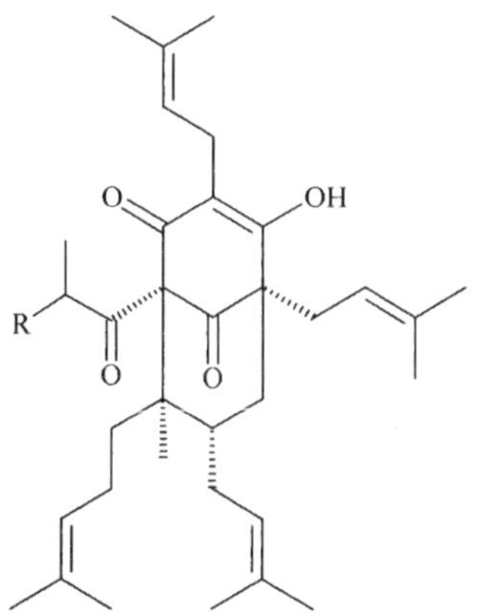

Hyperforin: R = CH$_3$
Adhyperforin: R = C$_2$H$_5$

icum extract are also effected by hyperforin. In rat, the plasma level for hyperforin, needed to inhibit synaptosomal uptake of several neurotransmitters, is about 700 nM. This level is reached at a dose of hypericum extract of 300 mg/kg, which is effective in most behavioural studies. Besides inhibiting the synaptosomal uptake of serotonin, norepinephrine and dopamine, like many other antidepressants, hyperforin also inhibits GABA and L-glutamate uptake, which indicates a different mechanism of action. There is evidence that the broad-spectrum effect on neuronal uptake is obtained by an elevation of the intracellular Na$^+$ concentration, probably due to the activation of sodium conductive pathways not yet finally identified but most likely ionic channels. This makes hyperforin the first member of a new class of compounds with antidepressant activity due to a completely novel mechanism of action.

Hypericum has been found to interact with many prescribed medicines such as warfarin, cyclosporin, theophylline, digoxin, HIV protease inhibitors, HIV non-nucleoside reverse transcriptase inhibitors, anticonvulsants, selective serotonin reuptake inhibitors, triptans and oral contraceptives. The main reason for many such interactions is the activation by hyperforin of the pregnane X receptor (PXR), an orphan nuclear receptor that regulates expression of cytochrome P450-3A4 (CYP3A4) monooxygenase. The activation of PXR induces expression of CYP3A4 which is involved in the oxidative metabolism of more than 50 % of all drugs. Increased levels of CYP3A4 causes a more rapid metabolism resulting in lower plasma levels of the prescribed medications. Induction of the cytochrome P450 isoenzymes CYP2C9 and CYP1A2 as well as the transport protein P-glycoprotein also seem to be involved.

Patients taking medicines metabolized by CYP3A4 or the other isoenzymes should consult their physician before taking hypericum as dose adjustment of conventional treatment might be necessary. When drugs such as warfarin and cyclosporin are prescribed, the doses of which have to be individually adjusted, it is also vital that the doctor knows that the patient is taking hypericum. If a patient is taking hypericum during the titration of such drugs, stopping the intake of hypericum can result in dangerous situations caused by increased blood levels of the drugs.

Biosynthesis of hyperforin

Biosynthetically hyperforin can be regarded as an acyl phloroglucinol moiety to which five isoprenoid units have been added. Labelling experiments with $1-^{13}C$ glucose and $U-^{13}C_6$ glucose followed by NMR analysis of the isolated hyperforin have shown that the acyl phloroglucinol moiety (A below) is formed via a polyketide mechanism and that the biosynthesis of the remaining part of the molecule involves a triple electrophilic substitution of the aromatic nucleus by two dimethylallyl diphosphate units and one geranyl-diphosphate formed via the non-mevalonate pathway (page 369ff). A ring closure is triggered by electrophilic attack of a third dimethylallyl diphosphate unit on the 2'/3' double bond of the attached geranyl chain. The sequence of the necessary steps is at present unidentified. In the reaction scheme below the bonds of the five isoprenoid units have been highlighted.

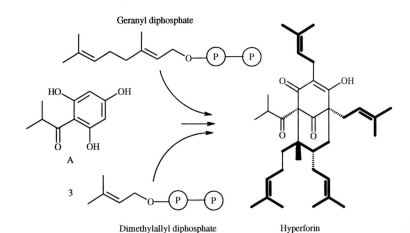

Geranyl diphosphate

Dimethylallyl diphosphate

Hyperforin

Polyketides of mixed biogenetic origin

The biosynthesis of most of the polyketides, discussed above, starts with acetate in the form of acetyl-CoA and the chain is lengthened stepwise by reaction with malonyl-CoA units. Other aliphatic acids, e.g. propionic acid, can replace acetic acid as starters. An aromatic acid derived from the shikimic acid route, e.g. cinnamic acid, can also function as a starter in a polyketide biosynthesis, thus giving a compound of mixed biogenetic origin. *Flavonoids* and *kava pyrones* are examples of such substances. *Flavanolignans* are derivatives of flavonoids, the structures of which also involve coniferyl alcohol which is derived from the shikimic acid pathway. *Mycophenolic acid* is a combination of a polyketide and a terpene structure derived from farnesyl diphosphate. The structures of the *ansamycin* group of antibiotics include a ring derived from the shikimic acid pathway. *Rotenoids* are examples of still more complicated polyketides in the biosynthesis of which cinnamic acid acts as a starter but where also a structural element derived from the isopentenyl diphosphate pathway (page 366) is incorporated at the last stage. *Khellin* is another example of a polyketide in which a fragment originating from isopentenyl diphosphate has been incorporated.

Flavonoids

Structurally flavonoids are derivatives of 1,3-diphenylpropane. As shown below (Fig. 85), one of the benzene rings – ring B – is derived from the shikimic acid pathway while the other – ring A – originates from acetate and is formed by ring closure of a polyketide. One hydroxyl group in ring A is practically always situated in *ortho*-position to the side chain. Ring closure involving the 3-carbon chain and this hydroxyl group can form a third 6-membered ring, characteristic of most of the flavonoids, or a 5-membered ring present in the *aurones*. Finally, movement of the β-ring to carbon 2 of the 3-carbon chain forms the *isoflavonoids*. Structurally all flavonoids can be referred to the four basic structures presented on the next page.

Flavonoids are found in ferns and higher plants where they occur both in the free state and as glycosides. They differ in their substituents – mostly hydroxyl or methoxyl groups – and in the nature and position of the sugar residues bound to the aglycones. About 6 000 compounds have so far been isolated from thousands of plant species and they form the largest group of naturally occurring phenols. The most common classes are the *flavones, flavonols, anthocyanidines, flavanones* and *isoflavonoids*. Together these classes represent more than 80 % of the known flavonoids.

The flavonoids are yellow compounds and contribute to the yellow colours of flowers and fruits where they are present as glycosides, dissolved in the cell sap. Exceptions are the anthocyanidins and their glycosides (known as anthocyanins) which are red, violet or blue depending on the pH of the cell sap, as the colour is associated with the distribution of positive charge throughout the arylsubstituted chroman ring system. Chelate formation with Fe^{3+} or Al^{3+} also influences the colour of the anthocyanidines. Free flavonoids occur dissolved in essential oils.

Biosynthesis of flavonoids

Flavonoids are biosynthesized via a combination of the shikimic acid and acylpolymalonate pathways. The starting material is

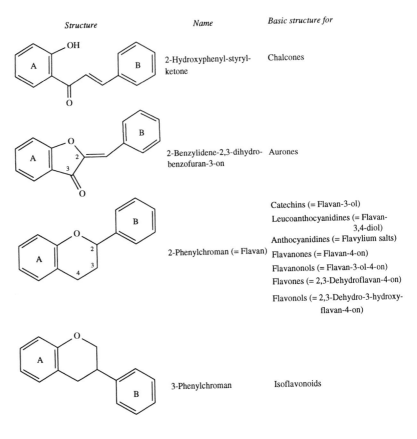

Structure	Name	Basic structure for
	2-Hydroxyphenyl-styryl-ketone	Chalcones
	2-Benzylidene-2,3-dihydro-benzofuran-3-on	Aurones
	2-Phenylchroman (= Flavan)	Catechins (= Flavan-3-ol) Leucoanthocyanidines (= Flavan-3,4-diol) Anthocyanidines (= Flavylium salts) Flavanones (= Flavan-4-on) Flavanonols (= Flavan-3-ol-4-on) Flavones (= 2,3-Dehydroflavan-4-on) Flavonols (= 2,3-Dehydro-3-hydroxy-flavan-4-on)
	3-Phenylchroman	Isoflavonoids

phenylalanine which is deaminated by PAL (phenylalanine ammonia lyase, page 208) to yield cinnamic acid which in a reaction catalysed by trans-*cinnamate 4-monooxygenase* (EC 1.14.13.11, also known as cinnamate 4-hydroxylase) is converted to *p*-coumaric acid. PAL and cinnamate 4-hydroxylase are closely associated. Cinnamic acid is not released from PAL into the surrounding environment but is directly transferred to the active centre of the second enzyme – cinnamate 4-hydroxylase. Such a process is called *channelling*. Control of the biosynthesis by assembly–disassembly of the channelling mechanism seems likely.

The next step is addition of CoA, catalysed by *4-coumarate-CoA ligase* (EC 6.2.1.12) yielding *p*-coumaroyl-CoA (Fig. 86). *p*-coumaryl-CoA functions as the starting compound in a polyketide synthesis, catalysed by *naringenin-chalcone synthase* (EC 2.3.1.74, also known as chalcone synthase) in which three malonyl-CoA react to form naringenin chalcone (4,2',4',6'-tetrahydroxychalcone). Chalcone synthase is a type III (page 270) polyketide synthase which has been cloned and sequenced from a number of plants. The enzyme is a dimer and the molecular mass of the subunits is 42 kDa. The enzyme is localized in the cytoplasm and is associated with endoplasmic reticulum membranes, where it may function as part of a complex.

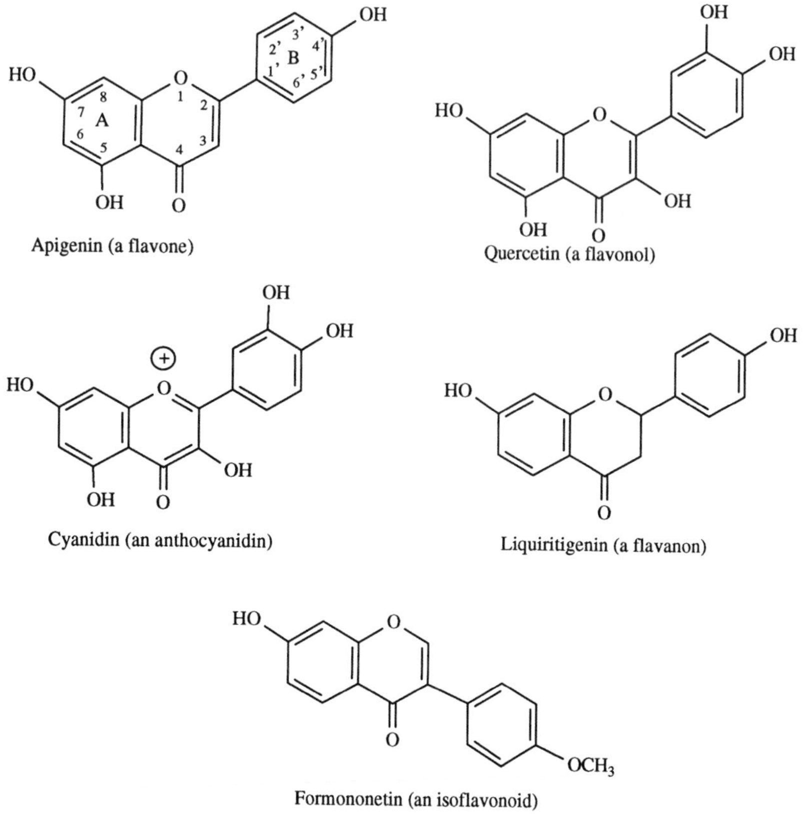

Fig. 85. Examples on the most common classes of flavonoid aglycones

The three-dimensional structure of crystallized recombinant chalcone synthase from *Medicago sativa* L. has been determined. The enzyme forms a symmetric dimer where each monomer has two distinct structural domains (upper and lower). The active site of each monomer lies in a cleft between the upper and lower domains and they are functionally mutually independent. Four conserved residues – Cys-164, His-303, Asn-336 and Phe-215 – have been

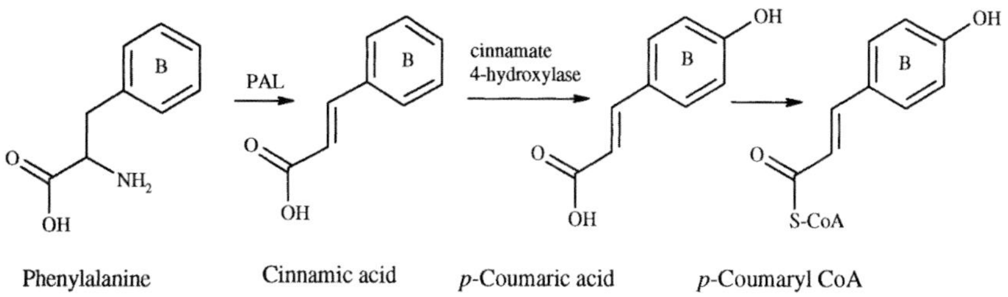

Fig 86. Biosynthesis of p-coumaryl-CoA

331

shown to participate in the catalytic cycle of chalcone synthase (for the three-letter code for amino acids, see page 509). Unlike other polyketide synthases, chalcone synthase does not have an acyl carrier protein with 4´-phosphopantetheine as a flexible arm for shuttling substrates and intermediate polyketides. Instead CoA thioesters are used for that purpose.

The proposed catalytic mechanism is presented in Fig. 87.

The assembly of the polyketide begins with the binding of *p*-coumaryl-CoA, which is then covalently attached to Cys-164 in a reaction mediated by His-303 (A in Fig. 87). Asn-336 hydrogen bonds with the thioester carbonyl, further stabilizing formation of the tetrahedral reaction intermediate. CoA then dissociates from the enzyme, leaving a coumaroyl-thioester at Cys-164. The first molecule of malonyl-CoA is then bound and undergoes decarboxylation, promoted by the action of Asn-336 and Phe-215 (B in Fig. 87). The role of Phe-215 in the decarboxylation process is proposed to orient the terminal carboxylate of malonyl-CoA and to favour conversion of the negatively charged carboxyl group to a neutral carbon dioxide molecule. The decarboxylation yields the enolate of acetyl-CoA, which then (C in Fig. 87) attacks the carbonyl of the enzyme-bound coumaryl thioester, releasing the thiolate anion of Cys-164 and transferring the coumaroyl group to the acetyl moiety of the CoA thioester. The elongated coumaroyl-acetyl-diketide-CoA is recaptured by Cys-164. Following release of CoA the diketide is subjected to two additional rounds of elongation, resulting in a tetraketide which undergoes an intramolecular Claisen condensation followed by aromatization to yield naringenin chalcone. The final step is ring closure, catalysed by *chalcone isomerase* (EC 5.5.1.6) to yield the flavanone naringenin.

Thus in flavonoids, ring A is formed from acetate, whereas ring B originates from shikimic acid and the three carbon atoms connecting rings A and B come from phosphoenol pyruvate (page 138). Subsequent hydroxylations, reductions and methylations lead to the different flavonoids. Numerous enzymes catalysing these reactions have been isolated and characterized. Additional hydroxylation of the B-ring occurs after formation of the C15 skeleton and not – as previously believed – by incorporation of properly substituted hydroxycinnamic acids. Specific oxygenases introducing the 3'- as well as the 5'-hydroxyl group have been characterized. The glycosidation is probably one of the last reactions in the biosynthesis of the various plant pigments.

Being relatively simple proteins, naringenin-chalcone synthase and other type III PKSs have been subject to genetic engineering aiming at finding more diverse and novel compounds with desirable pharmacological profiles.

Biological and pharmacological activity

Flavonoids are the major red, blue and purple pigments in plants

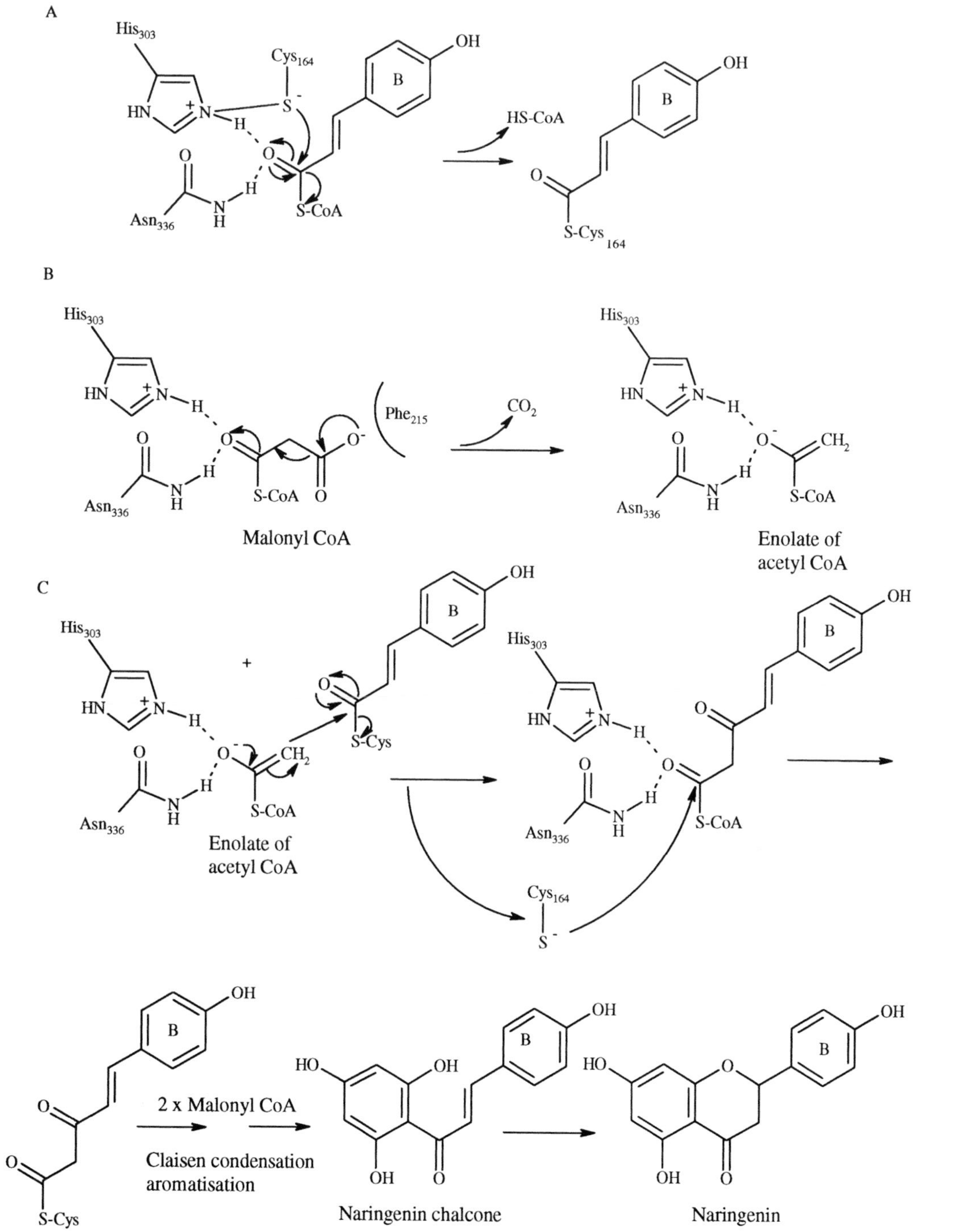

Fig. 87. Reaction mechanism of chalcone synthase. A. Attachment of p-coumaryl-CoA to Cys-164 in chalcone synthase. Dotted lines indicate non-covalent bonds. B. Decarboxylation of malonyl-CoA. C. Formation of a diketide intermediate and continuation of the biosynthesis

that are important in the recruitment of pollinators and seed dispersers. Besides providing beautiful pigmentation in flowers, fruits, seeds and leaves, flavonoids also have key roles in signalling between plants and microbes, in male fertility in some species, in defence as antimicrobial agents and feeding deterrents and in UV protection.

The first suggestion of pharmacological activity of flavonoids was presented by Szent-Gyorgyi in 1938, who reported that citrus peel flavonoids were effective in preventing capillary bleeding and fragility associated with scurvy. This effect has been called vitamin P activity, but the existence of the effect is not regarded as entirely proven. Since then a great number of pharmacological effects have been ascribed to flavonoids and certain individual members of the group have been found to exert a multiplicity of actions. The following activities have been described: anti-inflammatory, antihepatotoxic, antitumour, antimicrobial, antiviral, enzyme inhibiting, antioxidant and central vascular system effects. Most of the pharmacological investigations of flavonoids have been performed *in vitro*, and it is in most cases not known if the concentrations giving the observed effects are therapeutically realistic. As very few pharmacokinetic studies have been performed one does not know whether the studied flavonoids arrive at the alleged site of action in the native state and in sufficient concentrations. This is the main reason why so few flavonoid based drugs are used in professional medicine. Many herbal remedies contain crude drugs, the constituents of which are mainly flavonoids and whose alleged effect is ascribed to these compounds, a statement the truth of which remains to be proved in most cases.

At present the antioxidant and radical scavenging effects are subjects of great reseach interest. The presence of flavonoids in vegetables and fruit is thought to be one of the reasons for the beneficial influence on human health of these components of the diet. Especially coronary heart disease and cancer are areas where flavonoids may be of importance for prevention. The French have plasma cholesterol levels comparable with the Americans and the fat intakes are comparable as well, but in spite of this the French have a far lower incidence of coronary disease. This has been termed "the French paradox" and has been related to the high French consumption of red wine which contains flavonoids like quercetin and myricetin. Oxidation of LDL (Low Density Lipoproteins) is an important factor in the development of atherosclerosis, which, however, is a very complex process. *In vitro* experiments have shown that flavonoids inhibit the oxidation of LDL, triggered by cellular (e.g. macrophages) as well as non-cellular factors (e.g. copper ions). Scavenging of free radicals participating in oxidative processes may thus be an explanation for some of the observed effects of the flavonoids. Other mechanisms, such as protection of α-tocopherol, have also been suggested. In addition, quercetin reduces cytotoxic effects of oxidized LDL. The mechanism underlying this effect is not known. It is also of interest to note that 15-lipoxygenase has been suggested to participate in the oxidation of LDL and that several flavonoids inhibit this enzyme.

Epidemiological studies have been performed where the intake of flavonoids has been found to be inversely related to coronary heart mortality. Corresponding studies of effect of flavonoids for prevention of cancer have not given equally positive results.

There are considerable difficulties involved in the study of the importance of dietary flavonoids: (1). Flavonoids undergo structural changes in the gastrointestinal tract. Adsorbed metabolites may differ very much in structure from those ingested. It is therefore necessary to sharply distinguish between experiments where flavonoids are given orally and those where administration is performed via other routes. (2). The food also contains other active substances such as vitamin C. The possible effects of flavonoids may be dependent on interaction with other dietary components. (3). The effect seems to be mainly preventive and it is much more difficult to study prevention than treatment. (4). Flavonoids have a broad spectrum of effects and there seems to be no effect for which the flavonoids are solely responsible. (5). The metabolism and pharmacokinetics of flavonoids in humans are not very well understood. Depletion/repletion studies are difficult to perform as it is virtually impossible to compose a diet free of flavonoids.

Rutin

Rutin (Rutosid) is the 3-rhamnoglucoside of 5,7,3',4'-tetrahydroxy-flavonol (quercetin). Rutin is a component in multivitamin preparations because of its alleged so-called vitamin-P effect (see above). Rutin can be extracted from the buds of *Sophora japonica* L. (*Fabaceae*), an Indonesian tree. It can also be obtained from buckwheat, *Fagopyrum esculentum* Moench (*Polygonaceae*), which grows in Europe, or from leaves of the Australian tree *Eucalyptus macrorhyncha* F. Muell. (*Myrtaceae*).

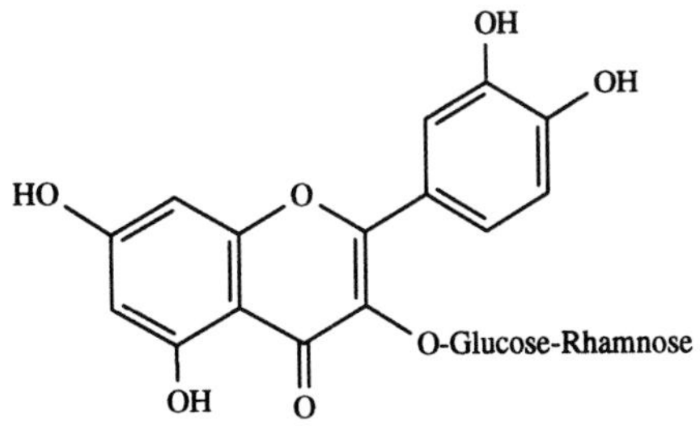

Rutin

Genistein and Daidzein

Isoflavones have also been called *phytoestrogens* as they exert a weak ability to act as agonists at oestrogen receptors. They have been extensively investigated for potential health effects in menopausal women, as considerable differences have been found between Asian and Western women in the occurrence of cardiovascular disease, osteoporosis, menopausal symptoms and breast cancer. In Asian women these conditions are rare and the levels of isoflavones in urine and plasma are high. The main dietary source of isoflavones in Asia are soyabean seeds (*Glycine max* Merrill., see page 251) with genistein and daidzein as the major components (up to 3 mg/g).

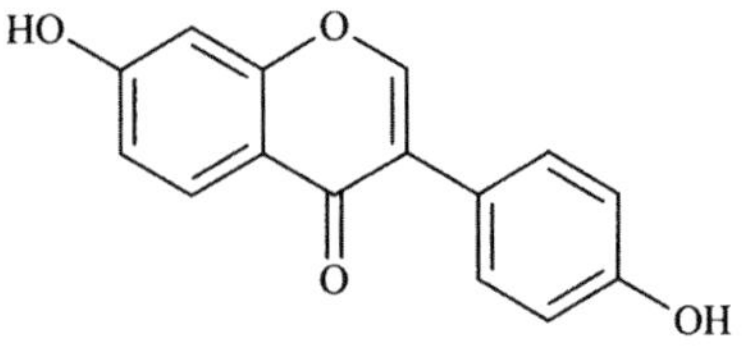

Daidzein

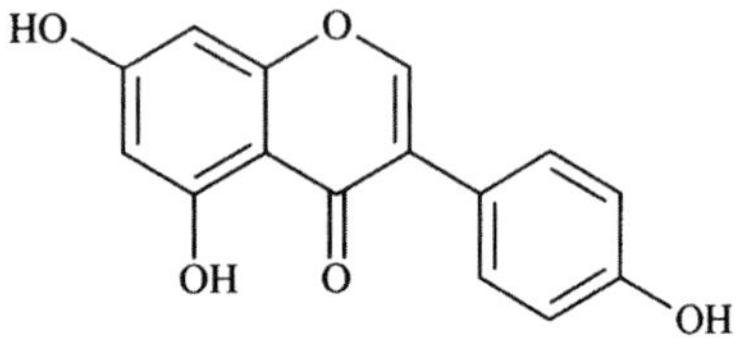

Genistein

Diet supplementation with soyabean phytoestrogens has been reported to ameliorate hot flashes and other symptoms of postmenopausal women. After menopause the ovaries stop producing the hormone oestrogen. Oestrogen has a positive influence on the metabolism of calcium and a lack of sufficient oestrogen can lead to bone loss (thinning) and/or the brittle-bone disease osteoporosis. Homone replacement therapy can reduce bone loss and the risk for osteoporosis but has been shown to induce an increased risk for breast cancer, heart disease, stroke and venous thromboembolism. There is therefore a need for oestrogen-like compounds that can selectively act against bone loss without causing negative oestrogenic function against the uterus (selective oestrogen receptor modulator = SERM). Genistein has shown SERM activity in ovariectomized mice and soyabean isoflavonoids have been shown to improve bone mass and reduce bone resorption. Dietary soyabean isoflavonoids also appear to possess cancer chemopreventive abilities and there is a correlation between the lower incidence and mortality from breast cancer in Asian women, compared with Western women, and the intake of soyabean isoflavones. Various biological activities have been demonstrated for soyabean isoflavones that may help to explain the chemopreventive properties, but the exact mechanism is not clear and is a hot topic for current research.

Numerous commercial phytoestrogen supplements are available, most of them including concentrated soyabean extracts. There are, however, considerable quality differences. There are limited data examining the relative clinical effectiveness of specific preparations. The most widely used soya products, soya oil, soya sauce and soya lecitin, do not have significant levels of isoflavones. This is also the case for alcohol-washed soya proteins. Recent research suggests significant differences in the bioavailability between foods rich in phytoestrogens and supplements. Current pharmacokinetic data indicate that divided doses of the soya food or supplement throughout the day are better than a single dose for maintenance of a steady-state serum level of isoflavones.

Kava pyrones

Kava pyrones occur in kava, the root of *Piper methysticum* Forst. (*Piperaceae*). This crude drug contains the three substances depicted below.

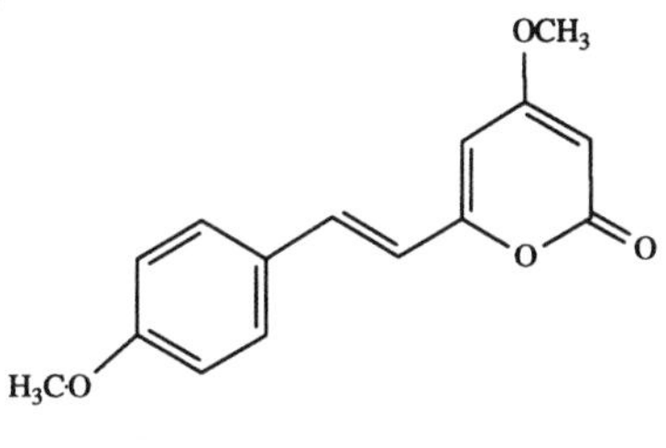

Yangonin

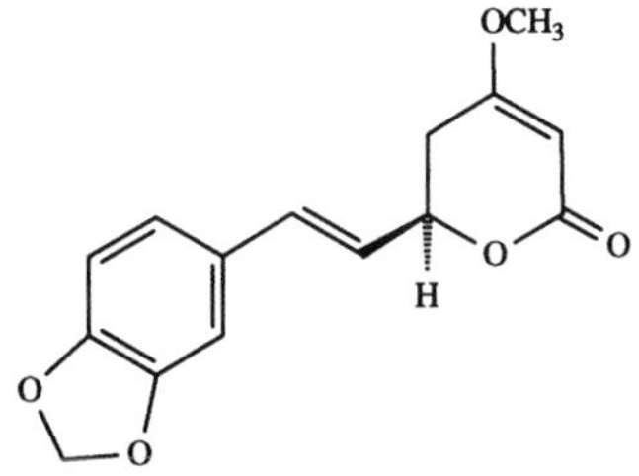

Methysticin

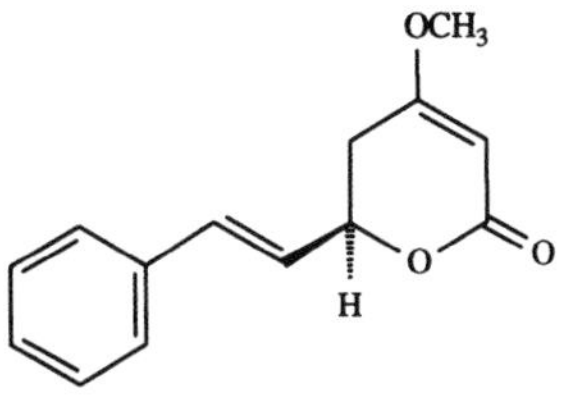

Kavain

Dihydrokavain and dihydromethysticin (both lacking the double bond in the bridge connecting the two rings), are also present, as well as desmethoxyyangonin, which lacks the methoxyl substituent on the benzene ring.

Biosynthesis Kava pyrones are biosynthesized via a combined shikimic acid–acylpolymalonate pathway. As with the flavonoids, a cinnamic acid derivative acts as the starting substance, giving the benzene ring, the bridge, and the bridgehead carbon atom of the pyrone ring, whereas the other four carbon atoms of the pyrone ring are acetate-derived.

Traditional use In Polynesia, Melanesia and Micronesia kava is used for the preparation of a drink which plays an important role in social life and which has been drunk ceremonially for thousands of years. Today kava is used in both traditional ceremonies and at informal social events. At the traditional ceremonies the procedure is carried out in accordance with a complex ritual, starting with preparation of the drink that is done by crushing the roots, which can be dried or fresh. Fresh roots give a more potent drink. The crushed roots are mixed with water and the mixture strained to remove the plant debris. The crushing of the roots was formerly done by young persons, often boys but occasionally virginal women, selected for their flawless teeth. They chewed the roots and then expectorated the macerated pulp into a bowl, where water was added. In Samoa the *taupou* or village virgin was traditionally the only person allowed to prepare and serve kava to assembled chiefs on ceremonial occasions. This custom continues in many villages today. But the chewing method of preparing kava has declined in modern Polynesia and instead kava roots are pulverized mechanically. The powder is then mixed with water and strained by hand with the traditional plant fibres (made from the bark of *Hibiscus tiliaceus* L., *Malvaceae*) or with porous cotton cloth.

In a traditional ceremony adult men sit cross-legged on the ground in a modified circle and the server sits in the centre next to the kava bowl. The ceremony begins with the recitation of ritual chants over guests, roots and the beverage. The server then ladles kava into cups made of a coconut shell or calabash. The cups are served to participants in descending order of social rank. Sounds of solemn clapping and elaborate chants fill the ceremony. Cups are continually refilled and the process repeated until the ceremony's conclusion, several to dozens of rounds later. In most kava societies, the participation by women in this ceremony was, and still is, unacceptable.

Formerly kava ceremonies were held only at very solemn occasions, e.g. negotiations between chiefs and in connection with religious events, and the drinking was in some islands restricted to royalty and priests. Ceremonial kava drinking today is performed to welcome visitors to the villages and to help the villagers reach consensus on potentially controversial decisions affecting the community. Both the beverage and the ceremony seem designed to increase friendly feelings and reduce the possibility of hostility.

In some islands e.g. Western Samoa, Fiji and Vanuatu, the use of kava has become more secularized and powdered kava is prepared commercially and sold like tea in muslin bags ready for preparation of "instant" kava by soaking the bag in water. It is also possible to drink kava informally in pubs and bars. Kava is also gaining a more important role in Western nations, e.g. in the USA where it is sold in many health food stores as a remedy against insomnia and nervousness.

Pharmacology, side effects   The kava drink has a relaxing effect, being at first mildly euphoric, and later hypnotic, due to the presence of kava pyrones. The principal pharmacological effect of these compounds is a centrally derived relaxation of the skeletal muscles. There are quantitative differences between the different pyrones, but if mixed, they have mutually synergistic effects. The kava pyrones also affect the pattern of movement and reflex excitability, and they have spasmolytic and anti-inflammatory effects. The kava pyrones have been tested clinically as remedies against epilepsy, but have been found to be unsuitable due to side-effects such as vomiting, diarrhoea and allergic skin reactions.

The use of kava for long periods or in high doses may cause changes of the skin, manifested as a peculiar scaly eruption. It is not known which constituent(s) of the kava is responsible for this effect, nor has the mechanism of it been elucidated. The condition reverses promptly with cessation of drinking kava and the new skin formed is smooth and clear. It is also said that skin conditions that existed before the outbreak of kava dermopathy are cured, but this allegation has never been formally studied. There are also reports of severe liver toxicity, allegedly caused by use of over-the-counter drugs containing kava. These reports have caused authorities in several countries to ban the sale of kava products.

Flavonolignans

Silybin Silybin is a main component of the complex *silymarin*, which can be extracted from the fruit of the St. Mary thistle, *Silybum marianum* Gaertn. (*Asteraceae*). Other compounds present in silymarin are silydianin and silychristin. They are all derived from the flavanonol taxifolin (2,3-dihydroquercetin) and coniferyl alcohol. Compounds of this type are called *flavonolignans*.

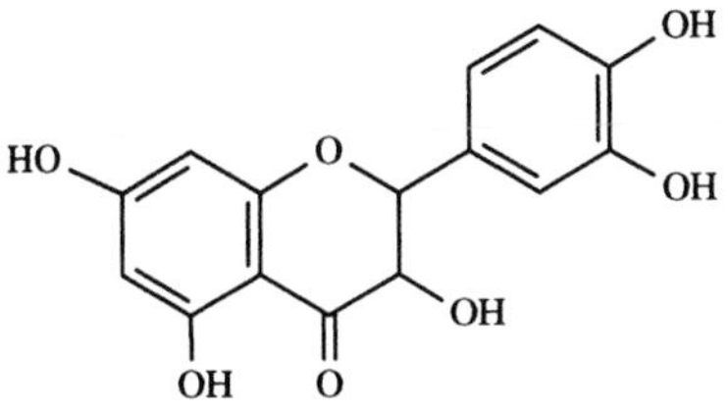

Taxifolin

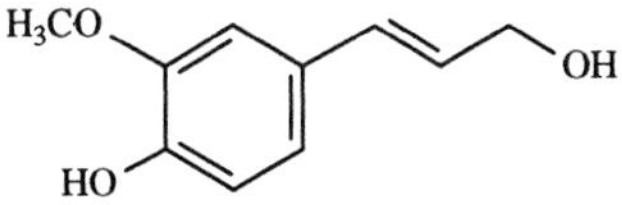

Coniferylalcohol

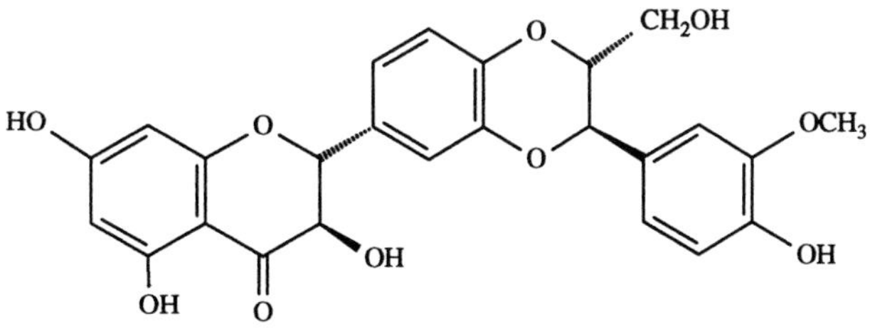

Silybin

In pharmacological experiments, silybin has been shown to protect the liver from the effects of the amatoxins, the poisonous principles of certain species belonging to the genus *Amanita*, a group of highly toxic fungi (page 519). Silybin, in combination with penicillin, has been used clinically, with good results, in the treatment of poisoning by the death cap *Amanita phalloides* Fr. Link. Drugs containing silymarin are used for the treatment of various types of liver disease.

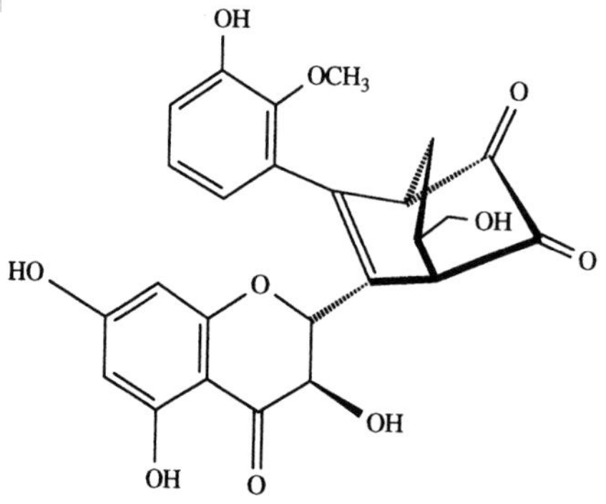

Silydianin

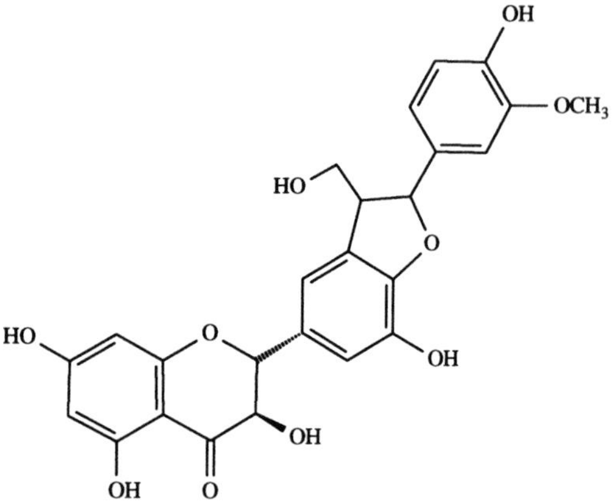

Silychristin

Crude drug **St. Mary Thistle** (Cardui mariae fructus) is the dried fruit, freed from its pappus (tuft of silky hairs), of the St. Mary thistle, *Silybum marianum* (L.) Gaertn. (*Asteraceae*), which is an about 1 m high thistle with purple tubulate florets and leaves with white areas. The plant originates from the Mediterranean area and is cultivated in North Africa and South America for production of the crude drug. The fruit contains fixed oil (20–30 %) and the above-mentioned mixture of flavonolignans – silymarin – as well as taxifolin and other flavonoids. Also free coniferyl alcohol is contained in the fruit. Cardui mariae fructus is used for extraction of silymarin and is also a component in herbal remedies for the treatment of diseases of the bile and the liver.

Mycophenolic acid

Mycophenolic acid is a polyketide derivative produced by the fungus *Penicillium brevicompactum*. The compound was isolated in 1896 and was shown to inhibit the growth of different bacteria. The structure was determined in 1952. Mycophenolic acid is too toxic to be used clinically as an antibiotic but it has been found to have immunosuppressant activities and is applied as the *N*-morpholinoethyl ester (**mycophenolate mofetil**), usually in combination with cyclosporine, to prevent rejection of transplanted organs, particularly kidneys. Since 1998 it has also been approved for use in heart transplants, in combination with other materials such as cyclosporin and corticosteroids.

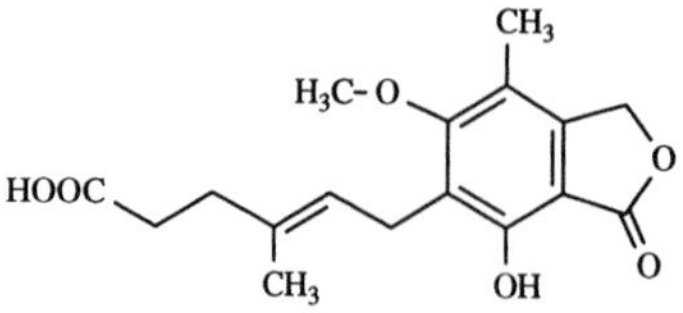

The body's normal rejection of transplanted organs is due to the action of "killer T-lymphocytes" which attack the foreign tissue. The rejection problem can be overcome by inhibition of lymphocyte proliferation, which requires DNA synthesis, which in turn is dependent

Mycophenolic acid

on the formation of purine monophosphates. The mechanism for the immunosuppressive action of mycophenolic acid is inhibition of the enzyme *IMP-dehydrogenase* (EC 1.1.1.205), also known as inosine 5′-monophosphate degydrogenase (IMPDH), an enzyme which is involved in the *de novo* biosynthesis of guanosine-5′-monophosphate, an important purine component of DNA. Lymphocytes are uniquely dependent on the *de novo* pathway, as they almost entirely lack the ability to reconvert purines formed in catabolic processes to purine monophosphates. Inhibition of IMPDH by mycophenolic acid results in an overall reduction of DNA synthesis in lymphocytes and thus to inhibition of proliferation of killer T-lymphocytes.

Biosynthesis   The biosynthetic precursor is a polyketide chain comprising 8 carbons, which is C-methylated and subjected to an aldol reaction and aromatization yielding 5-methylorsellinic acid (Fig. 88). Following oxidation of the methyl group *ortho* to the carboxyl, a lactone ring is formed, yielding a phthalide intermediate which is alkylated by farne-

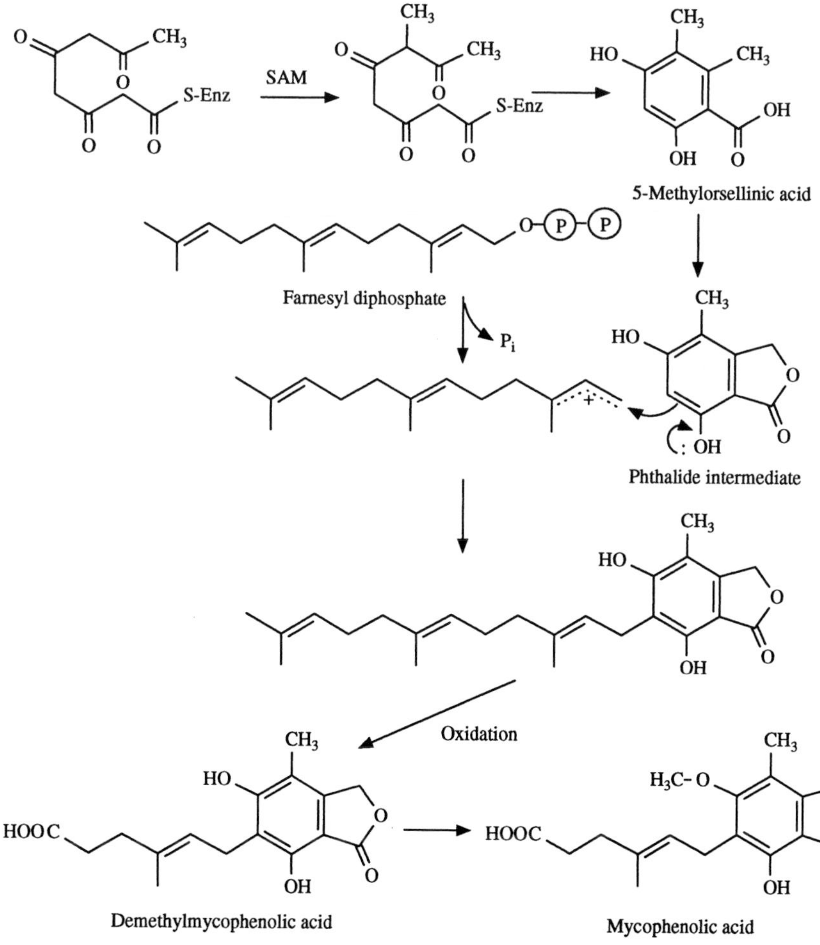

Fig. 88. Biosynthesis of mycophenolic acid

341

syl diphosphate. A double bond in the farnesyl residue is oxidized, leading to the shortening of the chain with formation of demethylmy-cophenolic acid which finally is *O*-methylated to yield mycophenolic acid.

The ansamycin group of antibiotics

The ansamycins constitute a class of antibiotics characterized by an aliphatic bridge (the "ansa bridge"; ansa = handle) linking two non-adjacent positions of an aromatic nucleus. The ansamycins can be divided into two subclasses depending on whether the aromatic nucleus is a naphthalene derivative or whether it is a benzene deriv-ative. The streptovaricins, the rifamycins, the halomicins, the toly-pomycins and naphthomycin belong to the naphthalene subclass. Representatives of the benzene subclass are geldanamycin and the maytansinoid compounds. Only the rifamycins are of medical importance and will be dealt with here.

The rifamycins were isolated in 1959 from cultures of *Amyco-latopsis mediterranei* (formerly *Nocardia mediterranea*), originat-ing from a soil sample collected in the Côte d'Azur. The original

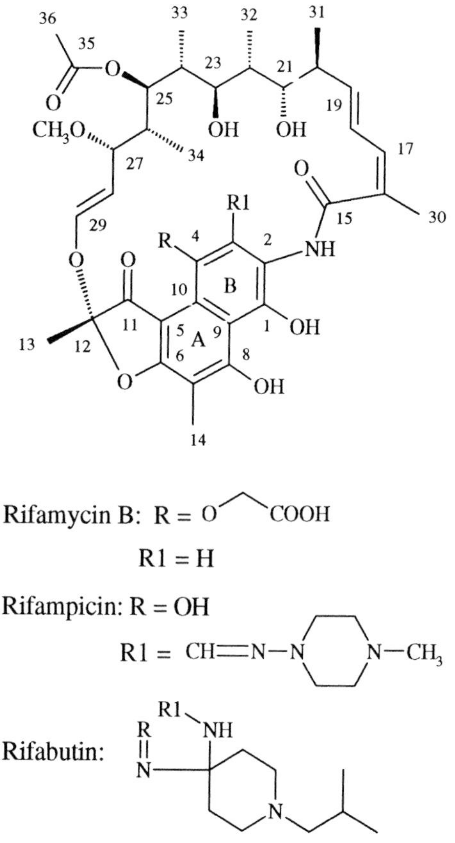

Rifamycin B: R = O⌢COOH

R1 = H

Rifampicin: R = OH

R1 = CH=N—N⌒N—CH₃

Rifabutin:

product, termed "rifamycin", was shown to be a mixture of five compounds called rifamycin A–E. The most stable of the rifamycins was rifamycin B and it was possible to change the conditions for the cultures so that only rifamycin B was obtained. Rifamycin B had only moderate antibiotic activity when administered to infected animals, but derivatives were prepared which turned out to be efficient remedies against tuberculosis and for prophylaxis against non-tuberculous mycobacterial infections such as the *Mycobacterium avium* complex (MAC) infection, a common and troublesome opportunistic infection in AIDS patients. The most important derivatives today are rifampicin (against tuberculosis) and rifabutin (against tuberculosis and prophylactic against MAC-infections). The structural differences between rifamycin B and these semisynthetic derivatives involve C-3 and C-4 in rifamycin B.

Biosynthesis of rifamycin

The starter for the biosynthesis of rifamycin is 3-amino-5-hydroxybenzoic acid (AHBA), which is biosynthesized via a variation of the

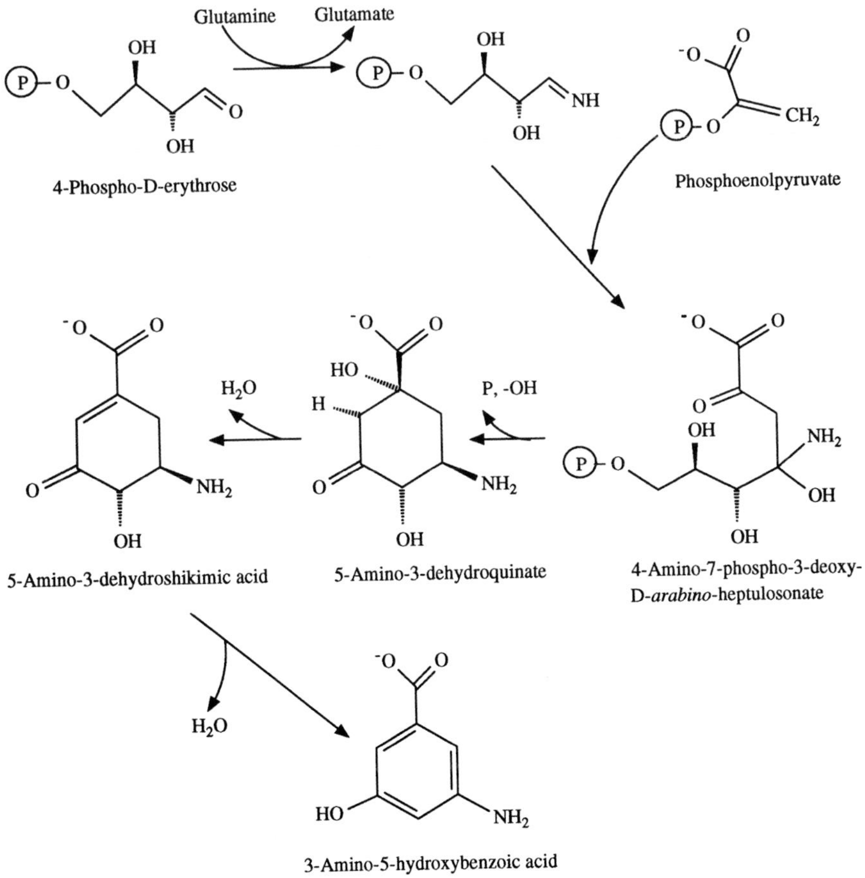

Fig. 89. Biosynthesis of 3-amino-5-hydroxybenzoic acid (AHBA)

shikimate pathway. Carbon 8 and the phenolic oxygen attached to this carbon atom originate from the carboxy group of AHBA.

The proposed pathway for the biosynthesis of AHBA is presented in Fig. 89. This pathway branches from the regular shikimic acid pathway (see page 196), already in the very first reaction. Glutamine is hydrolysed to form ammonia which immediately condenses with the aldehyde group of 4-phospho-D-erythrose to yield an imine which then reacts with phosphoenol pyruvate in the usual way to form 4-amino-7-phospho-3-deoxy-D-*arabino*-heptulosonate (aminoDAHP). (The role of glutamine as a specific nitrogen donor is not completely proved). AminoDAHP is transformed into 5-amino-3-dehydroquinate (aminoDHQ) in reactions analogous to those required for the formation of 3-dehydroquinate. 5-Amino-3-dehydroshikimic acid (aminoDHS) is formed from aminoDHQ by loss of one molecule of water and in the last reaction another molecule of water is removed to yield AHBA. This last reaction is catalysed by the enzyme *AHBA synthase* which has been isolated from *Amycolatopsis mediterranei* and shown to be a homodimer of about 39 kDa subunit molecular mass. The corresponding gene, designated *rifK*, corresponds to a protein of 388 amino acids with a predicted molecular mass of 42 281 Da. *rifK* has been overexpressed in *Escherichia coli* and the properties of the recombinant protein found to match those of the native enzyme. The enzyme uses PLP (= pyridoxal phosphate) as a cofactor and the presence of one mole of PLP per subunit of the protein has been confirmed.

The gene cluster for rifamycin biosynthesis in *Amycolatopsis mediterranei* has been sequenced. It contains genes homologous to those in the shikimate pathway, tailoring and regulatory genes and five large genes coding for a type I modular PKS ending with a translationally coupled amide-forming protein denoted as RifF. The PKS contains five proteins containing a total of 10 modules and a starting module. RifA (4 375 amino acid residues) comprises the starting module plus modules 1, 2 and 3. RifB (5 060 aa residues) contains modules 4, 5 and 6. RifC (1 763 aa residues) and RifD (1 728 aa residues) contain only one module each (no. 7 and 8, respectively). Finally, RifE (3 413 aa residues), comprises modules 9 and 10. The organization of modules and domains is illustrated in table 7.

The acytransferase in the loading module is a ligase with AMP binding motif which activates AHBA as its acyl adenate and transfers it to ACP in the same module for further transport to the KS of module 1. Module 2 is a very short module containing only KS, AT and ACP domains. There is no ER domain in any module. The DH domains in modules 1 and 5 are inactive because of large deletions in the active site motif and also the DH in module 8 has large deletions at both termini. The DHs in modules 6 and 7 have an apparently intact sequence but are not functional. The reason for this is not known. The active site motif of the DH in module 4 conforms with all other known DHs in modular PKSs, but the active motifs

Table 7. Modules and domains of RifA–RifE

Domain		RifA Module No.				RifB Module No.			RifC Module No.	RifD Module No.	RifE	
	0	1	2	3	4	5	6	7	8	9	10	
KS	-	+	+	+	+	+	+	+	+	+	+	
AT	CL	P	A	P	P	P	P	P	P	A	P	
DH	-	¤	-	-	+	¤	(+)	(+)	¤	+	+	
ER	-	-	-	-	-	-	-	-	-	-	-	
KR	-	+	-	¤	+	+	+	+	+	+	+	
ACP	+	+	+	+	+	+	+	+	+	+	+	RifF

Abbreviations: *KS* = β-ketoacyl synthase. *AT* = acyltransferase. *CL* = ATP-dependent carboxylic acid ligase. *A* = acetate specific acyltransferase. *P* = propionate specific acyltransferase. *DH* = dehydratase. *ER* = enoyl reductase. *KR* = β-ketoacyl reductase. *ACP* = acyl carrier protein. A + sign indicates a functioning domain. A - sign indicates absence of the domain. (+) indicates an inactive domain with apparently intact sequence. ¤ indicates an inactive domain with dysfunctional sequence.

of the DHs in modules 9 and 10 show differences involving two amino acids and the overall amino acid sequence of these domains displays considerable differences from other PKS functional DHs. The KR domain in module 3 is non-functional because of lack of an NADPH binding motif. Immediately downstream of the modular cluster there is a gene which codes for an enzyme – RifF – that catalyses the intramolecular cyclization of the alicyclic growing acyl chain through reaction of the AHBA amino group with the terminal thioester to form the amide bond in proansamycin X (Fig. 90). As in nystatin and amphotericin (pages 288, 290) there is a gene coding for an additional thioesterase, the function of which is unknown.

Figure 90 depicts the biosynthesis of rifamycin B. The biosynthesis starts with the reactions catalysed by RifA yielding compound A. Two genes – *rifS* and *rifT* – encode two P450s, one of which is thought to oxidize the ABHA aromatic ring to the corresponding quinoid (B), enabling a Michael attack by the first part of the polyketide chain to occur. The second P450 forms the dihydronaphthoquinone unit C. It is not completely established if these oxidations take place during chain assembly or afterwards, but it is attractive to assume them to take place before RifB starts its activity. Continuation of the synthesis yields the polyketide chain D and ring closure is catalysed by RifF, yielding proansamycin X. Conversion of the PKS product proansamycin to rifamycin B involves many steps, some of which are not yet clarified. Oxidation (P450) and oxidative aromatization yields rifamycin W. C-30 is lost via decarboxylation and an oxygen bridge is formed between C-29 and C-12. A five-membered ring involves the hydroxyl group at C-6 and

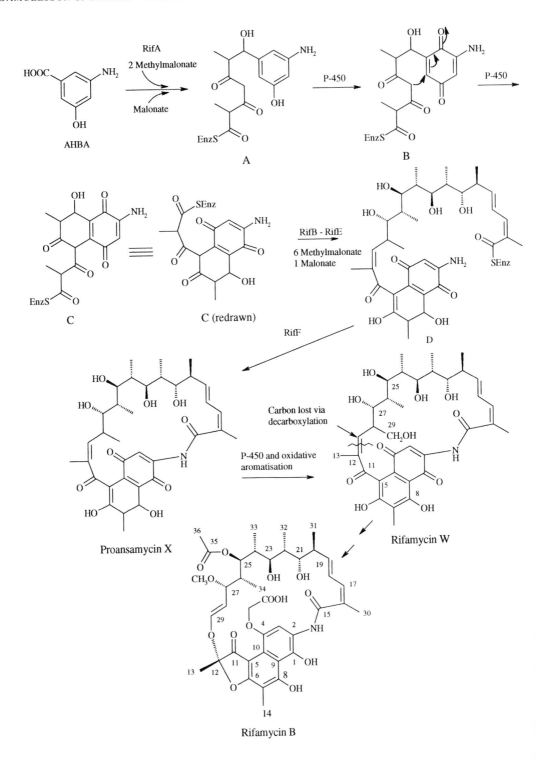

Fig. 90. Biosynthesis of rifamycin B

carbons 11, 12 and 5. Finally the B-ring is aromatized and a O-CH$_2$COOH group is formed at C-4.

Mechanism of the antibacterial activity

The rifamycins act by inhibiting the DNA-dependent RNA-polymerase in bacteria and thus also in chloroplasts and mitochondria but not in the cell nucleus of eukaryotes. This polymerase couples ribonucleosidetriphosphate molecules to each other on a DNA matrix and catalyses the transcription. The rifamycins inhibit the coupling of the first purine nucleotide. Elongation of a chain that is already growing is not hindered.

Rapamycin

Rapamycin (sirolimus) is a potent immunosuppressant which is produced by *Streptomyces hygroscopicus*. Rapamycin is a 31-membered macrolactone ring which also incorporates an *N*-heterocyclic ring. Attached to the large ring is a side chain composed of two carbons and a cyclohexane ring containing a hydroxyl and a methoxyl group. An internal ether bridge forms another six-membered ring.

Biosynthesis Rapamycin and the related tacrolimus (= FK-506, see below) are very interesting from a biosynthetic standpoint, because they incorporate both polyketide and polypeptide structures. Rapamycin is formed from a polyketide which is assembled from (4*R*,5*R*)-4.5-dihydroxycyclohex-1-enecarboxylic acid-CoA thioester as a starter unit, followed by seven acetate and seven propionate extender units, terminating with a pipecolic acid unit to form the large macrocyclic ring.

(4*R*,5*R*)-4,5-Dihydroxycyclohex-1-enecarboxylic acid is formed from shikimate according to the following scheme:

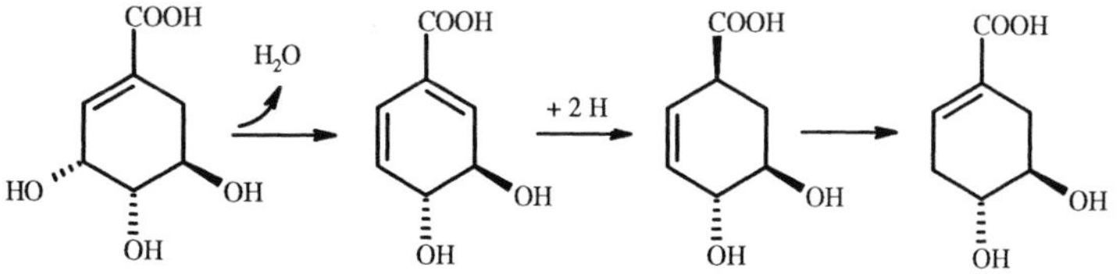

Shikimic acid

(4*R*,5*R*)-4,5-dihydroxy-cyclohex-1-enecarboxylic acid
(= DHCHC)

Cloning and sequencing of 107 000 bp from the genome of *Streptomyces hygroscopicus* has revealed the presence of the three PKS genes *rapA*, *rapB* and *rapC*, as well as the gene *rapP* coding for a *pipecolic acid inserting enzyme (PIE)* and 24 additional ORFs that may be involved in the biosynthesis of rapamycin.

The three PKS genes code for three colossal multifunctional polypeptides designated as RAPS1 (4 modules + a starter module),

347

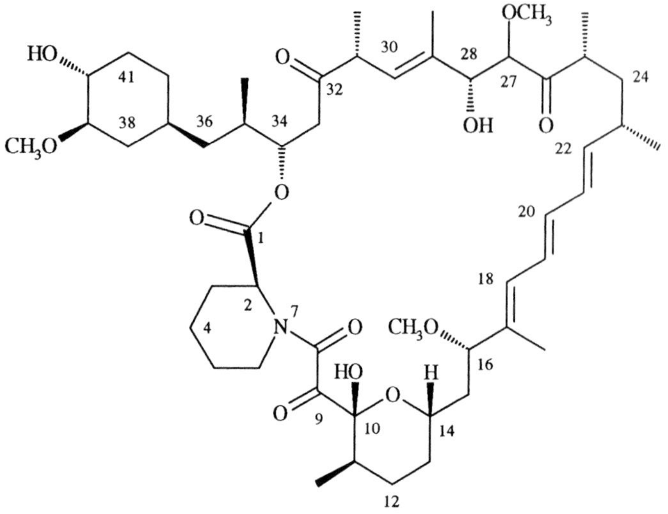

Rapamycin

RAPS2 (6 modules) and RAPS3 (4 modules). The predicted molecular masses are respectively 900, 1070 and 660 kDa, corresponding to 8 563, 10 222 and 6 260 amino acid residues.

The domains of all the modules have been identified. A loading module (module 0) precedes the four synthesizing modules in RAPS1. This loading module contains an ATP-dependent carboxylic acid ligase (CL), which might accept the CoA thioester of (4R,5R)-4,5-dihydroxycyclohex-1-enecarboxylic acid, the double bond of which is then reduced by the ER domain. The formed compound is transferred by the ACP domain to module 1 for further condensation with methylmalonate. These steps, however, have not been completely elucidated. An overview of the identified domains in the 14 modules is given in table 8.

These giant proteins possess a considerable number of superfluous domains. The superfluous peptide sequence may still be required for folding/spatial orientation (or other as yet unknown function) of the multifunctional protein, or it may not yet have been lost as a result of evolution.

Pipecolic acid is formed by deaminative cyclization of lysine (Fig. 91).

PIE (pipecolic acid inserting enzyme), which is connected to RAPS3, connects the amino group of pipecolic acid with the carbonyl bound to the ACP domain of module 14 of RAPS3 and also forms an ester bond with the first hydroxyl group in the polyketide chain (Fig. 92). These reactions close the macrolactone ring.

The deduced amino acid sequence of PIE (1 541 amino acid residues) has conserved sequence motifs comparable with those of non-ribosomal peptide synthases, including amino acid activation, ATP binding and a core sequence of LGGHS (one-letter abbreviations) for attachment of the phosphopantetheine moiety (see page

Table 8. Modules and domains of RAPS1 – RAPS3

Domain	RAPS1 Module No.					RAPS2 Module No.						RAPS3 Module No.			
	0	1	2	3	4	5	6	7	8	9	10	11	12	13	14
KS	-	+	+	+	+	+	+	+	+	+	+	+	+	+	+
AT	CL	P	A	P	P	A	P	P	A	A	P	A	A	P	P
DH	-	+	¤	(+)	+	¤	(+)	+	+	+	+	¤	¤	+	-
ER	+	+	-	(+)	-	-	~	+	-	-	-	-	-	+	-
KR	-	+	+	(+)	+	+	(+)	+	+	+	+	+	+	+	-
ACP	+	+	+	+	+	+	+	+	+	+	+	+	+	+	+

Abbreviations: KS = β-ketoacyl synthase. AT = acyltransferase. CL = ATP-dependent carboxylic acid ligase. A = acetatespecific acyltransferase. P = propionatespecific acyltransferase. DH = dehydratase. ER = enoyl reductase. KR = β-ketoacyl reductase. ACP = acyl carrier protein. A + sign indicates a functioning domain. A - sign indicates absence of the domain. (+) indicates an inactive domain with apparently intact sequence. ¤ indicates an inactive domain with dysfunctional sequence.

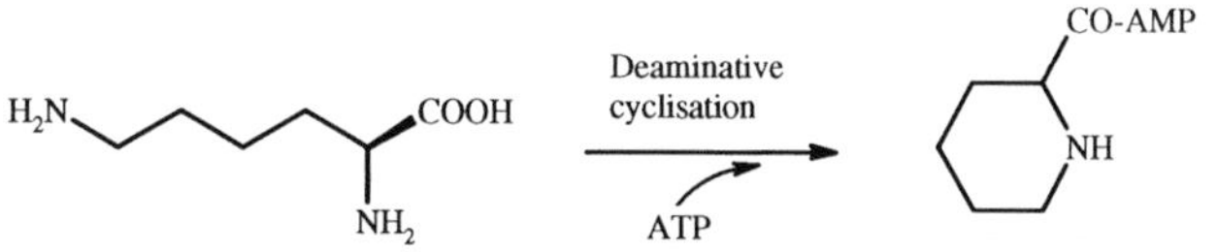

Fig. 91. Formation of pipecolyl adenosylmonophosphate from lysine

531 ff). There is also a peptide transfer motif (HHX3DGXS, where X = any amino acid) for epimerization and peptide bond formation. PIE thus forms a link between non-ribosomal polypeptide biosynthesis and polyketide biosynthesis.

RAPJ (386 amino acids) and RAPN (404 aa) are two P450 proteins presumed to catalyse oxidation of C-9 and C-27 in the product resulting from the synthesis and ring closure of the polyketide (Fig. 92). There is also a gene – *rapO* – which codes for an acidic protein, containing 78 amino acids, which is homologous with ferredoxin and which might serve for electron transfer in the P450 systems.

Rapamycin prevents T-cell proliferation by binding to immunophilins and is particularly effective at preventing graft rejection after organ transplantation. It acts at much lower concentrations than cyclosporin (page 546).

Tacrolimus (FK-506)

Tacrolimus is also an immunosuppressant and is obtained from cultures of *Streptomyces tsukubaensis*. The structure resembles that of

349

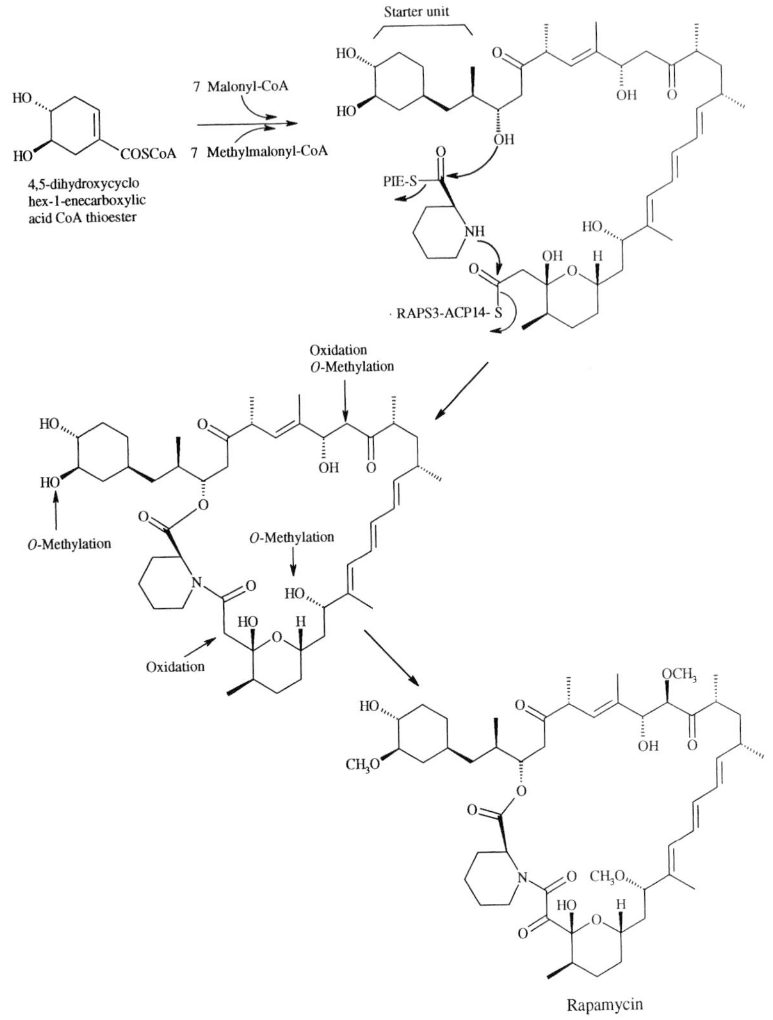

Fig. 92. Biosynthesis of rapamycin. PIES = pipecolic acid inserting enzyme. RAPS3-ACP14-S = ACP of module 14 of RAPS

rapamycin with a 23-membered macrolactone ring, incorporating an *N*-heterocyclic ring.

Biosynthesis The biosynthesis of tacrolimus (Fig. 93) resembles that of rapamycin and has also been elucidated by studies of the relevant genes. Similarly, the starter unit is (4*R*,5*R*)-4.5-dihydroxycyclohex-1-enecarboxylic acid-CoA thioester. The PKS (denoted as FKBS) consists of three proteins: FKBA containing four modules (6 420 amino acids), FKBB also with four modules (7 540 amino acids) and FKBC with two modules (3 592 amino acids). The domains have been identified and are schematically represented in table 9.

As for rapamycin, module 0 is the starter module which accepts

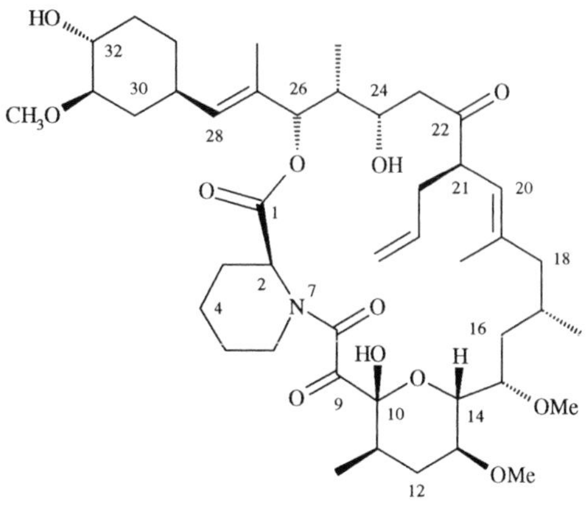

Tacrolimus (FK-506)

(4*R*,5*R*)-4.5-dihydroxycyclohex-1-enecarboxylic acid-CoA thio-ester, reduces it and transfers it to its ACP for further transport to module 1. These steps have not yet been fully clarified. The KR of module 2 has the opposite stereochemistry and the resulting OH group forms an ester bond with pipecolic acid, thereby closing the macrolactone ring. *Pipecolic acid inserting enzyme (PIE)* is the catalyst for the ring closure. A unique feature is the acyltransferase domain in module 4, which loads propylmalonate, thereby accounting for the propen-2-yl side-chain attached to C-21 in tacrolimus. Oxidation and dehydration of the saturated propylmalonate to form this chain are post-PKS reactions as are oxidation at C-9 and oxidation and methoxylation at C-13, C-15 and C-31.

Tacrolimus reduces peptidyl-prolyl isomerase activity by binding to the immunophilin FKBP-12, creating a new complex. This complex interacts with and inhibits calcineurin, thus inhibiting both T-lymphocyte signal transduction and IL-2 transcription.

Table 9. Modules and domains of FKBA–FKBC

Domain		FKBB Module No.				FKBC Module No.		FKBA Module No.			
	0	1	2	3	4	5	6	7	8	9	10
KS	-	+	+	+	+	+	+	+	+	+	+
AT	CL	P	P	A	N	P	P	A	A	P	A
DH	-	+	(+)	¤	(+)	+	+	+	¤	+	-
ER	+	-	-	-	-	-	+	+	-	+	-
KR	-	+	+	+	¤	+	+	+	+	+	-
ACP	+	+	+	+	+	+	+	+	+	+	+

Abbreviations: *KS* = β-ketoacyl synthase. *AT* = acyltransferase. *CL* = ATP-dependent carboxylic acid ligase. *A* = acetate specific acyltransferase. *P* = propionate specific acyltransferase. *N* = propionate specific acyltransferase, loading propylmalonate. *DH* = dehydratase. *ER* = enoyl reductase. *KR* = β-ketoacyl reductase. *ACP* = acyl carrier protein. A + sign indicates a functioning domain. A - sign indicates absence of the domain. (+) indicates an inactive domain with apparently intact sequence. ¤ indicates an inactive domain with dysfunctional sequence.

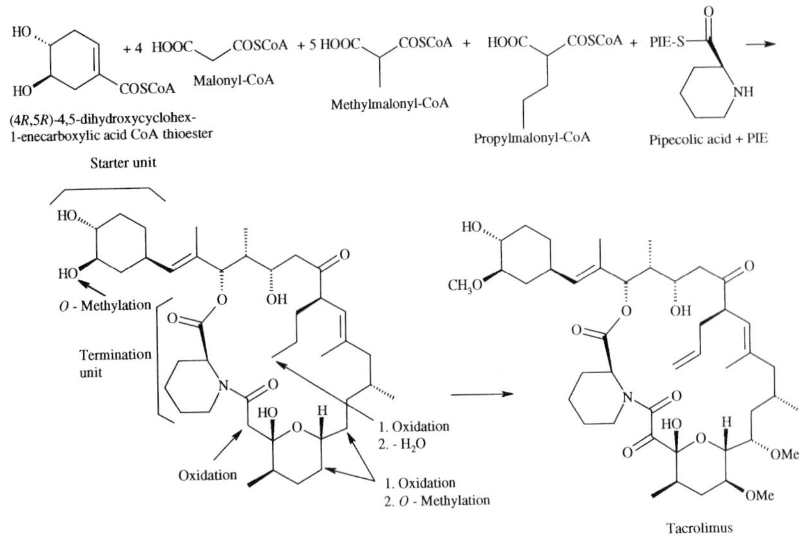

HO,,,
+ 4 HOOC⌒COSCoA + 5 HOOC⌒COSCoA + HOOC⌒COSCoA + PIE-S

HO⌒COSCoA   Malonyl-CoA

(4R,5R)-4,5-dihydroxycyclohex-
1-enecarboxylic acid CoA thioester   Methylmalonyl-CoA

Starter unit   Propylmalonyl-CoA   Pipecolic acid + PIE

O - Methylation

Termination
unit

1. Oxidation
2. - H₂O

Oxidation

1. Oxidation
2. *O* - Methylation

Tacrolimus

Fig. 93. Biosynthesis of tacrolimus

Pimecrolimus   Pimecrolimus (Elidel®) is a derivative of ascomycin, which differs from tacrolimus at C-21, where the side-chain is ethyl instead of allyl. Ascomycin is isolated from cultures of *Streptomyces hygroscopicus* var. *ascomyceticus*. In pimecrolimus the substituent at C-32 in ascomycetin is Cl instead of OH. Pimecrolimus acts as a calcineurin inhibitor and has anti-inflammatory and immunosuppresant activity. It is used for short-term and intermittent long-term treatment of mild to moderate atopic eczema in non-immunocompromised patients.

Rotenonoids

Rotenone and related compounds (rotenoids) are insecticides which are found in species of the genera *Derris*, *Lonchocarpus* and *Tephrosia* of the family *Fabaceae*.

Biosynthesis   Biosynthetically, rings A, C and D derive from the flavanone liquiritigenin, which is formed via a variant of the flavonoid biosynthesis catalysed not by chalcone synthase but by *6'-deoxychalcone synthase* (EC 2.3.1.170). Removal of the 6'-hydroxyl group takes place before cyclization of the polyketide in the chalcone synthase reaction. 6'-deoxychalcone synthase has been isolated and shown to be a polypeptide with a molecular mass of 34 kDa.

Liquiritigenin is then subjected to an aryl rearrangement followed by methylation to yield 7-hydroxy-4'-methoxyisoflavone. The mechanism for this rearrangement is not clear but a free 4'-hydroxy group seems to be required which means that methylation must occur subsequent to the rearrangement. The 7-hydroxy-4'-

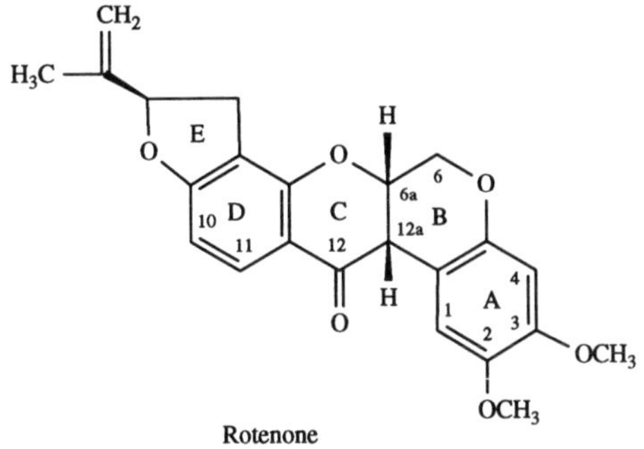

Rotenone

methoxyisoflavone then undergoes a series of hydroxylation and methylation reactions yielding 7-hydroxy-2',4',5'-trimethoxy-isoflavone. The 2'-methoxy group then forms a methoxyl radical which adds to the carbonyl-conjugated double bond, forming ring B. The final step in the biosynthesis of rotenone is prenylation at C-8 followed by a dehydrogenation process in which two hydrogen atoms are removed with attendant cyclization forming ring E. The oxidation is probably a radical process initiated by an enzyme similar to a P450 type or deguelin cyclase.

Crude drug **Derris** (Derridis radix) consists of the dried roots of *Derris elliptica* Benth. and *Derris malaccensis* Prain (*Fabaceae*), both of which occur in South-East Asia; they are also cultivated in Malaysia, Indonesia and the Philippines. The inhabitants have used derris as a fish poison. Its rotenone content may be as much as 13 % and it is used either in the form of a powder or as an extract of the root.

Rotenone and related compounds, the rotenoids, which are the

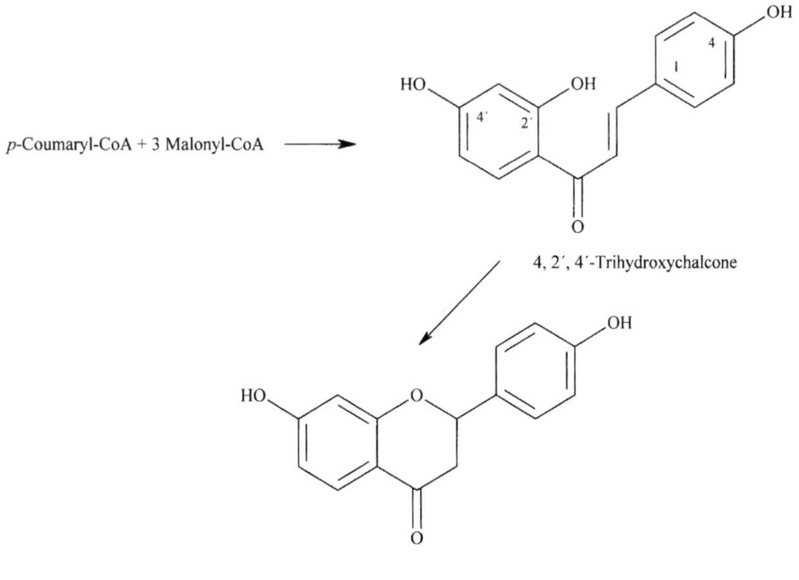

p-Coumaryl-CoA + 3 Malonyl-CoA ⟶

4, 2', 4'-Trihydroxychalcone

Liquiritigenin

353

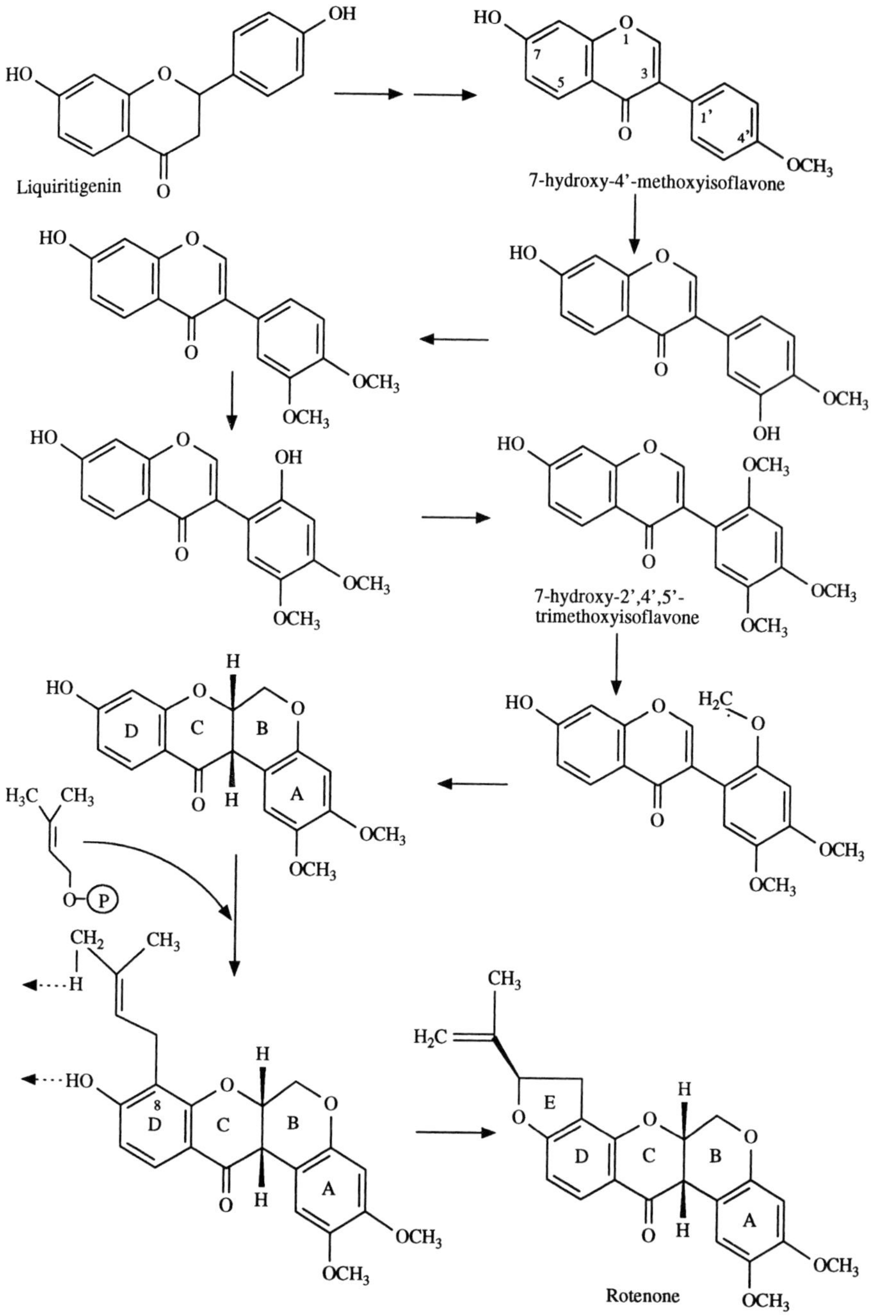

Fig. 94. Biosynthesis of rotenone

active insecticidal principles, have increased in importance, while the use of DDT has decreased. Rotenone and rotenoids also occur in other genera of the *Fabaceae*: *Lonchocarpus* (cubé, timbo), from the Amazonian region of Brazil, and *Tephrosia*, occurring in Equatorial Africa and many other places.

Khellin

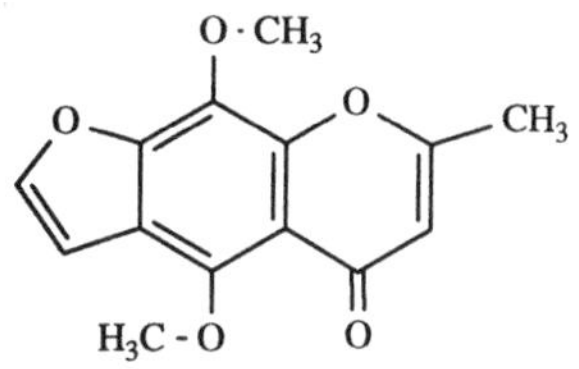

O·CH₃

Khellin

Biosynthesis Khellin is structurally related to the furanocoumarins (page 231), but is a derivative of 5,6-benzo-4-pyrone (γ-chromone), rather than 5,6-benzo-2-pyrone (α-chromone). Its biosynthesis is, however, very different. The benzene ring and the pyrone ring are formed from a C₁₀-polyketide whereas the furan ring derives from dimethylallyl diphosphate from the isopentenyl diphosphate pathway (page 366). This is an example showing that structurally similar natural compounds can have very different biosynthetic origin. The polyketide is a poly-β–keto ester, formed in the usual way from one acetate and four malonate units. The benzene ring is formed by Claisen reaction and enolization and the pyrone ring by nucleophilic attack of OH on to the enolic tautomer followed by loss of a leaving group yielding 5,7-dihydroxy-2-methylchromone. C-alkylation with dimethylallyl diphosphate adds a side-chain at C-6 which forms the furan ring combined with oxidative loss of three carbon atoms. O-methylation at C-5 with S-adenosylmethionine (SAM) yields visnagin and hydroxylation at C-8 followed by methylation of hydroxyls at C-5 and C-8 finally forms khellin.

Khellin has antispasmodic properties and was formerly used to ease cramps in the stomach, intestines, and urinary tract, and for the treatment of angina pectoris and asthma. Nowadays, it is little used because of its unpleasant side effects, such as nausea and vomiting. Attempts to find better substances led to the development of sodium cromoglycate, a prophylactic against allergic manifestations (page 356). This work is a good example of that branch of pharmacognostical research which uses the structure of a natural product as a model for the synthesis of drugs with improved properties.

Crude drug **Visnaga** (Ammeos visnagae fructus) is the schizocarp fruit of *Ammi visnaga* Lam. (*Apiaceae*). This plant grows in the Mediterranean region but is also found in Argentina, Chile, Mexico and the United States. The mericarp is 1.5–2.5 mm long and about 0.9 mm broad. It is very similar to the ammi fruit (page 232), but the two can be distinguished by anatomical differences in the pericarp. Visnaga contains 0.5–1 % khellin, 0.05–0.1 % visnagin, 0.3–1 % khellol and khellol glucoside, as well as the pyranocoumarin derivatives visnadin, samidin and dihydrosamidin. For structures see page 357.

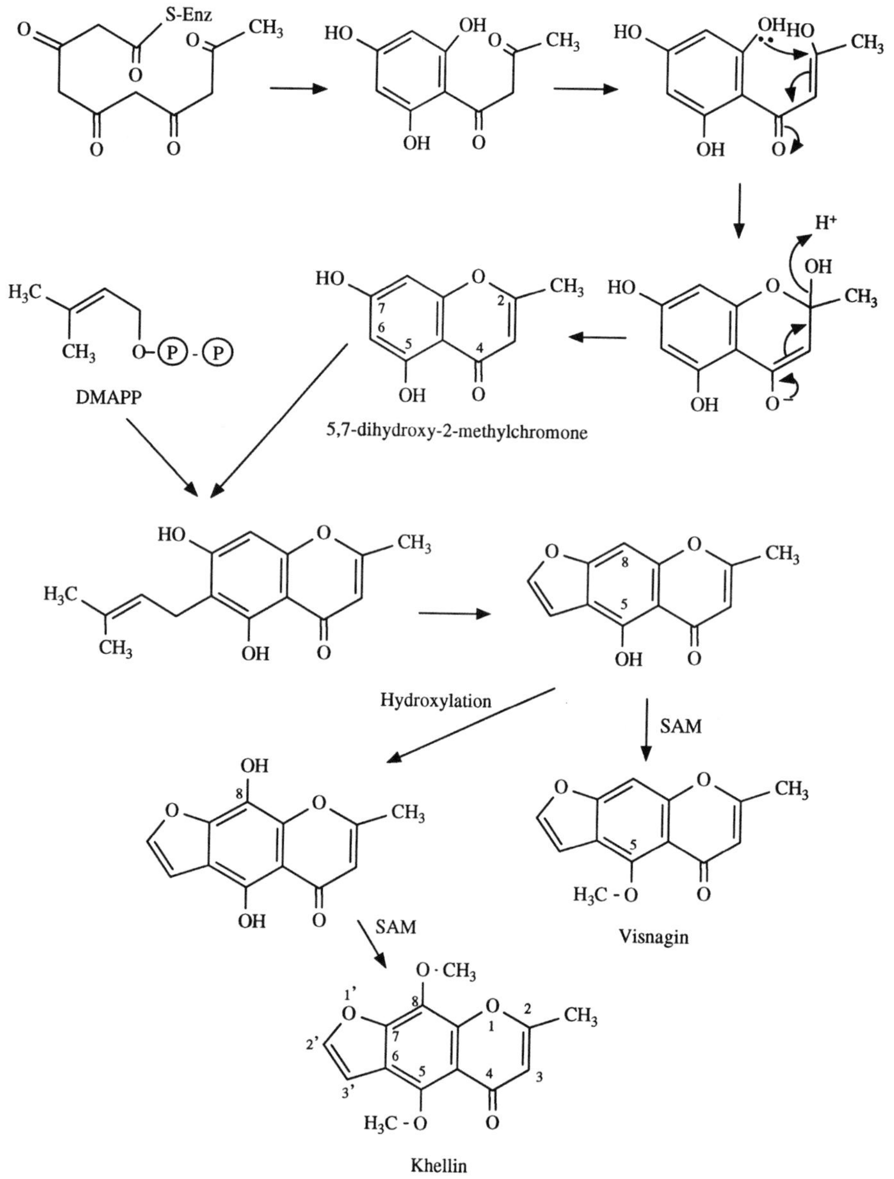

Fig. 95. Biosynthesis of khellin

Sodium cromoglycate

Sodium cromoglycate is a synthetic compound that was discovered during research involving analogues of khellin as potential bronchodilators for use against asthma. The compound is not a bronchodilator or histamine antagonist, but acts by preventing degranulation of mast cells by allergens and consequently the liberation of histamine and other substances causing asthma. The drug has no

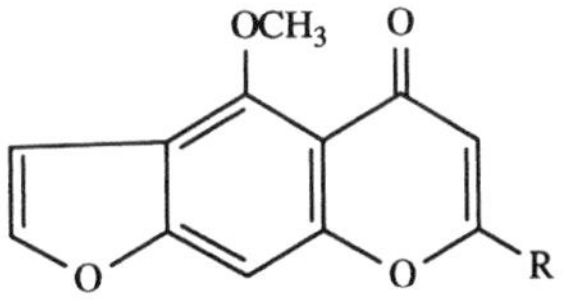

Visnagin: R = CH$_3$

Khellol: R = CH$_2$OH

Khellol glucoside: R = CH$_2$O-Glucose

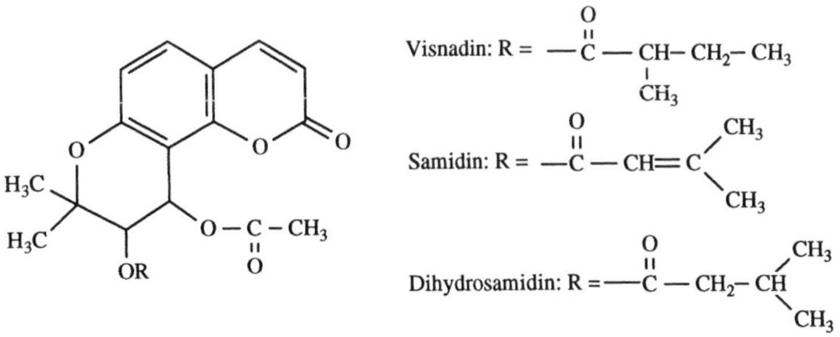

Visnadin: R = $-\overset{\overset{O}{\parallel}}{C}-\underset{\underset{CH_3}{|}}{CH}-CH_2-CH_3$

Samidin: R = $-\overset{\overset{O}{\parallel}}{C}-CH=C\overset{CH_3}{\underset{CH_3}{\diagup}}$

Dihydrosamidin: R = $-\overset{\overset{O}{\parallel}}{C}-CH_2-CH\overset{CH_3}{\underset{CH_3}{\diagdown}}$

effect on an acute asthmatic attack, but is used prophylactically when the patient expects exposure to an allergen, e.g. in the flowering season of certain plants.

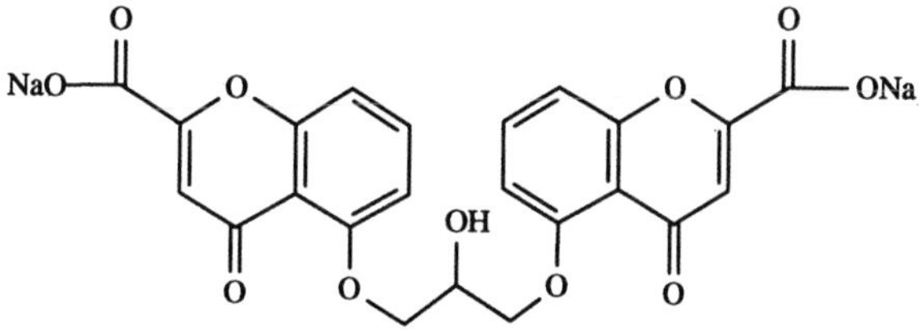

Sodium cromoglycate

Further reading

Biosynthesis of fatty acids. Type I FAS

Chang, S.-I. and Hammes, G.G. Structure and mechanism of action of a multifunctional enzyme: fatty acid synthase. *Accounts of Chemical Research*, **23**, 363–369 (1990).

Smith, S., Witkowski, A. and Joshi, A.K. Structural and functional organization of the animal fatty acid synthase. *Progress in Lipid Research*, **42**, 289–317 (2003).

Wakil, S.J. Fatty acid synthase, a proficient multifunctional enzyme. *Biochemistry*, **28** (11), 4523–4530 (1989).

Witkowski, A., Rangan, V.S., Randhawa, Z.I., Amy, C.M. and Smith, S. Structural organization of the multifunctional animal fatty-acid synthase. *European Journal of Biochemistry*, **198**, 571–579 (1991).

Type II FAS

White, S.W., Zheng, J., Zhang, Y.-M. and Rock, C.O. The structural biology of type II fatty acid biosynthesis. *Annual Review of Biochemistry*, **74**, 791–831 (2005)

Fats and waxes

Gaginella, T.S., Capasso, F., Mascolo, N. and Perilli, S. Castor oil: New lessons from an ancient oil. *Phytotherapy Research*, **12**, 128–130 (1998).

Prostaglandins

Smith, W.L., Marnett, L. and DeWitt, D.L. Prostaglandin and thromboxane biosynthesis. *Pharmacology and Therapy*, **49**, 153–179 (1991).

Leukotrienes

Funk, C.D., Rådmark, O., Fu, J.Y., Matsumoto, T., Jörnvall, H., Shimizu, T. and Samuelsson, B. Molecular cloning and amino acid sequence of leukotriene A4 hydrolase. *Proceedings of the National Academy of Science USA*, **84**, 6677–6681 (1987).

Haeggström, J.Z., Wetterholm, A., Shapiro, R., Vallee, B.L. and Samuelsson, B. Leukotriene A4 hydrolase: A zinc metalloenzyme. *Biochemical and Biophysical Research Communications,* **172** (3), 965–970 (1990).

Lam, B.K., Penrose, J.F., Freeman, G.J. and Austen, K.F. Expression cloning of a cDNA for human leukotriene C4 synthase, an integral membrane protein conjugating reduced glutathione to leukotriene A$_4$. *Proceedings of the National Academy of Sciences USA,* **91**, 7663–7667 (1994).

Minami, M., Minami, Y., Emori, Y., Kawasaki, H., Ohno, S., Suzuki, K., Ohishi, N., Shimizu, T. and Seyama, Y. Expression of leukotriene A4 hydrolase cDNA in *Escherichia coli*. *FEBS Letters*, **229** (2), 279–282 (1988).

Samuelsson, B., Dahlén, S.-E., Lindgren, J.Å., Rouzer, C.A. and Serhan, C.N. Leukotrienes and lipoxins: Structures, biosynthesis, and biological effects. *Science*, **237**, 1171–1176 (1987).

Samuelsson, B., Haeggström, J.Z. and Wetterholm, A. Leukotriene biosynthesis. *Annals of the New York Academy of Sciences*, **629**, 89–99 (1991).

Welsch, D.J., Creely, D.P., Hauser, S.D., Mathis, K.J., Krivi, G.G. and Isakson, P.C. Molecular cloning and expression of human leukotriene-C4 synthase. *Proceedings of the National Academy of Sciences USA*, **91**, 9745–9749 (1994).

Lipstatin

Schuhr, C., Eisenreich, W., Goese, M., Stohler, P., Weber, W., Kupfer, E. and Bacher, A. Biosynthetic precursors of the lipase inhibitor Lipstatin. *Journal of Organic Chemistry*, **67**, 2257–2262 (2002).

Weibel, E.K., Hadvary, P., Hochuli, E., Kupper, E. and Lengsfeld, H. Lipstatin, an inhibitor of pancreatic lipase, produced by *Streptomyces toxytricini*. *The Journal of Antibiotics*, **40** (8), 1081–1085 (1987).

Polyketides, general rewiews

Hopwood, D.A. Genetic contributions to understanding polyketide synthases. *Chemical Reviews*, **97**, 2465–2497 (1997).

Kaiz, L. Manipulation of modular polyketide synthases. *Chemical Reviews*, **97**, 2557–2575 (1997).

Khosla, C. Harnessing the biosynthetic potential of modular polyketide synthases. *Chemical Reviews*, **97**, 2577–2590 (1997).

Rawlings, B.J. Type I polyketide biosynthesis in bacteria (Part B). *Natural Products Reports*, **18**, 231–281 (2001).

Rawlings, B.J. Biosynthesis of polyketides (other than actinomycete macrolides). *Natural Products Reports*, **16**, 425–484 (1999).

Staunton, J. and Weissman, K.J. Polyketide biosynthesis: a millennium review. *Natural Products Reports*, **18**, 380–416 (2001).

Macrolides

Brautaset, T. *et al*. Biosynthesis of the polyene antifungal antibiotic nystatin in *Streptomyces noursei* ATCC 11455: analysis of the gene cluster and deduction of the biosynthetic pathway. *Chemistry & Biology*, **7**, 395–403 (2000).

Caffrey, P. *et al.* Amphotericin biosynthesis in *Streptomyces nodosus*: deductions from analysis of polyketide synthase and late genes. *Chemistry & Biology*, **8**, 713–723 (2001).

Douthwaite, S. Structure-activity relationships of ketolides vs. macrolides. *Clinical Microbiology and Infection*, **7** (Suppl. 3), 11–17 (2001).

Ikeda, H. and Omura, S. Avermectin biosynthesis. *Chemical Reviews*, **97**, 2591–2609 (1997).

Rawlings, B.J. Type I polyketide biosynthesis in bacteria (Part A– erythromycin biosynthesis). *Natural Products Reports*, **18**, 190–227 (2001).

Staunton, J. and Wilkinson, B. Biosynthesis of erythromycin and rapamycin. *Chemical Reviews*, **97**, 2611–2629 (1997).

White, R.L. Antibiotic resistance: Where do ketolides fit? *Pharmacotherapy*, **22**, 18s–29s (2002).

Epothilones

Altmann, K.-H. and Gertsch, J. Anticancer drugs from nature – natural products as a unique source of new microtubule-stabilizing agents. *Natural Products Reports*, **24**, 327–357 (2007).

Mulzer, J., Altmann, K.-H., Höfle, G., Müller, R. and Prantz, K. Epothilones – a fascinating family of microtubule stabilizing antitumor agents. *Comptes Rendus Chemie*, **11**, 1336–1368 (2008).

Aflatoxins

Yabe, K., Nakajima, H. Enzyme reactions and genes in aflatoxin biosynthesis. *Applied Microbiology and Biotechnology*, **64**, 745–755 (2004).

Mevastatin and lovastatin

Alberts, A.W. *et al.* Mevinolin: A highly potent competitive inhibitor of hydroxymethylglutaryl-coenzyme A reductase and a cholesterol-lowering agent. *Proceedings of the National Academy of Sciences USA*, **77**, 3957–3961 (1980).

Brown, A.G., Smale, T.C., King, T.J., Hasenkamp, R. and Thompson, R.H. Crystal and molecular structure of compactin, a new antifungal metabolite from *Penicillium brevicompactum*. *Journal of the Chemical Society. Perkin Transactions I*, 1165–1170 (1976).

Endo, A., Tsujita, Y., Kuroda, M. and Tanzawa, K. Inhibition of cholesterol synthesis *in vitro* and *in vivo* by ML-236A and ML-236B, competitive inhibitors of 3-hydroxy-3-methylglutaryl-coenzyme A reductase. *European Journal of Biochemistry*, **77**, 31–36 (1977).

Endo, A. Monacolin K, a new hypocholesterolemic agent that

specifically inhibits 3-hydroxy-3-methylglutaryl coenzyme A reductase. *The Journal of Antibiotics*, **33** (3), 334–336 (1980).

Yoshizawa, Y., Witter, D.J., Liu, Y. and Vederas, J.C. Revision of the biosynthetic origin of oxygens in Mevinolin (Lovastatin), a hypocholesterolemic drug from *Aspergillus terreus* MF 4845. *Journal of the American Chemical Society*, **116**, 2693–2694 (1994).

Leptospermone and nitisinone

Hall, M.G., Wilks, M.F., McLean Prowan, W., Eksborg, S. and Lumholtz, B. Pharmacokinetics and pharmacodynamics of NTBC (2-(2-nitro-4-fluoromethylbenzoyl)-1,3-cyclohexane-dione) and mesotrione, inhibitors of 4-hydroxyphenyl pyruvate dioxygenase (HPPD) following a single dose to healthy male volunteers. *Journal of Clinical Pharmacology*, **52**, 169–177 (2001).

Mitchell, G., Bartlett, D.W., Fraser, T.E.M., Hawkes, T.R., Holt, D.C., Townson, J.K. and Wichert, R.A. Mesotrione: a new selective herbicide for use in maize. *Pest Management Science*, **57**, 120–128 (2001).

Mupirocin

El-Sayed, A.K., Hothersall, J., Cooper, S.M., Stephens, E., Simpson, T.J. and Thomas, C.M. Characterization of the mupirocin biosynthesis gene cluster from *Pseudomonas fluorescens* NCIMB 10586. *Chemistry& Biology*, **10**, 419–430 (2003).

Hothersall, J., Wu, J., Rahman, A.S., Shields, J.A., Haddock, J., Johnson, N., Cooper, S.M., Stephens, E.R., Cox, R.J., Crosby, J., Willis, C.L., Simpson, T.J. and Thomas, C.M. Mutational analysis reveals that all tailoring region genes are required for production of polyketide antibiotic mupirocin by *Pseudomonas fluorescens*. Pseudomonic acid B biosynthesis precedes pseudomonic acid A. *Journal of Biological Chemistry*, **282** (21), 15451–15461 (2007).

Tetracyclines

Butler, M.J., Friend, E.J., Hunter, I.S., Kaczmarek, F.S., Sugden, D.A. and Warren, M. Molecular cloning of resistance genes and architechture of a linked gene cluster involved in biosynthesis of oxytetracycline by *Streptomyces rimosus*. *Molecular & General Genetics* **215**, 231–238 (1989).

Kim, E.-S., Bibb, M.J., Butler, M.J., Hopwood, D.A. and Sherman, D.H. Sequences of the oxytetracycline polyketide synthase-encoding otc genes from *Streptomyces rimosus*. *Gene*, **141**, 141–142 (1994).

Livermore, D.M. Tigecycline: What is it, and where should it be used? *Journal of Antimicrobial Chemotherapy*, **56** (4), 611–614 (2005).

Anthracyclines

Fujii, I. and Ebizuka, Y. Anthracycline biosynthesis in *Streptomyces galilaeus*. *Chemical Reviews*, **97**, 2511–2523 (1997).

Hutchinson, C.R. Biosynthetic studies of daunorubicin and tetra-cenomycin C. *Chemical Reviews*, **97**, 2525–2535 (1997).

Enediynes

Ahlert, J., Shepard, E., Lomovskaya, N., Zazopoulos, E., Staffa, A., Bachmann, B.O., Huang, K., Fonstein, L., Czisny, A., Whitwam, R.E., Farnet, C.M. and Thorson, J.S. The calicheamicin gene cluster and its iterative type I enediyne PKS. *Science*, **297**, 1173–1176 (2002).

Damle, N.K. and Frost, P. Antibody-targeted chemotherapy with immunoconjugates of calicheamicin. *Current Opinion in Pharmacology*, **3**, 386–390 (2003).

Galm, U., Hager, M.H, Van Lanen, S.G., Ju, J., Thorson, J.S. and Shen, B. Antitumor antibiotics: Bleomycin, enediynes, and mitomycin. *Chemical Reviews*, **105**, 739–758 (2005).

Giles, F., Estey, E. and O'Brien, S. Gemtuzumab ozogamicin in the treatment of acute myeloid leukemia. *Cancer*, **98**, 2095–2104 (2003).

Smith, A.L. and Nicolaou, K.C. The enediyne antibiotics. *Journal of Medicinal Chemistry*, **39** (11), 2103–2117 (1996).

Sthapit, B., Oh, T.-J., Lamichhane, R., Liou, K., Lee, H.C., Kim, C.-G. and Sohng, J.K. Neocarzinostatin naphtoate synthase: an unique iterative type I PKS from neocarzinostatin producer *Streptomyces carzinostaticus*. *Febs Letters*, **566**, 201–206 (2004).

Anthraquinones

Manitto, P., Monti, D., Speranza, G., Mulinacci, N., Vincieri, F.F., Griffini, A. and Pifferi, G. Conformational studies of natural products. Part 4. Conformation and absolute configuration of cascarosides A, B, C, D. *Journal of the Chemical Society. Perkin Transactions I*, 1577–1580 (1993).

Hypericin and *Hypericum perforatum*

Adam, P., Arigoni, D., Bacher, A. and Eisenreich, W. Biosynthesis of hyperforin in *Hypericum perforatum. Journal of Medicinal Chemistry*, **45** (21), 4786–4793 (2002).

Barnes, J., Anderson, L.A. and Phillipson, J.D. St John's wort

(*Hypericum perforatum* L.): a review of its chemistry, pharmacology and clinical properties, *Journal of Pharmacy and Pharmacology*, **53**, 583–600 (2001).

Bombardelli, E. and Morazzoni, P. Hypericum perforatum. *Fitoterapia*, **66**, (1), 43–68 (1995).

Henderson, L., Yue, Q.Y., Bergquist, C., Gerden, B. and Arlett, P. St. John's wort (*Hypericum perforatum*): drug interactions and clinical outcomes. *British Journal of Clinical Pharmacology* **54**, 349–356 (2002).

Moore, L.B. *et al.* St. John's wort induces hepatic drug metabolism through activation of the pregnane X receptor. *Proceedings of the National Academy of Science*, **97** (13), 7500–7502 (2000).

Müller, W.E. and Schäfer, C. Johanniskraut. In-vitro-Studie über Hypericum-Extrakt, Hypericin und Kämpferol als Antidepressiva. *Deutsche Apothekerzeitung*, **136** (13), 1015–1022 (1996).

Müller, W.E., Singer, A. and Wonnemann, M. Hyperforin – Antidepressant activity by a novel mechanism of action. *Pharmacopsychiatry*, **34** (Suppl. 1), S98–S102 (2001).

Flavonoids

Austin, M.B. and Noel, J.P. The chalcone synthase superfamily of type III polyketide synthases. *Natural Product Reports*, **20**, 79–110 (2003).

Coward, L., Barnes, N.C., Setchell, K.D.R. and Barnes, S. Genistein, Daidzein and their β-glycoside conjugates: Antitumor isoflavones in soybean foods from American and Asian diets. *Journal of Agricultural and Food Chemistry*, **41**, 1961–1967 (1993)

Ferrer, J.-L. *et al.* Structure of chalcone synthase and the molecular basis of plant polyketide biosynthesis. *Nature. Structural Biology* **6** (8), 775–784 (1999).

Forkmann, G. *Biosynthesis of Flavonoids. Polyphenolic Phenomena*, A. Scalbert, ed. INRA Editions, Paris, pp. 65–79 (1993).

Harborne, J.B. Advances in flavonoid research since 1992. *Phytochemistry*, **55** (6), 481–504 (2000).

Hooper, L. and Cassidy, A. A review of the health care potential of bioactive compounds. *Journal of the Science of Food and Agriculture*, **86**, 1805–1813 (2006).

Jez, J.M. *et al.* Dissection of malonyl-coenzyme A decarboxylation from polyketide formation in the reaction mechanism of a plant polyketide synthase. *Biochemistry* **39** (5), 890–902 (2000).

Jez, J.M., Bowman, M.E. and Noel, J.P. Structure-guided programming of polyketide chain-length determination in chalcone synthase. *Biochemistry* **40,** (49), 14829–14838 (2001).

McCue, P. and Shetty, P. Health benefits of soy isoflavonoids and strategies for enhancement: A review. *Critical Reviews in Food Science and Nutrition*, **44**, 361–367 (2004).

Meltzer, H.M. and Malterud, K.E. Can dietary flavonoids influence the development of coronary heart disease? *Scandinavian Journal of Nutrition/Näringsforskning*, **41** (2), 50–57 (1997).

Watanabe, K., Praseuth, A.P. and Wang, C.C.C. A comprehensive and engaging overview of the type III family of polyketide synthases. *Current Opinion in Chemical Biology*, **11**, 279–286 (2007).

Winkel-Shirley, B. Flavonoid biosynthesis. A colorful model for genetics, biochemistry, cell biology, and biotechnology. *Plant Physiology*, **126**, 485–493 (2001).

Kava pyrones

Norton, S.A. and Ruze, P. Kava dermopathy. *Journal of the American Academy of Dermatology*, **31** (1), 89–97 (1994).

Flavonolignans

Fraschini, F. Pharmacology of Silymarin. *Clinical Drug Investigation*, **22** (1), 51–65 (2002).

Wellington, K. and Jarvis, B. Silymarin: A review of its clinical properties in the management of hepatic disorders. *BioDrugs*, **15** (7), 465–489 (2001).

Mycophenolic acid

Bentley, R. Mycophenolic acid: A one hundred year odyssey from antibiotic to immunosuppressant. *Chemical Reviews,* **100**, 3801–3825 (2000).

Ansamycin antibiotics

Floss, H.G. Natural products derived from unusual variants of the shikimate pathway. *Natural Products Reports*, **14**, 433–452 (1997).

Ghisalba, O. Biosynthesis of rifamycins (ansamycins) and microbial production of shikimate pathway precursors, intermediates and metabolites. *Chimia*, **39** (4), 79–88 (1985).

Rinehart Jr, K.L. and Shield, L.S. Chemistry of the ansamycin antibiotics. *Fortschritte der Chemie Organischer Naturstoffe (Progress in the chemistry of organic natural products)*, **33**, 231–307 (1975).

Rotenone

Bhandari, P., Crombie, L., Kilbee, G.W., Pegg, S.J., Proudfoot, G., Rossiter, J., Sanders, M. and Whiting, D. Biosynthesis of rotenone and amorphigenin. Study of the origins of iso-propenyl-substituted dihydrofuran E-rings using isotopically labelled late precursors. *Journal of the Chemical Society. Perkin Transactions I*, 851–863 (1992).

7b. The isopentenyl diphosphate pathway Isoprenoids

481Many constituents of volatile oils are unsaturated hydrocarbons (terpenes), with the molecular formula $C_{10}H_{16}$, or alcohols and ketones, with similar skeletons.

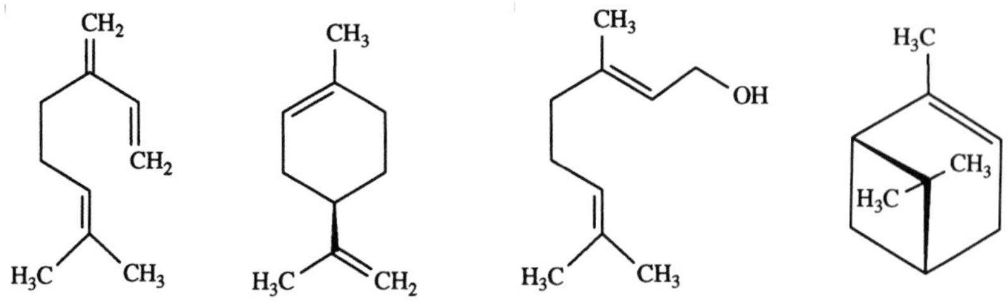

| Myrcene | (+)-Limonene | Geraniol | (+)-α-Pinene |

A close study of these structures shows that they could conceivably arise by a combination of two branched C_5 units, such as isoprene.

Other compounds of this type are larger molecules containing 15, 20, 25, 30 or 40 carbon atoms. All these can be regarded as products resulting from the condensation and hydrogenation of isoprene units. The compounds are called *monoterpenes* (C_{10}), *sesquiterpenes* (C_{15}), *diterpenes* (C_{20}), *sesterterpenes* (C_{25}), *triterpenes* (C_{30}) and *tetraterpenes* (C_{40}). Also the *steroids* can be fitted into this scheme if they are regarded as modifications of triterpenes.

All these substances, the basic skeleton of which may be conceived as derived from isoprene, are called *isoprenoids*. The hypothesis of their biogenesis was developed especially by Ruzicka, L. during the 1920s and 1930s and was named the *isoprene rule*. In order to prove this rule, there was a long search to find the so-called "active isoprene", i.e. the C_5 unit that could be incorporated into the isoprenoids. The solution came in 1956, when *mevalonic acid* was isolated and found to be very readily built into isoprenoids. It is true that mevalonic acid is a C_6 compound, but, before it takes part in biosynthetic reactions it is converted into *isopentenyl diphosphate,* which thus represents the long sought "active isoprene". It will be shown below that mevalonic acid is formed from acetate and is thus a key substance in the second main biosynthetic pathway from acetate which thus has been called *the mevalonic acid pathway*. However, since it has now been shown (page 369) that mevalonic acid is not the only precursor of isopentenyl diphosphate, it seems reasonable to change the name of this pathway to *the isopentenyl diphosphate pathway*. (Diphosphate was formerly called pyrophosphate and this name still remains in many abbreviations, e.g. IPP for isopentenyl diphosphate, relating to the previous name isopentenyl pyrophosphate.)

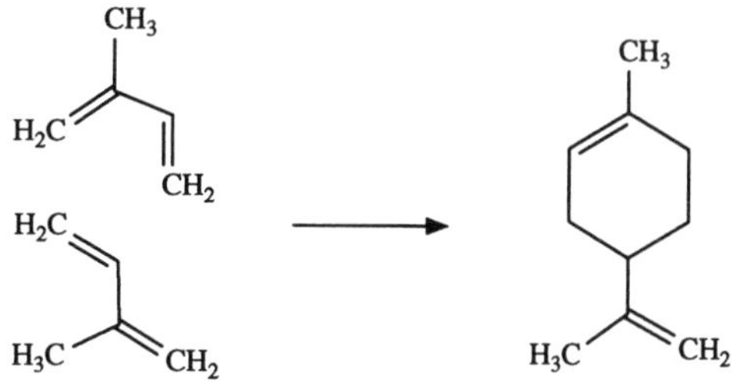

2 Isoprene   Limonene

Mevalonic acid and isopentenyl diphosphate

Like the acylpolymalonate pathway, the isopentenyl diphosphate pathway starts with acetyl CoA (Fig. 96). Catalysed by *acetyl-CoA C-acetyltransferase* (AACT, EC 2.3.1.9, also known as acetoacetyl-CoA thiolase), two molecules of this basic unit condense to yield acetoacetyl CoA. Aldol condensation with another molecule of acetyl CoA, followed by hydrolysis, gives 3-hydroxy-3-methylglutaryl CoA (HMG-CoA). The enzyme which catalyses this reaction is *hydroxymethylglutaryl-CoA synthase* (HMGS, EC 2.3.3.10). In animal and yeast systems these two enzymes can be separated and have been extensively studied. In plants, however, the two enzymes seem to be localized on a single polypeptide, which in radish (*Raphanus sativus* L.) is a soluble monomeric protein. Also in *Vinca rosea L.* (sometimes referred to as *Catharanthus roseus* G. Don) the two enzyme activities co-purify. The situation is, however, unclear as HMGS obtained from cDNA of *Arabidopsis thaliana* has been found to be devoid of thiolase activity.

Reduction of the thiol ester group in HMG-CoA, catalysed by *hydroxymethylglutaryl-CoA reductase (NADPH)* (HMGR, EC 1.1.1.34), yields (3*R*) mevalonic acid. This reaction is NADPH-dependent. The reduction step is irreversible and the mevalonate formed is used only for the further synthesis of terpenes and steroids. Only (3*R*) mevalonic acid is biologically active.

HMGR of animal origin has been extensively studied while there is much less information available on the plant derived enzyme. HMGR is membrane bound and has been isolated from several plants. The molecular size ranges from 110 kDa (potato, dimer) to 180 kDa (radish, tetramer). Sequence comparisons of HMGRs from plants, mammals and yeasts show high levels of sequence identity within the catalytic domain, but significant differences in the membrane domain. Mammalian enzymes are integral membrane proteins of the endoplasmic reticulum with eight membrane-spanning regions, whereas the membrane domains of plant enzymes contain only one or two transmembrane regions. The structures of the *N*-terminal membrane domains of two differentially regulated

HMGR-isoforms from tomato have been analysed showing that membrane insertion does not involve cleavage of an *N*-terminal targeting peptide. Both enzymes span the membrane twice; the *C*- and *N*-termini are both located in the cytosol and a short interconnecting peptide sequence is localized in the lumen. Genes encoding HMGR have been isolated and sequenced from wheat, rice, radish, the rubber tree (*Hevea brasiliensis*) and potato. cDNA for HMGR from radish encodes a polypeptide of 583 amino acids containing two membrane-spanning domains. Many plants seem to possess small multigene families rather than a single gene, as in animals. Thus radish and *Arabidopsis thaliana* are reported to have two genes, whereas the rubber tree has three. Four genes are reported for tomato and in potato seven genes have been encountered. Single genes are known from *Vinca rosea*, *Nicotiana sylvestris*, rice and wheat. There is some evidence that the individual genes may play a role in the biosynthesis of different classes of terpenoid metabolites.

The activity of higher eukaryote HMGRs can be regulated post-translationally by reversible phosphorylation, catalysed by *[hydroymethylglutaryl-CoA reductase (NADPH)] phosphatase* (EC 3.1.3. 47). HMGR is considered a key regulatory enzyme controlling isoprenoid metabolism in mammals and fungi, but the rate-limiting nature of this enzyme in plants is less well defined. Compartmentation, channelling or other rate-limiting enzymes may also be involved in determining relative accumulation of terpenoids.

The enzymes *mevalonate kinase* (EC 2.7.1.36) and *5- phospho-*

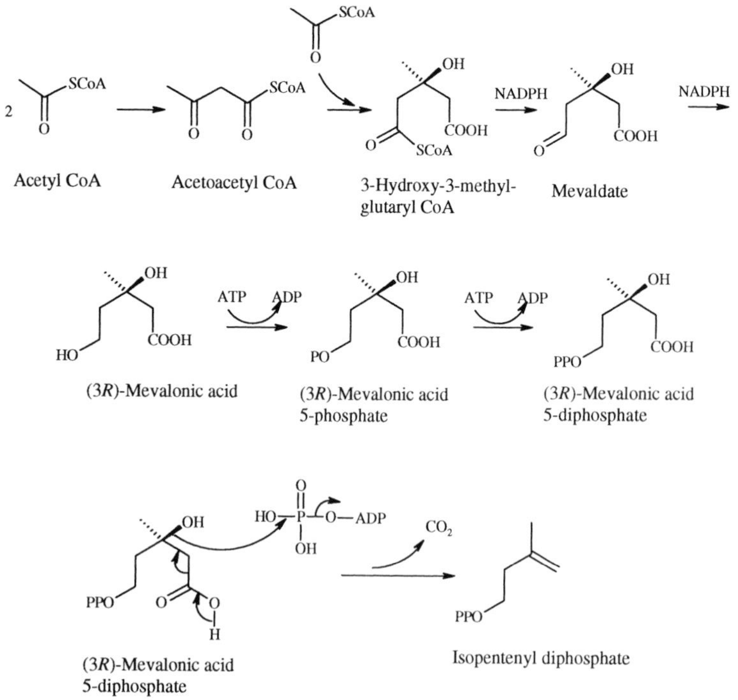

Fig. 96. Biosynthesis of isopentenyl diphosphate via mevalonic acid

mevalonate kinase (EC 2.7.4.2) catalyse phosphorylation of meval-onic acid in two steps yielding (3R)-mevalonic acid 5-diphosphate. This compound is finally transformed to isopentenyl diphosphate (IPP) by the elimination of a carboxyl and a hydroxyl group medi-ated by *diphosphomevalonate decarboxylase* (EC 4.1.1.33, also known as: mevalonate 5-diphosphate decarboxylase). The mecha-nism of this enzyme still awaits full clarification. Whilst a third molecule of ATP is required for the transformation there is little evi-dence for phosphorylation of the tertiary hydroxy group (as previ-ously believed). A mechanism in which an ATP molecule facilitates the decarboxylation-elimination is presented in Fig. 96.

Sites of isoprenoid biosynthesis

Within the plant cell three true plasmic phases are present: the cyto-plasm, the mitoplasm (the mitochondrial matrix) and the plasto-plasm (the chloroplast stroma). These are not directly connected, being separated from each other by non-plasmic phases between the inner and outer envelope membranes of the mitochondria and plas-tids. All three plasmic phases are sites of isoprenoid biosynthesis, often in cooperation with membranes to which enzymes might be bound. The membrane cooperating with the cytoplasm is the endo-plasmic reticulum, in the case of the mitoplasm it is the inner mito-condrial membrane, and in plastoplasm it is the inner membrane of the plastid envelope.

Monoterpenes are constituents of essential oils (see page 210) and are mainly synthesized in special anatomical structures of the plant, such as glandular hairs, secretory cavities, idioblasts and resin canals. It has been demonstrated that in these structures plastids are generally the site of monoterpene biosynthesis.

Sesquiterpenes, on the other hand, are not synthesized in plastids but in the cytoplasm/endoplastic reticulum compartments.

There are only a few reports on the site of diterpene biosynthesis but the formation of *ent*-kaurene has been assigned to plastids. The oxidation of gibberellins, on the other hand, is assigned to the endo-plastic reticulum.

The precursor of triterpenes and steroids – squalene – is formed in the endoplastic reticulum, while the precursor of tetraterpenes – is formed within plastids.

The non-mevalonate pathway for the biosynthesis of isoprenoids

Ever since the establishment of mevalonate as a precursor of cho-lesterol and the delineation of the intermediate steps and stereo-chemical features of the pathway, mevalonate has been accepted as the precursor of terpenoids in general and of meroterpenoids where terpenoid elements are combined with other biosynthetic building blocks. However, biosynthetic studies have often been hampered by the poor or negligible incorporation of labelled mevalonate into terpenoids or meroterpenoids, particularly in the plant system. Nobody, however, seriously challenged the fundamental belief that terpenoid biosynthesis proceeded via the same mevalonate pathway as demonstrated for mammalian steroids. The main reason for this was that research was predominantly focused on the exploration of post-mevalonate steps, particularly the utilization of polyprenyl diphosphates (see below), subsequent rearrangements and structural modifications. The poor utilization of mevalonate precursors therefore became an accepted inconvenience. The first indication of the existence of an alternative pathway came as early as 1989 with observations on incompatibilities in the labelling patterns from glucose precursors into triterpenoid hopanes produced by the bacterium *Zymomonas mobilis*. Further studies on the origins of the carbon atoms of ubiquinones and hopanoids in bacteria from labelled glucose, acetate, pyruvate and erythrose substrates suggested that the C_5 framework was most probably built up from a pyruvate-derived C_2 unit and a triose phosphate C_3 unit. Research reported since 1996 has clarified these steps and has also indicated how widespread this pathway is. This has been demonstrated by labelling studies in a range of organisms of different types and for a variety of different terpenoid classes. The results indicate that the mevalonate pathway is used in archaea, fungi and mammals. On the contrary, all eubacteria, including cyanobacteria and chloroplasts use the non-mevalonate pathway. **Plants utilize the mevalonate pathway in the cytosolic compartment, but in plastids the non-mevalonate pathway is used.** Putting these results in relation to the sites for isoprenoid biosynthesis discussed above, one could conclude that, as a rule, compounds which are formed in the cytoplasm, like **sesquiterpenes, sterols and triterpenes, are formed via mevalonate**, whereas those formed in plastids such as **monoterpenes, diterpenes and tetraterpenes derive from the non-mevalonate pathway**. There is very little information of a possible cooperation between cytosol and plastids. There is evidence that plastids cannot provide IPP or higher prenyl homologues for cytosolic sterol formation, but the possiblity of an import of IPP and related substances and/or their higher homologues from the cytosol into the plastid compartment has not yet been fully investigated.

This new pathway to isoprenoids has been termed *the glyceraldehyde phosphate/pyruvate pathway* or the Rohmer pathway (acknowledging Michel Rohmer, who first challenged the accepted beliefs). Other terms in use are: the mevalonate-independent pathway, the deoxyxylulose (DXP or DOXP) pathway, and the methylerythriol phosphate (MEP) pathway. In a review (1999) the terminology "*non-mevalonate pathway*" has been proposed and this term will be used in this book.

The non-mevalonate pathway for biosynthesis of isopentenyl

diphosphate is illustrated in Fig. 97. Genes coding for the enzymes which catalyse the reactions have been identified. The pathway starts with glyceraldehyde-3-phosphate (G3P) from the Calvin–Benson cycle (page 128) which reacts with pyruvate to form 1-deoxy-D-xylulose-5-phosphate (DXP). The reaction is catalysed by *1-deoxy-D-xylulose 5-phosphate synthase* (EC 2.2.1.7) and involves thiamine diphosphate. DXP is converted to 2-*C*-methyl-D-erythritol 4-phosphate (MEP) by *1-deoxy-D-xylulose-5-phosphate reducto-isomerase* (EC 1.1.1.267). DXP is also involved in the biosynthesis of pyridoxal, but MEP is not known to function anywhere else than in the non-mevalonate pathway and its formation might therefore be

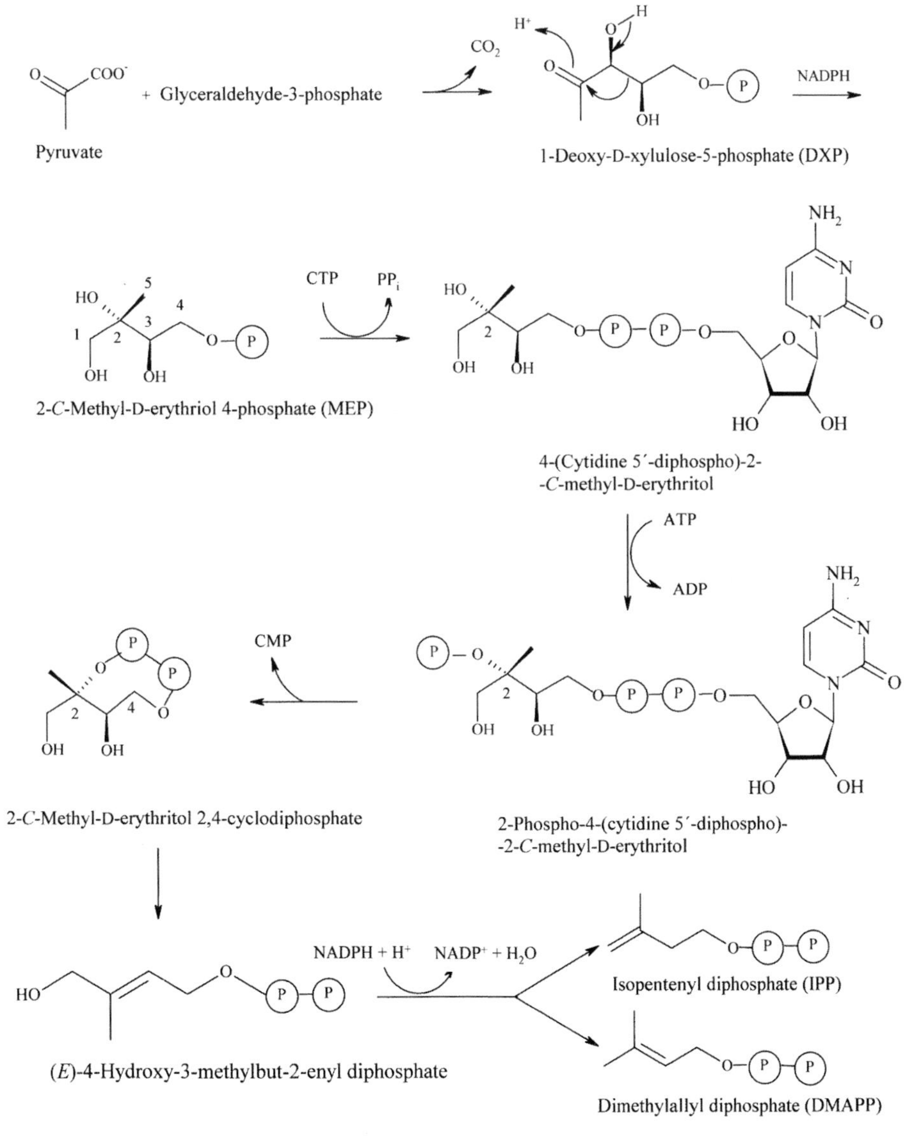

Fig. 97. Biosynthesis of IPP and DMAPP via the non-mevalonate pathway

371

the key step of this pathway. The enzyme *2-C-methyl-D-erythritol 4-phosphate cytidyltransferase* (EC 2.7.7.60), which requires divalent cations (Mg^{2+} or Mn^{2+}), catalyses a reaction between MEP and cytidine 5′-triphosphate to yield 4-(cytidine 5′-diphospho)-2-C-methyl-D-erythritol (CDP-ME).

CDP-ME is phosphorylated at the C-2 hydroxy group by *4-(cytidine 5′-diphospho)-2-C-methyl-D-erythritol kinase* (EC 2.7.1.148) with formation of 2-phospho-4-(cytidine 5′-diphospho)-2-C-methyl-D-erythritol, which is converted to 2-C-methyl-D-erythritol 2,4-cyclodiphosphate by *2-C-methyl-D-erythritol 2,4-cyclodiphosphate synthase* (EC 4.6.1.12). In the next reaction 2-C-methyl-D-erythritol 2,4-cyclodiphosphate is transformed to (*E*)-4-hydroxy-3-methylbut-2-enyl diphosphate by the enzyme *4-hydroxy-3-methylbut-2-enyl diphosphate synthase* (EC 1.17.4.3) in a complicated process in which two SH-groups in the enzyme form a disulphide bond which then has to be reduced for multiple turnover. Finally the enzyme *4-hydroxy-3-methylbut-2-enyl diphosphate reductase* (EC 1.17.1.2) catalyses a reaction in which isopentenylphosphate and dimethylallyldiphosphate are formed simultaneously in the ratio 5:1.

Monoterpenes

Many of the substances belonging to this group are constituents of volatile oils (see page 210). Most of them are of no medicinal use, but many are pharmaceutically important as means of improving the taste and smell of drugs. Many plant products containing volatile oils of the monoterpene type are used as spices. The great economic importance of essential oils in the perfume industry was mentioned earlier (page 212). Some oxidized monoterpene derivatives are of medicinal use as rubefacients, sedatives and bitters. The pyrethrins, important as insecticides, are also monoterpenes.

Biosynthesis of monoterpenes

As illustrated in Fig. 98, monoterpenes are formed from one molecule of IPP and one molecule of DMAPP, which are produced via the non-mevalonate pathway as indicated above.

The diphosphate group of DMAPP becomes ionized, generating an allylic cation which forms a new C-C bond by electrophilic attack on the double bond of a second molecule of IPP. This reaction is accompanied by loss of a proton from C-2 in IPP. The resulting product is *trans*-geranyl diphosphate[1] (GPP).

The enzymes which catalyse this reaction are members of a group of enzymes called *prenyltransferases*. Many prenyltransferases are relatively non-specific and catalyse the condensation of DMAPP with varying numbers of IPP units, building up polyprenyl

---

[1] At disubstituted double bonds the terms *cis* and *trans* denote the positions of the larger (longer) substituents in relation to each other.

diphosphate chains of differing lengths. However, chain-length specific *GPP synthases* have been found to be widely distributed in
plants and are different and separable from, for example, *farnesyl
diphosphate synthases* (see below) present in the same plants.

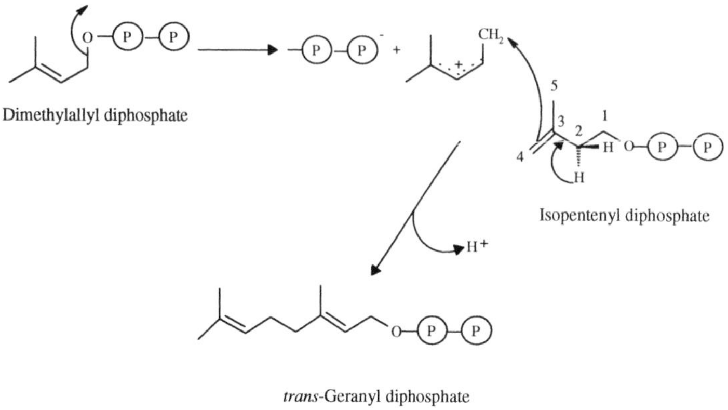

Dimethylallyl diphosphate

Isopentenyl diphosphate

trans-Geranyl diphosphate

Fig. 98. Biosynthesis of geranyl diphosphate

Hydrolysis of GPP yields geraniol, a monoterpene which is a component of many essential oils, e.g. oil of rose (page 378). Oxidation
of geraniol yields geranial, also a common component of volatile
oils. An NADP⁺-dependent *geraniol dehydrogenase* (EC 1.1.1.
183) has been characterized from leaves of lemongrass
(*Cymbopogon flexuosus*).

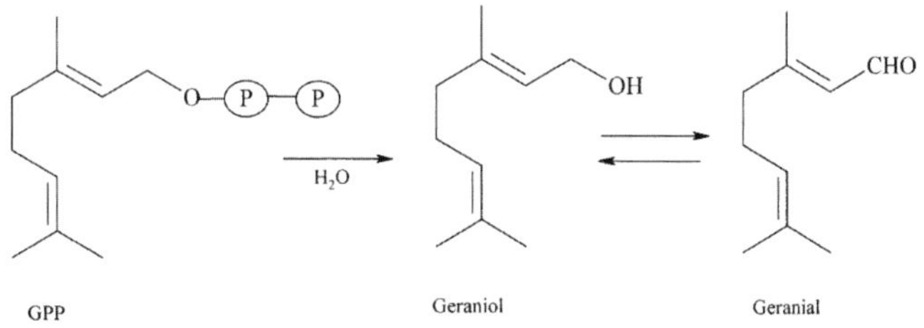

GPP

Geraniol

Geranial

The vast majority of the several hundred naturally occurring monoterpenes are cyclic, and they represent a relatively small number of
skeletal themes multiplied by a very large range of simple derivatives, positional isomers and stereochemical variants. The enzymes
which catalyse the cyclizations are called *monoterpene cyclases*.
Direct cyclization of GPP is not possible because the configuration of
the C_2-C_3 double bond of GPP prevents direct closure to form the
cyclohexene ring. The initial reaction must therefore be an isomerization to (3*R*)-(-)- or (3*S*)-(+)-linalyl diphosphate (LPP). These compounds have, however, never been isolated from the biosynthesizing
organisms. It has now been shown that the cyclase performs both the
isomerization and the cyclization reactions. LPP thus never leaves the
enzyme but is immediately converted to the cyclic compound.

A great number of monoterpene cyclases have been isolated both from microorganisms and higher plants. An example is *(4S)-limonene synthase* (EC 4.2.3.16), which has been isolated from the glandular trichomes of peppermint (*Mentha x piperita* L.) or spearmint (*Mentha x spicata* L.). This enzyme, which has a molecular mass of 56 kDa and is monomeric, catalyses the biosynthesis of (4S)-(-)-limonene. The proposed mechanism of this reaction, which proceeds via a series of carbocationic intermediates, is presented in Fig. 99. GPP first ionizes and then isomerizes to linalyl diphosphate, which remains bound to the enzyme. After rotation from the transoid to the cisoid conformer, linalyl diphosphate is itself ionized and then cyclized to the α-terpinyl cation, which is subsequently deprotonated to give limonene. Both the isomerization and cyclization steps are considered to take place at the same active site. The enzyme requires a divalent cation, Mn^{2+} being preferred over Mg^{2+}. The cation is thought to assist in the initial ionization of GPP by forming a complex with the diphosphate group. Treatment with special reagents has indicated the presence of an essential cysteine

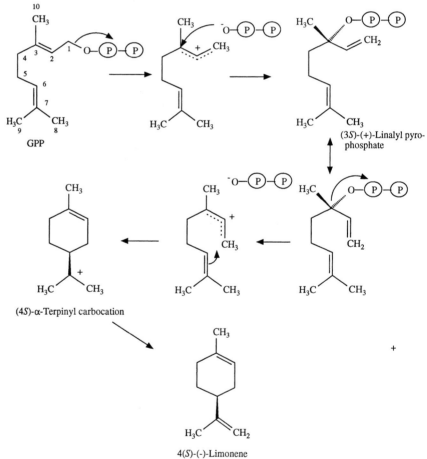

Fig. 99. Biosynthesis of (4S)(-)-limonene from GPP

residue and that a histidine residue is also important for the function of the enzyme.

Monoterpenes can be classified by their degree of oxidation, starting with hydrocarbons and continuing with alcohols, aldehydes, ketones, phenols and oxides.

Hydrocarbons

Some monoterpenes belonging to this group are:

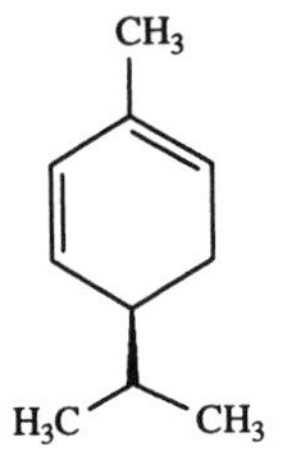

(+)-α-Phellandrene 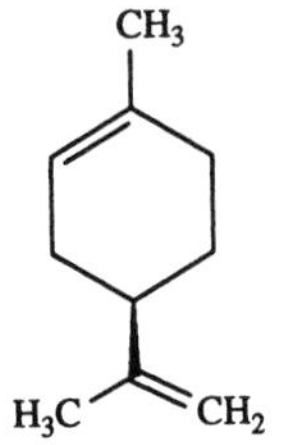   (+)-Limonene 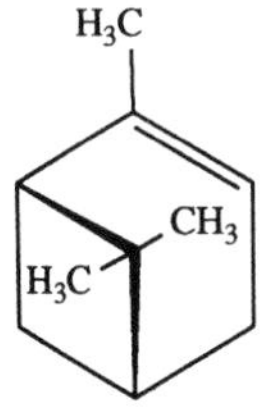   (+)-α-Pinene 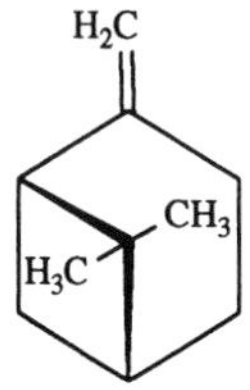   (+)-β-pinene

Crude drugs   **Black pepper** (Piperis nigri fructus) and **white pepper** (Piperis fructus albus) are obtained from *Piper nigrum* L. (*Piperaceae*) (Plate 8, page 377), indigenous in Vietnam and India but now cultivated in most tropical countries. The plant is a woody, perennial climber. Indonesia and India are leading exporters.

Pepper has spikes of small white flowers. After fertilization, they develop into drupes, which are at first green, then red, and finally, when completely ripe, yellow. Black pepper is the dried unripe (green) fruit, which on drying becomes shrivelled and black. White pepper is the ripe fruit, from which the outer, fleshy part of the pericarp has been removed a few days after the harvest. White pepper thus consists of the hard inner part of the pericarp and the seed. Both black and white pepper are important spices with an aromatic and sharp taste. The aromatic smell is due to the volatile oil (1–2.5 % of the weight of the fruit), the main constituents of which are α- and β-pinenes, limonene and phellandrene.

The sharp taste is due mainly to piperine, a nitrogenous substance.

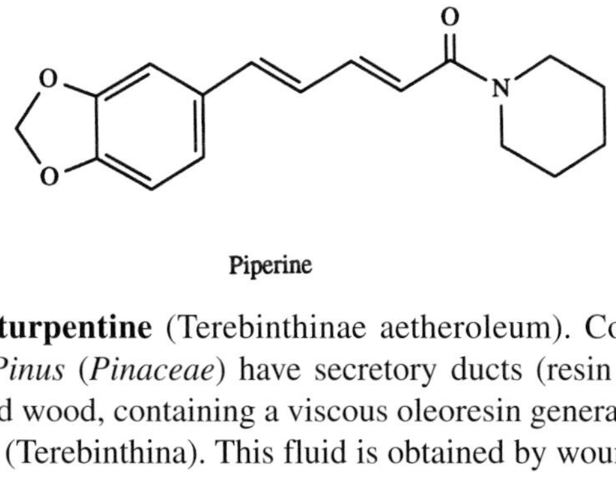

Piperine

Oil of turpentine (Terebinthinae aetheroleum). Conifers of the genus *Pinus* (*Pinaceae*) have secretory ducts (resin ducts) in the bark and wood, containing a viscous oleoresin generally called turpentine (Terebinthina). This fluid is obtained by wounding the tree so that the oleoresin oozes out. It is collected in small cups (mostly of earthenware) attached to the trunk below the wound. The initial turpentine flow soon ceases, but damage stimulates the tree to produce new tissue, richer in secretory ducts, from which further resin can be obtained. The turpentine flow can also be stimulated by spraying the wound with sulphuric acid.

Distillation of the turpentine gives *oil of turpentine*. The distillation residue is a solid resin, *colophony*. There are many grades of turpentine oil, depending on the *Pinus* species from which it has been obtained. Oil of turpentine for medicinal use comes mainly from *Pinus palustris* Miller, in the United States, and from *Pinus pinaster* Loud. ex Aiton, in France. After distillation, it is treated with alkali, to remove acidic impurities, and then redistilled. The main constituent of the product (60–96 %) is α-pinene. The rest is β-pinene and small amounts of other terpenes. Both α- and β-pinene are found in Nature in the dextrorotatory (+)-form as well as in the laevorotatory (-)-form. American oil of turpentine, from *P. palustris* Miller, contains mainly (+)-α-pinene, whereas (-)-α-pinene dominates in the French product.

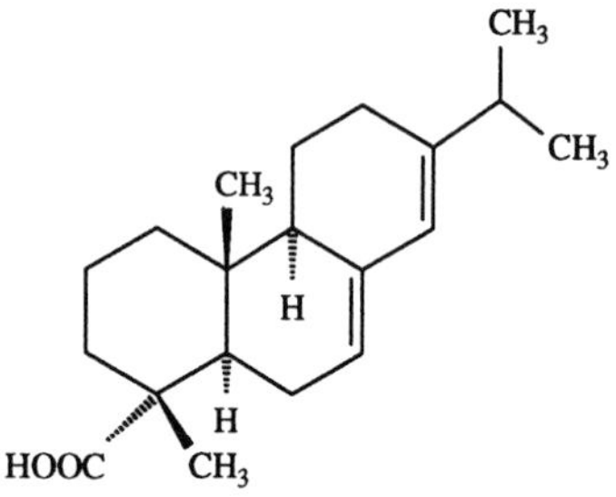

Colophony contains 80–90% diterpene acids, e.g. abietic acid.

Oil of turpentine is used as a rubefacient in liniments for rheumatic diseases. The irritation causes increased blood circulation through the skin, which gives a feeling of warmth and eases the rheumatic pains.

Abietic acid

Alcohols

The following monoterpene alcohols are constituents of essential oils and essential-oil drugs:

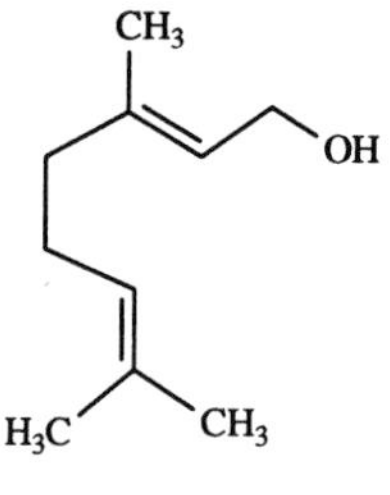

Geraniol

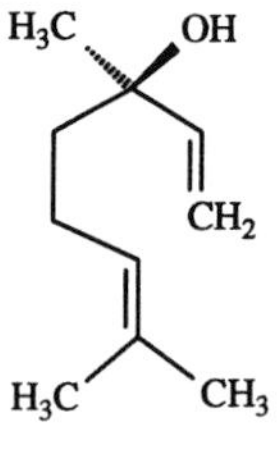

(−)-Linalool

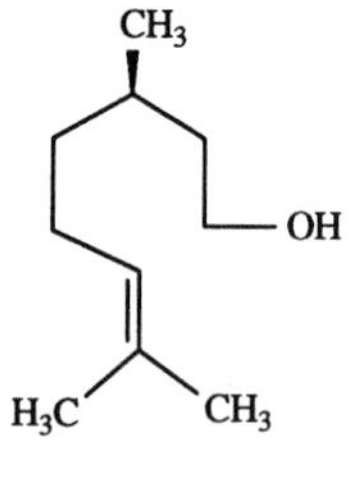

(-)-β-Citronellol

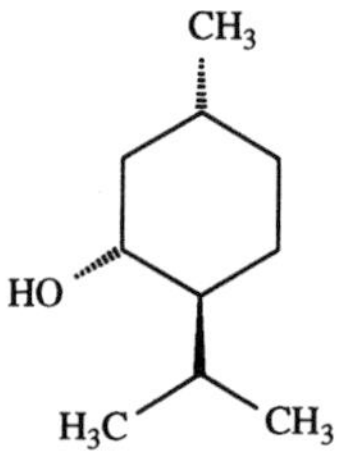

(-)-Menthol

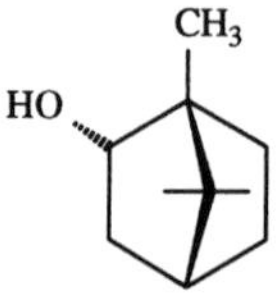

(+)-Borneol

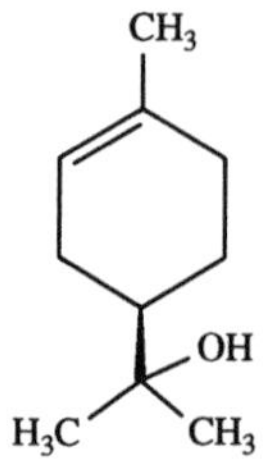

(+)-α-Terpineol

Plate 8.
Piper nigrum L. The fruit is processed to yield black and white pepper (page 375). The picture shows a plantation in Sri Lanka. Photograph by Gunnar Samuelsson.

Crude drugs

Jasmone

(+)-Menthofuran

Peppermint leaf (Menthae piperitae folium) and **Peppermint oil** (Menthae piperitae aetheroleum) are obtained from peppermint, *Mentha piperita* L. (*Lamiaceae*). This is a perennial European herb, which is also grown in the United States and Canada. In cultivation, it must be propagated vegetatively by runners or cuttings. Both the dried leaf and the flowering tops (Menthae piperitae herba) are available as crude drugs.

Peppermint contains approximately 1 % of volatile oil, which can be isolated by steam distillation. About 50 % of the oil is menthol, mostly as the free alcohol, but some is in the form of the acetate and valerate. The total ester content is 4–9 %. Other constituents of the oil are: 10 % menthone (the ketone corresponding to menthol) and small amounts of phellandrene, α-pinene, limonene, jasmone, menthofuran, acetaldehyde, amyl alcohol, acetic acid, valeric acid, etc.

The smell and taste – and hence the quality of the oil – is determined partly by the content of menthol esters, which should be high, and partly by the concentrations of other constituents, especially of jasmone and menthofuran. Jasmone is present only in a low concentration (less than 0.1 %), but it gives the oil a sweet and pleasant taste. Menthofuran has a sharp and disagreeable smell and the amount present should be as low as possible. Oil distilled from young plants has a high menthofuran content and is, therefore, a lower grade of oil than that obtained from older plants.

In pharmacy, peppermint and oil of peppermint are used mainly for taste improvement. Large quantities of oil of peppermint are consumed in making chocolates and sweets. Tea of peppermint leaf is used as a herbal remedy for treatment of indigestion and other stomach complaints.

Japanese peppermint oil is obtained from *Mentha arvensis* L. var. *piperascens* Malinv. ex Holmes (*Lamiaceae*) by steam distillation. The plant is a Japanese herb and is also cultivated in the United States and in Brazil. The oil contains 50–70, occasionally 90 % menthol, but its flavour is inferior to that of ordinary peppermint oil. The oil is therefore used almost entirely as a source of menthol. This compound crystallizes when the oil is cooled to about –22 °C and is purified by recrystallization. Natural menthol obtained in this way is laevorotatory. Menthol can also be synthesized using thymol (page 381) or α-pinene as a starting material. Owing to the presence of three asymmetric carbon atoms in the molecule, the synthetic product also contains stereoisomers with inferior organoleptic properties.

Menthol gives a chilling sensation when applied to the skin or to mucous membranes and it is used in ointments and liniments as a remedy against itching.

Oil of rose (Rosae aetheroleum) is obtained by steam distillation of flowers from species of *Rosa* (*Rosaceae*), especially *R. centifo-*

lia L., *R. gallica* L. and *R. damascena* Mill. The largest producer is Bulgaria, but oil of rose is also obtained from Turkey, France and Morocco. The oil content of the petals is low (0.01–0.04 %), so that 3–5 tons of petals are needed for the production of 1 kg of oil.

The main constituents of oil of rose are geraniol and citronellol. In addition, there are linalool, eugenol (page 214) and 5–10 % nerol, which is a stereoisomer of geraniol and of great importance as regards the scent of the oil. Genuine oil of rose also contains high-molecular-weight hydrocarbons (stearoptenes), which precipitate at temperatures below 25 °C. Oil of rose is a highly valued product and is therefore often adulterated. It is mainly used in perfumery and for high-class cosmetics. In pharmaceutical practice it is customary to use the substitute **rose geranium oil** (Pelargonii aetheroleum), which contains esters of geraniol and citronellol and has a scent that is very similar to that of authentic oil of rose. It is obtained by steam distillation of *Pelargonium* species (*Geraniaceae*), which are cultivated in the Mediterranean region as well as on Madagascar and Réunion.

CH₃

H₃C   CH₃

α-Terpinene

Cardamom fruit (Cardamomi fructus) is obtained from *Elettaria cardamomum* (L.) Maton (*Zingiberaceae*), a perennial herb, 2–3 metres high, which is grown in Guatemala, in Sri Lanka and on the Malabar Coast in South India. The fruit is a three-lobed capsule, which is harvested shortly before ripening, dried, and bleached with sulphur dioxide. The volatile oil, which is the reason for the use of the crude drug, occurs mainly in the seeds, but the drug is sold in the form of whole fruits, partly to facilitate identification and separation from inferior varieties, and partly in order to preserve the oil in the seeds.

Cardamom seeds contain 3–6 % of volatile oil, the main constituents of which are α-terpineol (free and acetylated) and the hydrocarbon α-terpinene. Low proportions of other monoterpene hydrocarbons and monoterpene alcohols are also present, as well as the terpene oxide cineole (page 390). Cardamom is used as a spice.

Lavender flower (Lavandulae flos) are the flowers of *Lavandula vera* DC. *(*occasionally known as *L. angustifolia* Mill. or *L. officinalis* Chaix) (*Lamiaceae*). *Lavandula* species grow naturally in southern Europe and north-western Africa and are also cultivated elsewhere in Europe and in the United States. *Lavandula vera* is an evergreen shrub, in cultivation up to 1 metre high, with small, narrow leaves. The flowering period is July–September. The inflorescence, a spike, comprises 6–10 blue flowers, which on steam distillation yields about 0.5 % volatile oil. Lavender growing in mountainous districts gives oil of a better quality than that growing at low altitudes. The oil contains 30–60 % esters (mainly linaloyl acetate), as well as geraniol, (-)-linalool, limonene and a small proportion of cineole. Lavender flowers and oil of lavender are used in perfumery.

Rosemary leaf (Rosmarini folium) is the leaf of *Rosmarinus offic-inalis* L. (*Lamiaceae*), an evergreen shrub from southern Europe, 1–2 metres in height. The branches are slender, with an ash-coloured bark, and the 3 cm long leaves are opposite and linear. The upper surface of the leaves is glossy and the lower surface is matt and grey owing to the presence of numerous hairs. The leaves contain 1–2 % of a volatile oil comprising 0.8–6 % esters, bornyl acetate and 8–20 % alcohols, mainly borneol. Other constituents present are ca. 20 % cineole and several terpene hydrocarbons, e.g. α-pinene and camphene. Rosemary is used as a spice. Rosemary is used as a herbal remedy for treatment of dyspepsia and headache. The volatile oil is used for topical treatment of rheumatic diseases.

Aldehydes

Volatile oils of *Citrus* fruits contain the terpene aldehydes geranial and citronellal, which correspond to the alcohols geraniol and citronellol, respectively. The aldehydes constitute only 5 % of the *Citrus* fruit oil, but they account for most of its flavour.

Crude drug **Bitter-orange peel** (Aurantii pericarpium) is the dried fruit peel of the bitter orange (= Seville or Bigarade orange), *Citrus aurantium* L. (*Rutaceae*). This tree is a close relative of the orange and lemon trees. Its home country is India and it is grown in most subtropical countries. Like the peel of other *Citrus* fruits, the fresh bitter-orange peel comprises two layers: an outer, firm, yellow one called *flavedo* and an inner, soft colourless one, the zest or *albedo*. As little as possible of the zest should be present in the crude drug. The yellow layer has large cavities containing a volatile oil, with up to 90 % of the terpene hydrocarbon (+)-limonene. The aroma of the oil is due essentially to geranial, which, together with its *cis*-isomer neral, constitutes about 5 % of the oil.

Bitter-orange peel is an aromatic bitter tonic, used as a remedy for poor appetite.

Geranial

Lemon oil (Limonis aetheroleum) is the volatile oil from the peel of the lemon, *Citrus limon* (L.) Burm.f. (*Rutaceae*). Lemons are grown in most subtropical countries. Lemon oil – like oils from other *Citrus* fruits – is very sensitive to hydrolysis and oxidation and therefore cannot be obtained by steam distillation. Instead, other means must be used to rupture the oil cavities of the peel and liberate the oil. The simplest method is to press the lemon peel by hand and collect the oil in a sponge, which is then squeezed out over a bucket. On a larger scale, machines are used to prick or cut the outermost layer of the peel, taking care not to touch the zest, since this contains enzymes that can degrade the constituents of the oil. In California, whole lemons are squeezed under very high pressure, and the juice and oil are then quickly separated in high-speed centrifuges at a low temperature. Lemon oil is easily oxidized and must be stored in a cool place

(+)-Citronellal

and in the dark. It contains 94 % limonene, about 4 % citral (geranial + neral), and citronellal, as well as geranyl acetate and other esters.

In pharmacy, oil of lemon is used for flavouring. The greater part of the production goes to the food and perfumery industries.

Ketones

Camphor

Camphor is the main constituent of the volatile oil obtained by steam distillation of the wood of *Cinnamomum camphora* T. Nees & Eberm. (*Lauraceae*). Camphor is solid at room temperature and separates from the volatile oil on cooling.

Cinnamomum camphora is a tall tree up to 40 m in height, with coriaceous leaves having an aromatic smell. It grows in the coastal regions of South-East Asia.

CH₃

H

(+)-Camphor

The greatest producer of natural camphor is Taiwan, followed by Japan. All parts of the tree contain oil cells, but the largest amount of camphor is obtained from the trunk and root of old trees (50–60 years old). The production of natural camphor has largely been replaced by the industrial synthesis from pinene which occurs in oil of turpentine. Natural camphor is optically active, with $[\alpha]_D$ +41–43, but the synthetic product is a racemate.

Traditionally, camphor is used in liniments against rheumatic pains and a 3% ethanolic solution sooths itching.

Carvone

CH₃

O

H₂C   CH₃

(+)-Carvone

Carvone is the main component of the volatile oil of **caraway fruit** (Carvi fructus), which is the schizocarp of *Carum carvi* L. (*Apiaceae*). This plant is a biennial herb growing in Europe and Asia; it is also cultivated in the United States and Canada. The fruit contains 5–7 % of volatile oil in special cavities. The oil comprises 50 % (+)-carvone and the rest is mainly (+)-limonene. Caraway is used as a spice and, in traditional medicine, as an expectorant for treatment of colds and as a spasmolytic and a carminative for treatment of stomach complaints.

Phenols

Crude drugs

CH₃

OH

H₃C   CH₃

Thymol

Thyme (Thymi herba) consists of the dried leaves and flowers of *Thymus vulgaris* L. (*Lamiaceae*), a small, shrubby, perennial herb from the Mediterranean region, also cultivated in France, Germany, Spain and the United States. The plant has glandular hairs on the leaves and stems and these contain a volatile oil, amounting to 1–3 % of the weight of the drug. The main component of the oil is thymol. Other constituents are carvacrol, *p*-cymene, borneol, linalool and β-pinene. The two phenols, thymol and carvacrol, which are both bactericidal, together make up 25–40 % of the oil; usually there is more thymol than carvacrol present.

Thyme is used as a herbal remedy for treatment of bronchitis, whooping cough and catarrh of the upper respiratory tract.

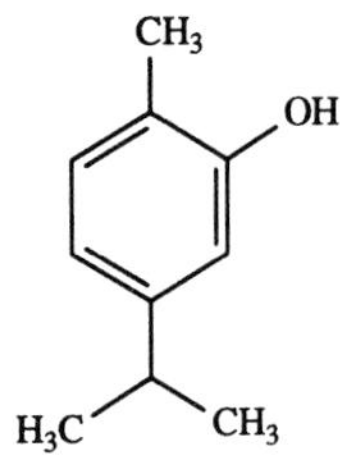

CH₃

OH

H₃C   CH₃

Carvacrol

Thyme oil (Thymi aetheroleum) is the volatile oil of *Thymus vulgaris* L. Thyme oil is toxic and should not be used internally. Externally it can cause dermal and mucous irritation and should be used only after dilution with a suitable carrier oil.

Ginger (Zingiberis rhizoma) is the peeled and dried rhizome of *Zingiber officinale* Rosc. (*Zingiberaceae*). Ginger is a very old Indian spice, nowadays grown in most tropical countries. The cultivated varieties of this are sterile and are thus propagated vegetatively. Ginger contains 1–2 % of a volatile oil, in which camphene, cineole, citral and the sesquiterpenes zingiberene, bisabolene and zingiberol are present.

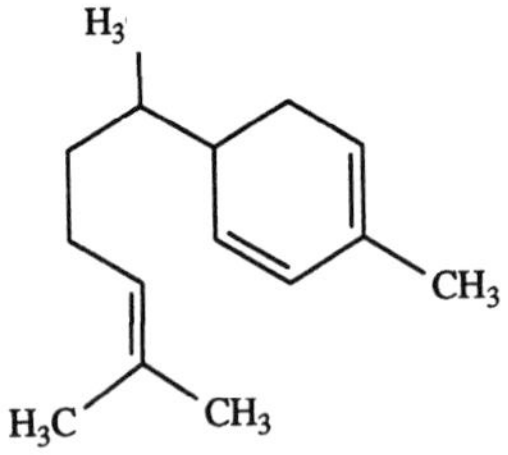

Ginger is used as a spice. The crude drug has a pungent taste due to the presence of gingerol, which is a mixture of homologous phenols with the following structures:

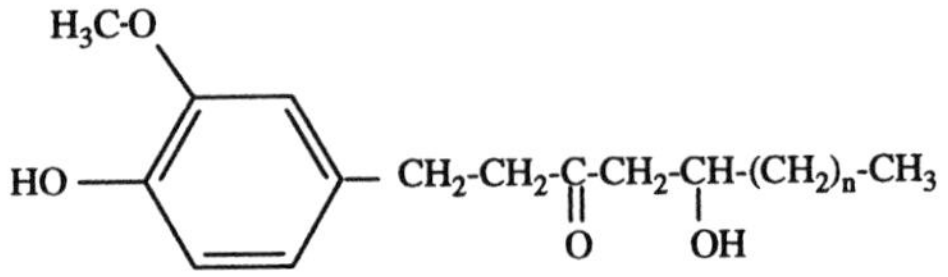

H₃

H₃C   CH₃

CH₃

Zingiberene

$H_3C\text{-}O$

HO — $CH_2\text{-}CH_2\text{-}C\text{-}CH_2\text{-}CH\text{-}(CH_2)_n\text{-}CH_3$

O OH

Gingerol (N = 3, 4 or 5)

Ginger is also used as a herbal remedy for treatment of dyspepsia, as a remedy against motion sickness and for treatment of osteoarthritis and rheumatoid arthritis.

Iridoids and secoiridoids

Iridoids are mainly C_{10} compounds containing a cyclopentanotetrahydropyran ring system. Secoiridoids can be regarded as being formed from iridoids by opening of the cyclopentane ring between C-7 and C-8. Most of these compounds are glucosides. The iridoids and secoiridoids can be classified as follows:

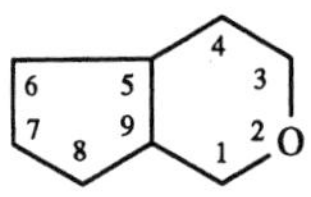

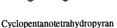

Cyclopentanotetrahydropyran

1. *Non-glycosidic iridoids*
Examples are:

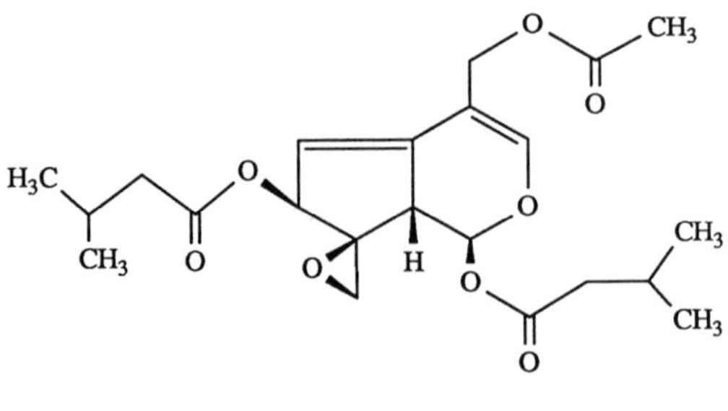

H₃C

CH₃   O

O

O   H

CH₃

O

CH₃

CH₃

O

O

O

CH₃

O

Valtrate

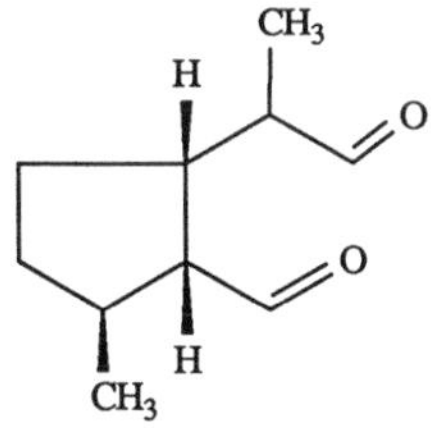

Iridodial

Iridodial is a precursor of the genuine iridoids (see below). Valtrate is one of the so-called valepotriates, which are sedative compounds occurring in all species of the *Valerianaceae* including *Valeriana* and *Centranthus* species. These compounds will be dealt with in greater detail further on page 386.

2. Iridoid glycosides
This group can be represented by the following compounds:

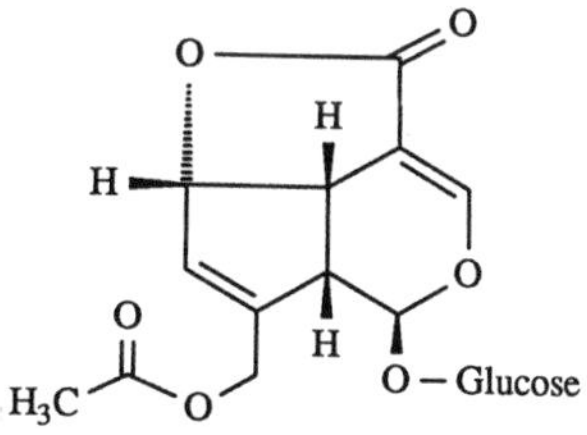

Asperuloside

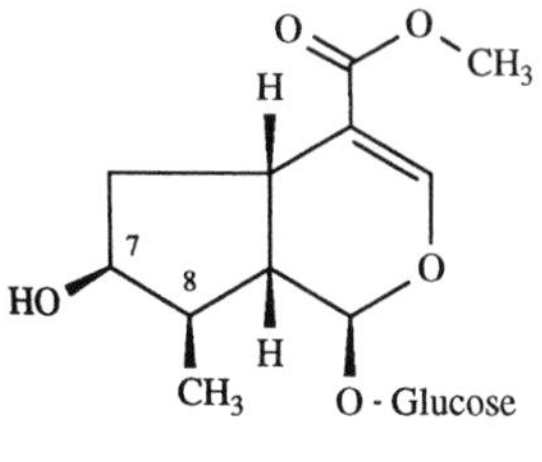

Loganin

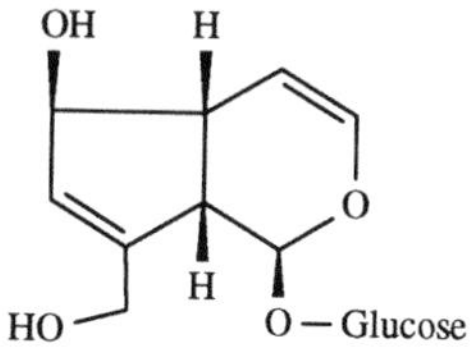

Aucubin

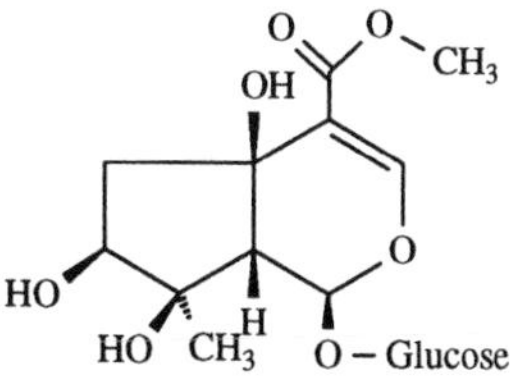

Lamiide

Asperuloside was the first iridoid to be isolated and was discovered in 1848. It is found in many plants of the family *Rubiaceae*, e.g. in *Asperula odorata* L. and in *Rubia tinctorum* L. Loganin is a bitter glucoside which is present in *Strychnos nux-vomica* L. (*Logani-aceae*, page 691) and *Menyanthes trifoliata* L. (*Menyanthanaceae*, page 387), among others. Aucubin occurs in many taxa in the *Lami-idae*, such as the genus *Aucuba* (*Aucubaceae*), *Plantago major* L. (*Plantaginaceae*) and *Scrophularia racemosa* Lowe (*Scrophulari-aceae s.str*). Lamiide is found, e.g. in *Lamium amplexicaule* L. (*Lamiaceae*).

3. Glycosidic secoiridoids
Examples:

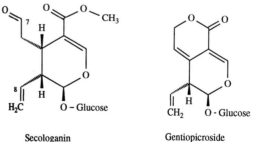

Secologanin

Gentiopicroside

Secologanin is a key intermediate in the biosynthesis of alkaloids containing a monoterpenoid skeleton (see pages 667 and 679). Gentiopicroside is a bitter compound, which is found in all species of the genus *Gentiana*, e.g. the medicinal plant *Gentiana lutea* L. (see below, page 388).

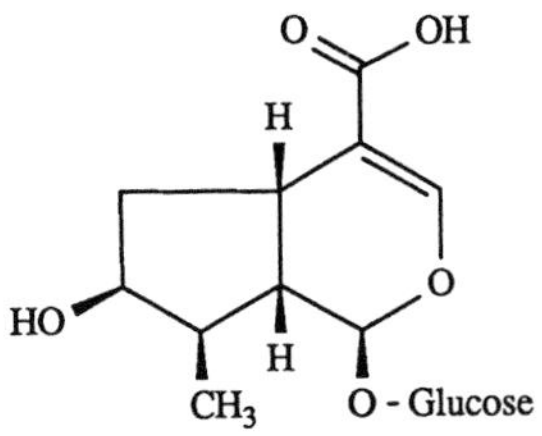

Loganic acid

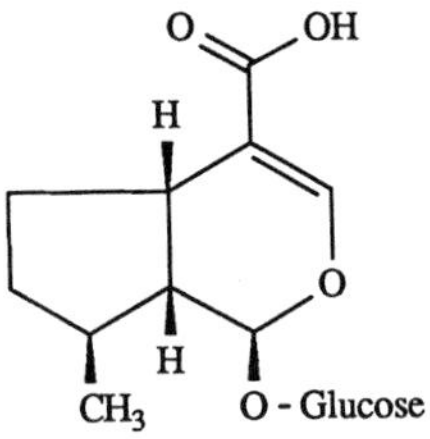

Deoxyloganic acid

Biosynthesis of iridoids and secoiridoids

Iridoids are biosynthesized from isopentenyl diphosphate (IPP) and dimethylallyl diphosphate (DMAPP) via 10-hydroxygeraniol or the stereoisomer 10-hydroxynerol. These compounds are oxidized to the corresponding aldehydes and the cyclopentane ring is then formed by cyclization, a process which divides the biosynthesis into two pathways depending on whether iridodial or the epimer 8-*epi*-iridodial is formed (Fig. 100).

The continuation of the pathways after formation of the cyclo-pentane ring is known in detail only in a few cases. Loganic acid, deoxyloganic acid or 8-*epi*-deoxyloganic acid are intermediates, formed by oxidation of C-9 in iridodial or 8-*epi*-iridodial.

Secologanin is formed from loganin by cleavage of the cyclopen-tane ring between C-7 and C-8 (see formula for cyclopentano-tetrahydropyran on page 382). This reaction is catalysed by the cytochrome P450 enzyme *secologanin synthase* (EC 1.3.3.9) and requires NADPH and molecular oxygen.

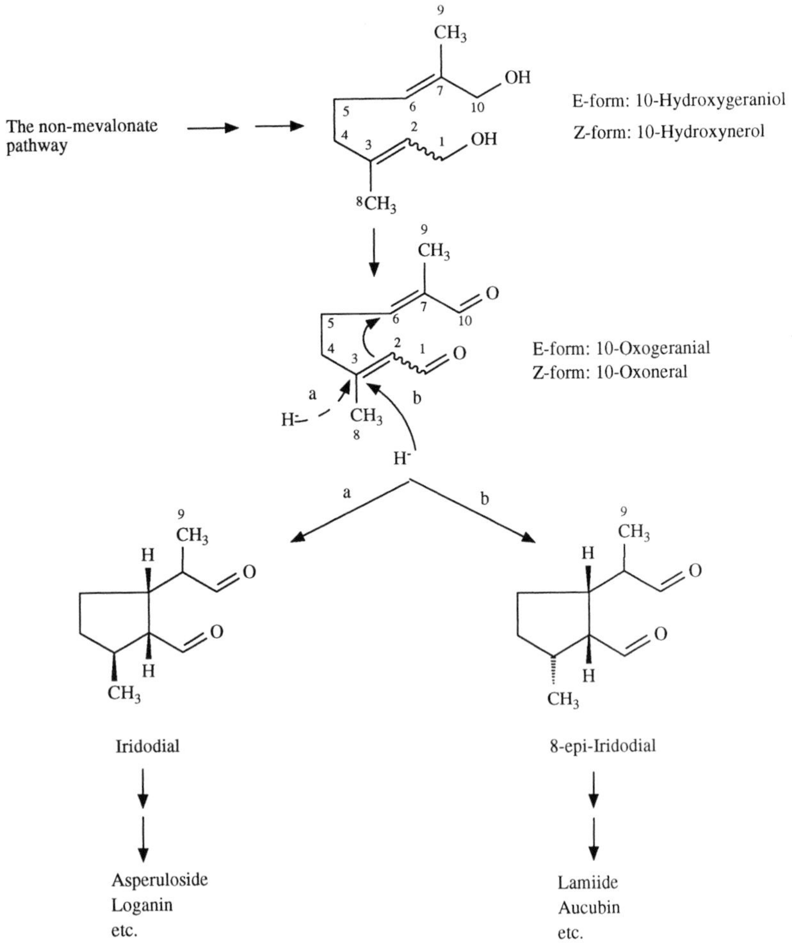

Fig. 100. Biosynthesis of iridoids and secoiridoids

Plate 9.
Valeriana officinalis L. The root is the source of the drug Valerian root (page 386). Kazanluk, Bulgaria. Photograph by Gunnar Samuelsson.

Plate 10.
Gentiana lutea L. The root contains extremely bitter compounds (page 388). The Botanical Garden, Uppsala. Photograph by Gunnar Samuelsson.

385

The medical importance of iridoids and secoiridoids is limited. Valepotriates have a sedative effect and loganin and some other glycosides with a bitter taste are ingredients of crude drugs used as *bitters* (see below).

Crude drugs   **Valerian root** (Valerianae radix) is the dried rhizome of *Valeriana officinalis* L. (*Valerianaceae*). This plant is a perennial herb growing all over Europe (Plate 9, page 385). The species is polymorphic and can be divided into several subspecies, varieties and forms. The crude drug is obtained from both wild and cultivated plants. Valerian is cultivated not only in Europe but also in the United States and Japan. The drug is best harvested at the end of September in the second year of growth. When the rhizome has been dug up, it is immediately washed free of adhering soil and dried in the air at 40° C. If prepared in this manner, the crude drug is light brown and has a faint, spicy scent. Careless or slow drying gives a darker product with a strong smell of valeric acid, indicating a drug of inferior quality. The valeric acid is formed through hydrolysis of the valepotriates, as well as of esters occurring in the volatile oil, which constitutes 1 % of the drug. Among the constituents of this oil are bornyl isovalerate, bornyl acetate, bornyl formate (see formula for borneol page 376), valeranone, valerenal and valerenic acid. In the rhizome, the oil is located in the hypodermis, which is made up of a single layer of large thin-walled cells. Because the oil is located near the surface of the rhizome and as at least some of the pharmacologically active compounds are present in the oil (see below), it is essential to collect and handle valerian rhizome with care, so that the oil cells are not damaged.

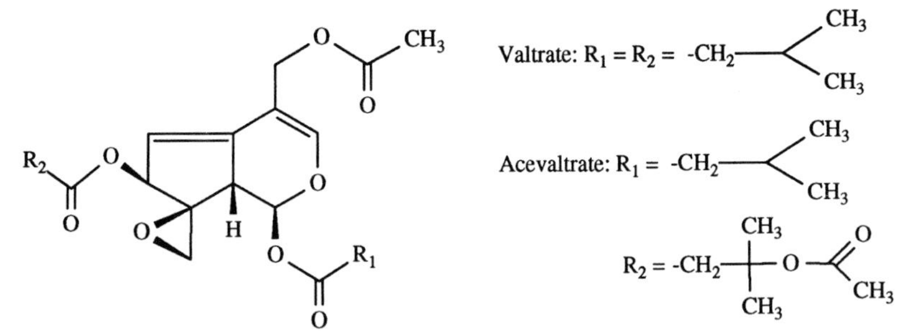

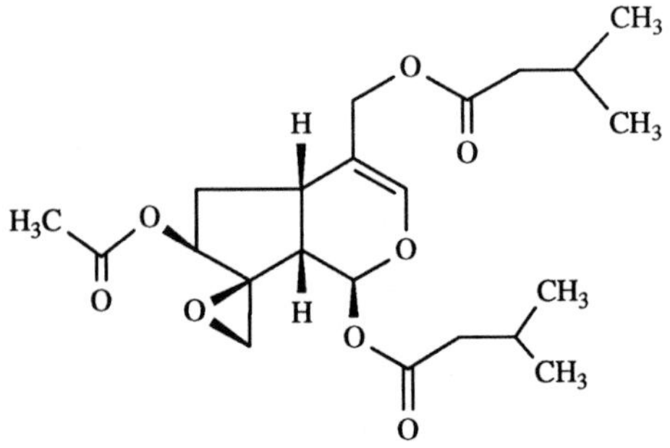

Valtrate: $R_1 = R_2 = -CH_2-$

Acevaltrate: $R_1 = -CH_2-$

$R_2 = -CH_2-$

Didrovaltrate

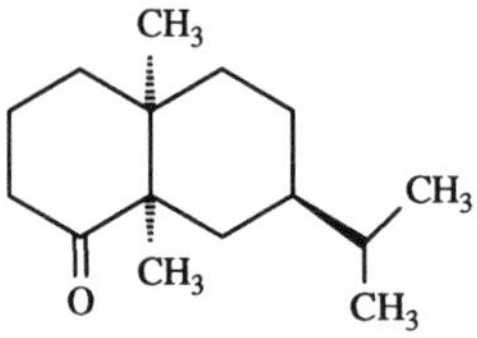

Valeranone 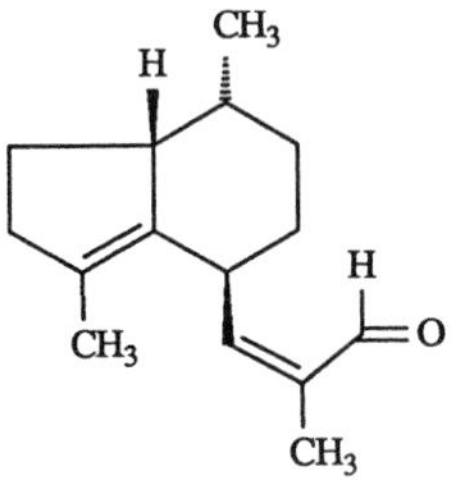   Valerenal 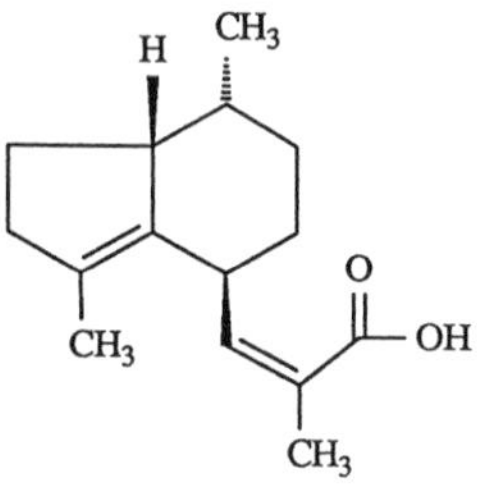   Valerenic acid

Carefully collected and dried Valerian rhizome contains about 0.8 % valepotriates, mainly valtrate (80–90 %). Acevaltrate and didrovaltrate are also present.

Preparations of valerian rhizome are used as mild sedatives. Numerous clinical studies on the effects of valerian preparations on subjective and/or objective sleep parameters have been performed, but the results obtained are inconclusive. The most recently published studies, which were also the most methodologically rigorous, have found no significant effects of valerian on sleep. Another problem is that no active constituent has been definitely identified. Valepotriates have been considered as the active constituents but their role is nowadays highly questioned. They are easily hydrolysed and decompose rapidly under the influence of moisture, heat (>40 °C) or acidity (pH < 3) to yield yellow unsaturated aldehydes (baldrinal, isopropylbaldrinal), which, however, have been shown also to possess sedative effect. *In vitro* valepotriates are highly cytotoxic, mutagenic and teratogenic, effects which have not been shown *in vivo*. Because of their sensitivity to low pH they are to a high degree destroyed in the stomach and thus may not present any danger. Modern preparations of valerian root are made to contain no or very low amounts of valepotriates.

Aqueous extracts of valerian rhizome have in some clinical trials been found to reduce the time taken to fall asleep. The quality of these trials are, however, highly questionable. As aqueous extracts do not contain any valepotriates, this alleged effect must be due to other compounds. Valeranone, valerenal and valerenic acid have shown a sedative effect in pharmacological tests. One investigation indicates that neither the valepotriates, nor the compounds of the essential oil, are responsible for the central depressant effect of valerian rhizome extracts, but that the main effect may be due to hitherto unknown substances.

Buckbean leaf (Menyanthis folium) is the dried leaf of *Menyanthes trifoliata* L. (*Menyanthaceae*). This is a herb growing in water and distributed all over the northern hemisphere. The leaf is three-lobed (similar to clover) and keeps its light colour even after drying. The crude drug has a bitter taste due to the presence of loganin and other iridoid glycosides. An extract of buckbean leaf is used as a *bitter* (amarum) to stimulate the appetite. The bitter taste increases the secretion and acidity of the gastric juice.

Bitters (amara) are usually not isolated compounds but aqueous or ethanolic extracts of crude drugs containing bitter-tasting but otherwise pharmacologically relatively inert substances. Bitters are not a uniform group with respect to their constituents. For this reason, their intrinsic quality is not determined chemically but biologically, utilizing the bitter taste. The lowest concentration of an aqueous extract of the drug at which a bitter taste is just noticeable is determined. However, the method is unreliable, because individual sensitivity to bitter taste differs widely. It is, therefore, advisable to use a panel of several tasters to obtain a better estimate. Attempts have been made to calibrate the tasters by means of a standard substance, e.g. the alkaloid brucine (page 690). However, this has not always been successful, because the bitter taste of the standard substance is often perceived as being different from that of the drug extract being tested. It is customary to report the result of the determination as the *bitterness value* of the drug. If the lowest concentration required for a bitter taste is that of an extract of 1 g of drug in 20 000 ml of water, the bitterness value of the drug is recorded as 1:20 000. It is common to quote only the denominator, i.e. the crude drug is said to have the bitterness value 20 000. In the case of the buckbean leaf discussed above, the value ranges between 1 500 and 9 000.

Gentian root (Gentianae radix) is the dried root of *Gentiana lutea* L. (*Gentianaceae*) (Plate 10, page 385). This plant is a more than 1

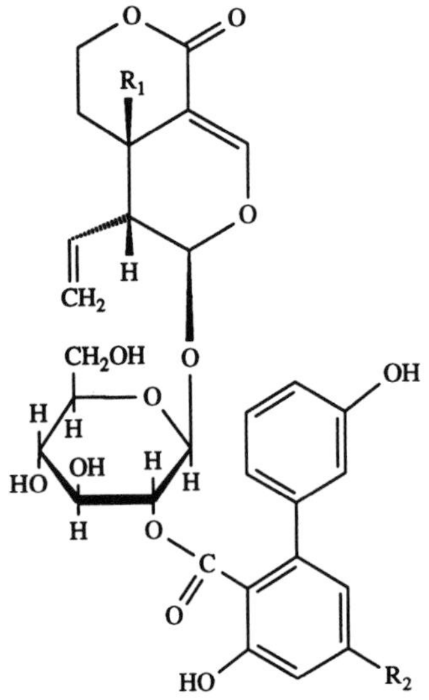

Amarogentin: R_1 = H, R_2 = OH

Amaroswerin: R_1 = R_2 = OH

Amaropanin: R_1 = R_2 = H

metre tall herb with broad, elliptic, decussate leaves and large yellow flowers. It grows in mountainous areas in central and southern Europe. The root, which can be over a metre long, is harvested when the plant is 2–5 years old. The drying process involves a certain fermentation, which makes the root darker and gives it a more distinctive aroma.

Gentian is widely used not only medicinally but also in the manufacture of alcoholic drinks. For the latter use, a more heavily fermented drug is preferred, with a pronounced aroma and a less bitter taste.

The crude drug is extremely bitter with a bitterness value of at least 20 000. This taste is due to the secoiridoid glycosides present in the drug. The fresh root contains about 2 % gentiopicroside (= gentiopicrin), which, however, is largely decomposed during the drying process. The bitter taste of the crude drug is attributed mainly to amarogentin (0.05–0.3 %) and amaroswerin (0.03–0.1 %), which both have a bitterness value of 58 000 000, and to amaropanin (0.05–0.2 %), which has a bitterness value of 20 000 000. The bitterness value of gentiopicroside is only 12 000.

A notable feature of the structures of amarogentin, amaroswerin and amaropanin is the presence of a glucose residue esterified with a derivative of 2-phenylbenzoic acid.

Gentian root is a herbal remedy against digestive disorders such as lack of appetite, fullness and flatulence.

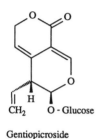

Gentiopicroside

Devil's claw root (Harpagophyti radix) is the secondary storage roots of *Harpagophytum procumbens* DC. (*Pedaliaceae*). The plant is native to the Karoo (semi-desert) of South Africa and Namibia (Kalahari). Several prostate stems, up to 2 m long, protrude from the primary root and form a star. The leaves are alternate or opposite and the bell-shaped, red-violet flowers project from the axils. The secondary roots branch off from the primary root and can reach a depth of 2 m in the ground. They are thick and store up to 90 % of water.

Devil's claw is used in South African traditional medicine for the treatment of a variety of disorders, such as fever, stomach upsets and rheumatic diseases. The drug was brought to Europe after the Second World War and has become widely used as a herbal remedy for rheumatic complaints. In pharmacological experiments, an aqueous extract of the plant has been shown to have an anti-inflammatory effect on experimentally induced oedema of the joints. The clinical efficacy has not been proved beyond doubt. The iridoid gly-

Harpagoside: R = O-C-CH=CH-C$_6$H$_5$
                                    ‖
                                    O

Harpagide: R = OH

cosides *harpagoside* and *harpagide* have been isolated and may contribute to the pharmacological effect of extracts of the crude drug, but other, not yet identified, compounds might also be responsible.

Other oxidized monoterpenes

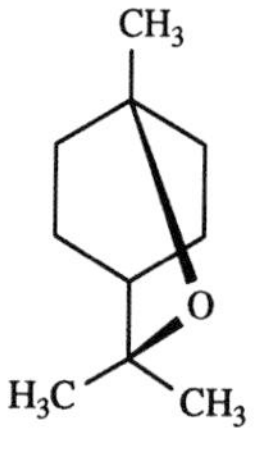

Eucalyptol (Cineole)

Eucalyptol is the chief constituent of **eucalyptus oil** (Eucalypti aetheroleum), which is obtained by steam distillation of the fresh leaves of *Eucalyptus globulus* Labill. and other *Eucalyptus* species (*Myrtaceae*). *Eucalyptus* species are Australian trees which have been introduced into southern Europe, the southern United States, and many other subtropical and tropical countries. Leaves of two different shapes are found on the trees: young leaves are ovate and sessile, whereas older leaves are longer, petiolate and scythe-shaped. The volatile oil is located in special oil glands in the leaves. Eucalyptus oil consists of up to about 70 % eucalyptol, which can be separated by freezing. Eucalyptol is a liquid at room temperature, but solidifies when the temperature falls below 0 °C.

Eucalyptus oil and eucalyptol are used in lozenges and inhalants to ease the discomfort of colds. Eucalyptol is preferable to the oil, because the latter contains aldehydes (butyraldehyde, valeraldehyde, caproaldehyde) which irritate mucous membranes.

CH$_3$

H$_3$C   CH$_3$

O

Eucalyptol (Cineole)

Pyrethroids

The dried flowers of *Chrysanthemum cinerariaefolium* contain compounds that are efficient contact insecticides but harmless to man and other warm-blooded animals. These compounds are esters of the monoterpenes pyrethric acid and chrysanthemic acid with a cyclopentenolone bearing an unsaturated side-chain.

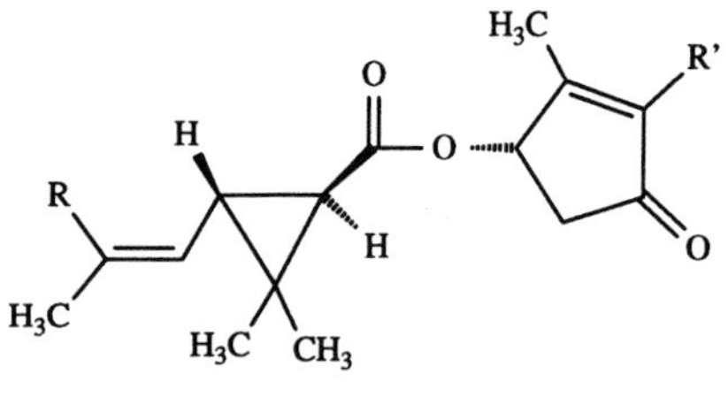

	Acid	R	R'
Pyrethrin I	Chrysanthemic acid	CH$_3$	CH$_2$-CH=CH-CH=CH$_2$
Pyrethrin II	Pyrethric acid	COOCH$_3$	CH$_2$-CH=CH-CH=CH$_2$
Cinerin I	Chrysanthemic acid	CH$_3$	CH$_2$-CH=CH-CH$_3$
Cinerin II	Pyrethric acid	COOCH$_3$	CH$_2$-CH=CH-CH$_3$

Like other monoterpenes, the acids are biosynthesized from isopentenyl diphosphate, but the cyclization leads to the formation of a cyclopropane ring. The alcohol moiety is also formed from acetate, but not via isopentenyl diphosphate.

390

Crude drug **Pyrethrum flowers** (Pyrethri flos) are the dried inflorescenses of *Chrysanthemum cinerariaefolium* Vis. (*Asteraceae*). The plant is a perennial herb, about one metre in height, naturally occurring in the [Dalmatian] Balkan peninsula, but now grown in many tropical countries. Every part of the plant contains pyrethrins, but the highest proportion is found in the flowers and only these are used as crude drugs in industrial processes. Selection and hybridization have led to clones with a pyrethrin content of up to 2 % of the dry weight of the flowers.

The world production of pyrethrum amounts to about 22 000 tons per annum. The main producer is Kenya (14 000 tons), but large quantities come from Tanzania, Ecuador and Japan. In Kenya, the production is spread over numerous small units and the number of families taking part is estimated at 85 000. The inflorescenses are collected by hand and this work goes on for 7–11 months of the year. The plantations are cooperatives and the entire production is delivered to the *Pyrethrum Marketing Board*, which is responsible for the analysis, extraction and export of the products.

Most of the crop is extracted with light petroleum. The solvent is removed by distillation, leaving a dark brown viscous extract, which is purified by extraction with methanol followed by treatment with charcoal. The product thus obtained is used in insecticidal sprays. It is customary to increase the effect of the extracts by adding compounds which act synergistically, e.g. piperonyl butoxide and sesamolin (page 251). A small part of the crop is used for the production of pyrethrum coils, which are burned in houses and which give an insecticidal smoke containing pyrethrins.

Pyrethrum has been used as an insecticide for over a century and toxicological studies have confirmed early experience that their toxicity to birds and mammals is low. The pyrethrum compounds are rapidly decomposed in Nature and there is no risk of accumulation in the nutrition chain, as with DDT. However, this rapid decomposition is also a disadvantage. Pyrethrum is essentially a poison that kills insects through contact and the main use is as a constituent of insect sprays for household use and as post-harvest insecticides.

Since the decision by many countries to prohibit the use of DDT, pyrethrum has become a very important insecticide. Synthetic pyrethroid analogues are now available which have better properties including a longer lifetime and greater toxicity towards insects. These compounds also have low mammalian toxicity and can be used as agricultural insecticides.

Cannabinoids

Cannabinoids are a group of C_{21} compounds occurring in the glandular hairs of *Cannabis sativa* L. (*Cannabidaceae*). The most important representative is Δ^1tetrahydrocannabinol (Δ^1-THC), which has hallucinogenic properties.

Several systems for numbering the carbon atoms of cannabinoids

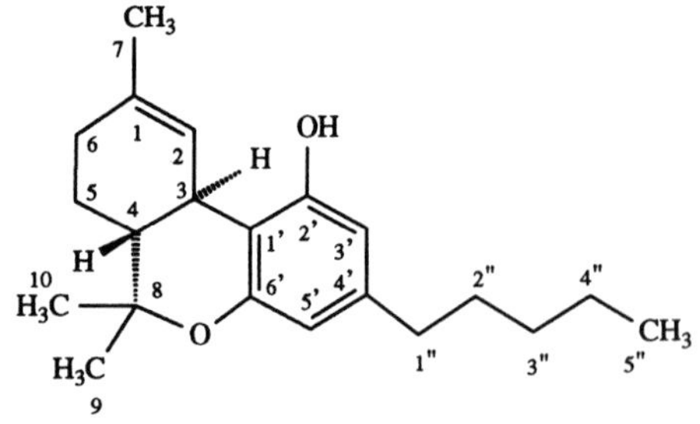

Δ¹-Tetrahydrocannabinol

monoterpenoid numbering

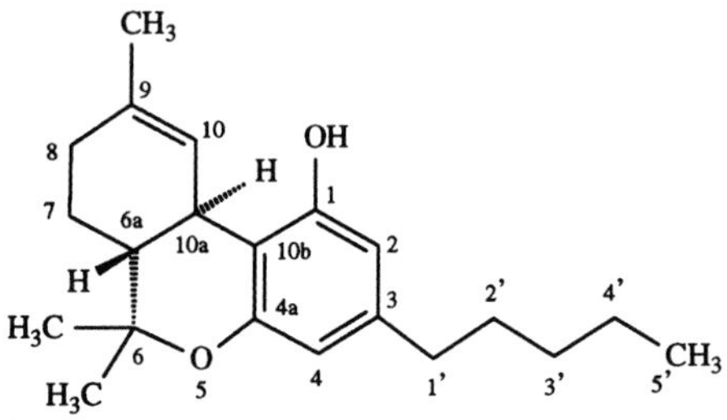

Δ⁹-Tetrahydrocannabinol

are used in the literature. The two most common systems are shown overleaf in the formula of tetrahydrocannabinol. In one of them the substance is regarded as a monoterpene derivative, in the other as a dibenzopyran derivative. The first system is consistent with the presumed biosynthesis of the compound (see below) and has the advantage of being applicable also to cannabinoids that are not dibenzopyran derivatives. For these reasons, the monoterpenoid numbering will be used here.

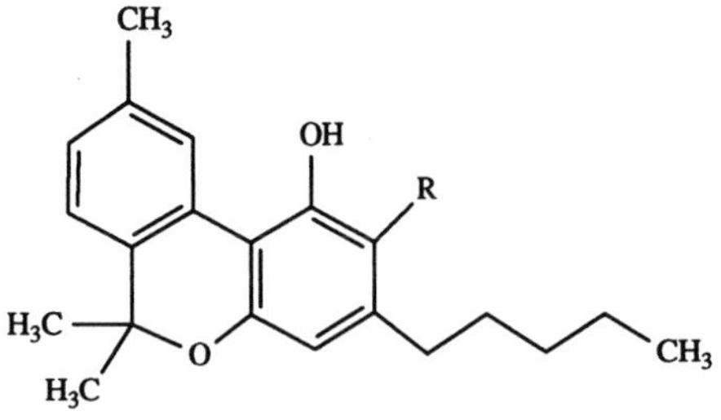

R = H: Cannabinol

R = COOH: Cannabinolic acid

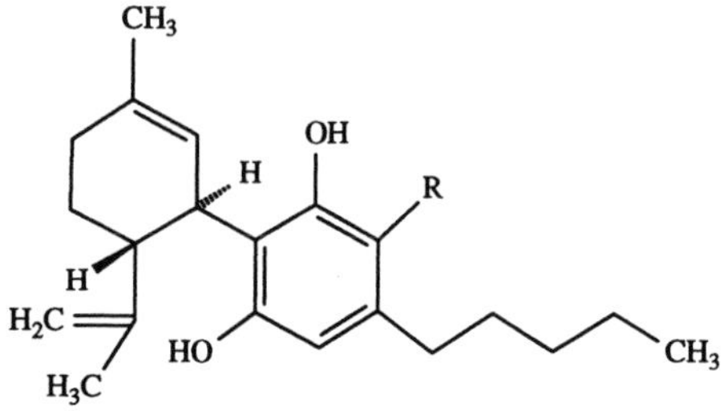

R = H: Cannabidiol

R = COOH: Cannabidiolic acid

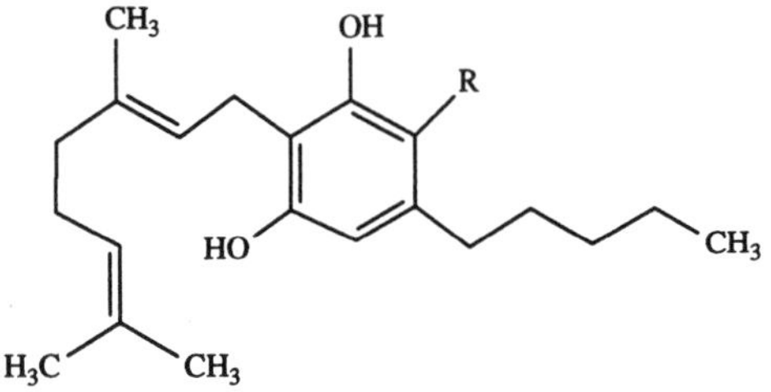

CH₃ OH R

HO CH₃

H₃C CH₃

R = H: Cannabigerol

R = COOH: Cannabigerolic acid

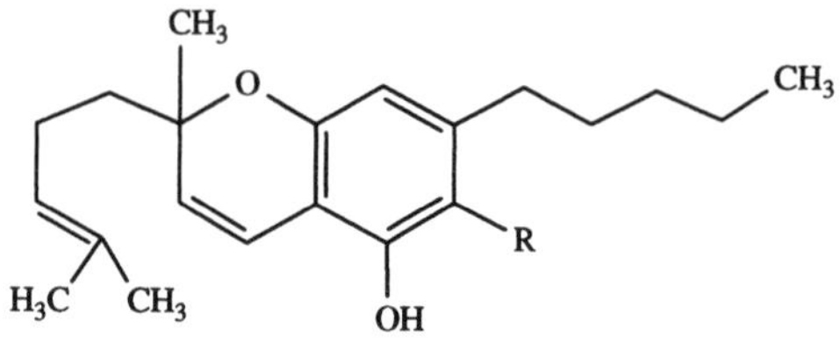

CH₃ O CH₃

H₃C CH₃ R

OH

R = H: Cannabichromene

R = COOH: Cannabichromenic acid

More than 60 cannabinoids have been identified in addition to Δ^1-tetrahydrocannabinol, but many of them are present only in very low concentrations. Some of the most abundant cannabinoids are presented above.

Most of the neutral cannabinoids have an acidic analogue, which differs only in the presence of a carboxyl group. These may occur in larger amounts than their neutral counterparts, especially in fresh

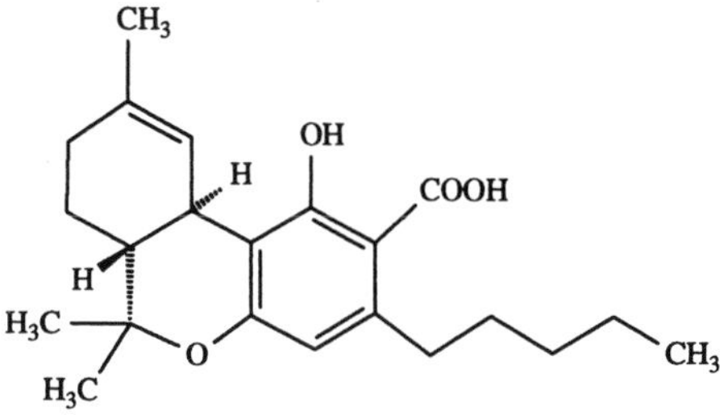

CH₃ OH COOH

H H₃C O CH₃

H₃C

Δ^1-THC-ic acid A

CH₃ OH H

H₃C O CH₃

H₃C COOH

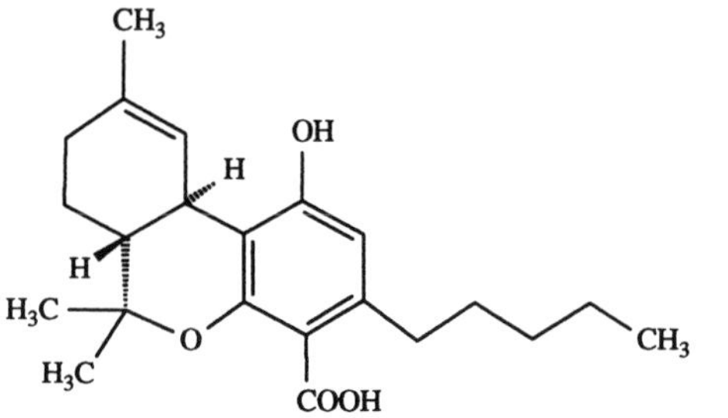

Δ^1-THC-ic acid B

plant material. Δ^1-THC has two acidic analogues: Δ^1-tetrahydro-cannabinolic acids A and B.

The acidic cannabinoids are regarded as being the primary compounds. The neutral substances, then, are secondary products, formed by decarboxylation, which occurs during growth of the plant and storage of the plant products.

Δ^1-THC is responsible for the greater part of the hallucinogenic effect of cannabis.

Biosynthesis of Δ^1-THC and cannabidiolic acid

The biosynthesis of Δ^1-THC is set out in Fig. 101. It starts with the condensation of geranyl diphosphate and olivetolic acid which is formed from a C_{12} polyketide. The resulting cannabigerolic acid undergoes oxidocyclization to yield Δ^1-tetrahydrocannabinolic acid A. The reaction is catalysed by Δ^1-tetrahydrocannabinolic-acid synthase (Δ^1-THCA synthase). This enzyme has a molecular mass of 74 kDa, its pI is 6.4 and it appears to be a monomeric enzyme. Ions such as Mg^{2+} and Mn^{2+} which are known to be cofactors for monoterpene cyclase have little influence on the Δ^1-THCA synthase activity.

393

Unlike other oxygenases and oxidases the enzyme does not require molecular oxygen for the oxidation reaction. Addition of hydrogen peroxide does not stimulate the enzyme indicating that it is not a peroxidase. Δ^1-THCA synthase should be regarded as a dehydrogenase, although coenzymes such as NAD, NADP, FAD and FMN do not stimulate the enzyme activity. Δ^1-THCA synthase appears to be a unique cyclase. Decarboxylation finally gives Δ^1-THC.

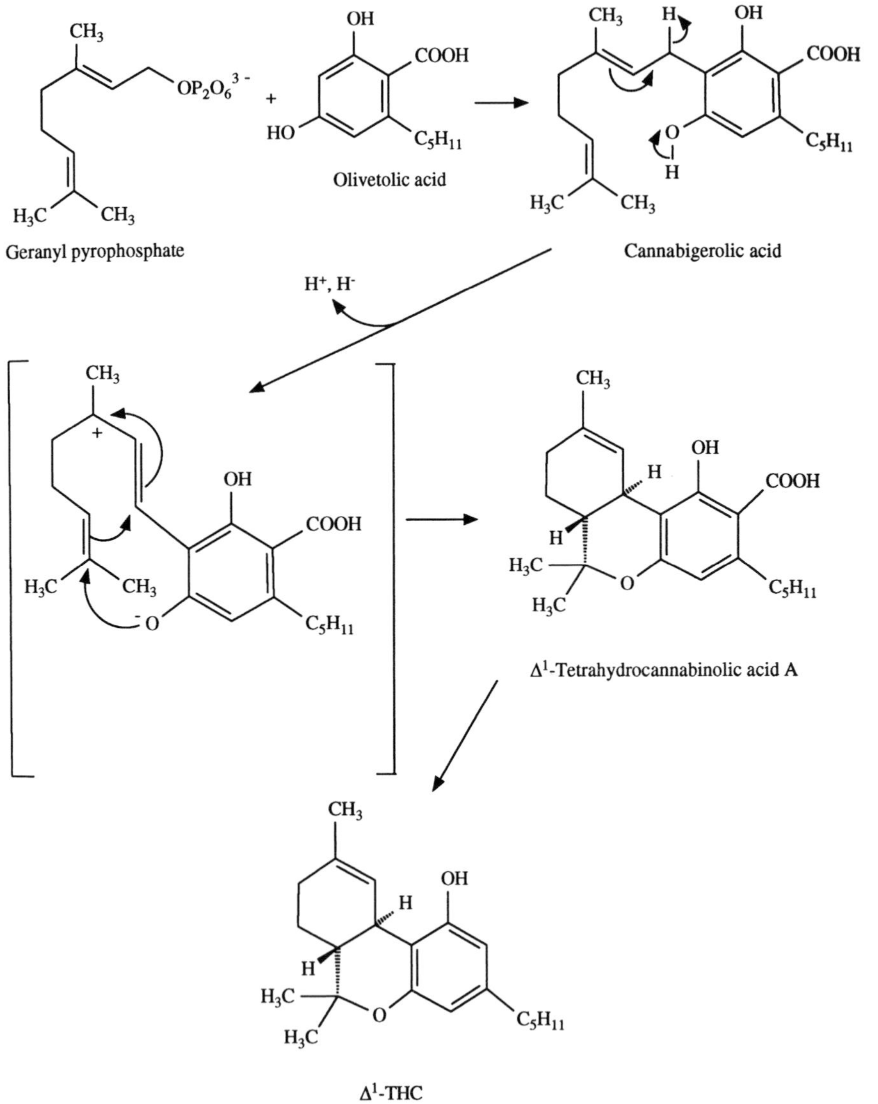

Fig. 101. Biosynthesis of Δ^1-THC

Previously, cannabidiolic acid was thought to be formed from cannabigerolic acid via hydroxylation and allylic rearrangement to then act as the immediate precursor of Δ^1-tetrahydrocannabinolic acid. As shown above, this hypothesis turned out to be false as Δ^1-tetrahydrocannabinolic acid is formed direct from cannabigerolic acid. In 1996 another enzyme which catalyses formation of cannabidiolic acid from cannabigerolic acid through oxidocyclization without hydroxylation – *cannabidiolic-acid synthase (CBDA synthase)* – was isolated (Fig. 102). The properties of this enzyme are very similar to those of Δ^1-THCA synthase. It is a monomeric enzyme with a molecular mass of 74 kDa and a pI of 6.1. Mg^{2+} and Mn^{2+} has no influence on the enzyme activity nor is molecular oxygen required. Hydrogen peroxide is also without influence. Like Δ^1-THCA synthase, CBDA synthase is neither an oxygenase, nor a peroxidase and should be regarded as a dehydrogenase which, however, does not require coenzymes such as NAD, NADP, FAD and FMN.

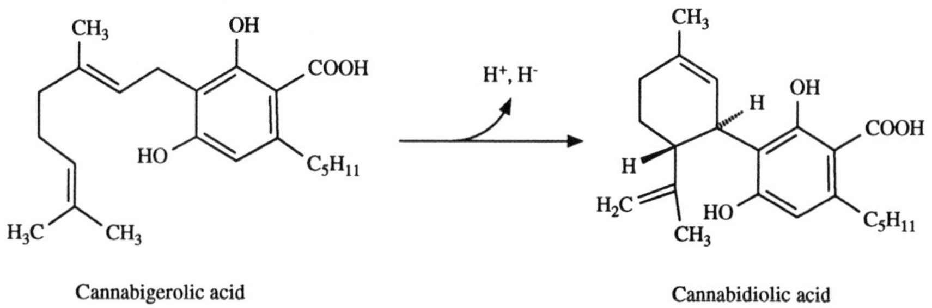

Fig. 102. Biosynthesis of cannabidiolic acid

Crude drug **Cannabis** (Indian hemp, Cannabis sativae herba) consists of the dried, aerial parts of *Cannabis sativa* L. (*Cannabidaceae*). The plant is an annual herb, which may reach 4–6 m in height. It is indigenous to Central and Western Asia and is grown in numerous tropical and temperate countries for fibre and seed production. The stem, which can be as much as 10 cm in diameter, contains fibres which are long and tough and are used for the production of ropes, carpets, etc. The seeds contain fixed oil which has a range of industrial uses.

The species is dioecious, i.e. it has male and female flowers on separate plants. The leaves and bracts on both types of plant have unicellular covering hairs with a pointed end and a wide base, containing a calcium-carbonate cystolith. The leaves and bracts also have glandular hairs which secrete a resin rich in cannabinoids.

Cannabis sativa is not a uniform species. Several hundred varieties (races) have been described. Regarding the content of cannabinoids, three chemical variants have been defined as the "drug type", the "intermediate type" and the "fibre type". Both male and female plants produce cannabinoids, but the female plants are preferred because they produce larger amounts of resin. In a hot, dry climate, the resin production in the flowering tops of the female plants can be so abundant that even under the midday sun the plants seem to be covered with dew. The presence of Δ^1-THC in the glandular hairs is the reason for the illegal use of Indian hemp as a hedonistic (hallucinogenic) drug. There are many different preparations of this crude drug. The most potent ones consist almost exclusively of the glandular hairs and are brown masses with a conspicuous smell. In Europe they are called hashish, in India charas (churrus). This quality is obtained in various ways. Harvesters dress in leather suits and walk through the plantation when the female plants are mature. The resin sticks to the leather and is then scraped off. Another method is to thrash the flowering tops against smooth concrete walls and then collect the powder and resin that sticks to the wall.

Lower grades of Indian hemp consist of flowering tops and are known under different names: in India as ganja, in North Africa as kief, in South and East Africa as dagga, in South America and the United States as marihuana.

The content of Δ^1-THC in Indian hemp varies a great deal. It depends partly on genetic factors and partly on the age of the drug, because Δ^1THC is transformed into cannabinol on storage. This transformation makes it doubtful whether cannabinol is a genuine constituent of Indian hemp. Good quality hashish contains 4–10 % and marihuana 0.1–2.7 % Δ^1-THC. Cannabidiol and cannabidiol-carboxylic acid are the main components of the glandular hairs (up to 15 %). The remaining cannabinoids occur in smaller amounts.

Cannabis smoking

The most common, and apparently efficient, method for administering cannabis is smoking, about 20 % of the Δ^1THC content of a cannabis cigarette being absorbed. Oral administration is less efficient and the bioavailability is more variable than when the drug is smoked.

Intoxication by Indian hemp is characterized by heightened awareness and perception, euphoria, sedation and sometimes hallucination. The sense of time and space is altered and the sensitivity to sounds is increased. Indian hemp is not as strongly habit-forming as narcotics such as morphine or heroin. The use and the possession of Indian hemp are illegal in most countries.

Contrary to the opinion of advocates of liberation of cannabis smoking, cannabis is not a harmless drug. Advanced head and neck cancer has been reported in young (19–38 years) cannabis smokers who had smoked daily since high school but did not smoke tobacco or use much alcohol. This type of cancer has otherwise been seen only in subjects 60 years of age or older who were heavy drinkers and tobaccco smokers for decades. Also tumours of the mouth, larynx, upper jaw and respiratory tract have been reported to result from extensive cannabis smoking. Smoking cannabis during pregnancy is associated with great risks for the foetus. There is a tenfold increased risk of leukaemia in the offspring of mothers who smoke cannabis just before or during pregnancy. Anomalies in newborn babies exposed to cannabis during gestation have been reported by several investigators. These were manifested as lower weight and head circumference. Impairment of mental performance by cannabis in man is well recognized. This impairment is present well beyond the period of acute intoxication. A well-controlled study showed short-term memory impairment to last for at least six weeks after exposure to cannabis. This finding might be correlated to binding of THC in the hippocampus, the relay centre which receives information during memory consolidation and which codes spatial and temporal stimuli and responses. The effects of cannabis on the brain constitute a risk in connection with the operation of complicated machinery, airplanes, trains and automobiles.

Metabolism of Δ^1-THC

The mammalian metabolism of Δ^1-THC has been studied in great detail and some eighty metabolites have been isolated. Thirty of these have been found in man. The most abundant metabolite found in human plasma, urine and faeces is Δ^1-THC-7-carboxylic acid. Δ^1-THC and other cannabinoids are transformed mainly in the liver by the cytochrome P450 enzyme system. Δ^1-THC is metabolized slowly. The elimination half-life of Δ^1-THC and its major metabolite Δ^1-THC-7-carboxylic acid has been found to be 3 days when the levels are followed for 10 days after smoking. Nanogram quantities of Δ^1-THC/g fat have been detected four weeks after smoking. The slow elimination results in storage of THC in the body following frequent consumption of cannabis.

Δ^1-THC-7-carboxylic acid

The endocannabinoid system

Studies of the pharmacological properties of synthetic, enantiomerically pure, THC, enabled the detection of cannabinoid receptors in animal organisms. Two receptors, CB_1 and CB_2, have been cloned and characterized. CB_1 is present preferentially in the brain and spinal cord but has also been found in the periphery, although at much lower concentrations. CB_2 is located mainly in cells of the immune and haematopoietic systems, but has recently been found also in the brain and other tissues. Δ^1-THC is an agonist at both the CB_1 and the CB_2 receptors. Following the detection of these receptors, endogenous ligands to them were also detected. The two best studied are anandamide (*N*-arachidonoylethanolamide) and its glycerol ester analogue 2-AG (2-arachidonoyl glycerol), but also other ligands such as noladin ether (2-arachidonoyl glycerol ether), arachidonoyl dopamine (NADA) and virodhamine (*O*-arachidonoyl

Anandamide

2-Arachidonoyl glycerol

ethanolamine) have been found, but their significance and bio-chemical characteristics are largely unknown.

Anandamide acts as an endogenous ligand for the CB_1 receptor but has very low efficiency for the CB_2 receptor. 2-AG acts as an agonist at both receptors. As it is active at the CB_2 receptor, to which anandamide binds poorly, some authors consider 2-AG as the true ligand for this receptor. In the brain the basal levels of 2-AG are much higher than those of anandamide which suggests that only a fraction of the total is involved in endocannabinoid signalling.

The cannabinoid receptors and their endogenous ligands are referred to as the endocannabinoid system which has been found to be involved in a number of physiological functions both in the central and in the peripheral nervous systems and in peripheral organs. The endocannabinoids are produced from membrane phosphoglycerides via Ca^{2+} sensitive biosynthetic pathways, triggered on demand and are not stored in secretory vesicles. They do not function like hormones but as local mediators. In the brain, postsynaptic neuron depolarization or stimulation of postsynaptic neurotransmitter receptors, coupled to Ca^{2+} mobilization from intracellular stores, causes elevation of intracellular Ca^{2+}, which stimulates enzymes catalysing the biosynthesis of anandamide and 2-AG. These compounds are then released from the neuron to activate presynaptic CB_1 receptors that reduce the release of both excitatory (e.g. glutamate) and inhibitory (e.g. GABA) neurotransmitters, thereby producing various effects on neuronal synapses. This mode for endocannabinoid action has been implicated in several physiopathological functions of the brain, including the control of food intake, habit-forming and mnemonic processes, neuronal plasticity and excitotoxicity. The reciprocal pattern of expression of endocannabinoid-synthesizing enzymes and cannabinoid receptors determines the specificity of endocannabinoid action, whereas the localization of degrading enzymes sets its duration. Research is going on to prove this mode of action also for CB_2 receptors and other areas where endocannabinoid activity has been detected.

The endocannabinoid system is currently an extensive target for research aiming at the discovery of new drugs in the areas of diseases of energy metabolism, pain and inflammation, central nervous system disorders and cardiovascular and respiratory disorders.

Drugs acting on the endo-cannabinoid system

Cannabis sativa is one of the oldest cultivated plants and has been used for thousands of years as a source of fibre and as a medicament. In the 19th century, cannabis was widely used as an analgesic for rheumatic pain and toothache and also for other painful conditions. This use was especially prominent in England and probably originated in India, where cannabis has been in use as a painkiller for a very long time. Δ^1-THC has been investigated for its analgesic activity, but although such activity can be demonstrated, the substance has not been found suitable for modern use.

In India cannabis has also been used for hundreds of years as an anti-emetic drug. In 1975 it was discovered that cannabis smokers who were undergoing anticancer chemotherapy did not suffer from emesis, which is usually a serious side effect of the treatment. This anti-emetic effect was intensively studied and it was concluded that Δ^1-THC is a potent anti-emetic agent, which is very effective against nausea and vomiting caused by radiation therapy, as well as by many – but not all – anticancer drugs. The oral dose is ca. 15 mg and it causes side effects such as somnolence and sedation in many patients. A psychological "high" has also been reported in some cases. As a result of these investigations, in 1986, the United States Food and Drug Administration approved the use of Δ^1-THC (=Dronabinol, tradename: Marinol) in cases of nausea and vomiting caused by anticancer treatments. In 1996 Marinol was approved in the USA for treatment of anorexia in AIDS patients.

The discovery of the anti-emetic effect of Δ^1-THC initiated a search for synthetic analogues with less pronounced cannabimimetic effects. One synthetic cannabinoid, *nabilone*, is about 10 times more potent than Δ^1-THC. However, it is not devoid of cannabimimetic effects although there is a considerable divergence of opinion regarding the degree of euphoria elicited. Nabilone is on the market in several countries, including the United Kingdom and Switzerland, as a remedy for nausea and vomiting arising during anticancer therapy.

The discovery, in 1971, that marihuana smoking decreases lachrymation and lowers the intra-ocular pressure of the eye in normal subjects, led to an interest in Δ^1-THC as a remedy for glaucoma. This disease is characterized by a progressive increase in the intra-ocular pressure, which leads to deterioration of vision and eventually to complete blindness. Glaucoma is usually treated with parasympathicomimetics such as pilocarpine (see page 699), which induce miosis and drain the excess liquid from the eye, thus reducing the pressure. Δ^1-THC is an effective agent with a duration of action as long as or longer than that of currently used antiglaucoma drugs. The side effects have, however, prevented its use as a drug.

In Canada, the preparation **Savitex** which is a 1:1 mixture of Δ^1-THC and cannabidiol (page 392) has been approved for treatment of neuropathic pain associated with multiple sclerosis. The extensive research on the endocannabinoid system has resulted in the marketing in EU of the CB_1 receptor antagonist **Rimonabant** for the therapy of obesity.

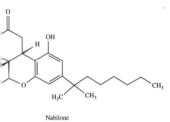

Nabilone

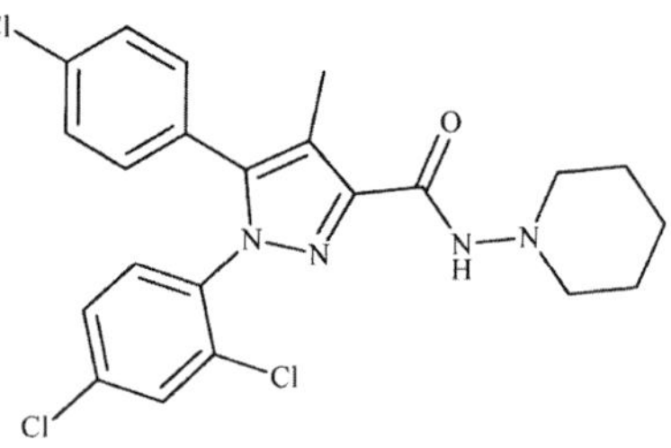

Rimonabant

Sesquiterpenes

Sesquiterpenes are terpenes containing 15 carbon atoms. Many natural products are sesquiterpenes, but only a few of them are of medicinal and pharmaceutical interest.

Biosynthesis of sesquiterpenes

As mentioned on page 369, sesquiterpenes are formed in the cytoplasm and the IPP from which they derive is consequently formed via the mevalonate pathway. The biosynthesis of sesquiterpenes starts with formation of geranyl diphosphate in the same type of reaction already described for monoterpenes (page 372). The DMAPP needed for this reaction is not synthesized directly in the mevalonate pathway, but is produced by rearrangement of IPP. The reaction is catalysed by *isopentenyl-diphosphate Δ-isomerase* (EC 5.3.3.2, also known as IPP isomerase), an enzyme which has been isolated from bacteria, fungi including yeast, insects and several plants. The molecular masses of various IPP isomerases vary considerably but are usually in the range 22–40 kDa. Genes coding for the enzyme have been studied in yeast, the plants *Clarkia breweri* and *Arabidopsis thaliana* as well as in man. Cysteine 139 and glutamic acid 207 have been identified as being involved in the active site of the yeast enzyme. The 1,3-allylic rearrangement reaction which converts IPP into DMAPP is postulated to proceed via a two-base cationic mechanism as illustrated in Fig. 103.

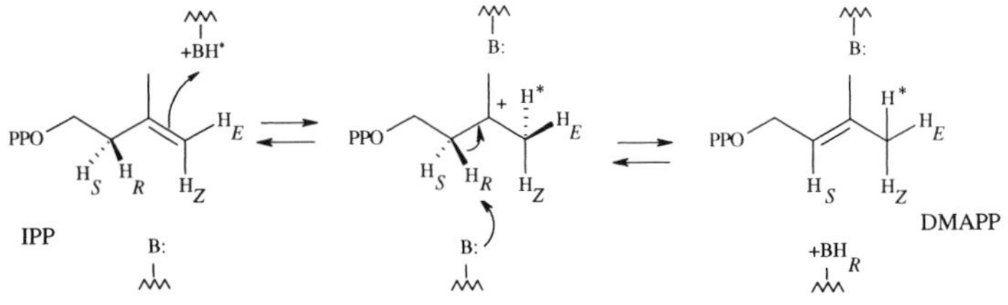

Fig. 103. Rearrangement of IPP into DMAPP. Enzyme: IPP isomerase

Geranyl diphosphate (page 373) can condense further with one molecule of isopentenyl diphosphate, giving *trans*, *trans*-farnesyl diphosphate (FPP). As in the biosynthesis of GPP, the diphosphate group of GPP is ionized with formation of an allylic cation which in an electrophilic attack on the double bond of IPP forms a new C-C bond accompanied by loss of a proton (Fig. 104). The condensation is catalysed by *geranyl*trans*transferase* (EC 2.5.1.10, also known as farnesyl diphosphate synthase or FPP synthase), a prenyl-transferase, which appears to occur in all living organisms studied thus far. FPP synthases have been isolated and purified from animals, plants, fungi and bacteria. They are all homodimers of subunit molecular mass 38–43 kDa, requiring the divalent cations Mg^{2+} or Mn^{2+} for activity. Most of the isolated synthases can accept both DMAPP and GPP as primers in the presence of IPP. The FPP synthase gene has been cloned and sequenced from rat liver, human liver, the yeast *Saccharomyces cerevisiae* and the higher plants *Artemisia annua, Arabidopsis thaliana, Lupinus albus* and *Hevea brasiliensis*.

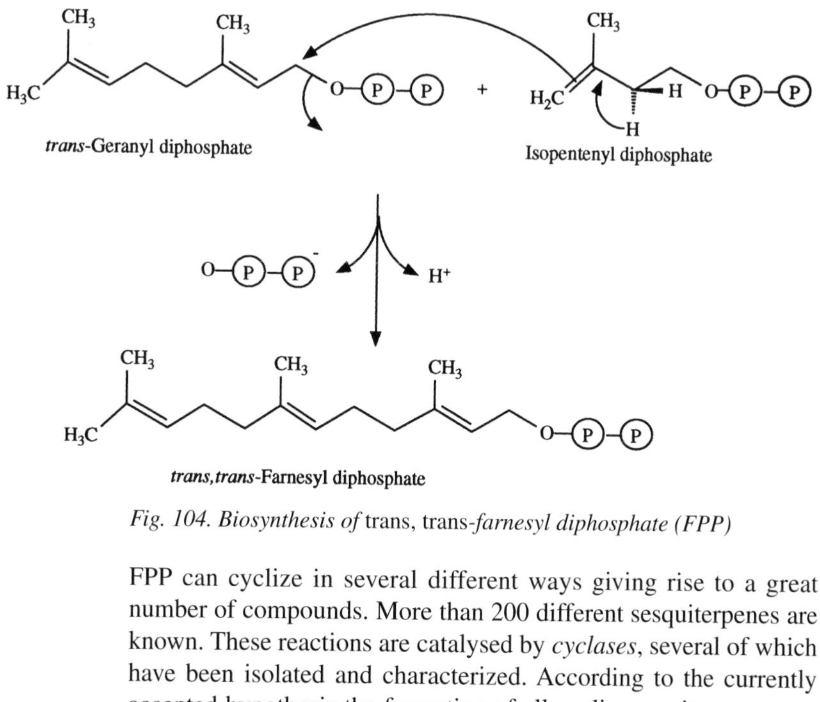

trans-Geranyl diphosphate

Isopentenyl diphosphate

trans,trans-Farnesyl diphosphate

Fig. 104. Biosynthesis of trans, trans-*farnesyl diphosphate (FPP)*

FPP can cyclize in several different ways giving rise to a great number of compounds. More than 200 different sesquiterpenes are known. These reactions are catalysed by *cyclases*, several of which have been isolated and characterized. According to the currently accepted hypothesis the formation of all cyclic sesquiterpenes can be accounted for by ionization of the diphosphate group of FPP, followed by an electrophilic attack of the resultant allylic cation on either the central or distal double bond, followed by well-precedented cationic transformations involving further cyclizations and rearrangements, including methyl migrations and hydride shifts, culminated by quenching of the positive charge by loss of a proton or capture of an external nucleophile such as water or the original diphosphate anion. Figure 105 illustrates the possible electrophilic attacks of the allylic cation on the double bonds of FPP.

Only the distal double bond can be directly attacked by the pri-

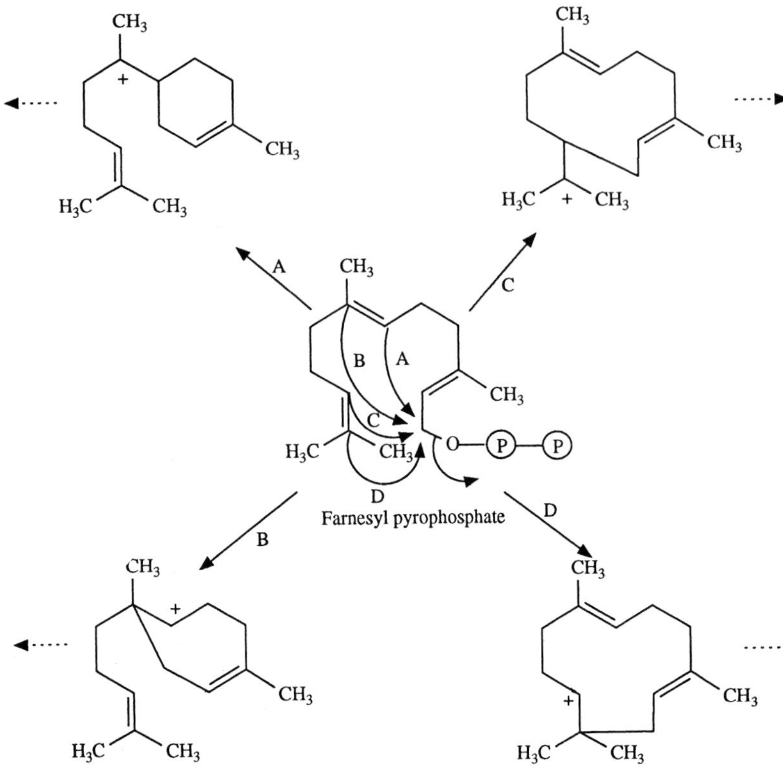

Fig. 105. Possibilities for formation of cyclic sesquiterpenes from farnesyl diphosphate

marily formed allylic cation (route C and D in Fig. 105). Attack on the central double bond is stereochemically hindered. This hindrance is overcome by isomerization to the corresponding tertiary allylic isomer, nerolidyl diphosphate (NPP) (Fig. 106). NPP can undergo a simple rotation about the newly generated 2,3 single bond, followed by reionization to the corresponding cisoid allylic cation-diphosphate ion pair. The cation can then launch an electrophilic attack on either the central or distal double bonds with net antidisplacement of the diphosphate moiety. Thus all the routes depicted in Fig. 105 can be realized after isomerization of FPP to NPP. The various skeletons are further transformed to the individual sesquiterpenes by various reactions such as oxidation, dehydrogenation and rearrangement.

The ion-pair model for the enzymatic cyclization of FPP is supported by related studies of terpene biosynthesis carried out at both the intact cell and the enzyme level and is consistent with a substantial body of chemical model reactions. It is also interesting to note the resemblance to the formation of cyclic monoterpenes from GPP involving isomerization to linalyl diphosphate and subsequent ion-pair formation (Fig. 99). In order for cyclization of either FPP

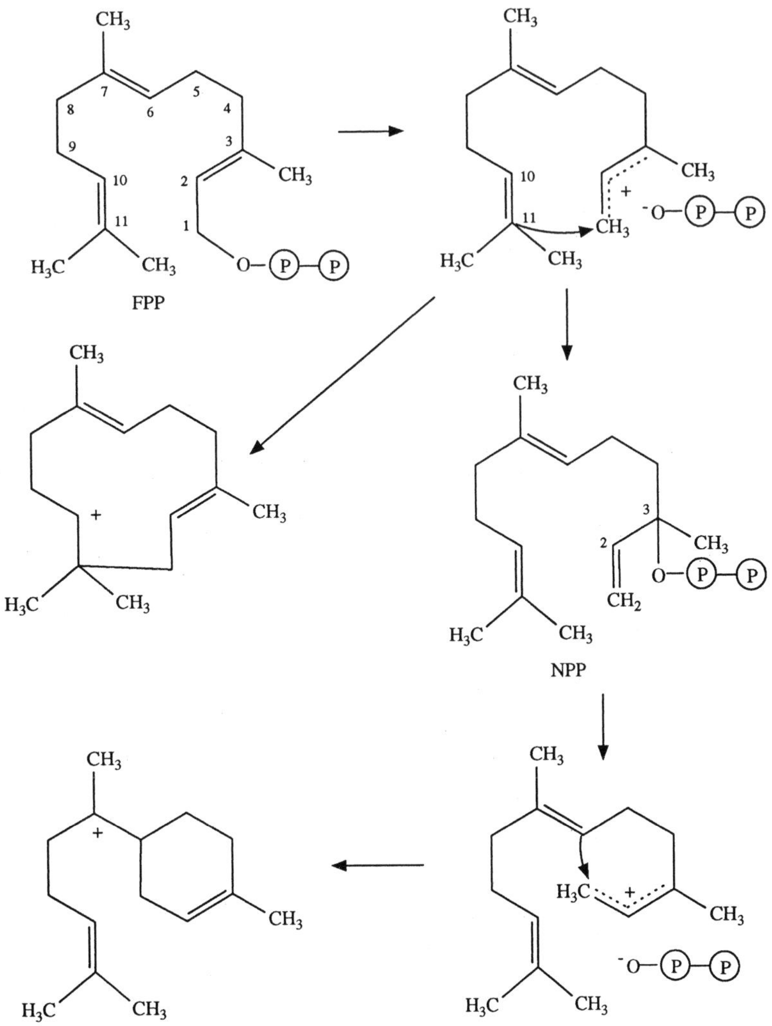

Fig. 106. Biosynthesis of cyclic sesquiterpenes from FPP

or NPP to occur, the π-orbitals of the relevant double bonds must be properly aligned so as to achieve the required geometry for interaction. A systematic analysis has shown that only a limited number of conformations of the substrate satisfy this condition.

So far only a handful of the some 200 possible sesquiterpene synthases have been isolated and characterized. All the enzymes so far examined are operationally soluble proteins of molecular masses ranging from 40 to 44 kDa. The enzymes require no cofactors other than a divalent cation, Mg^{2+} usually being preferred (see the monoterpene cyclases).

Cnicin The bitter compound cnicin is a sesquiterpene lactone, esterified with a substituted acrylic acid.

Crude drug   **Holy thistle** (Cardui benedicti herba) consists of the dried aerial parts of *Cnicus benedictus* L. (*Asteraceae*), an annual, thistle-like herb with sharp thorns on the leaves and yellow flowers. It is naturally occurring in the Mediterranean region but widely cultivated in tempered areas. The crude drug contains about 0.2 % cnicin and is used as a bitter tonic. The bitterness value is approximately 1 500.

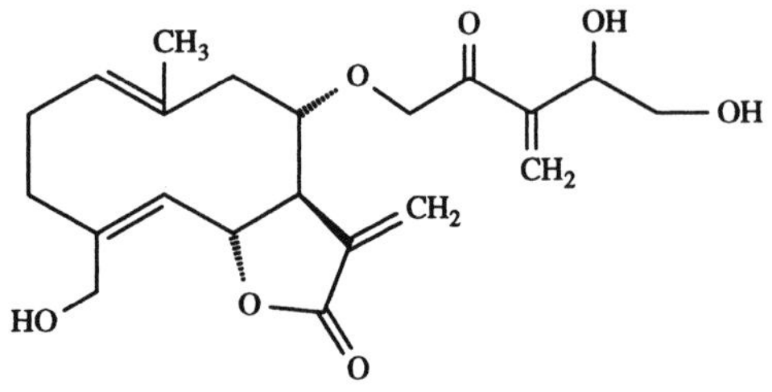

Cnicin

Absinthin   Absinthin is a dimeric sesquiterpene lactone with a strong bitter
Wormwood (Absinthii herba) consists of the dried aerial parts of

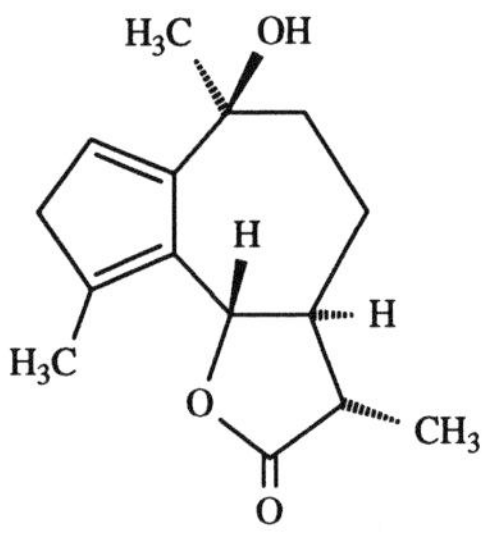

Artabsin

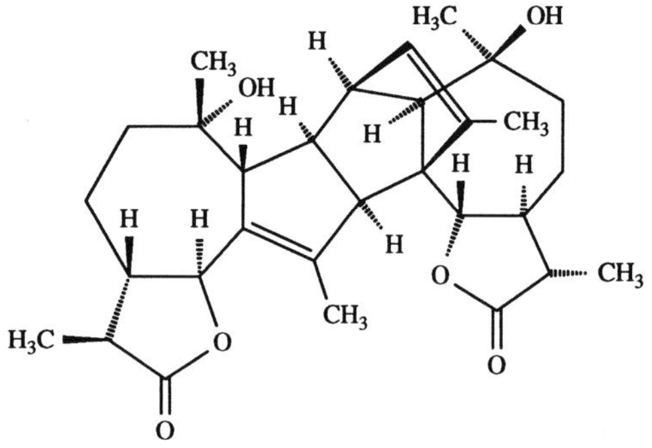

Absinthin

Crude drugs   *Artemisia absinthium* L. (*Asteraceae*). This is a perennial herb, 60–120 cm tall, densely covered with hairs on the leaves and stems. The plant has an aromatic smell and a very bitter taste. It grows in Asia, Europe and North Africa.

The bitter taste of the crude drug is due to its content of absinthin. The bitterness value is 10 000. Besides absinthin, wormwood herb contains at least 0.5 % of a volatile oil, which gives the drug its aroma and which can be obtained by steam distillation.

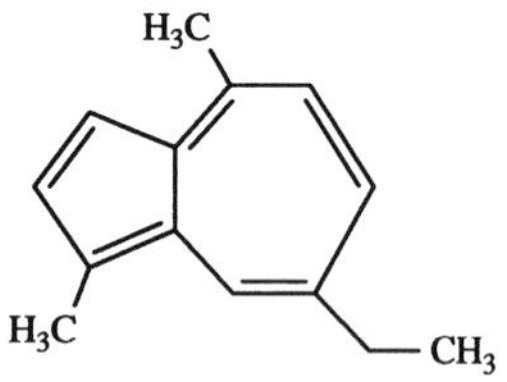

The oil is often coloured blue due to the presence of the sesquiterpene chamazulene, which arises during the distillation through decomposition of sesquiterpenes, such as artabsin. The oil also contains the monoterpene ketone thujone and the corresponding alcohol thujol. These are poisonous substances, which affect the central nervous system and cause dizziness, convulsions and delirium.

Chamazulene

CH₃

Thujone

Bisabolol

Wormwood herb was formerly used in France, Switzerland, Belgium and Germany as one of the constituents for the preparation of absinth, a liqueur whose manufacture is now prohibited because of the toxic effects due to thujone and thujol. Medicinally wormwood herb is used as an aromatic bitter tonic.

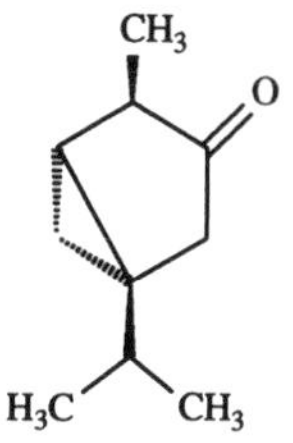

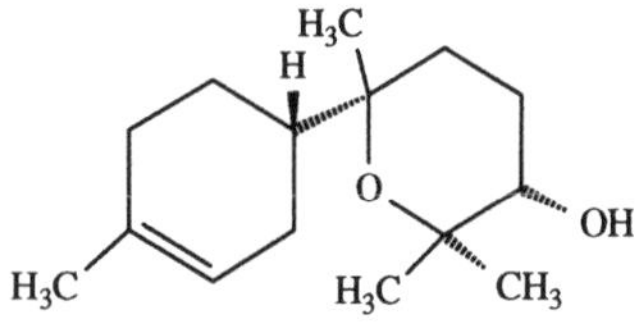

(-)-α-Bisabololoxide A

HO CH₃

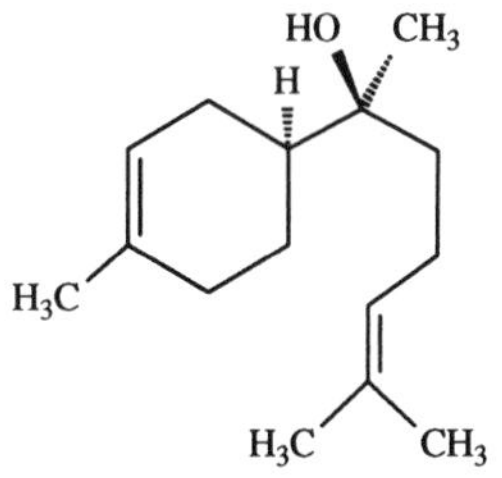

(-)-α-Bisabolol

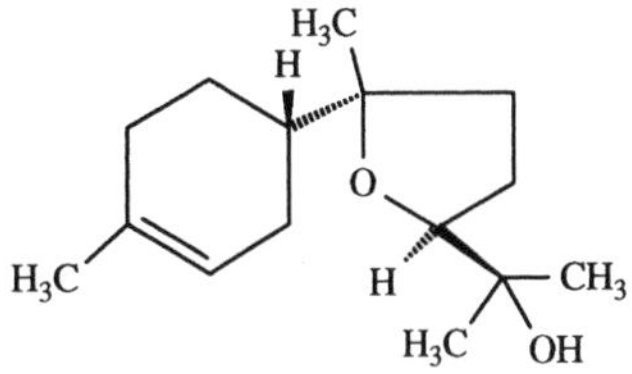

(-)-α-Bisabololoxide B

(-)-α-Bisabolol and (-)-α-bisabolol oxides A and B are present in the volatile oil of German chamomile.

Bisabolol and its oxides have anti-inflammatory and spasmolytic activities combined with very low toxicity.

Crude drugs   **Matricaria flower** (German chamomile, Matricariae flos), consists of the inflorescence of *Matricaria chamomilla* L. (also known as *M. recutita* (L.) Rausch.) (*Asteraceae*). The plant originates from southern Europe and is cultivated in many parts of the world. Argentina, Egypt and Hungary are big producers of German chamomile.

The inflorescence is 10–17 mm in diameter and consists of a receptacle with 12–20 marginal ligulate florets and numerous central tubulate florets. The receptacle is hollow, which distinguishes German chamomile from a common adulterant – *Matricaria inodora* L. (scentless false chamomile) – the inflorescence of which is solid.

The drug contains 0.25–1 % volatile oil, the main constituents of which (up to 50 %) are bisabolol and its oxides. In the fresh plant the oil is contained in glandular hairs. Besides bisabolol and its oxides the oil in the hairs also contains proazulenes, e.g. matricin. These are labile compounds and are easily transformed to chamazulene which has a characteristic blue colour. Chamomile oil produced by steam distillation is blue due to this reaction. Also present in the oil are the acetylene derivatives *E*- and *Z*-en-in-dicycloether.

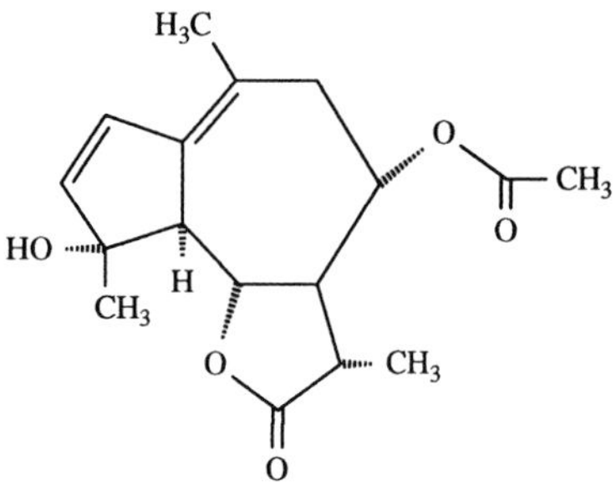

Matricin

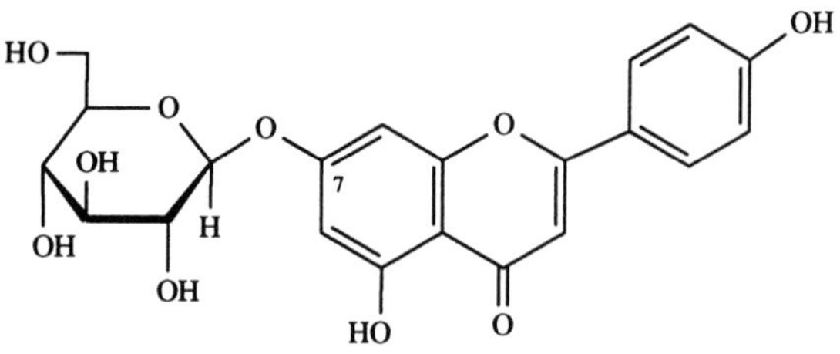

Apigenin 7-glucoside

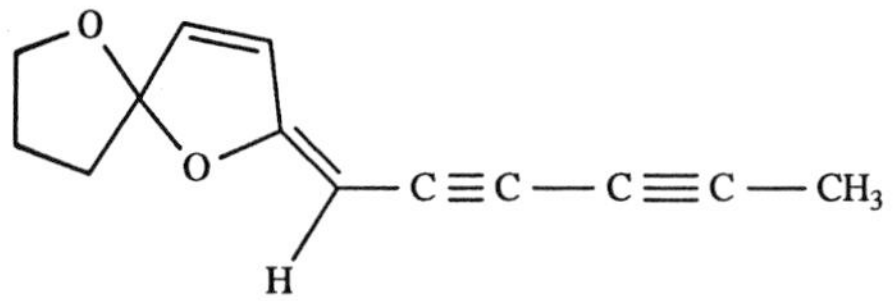

(*E*)-En-in-dicycloether

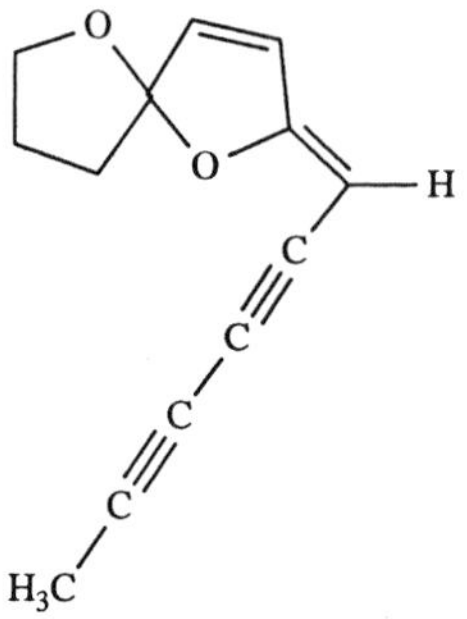

(*Z*)-En-in-dicycloether

German chamomile also contains a number of flavonoids, several of which have a spasmolytic effect, the most important of which is apigenin 7-glucoside. These flavonoids are present only in the ligulate florets. German chamomile is widely used as a herbal remedy for the treatment of various gastric upsets like dyspepsia and colic. The drug is mostly administered as a tea, prepared by pouring boiling water over the dried flowers. Extracts of German chamomile are also used externally for the treatment of inflammation of the skin and in the nose and the mouth, as well as in bath preparations, hair dyes, mouthwashes, shampoos and sunburn preparations. Bisabolol and apigenin are the main anti-inflammatory constituents.

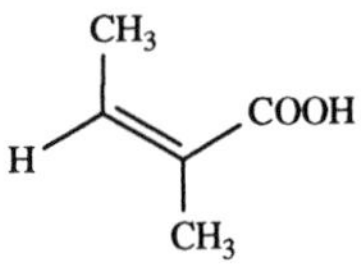

Roman chamomile flower (Chamomillae romanae flos) is the inflorescense of *Anthemis nobilis* L. (sometimes known under the name *Chamaemelum nobile* (L.) All.). The plant grows wild in southern and western Europe and also in North Africa. It is cultivated in several European countries, as well as in Egypt and in Argentina.

The inflorescence is white to yellowish grey, with a diameter of 8–20 mm. The receptacle is solid. Most florets are ligulate but a few pale-yellow tubulate florets may occur in the centre.

Roman chamomile flowers contain 0.6–2.4 % volatile oil, the main components of which are esters, in particular the isobutyl

Angelic acid

ester, of angelic acid. Also present in the drug are sesquiterpene derivatives, e.g. the lactone nobilin, which has a bitter taste, and flavonoids with a spasmolytic effect, e.g. apigenin 7-glucoside.

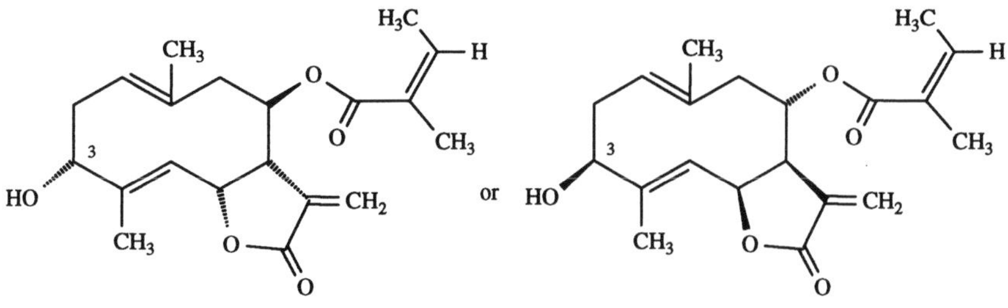

Nobilin

Roman chamomile flower – usually taken as a tea – is a herbal remedy for colic and several other complaints.

Parthenolide

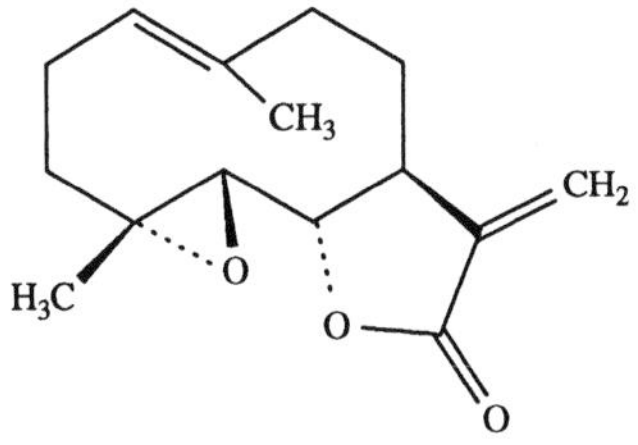

Parthenolide

Parthenolide is a sesquiterpene derivative which is contained in leaves of feverfew, *Chrysanthemum parthenium* Bernh. (sometimes known as *Tanacetum parthenium* Sch. Bip.), a herbal remedy used for prophylaxis of migraine (see *crude drug* below). This activity has been proved clinically. Parthenolide is an inhibitor of blood platelet aggregation. Release of 5-hydroxytryptamine (serotonin) accompanies platelet aggregation and has been linked to the onset of migraine. Inhibition of aggregation by parthenolide could therefore be a mechanism for the efficacy of the leaves of feverfew as a migraine prophylactic.

The α-methylenebutyrolactone function of parthenolide is a Michael acceptor of thiols (see below). It has been suggested that this reaction is responsible for the inhibitory effect of the compound on blood platelet aggregation. Three pieces of evidence support this theory. Firstly, addition of cysteine or 2-mercaptoethanol to crude feverfew extracts or to pure parthenolide completely suppresses their ability to inhibit platelet aggregation. Secondly, the inhibitory effects are dose- and time-dependent, and thirdly, treatment of platelets with feverfew extracts or parthenolide causes a dramatic reduction in the number of acid-soluble thiol groups present. On the

407

other hand doubts have been raised as to the credibility of this explanation in the clinical situation as parthenolide entering the bloodstream would be rapidly "neutralized" by Michael addition of the thiol residue in glutathione, which is one of the body's main defences against such compounds.

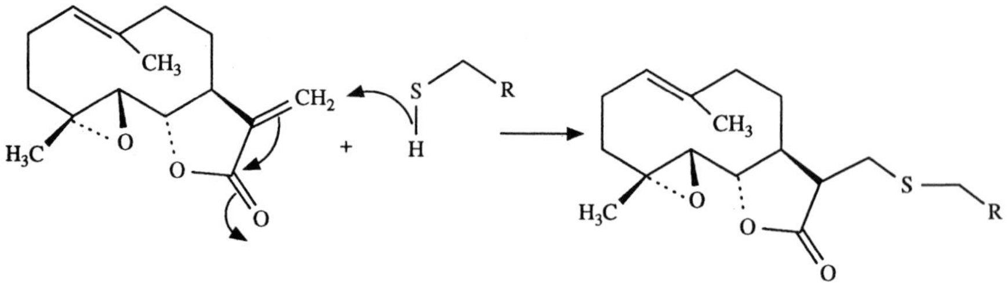

Crude drug   **Feverfew** (Chrysanthemi parthenii folium) is the dried leaves of *Chrysanthemum parthenium* Bernh. (sometimes known as *Tanacetum parthenium* Sch. Bip.), *(Asteraceae)*. This plant is native to southern and eastern Europe. The height is 20–60 cm and the leaves are yellow-green, divided into 2–5 lobes. The upper part of the stem is branched with 5–20 inflorescenses, 1.2–2 cm in diameter. The ligulate florets are broad and white and the bracts downy and green. The plant has a strong, aromatic odour. The drug contains a rich mixture of mono- and sesquiterpenes. The abundant monoterpene is camphor. Among the sesquiterpenes, farnesene and germacrene D can be mentioned. The most significant components are a complex series of sesquiterpene α-methylenebutyrolactones of which parthenolide is present in the highest amount. It should be observed that at least three different chemotypes of *Chrysanthemum parthenium* have been discovered, two of which do not contain parthenolide. Control of the crude drug must therefore include quantitative determination of this compound, the content of which should be at least 0.2 %.

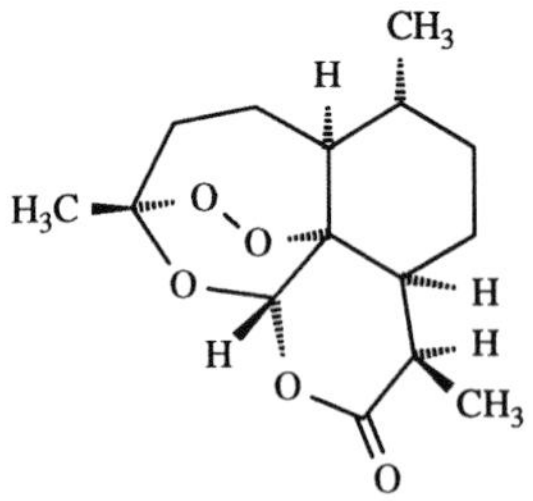

Artemisinin (Qinghaosu)

Artemisinin   This oxidized sesquiterpene derivative is the active compound of an ancient Chinese antimalarial remedy – "qinghao" – which consists of the aerial parts of *Artemisia annua* L. *(Asteraceae)*. (For a discussion of the cause of malaria and other remedies against this disease, see page 692.) In clinical tests, performed in China, artemisinin has shown very good activity against different forms of malaria, including the dangerous cerebral malaria caused by *Plasmodium falciparum*. It is also active against strains of *Plasmodium* which have become resistant to chloroquine and other synthetic antimalarials.

Artemisia annua is cultivated in China and several other East Asian countries. Artemisinin is found in the leaves and flowering branches of the plant but not in the roots. Wild plants contain 0.06–0.5 % artemisinin but breeding has yielded plants which in cultivation may contain up to 2 %. Artemisinin can be produced inexpensively. Formulations (250 mg tablets, 250 mg capsules, 100 mg suppositories) are manufactured in Vietnam and China and are registered in both countries. The suppository formulation is unique among antimalarial agents and represents a promising route of administration as it can be given to patients who are too ill to take medication orally.

Artemisinin is poorly soluble in water and oil-soluble and water-soluble derivatives have been synthesized. They are derivatives of dihydroartemisinin which is also used as a drug.

Artemether is an oil-soluble compound that was developed in China and since 1992 used in combination with a synthetic anti-malarial (lumefantrin, also called benflumetol) for treatment of *Plasmodium falciparum* infections. Cooperation with a Western pharmaceutical company resulted in marketing of a fixed combination of arthemether and lumefantrin (1:6) as coartemether. Coarthemether is available under the trade name Riamet® in developed, non-endemic countries at regular prices while (as the result of an agreement with WHO) the same compound is marketed as Co-artem® in malaria-endemic countries at prices comparable with locally available products.

Artesunate is a water-soluble compound which is used in combination with synthetic drugs (mefloquine, amodiaquine, sulfadoxin/pyrimethamine). These combinations of derivatives of artemisinin with synthetic drugs have been found to be effective for treatment of drug-resistant malaria.

In pharmacological experiments artemisinin and its derivatives –

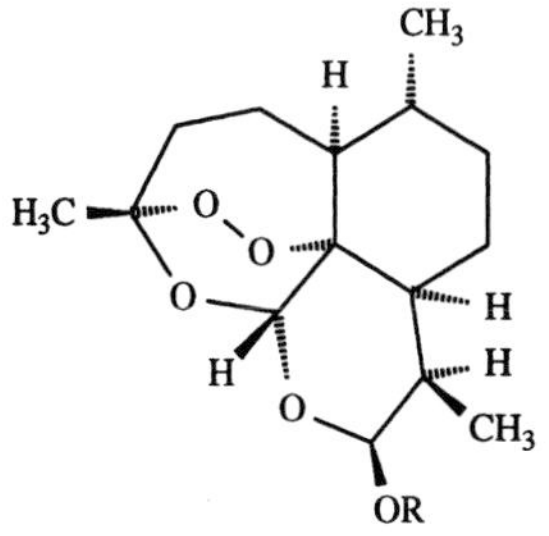

Dihydroartemisinin: R = H
Artemether: R = CH_3
Artesunate: R = $OCO(CH_2)_2COONa$

particularly the principal metabolite dihydroartemisinin – have been found to be neurotoxic, a fact which has caused concern regarding possible risks in the clinical use of the drug. However, so far no side-effects attributable to drug-induced neurotoxicity have been noticed in patients.

The complex ring structure of artemisinin is not required for antimalarial activity; only the endoperoxide bridge is necessary. Intense synthetic studies are presently conducted in many countries aiming at finding a simpler molecule with equal or better properties than those of artemisinin.

Artemisinin and its derivatives are toxic to malaria parasites at nanomolar concentrations, whereas micromolar concentrations are needed to produce toxicity to mammalian cells. The killing of the parasites is believed to be mediated by free artemisinin radicals produced by activation of the compounds by haem or molecular iron. In a second step these radicals react with and damage specific membrane-associated proteins in the malaria parasite. Thus artemisinin and its derivatives are free-radical generators but they differ from "oxidant" free-radical generating drugs in that the free radicals are carbon centred and that the activity is directed towards specific target proteins. This hypothesis for the mode of action has recently been challenged, but a definite explanation of the mechanism for the antimalarial activity has not yet been produced.

Biosynthesis   The first step in artemisinin biosynthesis is the formation of amorpha-4,11-diene from farnesyl diphosphate (Fig. 107). The reaction is catalysed by *amorpha-4,11-diene synthase* (EC 4.2.3.24), an enzyme which has been isolated from the leaves of *Artemisia annua*. Like most sesquiterpene synthases it has a broad pH optimum (6.5–7.0) and its molecular mass is 56 kDa. From young leaves a cDNA clone has been isolated which contains a 1 641-bp open reading frame coding for 546 amino acids. The clone was expressed in *Escherichia coli*, yielding a product with the properties of amorpha-4,11-diene synthase. In the next step, amorpha-4,11-diene is oxidized to artemisinic alcohol. An enzyme that catalyses this reaction has not yet been identified, but leaf microsomes from *A. annua* perform the reaction in the presence of NADPH. The following steps to produce artemisinin are not entirely clear. Theoretical pathways are presented in Fig. 107, which are supported by identification of a cDNA clone, which, when expressed in yeast, yielded a multifunctional cytochrome P450-enzyme, designated CYP71AV1, which oxidized amorpha-4-11-diene to artemisinic alcohol and subsequently to the corresponding aldehyde and to artemisinic acid. That reduction of the C11-C13 double bond occurs after formation of artemisinic acid has not been proved. Theoretically, the oxidation of artemisinic alcohol to artemisinic acid might represent a branch in the pathway to artemisinin and, as indicated in Fig. 107, reduction of the C11-C13 double bond could occur before further oxidation of artemisinic alcohol.

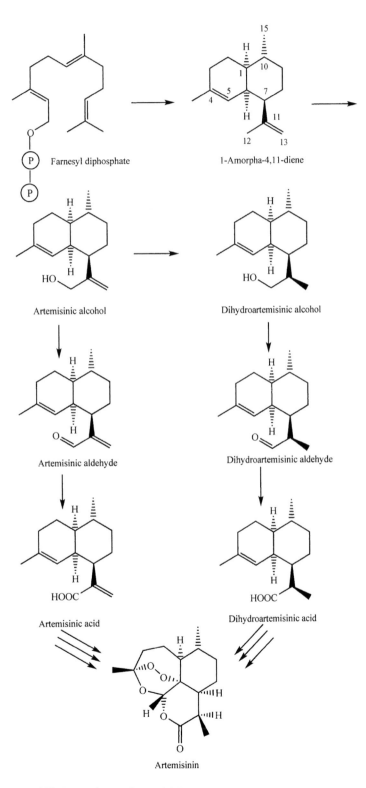

Farnesyl diphosphate

1-Amorpha-4,11-diene

Artemisinic alcohol

Dihydroartemisinic alcohol

Artemisinic aldehyde

Dihydroartemisinic aldehyde

Artemisinic acid

Dihydroartemisinic acid

Artemisinin

Fig. 107. Biosynthesis of artemisinin

411

Cantharidin

Cantharidin was long regarded as a monoterpene, but later investigations have shown that it is formed from farnesol by elimination of carbon atoms 1, 5, 6, 7, and 7'. This leads to a C_{10} compound which is formally a monoterpene, but which biosynthetically must be regarded as a sesquiterpene derivative.

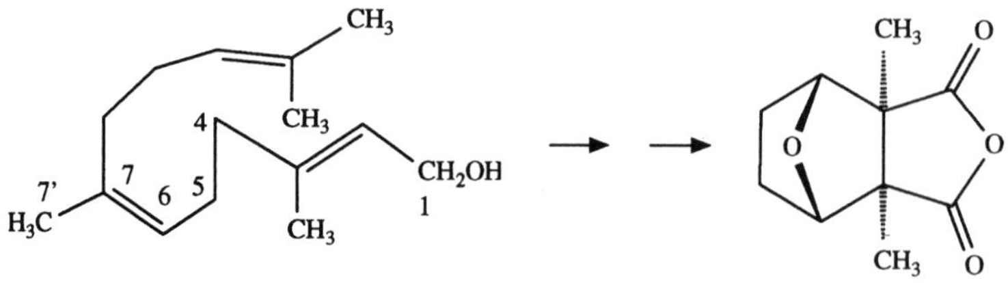

Farnesol

Cantharidin

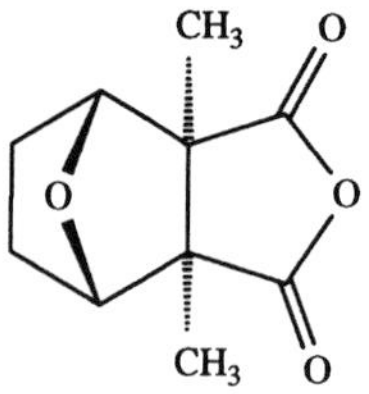

Cantharidin

Cantharidin is a strong irritant and can cause blisters on the skin. It is used in small amounts in liniments for the topical treatment of rheumatic pains. Cantharidin is very toxic and doses of 5 mg can seriously damage the kidneys. The compound is excreted unchanged through the kidneys and irritates the urogenital organs. This has given rise to the abuse of cantharidin and cantharides (see below) as abortifacients and aphrodisiacs. This improper use has caused severe poisoning and death.

Crude drug

Cantharides (Cantharis) consists of the dried beetle *Lytta vesicatoria*. This beetle, the so-called Spanish fly or blister beetle, is found all over southern and central Europe and also in Asia. It feeds mostly on the leaves of plants belonging to the family *Oleaceae*, e.g. *Fraxinus*, *Olea*, *Syringa* and *Ligustrum* species. These beetles often appear in large numbers in June and July and are then collected from the trees attacked. This is done early in the morning, when the beetles are still numb from the cold of night, so that they can be shaken from the trees on to cloths spread on the ground. The insects are killed by being poured into vessels containing carbon disulphide, chloroform or petrol, and then dried.

The metallic green, yellowish green, or bluish beetles are 1.5–2.5 cm long and 5–8 mm broad. The head is heart-shaped and carries two black antennae. Each of the three pairs of legs is borne on a separate segment of the thorax. The legs are covered with hairs. The elytra, on the second thoracic segment counting from the head, are lustrous metallic green in the male and yellow-green in the female. The wings are brown, reticulate-veined, and situated on the third thoracic segment. The abdomen comprises eight segments and is covered with hairs. The underside of head and thorax is yellow.

Spanish fly contains 0.6–1.0 % cantharidin, which it secretes for defence.

Diterpenes

Many important natural products are diterpenes. The alcohol phytol, which constitutes part of the chlorophyll molecule (page 118) and of the vitamin E and K_1 molecules (see below), is a diterpene with a straight carbon chain. The gibberellins, which are plant growth regulators, and abietic acid, one of the components of colophony (page 376), are examples of cyclic diterpenes. Several alkaloids contain diterpenoid moieties in their skeletons (page 704). Vitamin A is another C_{20} compound and can be regarded as a diterpene, but it is formed by cleavage of tetraterpenes, which means that it has a different biosynthetic origin. For this reason it will be treated in connection with the tetraterpene group.

Biosynthesis of diterpenes

As a rule the IPP required for the biosynthesis of diterpenes is produced via the non-mevalonate pathway. The formation of geranyl diphosphate thus proceeds as described for monoterpenes (page 372) and farnesyl diphosphate is obtained via reaction between geranyl diphosphate and IPP as described for sesquiterpenes (page 400). Farnesyl diphosphate can react with another molecule of isopentenyl diphosphate, giving *all trans*-geranylgeranyl diphosphate, from which diterpenes are formed. The condensation is catalysed by *farnesyl*trans*transferase* (EC 2.5.1.29, also known as geranylgeranyl diphosphate synthase or GGPP synthase), a prenyltransferase analogous to GPP synthase and FPP synthase. Most eubacteria and eukaryotes seem to possess distinct FPP and GPP synthases. GGPP synthase isolated from chromoplasts of *Capsicum annuum* L. has a dimeric structure with two 37 kDa subunits. This enzyme is not chain-length specific and can use DMAPP, GPP and FPP equally well. Also GGPP synthase from white mustard (*Sinapis alba* L.) is a homodimer. In contrast GGPP synthase from bovine brain is a homo-oligomer with 4–5 subunits. Gene sequences

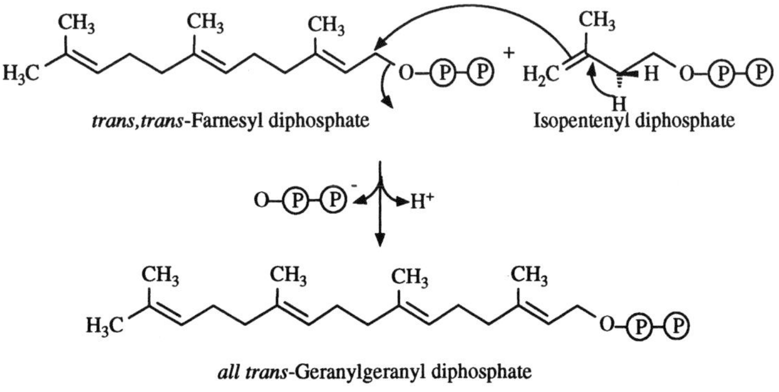

Fig. 108. Biosynthesis of GGPP

413

encoding for GGPP synthase have been characterized from a variety of sources including *Capsicum annuum* L., *Arabidopsis thaliana* (L.) Heynh. and white lupin, *Lupinus albus* L.

The mechanism for the reaction between FPP and IPP is the same as described above (see pages 372 and 400) for formation of GPP and FPP, i.e. ionization of FPP to form an ion pair and electrophilic attack of the allylic cation on the double bond of IPP followed by loss of a proton, (Fig. 108).

Biosynthesis of diterpenes from GGPP is initiated by electrophilic attack on the terminal double bond. A series of cyclizations then follows, leading to mono- (rare), di-, tri- and tetracyclic derivatives which can be further modified by other reactions. As an example the biosynthesis of *ent*-kaurene is illustrated in Fig. 109. *Ent*-kaurene is a precursor to gibberellins, which are important plant growth hormones (se below), and to stevioside and other sweet glucosides from *Stevia rebaudiana* (page 422). The electrophilic attack on the terminal double bond of GGPP triggers cyclization to the bicyclic compound copalyl diphosphate. In analogy with the mechanisms described above for cyclization of mono- and sesquiterpenes, the diphosphate group is then ionized followed by an electrophilic attack on the resultant allylic cation resulting in the tricyclic cation A (Fig. 109). Further cationic transformations via ions B and C finally yield *ent*-kaurene. This series of reactions is catalysed by two enzymes: ent-*copalyl diphosphate synthase* (EC 5.5.1.13) and ent-*kaurene synthase* (EC 4.2.3.19). *Ent*-copalyl diphosphate synthase catalyses formation of copalyl diphosphate and *ent*-kaurene synthase accomplishes the further cyclization and subsequent rearrangement. Biosynthesis of *ent*-kaurene has been observed using cell-free extracts from wheat (*Triticum aestivum*) seedlings, pea (*Pisum sativum*) shoots and pumpkin (*Curcubita maxima*) endosperm. In each case the activity was located in the plastids. A gene encoding *ent*-copalyl diposphate synthase has been isolated from *Arabidopsis thaliana* and shown to express an active protein in *Escherichia coli*. A full-length cDNA encoding *ent*-kaurene synthase has been isolated from developing cotyledons in immature seeds of pumpkin. This was expressed as a fusion protein in *E. coli* and shown to possess *ent*-kaurene synthase activity by the cyclization of copalyl PP to *ent*-kaurene. The deduced amino acid sequence of the protein shared significant homology with other terpene cyclases including a conserved motif comprising a putative binding site for a divalent metal ion-diphosphate complex. A putative transit peptide sequence that may target the translated product into the plastids is present in the *N*-terminal region.

Gibberellins

The gibberellins are acids which were first isolated from the fungus *Gibberella fujikuroi* (Sawada) Wollenweber and which were found to stimulate growth in seedlings. Subsequently, gibberellins have also been encountered in higher plants and more than 90 have been isolated and elucidated structurally. Together with auxins, they regulate the growth processes of plants, especially enhancement of

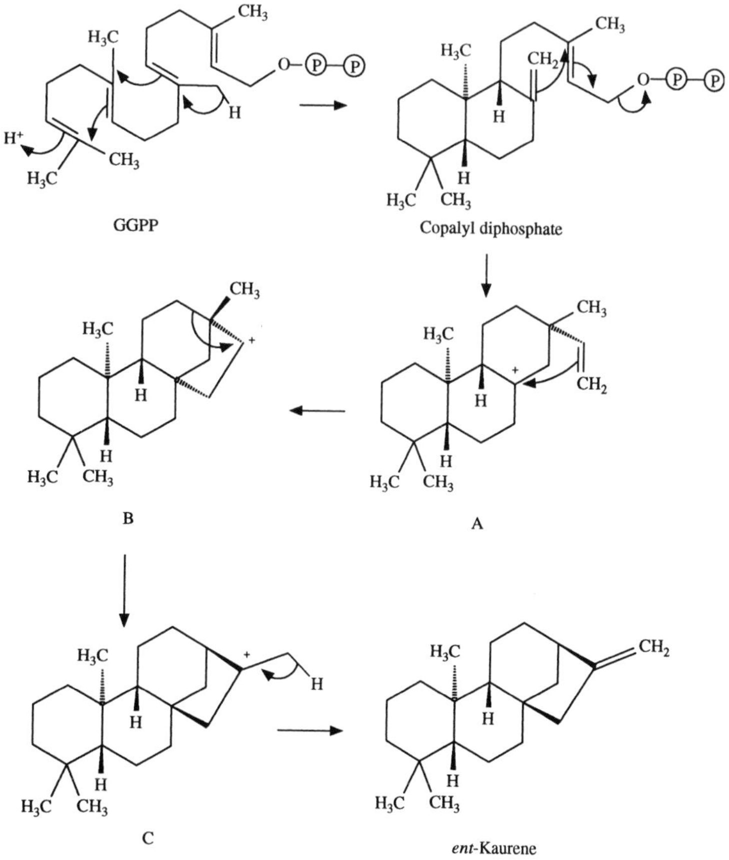

Fig. 109. Biosynthesis of ent-*kaurene*

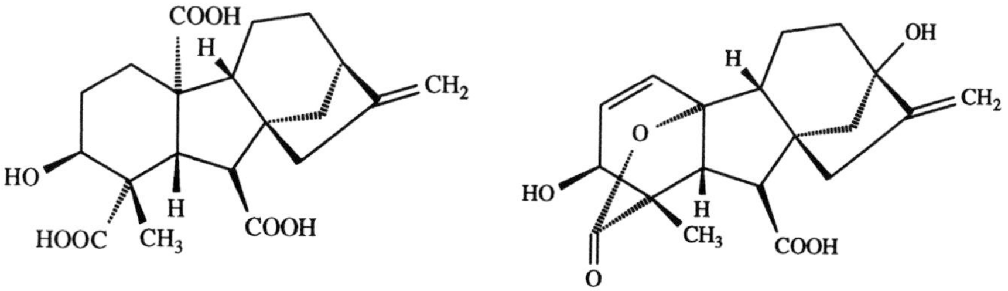

Gibberellin A₁₃

Gibberellic acid

stem growth. Two groups of gibberellins can be distinguished: C_{20} and C_{19} gibberellins. One example from each group is shown above: gibberellin A_{13}, which is a C_{20} gibberellin and gibberellic acid, which is a C_{19} gibberellin.

415

Biosynthesis   *ent*-Kaurene is the precursor of all gibberellins and the biosynthesis of this important group of diterpenes has been extensively studied, including isolation and characterization of several enzyme systems. Genes and c-DNA clones have also been isolated and expressed in *Escherichia coli* yielding biosynthetically active proteins.

Gibberellins are applied in agri- and horticulture for various purposes, e.g. thinning of grape flowers which yields increased berry size in seedless table grapes, applications to *Citrus* fruits to strengthen the rind and inducement of off-season flowering in ornamental plants. Gibberellic acid is produced commercially in ton quantities by cultivation of *Gibberella fujikuroi* in bioreactors.

Tocochromanols (Vitamin E)   The general structure of tocochromanols is a chromanol ring connected to a 16-carbon chain. The tocochromanols can be divided into tocopherols and tocotrienols which differ from each other in the composition of the side-chain. Tocopherols have a fully saturated tail while the tocotrienols have a tail with double bonds at carbons

Tocopherols

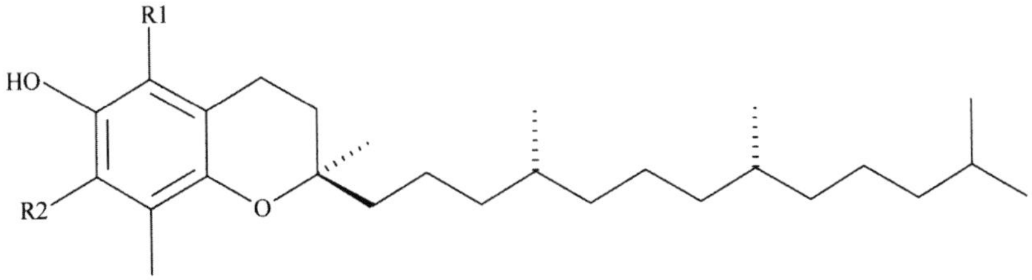

	R_1	R_2
α-Tocopherol	CH_3	CH_3
β-Tocopherol:	CH_3	H
γ-Tocopherol:	H	CH_3
δ-Tocopherol:	H	H

Tocotrienols

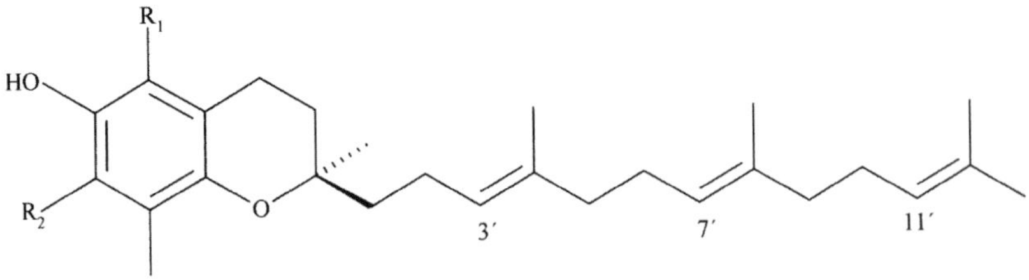

	R_1	R_2
α-Tocotrienol:	CH_3	CH_3
β-Tocotrienol:	CH_3	H
γ-Tocotrienol:	H	CH_3
δ-Tocotrienol:	H	H

Fig. 110. Structures of tocopherols and tocotrienols

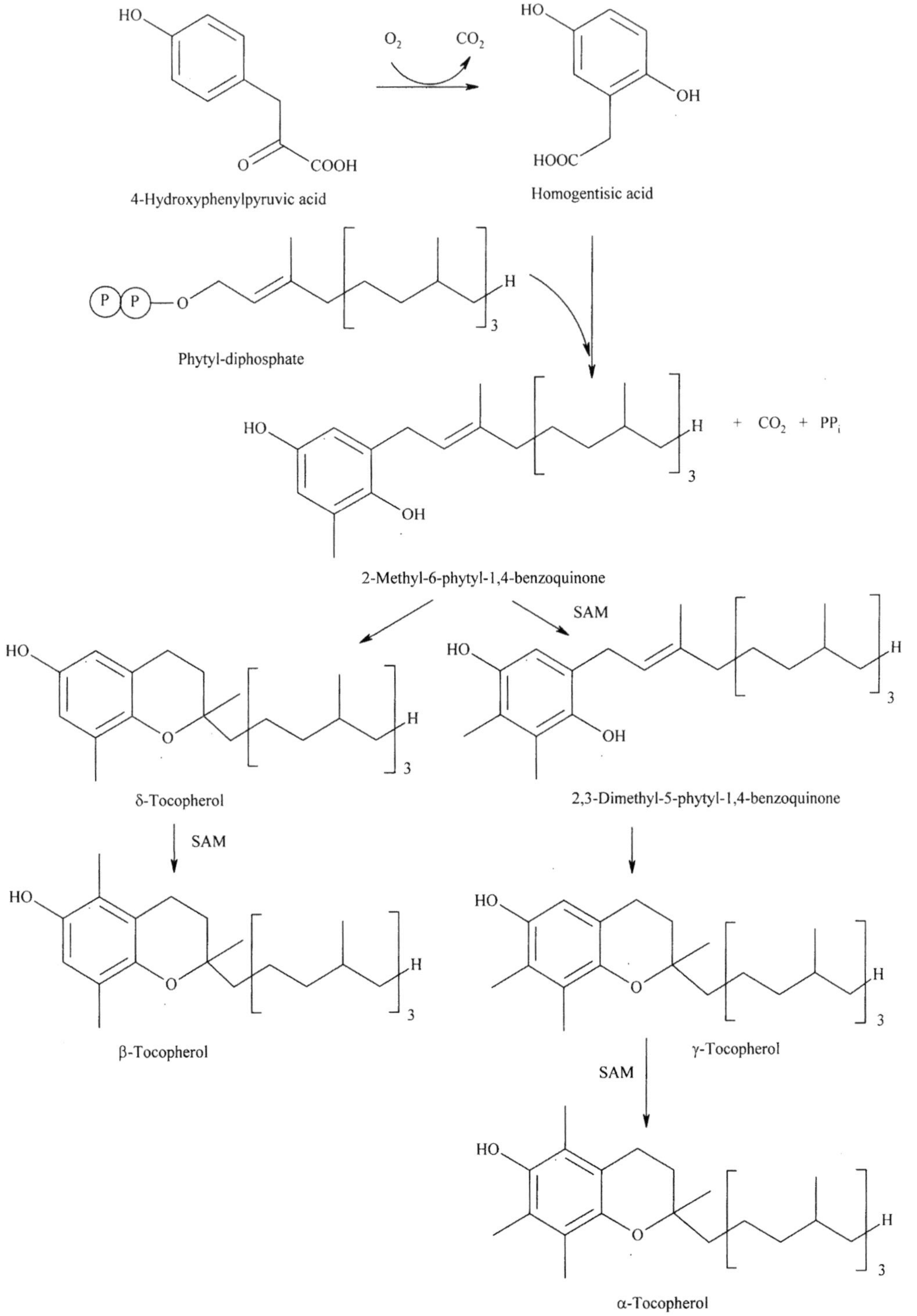

Fig. 111. Biosynthesis of tocopherols. Tocotrienols are synthesized in the same way but with substitution of geranygeranyl diphosphate for phytyl-phosphate in step 2

417

3′, 7′and 11′. The tocochromanols also differ in the number of methyl substituents on the aromatic ring. α-Tocohromanols have three methyl substituents, β- and γ- tocochromanols have two and δ- tocochromanols have one methyl substituent. Fig. 110 illustrates the structures of the eight tocochromanols.

Biosynthesis As illustrated in Fig. 111 tocochromanols emanate from two biosynthetic pathways – the shikimic acid pathway for synthesis of the tocopherol head group and the plastidic non-mevalonate pathway for synthesis of the hydrophobic tail. The first step is formation of homogentisic acid from *p*-hydroxyphenylpyruvic acid, catalysed by *4-hydroxyphenylpyruvate dioxygenase* (EC 1.13.11.27). This is a complex, irreversible, reaction that catalyses addition of two oxygen molecules, a decarboxylation and rearrangement of the side-chain of 4-hydroxyphenylpyruvic acid. The second step is prenylation of homogentisic acid with phytyl diphosphate or geranylgeranyl diphosphate to yield 2-methyl-6-phytyl-1,4-benzoquinone (in the tocopherol group) or 2-methyl-6-geranylgeranyl-1,4-benzoquinone (in the tocotrienol group). The enzyme that catalyses this reaction is *homogentisate phytyltransferase*. The continuation of the biosynthesis of the two groups of tocochromanols involves cyclization to form the oxane ring and further methylation of the aromatic ring as indicated in Fig. 111. The genes that code for the enzymes that catalyse these reactions have been identified in the model organisms *Arabidopsis thaliana* (L.) Heynh. and the cyanobacterium *Synechocystis PCC6803*. The corresponding enzymes have been cloned and characterized.

Tocochromanols can be obtained from wheat-germ oil and they also occur in cotton-seed oil, leafy vegetables, egg yolk and meat. They are lipophilic and are important as antioxidants for fats. Biologically, they are classified as antisterility vitamins, or vitamin E. All tocopherols have vitamin E-activity. α-Tocopherol is most active while β-tocopherol has half the activity and γ- and δ- tocopherols have only 10 and 3 %, respectively, of the activity of α-tocopherol. The tocotrienols are less active. α-Tocotrienol has 30 % of the activity of α-tocopherol, β-tocotrienol has 5 % activity and γ- and δ- tocotrienols are inactive.Vitamin E is necessary for normal foetal development and growth in rats, but its role in human physiology has not been determined. Vitamin E is used for the treatment of neuromuscular disorders, though its effects are questionable.

Synthetic α-tocopheryl acetate is the main commercial form of vitamin E which is used for food supplementation and for medicinal purposes.

Vitamin K Vitamin K is involved in the processes leading to the coagulation of blood. Like vitamin E, it is not a single chemical compound. In Nature, there are two substances with this biological property, viz. vitamins K_1 and K_2: (phylloquinone and menaquinone, respec–tively).

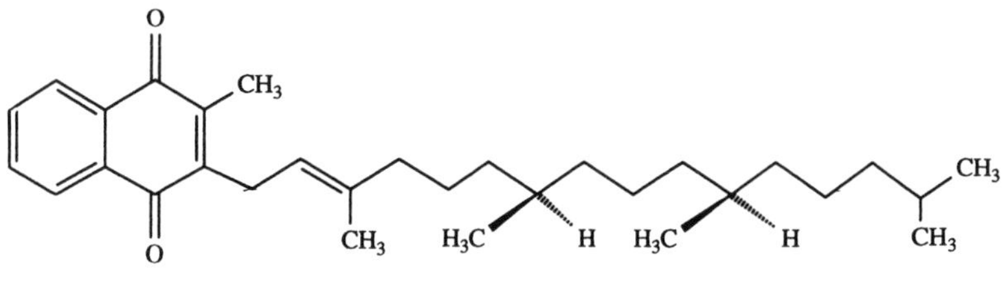

Vitamin K₁

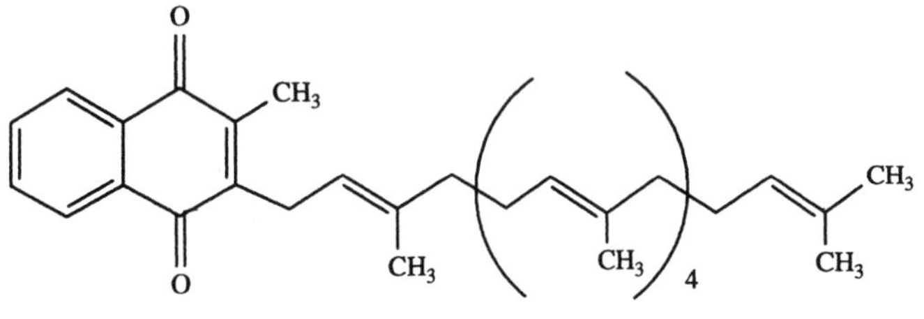

Vitamin K₂

The side-chain of vitamin K_1 is derived from phytol, which means that the vitamin can be regarded as a diterpene derivative. The longer side-chain of vitamin K_2 can vary in length and is derived from squalene (page 434), so that this vitamin can also be said to be a triterpene derivative. The side-chain is not necessary for the biological activity and synthetic 2-methyl-1,4-naphthoquinone has been marketed as a vitamin-K effective substitute.

Biosynthesis   The naphtoquinone skeleton of the K vitamins emanates from the shikimate acid pathway. Chorismate (page 202 is transformed to isochorismate in a reaction (Fig. 112) catalysed by *isochorismate synthase* (EC 5.4.4.2). In a complicated reaction, catalysed by *2-succinyl-6-hydroxycyclohexa-2,4-diene-1-carboxylate synthase* (EC 2.5.1.64) isochorismate reacts with 2-oxoglutarate to yield (1R,2S,5S,6S)-2-succinyl-5-enolpyruvyl-6-hydroxy-3-cyclohex-ene-1-carboxylate. A third enzyme converts this compound to 6-hydroxy-2-succinylcyclohexa-2,4-dienecarboxylate. This compound loses water with formation of 2-succinylbenzoate, thus establishing the aromatic ring of the naphtoquinone skeleton. 2-Succinylbenzoate is transformed into an active CoA ester in a reaction catalysed by *o-succinylbenzoate-CoA ligase* (EC 6.2.1.26). The enzyme *naphtoate synthase* (EC 4.1.3.36) then catalyses ring-closure to yield 1,4-dihydroxy-2-naphtoate. C-Alkylation with phytyl diphosphate, followed by decarboxylation of the resultant β-keto acid and oxidation, yields a *p*-quinone which is subsequently methylated to form vitamin K_1. Vitamin K_2 is formed similarly by alkylation with a longer isoprenoid phosphate.

Natural vitamin K is found in green leaves, e.g. spinach, as well as in tomatoes and in vegetable oils. Vitamin K_1 can be obtained from spinach and vitamin K_2 from decaying fishmeal.

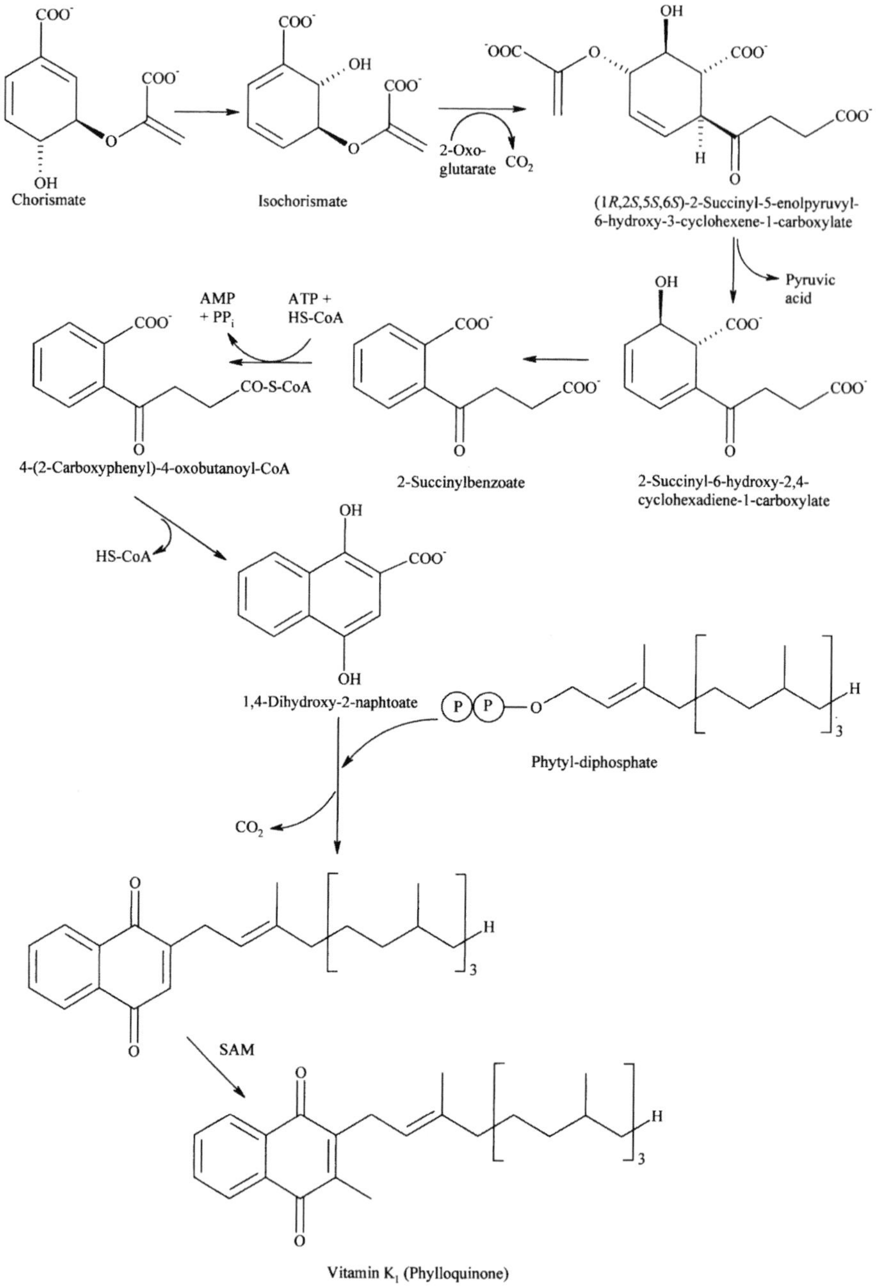

Fig. 112. Biosynthesis of Vitamin K₁

Vitamin K is important for the formation of prothrombin, which is needed for the normal coagulation of blood and lack of it increases the risk of bleeding. Shortage of vitamin K is rare, since it is normally formed by bacteria the intestine. Prolonged diarrhoea and excessive use of purgatives, as well as treatment with broad-spectrum antibiotics, may, however, bring about a deficiency of vitamin K. Therapeutically, vitamin K is mainly used for the treatment of haemorrhage in newborn children.

Forskolin Forskolin is a diterpene obtained from the roots of *Coleus barbatus* Benth. (*Lamiaceae*), a plant which has been described in Ayurvedic materia medica and in ancient Hindu medicinal texts as a remedy for several complaints, including heart diseases and CNS disorders such as insomnia and convulsions.

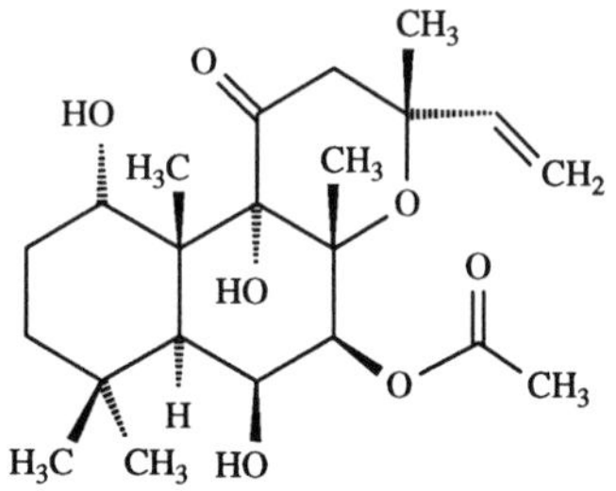

Forskolin

Forskolin has a strong positive inotropic effect on the heart muscle and is also an antihypertensive. It has attracted considerable interest because the mode of action is different from that of other compounds with a positive inotropic effect, such as the cardiac glycosides (page 459) and xanthine derivatives. Forskolin activates membrane-bound adenylate cyclase and causes an increase in cellular cAMP levels. Hormone receptors are not needed for this activation, but in hormone-mediated adenylate cyclase reactions the hormonal stimulation is markedly potentiated at extremely low doses of the activating agent. Forskolin causes a rapid and reversible, eightfold activation of adenylate cyclase in membranes from rat cerebral cortex.

Forskolin is an example of a pharmacologically active natural product with a unique mode of action. It is useful as a tool in pharmacology and is also of interest for the development of, for example, antihypertensive drugs based on a novel mechanism of action.

Andromedotoxin Andromedotoxin is a toxic compound which occurs in the gen-

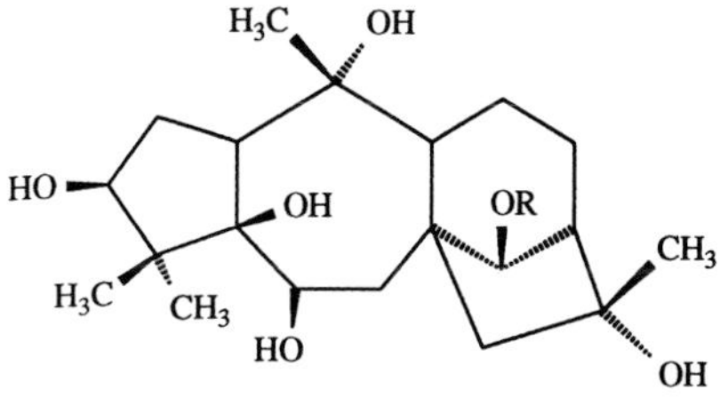

Andromedotoxin (Grayanotoxin I) R = COCH₃

era *Kalmia, Leucothoe* and *Rhododendron* of the family *Ericaceae*. It is present mainly in the leaves but can also occur in the nectar and the pollen grains of the flowers. Honey produced by bees collecting nectar from such species has a bitter taste and is poisonous. The symptoms of poisoning are irritation of the mucous membranes of the mouth and stomach, vomiting and diarrhoea, dizziness, headache, fever and dyspnoea.

Stevioside

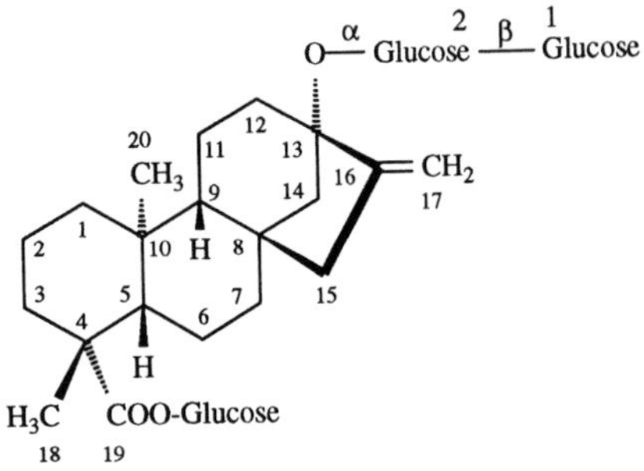

Stevioside

Stevioside is a sweet-tasting compound (about 300 times sweeter than saccharose) which can be obtained from the leaves of *Stevia rebaudiana* Bertoni (*Asteraceae*). The plant originates from Northern Paraguay and parts of Brazil. It is now cultivated commercially in Brazil, Israel, Korea, China and Japan. It is a 40-cm tall herb with lanceolate, 2–3 cm long, pinnatifid leaves which have been used for centuries by the Paraguayan Indians for sweetening of maté tea (page 734).

The content of stevioside in the leaves is high (up to 10%). The compound is obtained commercially by extraction of a concentrated, aqueous extract of the leaves with *n*-butanol. The butanol extract is clarified with charcoal, evaporated and the residual stevioside crystallized from methanol. A number of related sweet glycosides known as dulcosides and rebaudosides have also been extracted from the leaves. They all have the same aglycone (steviol) but differ in their sugar composition. Chromatographic methods for separation of stevioside from these glycosides have been developed.

Biosynthesis   As expected, steviol, the aglycone of stevioside, has been shown to be biosynthesized via the non-mevalonate pathway. The biosynthesis proceeds via *ent*-kaurene (page 414) which is oxidized in a three-step reaction, catalysed by ent-*kaurene oxidase* (EC 1.14.13. 78) to ent-kaurenoic acid (Fig. 113) which is further oxidized to steviol. The enzyme for this latter reaction – ent-*kaurenoic acid 13-hydroxylase* – has been partially purified and an *N*-terminal sequence described. The identity of this enzyme has, however, been questioned and it is suggested that it actually is *fructose biphosphate aldolase*. The glycosides are produced in the leaves, where all the steps up to *ent*-kaurene occur in the plastids, one of the two oxidation steps is located on the surface of the ER and glycosylation takes place in the cytoplasm. Enzyme systems have been isolated which will glucosylate steviol at C-13 and C-19. The sequence is steviol –> steviol 13–*O*-glucoside –> steviol-bioside –> stevioside.

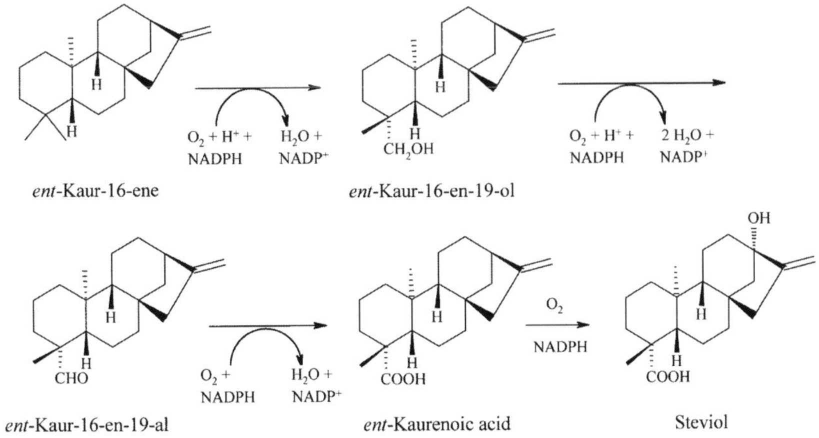

ent-Kaur-16-ene $O_2 + H^+ +$ NADPH $H_2O +$ NADP$^+$ CH$_2$OH   *ent*-Kaur-16-en-19-ol $O_2 + H^+ +$ NADPH $2 H_2O +$ NADP$^+$

ent-Kaur-16-en-19-al   CHO $O_2 +$ NADPH $H_2O +$ NADP$^+$ COOH   *ent*-Kaurenoic acid O_2 NADPH   OH   COOH   Steviol

Fig. 113. Biosynthesis of steviol from ent-*kaurene*

Stevioside is used in Japan and some other countries as a sweetener for beverages and foods, but there is some hesitation about possible mutagenic and kidney-damaging effects.

Recently a product called rebiana has been proposed to be a safe general purpose sweetener (see the reference Carakostas 2008 in *Further reading*). Rebiana contains not less than 97 % of the glycoside rebaudioside A and not more than 3 % of other steviol glycosides. Rebaudioside A has the same structure as stevioside but contains an additional glucose residue in the chain attached to C-13 in steviol. This residue is connected to the first glucose residue in the chain via a 3→1-α bond.

Ginkgolides Ginkgolides are diterpenes with a unique structure and are characteristic constituents of the leaves of *Ginkgo biloba* L. (*Ginkgoaceae*), the only living representative of a family which once contained at least six genera of which 200 million-year-old fossils have been found. *Ginkgo biloba* is a tree with an average height of 20 m and originated in east Asia where it is cultivated for its seeds, which are used as food and for medicinal purposes. *Ginkgo biloba* is also widely cultivated in Japan, the United States and in Europe both as an ornamental and for pharmaceutical purposes. The species is dioecious and thus female plants are preferred for the seed production, but males as ornamentals since the fruit pulp deteriorates under fruit maturation and produces a remarkable smell from butyric acid. Besides the ginkgolides the leaves also contain a sesquiterpene

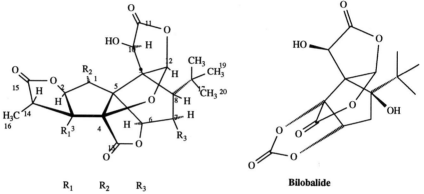

Flavonoid glycosides

Kaempferol derivatives: R = H
Quercetin derivatives :   R = OH
Isorhamnetin derivatives: R = OCH₃

Example of a biflavone (amentoflavone)

	R₁	R₂	R₃
Ginkgolide A:	OH	H	H
Ginkgolide B:	OH	OH	H
Ginkgolide C:	OH	OH	OH
Ginkgolide J:	OH	H	OH
Ginkgolide M:	H	OH	OH

Bilobalide

Fig. 114. Constituents of the leaves of Ginkgo biloba

derivative – *bilobalide* – and a great number of flavanoids present as mono-, di- and triglycosides of kaempferol, quercetin and isorhamnetin. Non-glycosidic biflavonoids, cathechins and proanthocyanidins have also been isolated. Structurally interesting are flavonoid glycoside esters with coumaric acid. The leaves also contain potent allergens (ginkgolic acids) which must be removed before an extract can be used as a drug. Another potentially toxic constituent is 4´-O-methylpyridoxine, which occurs mainly in the seeds, but which has been found also in some batches of leaves. Excessive consumption of seeds causes convulsions, loss of consciousness and even death.

Biosynthesis   The biosynthesis of the ginkgolides has been only partly elucidated. As outlined in Fig. 115 the ultimate precursor is geranylgeranyl diphosphate which through proton-initiated carbocation formation is transformed to labdadienyl diphosphate (=copalyl diphosphate).

Further cyclization with loss of the phosphate ion yields a cation which undergoes a special case of six-electron electrocyclic rearrangement resulting in migration of a methyl group and formation of levopimaradiene (= 8,12-abietadiene) via a second cation. The cyclization of geranylgeranyl diphosphate to levopimaradiene is catalysed by the multifunctional enzyme *levopimaradiene synthase* which has been cloned and functionally characterized. The nucleotide sequence of the corresponding cDNA predicts the enzyme to contain 873 amino acids. Dehydrogenation of ring C in levopimaradiene yields (+)-dehydroabietane which is further processed to the ginkgolides in a complex set of reactions involving several oxidation steps. There is evidence that dehydroabietane is formed in the plastids and then transported into the cytoplasm, where the oxidation reactions occur. Bilobalide is supposed to be formed by partial degradation of the ginkgolides.

Medical use   In China the medicinal use of ginkgo leaves has a very long history. The plant is recorded in herbals, which are almost 5 000 years old, as a remedy against asthma and "to benefit the brain". In the

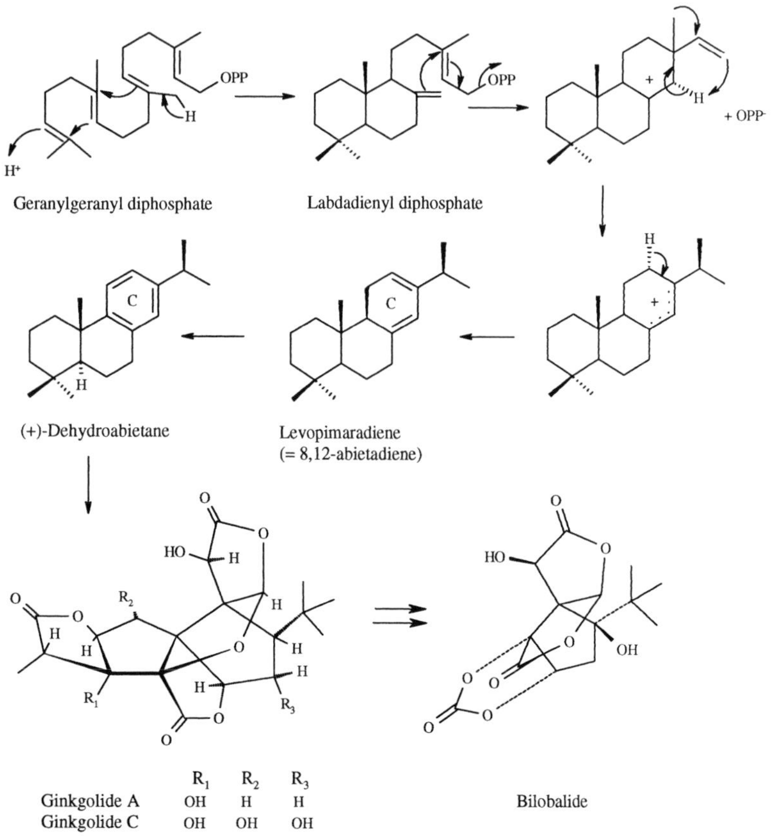

Geranylgeranyl diphosphate

Labdadienyl diphosphate

(+)-Dehydroabietane

Levopimaradiene
(= 8,12-abietadiene)

	R₁	R₂	R₃
Ginkgolide A	OH	H	H
Ginkgolide C	OH	OH	OH

Bilobalide

Fig. 115. Biosynthesis of ginkgolides

1970s a highly purified, patented extract of the leaves (EGb 761) was introduced in Western countries as a herbal remedy for the treatment of diseases appearing with advanced age and with symptoms such as vertigo, tinnitus, headache, impaired short-term memory, hearing loss, decreased vigilance and mood disturbance. These symptoms are associated with peripheral circulatory insufficiency due to degenerative angiopathy and cerebrovascular insufficiency. Flavonoids and the ginkgolides are regarded as the pharmacologically active principles.

The flavonoids inhibit cyclo-oxygenase and lipoxygenase and could thus decrease production of thromboxane A_2, which is a potent platelet aggregator involved in clot formation and subsequent thrombosis (page 262). Ginkgo flavonoids are also free radical scavengers and could thus diminish the damage to cell membranes which causes malfunction and death of cells in the brain.

Most attention has been given to the ginkgolides which have potent antagonist activity against platelet-activating factor (PAF). PAF is a phospholipid with an ether linkage to a long-chain fatty alcohol. It plays an important role in several pathological conditions such as asthma, shock, ischaemia, anaphylaxis, graft rejection, renal disease, CNS-disorders and numerous inflammatory conditions. The ginkgolides are highly specific inhibitors of the binding of PAF to its cellular receptors. Ginkgolide B is the most powerful antagonist. PAF stimulates the conversion of phospholipids in cells to arachidonic acid and has thus an influence on the biosynthesis of prostaglandins and leukotrienes which, as mentioned above, are associated with thrombosis and inflammation especially in connection with cerebral oedema. Several clinical trials have been performed particularly on EGb 761 which indicate that ginkgo extract could be of benefit in cases of cerebral insufficiency in elderly patients. EGb 761 is standardized with respect to ginkgolides and flavonoids, a quality standard which must be required for *Ginkgo* extracts. Not all extracts on the market fulfil this requirement.

Extracts of *Ginkgo biloba* leaf inhibit cytochrome P450 enzyme activity and consequently may produce CYP-mediated herb–drug interactions

e.g. R = $(CH_2)_{15}CH_3$

PAF

Paclitaxel (Taxol®)   Paclitaxel is a diterpene which was first isolated from the bark of the Western yew *Taxus brevifolia* Nutt. (*Taxaceae*), a tree with a height of 4–15 m and native to western North America. Its isolation was a result of a large research project under the auspices of the *NIH* (National Institutes of Health, USA) in which thousands of plant extracts were screened for tumour-inhibiting activity. Bioassay-guided fractionation of an ethanol extract of the bark yielded a new compound, denominated as Taxol, a name which later became a registered trademark.

Paclitaxel has a taxane nucleus consisting of two six-membered rings (A and C) fused at an almost perpendicular angle with an eight-membered ring (B). The C-13 hydroxyl group of this nucleus

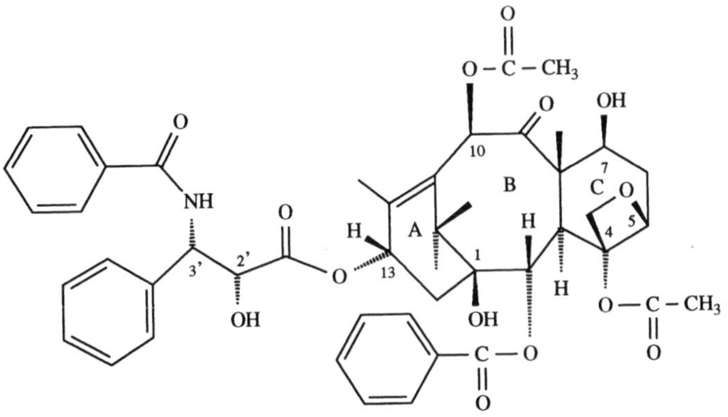

Paclitaxel

is esterified with (2'*R*-3'*S*)-*N*-benzoyl-3'-phenylisoserine. Other ester bonds involve hydroxyl groups at C-2 (benzoic acid) as well as C-4 and C-10 (acetic acid). A remarkable feature is the oxetane ring involving the bond between C-4 and C-5 of the taxane nucleus. This ring is a crucial component with respect to the biological and pharmacological activity of paclitaxel. Also the side-chain at C-13 and the benzoyloxy group at C-2 are of great importance.

Biosynthesis   In analogy with other diterpenes paclitaxel is formed from geranylgeranyl diphosphate (GGPP). This, however, does not originate from mevalonate but is presumably biosynthesized from isopentenyl diphosphate, which is formed via the non-mevalonate pathway (page 369) in analogy with other diterpenes. The enzyme *taxadiene synthase* (EC 4.2.3.17) converts geranylgeranyl diphosphate to taxa-4(5),11(12)-diene, a reaction involving formation of three C-C bonds (**5** in Fig. 116). A cDNA fragment from a cDNA library of *Taxus brevifolia* has been expressed in *Escherichia coli* resulting in a protein with the properties of the expected taxadiene synthase. The deduced amino acid sequence comprises 862 amino acids and the molecular mass is 98 kDa.

The biosynthesis (Fig. 116) starts with ionization of the diphosphate group of GGPP followed by cyclization to a transient verticillyl intermediate (**2**), proposed to have the 11(*R*) configuration (see the mechanisms for cyclization of farnesyl phosphate page 402). This allows intramolecular transfer of the C-11 proton of **2** to C-7 (**3**), initiating transannular B/C-ring closure to the taxenyl cation (**4**) followed by deprotonation at C-5 to yield taxa-4,11-diene (**5**). A detailed study of the electrophilic cyclization reaction cascade has been performed using recombinant taxadiene synthase and deuterium labelled substrates. (4*R*)-[4-^{2}H$_1$]GGPP yielded taxa-4,11-diene which bore no detectable deuterium compared with the product obtained from the unlabelled precursor, indicating that the

β-hydrogen at C-4 of GGPP (= C-5 of taxa-4,11-diene) is selectively and entirely lost in the terminating deprotonation step. Feeding of
[10-^{2}H$_1$]GGPP resulted in formation of taxa-4,11-diene with 99 %
of the deuterium originally present at C-10 of GGPP at position 7 of
the taxane skeleton, indicating migration of the hydrogen at C-11 in
2 (= C-10 in GGPP) to C-7 (**3**). NMR studies of the labelled taxa-
4,11-diene obtained in the experiment and unlabelled taxa-4,11-
diene, showed that the migrating hydrogen becomes positioned in

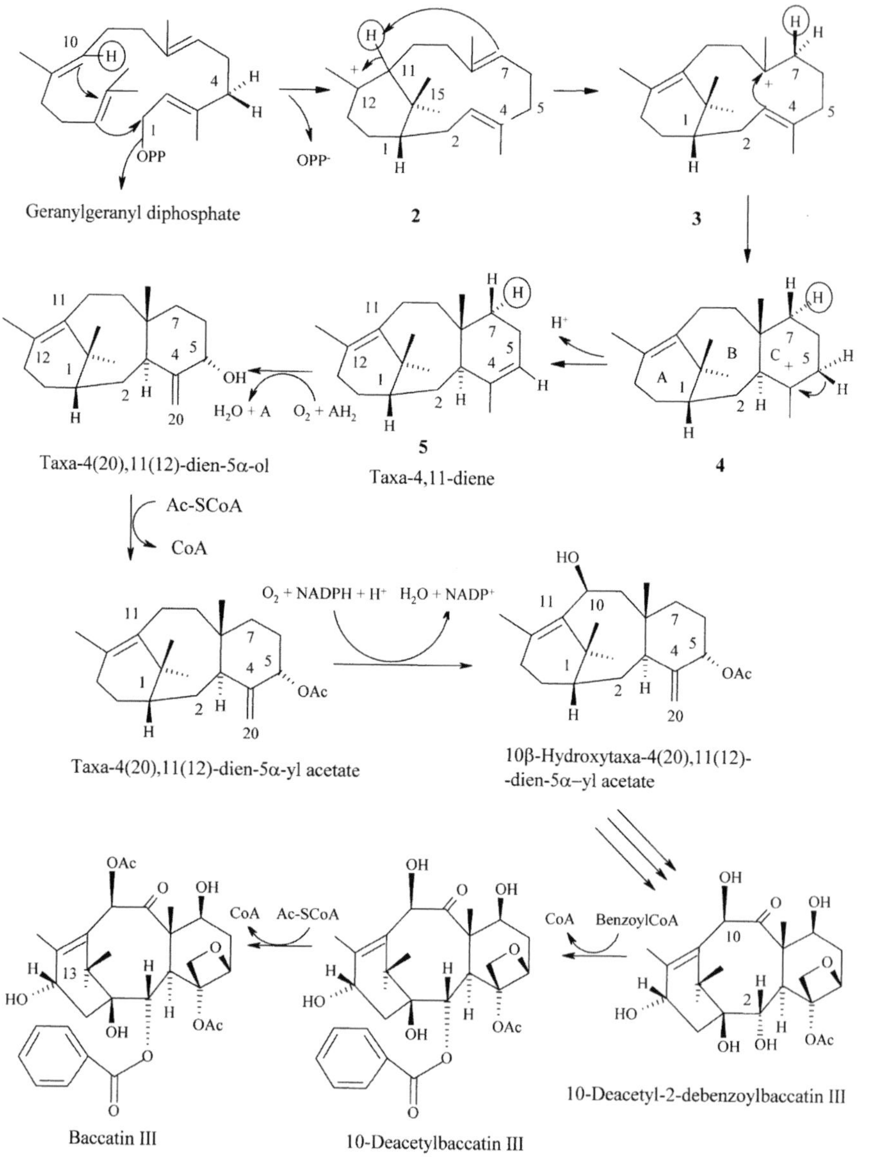

Fig. 116. Biosynthesis of baccatin III. The migrating hydrogen is marked
with a circle

Plate 11.
Taxus baccata L. The leaves (needles) are a source of 10-deacetyl-baccatin III which is a raw material for the semisynthesis of paclitaxel (page 426). Rimforsa, Östergötland, Sweden. Photograph by Gunnar Samuelsson.

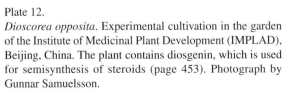

Plate 12.
Dioscorea opposita. Experimental cultivation in the garden of the Institute of Medicinal Plant Development (IMPLAD), Beijing, China. The plant contains diosgenin, which is used for semisynthesis of steroids (page 453). Photograph by Gunnar Samuelsson.

the α-position at C-7. Molecular modelling of the C-12 carbocation in **2** (resulting from the C-11 to C-15 closure) revealed that the C-11 proton is close to the C-7 carbon and seemingly perfectly poised for the transannular migration to the *re*-face at C-7. Furthermore, inspection of the conformation of the C-12 carbocation strongly suggested that it is unlikely that an amino acid at the active site of

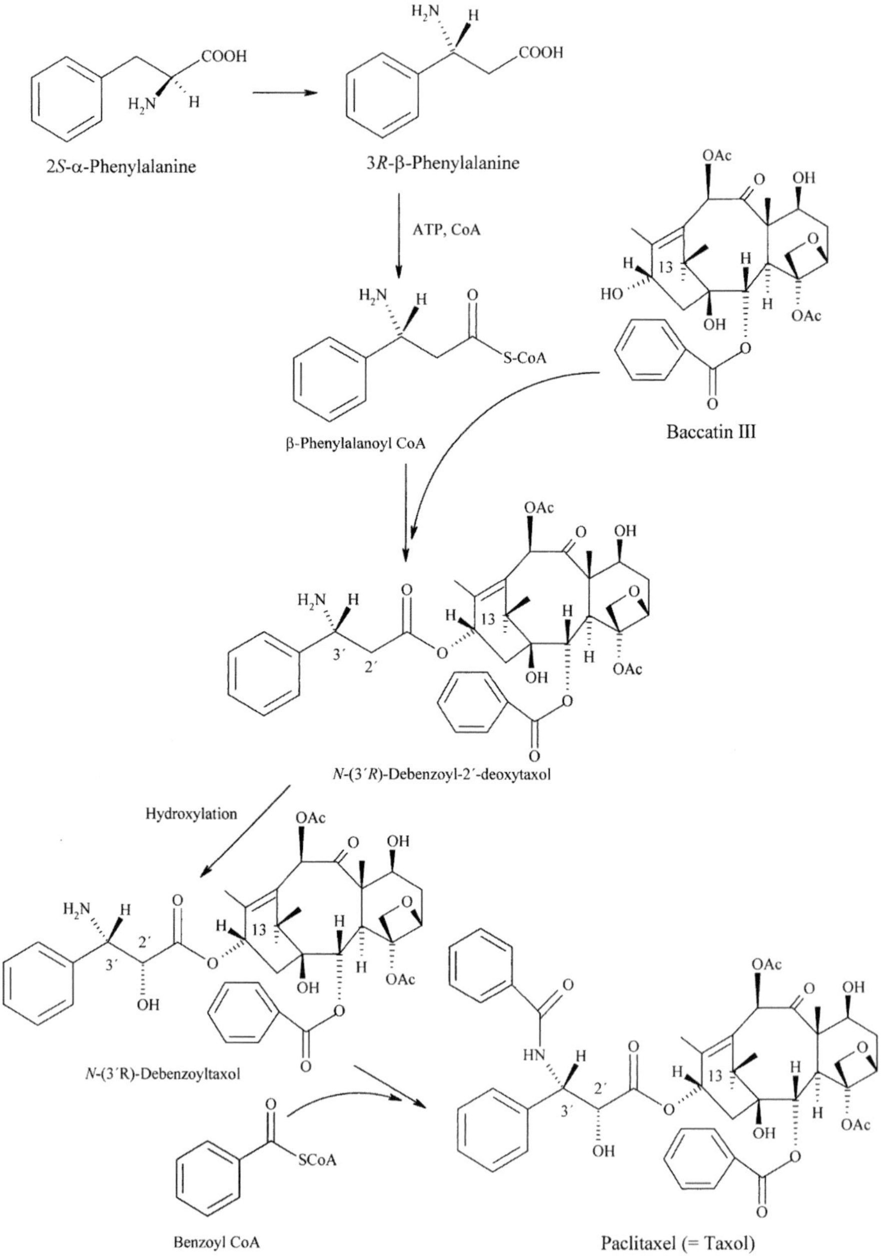

Fig. 117. Biosynthesis of paclitaxel from baccatin III

the enzyme could gain access to the C-11 proton to mediate the C-11 to C-7 transfer, since this proton is buried deep within the concave face of the 12-membered macrocycle. It thus appears that an **unassisted intramolecular transfer** is the most plausible mechanism for the initiation of the final ring closure step, mediated by taxadiene synthase. Thus the C-7 to C-8 olefin serves as the Brönsted base that quenches the incipient carbocation at C-12 and no active site enzyme base is needed for this process. Taxadiene synthase is a remarkable enzyme which is able to bring about an enantio- and face-selective polyolefin cation cascade involving formation of three carbon-carbon bonds, three stereogenic centres and the loss of hydrogen in a single step. The unassisted intramolecular proton transfer mechanism is very unusual and suggests that this enzyme is capable of performing complex olefin cation cyclizations with absolute stereochemical fidelity by conformational control alone.

The next step in the biosynthetic pathway is oxidation at C-5, accompanied by a shift of the double bond C-4(5) to C-4(20), yielding taxa-4(20),11(12)-dien-5α-ol. This reaction is catalysed by *taxadiene 5-hydroxylase* (EC 1.14.99.37, also known as taxa-4(5),11(12)-dien-5α-hydroxylase), which is a membrane-bound, NADPH-dependent P450 oxygenase that uses molecular oxygen.

The enzyme *taxadienol O-acetyltransferase* (EC 2.3.1.162) catalyses acetylation of taxa-4(20),11(12)-dien-5α-ol, yielding taxa-4(20),11(12)-dien-5α-yl acetate, which is hydroxylated at C-10 by the enzyme *taxane 10β-hydroxylase* (EC 1.14.13.76) with formation of 10β-hydroxytaxa-4(20),11-dien-5α-yl acetate. This compound is then oxidized in several, not yet elucidated, steps to yield 10-deacetyl-2-debenzoylbaccatin III, which is benzoylated in a reaction catalysed by *2α-hydroxytaxane 2-O-benzoyltransferase* (EC 2.3.1.166), resulting in formation of 10-deacetylbaccatin III. Finally baccatin III is formed by acetylation at C-10, catalysed by *10-deacetylbaccatin III 10-O-acetyltransferase* (EC 2.3.1.167). The side-chain in paclitaxel derives entirely from phenylalanine that is subjected to a reaction which has never before been detected in plants involving a migration of the amino group from the α to the β position (Fig. 117). This is catalysed by *phenylalanine aminomutase*, an enzyme, the activity of which was first demonstrated in the soluble fraction of crude cell-free extracts of *Taxus brevifolia* stems in 1998. Analysis of the mechanism of the enzymatic activity showed that the migration of the amino group in phenylalanine is strictly intramolecular. Also the steric course of the reaction was determined. The enzyme has since been purified from cell cultures of *Taxus chinensis* and the corresponding gene has been identified, cloned and expressed in *Exherichia coli*. Identification of the gene, cloning and expression has also been performed starting with cell cultures of *Taxus cuspidata*.

The β-phenylalanine formed is esterified with the C-13 hydroxyl group of baccatin III yielding β-phenylalanoyl baccatin III. An

431

enriched cDNA library has been constructed from mRNA isolated from *Taxus cuspidata* cells and from this library a cDNA clone has been isolated and used to prepare the recombinant O-*(3-amino-3-phenylpropanoyl) transferase* which catalyses the esterification.

The final steps in the construction of the side-chain of paclitaxel involves hydroxylation at C-2′ of the β-phenylalanoyl moiety and benzamidation of the aminogroup. Microsomal preparations from *Taxus* suspension cells were found to enable cytochrome P450 mediated hydroxylation of β-phenylalanoyl baccatin III to phenylisoserinoyl baccatin III and a recombinant enzyme, which catalyses the benzamidation of this compound, has been obtained by expression of a cDNA clone from the above-mentioned cDNA library of *Taxus cuspidata*. The benzoyl CoA needed for the benzamidation is derived from β-phenylalanine and not from cinnamic acid which is the usual route in plants.

Elucidation of the biosynthesis of taxol is important for development of gene technological methods for the production of paclitaxel in cell cultures, and for the attachment of modified aroyl groups to taxoid precursors for the purpose of improving drug efficacy.

Production of
paclitaxel

As mentioned above, paclitaxel was first isolated from the bark of the Western yew tree, *Taxus brevifolia*. However, the content of paclitaxel is very low, only 0.01–0.04 % of the dried inner bark (phloem-cambial tissue), which is the main location of paclitaxel. This means that production of 1 kg of paclitaxel requires 9 000 kg of bark from 2 000–3 000 trees. As *Taxus brevifolia* is slow-growing, production from wild growing trees involves a risk of extinction of this plant. Paclitaxel has since been found in the leaves (needles) of several *Taxus* species in yields comparable with those from the bark. As needles are a renewable source the environmental objections to bark collection for production of paclitaxel are avoided. Another source of paclitaxel is semisynthesis from 10-deacetylbaccatin III, which is available from the leaves of *Taxus baccata* L. (Plate 11, page 429) and other yews in yields of at least 0.1 %.

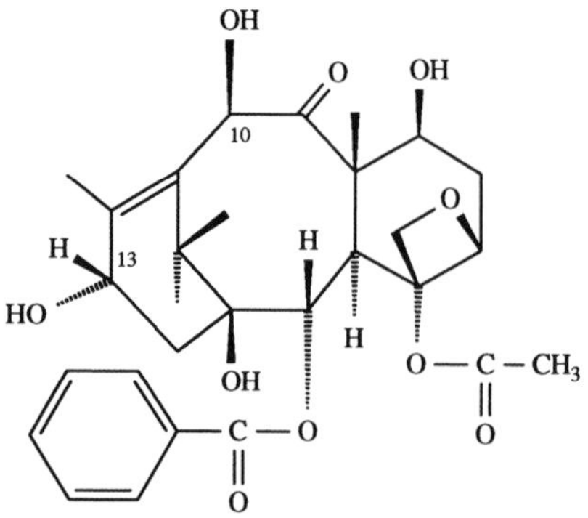

10-Deacetylbaccatin III

As seen from the structures, paclitaxel can be obtained from 10-deacetyl-baccatin III by acetylation of the hydroxyl at C-10 and esterification with (2R-3S)-N-benzoyl-3-phenylisoserine, for which cost-effective methods of synthesis are available. As a result of these developments *Taxus brevifolia* bark is no longer used as a source of paclitaxel.

Important progress has been made in the development of paclitaxel production by plant tissue-culture methods and it has been announced that paclitaxel is likely to become the first plant-derived drug, commercially produced by this technique.

Total synthesis of paclitaxel has been achieved but is too complex to become a commercial source of the drug.

Medical use Paclitaxel is used as a chemotherapeutic agent for ovarian cancer. The mechanism of action is unique in that the substance – in contrast to other inhibitors of mitosis such as colchicine and the *Vinca* alkaloids – does not bind to tubulin but instead to the microtubules, thereby stabilizing these and interfering in the equilibrium between tubulin and the microtubules (see page 219). The binding of paclitaxel to microtubules results in two distinct morphological effects. During all phases of the cell cycle paclitaxel causes reorganization of the cellular microtubule cytoskeleton to give abundant arrays of disorganized microtubules which are often aligned in parallel bundles. During cell division (mitosis) microtubules usually grow from centrioles that migrate to opposite poles of the cell, forming a spindle aster for daughter chromatid separation. In contrast, paclitaxel-treated cells form large numbers of abnormal asters that do not require centrioles for formation and arrest the cell cycle during mitosis.

Paclitaxel is very poorly soluble in water and is therefore formulated as a 6 mg/ml solution in a solvent consisting of 50 % polyoxyethylated castor oil and 50 % dehydrated ethanol. It is administered as an infusion after dilution with 5 % glucose or 0.9 % NaCl. Paclitaxel may occasionally crystallize in these aqueous dilutions and in-line filtration is used in administration of the drug for elimination of potential microcrystals. Hypersensitivity reactions (presumably due to the solvent) can occur and premedication with antiallergic drugs is therefore required.

Docetaxel (Taxotere®) Docetaxel is a semisynthetic analogue of paclitaxel where an *N*-tert-butoxycarbonyl group is substituted for the *N*-benzoyl group in the side-chain and the C-10 hydroxyl group is free. The starting material for synthesis of docetaxel is 10-deacetylbaccatin III, obtained from the needles of *Taxus baccata* L. Docetaxel is also a stabilizer of microtubules and is cytotoxic against murine and human tumour cells. It is used as a single agent against locally advanced or metastatic breast cancer which resists other treatment or which has recurred after treatment with cytostatics. It has also been clinically tested against lung cancer.

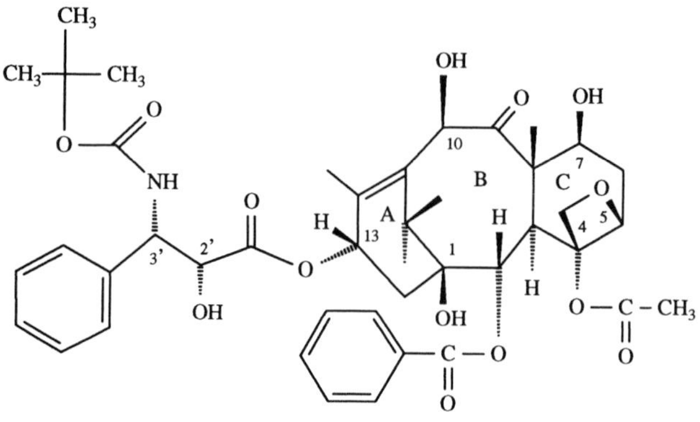

Docetaxel

Triterpenes and steroids

These two groups of isoprenoids are treated together because of their biosynthetic origin. There are many triterpenes found in Nature, but only a few are of medicinal interest. These include the antibiotic fusidic acid, a group of glycosides, saponins, in which the aglycone is a triterpene, and the bitter principle quassin (page 448), which has lost ten carbon atoms by oxidation. Two very important groups of natural products, steroidal hormones and cardiac glycosides, belong to the steroids.

Biosynthesis of triterpenes and steroids

In plants, triterpenoids and steroids are generally synthezised in the endoplasmic reticulum and the IPP required for the primary stage of their biosynthesis is therefore obtained via the mevalonate pathway. Consequently the DMAPP for formation of geranyl diphosphate is obtained via rearrangement of IPP as described for sesquiterpenes (page 400).

It has been shown that the direct precursor for triterpenes and steroids is squalene – a rare C_{30}-hydrocarbon which was first isolated from shark liver but was later found to be ubiquitously distributed. Squalene can be folded in many ways including conformations in which the angular methyl groups and side-chains of cyclic triterpenoids are in the correct positions. Squalene is formed through the coupling of two molecules of farnesyl diphosphate which formally joins carbon atom 1 of one unit to carbon atom 1 of the second unit. However, this coupling is not achieved directly but proceeds via presqualene diphosphate, a compound which includes a cyclopropane structure (Fig. 118). In analogy with the start of the reactions leading to GPP, FPP and GGPP, the first reaction in the biosynthesis of presqualene diphosphate is ionization of one of the two FPP molecules with formation of an allylic cation. This cation then launches an electrophilic attack on the double bond between

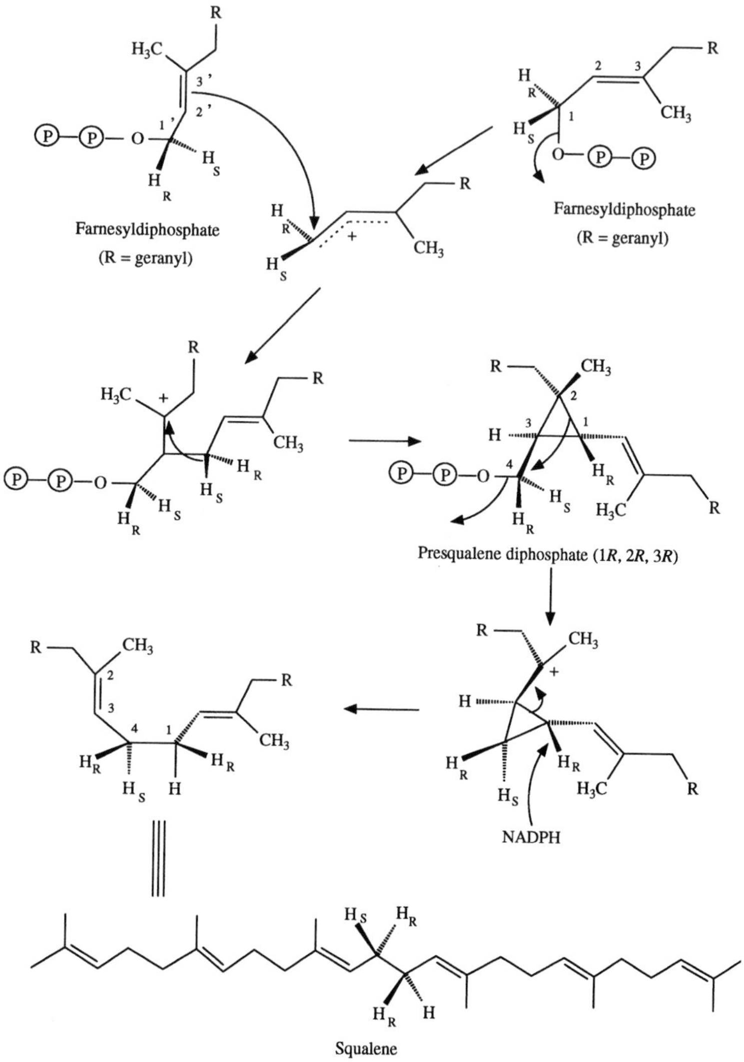

Fig. 118. Biosynthesis of all-trans-squalene

C-2' and C-3' of the second FPP molecule. The alkylation proceeds with inversion of configuration. Stereospecific elimination of one of the hydrogens at C-1 of the first molecule then forms the cyclopropane ring of presqualene diphosphate. Ionization of this molecule with loss of the diphosphate group at C-4 yields a C-2 cation which rearranges under acceptance of a proton from NADPH and inversion of configuration at C-4 to yield the linear all-*trans*-squalene.

The reactions are catalysed by *squalene synthase* (EC 2.5.1.21). This enzyme has been isolated from yeast and is a single 53 kDa protein. The enzyme has been cloned from yeast, from rat liver and from man. cDNAs for the enzyme have been isolated from several plants such as *Arabidopsis thaliana*, *Nicotiana tabacum* and *Glycyrrhiza glabra*. Prokaryotic expression of these cDNAs has yielded the corresponding squalene synthases. cDNA clones for squalene synthase have been isolated from several plants, e.g. *Glycine max* and the monocots *Oryza sativa* and *Zea mays*.

Squalene synthase requires Mg^{2+} for activity and as shown above, reduction of presqualene diphosphate is dependent on NADPH, but NADH can function as well.

Theoretically squalene can fold in a number of ways and following cyclization reactions could yield products with great variations in size and number of rings. However, there are few skeletal variations among the triterpenes. The cyclization is always initiated at the terminal double bond and with few exceptions the first three rings are six-membered (usually denoted A, B and C). Most triterpenes contain either four or five rings. Practically all of them contain a β-hydroxy group at position 3 in the A-ring. This group derives from oxidation of squalene to the epoxide 2,3-oxidosqualene, a reaction mediated by *squalene monooxygenase* (EC 1.14. 99.7, also known as squalene epoxidase) that requires molecular oxygen, NADPH, FAD and NADPH-cytochrome P450-reductase. Squalene epoxidase is the only flavoprotein known to catalyse the epoxidation of an olefin. In humans, animals and higher plants, the product of the oxidation is the (3*S*) enantiomer. Formation of 2,3-oxidosqualene is one of the rate-limiting steps in the pathway to sterols and triterpenes.

The epoxide can assume several different conformations and subsequent cyclization reactions yield different series of triterpenes.

Biosynthesis of pentacyclic triterpenes

Pentacyclic triterpene derivatives are present in many plants especially as aglycones to saponins (page 440). The cyclase catalysing the biosynthesis of these compounds folds the (*S*) 3-oxidosqualene-2,3-epoxide in a chair-chair-chair conformation and the cyclization is initiated by a proton-catalysed opening of the epoxide. The first product formed is the tetracyclic dammarenyl C-20 cation, and the subsequent rearrangements lead to the pentacyclic oleanyl cation via the baccharenyl and lupenyl cationic intermediates. Finally, a series of hydride shifts with elimination of a proton at C-12 gives the α-amyrin or β-amyrin framework with Δ^{12} double bond. It has been verified that elimination of the α-proton at C-12 is involved in formation of β-amyrin while α-amyrin is formed with elimination of the β-positioned proton. The enzymes catalysing these reactions are known only to a limited extent. *2,3-Oxido-β-amyrin cyclase* has been isolated from peas.

α-Amyrin and β-amyrin represent two of the three different groups of pentacyclic triterpene skeletons. The third group contains four six-membered rings and one five-membered ring and is derived from lupeol which is formed from the lupenyl cation by loss of a

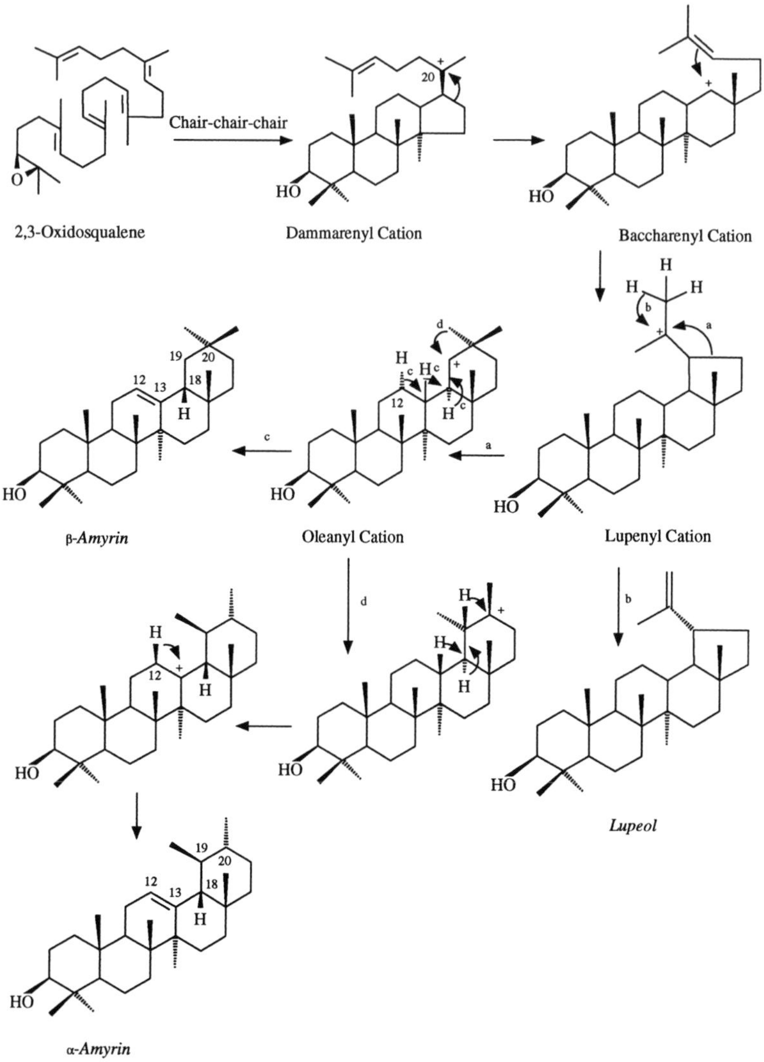

Fig 119. Biosynthesis of pentacyclic triterpenes from oxidosqualene

proton from the side-chain. The different triterpenes arise by further reactions including oxidation to form hydroxyl and carbonyl groups, backbone rearrangements by Wagner–Meerwein shifts, dehydrogenations and side-chain alkylations.

Most tetracyclic triterpene derivatives are biosynthesized from *cycloartenol* (see below) but some are derived from the dammarenyl C-20 cation. One example is the aglycones of the ginsenosides (page 446).

Biosynthesis of tetracyclic triterpenes and steroids

For the biosynthesis of these derivatives the catalysing enzyme folds (3S)-2,3-oxidosqualene in a chair-boat-chair-boat-endo-conformation. The epoxide is subsequently opened under catalytic influence by a protonated group on the enzyme. Fourfold cyclization then yields the protosterol carbonium ion I with the side-chain at C-17 in the β position. Elimination of H at C-9 forms the double

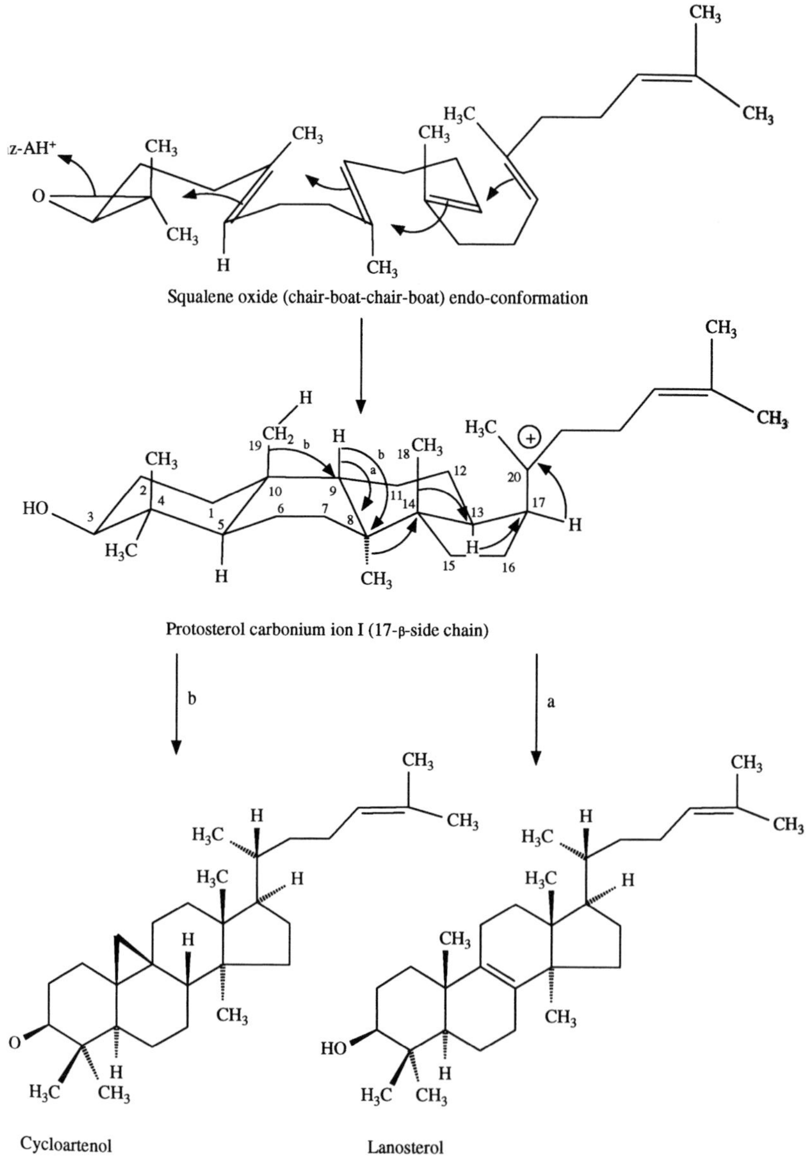

Squalene oxide (chair-boat-chair-boat) endo-conformation

Protosterol carbonium ion I (17-β-side chain)

Cycloartenol

Lanosterol

Fig.120. Biosynthesis of lanosterol and cycloartenol from oxidosqualene

bond C-8–C-9 in lanosterol (route a in Fig. 120) while in the formation of cycloartenol C-9 H is shifted to C-8 (with retention of the β-configuration) and a new bond is formed between C-19 and C-9 (route b in Fig. 120). Further shifts are the same in the formation of lanosterol and cycloartenol. The α-methyl group attached to C-8 is shifted to C-14 and at the same time the β-methyl at C-14 is moved to C-13. Both shifts occur with retention of the configurations of the groups. The α-H at C-13 is shifted to C-17 and β-H at C-17 is moved to C-20. In both shifts the configurations of the shifted atoms are retained.

The enzymes mediating these reactions have been difficult to isolate because they are membrane-bound and have low stability and solubility. The breakthrough in purification of the oxidosqualene cyclases was first achieved by the purification of several plant triterpene cyclases in 1988–1992.

Lanosterol synthase (EC 5.4.99.7, also known as oxidosqualene: lanosterol cyclase or sterolcyclase) which is responsible for the biosynthesis of lanosterol, was isolated for the first time from rat liver in 1991. It has a molecular mass of 75 kDa and an isoelectric point of 5.5. This enzyme has been the subject of extensive studies of the mechanisms involved.

Cycloartenol synthase (EC 5.4.99.8, also known as cycloartenol cyclase) has *i.a.* been isolated from pea seedlings (*Pisum sativum* L.). It has a molecular mass of 55 kDa.

Lanosterol and cycloartenol are precursors for steroids which are natural products of great medicinal importance. The group comprises hormones, vitamins, saponins and cardiac glycosides (page 459). The steroids are formed from the triterpene precursors by the loss of three methyl substituents from the skeleton and of a shorter or longer portion of the side-chain. Further reactions such as oxidation, dehydrogenation and rearrangements form the individual steroids.

Steroids from mammals and fungi such as yeast are derived from lanosterol, while plant steroids are formed from cycloartenol. Cycloartenol is also the main precursor of tetracyclic triterpenes.

Triterpenes

This tetracyclic triterpene has lost carbon 30 and has thus only 29 carbon atoms. The β-positioned hydrogen at C-9, the α-positioned methyl group at C-8 and the β-positioned methyl group at C-14 are unusual structural features.

Fusidic acid Fusidic acid is an antibiotic produced by the fungus *Acremonium fusidioides* (formerly *Fusidium coccineum)* and is isolated from culture filtrates of this organism. It is used for the treatment of infections caused by staphylococci and it is also active against strains which have developed resistance to penicillin. Its mode of action is inhibition of the ribosomal translocation process during bacterial protein synthesis.

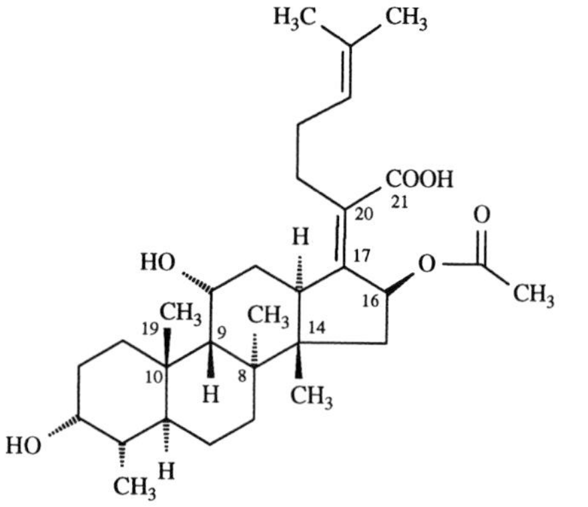

Fusidic acid

Saponins

The term saponin is applied to a group of glycosides which have the ability to lower the surface tension of aqueous solutions and to bring about haemolysis of red blood corpuscles. The word saponin refers to the first property: an aqueous saponin solution foams just like a soap solution. Saponins are used technically to decrease surface tension, e.g. as emulsifiers and in fire extinguishers.

The most important biological property of saponins is their ability to cause haemolysis, a process in which changes in the walls of the red blood corpuscles allow the haemoglobin to pass into the surrounding fluid. This is easily demonstrated: a suspension of red blood corpuscles in physiological saline is cloudy, but if a saponin is added, it becomes transparent and bright red. This effect can be used to assess saponin-containing plant extracts, by determining the lowest concentration required for the complete haemolysis of a suspension with a known content of red blood corpuscles. The result is reported as the *haemolytic index.* If the lowest concentration of a saponin drug extract needed for complete haemolysis is obtained from 1 g of the drug in 100 000 ml of water, the haemolytic index of the drug is said to be 1:100 000. Two groups of saponins can be distinguished, differing in the aglycone moiety: *triterpenoid saponins* and *steroidal saponins,* the aglycone being a triterpene and a steroid, respectively. These types behave somewhat differently in the haemolysis reaction. Steroidal saponins cause rapid haemolysis, whereas triterpenoid saponins give a slower effect, which makes it more difficult to determine the concentration limit that defines the haemolytic index.

In man and other warm-blooded animals, saponins are absorbed by the intestines only to a small extent. For this reason, they are normally not very toxic on oral administration. In contrast, they are very poisonous to fish. Saponin-containing plants are often used for fishing in developing countries. Part of a stream is dammed and crushed material of a saponin-containing plant is thrown into the water. The fish die, rise to the surface, and can easily be caught. They are perfectly fit for human consumption, as saponins are not toxic to man.

A great number of *in vitro* and *in vivo* pharmacological studies have been performed on saponins. Antitumour, chemopreventive, antiphlogistic, immunomodulating, antihepatotoxic, antiviral, hypoglycaemic, antifungal and molluscicidal activities have been found. Saponins have also shown activity on the cardiovascular system, the central nervous system and on the endocrinal system. However, very few clinical studies have been carried out to promote saponins as drugs. This may be due to the high toxicity of most of the compounds when given intravenously to higher animals, although the toxicity – as mentioned above – is very low when they are given orally.

As mentioned before, the aglycone of a saponin can be either a triterpene or a steroid. Saponins have been found to contain up to 12 sugar residues (monosaccharides or uronic acids). The numerous sugar residues make the saponins highly polar and hence difficult to isolate and purify.

Triterpenoid saponins will be treated here, while steroidal ones, which are important raw materials for the partial synthesis of steroidal hormones, will be discussed later, together with the steroids (page 449).

Triterpenoid saponins

Most aglycones of triterpenoid saponins are pentacyclic compounds derived from one of the three basic structures represented by α-amyrin, β-amyrin and lupeol (page 437). The β-amyrine type is the most common one and among the substituents present are hydroxyl, ketone and carboxyl functions. The saponins can be either *monodesmosides* with sugar residues attached to one OH-group, usually at C-3, or *bisdesmosides* with sugar residues connected to two carbon atoms, usually C-3 and C-28, e.g. as in quillaja saponin (page 443). Generally bisdesmosides have a weak haemolytic activity and their foam-forming ability is usually also weak. Most triterpenoid saponins contain at least one carboxyl group.

Crude drugs containing triterpenoid saponins are used as expectorants. The saponin stimulates the secretion of a thin mucus, which facilitates the expectoration and discharge of phlegm in infections of the respiratory organs. As the saponin is not absorbed, the effect is presumed to be due to irritation of the mucous lining of the stomach, which, by reflex action, causes a general increase in glandular secretion, especially of the bronchi. Another suggested explanation is that the saponin, when taken, comes into direct contact with the outermost layer of phlegm and through its ability to lower the surface tension makes the phlegm less viscous and easier to discharge. The value of saponin-containing drugs as expectorants has been questioned.

Glycyrrhizin Glycyrrhizin is the name of the potassium and/or calcium salt of glycyrrhizic acid, (also called glycyrrhizinic acid) a triterpenoid saponin occurring in the root of *Glycyrrhiza glabra*. Its constituent parts are the aglycone glycyrrhetic acid (= glycyrrhetinic acid), an α-D-glucuronic acid residue which is attached to the aglycone, and a β-D-glucuronic acid residue which is attached to C-2 of the first glucuronic acid residue.

Glycyrrhizinic acid: R =

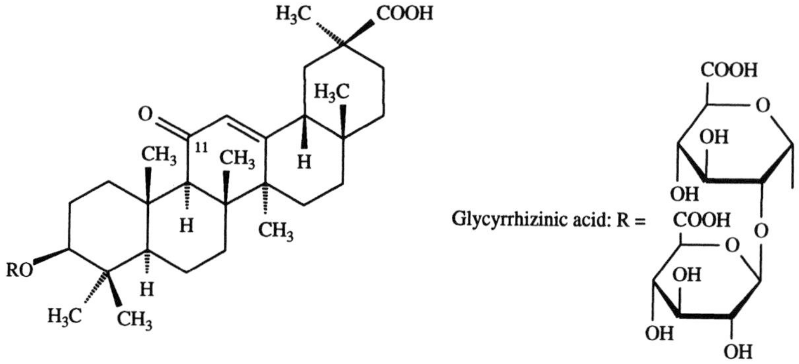

Glycyrrhetic acid: R = H

Crude drug   **Liquorice root** (Liquiritiae radix) and **liquorice extract** (Glycyrrhizae extractum crudum, Succus liquiritiae) are obtained from two species of the genus *Glycyrrhiza* (*Fabaceae*). Liquorice root comprises both roots and underground stems (stolons). These plants are perennial herbs, 1–2 metres high, growing naturally in southern and central Europe (*Glycyrrhiza glabra* L.), in central and southern Russia (*Glycyrrhiza glandulifera* Waldst. & Kit.), and in Iran and Iraq (*Glycyrrhiza glabra* L., sometimes in this respect known as *Glycyrrhiza violacea* Boiss). *Glycyrrhiza glabra* is cultivated in Spain, Italy, France, Germany, England and the United States. The crude drug from Russia and the Middle East is derived from plants growing in the wild. The roots and stolons are harvested when the plant is 3–4 years old.

Liquorice extract is largely prepared in southern Italy. Liquorice root is extracted with water and the extract is filtered and concentrated in a vacuum to a viscous mass, which is cast into blocks or sticks and gradually solidifies. Liquorice extract is brownish black and has a shiny fracture and a sweet taste.

The taste of glycyrrhizin is some 50 times sweeter than that of sucrose. The content of glycyrrhizin is 6–13 % in the root and 20–25 % in the extract. Extracts that have not been concentrated in a vacuum are also available. Their glycyrrhizin content is 10–15 %. Liquorice extract is used as an expectorant in cough syrups. It is also used for the treatment of stomach complaints. Because of its sweet taste, liquorice extract is employed in the confectionery industry. When taken in large amounts, liquorice extract has been shown to affect the carbohydrate and mineral metabolism of the body, resulting in sodium and water retention, hypertension, hypokalaemia and suppression of the renin–aldosterone system. The aglycone glycyrrhetic acid has a keto group at position 11, a structural feature shown also by the adrenocortical hormone cortisone. For years it was thought that liquorice produces these effects through the binding of glycyrrhizic acid and its aglucone to mineralocorticoid receptors. It has now been shown, however, that the affinity of glycyrrhetic acid for mineralocorticoid receptors is only 0.01 % of that of aldosterone. Furthermore, liquorice or glycyrrhetic acid does not have mineralocorticoid effects in patients with Addison's disease or in adrenalectomized rats unless accompanied by administration of cortisone or hydrocortisone. The effects of liquorice on carbohydrate and mineral metabolism can instead be explained by the action of glycyrrhetic acid and glycyrrhizic acid as inhibitors of important steroid-metabolizing enzymes. Glycyrrhetic acid is a potent inhibitor of the enzyme *11β-hydroxysteroid dehydrogenase* (11β-OHSD, EC 1.1.1.146), which converts cortisol to cortisone (for structures of the steroids see pages 449-450). Cortisol has the same binding affinity for mineralocorticoid receptors as aldosterone, while that of cortisone is much less. 11β-OHSD is found in high concentrations in the kidney, which is an aldosterone-responsive tissue. By inhibiting the enzyme in this organ, liquorice produces high renal levels of cortisol which then binds to and activates mineralocorticoid receptors thus promoting sodium reabsorption. Glycyrrhetic acid is also a potent inhibitor of *Δ⁴-3-oxosteroid 5β-reductase* (EC 1.3.1.3), an enzyme which inactivates both glucocorticoids and mineralcorticoids by reduction of the steroid ring A. Liquorice derivatives thus reroute the metabolism of aldosterone, deoxycorticosterone and glucocorticoids resulting in the accumulation of unmetabolized hormones and their corresponding 5α-dihydro derivatives. The amount of liquorice confectionery which can be consumed daily without untoward effects depends on the amount of liquorice extract present and the content of glycyrrhetic and glycyrrhizic acid in the extract. The acceptable daily intake (ADI) for glycyrrhizin has been proposed to be 0.2 mg/kg/day, which for a person weighing 60 kg amounts to 12 mg, approximately corresponding to 60 mg of liquorice extract. It should be noted that the content of glycyrrhizin in confectionary is highly variable.

Besides having effects on carbohydrate and mineral metabolism, liquorice extract can also interfere with the metabolism of conventional drugs, causing increased or decreased pharmacological activity of these. This is due to interaction by glycyrrhizin and possibly also other components in the extract with cytochrome P450 *iso*-enzymes. Patients should therefore inform their doctors of regular use of liquorice preparations, taken either as herbal remedies or as confectionary.

Quillaja saponin   The bark of *Quillaja saponaria* contains a mixture of saponins consisting of a triterpene aglycone with a number of sugar residues attached at C-3 and C-28. A complex acyl moiety is also present on one of the sugar residues. The complete structure of a component from the mixture is as shown.

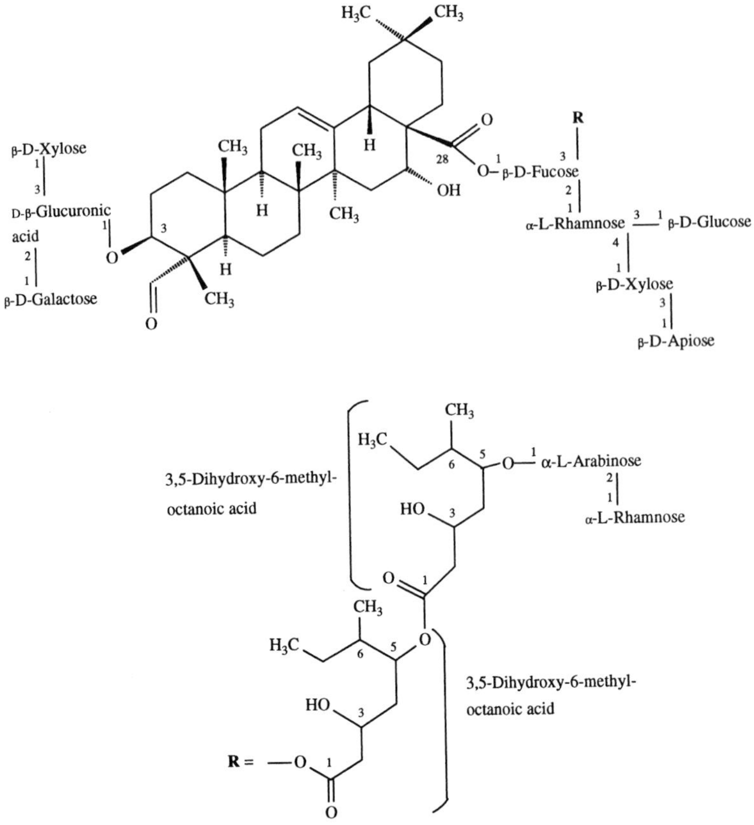

β-D-Xylose

D-β-Glucuronic acid

β-D-Galactose

O–β-D-Fucose

α-L-Rhamnose — β-D-Glucose

β-D-Xylose

β-D-Apiose

3,5-Dihydroxy-6-methyl-octanoic acid

O— α-L-Arabinose

α-L-Rhamnose

3,5-Dihydroxy-6-methyl-octanoic acid

R =

A major component of quillaja saponin

Crude drug **Quillaja bark** (Quillajae cortex) is the cork-free bark of *Quillaja saponaria* Molina (*Rosaceae*). This is a tree, about 18 metres high, native to Chile, Peru and Bolivia. The drug consists of the inner parts of the bark, which are generally cut before marketing. It has an acrid taste and causes sneezing. Quillaja bark contains about 10 % saponins and an extract of it is used as an expectorant. The drug also finds technical application as a mild detergent for washing fine textiles.

A saponin preparation called Quil A is used as an adjuvant in veterinary vaccines, particularly those against foot-and-mouth disease. Quil A augments antibody responses and induces Ag-specific helper T-lymphocyte memory. The toxicity of the saponins can be reduced by incorporation in a so-called immunostimulating complex (ISCOM) containing quillaja saponins, cholesterol, antigens and phospholipids. This complex forms cage-like structures with a diameter of 35 nm. The use of ISCOMs as adjuvants has been successfully applied in experimental vaccines against influenza, herpes simplex, respiratory syncytial virus and HIV. The saponins in the ISCOMs activate cytotoxic T-cells or T-helper cells (Th1- and Th2-

Senegin II

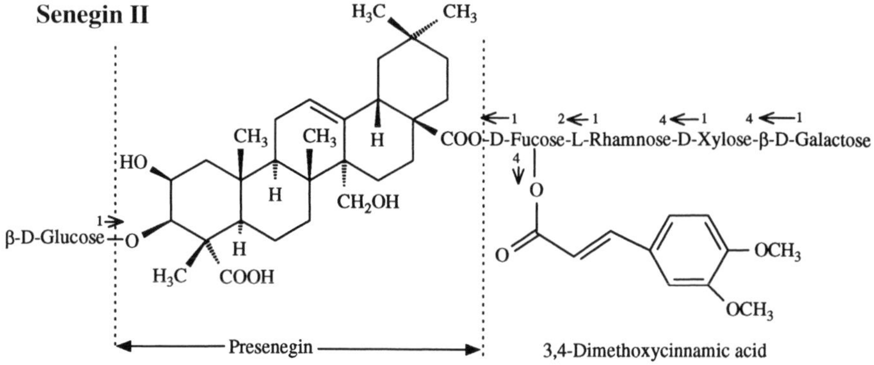

Senegin II

cells) which results in secretion of interleukin 5 and γ-interferon. These cytokines stimulate formation of specific antibodies (IgG_1, $IgG_{2'a}$).

Polygala senega contains several saponins related to senegin II, the major component of the mixture present in the var. *latifolia*. They all have glucose attached at C-3, but they differ in the oligosaccharide chain that is ester-bound at C-28. It is also worth noting that the fucose residue in this chain is esterfied with 3,4-dimethoxycinnamic acid.

Crude drug **Senega root** (Polygalae radix, Senegae radix) is the dried root and rhizome of *Polygala senega* L. var. *senega* and/or var. *latifolia* Torrey & A. Gray (*Polygalaceae*). The plants are perennial herbs, growing in the United States and Canada and also cultivated in Japan. Senega root has a typical smell due to its methyl-salicylate content. It contains about 10 % triterpenoid saponins, the aglycone of which is presenegenin.

Senega-root extract is used as an expectorant.

Aescin Aescin (escin) is a mixture of triterpenoid saponins present in the seed cotyledons of the horse chestnut *Aesculus hippocastanum* L. (*Sapindaceae*) (Plate 2 page 211). The saponins are derived from two aglycones – protoaescigenin and barringtogenol C – which are members of the β-amyrin series of triterpenes (page 436). Both aglycones are polyhydroxylated. The hydroxyl group at C-21 is esterified with various low-molecular-weight aliphatic acids, while that at C-22 forms an ester with acetic acid. The sugar part is attached to C-3 and consists of a trisaccharide containing D-glucuronic acid, D-glucose and a third sugar which can be either glucose, galactose or xylose. D-Glucuronic acid forms the glucoside link.

Extracts of horse chestnut seeds, standardized with respect to the content of aescin (16–20 %), are used for treatment of chronic venous insufficiency which is a common circulation problem, caused by veins that do not function properly. The symptoms are

445

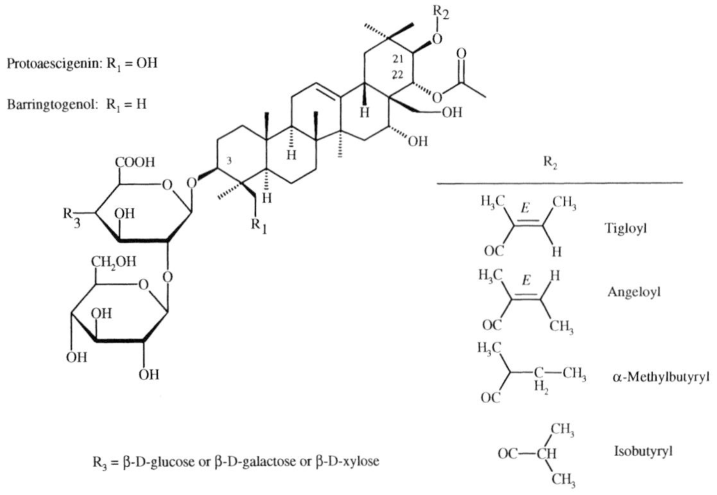

Protoaescigenin: R_1 = OH

Barringtogenol: R_1 = H

COOH

R_3
OH

R_1

CH$_2$OH

OH

OH

OH

R_3 = β-D-glucose or β-D-galactose or β-D-xylose

R_2

H$_3$C, E, CH$_3$ / OC, H	Tigloyl
H$_3$C, E, H / OC, CH$_3$	Angeloyl
H$_3$C / C—CH$_3$, H$_2$ / OC	α-Methylbutyryl
OC—CH, CH$_3$, CH$_3$	Isobutyryl

swelling, pain, tension and hardening of the skin, particularly in legs. This disease is generally treated by application of compression stockings which is uncomfortable. Randomized, double-blind clinical trials have indicated that treatment with standardized extracts of horse chestnut seeds might be an effective treatment of chronic venous insuffiency (see the Cochrane Review by Pittler and Ernst cited in *Further Reading* page 485). Aescin inhibits the activity of *elastase* and *hyaluronidase in vitro*. Both enzymes are involved in proteoglycan degradation. An important pathophysiological mechanism of chronic venous insuffiency is considered to be the accumulation of white blood cells (leucocytes) in the affected limbs and subsequent activation and release of elastase and hyaluronidase. Aescin is thought to shift the equilibrium between degradation and synthesis of proteoglycans towards a net synthesis, thus preventing vascular leakage.

Ginsenosides
A series of tetracyclic triterpenoid saponins, ginsenosides R_b–R_h has been isolated from the root of *Aralia quinquefolia* Decne & Planch. (formerly known as *Panax ginseng* C. A. Mey.) (*Araliaceae*). R_b–R_d are derivatives of 20(S)-protopanaxadiol, and R_e–R are derivatives of 20(S)-protopanaxatriol, which biosynthetically are derived from the dammarenyl cation (page 437) by an electrophilic attack on water forming the 20(S)-hydroxyl group. Theoretically both 20(S)- and 20(R)-OH derivatives can be formed but only the 20(S)-derivative is biologically active.

The ginsenosides differ in the location and nature of the sugar moieties. It is noteworthy that in the triol-group the sugars are linked via the hydroxyls at C-6 and C-20. In the diol-group the more common linkage at C-3 is represented besides the linkage at C-20 as in the triol-group. About 30 ginsenosides have been isolated. The most important are R_{b1}, R_{b2}, R_c, R_d, R_e, R_f, Rg_1 and R_{g2}. Most abundant are R_{b1}, R_{b2}, R_e and R_{g1}. The main component of the diol-group

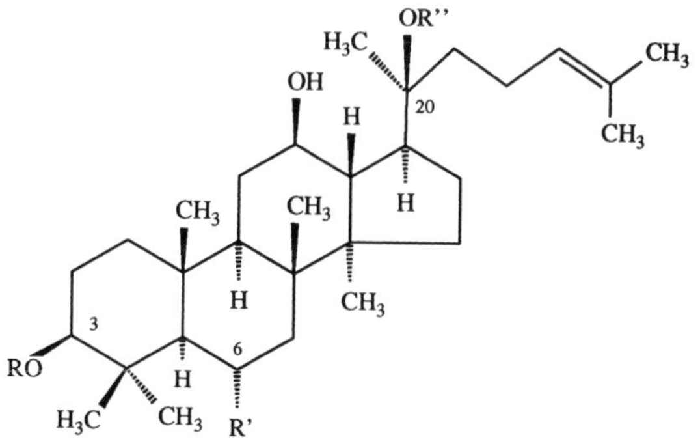

OR''
H₃C,
OH
H
20
CH₃
CH₃
CH₃
CH₃
H
H
CH₃
H
H
3
6
RO
H
H₃C
CH₃
R'

	R	R'	R''
20(S)-protopanaxadiol	H	H	H
Ginsenoside R$_{b1}$	β-D-Glc(1→2)-β-D-Glc	H	β-D-Glc(1→6)-β-D-Glc
Ginsenoside R$_c$	β-D-Glc(1→2)-β-D-Glc	H	α-L-Araf(1→6)-β-D-Glc
20(S)-protopanaxatriol	H	OH	H
Ginsenoside R$_e$	H	O-α-L-Rha(1→2)-β-D-Glc	β-D-Glc
Ginsenoside R$_{g1}$	H	O-β-D-Glc	β-D-Glc

Araf = Arabinofuranose. Glc = Glucopyranose. Rha = Rhamnopyranose.

is R$_{b1}$, and R$_{g1}$ is the principal representative of the triol-group. Also R$_c$ and R$_e$ do occur in appreciable amounts.

Crude drug **Ginseng** (Ginseng radix) is the root of *Aralia quinquefolia* Decne & Planch. (formerly known as *Panax ginseng* C. A. Mey.) (*Araliaceae*). The plant originates from the highlands of the Chinese region of Manchuria and Korea. It is an about 50 cm tall bush with a crown of dark green verticillate leaves and small green flowers which develop into clusters of bright red berries. Most of the crude drug is obtained from South Korea where it is cultivated under thatched covers. The roots are harvested when the plant is 4–6 years old. They are then branched, 8–20 cm long and about 2 cm thick. Thin parts of the tap-root and the branches are cut off, the root is washed and the outer layer removed. Sun-drying yields white ginseng whereas red ginseng is obtained by first steaming the root, followed by artificial drying and then sun-drying, yielding a red product with a glassy surface.

Besides the ginsenosides about 170 other compounds have been isolated including polysaccharides, glycans, lipids, fatty acids and volatile compounds.

Ginseng is an ancient Chinese medicine, employed as a kind of panacea. In Western countries white ginseng is a herbal remedy which is used as a so-called *adaptogen*, a Russian term which is defined as an agent that increases resistance to physical, chemical and biological stress and builds up general vitality, including the physical and mental capacity for work (see page 226 for more information on adaptogens). This effect sets in slowly, and several weeks' treatment is needed to obtain the full benefit of the drug. It is generally considered that the ginsenosides are responsible for the effects and many investigations of their pharmacological properties have been published. Generally, however, these studies have been performed with concentrations of the ginsenosides, much higher than those that can be expected in the body at normal dosage levels of ginseng extracts. Several clinical investigations appear to support the concept of adaptogenic activity and the opinion that the drug can be of value especially in geriatric practice. Control of the crude drug should comprise determination of the ginsenoside content. A minimum of 1.5 % of total ginsenosides, calculated as ginsenoside Rg_1 is required.

American ginseng is also the root of *Aralia quinquefolia* Decne & Planch.) (*Araliaceae*) but grown in United States and Canada. It was previously regarded as a different species – *Panax quinquefolium* L. – but now the American plant is considered to be the same species as that grown in Asia. An estimated 90 % of the USA-cultivated drug is produced by growers in north-central Wisconsin American ginseng is not used in Western countries but is exported to Asia via Hong Kong. The drug contains the same ginsenosides as the root grown in Asia with the exception of Rb_2, which is missing The content of Rb_1 is about twice as high.

Modified triterpenes

Quassin Quassin is an intensely bitter substance (bitterness value: 17 000 000) found in the wood of *Picraena* and *Quassia* species. Quassin and other quassonoids from the *Simaroubaceae* family have attracted attention because of their cytotoxic, antimalarial and amoebocidal properties. Biosynthetically, quassin is derived from a triterpene precursor – triterpene A – by oxidative removal of the side-chain and loss of a methyl substituent from ring A. Other reactions involving oxidation and methylation, give rise to the keto and methoxy groups.

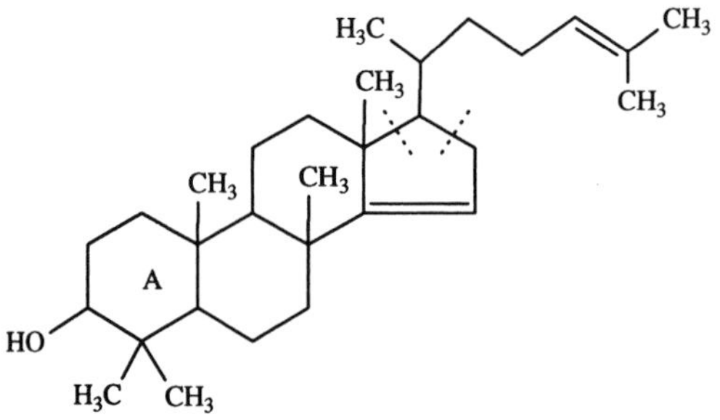

Triterpene A

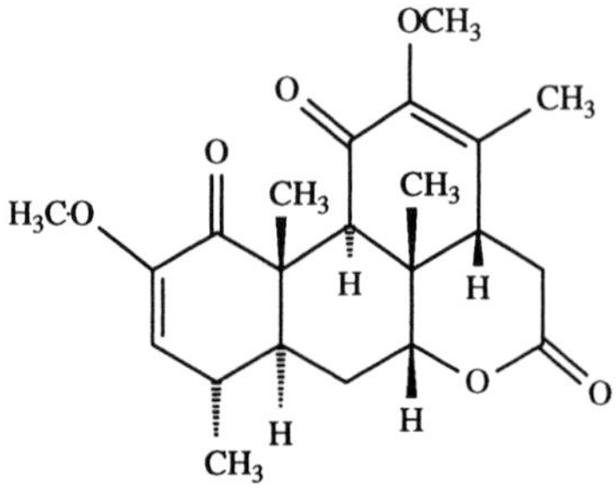

Quassin

Crude drug   **Quassia wood** (Quassiae lignum) is the wood of *Quassia amara* L. or *Picraena excelsa* Lindl. (previously known as *Picrasma excelsa* (Schwartz.) Planch.) (*Simaroubaceae*). *Q. amara* is a small tree, only 2–3 metres in height, occurring in Venezuela, northern Brazil and Guyana. *P. excelsa* is much taller, up to 25 metres, growing in the West Indies, especially on Jamaica. The crude drug Quassiae lignum is usually marketed in the form of chips. It was formerly used as a bitter (page 388), bitterness value 40 000, because of its content of quassin. Today an aqueous extract of the drug or isolated quassin is used as an insecticide which is non-toxic to man.

Steroidal hormones

Two classes of steroidal hormones can be distinguished: sex hormones and adrenocortical hormones. Sex hormones are produced in the gonads and control the growth, development and function of the sexual organs. Closely connected with this group of steroids are synthetic contraceptives (the "Pill"). As their name indicates, adrenocortical hormones are produced in the cortex of the adrenal glands. These hormones regulate protein and carbohydrate metabolism, as well as the salt and water balance in the body. The structures of the most important steroidal hormones are given here and on the following page to facilitate the dicussion.

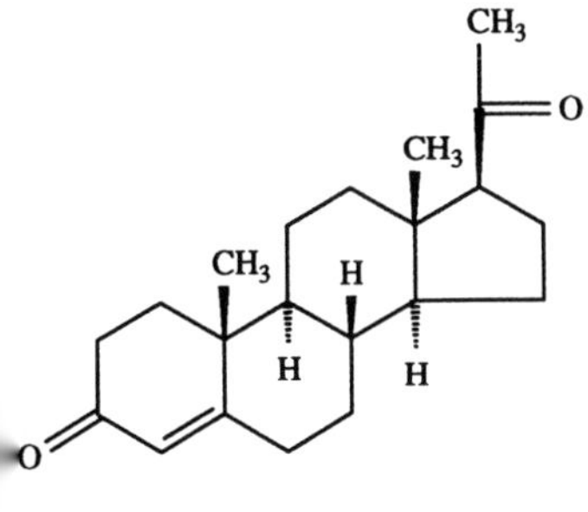

Progesterone

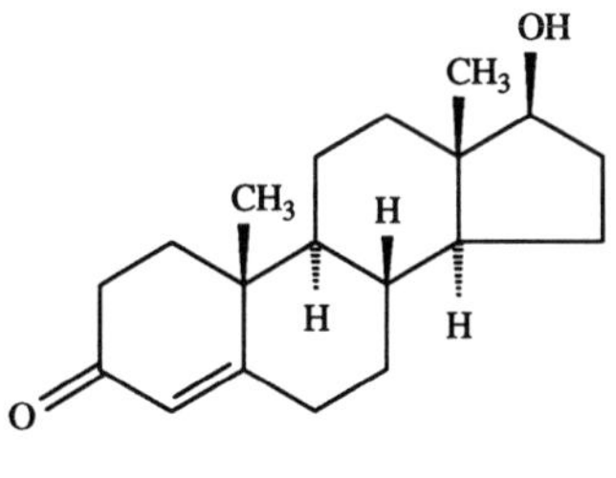

Testosterone

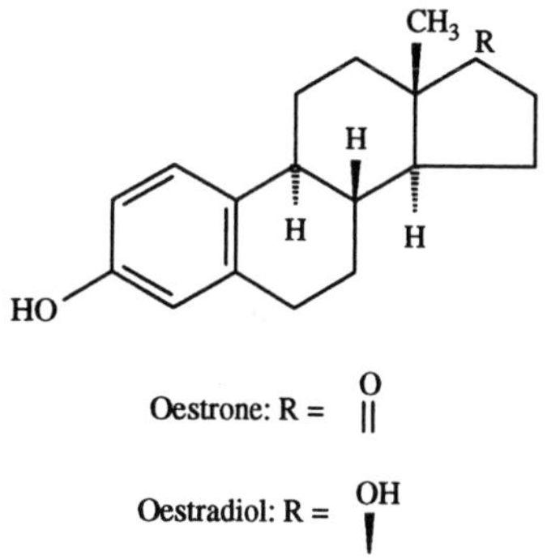

Oestrone: R = $\overset{O}{\underset{||}{}}$

Oestradiol: R = $\overset{OH}{\underset{|}{}}$

Development of the steroid industry

The first steroidal hormone to be isolated was the sex hormone oestrone. It was obtained in 1929 from the urine of pregnant women. The compound was isolated by two independent groups, one American the other German. The isolation of progesterone from pigs' ovaries was reported in 1934 by four different groups, two in Germany, one in Switzerland, and one in the United States. In 1936, cortisone was isolated by the Americans C.S. Myers and E.C. Kendall and by the Swiss T. Reichstein. The quantities of steroidal hormones obtained were, however, exceedingly small and not sufficient for detailed pharmacological studies and clinical testing. The following examples may be mentioned: ovaries from 50 000 sows were needed for the isolation of 20 mg of proges-

Contraceptive

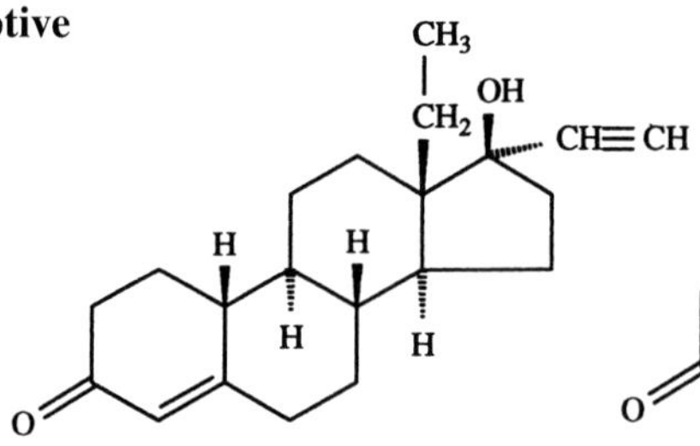

Norgestrel

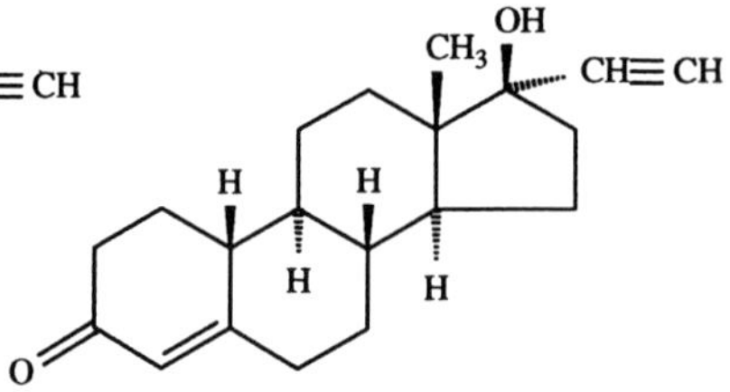

Norethisterone

Glucocorticoids

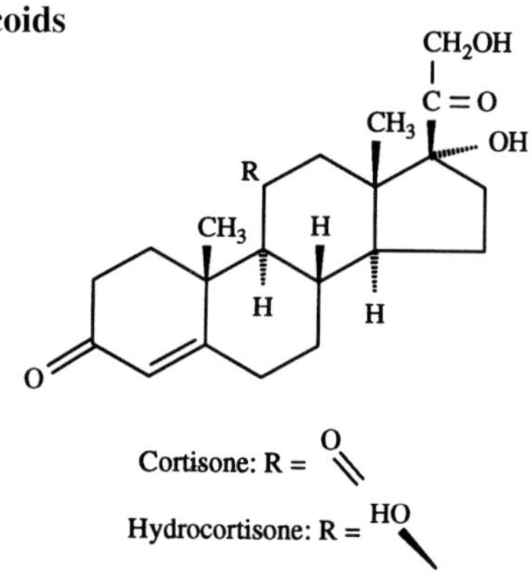

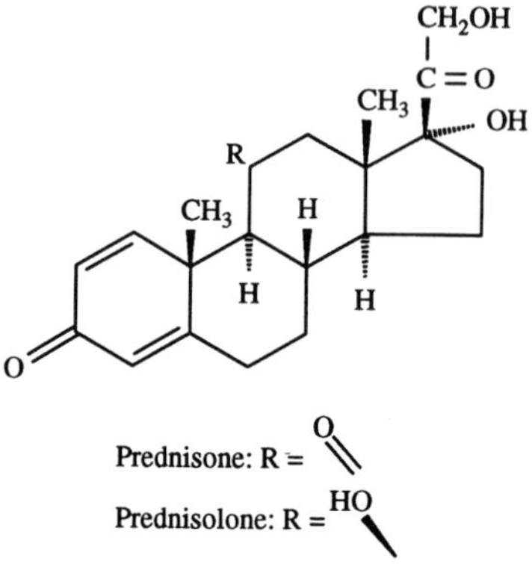

Cortisone: R =

Hydrocortisone: R =

Prednisone: R =

Prednisolone: R =

Mineralocorticoids

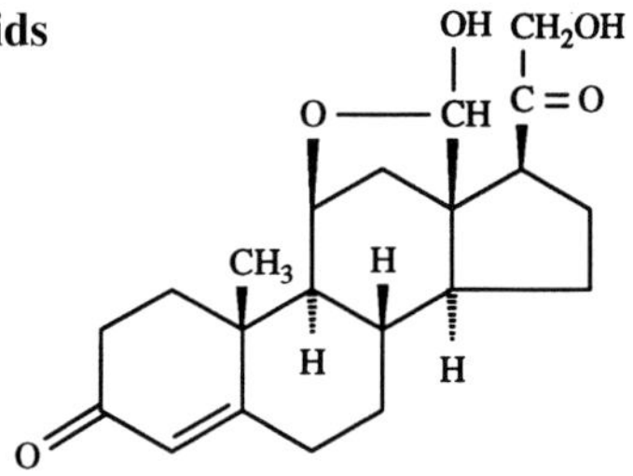

Aldosterone

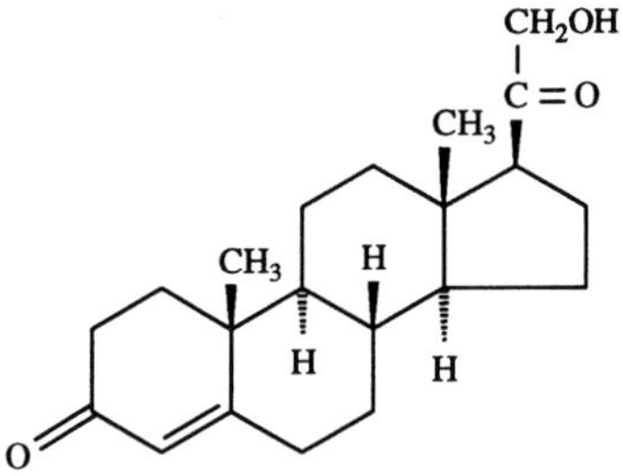

Deoxycorticosterone

**Aldosterone antagonist
(diuretic, antihypertensive)**

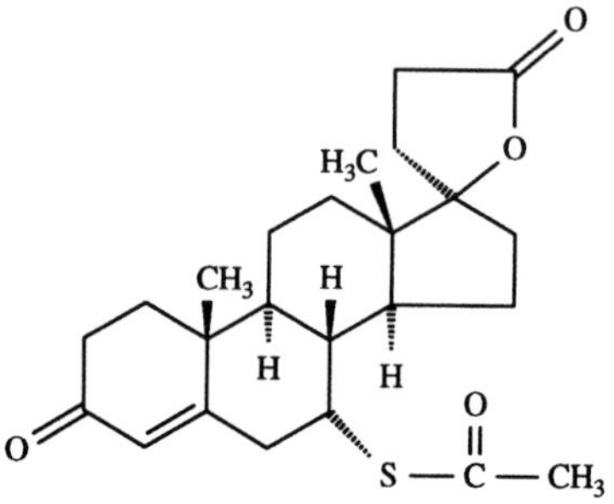

Spironolactone

terone, and adrenal glands from 20 000 head of cattle gave 200 mg of cortisone.

It took about 10 years from the first isolation of cortisone until it became available in larger quantities through partial synthesis starting with deoxycholic acid. In 1949, it was discovered in the United States that cortisone was a dramatic remedy for rheumatoid arthritis and this led to a great demand for cortisone, which could not be met by the synthetic processes then available. New, suitable starting materials were sought. One of the difficult stages in the cortisone synthesis was the introduction of the keto group at position 11 in ring C. A scrutiny of the steroid structures known at that time showed that *sarmentogenin*, an aglycone from a cardiac glycoside isolated from the seed of an African liana, *Strophanthus sarmentosus* DC. (*Apocynaceae*), had a hydroxyl group in this position and might, therefore, be a suitable starting material for the partial synthesis of cortisone. Several expeditions were sent to Africa to obtain supplies of this raw material, but the plant was soon found to be useless as a starting material for the industrial production of steroids, because the glycoside content varied too much and there were chemical races within the species.

Sarmentogenin

An American scientist, *Russel E. Marker*, made one of the most important contributions to the development of the steroid industry. He was a professor at Pennsylvania State College from 1935 to 1943 and his studies on steroid chemistry during that period resulted in 147 scientific reports and some 70 patents. During this work Marker became convinced that steroidal saponins would be better starting substances for the partial synthesis of steroidal hormones than the compounds that had been used so far, e.g. deoxycholic acid and cholesterol. In 1940, Marker had succeeded in determining the structure of the sapogenin ring system attached at positions 16 and 17 of the steroid nucleus (for numbering of the steroid skeleton see page 459), and he worked out an efficient method for degrading this ring system and transforming the steroidal sapogenin into progesterone. He also found that plants producing suitable saponins in large quantities grew in Mexico. He was unable, however, to interest any American pharmaceutical firm in using this method, and in 1944 he left his position at Pennsylvania State College for Mexico City. Here, he began to collaborate with the Mexican firm Hormona Laboratories in developing a process for producing progesterone from diosgenin. Later the Syntex company was formed to exploit the technology and the method proved to be so good that in the first year of production several kilograms of progesterone were obtained. A year or two later, Marker left the company after a disagreement. This was a serious set-back for Syntex, because Marker himself had carried out all the key operations in the production and had not left any notes behind. However, the problem was soon solved by the Hungarian-born chemist *George Rosenkranz* and Syntex has now become one of the world's largest producers of steroidal hormones. Methods were also found for transforming the progesterone originally produced into other steroidal hormones including cortisone. The access to a relatively cheap starting material soon caused drastic price reductions, as illustrated by the changing price of progesterone. In 1944, before Marker introduced his process, progesterone cost 80 dollars per gram. In the early 1950s the price had fallen to 48 cents per gram.

Raw materials for the partial synthesis of steroidal hormones

Diosgenin

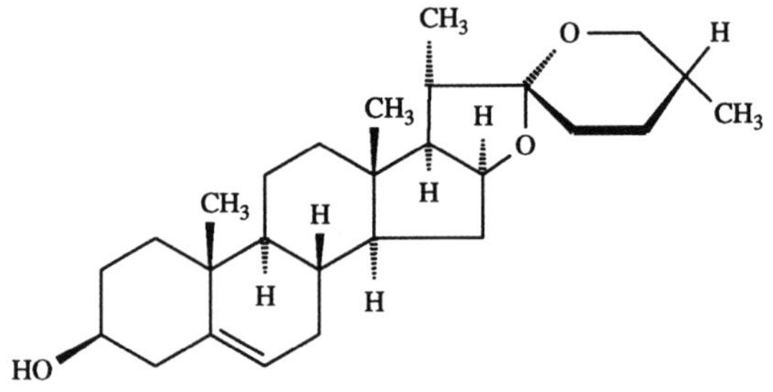

Diosgenin

Mexican diosgenin was up to 1970 the dominant raw material for the synthesis of steroids. Since then other sources of diosgenin and other steroidal raw materials have become available. As a consequence, the importance of the Mexican product has decreased.

Diosgenin is a saponin aglycone, which can be obtained from the roots of several species of plants belonging to the genus *Dioscorea* (*Dioscoreaceae*) (Plates 12, page 429 and 13, page 455). The most important Mexican species used are *Dioscorea composita* Hemsl. and *D. mexicana* Scheidw. In India, diosgenin is obtained from *D. deltoidea* Wall. and *D. floribunda* Mart. & Gal. while in China, *D. nipponica* Makino is used. *Dioscorea* roots from 4-year-old plants contain about 5 % diosgenin and older roots have even more. The various *Dioscorea* species grow in tropical climates with alternating rainy and dry seasons. They are climbers and have cordate leaves with a drip tip, from which the rain runs off. The roots are thick, food-storage organs. Some species, e.g. *D. divaricata* Blanco (previously known as *D. batatas* Decne), are important sources of food (yams) because of their content of starch.

In Mexico, the *Dioscorea* roots are collected from wild plants. When they are dug up they get broken and some pieces are left in the soil and develop into new plants which can be harvested 4–5 years later. Cultivation of the plants has been tried but is unprofitable, partly because the plants can only be harvested every 4–5 years. (In India, however, the cultivation of *D. floribunda* and harvesting the roots every second year has proved to be economically successful.) The roots collected are transported to depots, where they are crushed in a hammer-mill. The product is then left to ferment in tanks for about two days. The fermented mass is air-dried and sent to the steroid factory for acid hydrolysis and extraction of the diosgenin. Fermented roots give a higher yield of diosgenin than unfermented ones.

The supply of diosgenin from *Dioscorea* roots cannot satisfy the demands of the ever-growing steroid industry, and other sources have been sought. A useful raw material is the rhizome of *Costus speciosus* Sm. (*Zingiberaceae*). This plant is a subshrub, common in the tropical rain forests of India and South-East Asia. Its rhizomes contain 2–3 % diosgenin. Economically sound processes for growing the plant and isolating the diosgenin have been developed in India and the compound is now produced from this source.

Synthesis of steroids from diosgenin

Marker's method for degrading rings E and F of diosgenin (see Fig. 121) is still used with only minor changes in the original process.

Diosgenin is first heated with acetic anhydride (or isobutyric anhydride, which gives a better yield) at 200 °C. Ring F is opened and the double bond introduced into ring E is then split by oxidation with chromium trioxide. The oxidation product is boiled with acetic acid to remove what is left of ring F and yields 16-dehydropregnenolone acetate. In its present form the process gives an 80 % yield of the end product, calculated on the amount of diosgenin used, and the intermediates need not be isolated. 16-Dehy-

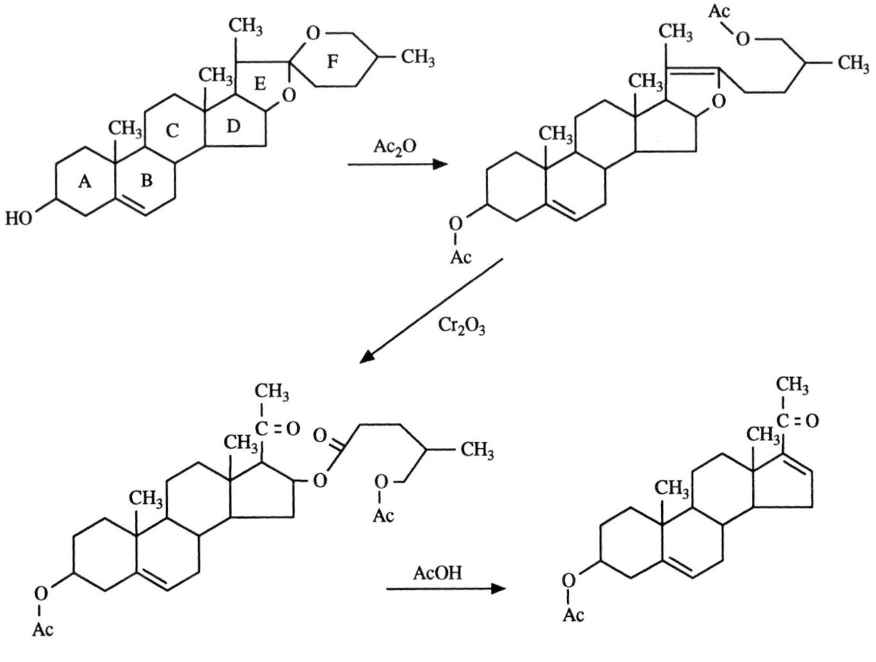

16-Dehydropregnenolone acetate

Fig. 121. Marker degradation of diosgenin

dropregnenolone acetate is the starting compound for the synthesis of progesterone, testosterone, cortisone, and contraceptives and other steroids. In the synthesis of cortisone, the keto group in position 11 is introduced with the aid of special microorganisms, e.g. varieties of *Rhizopus nigricans*. Nowadays, there are several microorganisms that are able to introduce hydroxyl groups at specific positions in the steroid skeleton.

Hecogenin

Hecogenin is a steroidal sapogenin that can be obtained from the waste left over from the production of textile fibres from sisal, the leaves of *Agave rigida* Mill. (previously known as *Agave sisalana* Perrine. ex. Engelm. (*Agavaceae*) (Plate 14, page 455). Hecogenin serves as the starting compound for about 5 % of the world production of steroids. The plant is a succulent with long, thick leaves that store water and are tipped with a spine. The leaves are rich in very long (over 1 m) tough fibres which are used for making ropes, sacks, etc. The plant is grown on a relatively large scale in East Africa, especially Tanzania and in Central America. For fibre production, the leaves are crushed and the fibres separated from the rest of the leaf material. To obtain the hecogenin, the leaf waste is fermented. A mixture of hecogenin and partially hydrolysed saponins then separates as a suspension, which is coagulated by addition of phenol in order to facilitate filtration. Hecogenin can be used for the synthesis of glucocorticoids and mineralocorticoids, because it has a keto group in ring C. This function is moved to position 11 by chemical or microbiological methods. The spiro ring system is removed by Marker degradation in the same way as in the diosgenin process.

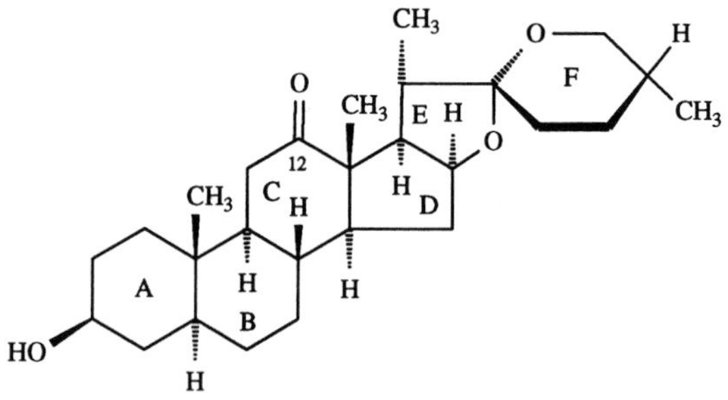

Hecogenin

The stereoisomers smilagenin and sarsasapogenin occur in plants of the genera *Smilax* (*Liliaceae*) and *Yucca* (*Agavaceae*). *Smilax* species grow in Central America and the West Indies. They are climbers with long, slender roots, which used to be sold under the drug name Sarsaparillae radix for use against syphilis and various skin diseases. *Yucca* species are succulents, growing in Mexico and the southern United States. They have a woody trunk with strap-

Plate 13.
Roots of *Dioscorea composita* Hemsley. Raw material for diosgenin (page 452). Photograph by F. Sandberg.

Plate 14.
Agave rigida Mill. A plantation in Tanzania. The leaves yield a fibre which is made into ropes, etc. The pressjuice, obtained as a by-product, is a source of hecogenin (page 454). Photograph by Gunnar Samuelsson.

Smilagenin and sarsasapogenin

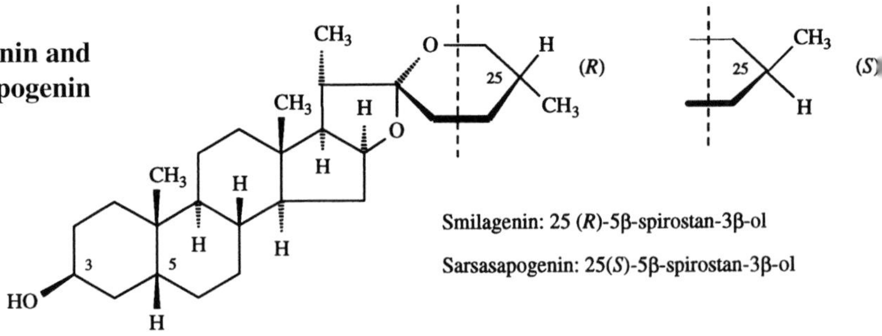

Smilagenin: 25 (R)-5β-spirostan-3β-ol

Sarsasapogenin: 25(S)-5β-spirostan-3β-ol

like, fleshy leaves. Some species can look like trees. The stem, as well as the leaves and the seeds, contain saponins. Plants of these two genera are used to a limited extent as raw materials for the pro duction of steroids.

Solasodine

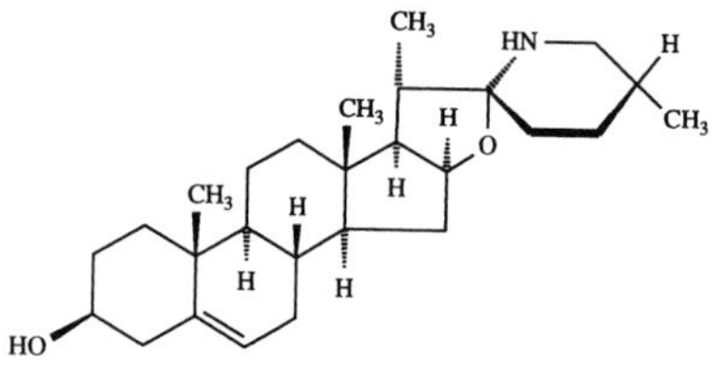

Solasodine

Solasodine is an alkaloidal analogue of diosgenin with a nitrogen atom in ring F. It occurs in the form of a glycoside in *Solanum avic ulare* Forst. f. (previously known as *Solanum laciniatum* Ait. (*Solanaceae*), which is grown for steroid production in the former USSR, Hungary and Australia. The unripe fruits contain 2–3 % solasodine, but the content decreases on ripening. In India, a spine free mutant of *S. khasianum* C.B. Clarke has been developed from which the solasodine-containing fruits can easily be harvest ed. Rings E and F of solasodine can be degraded in a Marker reac tion, like the corresponding rings in the steroidal sapogenins. How ever, the yield is a good deal lower than in the degradation of dios genin.

Stigmasterol

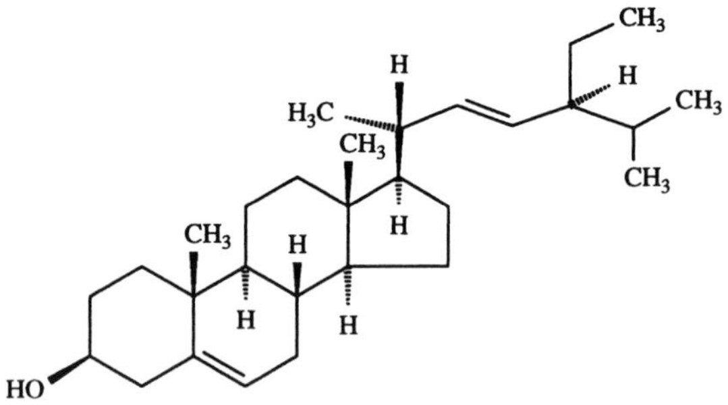

Stigmasterol

Stigmasterol was first isolated from calabar beans (page 671). The commercial source of the sterol is the unsaponifiable fraction of soya bean oil, which contains 12–25 % stigmasterol (page 251).

Stigmasterol has now overtaken diosgenin as the principal start ing material for the industrial semisynthesis of steroids. In the syn thesis of progesterone (see Fig. 122), stigmasterol is subjected to

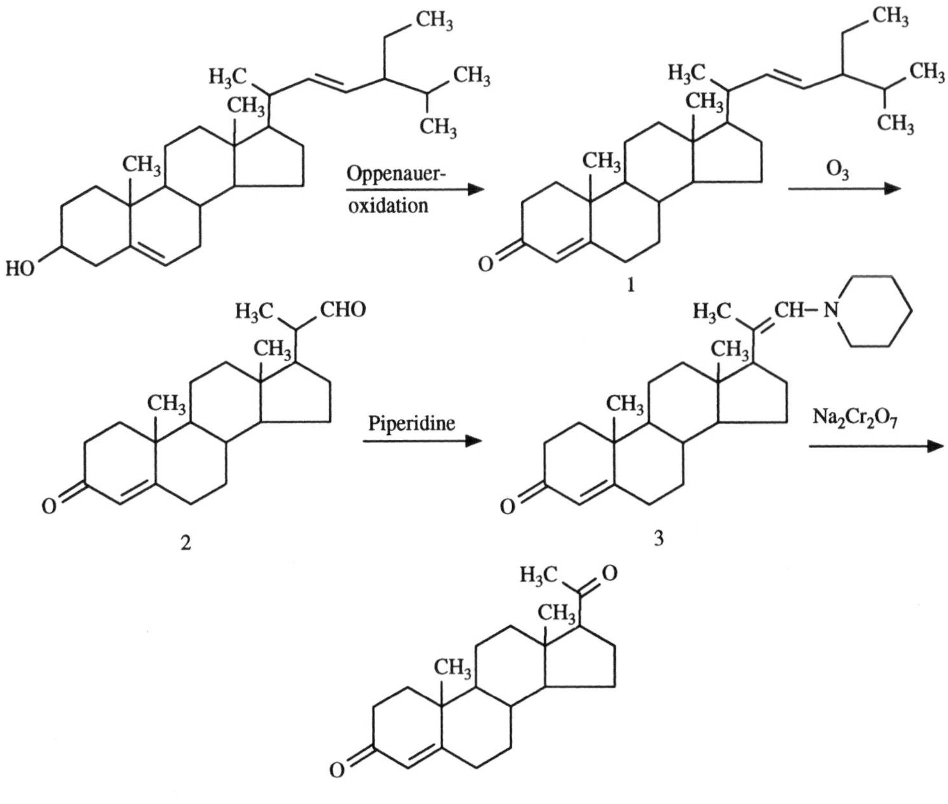

Fig. 122. Synthesis of progesterone from stigmasterol

Oppenauer oxidation to give stigmastadienone (1), which in turn is oxidized with ozone to the aldehyde (2). Distillation of this aldehyde with piperidine in benzene gives the enamine (3), which is oxidized to progesterone with sodium dichromate in a mixture of anhydrous acetic acid and benzene. The overall yield is approximately 60 %.

Progesterone is used as a starting material for the synthesis of other steroids.

Sitosterol   Sitosterol, an analogue of stigmasterol with a saturated side-chain, occurs together with stigmasterol in soya bean oil and can also be isolated from other oils, e.g. maize oil. Since the side-chain of sitosterol does not contain a double bond, it cannot be degraded by chemical methods as readily as that of stigmasterol. However, industrial microbiological methods for this degradation have been worked out. This means that nowadays even sitosterol can be used as a starting material for the synthesis especially of sex hormones, 19-norsteroids and spironolactone.

457

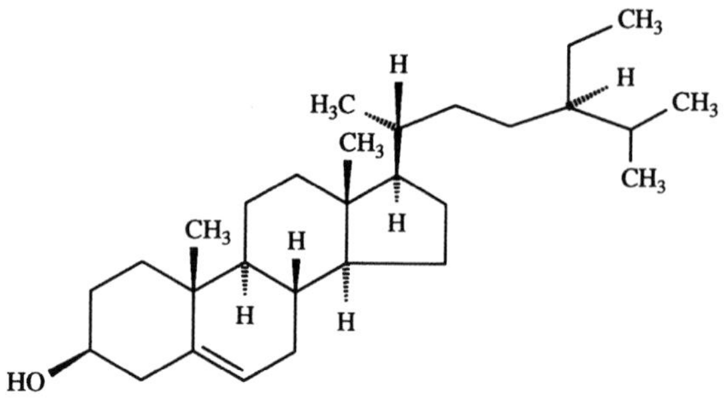

Sitosterol

Cholesterol and bile acids

Cholesterol and bile acids are the starting materials for about 10 % of the world production of steroids. Cholesterol is one of the main constituents of wool fat, but the main source of cholesterol is the backbone of slaughtered animals. Bile acids, likewise, are obtained from abattoirs.

Bile acids differ from other steroids in that the hydroxyl group at C-3 is α-oriented.

Microbiological processes for removal of the side-chain have become more and more important in the synthesis of steroids from these starting substances.

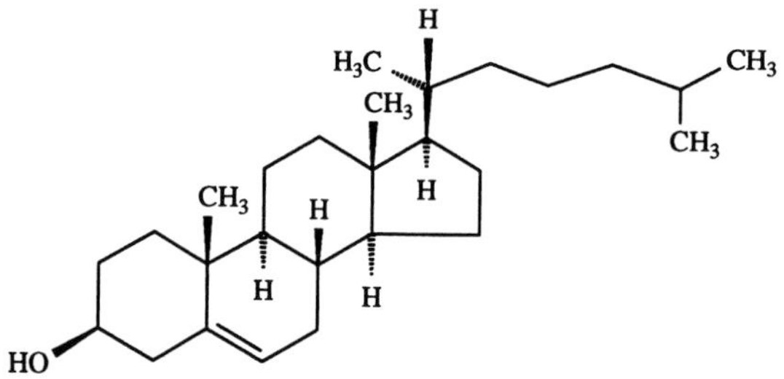

Cholesterol

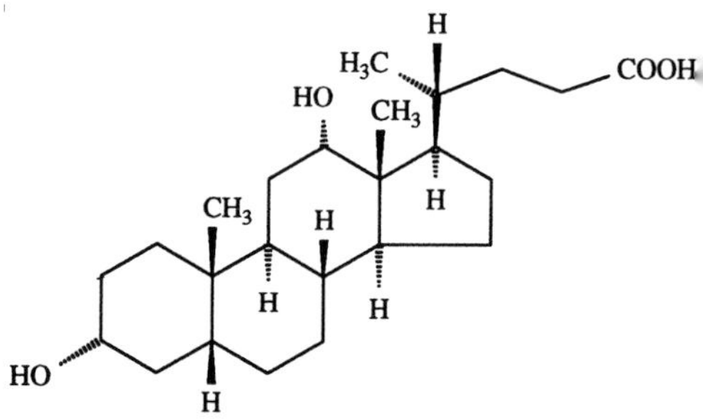

Deoxycholic acid

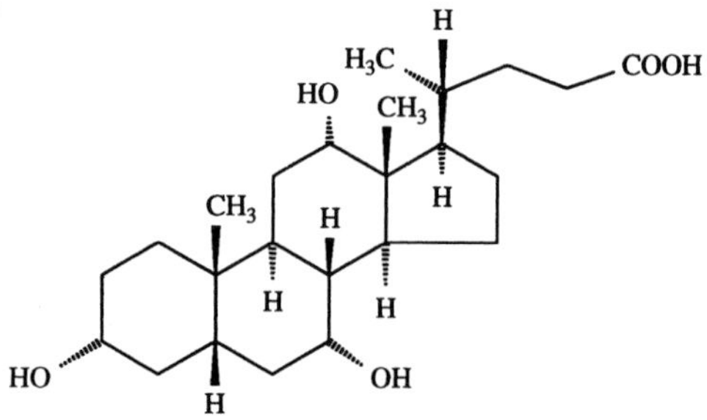

Cholic acid

Crude drugs **Ox bile** (Fel tauri, Fel bovis) is the gall of cattle (*Bos taurus*). It contains the sodium salts of some 20 bile acids, the main ones being cholic and deoxycholic acids.

Bile acids were the first starting materials for the industrial preparation of steroidal hormones. The original process comprised about 30 reaction steps and gave an overall yield of only a few tenths of a per cent. However, improvements have been introduced and the process can now compete with those methods which use plant steroids as starting materials.

A small quantity of ox bile is used medicinally in cases of insufficient secretion of bile.

Wool fat (Adeps lanae) is a wax-like product extracted from the wool of the sheep, *Ovis aries.* The extraction can be carried out either with organic solvents, e.g. light petroleum, or by washing with soapy water. The product is purified and bleached. Wool fat contains at least 30 % cholesterol, partly free and partly esterfied with fatty acids. Other constituents are isocholesterol and free, as well as esterfied, wax alcohols.

Wool fat is used as an ingredient in ointments. This use is based on the fact that it is a very good emulsifier for water and makes it possible to incorporate water-soluble drugs into ointments. Dry wool fat, obtained by heating the molten fat on a steam-bath while stirring until the water has evaporated, is preferred for this purpose.

Cardiac glycosides

Cardiac glycosides are a group of natural products characterized by their specific effect on myocardial contraction and atrioventricular conduction. In larger doses they are toxic and bring about cardiac arrest in systole, but in lower doses they are important drugs in the treatment of *congestive heart failure*. This disease is characterized by impaired blood circulation, due to a decrease in the force with which the heart muscle contracts. Congestive heart failure is often accompanied by *atrial fibrillation,* due to irregular and poorly coordinated contraction of the auricle and ventricle of the heart. An obvious symptom of congestive heart failure is *oedema*, i.e. accumulation of fluid especially in the extremities, caused by poor circulation of the blood. This explains the old name of the disease: *dropsy* (= hydropsy).

Cardiac glycosides have a direct cardiotonic action on the myocardium, resulting in an increase in the force of contraction. The increased contractility is caused by inhibition of the membrane-bound sodium-potassium-activated ATPase, leading to an increase in the intracellular stores of calcium. When cardiac glycosides are given to a patient suffering from congestive heart failure, the stroke volume of the heart is increased, causing a more effective emptying of the ventricles and a lowering of the diastolic ventricular pressure, which in turn causes a decrease in pulmonary and central venous

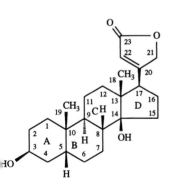

Card-20(22)-enolide

pressures. The increased cardiac output improves renal blood flow and glomerular filtration, which in turn leads to a decrease in juxtaglomerular renin secretion. Renal absorption of sodium and water are thus decreased and oedema diminished. Diuresis is promoted. Raised hepatic blood flow increases clearance of aldosterone, further reducing the oedema. The improvement in cardiac output also reduces the compensating increased cardiac rate.

In higher doses, cardiac glycosides also have a direct inhibiting action on atrioventricular conduction and are employed in the treatment of *atrial flutter, atrial fibrillation* and *paroxysmal atrial tachycardia*. The effect of cardiac glycosides is particularly dramatic in patients suffering from a combination of congestive heart failure and atrial fibrillation.

The aglycones of cardiac glycosides are steroids, in which the side-chain is an unsaturated lactone ring, which can be either 5- or 6-membered. The various aglycones differ mainly in the numbers and positions of the hydroxyl substituents and in the extent of oxidation of the C-19 methyl group. Almost all the known aglycones however, have two β-oriented hydroxyl groups, one at C-3 and the other at C-14. The sugar moiety is attached via the C-3 hydroxyl group.

Cardiac glycosides contain a wide variety of sugars, many of which are rare:

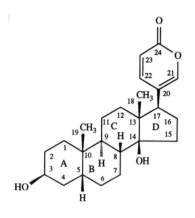

Bufa-20,22-dienolide

CHO	CHO	CHO	CHO	CHO
CH₂	CH₂	H-C-OH	H-C-OH	CH₂
H-C-OH	H-C-OCH₃	CH₃O-C-H	H-C-OH	H-C-OCH₃
H-C-OH	H-C-OH	HO-C-H	HO-C-H	HO-C-H
H-C-OH	H-C-OH	H-C-OH	HO-C-H	H-C-OH
CH₃	CH₃	CH₃	CH₃	CH₃
D-Digitoxose	D-Cymarose	D-Digitalose	L-Rhamnose	D-Sarmentose

Digitoxose, cymarose and sarmentose are 2-deoxysugars and lack the hydroxyl group at C-2. In some sugars the C-3 hydroxyl group is methylated, e.g. cymarose, digitalose and sarmentose.

The pharmacological effects of cardiac glycosides are due mainly to the aglycone. The sugar moiety increases the water solubility of the compounds and modifies the intensity and duration of the effect. The unsaturated lactone ring is essential for the pharmacological action. If the ring is saturated, the activity is substantially decreased, and opening the ring renders the glycoside inactive. Stereochemical factors are also significant. If the C-3 hydroxyl group is α-oriented, the effect is weaker than that of the β-isomer. The activity is strongest if the 17-lactone ring is β-oriented and the A/B and C/D ring junctions are both *cis*. Variation in the substituents on the steroid skeleton influences the intensity and duration of the effect to only a minor extent.

Cardiac glycosides occur in a limited number of plant families. Medicinally important glycosides are obtained from plants belonging to the *Liliaceae* (*Drimia*), *Ranunculaceae* (*Adonis*), *Apocynaceae* (*Strophanthus, Thevetia*) and *Plantaginaceae s. lat* (*Digitalis*). In addition, cardiac glycosides have been encountered in plants belonging to the *Iridaceae, Moraceae, Brassicaceae, Fabaceae, Euphorbiaceae, Meliaceae, Melianthaceae, Celastraceae, Tiliaceae, Sterculiaceae* and the subfamily *Asclepioideae* of the *Asclepiadaceae*. The *Apocynaceae* is especially rich in plants containing these substances, several of which have been, and still are, used by primitive peoples for the preparation of arrow poisons.

Extraction and isolation of cardiac glycosides

Cardiac glycosides are neutral, polar compounds and, for that reason, difficult to isolate from plant materials and to obtain in a pure state. Moreover, glycosides with a large sugar moiety are hard to obtain crystalline. Plants containing cardiac glycosides usually produce a group of similar compounds with relatively small differences in the aglycone or sugar part. When the plant material is dried, some of the original, sugar-rich glycosides lose one or more sugar residues as a result of hydrolysis by glycosidases, which always occur together with the glycosides.

The polarity of the glycosides makes it necessary to use a polar solvent for their extraction, in more recent work usually ethanol or mixtures of ethanol and water. The following method, once worked out by *Arthur Stoll* for the isolation of the naturally occurring, primary (genuine) cardiac glycosides from *Digitalis lanata,* is an example of how these difficulties can be circumvented. The fresh leaves were ground together with solid ammonium sulphate to precipitate the enzymes present, so that they could not attack the glycosides. These were then extracted from the mixture with ethyl acetate or a mixture of chloroform and ethanol. The glycoside-phenol complex obtained was purified by the addition of a soluble lead salt, the excess lead being later precipitated as sulphate or sulphide. The mother liquors were concentrated to give a crude crystalline glycoside mixture, from which the genuine glycosides were separated chromatographically. The principles of this method are still applicable, purification with lead salts is, however, not used anymore.

Glycosides from *Digitalis purpurea*

The leaf of *Digitalis purpurea* L. (foxglove, *Plantaginaceae*) (Plate 15, page 465) contains a large number of glycosides. Most of these are cardiac glycosides, but others are of the saponin type and still others are digitanol glycosides, i.e. a group of pharmacologically inactive glycosides based on steroidal aglycones. The enzymes present in the leaves of the foxglove degrade the sugar moieties of the glycosides, so that the dried leaves (the crude drug) may contain both sugar-rich primary glycosides and secondary glycosides with a smaller number of sugar residues.

The cardiac glycosides of *Digitalis* can be divided into three groups, according to the structure of the aglycone.

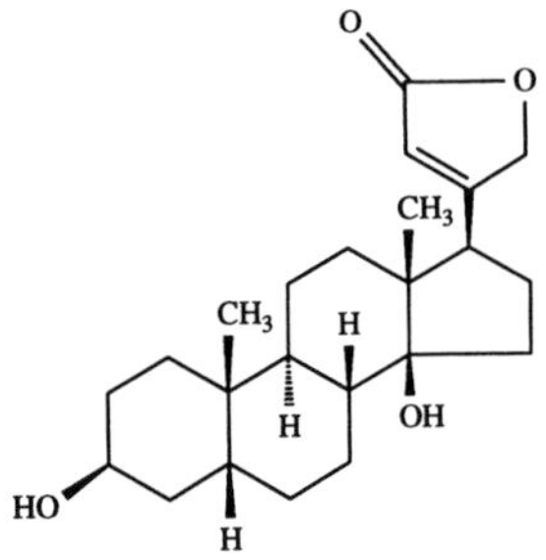

Digitoxigenin

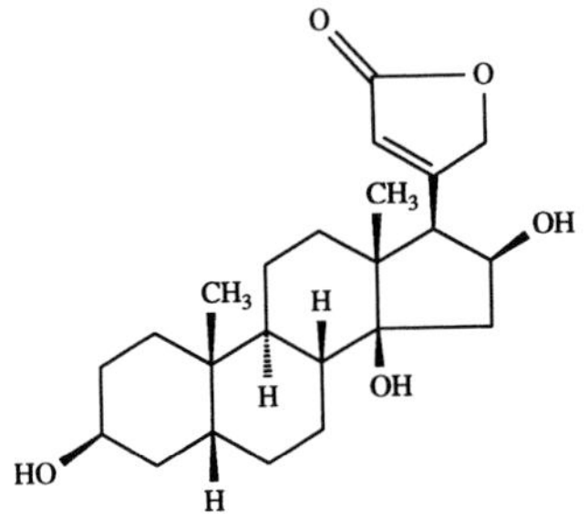

Gitoxigenin

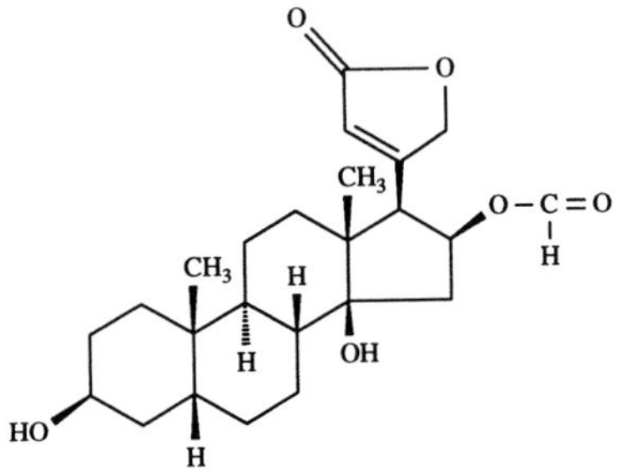

Gitaloxigenin

Table 10 includes only the main glycosides and the more important minor ones. In all, some 30 glycosides have been isolated from dried foxglove leaves, most of which are present only in very small amounts. The total cardiac-glycoside content of the dried leaves is about 0.3 %.

Biosynthesis of digitoxin

As previously mentioned (page 439), plant steroids are derived from cycloartenol, which, in turn, originates from squalene. Demethylation of cycloartenol results in cholesterol, which is oxidized to 20α, 22β-dihydroxycholesterol (Fig. 123). The side-chain of this compound is cleaved to isocaproic aldehyde and pregnenolone. These two reactions are catalysed by the enzyme *cholesterol monooxygenase (side-chain-cleaving)* (EC 1.14.15.6, also known as cholesterol side-chain cleaving enzyme (SCCE) which has been detected in protein extracts from leaves of *Digitalis purpurea*. The next reaction – oxidation and isomerization to progesterone – is catalysed by Δ5-3β-*hydroxysteroid dehydrogenase/Δ5-Δ4-ketosteroid isomerase* (3β-HSD). It has been detected in extracts from suspension-cul-

Table 10. Cardiac glycosides of Digitalis purpurea L

	Glycoside	Sugar moiety
A. Digitoxigenin group		
Aglycone: Digitoxigenin		
Main glycosides		
	Purpurea glycoside A	-digitoxose-digitoxose
		-digitoxose-glucose
	Digitoxin	-digitoxose-digitoxose
		-digitoxose
Minor glycosides		
	Gluco-evatromonoside	-digitoxose-glucose
	Gluco-digifucoside	-fucose-glucose
B. Gitoxigenin group		
Aglycone: Gitoxigenin		
Main glycosides		
	Purpurea glycoside B	-digitoxose-digitoxose
		-digitoxose-glucose
	Gitoxin	-digitoxose-digitoxose
		-digitoxose
Minor glycosides		
	Gluco-gitoroside	-digitoxose-glucose
	Digitalinum verum	-digitalose-glucose
C. Gitaloxigenin group		
Aglycone: Gitaloxigenin		
Main glycosides		
	Glucogitaloxin	-digitoxose-digitoxose
		-digitoxose-glucose
	Gitaloxin	-digitoxose-digitoxose
		-digitoxose
Minor glycosides		
	Gluco-verodoxin	-digitalose-glucose
	Verodoxin	-digitalose

tured cells and intact leaves of *Digitalis lanata*. Reduction of the double bond in ring A of progesterone (for the numbering of the steroid skeleton, see page 459) is catalysed by *Δ⁴-3-oxosteroid 5β-reductase* (EC 1.3.1.3, also known as progesterone 5β-reductase*)* which has been isolated from photomixotrophic shoot cultures of *Digitalis purpurea*. The same source yielded *3β-hydroxy-5β-steroid dehydrogenase* (EC 1.1.1. 277, also known as 3β-hydroxysteroid-5β-oxidoreductase) which transforms 5β-pregnane-3,20-dione to 5β-pregnane-3β-ol-20-one. The enzyme catalysing the hydroxyla-

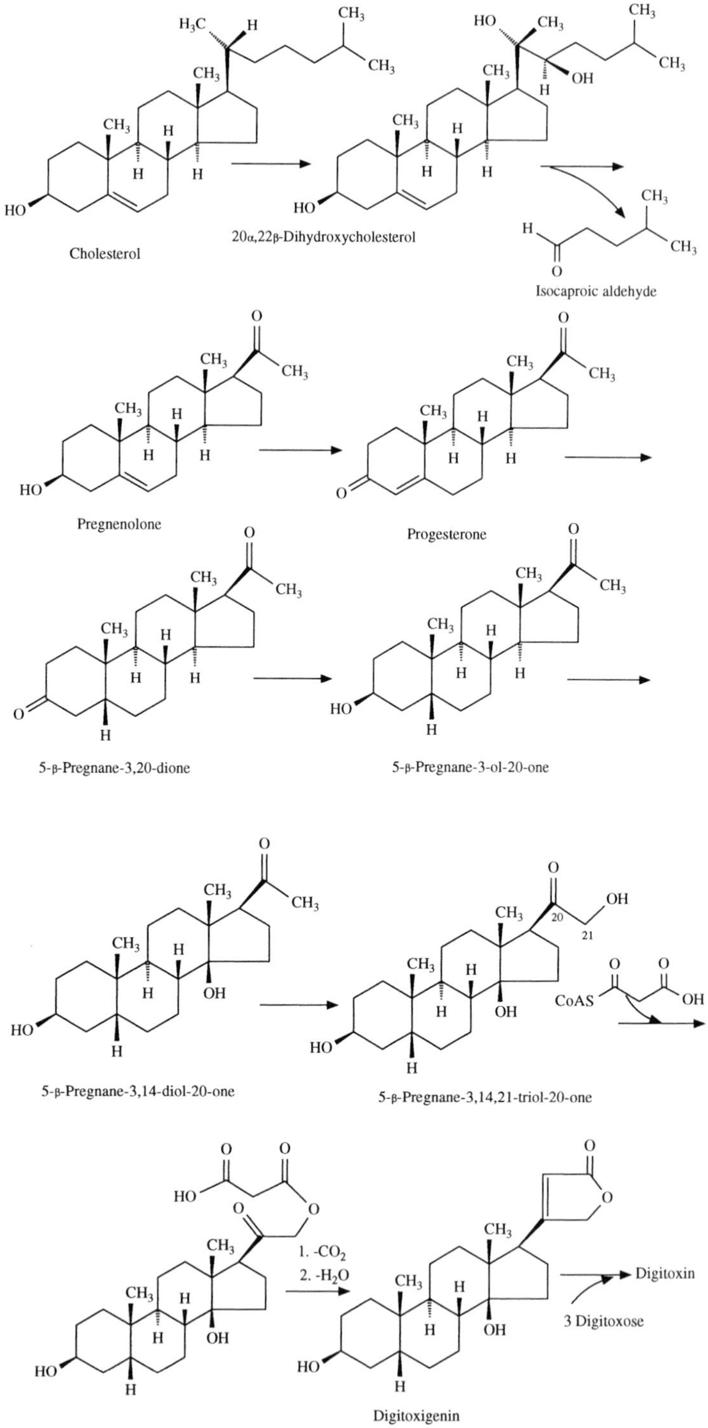

Fig. 123. Biosynthesis of digitoxin

Plate 15.
Digitalis purpurea L. Purple fox-glove. The leaves contain cardiac glycosides, particularly digitoxin, and are the source of the crude drug Digitalis leaf (page 467). Photographed by Gunnar Samuelsson in a museum garden, dedicated to "Kisa Mor", a famous traditional healer who lived in the 19th century in Horn, Östergötland, Sweden.

Plate 16.
Digitalis lanata Ehrh. The Botanical Garden, Uppsala. Photograph by Gunnar Samuelsson.

Plate 17.
First year rosette leaves of *Digitalis lanata* Erh. These leaves are the industrial source of lanatoside C and other cardiac glycosides (page 469). The Botanical Garden, Uppsala. Photograph by Gunnar Samuelsson.

465

tion of C-14 in 5β-pregnane-3-ol-20-one with formation of 5β-pregnane-3,14-diol-20-one has not yet been characterized, but it has been concluded that 14β-hydroxylation must occur prior to the formation of the butenolide ring, which involves transfer of a malonyl moiety to the 21-hydroxy group to form a malonyl hemi-ester. This reaction is catalysed by *malonyl-coenzyme A: 21-hydroxypregnane 21-hydroxy-malonyltransferase*. A subsequent Claisen-type condensation of the malonyl moiety with the C-20 ketone, accompanied by decarboxylation yields digitoxigenin, to which three digitoxose residues are added, forming digitoxin.

Alternative pathways to the above-described biosynthesis of digitoxin have been proposed and evidence is accumulating that cardenolides are not assembled on a straight conveyor belt but rather are formed via a complex multidimensional metabolic grid.

Saponins

Foxglove leaves also contain steroidal saponins, in whose aglycones the side-chain is transformed into a spiroketal system of the type seen earlier in diosgenin and other steroidal saponins used for the synthesis of steroidal hormones (page 452). The sugar moiety is linked to the hydroxyl group at C-3.

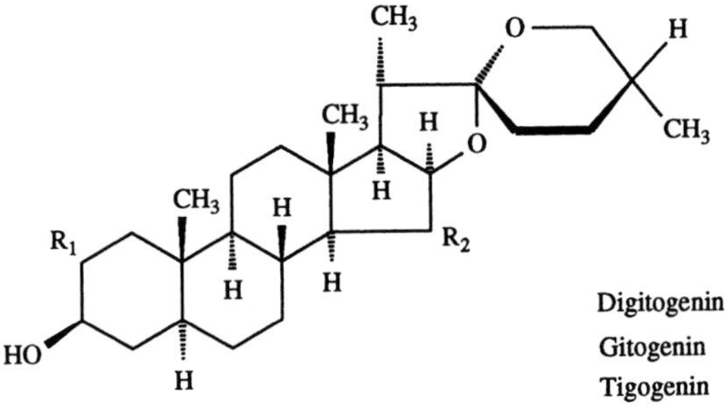

	R_1	R_2
Digitogenin	⁗OH	◂ OH
Gitogenin	⁗OH	-H
Tigogenin	- H	-H

Saponin	Aglycone	Sugar
Digitonin	Digitogenin	2 glucose, 2 galactose, xylose
Gitonin	Gitogenin	glucose, 2 galactose, xylose
Tigonin	Tigogenin	2 glucose, 2 galactose, xylose

Digitonin is the main saponin and a considerable amount of it occurs in the seeds of *D. purpurea,* from which it is isolated commercially. Digitonin is used as a reagent for cholesterol, with which it forms an insoluble complex. The saponins are also believed to have a favourable effect on the absorption of the cardiac glycosides, when the crude drug is administered as a medicine.

Digitanol glycosides

The digitanol glycosides are another group of glycosides isolated from foxglove leaves. These compounds contain a steroidal agly-cone, but they do not affect the heart, nor do they show any other marked pharmacological effects. Digitalonin is an example.

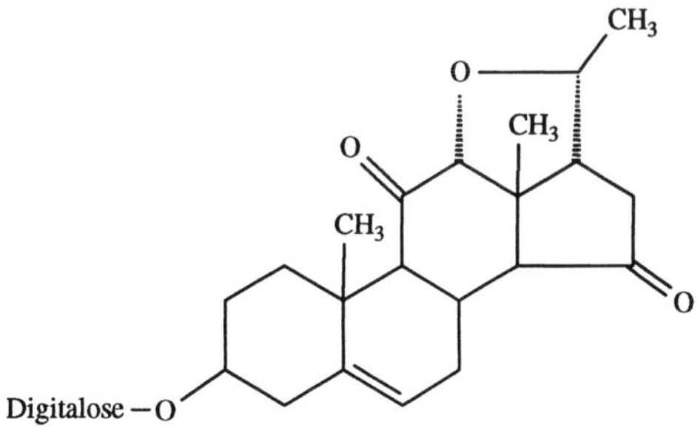

Digitalonin

Crude drug **Digitalis leaf** (Digitalis purpureae folium) is the dried leaf of the purple foxglove, *Digitalis purpurea* L. (*Plantaginaceae*). The plant is a biennial but can also become a perennial, expecially in cultivation. In the first year, a rosette of leaves is formed. These leaves are very hairy and up to 0.5 m long. In the following year, a stem develops, which carries scattered, smaller leaves and a raceme of pink or white flowers at the top. The herb is often grown for ornamental purposes in gardens. It is also cultivated for drug production in Europe as well as in the United States and Canada. It is harvested in the first year (the leaf rosette), when the yield of the leaf drug is the highest.

The English physician William Withering of Birmingham start-ed to use digitalis leaf in 1775. Withering found that patients suf-fering from dropsy improved when they took an extract of a plant mixture supplied by a Mrs Hutton from Shropshire. The mixture was made up from some 20 different plants and Withering succeed-ed in showing that digitalis leaf was the effective ingredient. His observations were published 10 years later in the classical work "An account of the Foxglove and Some of its Medical Uses: with Prac-tical Remarks on Dropsy and other Diseases".

The use of digitalis leaf and of isolated glycosides

Powdered digitalis leaf, given as pills or tablets, is used as a reme-dy against congestive heart failure. In many countries, however, medicinal preparations containing the crude drug **Digitalis leaf** are no longer official. There are several reasons for this. One difficulty in using the crude drug is the fact that the glycoside content varies a great deal. The content of each lot must therefore be determined, and batches with high and low contents must be mixed so that a drug with constant strength is obtained.

The analysis can be performed biologically and involves a toxicity determination on guinea pigs as experimental animals. This was previously the principal method. In this method, the amount of leaf extract that is needed to cause cardiac arrest on intravenous administration is determined. In order to compensate for differences in sensitivity among the experimental animals, the sample is compared with a standard preparation, and the determination is carried out on at least eight animals for the standard and the same number for the sample.

Even if two batches of digitalis leaf have the same glycoside content as determined biologically, differences in the relative proportions of the individual glycosides may give rise to differences in the pharmacological effect. Thus, on oral administration, digitoxin is more readily absorbed than gitoxin, whereas with the intravenous administration used in the toxicity test, the two glycosides are almost equivalent. A drug containing much digitoxin and little gitoxin thus has a stronger therapeutic effect than a drug with the reverse proportions, although the two drugs give the same result on biological testing.

The same objection can be raised about chemical methods, which only determine the total content of cardiac glycosides. Determination of individual glycosides in an extract by HPLC is preferable to the biological as well as to the previous chemical methods. Another reason why digitalis leaf is not used very much nowadays is that this crude drug is usually heavily contaminated with bacteria. The reason for this is that the leaves are densely covered with hairs, to which soil and dirt readily adhere. Digitalis leaf should therefore be sterilized before use (see page 64).

Digitoxin is the main secondary glycoside of *Digitalis purpurea*. The effect of digitoxin starts very slowly and reaches its maximum 8–12 hours after being taken orally. The excretion of the compound is also slow, so that the risk of overdosage through cumulation is high. As the safety margin is low, the dosage must be determined individually. To prevent cumulation, it is often indicated that the drug should not be taken during 1–2 days a week. Signs of overdosage are: nausea, vomiting, diarrhoea, abdominal pain, headache, drowsiness, fatigue, malaise, backache, trigeminal neuralgia, visual disturbances, convulsions and mental disturbances. Cardiac arrhythmias of all types are relatively common. Heart block and premature ventricular contractions are the most frequent, ventricular tachycardia the most ominous.

Glycosides from *Digitalis lanata*

The leaves of *Digitalis lanata* contain five main glycosides, named lanatosides A–E. In addition, a number of other glycosides are present as well as (in the dried leaves) enzymic degradation products of lanatosides. Three of the main glycosides have the same aglycones as the three genuine glycosides of *D. purpurea*, but differ from these in that one of the digitoxose residues in the sugar moiety is acetylated.

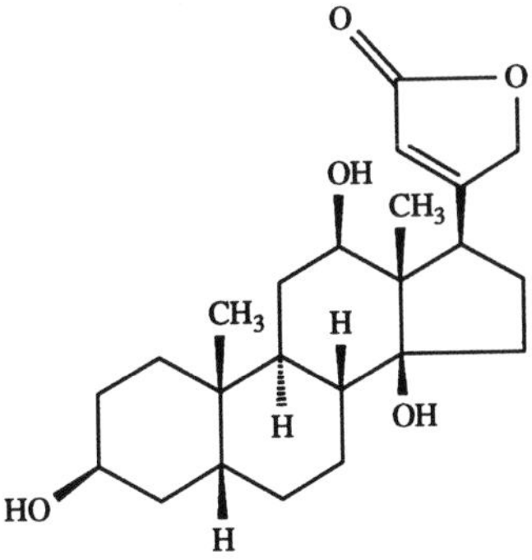

Digoxigenin

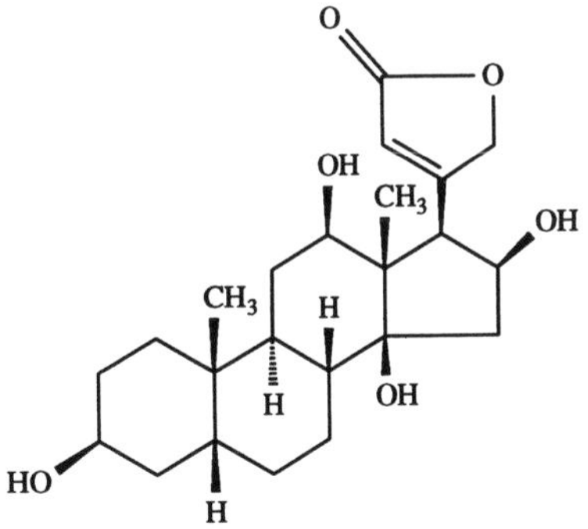

Diginatigenin

Lanatoside A: Aglycone: Digitoxigenin
Sugar: -digitoxose-digitoxose-acetyldigitoxose-glucose

Lanatoside B: Aglycone: Gitoxigenin
Sugar: -digitoxose-digitoxose-acetyldigitoxose-glucose

Lanatoside C: Aglycone: Digoxigenin
Sugar: -digitoxose-digitoxose-acetyldigitoxose-glucose

Lanatoside D: Aglycone: Diginatigenin
Sugar: -digitoxose-digitoxose-acetyldigitoxose-glucose

Lanatoside E: Aglycone: Gitaloxigenin
Sugar: -digitoxose-digitoxose-acetyldigitoxose-glucose

The only one of the genuine glycosides that is used therapeutically is *lanatoside C*. It has a rapid and short-lasting effect, so that the risk of cumulation is small. The following derivatives of *D. lanata* glycosides are also used therapeutically:

Acetyldigitoxin is obtained from lanatoside A by removal of the terminal glucose residue. The duration of action is shorter and the risk of cumulation is smaller than with digitoxin but greater than with lanatoside C.

Digitoxin can be prepared from lanatoside A by removal of the glucose residue and acetyl group from the acetyldigitoxose residue. For its therapeutic use, see the *D. purpurea* glycosides (page 468).

Deacetyllanatoside C (deslanoside) is formed by splitting off the acetyl group from the third digitoxose residue of lanatoside C. It is used especially in injections for rapid digitalization.

Digoxin is formed from lanatoside C by loss of the terminal glucose and removal of the acetyl group from the third digitoxose residue. Digoxin works relatively rapidly, the maximum effect being reached 5–7 hours after oral administration. The half-life is normally 1 1/2 days (36 h), as compared with about 6 days for digitoxin.

Crude drug **Digitalis lanata leaf** (Digitalis lanatae folium) is the dried leaf of *Digitalis lanata* Ehrh. (*Plantaginaceae*) (Plates 16 and 17, page 465). The plant is a biennial or perennial herb from central and southeastern Europe, which is cultivated elsewhere in Europe and in the United States. Like its relative *D. purpurea,* it only develops a rosette of leaves during the first year. In the second (and possibly subsequent) years it forms a one-metre tall herb with light brown flowers. The inflorescence is woolly because of the presence of numerous trichomes, whereas the leaves, as distinct from those of *D. purpurea,* are smooth and lanceolate in shape.

In plantations of *D. lanata,* it is the rosettes of leaves – formed during the first year – that are harvested, as they give the highest yield. The crude drug is used exclusively as a raw material for the extraction of glycosides, and as such it is more important than foxglove leaves. The total glycoside content is higher and can be up to 1 %. The cultivated crude drug is harvested from mid-September to the end of October. The leaves are quickly dried at a temperature of 50–60 °C, the rapid drying at a relatively high temperature being essential for preventing enzymic degradation of the glycosides. However, the temperature must not be too high, as this may lead to loss of the hydroxyl group at C-14 of the steroid nucleus, with formation of inactive anhydro-compounds.

Other cardiac glycosides

G-Strophanthin (=ouabain) This glycoside can be obtained from the seeds of *Strophanthus gratus* or from the bark of *Acokanthera schimperi* (A.DC.) Schweinf., both of which belong to the family *Apocynaceae.* Like the digitalis glycosides, g-strophanthin is used as a remedy for congestive heart failure. Its effect is rapid but of short duration; and there is little risk of cumulation. The glycoside is mainly administered by injection, because it is poorly absorbed orally.

Crude drug **Strophanthus gratus seed** (Strophanthi grate semen) is the seed of *Strophanthus gratus* Baill. (synonymus to *Roupellia grata* Hook.) (*Apocynaceae*). The genus *Strophanthus* comprises over 100 species, mostly lianas, occurring mainly in Equatorial Africa and, to a lesser extent, East Asia. In cultivation, it is possible to keep the plant in the form of a bush by pruning. The fruit consists of two follicles, joined at the base, 20–35 cm long and 2–2.5 cm in diameter. The follicles split open on ripening and the numerous seeds are set free. Each seed carries a long, feathery awn which aids its dispersal. This awn should not be present in the crude drug. The seed of *S. gratus* is flat-

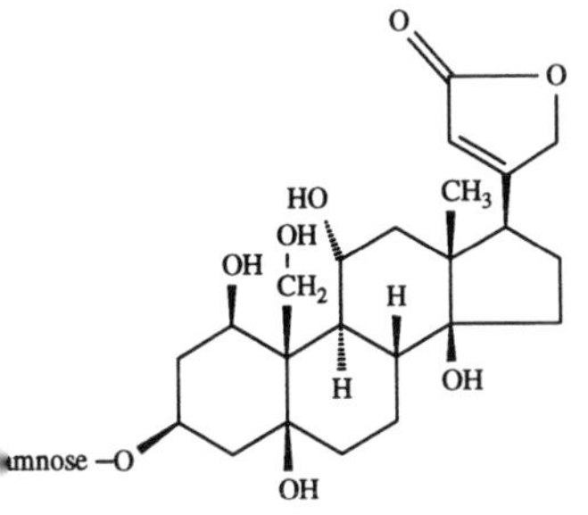

G-Strophanthin

tened, 10–20 mm long and, unlike other *Strophanthus* seeds, almost glabrous. The seeds of *Strophanthus* species have been widely used for preparing arrow poison, because of the cardiac glycosides present in them. *S. gratus* seeds contain 4–8 % glycosides, 90–95 % of which is g-strophanthin. The crude drug is used exclusively for the extraction of this glycoside.

Glycosides of the aglycone strophanthidin occur in the seeds of *Strophanthus kombe* Oliver and *S. hispidus* DC, as well as in the rhizome and aerial parts of the lily-of-the-valley, *Convallaria majalis* L. (*Liliaceae*). The *Strophanthus* seeds contain three principal glycosides, which differ in the sugar moiety:

Strophanthidin glycosides

Glycoside	Sugar
K-Strophanthoside	-Cymarose-β-glucose-α-glucose
K-Strophanthin-β	-Cymarose-β-glucose
Cymarin	-Cymarose

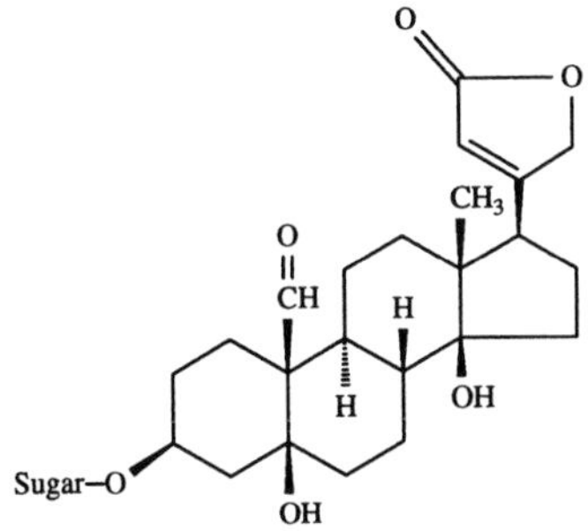

Only k-strophanthoside is to be regarded as a genuine glycoside. It is split by an α-glycosidase present in the seed to give k-strophanthin-β, which is then attacked by the enzyme strophanthobiase, also occurring in the seed, to give cymarin. A mixture of these glycosides is marketed under the name of strophanthin or k-strophanthin and has more or less the same use as g-strophanthin.

The glycoside cymarin also occurs in the lily-of-the-valley, which contains about 20 cardiac glycosides. Convalloside and convallatoxin are two of the main ones, and both contain strophanthidin as the aglycone. In convalloside, the sugar part is rhamnose-glucose, in convallatoxin it is rhamnose alone. Glycoside concentrates from lily-of-the-valley can be used medicinally in the same way as k- and g-strophanthin. The glycoside content of the plant is also of toxicological interest.

Strophanthidin glycoside

Crude drug **Strophanthus seed** (Strophanthi kombe semen) is the seed of *Strophanthus kombe* Oliver (*Apocynaceae*). The plant is a liana of East African origin (Tanzania, Kenya, etc.). The seed carries a feathery awn and develops follicles of the same type as those of *S. gratus*. The crude drug should be free of awns. The seed of *S. kombe* is yellowish green, covered by a mat of pale hairs, flattened, and 10–15 mm long. It is used only for glycoside extraction. The same glycosides can be obtained from the seeds of *S. hispidus* DC. (*Apocynaceae*), a species growing in West and Central Africa (Senegal, Guinea, Zaïre, etc.). The seeds of this species are brown, hairy and usually smaller than those of *S. kombe*.

Adonitoxin The aglycone of this glycoside – adonitoxigenin – is an isomer of strophanthidin, carrying a hydroxyl group at C-16 rather than C-5. Adonitoxin is the main component of the water-soluble glycoside mixture that can be obtained from the aerial parts of *Adonis ver-*

471

Adonitoxin

Crude drug

Proscillaridin A

Proscillaridin A

nalis. This glycoside has approximately the same pharmacological effect as the strophanthidin glycosides.

Adonis (Adonidis herba) consists of the aerial parts of *Adonis vernalis* L. (*Ranunculaceae*). The plant is a 10–30 cm high herb with large yellow flowers and laciniate leaves. It grows on calcareous ground especially in south-eastern and central Europe. The crude drug contains several glycosides, the main ones being cymarin and adonitoxin. The crude drug is chiefly used for making a standardized glycoside preparation.

The aglycone of this glycoside – scillarenin – contains a 6-membered lactone ring and a double bond between C-4 and C-5 in ring A. The glycoside must be regarded as a secondary glycoside, as other glycosides, richer in sugar, also occur in squill, *Drimia maritima*. Proscillaridin A has a rapid effect of short duration, with little risk of cumulation. It lowers the pulse less than other cardiac glycosides.

Crude drug

Squill (Scillae bulbus) consists of the dried intermediate scales of the bulb of *Drimia maritima* (L.) Stearn (*Urginea maritima* Baker = *Urginea scilla* Steinh.) (*Liliaceae*). This bulbous plant grows on the shores of the Mediterranean and on the Canary Islands. The bulb can reach the size of a child's head and weigh almost 2 kg. Leaves grow out from the bulb in the spring and wither away in the autumn, but a one-metre high scape then develops, carrying a raceme of numerous white flowers. The bulbs are collected in August, when the leaves have withered, and the intermediate scales are cut out and dried. The innermost parts of the bulb cannot be processed, because their mucilage content is too high.

Squill contains a large number of cardiac glycosides. Glucoscillaren A is a genuine glycoside, in which scillarenin is combined with rhamnose-glucose-glucose. Removal of one of the glucose residues gives scillaren A, and if both the glucose residues are split off, proscillaridin A (see above) remains. Other squill glycosides contain the aglycone scilliglaucosidin, which differs from scillarenin in that the C-19 methyl group is oxidized to an aldehyde group. The crude drug is used for the extraction of scillarenin glycosides, which are hydrolysed to proscillaridin A.

Bufotoxin

Bufotoxin is an ester of the steroid bufotalin with suberylarginine. The aglycone bufotalin belongs to the group of bufa-20,22-dienolides (page 460). The suberylarginine residue was long believed to be attached to the C-14 hydroxyl group, but later investigations have shown it to be ester-bound via the C-3 hydroxyl group, in analogy with the sugar moiety in the cardiac glycosides.

Crude drug

Bufotoxin

Toad venom. The parotoid glands located behind the eyes of toads produce a secretion which can be expressed without hurting the animal. This secretion is poisonous and contains steroids with a pharmacological effect similar to that of cardiac glycosides. Dried and powdered toad skins were used against dropsy before Withering introduced foxglove leaf. In Japan and China, dried toad skins – the crude drug "chan su" (Chinese) or "senso" (Japanese) – have been in use for centuries. Secretions of this kind are produced by many toad species. The main active constituent of the secretion of the European toad *Bufo bufo* (= *Bufo vulgaris*) is bufotoxin.

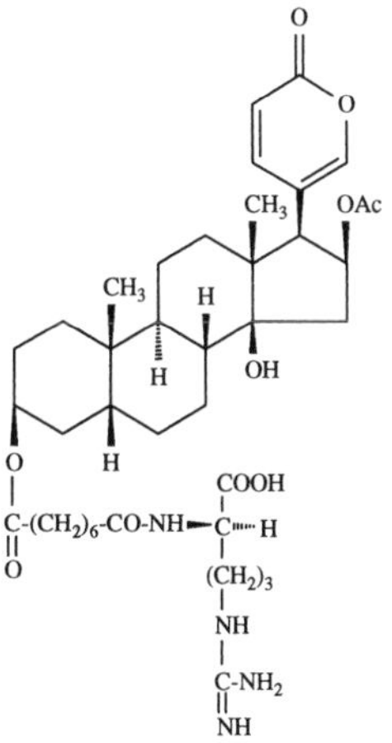

Vitamin D

The name vitamin D was originally used to denote the constituent of liver oil from cod (*Gadus morhua*) that was active against rickets (= rachitis). Nowadays, several substances with a vitamin D effect are known. They are chemically related to the steroids, but are not true steroids, because ring B of the steroid skeleton is open. Biosynthetically, they are derived from $\Delta^{5,7}$-sterols through complex photochemical reactions. Two D-vitamins are of therapeutic importance: vitamin D_3 (= cholecalciferol), occurring abundantly in fish-liver oils and in milk, butter and liver, and vitamin D_2 (= ergocalciferol), which can be prepared by UV irradiation of the steroid ergosterol.

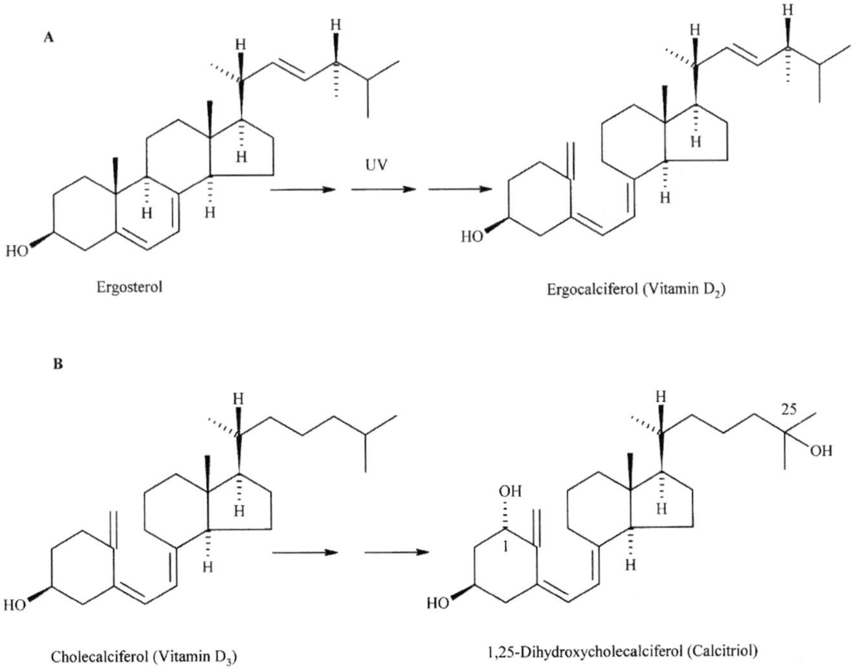

Fig. 124. A. Formation of vitamin D_2 by UV-irradiation of ergosterol. B. Metabolism of vitamin D_3 in the liver and kidney

473

Ergosterol occurs in yeast, from which it can be extracted. Its presence in ergot gave rise to the name ergosterol.

UV irradiation of ergosterol gives a series of products: lumisterol, tachysterol, vitamin D_2, toxisterol and suprasterols 1 and 2. Toxisterol and the suprasterols are poisonous and consequently their formation must be prevented by adjusting the irradiation conditions.

In human skin vitamin D_3 is synthesized from 7-dehydrocholesterol under influence of UV-light from the sun. It is also available from the food, especially fatty fish. Vitamin D_2 and D_3 are not active as such, but are oxidized in the liver to 25-hydroxyergosterol and 25-hydroxycholecalciferol, respectively. Further oxidation occurs in the kidneys with formation of 1,25-dihydroxyergosterol and 1,25-dihydroxycholecalciferol (=calcitriol), respectively, which are the active metabolites of the vitamins. These molecules bind to a vitamin D receptor to modulate gene transcription and regulate mineral ion homeostasis. Vitamin D is of great importance for the body's ability to absorb calcium from the intestine and provide a proper balance of calcium and phosphorus to support mineralization. Lack of vitamin D in children leads to rickets, which is characterized by malformation of the bones due to lack of calcium.

Crude drug **Cod-liver oil** (Jecoris aselli oleum) is the oil from the liver of the cod, *Gadus morhua,* and other species of the genus *Gadus.* The cod is caught in the Atlantic. The fish to be used for cod-liver oil production are landed as soon as possible and the livers are removed. The oil is separated from the livers by heating with steam in closed containers under carbon dioxide, to prevent oxidation. The drained oil is cooled to –5 °C and the solid precipitate formed is then removed. The oil is finally analysed for its vitamin content and different batches are mixed, so that a product with a uniform vitamin content is obtained. Apart from vitamin D, cod-liver oil also contains vitamin A (page 477). The minimum vitamin content required is 850 IU vitamin A and 85 IU vitamin D per gram oil.

Vitamins A and D are also present in high amounts (30 000–50 000 IU vitamin A and 600 IU vitamin D per gram) in **halibut-liver oil** (Jecoris hippoglossi oleum) which is obtained from the liver of the Atlantic halibut, *Hippoglossus hippoglossus.* This fish is also caught in the Atlantic, and the oil is produced in the same way as cod-liver oil.

Both oils are used as sources of vitamins A and D and for making concentrates of these vitamins.

Tetraterpenes

Tetraterpenes contain 40 carbons and the most important group is the *carotenoids*, which are yellow-red pigments of great importance in photosynthesis where they serve as collectors of light-energy in the antennas of the two photosystems. They are also common pigments in products such as egg yolk, carrots and tomatoes and they occur frequently in algae. Carotenoids can be divided into two groups: *carotenes* and *xanthophylls* of which carotenes are hydrocarbons whereas xanthophylls contain one or more oxygen-containing functional groups.

Biosynthesis of tetraterpenes

As a rule, the IPP and DMAPP, which are required in the early steps of the biosynthesis of tetraterpenes, are derived from the non-mevalonate pathway. The immediate precursor of carotenoids is *phytoene*, a colourless C_{40}-hydrocarbon which is formed by coupling of two geranylgeranyl diphosphate molecules, formally by joining carbon atom 1 of one unit to carbon atom 1 of the other. As in the formation of squalene from two molecules of farnesyl diphosphate (page 434) this coupling does not occur directly but proceeds via an intermediate containing a cyclopropane ring. Ionization of one of the two GGPP molecules yields an allylic cation which launches an electrophilic attack on the 2'–3' double bond of the second GGPP molecule. Stereospecific elimination of a proton

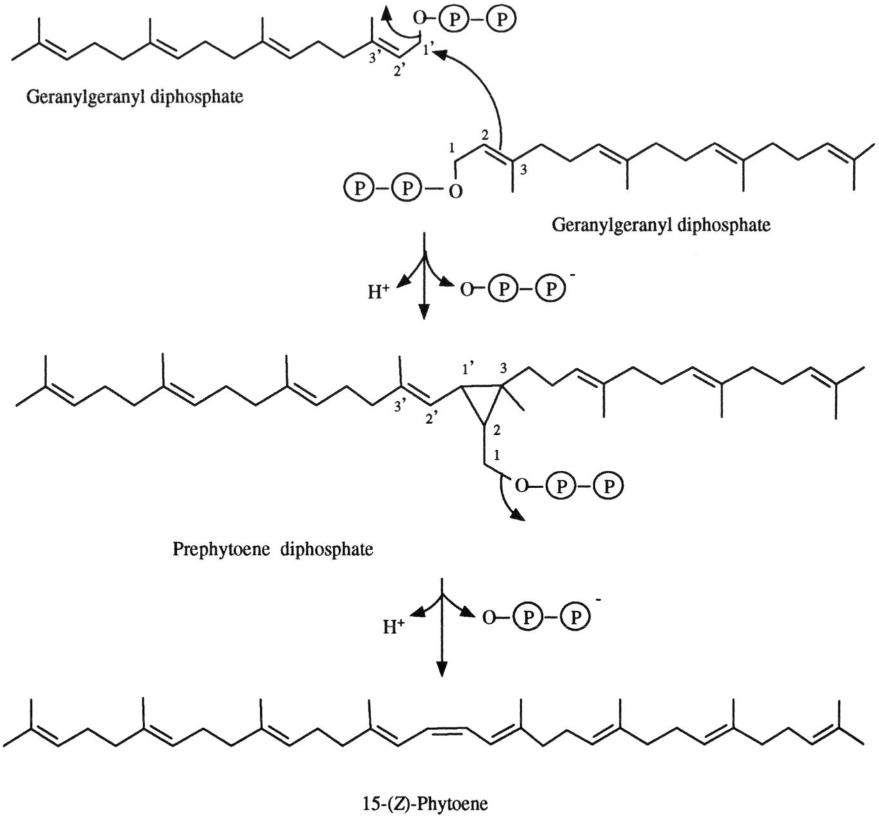

Geranylgeranyl diphosphate

Geranylgeranyl diphosphate

Prephytoene diphosphate

15-(Z)-Phytoene

Fig. 125. Biosynthesis of phytoene

from C-1 of the first molecule results in formation of the cyclo-propane ring of the resulting prephytoene diphosphate (Fig. 125). Ionization of this molecule with loss of the diphosphate group at C-1 yields a C-2 cation which undergoes a 1–1' rearrangement via a cyclobutyl cation. Depending on the stereochemistry of the subsequent hydrogen abstraction 15-(Z)- or all-*trans*-phytoene is formed. Most organisms, particularly higher plants, algae and fungi, synthesize 15-(Z)-phytoene.

The enzyme catalysing these reactions is *phytoene synthase*, (EC 2.5.1.32) which has been isolated from *Capsicum* chloroplasts. The corresponding gene has also been identified. The molecular mass of phytoene synthase is 47.5 kDa and the enzyme is dependent on Mn^{2+}. Its presence in chloroplasts has also been demonstrated.

A series of desaturation and cyclization reactions converts phy-

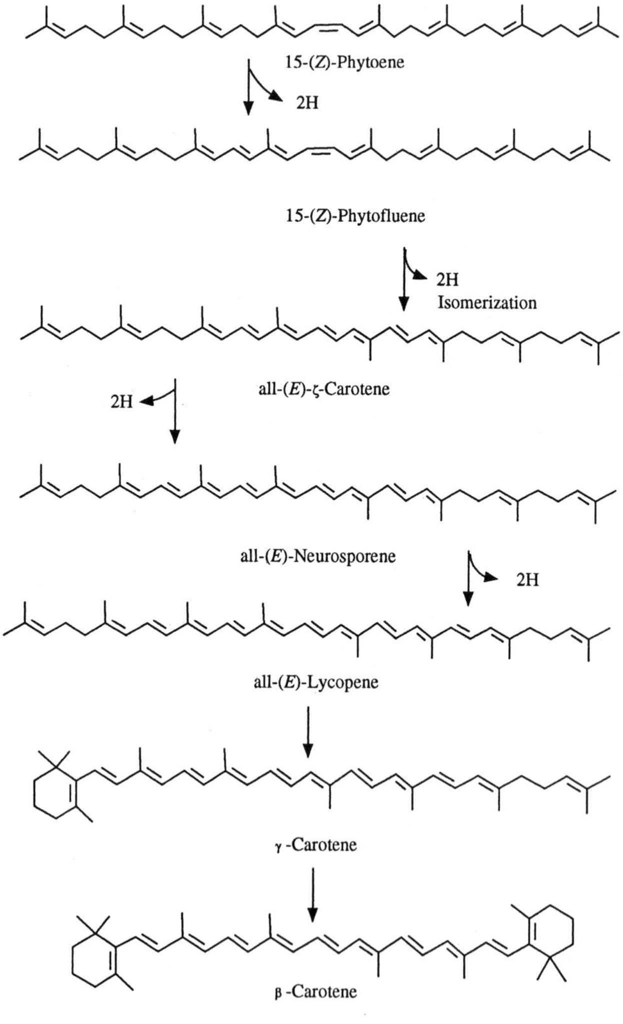

Fig. 126. Biosynthesis of cyclic carotenes from phytoene

toene into cyclic carotenes, such as β-carotene (Fig. 126). The successive introduction of conjugated double bonds during this process lengthens the chromophore, producing coloured carotenoids beginning with ζ-carotene. Lycopene, cyclic carotenes and xanthophylls usually exist in the all-*trans* configuration, indicating that at least one isomerization step must occur. The *cis* to *trans* conversion does not, however, appear to require a distinct isomerase activity.

Tetraterpenes contain 40 carbon atoms and the largest group is the carotenoids. They have been named after β-carotene, the main pigment of the carrot, *Daucus carota* L. (*Apiaceae*). Carotenoids are widespread in Nature, but they often occur only in small amounts. They are found in plants, insects, birds' feathers and egg yolk. Some carotenoids are among the pigments active in photosynthesis (page 119). Their system of conjugated double bonds strongly absorbs visible light, so that most carotenoids are yellow or orange in colour.

Vitamin A (retinol

Vitamin A is a C_{20}-compound which is formed in the liver from carotenoid precursors, ingested with the food. Lack of vitamin A causes damage to the eyesight through xerophthalmia, a turbidity in the outermost membrane of the eye (the conjunctiva). Another effect is night-blindness (nyctalopia). These conditions can be cured by administration of vitamin A. Severe lack of vitamin A causes hyperkeratosis, which is a change in the skin characterized by abnormal thickening of the epidermis. Lack of vitamin A also increases the susceptibility to infections.

Vitamin A (Retinol)

Crude drug
Cod-liver oil (Jecoris aselli oleum). See page 474.

Protocrocin
The pigments, as well as the aroma-yielding constituents, of saffron are believed to be formed when the drug is being dried, by decomposition of a hypothetical C_{40} carotenoid called protocrocin. The picrocrocin first formed is then further decomposed with the formation of safranal.

In crocin, the dibasic acid crocetin is esterfied with two molecules of gentiobiose. This is a disaccharide consisting of two glucose residues joined via a 1,6-β-glycosidic linkage. Crocin is an intensely yellow-coloured pigment and is the main colouring matter of saffron. Other glycosides of crocetin are present, and so also are other carotenoids. Picrocrocin and safranal account for most of the taste and flavour of saffron.

Crude drug
Saffron (Croci stigma) consists of the dried stigmas of *Crocus sativus* L. (*Iridaceae*). The plant is bulbous and has been grown for thousands of years. It cannot set fruit but is propagated vegetatively by means of the bulbs. These are planted in July–August, and the flowers appear in September–October of the following year. The flowers are pale violet, and the orange-coloured pistil has three long (3–3.5 cm) lobes. It is these stigma lobes that constitute the drug. Saffron originates from Greece and the Middle East. It is grown

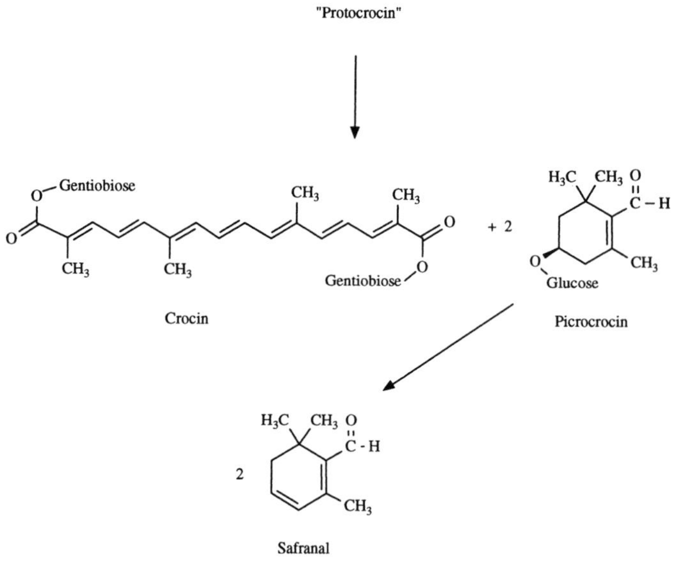

Fig. 127. Formation of pigments and aroma-yielding compounds in saffron

especially in Spain, which is the main producer, with an annual production of some 25 000 kg. Other exporters are India, France and Italy. Growing saffron requires considerable effort, both in running the plantation and above all in harvesting the product. The flowers are picked by hand and the stigma lobes are separated, likewise by hand, spread out on silk nets, and carefully dried over a charcoal fire; 100 000 flowers are needed to produce 1 kg of saffron. Hence, saffron is a very expensive product, which is often adulterated in various ways. It is mainly used as a spice, but is also of some toxicological interest, as it is believed to be an abortifacient, a belief which is supported by the demonstration of a stimulating effect on pregnant and non-pregnant uteri of guinea pigs, rabbits and dogs. At least one death has been attributed to the ingestion of 1.5 g of saffron, but the credibility of this report can be doubted as there are a number of studies published in 1980–1990 which clearly demonstrate the non-toxicity of saffron. Crocin and crocetin have radical scavenging properties and are used in sperm cryoconservation for their superoxide scavenger capacity.

In traditional medicine saffron has been used as an antispasmodic, eupeptic, gingival sedative, nerve sedative, carminative, diaphoretic, expectorant, stimulant, stomachic, aphrodisiac and emmenagogue. Although these uses have declined considerably, there is still an interest for investigations of the pharmacological properties of saffron, crocin and crocetin. Thus it has been shown that saffron has a hypolipidaemic effect which is considered to be related to the low incidence of cardiovascular disease in the regions of Spain in which saffron is consumed daily. A number of papers have been published concerning a tumour-inhibiting activity of crocetin and this compound has also been shown to promote diffusion of oxygen in different tissues, particularly in the brain.

Further reading

Biosynthesis of isoprenoids

Chappel, J. Biochemistry and molecular biology of the isoprenoid biosynthetic pathway in plants. *Annual Review of Plant Physiology and Plant Molecular Biology*, **46**, 521–547 (1995).

Dewick, P.M. The biosynthesis of C$_5$-C$_{25}$ terpenoid compounds. *Natural Products Reports*, **19**, 181 respectively,222 (2002), **16**, 97–130 (1999), **14** (2), 111–144 (1997), **12**, 507–534 (1995) and older reviews.

Kleinig, H. The role of plastids in isoprenoid biosynthesis. *Annual Review of Plant Physiology and Plant Molecular Biology*, **40**, 39–59 (1989).

Lichtenthaler, H.K., Rohmer, M. and Schwender, J. Two independent biochemical pathways for isopentenyl diphosphate and isoprenoid biosynthesis in higher plants. *Physiologia plantarum*, **101**, 643–652 (1997).

Ramos-Valdivia, A.C., van der Heijden, R. and Verpoorte, R. *Iso*pentenyl diphosphate isomerase: a core enzyme in isoprenoid biosynthesis. A review of its biochemistry and function. *Natural Products Reports*, **14**, 591–603 (1997).

Rohdich, F., Kis, K., Bacher, A. and Eisenreich, W. The non-mevalonate pathway of isoprenoids: genes, enzymes and intermediates. *Current Opinion in Chemical Biology*, **5** (5), 535–540 (2001).

Rohmer, M. The discovery of a mevalonate-independent pathway for isoprenoid biosynthesis in bacteria, algae and higher plants. *Natural Products Reports*, **16**, 565–574 (1999).

Monoterpene biosynthesis

Alonso, W.R., Rajaonarivony, J.I.M., Gershenzon, J. and Croteau, R. Purification of 4*S*-limonene synthase, a monoterpene cyclase from the glandular trichomes of peppermint (*Mentha x piperita*) and spearmint *(Mentha spicata)*. *The Journal of Biological Chemistry*, **267**, 7582–7587 (1992).

Croteau, R. Biosynthesis and catabolism of monoterpenoids. *Chemical Reviews*, **87**, 929–954 (1987).

Gershenzon, J. and Croteau, R. Regulation of monoterpene biosynthesis in higher plants. *Phytochemical Society of North America. Meeting 29, 1989. Biochemistry of the mevalonic acid pathway to terpenoids.* G.H.N. Towers and H.A. Stafford (Eds.) Plenum Press. pp. 99–160 (1990).

Rajaonarivony, J.I.M., Gershenzon, J. and Croteau, R. Characterization and mechanism of 4(*S*)-limonene synthase, a monoterpene cyclase from the glandular trichomes of peppermint *(Mentha x piperita)*. *Archives of Biochemistry and Biophysics*, **296** (1), 49–57 (1992).

Rajaonarivony, J.I.M., Gershenzon, J., Miyazaki, J. and Croteau, R. Evidence for an essential histidine residue in 4*S*-limonene synthase and other terpene cyclases. *Archives of Biochemistry and Biophysics*, **299** (1), 77–82 (1992).

Valerian root

Taibi, D.M., Landis, C.A., Petry, H. and Vitiello, M.V. A systematic review of valerian as a sleep aid: Safe but not effective. *Sleep Medicine Reviews*, **11**, 209–230 (2007).

Cannabis

Di Marzo, V., Bisogno, T. and Petrocellis, D. Endocannabinoids and related compounds: walking back and forth between plant natural products and animal physiology. *Chemistry and Biology*, **14** (July), 741–756 (2007).

Formukong, E.A., Evans, A.T. and Evans, F.J. The medicinal uses of cannabis and its constituents. *Phytotherapy Research*, **3** (6), 219–231, 1989.

Gómez-Ruiz, M., Hernández, M., de Miguel, R. and Ramos, J.A. An overview on the biochemistry of the cannabinoid system. *Molecular Neurobiology*, **36**, 3–14 (2007).

Iversen, L.L. Medical uses of marijuana? *Nature*, **365** (2 Sept.), 12–13 (1993).

Kulkarni, S.K., Ninan, I. and George, B. Anandamide: an endogeneous cannabinoid. *Drugs of Today*, **32** (4), 275–285 (1996).

Nahas, G. and Latour, C. The human toxicity of marijuana. *The Medical Journal of Australia*, **156**, 495–497 (1992).

Pacher, P., Bátkai, S. and Kunos, G. The endocannabinoid system as an emerging target of pharmacotherapy. *Pharmacological reviews*, **58** (3), 389–462 (2006).

Taura, F., Morimoto, S. and Shoyama, Y. First direct evidence for the mechanism of Δ^1-tetrahydrocannabinolic acid biosynthesis. *Journal of the American Chemical Society*, **117**, 9766–9767 (1995).

Taura, F., Morimoto, S. and Shoyama, Y. Purification and characterization of cannabidiolic-acid synthase from *Cannabis sativa*. *The Journal of Biological Chemistry*, **271** (29), 17411–17416 (1996).

Biosynthesis of sesquiterpenes

Cane, D.E. Enzymatic formation of sesquiterpenes. *Chemical Reviews*, **90**, 1089–1103 (1990).

Feverfew

Hewlett, M.J., Begley, M.J., Groenewegen, W.A., Heptinstall, S., Knight, D.W., May, J., Salan, U. and Toplis, D. Sesquiterpene lactones from feverfew, *Tanacetum parthenium*: isolation, structural revision, activity against human platelet function and implications for migraine therapy. *Journal of the Chemical Society. Perkin Transactions I*, 1979–1986 (1996).

Knight, D.K. Feverfew: chemistry and biological activity. *Natural Products Reports*, **12**, 271–276 (1995).

Artemisinin

Agtmael, M.A., Eggelte, T.A. and v. Boxtel, C.J. Artemisinin drugs in the treatment of malaria: from medicinal herb to registered medication. *Trends in Pharmacological Sciences*, **20** (5), 199–205 (1999).

Covello, P.S., Teoh, K.H., Polichuk, D.R., Reed, D.W. and Nowak, G. Functional genomics and the biosynthesis of artemisinin. *Phytochemistry*, **68**, 1864–1871 (2007).

Haynes, R.K. Artemisinin and derivatives: the future for malaria treatment? *Current Opinion in Infectious Diseases*, **14**, 719–726 (2001).

Jefford, C.W. Why artemisinin and certain synthetic peroxides are potent antimalarials. Implications for the mode of action. *Current Medicinal Chemistry*, **8**, 1803–1826 (2001).

Mercke, P. *et al.*, Molecular cloning, expression and characterization of amorpha-4,11-diene synthase, a key enzyme of artemisinin biosynthesis in *Artemisia annua* L. *Archives of Biochemistry and Biophysics*, **381** (2), 173–180 (2000).

Meshnick, S.R., Taylor, T.E. and Kamchonwongpaisan, S. Artemisinin and the antimalarial endoperoxides: from herbal remedy to targeted chemotherapy. *Micobiological Reviews*, **60** (2), 301–315 (1996).

Olliaro, P.L. and Taylor, R.J. Antimalarial compounds: from bench to bedside. *The Journal of Experimental Biology*, **206**, 3753–3759 (2003).

German chamomile

Ammon, H.P.T., Sabieraj, J. and Kaul, R. Kamille. Mechanismus der antiphlogistischen Wirkung von Kamillenextrakten und -inhaltstoffen. *Deutsche Apothekerzeitung*, **136** (22), 1821–1834 (1996).

Gibberellins

Mander, L.N. The chemistry of gibberellins: an overview. *Chemical Reviews*, **92**, 573–612 (1992).

Tocochromanols

DellaPenna, D. A decade of progress in understanding vitamin E synthesis in plants. *Journal of Plant Physiology*, **162**, 729–737 (2005).

Vitamin K

Jiang, M., Chen, M., Cao, Y., Yang, Y., Sze, K.H., Chen, X. and Guo, Z. Determination of the stereochemistry of 2-succinyl-5-enolpyruvyl-6-hydroxy-3-cyclohexene-1-carboxylate, a key intermediate in menaquinone biosynthesis. *Organic Letters*, **9** (23), 4765–4767 (2007).

Stevioside

Brandle, J.E. and Telmer, P.G. Steviol glycoside biosynthesis. *Phytochemistry*, **68**, 1855–1863 (2007).

Carakostas, M.C., Curry, L.L., Boileau, A.C., Brusick, D.J. Overview: The history, technical function and safety of rebaudioside A, a naturally occurring steviol glycoside, for use in food and beverages. *Food and Chemical Toxicology*, **46**, S1–S10 (2008).

Hanson, J.R. and De Oliveira, B.H. Stevioside and related sweet diterpenoid glycosides. *Natural Products Reports*, **10** (3), 301–309 (1993).

Kim, K.K., Sawa, Y. and Shibata, H. Hydroxylation of *ent*-kaurenoic acid to steviol in *Stevia rebaudiana* Bertoni – purification and partial characterization of the enzyme. *Archives of Biochemistry and Biophysics*, **332** (2), 223–230 (1996).

Paclitaxel

Foa, R., Norton, L. and Seidman, A.D. Taxol (paclitaxel): a novel anti-microtubule agent with remarkable anti-neoplastic activity. *International Journal of Clinical & Laboratory Research*, **24**, 6–14 (1994).

Joel, S.P. Taxol and taxotere: from yew tree to tumour cell. *Chemistry & Industry*, 172–175 (1994).

Kingston, D.G.I. Taxol: the chemistry and structure-activity relationships of a novel anticancer agent. *Trends in Biotechnology* **12**, 222–227 (1994).

Kingston, D.G. Recent advances in the chemistry of taxol. *Journal of Natural Products*, **63,** 726–734 (2000).

Long, R.M. and Croteau, R. Preliminary assessment of the C13-side chain 2´-hydroxylase involved in Taxol biosynthesis. *Biochemical and Biophysical Research Communications*, **338**, 410–417 (2005).

Rohr, J. Biosynthesis of taxol. *Angewandte Chemie International Edition in English*, **36** (20), 2190–2195 (1997).

Srinivasan, V., Pestchanker, L., Moser, S., Hirasuna, T.J., Taticek, R.A. and Shuler, M.L. Taxol production in bioreactors: kinetics of biomass accumulation, nutrient uptake, and taxol production by cell suspensions of *Taxus baccata. Biotechnology and Bioengineering*, **47**, 666–676 (1995).

Steele, C.L., Chen, Y., Dougherty, B.A., Li, W., Hofstead, S., Lam, K.S., Xing, Z. and Chiang, S.-J. Purification, cloning and functional expression of phenylalanine aminomutase: The first committed step in Taxol side-chain biosynthesis. *Archives of Biochemistry and Biophysics*, **438**, 1–10 (2005).

Walker, K., Long, R. and Croteau, R. The final acylation step in Taxol biosynthesis: Cloning of the taxoid C13-side-chain N-benzoyltransferase from *Taxus. Proceedings of the National Academy of Sciences*, **99** (14), 9166–9171 (2002).

Walker, K., Fujisaki, S., Long, R. and Croteau, R. Molecular cloning and heterologous expression of the C-13 phenylpropanoid side chain-CoA acyltransferase that functions in Taxol biosynthesis. *Proceedings of the National Academy of Sciences*, **99** (20), 12715–12720 (2002).

Walker, K.D., Klettke, K., Akiyama, T. and Croteau, R. Cloning, heterologous expression, and characterization of a phenylalanine aminomutase involved in Taxol biosynthesis. *The Journal of Biological Chemistry*, **279** (52), 53947–53954 (2004).

Wheeler, N.C., Jech, K., Masters, S., Brobst, S.W., Alvarado, A.B., Hoover, A.J. and Snader, K.M. Effects of genetic, epigenetic, and environmental factors on taxol content in *Taxus brevifolia* and related species. *Journal of Natural Products*, **55** (4), 432–440 (1992).

Williams, D.C. *et al*. Intramolecular proton transfer in the cyclization of geranylgeranyl diphosphate to the taxadiene precursor of taxol catalyzed by recombinant taxadiene synthase. *Chemistry & Biology*, **7**, 969–977 (2000).

Ginkgo biloba

Braquet, P. and Hosford, D. Ethnopharmacology and the development of natural PAF antagonists as therapeutic agents. *Journal of Ethnopharmacology*, **32**, 135–139 (1991).

Chan, P.-C., Xia, Q. and Fu, P.P. *Ginkgo biloba* leave extract: biological, medicinal and toxicological effects. *Journal of Environmental Science and Health Part C*, **25**, 211–244 (2007).

Houghton, P. Herbal products. 2. Ginkgo. *The Pharmaceutical Journal*, **253**, 122–123 (1994).

Schepmann, H.G., Pang, J. and Matsuda, P.T. Cloning and characterization of *Ginkgo biloba* levopimaradiene synthase, which catalyzes the first committed step in ginkgolide biosynthesis. *Archives of Biochemistry and Biophysics*, **392** (2), 263–269 (2001).

Schwarz, M. and Arigoni, D. Ginkgolide biosynthesis. *Comprehensive Natural Product Chemistry*. Ed. D.E. Cane. Vol. 2, 367–400. Pergamon, Oxford (1999).

Sticher, O. Quality of Ginkgo preparations. *Planta Medica*, **59**, 2–11 (1993).

Biosynthesis of triterpenoids and steroids

Abe, I., Rohmer, M. and Prestwich, G.D. Enzymatic cyclization of squalene and oxidosqualene to sterols and triterpenes. *Chemical Reviews.* **93**, 2189–2206 (1993).

Bohlmann, R. The folding of squalene; an old problem has new results. *Angewandte Chemie International Edition in English*, **31** (5), 582–584 (1992).

Brown, G.D. The biosynthesis of steroids and triterpenoids. *Natural Products Reports,* **15**, 653–696 (1998).

Harrison, D.M. The biosynthesis of triterpenoids, steroids and carotenoids. *Natural Products Reports*, **7**, 459–484 (1990).

Ourisson, G. Pecularities of sterol biosynthesis in plants. *Journal of Plant Physiology*, **143**, 434–439, (1994).

Spencer, T.A. The squalene dioxide pathway of steroid biosynthesis. *Accounts of Chemical Research*, **27**, 83–90 (1994).

Saponins

Lacaille-Dubois, M.A. and Wagner, H. A review of the biological and pharmacological activities of saponins. *Phytomedicine*, **2** (4), 363–386 (1996).

Marston, A. and Hostettmann, K. Plant molluscicides. *Phytochemistry*, **24** (4), 639–652 (1985).

Thilborg, S.T., Christensen, S.B., Cornett, C., Olsen, C.E. and Lemmich, E. Molluscicidal saponins from *Phytolacca dodecandra. Phytochemistry*, **32** (5), 1167–1171 (1993).

Liquorice

Davis, E.A. and Morris, D.J. Medicinal uses of licorice through the millennia: the good and plenty of it. *Molecular and Cellular Endocrinology*, **78**, 1–6 (1991).

Farese Jr, R.V., Biglieri, E.G., Shackleton, C.H.L., Irony, I. and Gomez-Fontes, R. Licorice-induced hypermineralocorticoidism. *The New England Journal of Medicine*, **325** (17), 1223–1227 (1991).

Isbrucker, R.A. and Burdock, G.A. Risk and safety assessment on the consumption of Licorice root (*Glycyrrhiza* sp.), its extract and powder as a food ingredient, with emphasis on the pharmacology and toxicology of glycyrrhizin. *Regulatory Toxicology and Pharmacology*, **46**, 167–192 (2006).

Morris, D.J. Liquorice: New insights into mineralocorticoid and glucocorticoid hypertension. *Rhode Island Medicine*, **76**, 251– 254 (1993).

Quillaja saponins

Morein, B. The iscom antigen-presenting system. *Nature*, **332**, 287–288 (1988).

Aescin

Pittler, M.H. and Ernst, E. Horse chestnut seed extract for chronic venous insufficiency (Cochrane Review). *The Cochrane Library, Issue* **4,** 2002. Oxford: Update Software.

Cardiac glycosides

Gärtner, D.E. and Seitz, H.U. Enzyme activities in cardenolide-accumulating, mixotrophic shoot cultures of *Digitalis purpurea* L. *Journal of Plant Physiology*, **141**, 269–275 (1993).
Kreis, W., Hensel, A. and Stuhlemmer, U. Cardenolide biosynthesis in Foxglove. *Planta Medica*, **64**, 491–499 (1998).

Biosynthesis of carotenoids

Armstrong, G.A. and Hearst, J.E. Genetics and molecular biology of carotenoid pigment biosynthesis. *The FASEB journal*, **10**, 228–237 (1996).
Sandmann, G. Carotenoid biosynthesis in microorganisms and plants. *European Journal of Biochemistry*, **223**, 7–24 (1994).

Saffron

Rios, J.L., Recio, M.C., Giner, R.M. and Máñez, S. An update review of saffron and its active constituents. *Phytotherapy Research*, **10**, 189–193 (1996).

8. Amino Acids

The biosynthetic overview presented in Fig. 21 (page 135) shows that amino acids are the primary products formed when nitrogen is incorporated into organic substances. The nitrogen contained in natural products all comes from the atmosphere. Higher plants, however, are unable to assimilate atmospheric nitrogen, but cyanobacteria and other bacteria (especially of the genera *Azotobacter*, *Aerobacter* and *Clostridium*) are able to reduce nitrogen to ammonia. The ammonia can either be utilized by the bacteria for their own metabolism or it can be emitted to the surroundings, where it can be absorbed by plants or be oxidized to nitrate by other bacteria. The latter reaction is the predominant one.

Some plants belonging to the families *Fabaceae* and *Cycadaceae* live in symbiosis with nitrogen-fixing bacteria of the genus *Rhizobium*. The plant develops nodules on its roots, or a layer of specific tissue below the root cortex where the bacteria live. These reduce atmospheric nitrogen to ammonia, which is absorbed by ammonia-acceptors supplied by the plant, and the amino acids subsequently formed are utilized by both organisms.

Besides fixation by bacteria, atmospheric nitrogen is also made accessible to plants through electric discharges during thunderstorms. Every flash of lightning causes oxidation of atmospheric nitrogen to nitrogen oxides, which dissolve in rain and end up in the soil, where bacteria transform them into nitrate and other nitrogenous compounds.

Incorporation of nitrogen into amino acids

The above-described plants that live in symbiosis with nitrogen-fixing bacteria are exceptions as regards the uptake of nitrogen. Most higher plants have to use nitrate as their source of nitrogen. The nitrate can be absorbed from the soil by the roots of the plant but cannot react with a nitrogen acceptor without previous reduction to ammonia. This reduction takes place in two steps. The enzyme *nitrate reductase (NADH)* (EC 1.7.1.1) reduces the nitrate to nitrite, which is then reduced to ammonia by *ferredoxin-nitrite reductase* (EC 1.7.7.1)

The main acceptor of ammonia in higher plants is 2-oxoglutaric acid, one of the acids formed in the citric-acid cycle which is part of carbohydrate metabolism (page 137). The enzyme *glutamate dehydrogenase [NAD(P)+]* (EC 1.4.1.3) catalyses the reaction in plants.

At the pH of the cytosol, acids are ionized (COO-) and the amino acids are present as zwitterions (NH_3^+ and COO-). However, in the reaction scheme overleaf and in the following schemes they are presented as non-ionized structures.

Other amino acids are formed through transamination, whereby

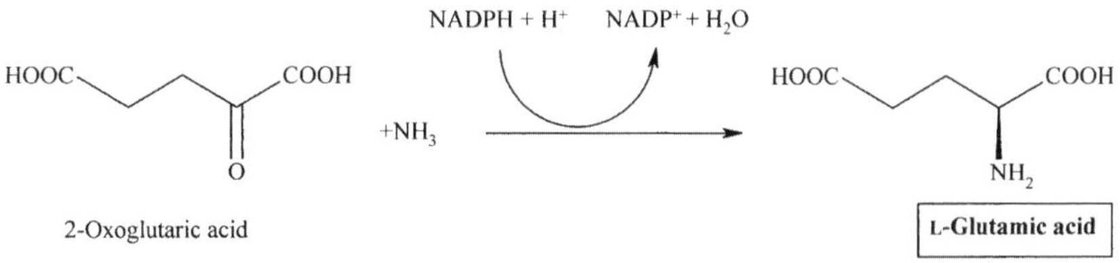

$$NADPH + H^+ \qquad NADP^+ + H_2O$$

2-Oxoglutaric acid

$+NH_3$

L-Glutamic acid

Fig. 128. Ammonia fixation in higher plants. Biosynthesis of glutamic acid.

the amino group of glutamic acid or some other amino acid is transferred to a keto-acid derived from carbohydrate metabolism. The enzyme *glutamate dehydrogenase* [NAD(P)$^+$] (EC 1.4.1.3) catalyses the reaction in plants.

Amino acids are structurally characterized by the presence of a carbon atom to which an amino- and a carboxyl group are attached. This carbon atom – the α carbon – is a member of a chain or is incorporated in a ring-structure and is thus asymmetric, making two configurations possible. By convention these are related to D- and L-glyceraldehyde and thus termed D- and L-amino acids. In Nature amino acids occur free or incorporated into peptides and proteins. Most natural amino acids, particularly those which are part of protein structures, have the L-configuration, but examples of peptides containing D-amino acids are presented in Chapter 9.

There are over 100 variations of the side-chain attached to the α-carbon, but only 20 of these are present in proteins. They are called proteinogenic amino acids and can be grouped into "families" on the basis of the carbon skeleton to which the amino function is attached. The biosynthesis of the amino acids has been studied most extensively in bacteria but a lot of knowledge is also available from plants. Most of the pathways described below have been confirmed also on the genetic level, with genes encoding the enzymes identified.

2-oxoglutaric acid group

The carbon skeleton of the amino acids belonging to this group comes from 2-oxoglutaric acid. The group comprises the following amino acids: glutamic acid, glutamine, ornithine, citrulline, arginine, proline and hydroxyproline. As described above, glutamic acid is the primary product of ammonia fixation in higher plants. This acid then functions directly or indirectly as the NH$_2$ donor in the biosynthesis of the other amino acids. In some cases, keto-acids other than 2-oxoglutaric acid, chiefly oxalacetic acid, can function as ammonia acceptors, but the route via glutamic acid generally predominates.

Glutamine   The enzyme *glutamate-ammonia ligase* (EC 6.3.1.2, also known as

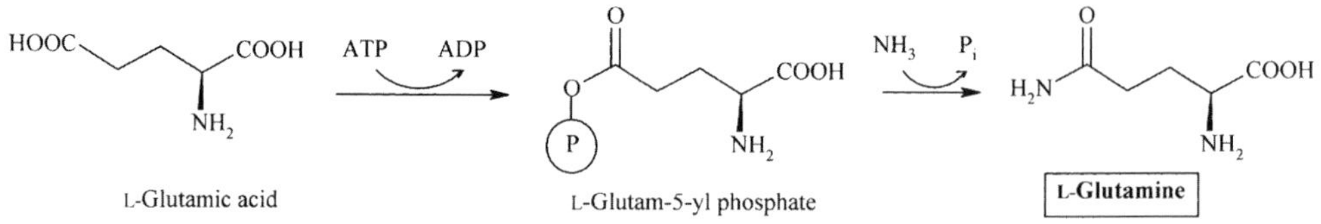

Fig. 129. Biosynthesis of glutamine

glutamine synthetase) catalyses the reaction of glutamic acid with a second molecule of ammonia to form glutamine. This reaction proceeds in two steps, via the intermediate L-glutam-5-yl phosphate.

Proline and hydroxyproline   Proline is a cyclized derivative of glutamic acid. The biosynthesis is presented in Fig. 130 and starts (as in the biosynthesis of glutamic acid) with phosphorylation of glutamic acid to form L-glutam-5-yl phosphate, but the reaction is catalysed by another enzyme – *glutamate 5-kinase* (EC 2.7.2.11). The γ-carboxyl group of this compound is reduced to a carbonyl group by the enzyme *glutamate-5-semialdehyde dehydrogenase* (EC 1.2.1.41) with formation of L-glutamic acid 5-semialdehyde. Spontaneous cyclization yields (*S*)-3,4-dihydro-2*H*-pyrrole-2-carboxylic acid which is transformed to proline by *pyrroline-5-carboxylate reductase* (EC 1.5.1.2).

Alternatively, proline can be formed from ornithine (see below) by cyclization, catalysed by *ornithine-cyclodeaminase* (EC 4.3.1.12).

The relatively rare amino acid hydroxyproline arises by oxidation of proline. This is a complex process, which has been demonstrated for the formation of hydroxyproline in collagen, the major connective tissue in vertebrates. In nascent collagen chains, proline residues on the

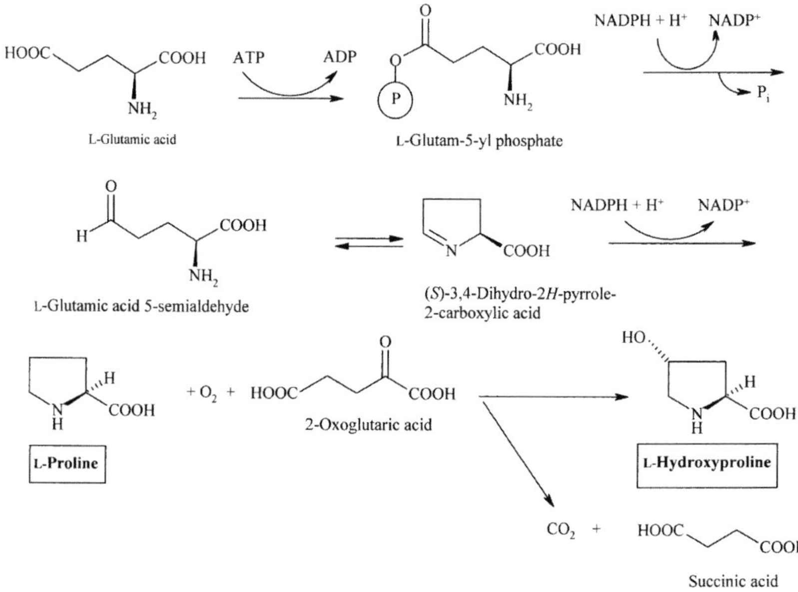

Fig. 130. Biosynthesis of proline and hydroxyproline

amino side of glycine residues are hydroxylated at C-4 by the enzyme *procollagen-proline dioxygenase* (EC 1.14.11.2, also known as proline-4-hydroxylase). The required oxygen atom comes from molecular oxygen – O_2. The other oxygen atom of O_2 is used for transformation of 2-oxoglutaric acid to succinic acid and CO_2 as illustrated in Fig. 130. Proline-4-hydroxylase contains a tightly bound Fe^{2+} ion, which is needed to activate O_2. The enzyme can also convert 2-oxoglutaric acid to succinic acid without hydroxylating proline. In this reaction an oxidized iron complex is formed, which inactivates the enzyme. The ferric ion of the inactivated enzyme can be reduced by ascorbate (vitamin C), whereby the activity of the enzyme is restored.

Ornithine and arginine Also ornithine is derived from glutamic acid. The pathway (Fig. 131) parallels some steps of the proline biosynthesis, but the first reaction is acetylation of glutamic acid, mediated by the enzyme *amino-acid* N-*acetyltransferase* EC 2.3.1.1). This is necessary to prevent cyclization of the glutamic acid γ-semialdehyde formed in the third reaction. The *N*-acetylglutamic acid is phosphorylated by *acetylglutamate kinase* (EC 2.7.2.8) to yield *N*-acetyl-L-glutamyl 5-phosphate, which is reduced by N-*acetyl-γ-glutamyl-phosphate reductase* (EC 1.2.1.38) with formation of *N*-acetyl-L-glutamic acid 5-semialdehyde. Transamidation, mediated by N²-*acetylornithine 5-transaminase* (EC 2.6.1.11) yields *N*-acetylornithine. The blocking acetyl group is finally removed by *acetylornithine deacetylase* (EC 3.5.1.16) yielding ornithine.

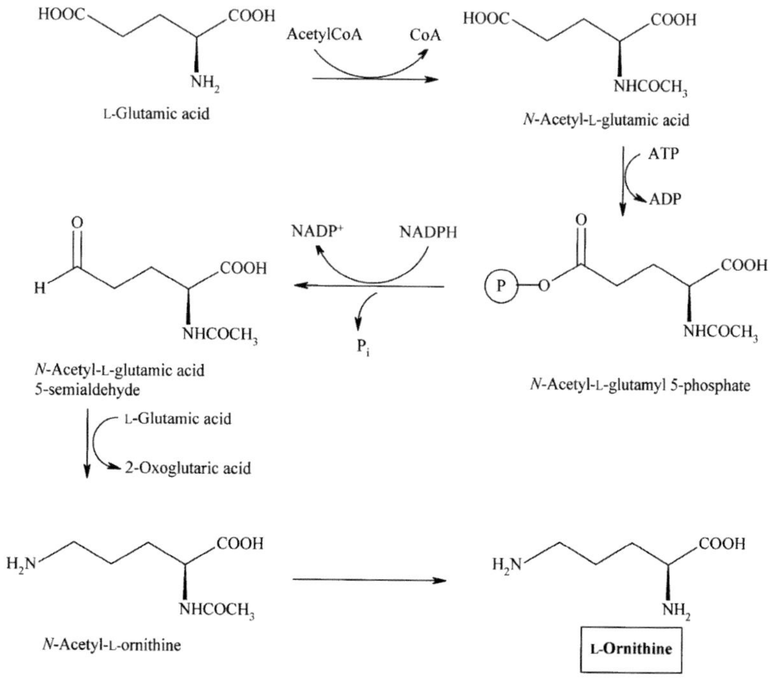

Fig. 131. Biosynthesis of ornithine

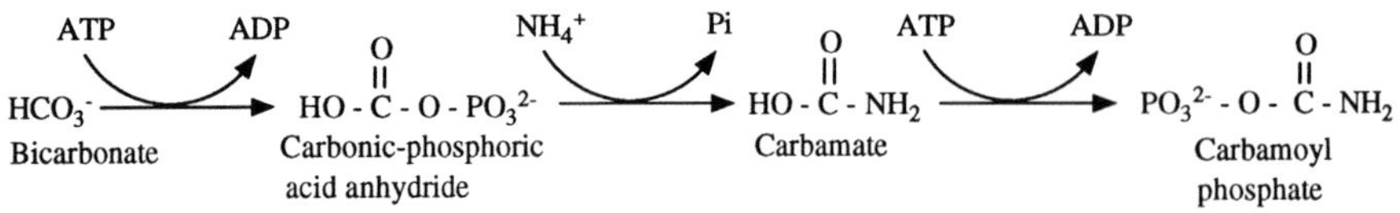

Fig. 132. Biosynthesis of carbamoyl phosphate

Ornithine is an important intermediate in the biosynthesis of tropane and other alkaloids. It is normally built into peptides and proteins but also plays an important role in the biosynthesis of arginine (Fig. 133).

The first step in the transformation of ornithine into arginine is a reaction with carbamoyl phosphate. This compound is formed from carbon dioxide and ammonia in an ATP-dependent reaction which is catalysed by the enzyme *carbamoyl phosphate synthase (ammonia)* (EC 6.3.4.16). Another variety of the enzyme – *carbamoyl-phosphate synthase (glutamine hydrolysing)* (EC 6.3.5.5) – first produces ammonia by hydrolysis of glutamine and then uses the liberated ammonia in the reaction with carbonic phosphoric acid anhydride. Carbamoyl phosphate is also used in the biosynthesis of uridine phosphate (UMP, page 738).

Ornithine reacts with carbamoyl phosphate to give citrulline in a reaction catalysed by *ornithine carbamoyltransferase* (EC 2.1.3.3).

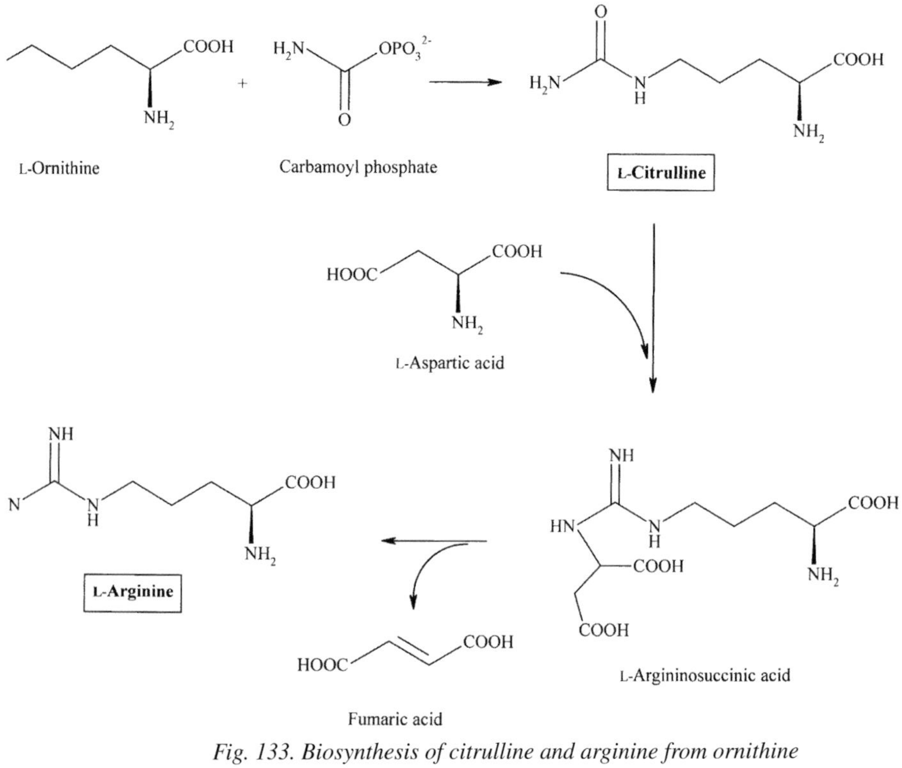

Fig. 133. Biosynthesis of citrulline and arginine from ornithine

490

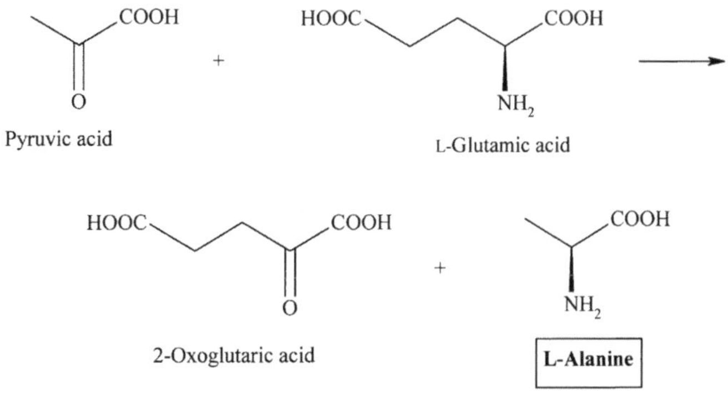

Fig. 134. Biosynthesis of alanine

Citrulline contains three nitrogen atoms. The fourth nitrogen atom needed for arginine comes from aspartic acid, the formation of which is described on page 493. In a reaction catalysed by *argininosuccinate synthase* (EC 6.3.4.5), aspartic acid and citrulline form argininosuccinic acid, which is finally split by the enzyme *argininosuccinate lyase* (EC 4.3.2.1) into arginine and fumaric acid.

Pyruvic acid group

This group of amino acids is derived from pyruvic acid, which is formed biosynthetically through glycolysis of carbohydrates. The members are alanine, valine and leucine. Pyruvic acid also enters the pathway for formation of isoleucine (see the oxalacetic acid group below).

Alanine   Alanine is formed from pyruvic acid by transamination, catalysed by *alanine transaminase* (EC 2.6.1.2) as illustrated in Fig. 134.

Some plants are able to produce alanine directly from pyruvic acid and ammonia by reductive amination.

Valine and   The biosynthesis of valine and leucine has been studied mainly in
leucine   bacteria, but experiments have indicated that it follows a similar course in higher plants. The two amino acids have a common precursor – 3-methyl-2-oxo-butanoic acid – which is synthesized in three steps, starting with pyruvic acid (Fig. 135). In the first reaction, catalysed by *acetolactate synthase* (EC 2.2.1.6), two molecules of pyruvic acid form α-acetolactic acid (= (S)-2-hydroxy-2-methyl-3-oxobutanoic acid) and CO_2. The enzyme requires thiamine pyrophosphate, which reacts with one of the pyruvic acid molecules with loss of CO_2 under formation of "active acetaldehyde", which then reacts with the second molecule of pyruvic acid. The enzyme *ketol-acid reductoisomerase* (EC 1.1.1.86) then

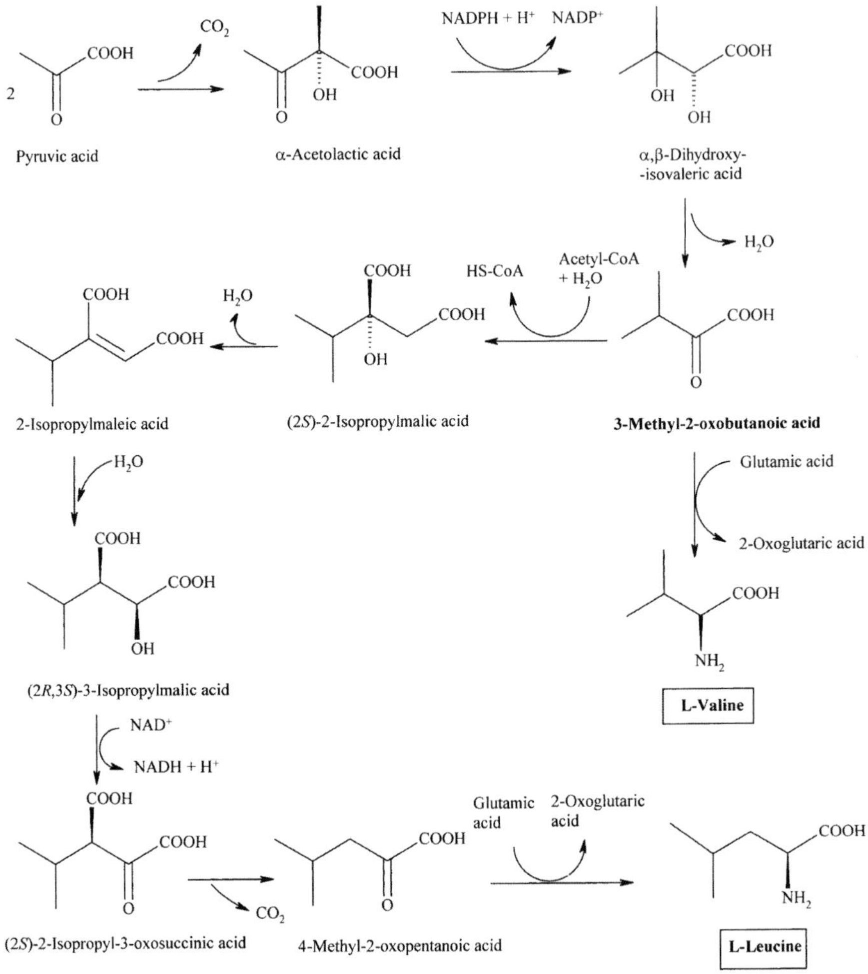

Fig. 135. Biosynthesis of valine and leucine

reduces and isomerizes α-acetolactic acid to α,β-dihydroxy-isovaleric acid (= (2*R*)-2,3-dihydroxy-3-methylbutanoic acid), which loses water with formation of 3-methyl-2-oxobutanoic acid in a reaction catalysed by *dihydroxy-acid dehydratase* (EC 4.2.1.9). The enzyme *branched-chain-amino-acid transaminase* (EC 2.6.1.42) then catalyses transamination with glutamic acid which yields valine.

The biosynthesis of leucine starts with condensation of 3-methyl-2-oxobutanoic acid with acetyl-CoA to yield, (2*S*)-2-isopropylmalic acid, a reaction which is catalysed by *2-isopropylmalate synthase* (EC 2.3.3.13). The next step is a two-stage reaction in which *3-isopropylmalate dehydratase* (EC 4.2.1.33) first catalyses loss of water from (2*S*)-2-isopropylmalic acid with formation of 2-isopropylmaleic acid, followed by addition of water to this compound, yielding (2*R*,3*S*)-3-isopropylmalic acid. The 3-isopropyl malic acid is oxidized by *3-isopropylmalate dehydrogenase* (EC

1.1.1.85) to (2S)-2-isopropyl-3-oxosuccinic acid, which spontaneously decarboxylates to 4-methyl-2-oxopentanoic acid (= α-ketoisocaproic acid). Transamination with glutamic acid finally yields leucine. The reaction is mediated by the same enzyme as in the valine biosynthesis – *branched-chain-amino-acid transaminase* (EC 2.6.1.42) – but can also be accomplished by the more specific *leucine transaminase* (EC 2.6.1.6), which does not act in the formation of valine.

Oxalacetic acid group

Oxalacetic acid is one of the oxidation products formed in the citric-acid cycle (page 137). It can also arise through direct carboxylation of pyruvic acid. This family comprises the amino acids aspartic acid, asparagine, lysine, threonine, methionine and isoleucine.

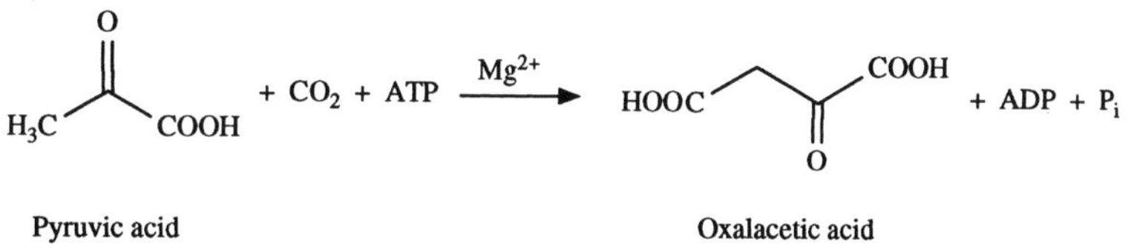

Pyruvic acid   Oxalacetic acid

Aspartic acid and asparagine As illustrated in Fig. 136, aspartic acid is formed in a transamination reaction between glutamic acid and oxalacetic acid, catalysed by *aspartate transaminase* (EC 2.6.1.1). Amidation of aspartic acid yields asparagine. In microorganisms the amidation can be achieved by fixation of an ammonium ion, in analogy with the biosynthesis of glutamine. Two enzymes – *aspartate-ammonia ligase* (EC 6.3.1.1) and *aspartate-ammonia ligase (ADP-forming)* (EC 6.3.1.4) – are known to catalyse this reaction. Formation of asparagine can also be achieved by transamination between aspartic acid and glutamine, a process that dominates in plants and animals. The catalysing enzyme for this process is *asparagine synthase (glutamine-hydrolysing)* (EC 6.3.5.4). The reaction is ATP dependent.

Lysine One biosynthetic route to lysine is known to occur in both bacteria and higher plants and a different one in fungi. In bacteria and higher plants the pathway (Fig. 137) starts with aspartic acid which is phosphorylated to L-aspart-4-yl phosphate in a reaction catalysed by the enzyme aspartate kinase (EC 2.7.2.4). This compound is reduced to L-aspartic acid 4-semialdehyde by *aspartate-semialdehyde dehydrogenase* (EC 1.2.1.11). Aldol condensation with pyruvic acid, catalysed by *dihydropicolinate synthase* (EC 4.2.1.52) yields a compound, which spontaneously forms an intermediate cyclic Schiff base under elimination of water. Loss of another molecule of water yields 2,3-dihydrodipicolinic acid, which is transformed to 2,3,4,5-tetrahydrodipicolinic acid by *dihydropicolinate reductase* (EC1.3.1.26). The enzyme *2,3,4,5-tetrahydropyridine-2,6-dicarboxylate N-succinyltransferase* (EC 2.3.1.117) catalyses opening of the tetrahydropyridine ring and subsequent succinylation of the aminogroup, yielding *N*-succinyl-L-2-amino-6-oxopimelic acid. Transamination with glutamic acid, catalysed by *succinyldiaminopimelate transaminase* (EC 2.6.1.17) results in formation of *N*-succinyl-L,L-2,6-diaminopimelic acid from which succinic acid is split off by *succinyl-diaminopimelate desuccinylase* (EC 3.5.1.18), yielding L,L-2,6-diaminopimelic acid. *Diaminopimelate epimerase* (EC 5.1.1.7) catalyses isomerization of L,L-2,6-diaminopimelic acid to *meso*-2,6-diaminopimelic acid. The final step in the biosynthesis of lysine is decarboxylation of this compund by *diaminopimelate decarboxylase* (EC 4.1.1.20) .

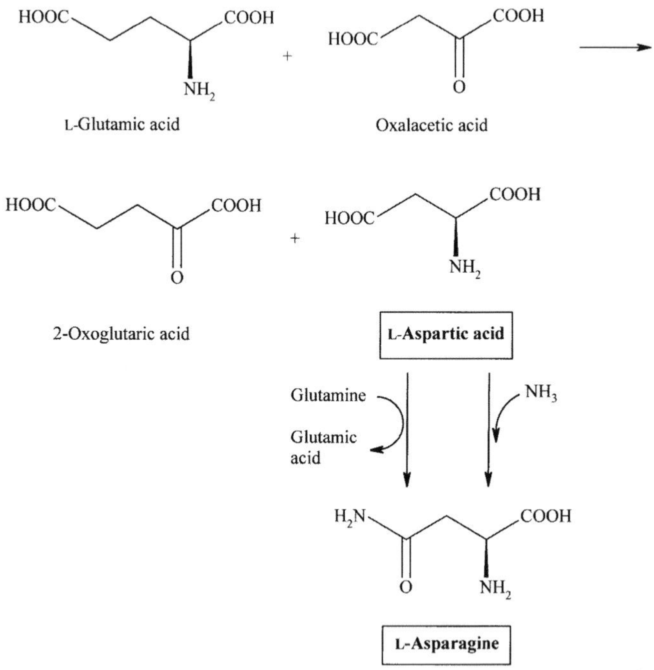

Fig. 136. Biosynthesis of aspartic acid and asparagine

In Euglenoids and higher fungi (Ascomycetes and Basidiomycetes) lysine biosynthesis proceeds through the intermediacy of L-α-aminoadipate in a pathway which is completely unrelated to the diaminopimelate route of bacteria and plants. This pathway, which is illustrated in Fig. 138, has been elucidated both at the enzymatic and genetic level and proceeds in eight steps involving seven free intermediates. The first step is a condensation between 2-oxoglutaric acid and acetyl-CoA, catalysed by *homocitrate synthase* (EC

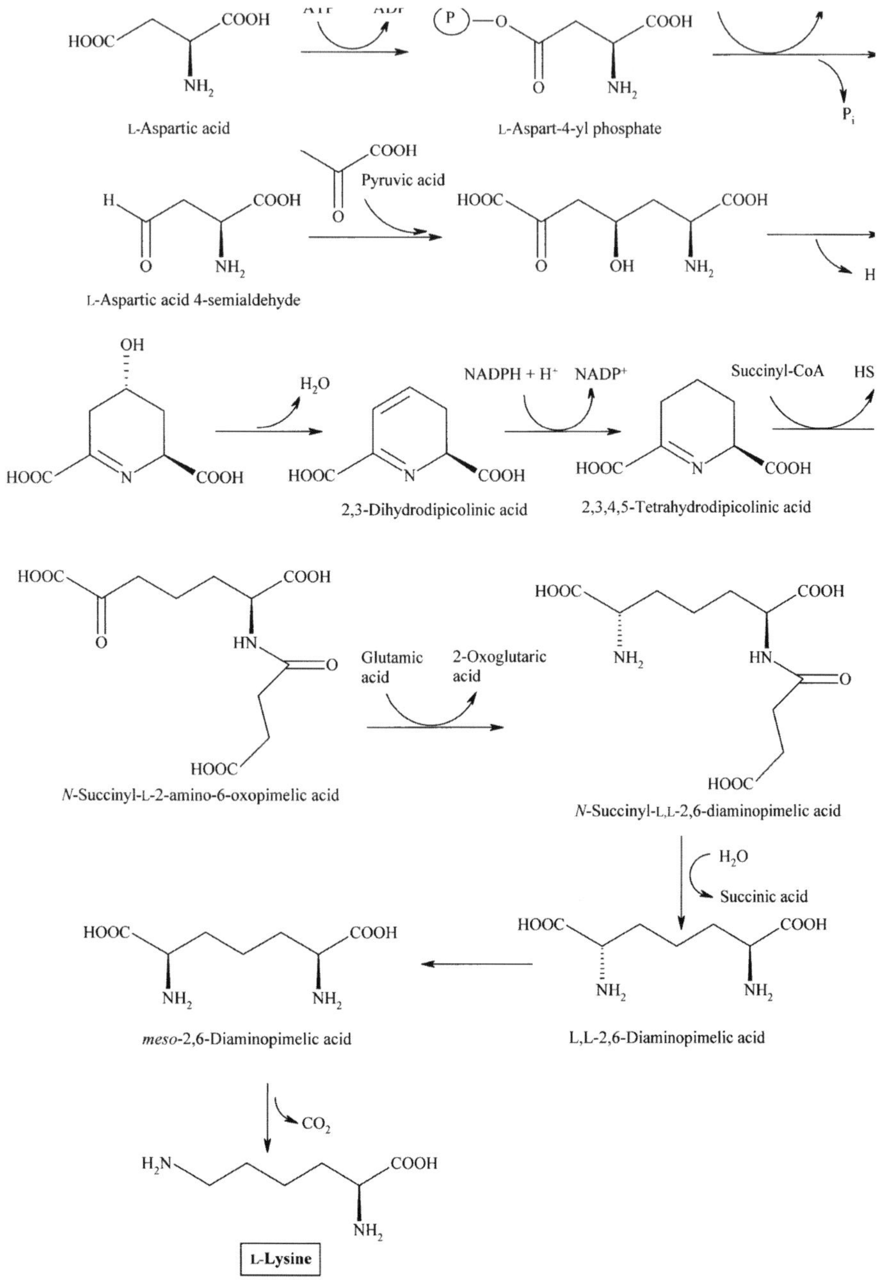

Fig. 137. Biosynthesis of lysine in bacteria and higher plants

2.3.3.14), yielding homocitric acid. *Homoaconitate hydratase* (EC 4.2.1.36) catalyses a two-step conversion of homocitric acid to homoisocitric acid with *cis*-homoacotinic acid as an intermediary product. In the next reaction, catalysed by *homoisocitrate dehydrogenase* (EC 1.1.1.87) homoisocitric acid is oxidized with loss of carbon dioxide to form 2-oxoadipic acid which is transaminated to

L-α-aminoadipic acid in a reaction involving glutamic acid, catalysed by *2-aminoadipate transaminase* (EC 2.6.1.39). In β-lactam-producing fungi the pathway forms a branch here, and part of the α-aminoadipic acid is used for the biosynthesis of β-lactams (page 563). In the pathway for the biosynthesis of lysine, the 6-COOH group of α-aminoadipic acid is reduced by *L-aminoadipate-semi-aldehyde dehydrogenase* (EC 1.2.1.31) to form L-α-aminoadipic acid-6-semialdehyde in a process requiring ATP and NADPH. In the next reaction, catalysed by *saccharopine dehydrogenase (NADP+, L-glutamate-forming)* (EC 1.5.1.10), L-α-aminoadipic acid-6-semialdehyde is condensed with glutamic acid, forming an imine, which is reduced with formation of L-saccharopine. In the final step, catalysed by *saccharopine dehydrogenase (NAD+, L-lysine-forming)*, (EC 1.5.1.7), the C-N bond within the glutamic acid moiety is cleaved, yielding L-lysine and 2-oxoglutaric acid.

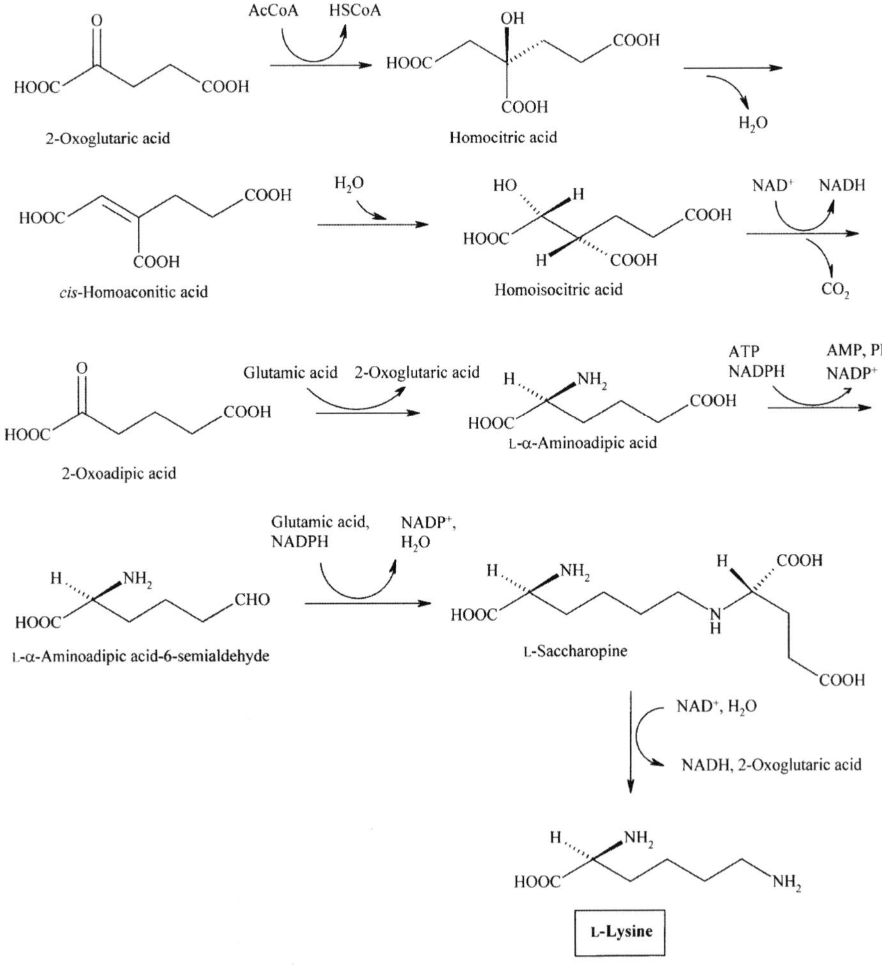

Fig. 138. Biosynthesis of lysine in higher fungi

Threonine and methionine

These two amino acids also have their origin in aspartic acid. The first two steps in the biosynthesis – aspartic acid $\rightarrow$ L-aspart-4-yl phosphate $\rightarrow$ L-aspartic acid 4-semialdehyde – are the same as in the biosynthesis of lysine (Fig.137) and are catalysed by the same enzymes. Aspartic acid 4-semialdehyde is transformed to L-homoserine by the enzyme *homoserine dehydrogenase* (EC 1.1.1.3). In plants this enzyme requires the presence of NADPH. The next reaction is catalysed by *homoserine kinase* (EC 2.7.1.39) which in the presence of ATP converts homoserine to 4-phospho-L-homoserine. The final step in the biosynthesis of threonine is an irreversible hydrolysis of 4-phospho-L-homoserine to yield threonine in a reaction catalysed by *threonine synthase* (EC 4.2.3.1).

In plants the biosynthesis of methionine starts with 4-phospho-L-homoserine (Fig. 140) which reacts with cysteine (page 501) to form L,L-cystathionine in a process catalysed by the enzyme *cystathionine γ-synthase* (EC 2.5.1.48). L,L-Cystathionine is subjected to β-cleavage by the enzyme *cystathionine-β-lyase*, with formation of homocysteine, pyruvic acid and ammonia. Methionine is finally formed by methylation of homocysteine in a reaction catalysed by *methionine synthase* (EC 2.1.1.13) which takes the methyl group from 5-methyl-tetrahydrofolate (page 725) in a cobalamin dependent reaction.

Methionine is an important donor of methyl groups in many biosynthetic reactions (*O*-, *N*- and *C*-methylations). In these, methionine is first converted to *S*-adenosylmethionine (SAM), in a reaction with ATP (Fig.141), catalysed by *methionine adenosyltransferase* (EC 2.5.1.6). The sulphur atom in SAM has a positive charge which facilitates nucleophilic substitution (S_N2) type mechanisms.

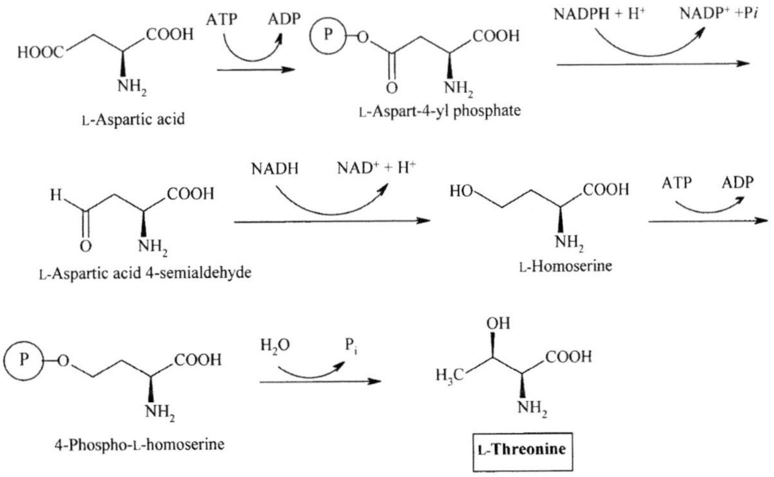

Fig. 139. Biosynthesis of threonine

497

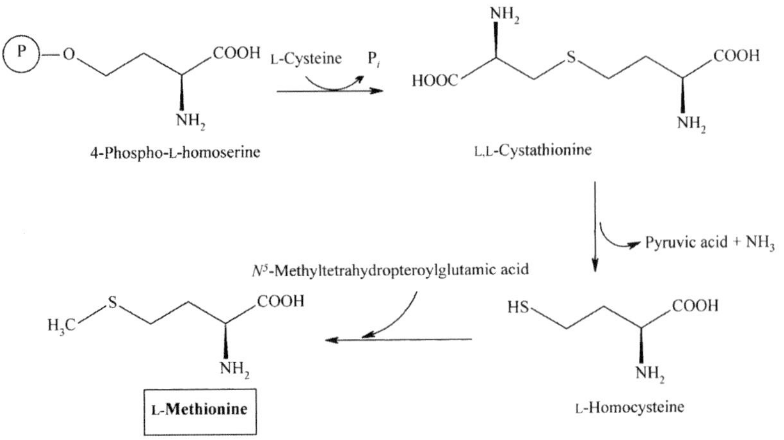

Fig. 140. End of the biosynthetic pathway to methionine

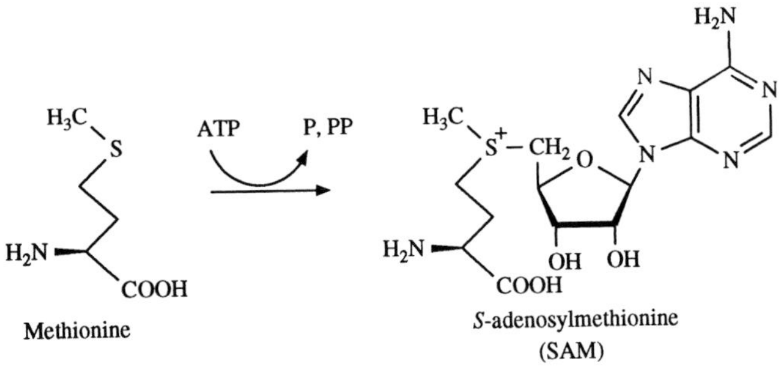

Fig. 141. Biosynthesis of S-adenosylmethionine (SAM)

Isoleucine   The biosynthesis of isoleucine is a continuation of the biosynthetic path that leads from aspartic acid to threonine. The first step (Fig. 142) is a deamination of threonine, catalysed by *threonine ammonia-lyase* (EC 4.3.1.19), with formation of 2-oxobutanoic acid. Condensation of 2-oxobutanoic acid with pyruvic acid yields (S)-2-ethyl-2-hydroxy-3-oxobutanoic acid, a reaction which is catalysed by *acetolactate synthase* (EC 2.2.1.6), the same enzyme that catalyses condensation of two molecules of pyruvic acid to α-acetolactic acid in the biosynthesis of valine and leucine (page 491). Also the next three steps are catalysed by the same enzymes as corresponding steps in the valine biosynthesis. (S)-2-ethyl-2-hydroxy-3-oxobutanoic acid is reduced to (2R,3R)-2,3-dihydroxy-3-methyl-pentanoic acid by *ketol-acid reductoisomerase* (EC 1.1.1.86). The reaction involves an alkylmigration and a NADPH-dependent reduction. Dehydration of (2R,3R)-2,3-dihydroxy-3-methyl-pentanoic acid to form (S)-3-methyl-2-oxopentanoic acid is catalysed by *dihydroxy-acid dehydratase* (EC 4.2.1.9). Finally isoleucine is formed by transamination with glutamic acid, catalysed by *branched-chain-amino-acid transaminase* (EC 2.6.1.42).

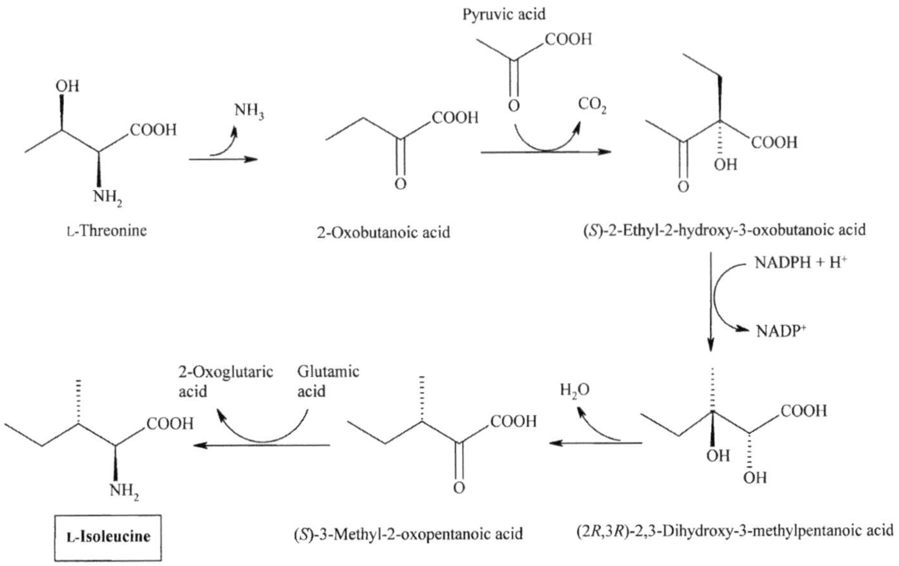

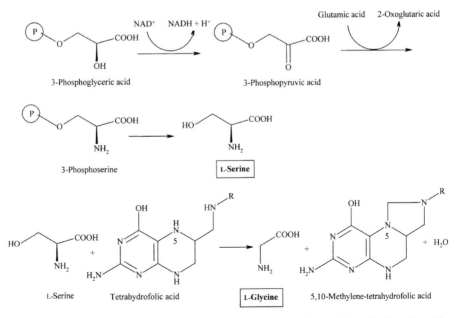

Fig. 142. Isoleucine biosynthesis from threonine

Serine group

This group of amino acids includes serine, glycine, cysteine and cystine. Serine, which is formed from 3-phosphoglyceric acid, a

Fig. 143. Formation of serine and glycine from 3-phosphoglyceric acid

product of the Calvin–Benson cycle (page 128), is a precursor of the other amino acids of this family. The hydroxymethyl group of serine can also be split off to form an important "active" one-carbon-fragment in biosynthesis. The role of serine in tryptophan formation has already been discussed (page 208).

Serine and glycine The first step in the serine biosynthesis (Fig. 143) is oxidation of the hydroxyl group of 3-phosphoglyceric acid to yield 3-phosphopyruvic acid in a reaction with NAD+, catalysed by *phosphoglycerate dehydrogenase* (EC 1.1.1.95). *Phosphoserine transaminase* (EC 2.6.1.52) then mediates transamination with glutamic acid forming 3-phosphoserine, which is hydrolysed by *phosphoserine phosphatase* (EC 3.1.3.3) to form serine.

Serine is transformed into glycine by *glycine hydroxymethyl transferase* (EC 2.1.2.1) in a reaction with tetrahydrofolic acid, which accepts the hydroxymethyl group of the serine side-chain with formation of 5,10-methylene-tetrahydrofolic acid.

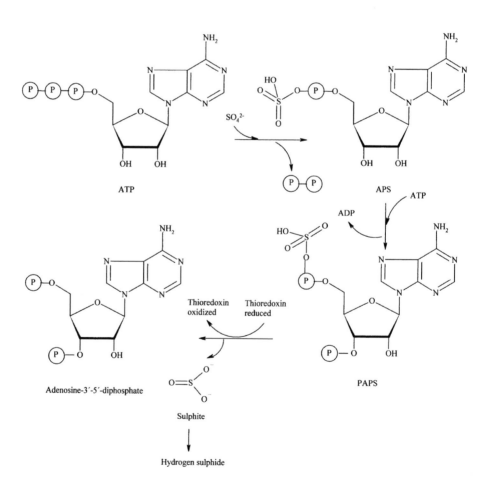

Fig. 144. Activation of sulphur via APS and PAPS

Alternative pathways for serine and glycine biosynthesis have also been discovered. In leaves, 3-phosphoglycerate can lose phosphate, forming glyceric acid which can be oxidized to hydroxypyruvic acid. Transamination with alanine can then give serine.

The conversion of serine into glycine is reversible. Moreover, glycine can be formed from glyoxylic acid in connection with an alternative biosynthetic pathway to hexoses.

Cysteine and cystine Plants absorb sulphur in the form of sulphate. To enable the sulphur to be built into organic compounds, the sulphate must first be reduced. This process begins with a two-step reaction between sulphate and ATP (Fig. 144). In the first step, catalysed by *sulphate adenylyltransferase* (EC 2.7.7.4), ATP reacts with sulphur yielding adenylyl sulphate (APS) and inorganic diphosphoric acid. APS then reacts with a second molecule of ATP with formation of 3´-phosphoadenylyl sulphate (PAPS) and ADP. This reaction is catalysed by *adenylyl-sulphate kinase* (EC 2.7.1.25). PAPS, also called active sulphate, yields sulphite in a reaction with thioredoxin, catalysed by the enzyme *phosphoadenylyl-sulphate reductase (thioredoxin)* (EC 1.8.4.8). Sulphite is reduced to sulphide by *sulphite reductase (ferredoxin)* (EC 1.8.7.1) for incorporation of sulphur into biomolecules.

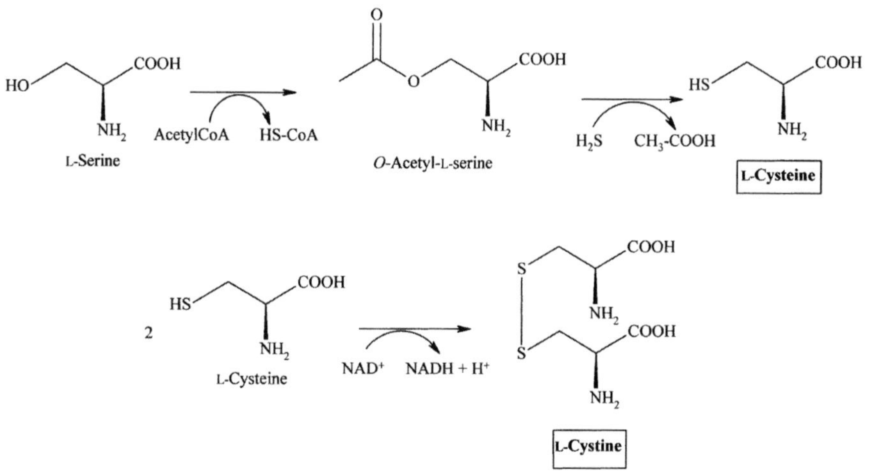

Fig. 145. Biosynthesis of cysteine and cystine

The biosynthesis of cysteine and cystine from serine is outlined in Fig. 145. Serine is acetylated by the enzyme *serine-O-acetyltransferase* (EC 2.3.1.30) and the *O*-acetylserine formed is condensed with sulphide to yield cysteine and acetic acid in a reaction catalysed by *cysteine synthase* (EC 2.5.1.47). Cystine is formed from two molecules of cysteine by oxidative coupling with formation of a disulphide bridge. The reaction involves NAD^+ and is catalysed by

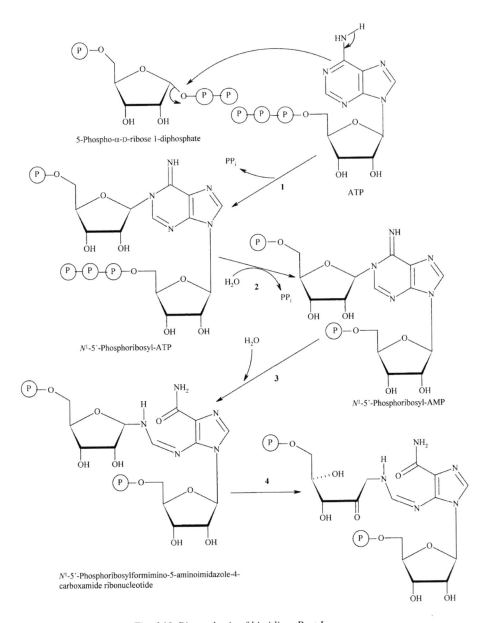

Fig. 146. Biosynthesis of histidine. Part I

Participating enzymes: Reaction 1: ATP phosphoribosyl transferase *(EC 2.4.2.17). Reaction 2:* phosphoribosyl-ATP diphosphatase *(EC 3.6.1.31). Reaction 3:* phosphoribosyl-AMP cyclohydrolase *(EC 3.5.4.19). Reaction 4–6:* 1-(5-phosphoribosyl)-5-[(5-phosphoribosylamino)methylidenamino] imidazole-4-carboxamide isomerase *(EC 5.3.1.16). Reaction 7 and 8:* imidazoleglycerol-phosphate dehydratase *(EC 4.2.1.19). Reaction 9:* histidinol-phosphate transaminase *(EC 2.6.1.9). Reaction 10:* histidinol-phosphatase *(EC 3.1.3.15). Reaction 11:* histidinol dehydrogenase *(EC 1. 1. 1. 23).*

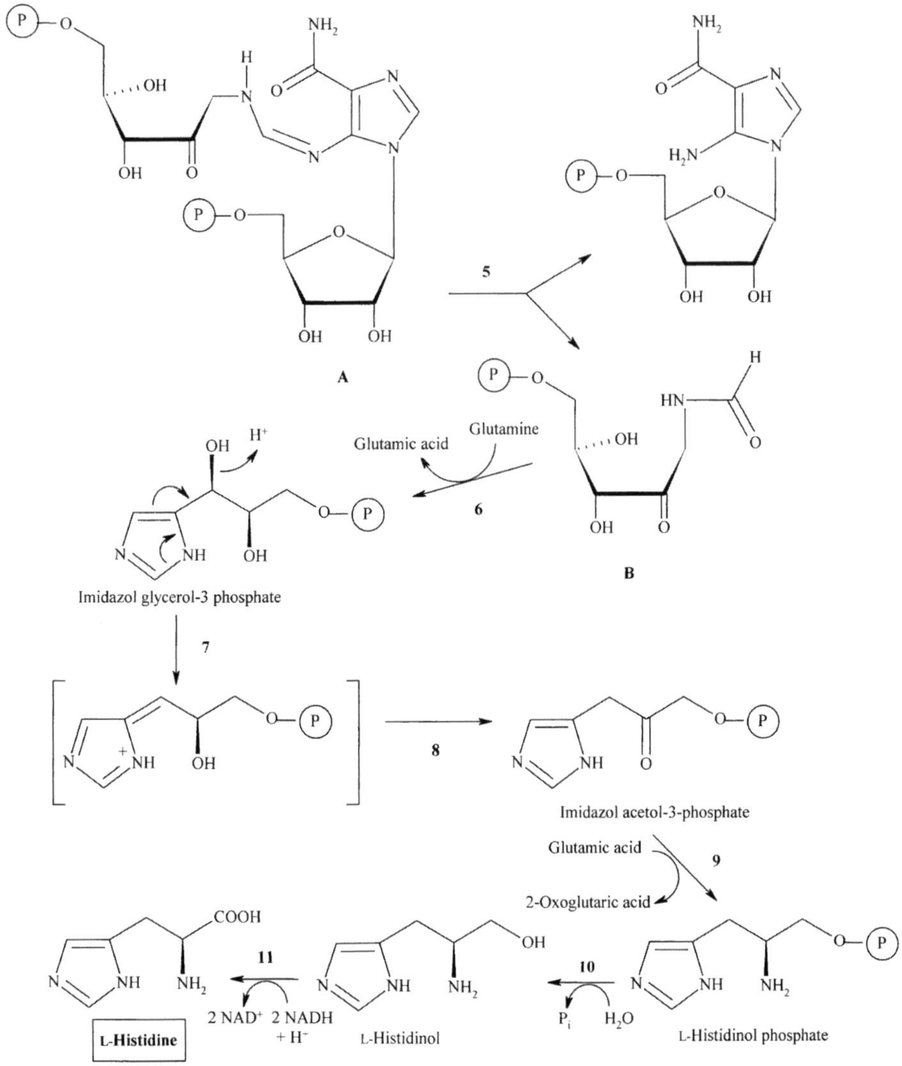

Fig. 146. Biosynthesis of histidine. Part II

cystine reductase (EC 1.8.1.6). The reacting cysteine residue is usually part of a peptide chain and the disulphide bridge formed is essential for protein structure and can form a bridge between two peptide chains – as in insulin – or form a loop in the same chain as in snake venom neurotoxins (page 524).

Histidine

The biosynthetic route to histidine differs entirely from the routes to the other amino acids. It has been studied mainly in bacteria and is intimately associated with the biosynthesis of purine (page 712).

5-Phospho-α-D-ribose 1-diphosphate reacts with ATP with loss of the diphosphate group and formation of an *N*-glycosidic linkage to the *N*(1) of the purine ring in ATP (Fig. 146 reaction **1**). The ATP moiety of the product then loses a diphosphate residue (reaction **2**) and its pyrimidine ring is opened (reaction **3**) forming N^1-5´-phosphoribosylformimino-5-aminoimidazole-4-carboxamide ribonucleotide in which the ribose-ring attached to N^1 is opened (reaction **4**) yielding the intermediate compound **A**, from which the entire ATP fragment, except the –NH-CH= bridge to the ribose moiety, is split off in reaction **5** (Fig. 146, part II), yielding compound **B**.

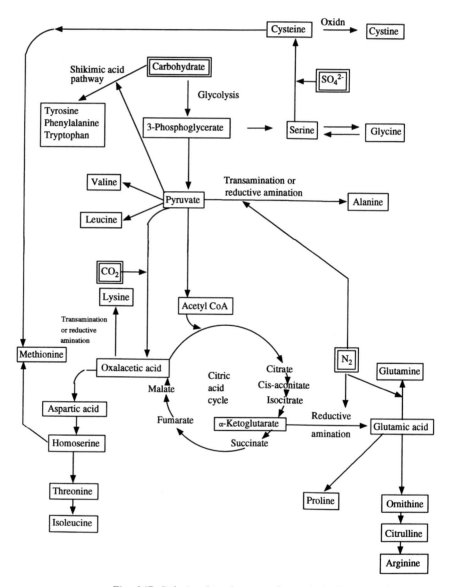

Fig. 147. Relationships between the carbohydrate metabolism and the biosynthesis of amino acids

Reaction **6** involves addition of ammonia from glutamine to compound **B** yielding imidazol glycerol-3-phosphate. Reactions 4–6 are catalysed by the same enzyme. In reactions **7** and **8** imidazol glycerol-3-phosphate loses water to form imidazol acetol-3-phosphate, which in reaction **9** is transaminated with glutamic acid to yield histidinol phosphate. In reactions **10** and **11**, histidinolphosphate is dephosphorylated and the histidinol formed is oxidized to yield histidine.

Aromatic amino acids

The biosynthesis of the aromatic amino acids tyrosine, phenylalanine and tryptophan via shikimic acid has been discussed earlier (page 202). Figure 147 shows the relationships between the biosynthesis of amino acids and the metabolism of carbohydrates. The biosynthesis of histidine has been omitted, because it proceeds along quite different lines (see page 503).

Essential amino acids

In the discussion of the shikimic acid pathway (page 194) it was mentioned that this pathway is not operative in animals which thus depend on the diet for supply of the aromatic amino acids. Also some other amino acids cannot be synthesized by animals. For human beings the following ten *essential amino acids* must be obtained from food: arginine, histidine, isoleucine, leucine, lysine, methionine, phenylalanine, threonine, tryptophan and valine. Tyrosine is produced from phenylalanine in the liver and is thus not an essential amino acid.

Toxic, non-proteinogenic amino acids

β-*N*-Methylamino-L-alanine
(BMAA)

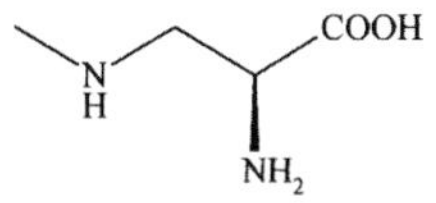

β-*N*-Methylamino-L-alanine

This amino acid has been found to cause amyotropic lateral sclerosis-Parkinsonism dementia complex (ALS-PDC) in the Chamorro people, who live in the island of Guam in the Pacific Ocean. ALS-PDC is a progressive neurodegenerative disease which has clinical and histological characteristics common to ALS (= Amylotrophic Lateral Sclerosis), Parkinson's disease and Alzheimer's disease. The occurrence of this disease in Guam was found to be related to the consumption of flying foxes *(Pteropus mariannus mariannus)*. Flying foxes forage on seeds of *Cycas micronesica* Hill *(Cycadaceae)*, which contain BMAA, produced by cyanobacteria, living as root symbionts of *Cycas* trees. BMAA has been found to be accumulated in the tissues of the animals and thus becomes a part of the diet of the Chamorro people, which are also subjected to exposure for BMAA contained in other parts of their diet, such as flour of gametophytes of seeds from *Cycas micronesica*. This flour has been found to contain protein-bound BMAA, which is not removed by the preparation technique.

Under the right conditions BMAA has been found to be produced by 97 % of all cyanobacteria species, including freshwater species, marine species as well as symbionts. These findings have raised concern that cyanobacteria may contribute to development of neurodegenerative disease also in other parts of the world, particularly as blooms of cyanobacteria have become more common due to water pollution. BMAA could accumulate in the tissues of fish, shellfish and other animals. Further research is, however, needed before it can be decided that BMAA from cyanobacteria poses a real risk for public health.

β-*N*-Oxalylamino-L-alanine (L-BOAA)

Seeds of *Lathyrus sativus* L. (chickling pea or grass pea, family: *Fabaceae*) are consumed in India, Bangladesh and Ethiopia during periods of famine, caused by flood or drought. Prolonged consumption can cause neurolathyrism, which is a disease characterized by muscle spasms and weakness in the legs, progressing to spastic paraparesis, which in the most severe cases leads to a crawling stage, where individuals are unable to move their legs. Neurolathyrism is caused by the presence in the seeds of the non-proteinogenic amino acid β-*N*-oxalylamino-L-alanine (L-BOAA,)also called γ-*N*-oxalyl-L-α,β-diaminopropionic acid (ODAP).

L-Glutamate is the major excitatory neurotransmitter in the mammalian CNS and as such the glutamate receptors play a vital role in the mediation of excitatory synaptic transmission. The glutamate receptors act both through ligand gated ion channels (ionotropic receptors) and G-protein coupled (metabotropic) receptors. The ionotroic glutamate receptors are divided into three subtypes – NMDA receptors, AMPA receptors and kainate receptors – named for their selective specific agonists, i.e. *N*-methyl-D-aspartate, α-amino-3-hydroxy-5-methyisoxazole-4-propionic acid and kainate, respectively. The AMPA receptors mediate fast synaptic transmission and are found in many parts of the brain and are the most commonly found receptors in the nervous system. The neurotoxicity of L-BOAA is mediated by this subtype of glutamate receptors. L-BOAA is structurally related to AMPA and acts as an agonist.

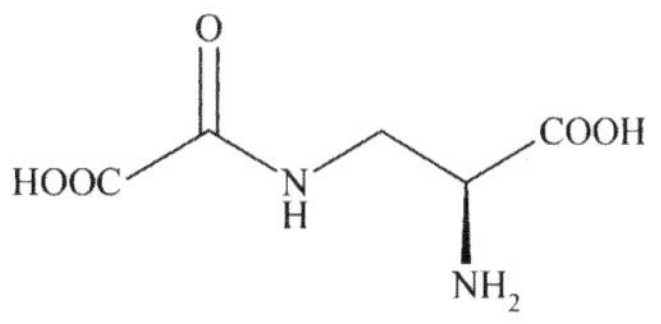

β-*N*-Oxalylamino-L-alanine (L-BOAA)

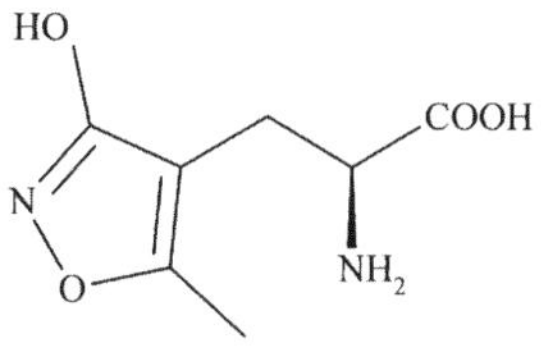

α-Amino-3-hydroxy-5-methyl-4-isoxazolepropionic acid (AMPA)

Further reading

Alifano, P., Fani, R., Lio, P., Lazcano, A., Bazzicalupo, M., Carlo-magno, M.S. and Bruni, C.B. Histidine biosynthetic pathway and genes: structure, regulation, and evolution. *Microbiological reviews*, **60** (1), 44–69 (1996).

Azevedo, R.A., Arruda, P., Turner, W.L. and Lea, P.J. The biosynthesis and metabolism of the aspartate derived amino acids in higher plants. *Phytochemistry*, **46** (3), 395–419 (1997).

Bannack, S.A., Murch, S.J. and Cox, P.A. Neurotoxic flying foxes as dietary items for the Chamorro people, Marianas Islands. *Journal of Ethnopharmacology*, **106**, 97–104 (2006).

Costa, L.G., Guizzetti, M. and Vitalone, A. Diet-brain connections: role of neurotoxins. *Environmental Toxicology and Pharmacology*, **19**, 395–400 (2005).

Cox, P.A. *et al*. Diverse taxa of cyanobacteria produce β-*N*-methylamino-L-alanine, a neurotoxic amino acid. *Proceedings of the National Academy of Sciences of the United States of America*, **102**, 5074–5078 (2005).

Cox, R.J. The DAP pathway to lysine as a target for antimicrobial agents. *Natural Products Reports*, **13**, 29–43 (1996).

Kuehn, B.M. Environmental neurotoxin may pose health threat. *JAMA*, **293** (20), 2460–2462 (2005).

Velíšek, J. and Cejpek, K. Biosynthesis of food constituents: Amino acids: 1. The glutamic acid and aspartic acid groups – a review. *Czech Journal of Food Sciences*, **24** (1), 1–10 (2006).

Velíšek, J. and Cejpek, K. Biosynthesis of food constituents: Amino acids: 2. The alanine-valine-leucine, serine-cysteine-glycine, and aromatic and heterocyclic amino acids groups – a review. *Czech Journal of Food Sciences*, **24** (2), 45–58 (2006).

Velíšek, J. and Cejpek, K. Biosynthesis of food constituents: Amino acids: 3. Modified proteinogenic amino acids – a review. *Czech Journal of Food Sciences*, **24** (2), 59–61 (2006).

Zabriskie, T.M. and Jackson, M.D. Lysine biosynthesis and metabolism in fungi. *Natural Products Reports*, **17**, 85–97 (2000).

9. Natural Products Derived Biosynthetically from Amino Acids

The bulk of the amino acids formed is used for building peptides and proteins, e.g. the thousands of enzyme systems needed to catalyse the many biochemical reactions in living organisms. Some amino acids are the precursors of other, so-called secondary, metabolites. From the pharmaceutical and medicinal point of view, the alkaloids constitute a very important group of such metabolites, but also other groups, such as non-ribosomal polypeptides, glycopeptides and β-lactams, are of great interest.

Peptides and proteins

Many proteins and peptides are used in medicine. A big group constitutes compounds which are part of the human body such as blood-protein fractions, enzymes and hormones. Another group is sera and vaccines. For such compounds the reader is referred to textbooks of biochemistry and medicine. In this book we will focus on compounds derived from plants, fungi and bacteria.

Symbols for the amino acids in the primary structures of peptides and proteins

When writing the primary structure of peptides and proteins (the amino acid sequence) it is too cumbersome to write the complete structure of each amino acid. Such structures would also be very difficult to read. Instead one uses symbols. Originally the first three letters in the names of the amino acids were used, but for large molecules even these abbreviations have become too long. One has therefore also constructed a complementary one-letter code.

Amino acid	Three-letter code	One-letter code
Alanine	Ala	A
Arginine	Arg	R
Asparagine	Asn	N
Aspartic acid	Asp	D
Cysteine	Cys	C
Glutamic acid	Glu	E
Glutamine	Gln	Q
Glycine	Gly	G
Histidine	His	H

Amino acid	Three-letter code	One-letter code
Isoleucine	Ile	I
Leucine	Leu	L
Lysine	Lys	K
Methionine	Met	M
Phenylalanine	Phe	F
Proline	Pro	P
Serine	Ser	S
Threonine	Thr	T
Tryptophan	Trp	W
Tyrosine	Tyr	Y
Valine	Val	V

In amino acid sequences, written with these symbols, the chain starts with the amino-terminal amino acid (the acid which has a free α-amino group) and ends with the acid which has a free α-carboxyl group (the carboxy-terminal amino acid). Two symbols are assumed to be connected by the peptide bond -CO-NH- formed between the carboxyl group of one amino acid and the amino group of the following acid. Example: Ala-Leu-Arg is a tripeptide in which alanine is the amino-terminal amino acid and arginine is the carboxy-terminal amino acid. The hyphen (-) is the peptide bond. With one-letter symbols the sequence would be written as ALR with no hyphens.

Proteolytic enzymes

Papain   Papain is the dried and purified latex from the fruit of *Carica papaya* L. (*Caricaceae*), the pawpaw. This is a tree, native to South and Central America and now grown in most tropical countries (Plate 18, page 511). The tree reaches a height of 5–10 metres and carries male and female flowers on separate trees, a dioecius habit. The fruit can be 30 cm long and weigh up to 5 kg. The outer, firm part of the pericarp covers a thick pulp and in the middle of the fruit there is a cavity (locule), filled with numerous black seeds.

In order to collect the latex, shallow cuts are made in the outer part of the pericarp of the full-grown, but still green and immature fruit. The latex exudes from the cuts and quickly solidifies. It is scraped off and dried either in the sun or under more controlled conditions in an oven or under reduced pressure. Drying in the sun destroys part of the enzyme activity. Latex can be repeatedly drawn from the same fruit, at intervals of about a week. Latex can also be obtained from the cortex of the stem and the petioles.

The crude papain is purified by dissolution in water and precipitation with ethanol. The product obtained contains a number of enzymes with proteolytic, amylolytic, lipolytic and rennin-like activity. The most important one is the proteolytic enzyme, which can be isolated in a pure state. Rather confusingly, this pure enzyme is also called papain (EC 3.4.22.2).

Plate 18.

Carica papaya L. Unripe fruits are the source of papain (page 510). The ripe fruit is an important food in tropi-al countries. The picture shows a plantation in Tanzania. Photograph by Gunnar Samuelsson.

Papain has a molecular weight of 23 kDa and is composed of 212 amino acid residues of known sequence. The enzyme contains three disulphide bridges and one free thiol group, essential for the proteolytic activity. In medical practice, papain is used to attenuate viscous mucilage in the mouth and stomach. It is also used in blood group diagnosis, as treatment of the blood cells with the enzyme destroys some blood group antigens, while the reactions with antibodies are enhanced for others. Thus it is often possible to identify additional Rh antibody specificities. Reactions of Rh, P_1, I, i, Lewis, Kidd, Colton and Dombrock antibodies are usually stronger with enzyme-treated than with untreated cells. Large amounts of papain are also used industrially to tenderize meat.

Ficin Ficin is a mixture of proteolytic enzymes and other constituents of the latex from several species of trees belonging to the genus *Ficus* (*Moraceae*), e.g. *Ficus carica* L., which also gives edible figs, and *Ficus anthelmintica* Mart. The latex, obtained by incisions in the trunk, is filtered and dried to give the commercial product. One of the main components of the preparation from *Ficus anthelmintica* has been isolated and found to be a glycoprotein with a molecular mass of 25 kDa. The pure substance is called *ficain* (EC 3.4.22.3). Like papain, this enzyme has a free thiol group which is essential for the proteolytic activity. The amino acid sequence has not been completely elucidated, but the zone surrounding the thiol group shows great similarities to the corresponding part of papain. Ficin can be used as a remedy against roundworms (*Ascaris*) and whipworms (*Trichocephalus*). It can be used in blood group diagnosis in the same way as papain. Ficin is also used in the food industry for tenderizing meat.

Bromelain Bromelain is the name of another preparation containing a number of proteolytic enzymes. It is prepared from the pineapple, *Ananas comosus* (L.) Merr. (*Bromeliaceae*), which contains several distinct cysteine proteinases. The major enzyme in the stem is *stem bromelain* (EC 3.4.22.32), while *fruit bromelain* (EC 3.4.22.33) is the major proteinase in the juice. Stembromelain is a glycoprotein with a molecular mass of 24 kDa. The amino acid sequence has been determined and shows that the protein part of this glycoprotein consists of a single polypeptide chain containing 211 or 212 amino acid residues. The carbohydrate moiety has a molecular weight of about 1 000 and is attached to an asparagine residue (no. 117). Like the main proteolytic enzymes in papain and ficin, bromelain contains a thiol group which is necessary for activity. The main component has been shown to be a mixture of two similar glycoproteins with small differences in size and charge. Six additional proteases have also been detected and partially characterized. One of these is about 15 times more potent than the main component(s) and contains no sugars.

Bromelain is used medicinally against inflammation and oedema, but the main consumer with respect to quantity is the food industry, just as with papain and ficin. It can also be used in blood group diagnosis (see papain). The medical interest in bromelain has recently been enhanced owing to the discovery that bromelain can act as an immunomodulator by raising the impaired immunotoxicity of monocytes against tumour cells from patients and by inducing the production of distinct cytokines such as tumour necrosis factor *alpha*, interleukin (Il)-1*beta*, Il-6 and Il-8. There are also reports on animal experiments claiming an antimetastatic efficacy and inhibition of metastasis-associated platelet aggregation as well as inhibition of growth and invasiveness of tumour cells. Apparently the anti-invasive activity does not depend on the proteolytic activity of bromelain.

Ribosome inactivating proteins (RIP toxins)

Some angiosperm plants contain toxic proteins which act as site-specific RNA *N*-glycosidases and inactivate eukaryotic ribosomes. Such toxins, called *Ribosome Inactivating Proteins* (RIPs or RIP toxins) are of two types. Type I consists of a single protein of approximately 30 kDa which is frequently, but not always, *N*-glycosylated. Type II RIPs are heterodimers composed of two chains of 32 kDa joined by a disulphide bond. These two chains are usually also *N*-glycosylated. In spite of their ribosome-inactivating activity, type I RIPs are not cytotoxic because they have no means of entering an intact eukaryotic cell. In fact type I RIPs are present in foods such as wheat germ and barley grain and are thus widely consumed by both humans and animals. Type II RIPs, on the other hand, are potent cytotoxins. In these, one of the chains (the B-chain) acts as a lectin (see page 518), binding to galactosides on the surface of eukaryotic cells and triggering toxin uptake by endocytosis. Once inside the cell the disulphide bond between the A- and B-chains is reduced, thus liberating the A-chain which inactivates cellular ribosomes by *N*-glycosidase activity. The A-chain corresponds to a type I RIP. Ricin and abrin are two very potent type II RIPs which have been extensively studied and which are of medical interest because they can be conjugated to antibodies and used as cell-specific therapeutic agents, e.g. for cancer treatment.

Ricin Besides a fixed oil (page 252) the seeds of the castor bean, *Ricinus communis* L. (Plate 4, page 253), contain a type II RIP called ricin. There are several isoforms of ricin including ricin D, ricin E and the closely related *Ricinus communis* agglutinin (RCA). Ricin is a potent cytotoxin but a weak haemagglutinin, whereas RCA is only weakly toxic to intact cells but is a strong haemagglutinin. The A-chain of ricin is composed of 267 amino acid residues and has a molecular size of 32 kDa. The B-chain contains 262 residues and is 34 kDa in size. Both chains are glycoproteins. The A-chain has one oligosaccharide chain attached to asparagine residue no. 10. The B-chain contains two oligosaccharide chains attached to asparagine residues nos. 93 and 133, respectively. The disulphide bond connecting the two chains involves cysteine 259 in the A-chain and cysteine 4 of the B-chain (the figures indicate the position of the amino acid residue in the sequence). In contrast to the dimeric ricin, RCA is a tetramer composed of two ricin-like heterodimers, each of which contains an A-chain (32 kDa) and a galactose-binding B-chain (36 kDa). The complete amino acid sequences of the ricin and RCA subunits have been determined chemically or deduced from the nucleotide sequence of cloned cDNAs and genes.

Formation of ricin   Ricin and its homologues are synthesized in the endosperm cells of maturing *Ricinus* seeds and in the mature seed are stored in an organelle, called *the protein body*, which is analogous to a vacuole compartment. The ricin genes encode a preproprotein containing both the A- and B-chain sequences. Preproricin consists of 576 amino acid residues, the first 35 of which include, but do not entirely consist of, an NH_2-terminal signal sequence, followed by the mature A-chain sequence (267 residues) joined by a 12 residue linker peptide to the mature B-chain sequence (262 residues). Following synthesis of preproricin in a ribosome, the NH_2-terminal signal sequence directs the nascent protein molecule across the membrane of the endoplasmic reticulum (ER) into the ER lumen. During this translocation three major modifications occur. First the signal sequence is split off by *ER luminal signal peptidase*. This probably occurs after serine 22 in the 35-residue preprotein leader sequence, yielding a proricin which has an N-terminal sequence of 12 amino residues before the sequence of the A-chain begins. The second event is *N*-glycosylation of proricin at four sites, two within the A-

Fig. 148. The three-dimensional structure of ricin. The alpha carbon backbone of the protein is shown as ribbons. Chain-A is shown in the upper right as the multistranded ribbon, while chain-B is the solid ribbon. The spheres between the chains mark the positions of solvent waters trapped in the interface. (From J. M. Lord, L.M. Roberts and J. Robertus, The FASEB Journal, 8 (February), 201–208 [1994]. Reproduced with permission from the publishers)

chain sequence and two within the B-chain. The third modification is the formation of five disulphide bonds within the folding proricin, a reaction which is catalysed by *protein disulphide isomerase*.

Vesicular flow transports the core-glycosylated, disulphide-bonded proricin from the ER lumen via the Golgi complex to the protein bodies. During the transport through the Golgi complex, enzymatic modification of the oligosaccharide chains of proricin takes place. Finally, Golgi-derived vesicles carrying the modified proricin fuse with the vacuolar membrane and deliver their contents into the protein body matrix. Here proricin is processed by an *endopeptidase* which removes the 12-residue peptide connecting the A- and B-chains. Also the 12-residue-long peptide, attached at the N-terminal end of the A-chain, is split off in the final formation of the mature ricin molecule.

Three-dimensional structure and mechanism of action

The three-dimensional structure of ricin has been determined by X-ray crystallography at a resolution of 2.5 Å. Chain-A is a globular protein with a pronounced binding site cleft while the B-chain is an elongated dumbbell with galactose binding sites at both ends. The active centre of the A-chain is a pronounced cleft where the amino acid residues tyrosine 80, tyrosine 123, glutamic acid 177 and arginine 180 are in close proximity to each other. These amino acid residues are invariant in the structures of all RIPs and are operative in the *N*-glycosidase activity of the A-chain (see below). The B-chain is made of two globular domains each one with a lactose binding site. Domain 1 is the amino-terminal half of the molecule and domain 2 forms the carboxy-terminal half. Each domain is composed of four subdomains designated λ, α, β and γ. The α, β and γ

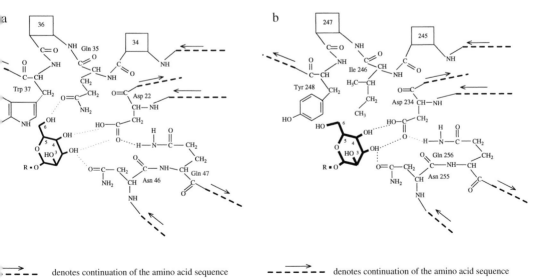

Fig. 149. Binding of galactose (as the terminal part of a cell wall structure) to ricin B-chain. (a) in the site of domain 1, (b) in the site of domain 2

515

subdomains are homologous and each is approximately 40 residues in length, while the λ-subunit is a 17-residue-long peptide. Subunit α of domain 1 and subunit γ of domain 2 are the lactose binding sites of the B-chain. Amino acid residues tryptophan 37 and aspartic acid 22 are operative for the binding of the galactose moiety of lactose at site 1, and at site 2, tyrosine 248 and aspartic acid 234 fulfil this function (Fig. 149). The aromatic amino residues contact the hydrophobic side of the sugar and the aspartic residues make the primary interaction and account for the epimeric specificity of binding, forming hydrogen bonds to the C(3)- and C(4)-hydroxyls of galactose. Hydrogen bonds from glutamine 47 in site 1 and glutamine 256 in site 2 lock the crucial aspartates in position. The C(3)-hydroxyl of galactose also forms a strong hydrogen bond with asparagine 46 in site 1 and asparagine 255 in site 2. At site 1 glutamine 35 forms a hydrogen bond to the C(6)-hydroxyl of galactose, while at site 2 the C(6)-OH is turned towards the solvent as there is a hydrophobic contact between C(6) of galactose and isoleucine 246 which in domain 2 is the analogue of Gln 35 in domain 1. The interactions between galactose and the B-chain are few in number compared with other sugar-binding proteins.

When ricin comes into contact with a eukaryotic cell the B-chain attaches the toxin to structures on the cell wall which contain ter-

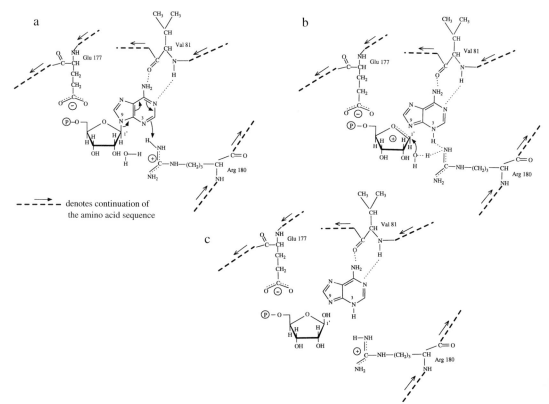

Fig. 150. Mechanism of action for the ricin A-chain

minal galactose residues. This attachment triggers endocytosis whereby the toxin enters the cell and the A-chain is released into the cytoplasm by reduction of the disulphide bond between the chains.

In the cell the A-chain causes release of adenine 4324 from the 28S rRNA found in the 60S subunit of eukaryotic ribosomes. Adenine 4324 is situated in a hairpin containing the tetranucleotide loop GAGA. The proposed mechanism for the action of the A-chain is illustrated in Fig. 150, a–c. The substrate (adenosine) binds in the above-described active cleft and adenine is sandwiched between tyrosines 80 and 123. Specific hydrogen bonds are formed from adenine to valine 81 as indicated by the dotted lines in the figure (a). N(3) of adenine is protonated by arginine 180 which promotes breaking of the bond between N(9) of adenine and C(1') of ribose. An oxycarbonium ion is formed on ribose as the bond is broken and this is stabilized by ion pairing to glutamic acid 177. A water molecule opposite to the ribose is hydrogen bonded to arginine 180. As arginine 180 transfers a proton to adenine it removes one from the water and thereby activates it (b). The activated water attacks the oxycarbonium ion, releasing the adenine leaving group (c).

Therapeutic applications   Several toxic proteins have been used in attempts to kill unwanted cells selectively, in particular malignant cells. The toxin is delivered to the target cell by linkage to an antibody or growth factor that specifically interacts with the cell in question. In attempts to use ricin for this purpose the toxin has generally been linked to monoclonal antibodies by a disulphide bond formed using heterobifunctional crosslinkers. Such conjugates are called *immunotoxins*. Ricin immunotoxins are preferentially made from the A-chain, thereby avoiding non-specific cell interactions mediated by the B-chain in the holotoxin. The A-chain can be produced by expression of the corresponding cDNA in *Escherichia coli*. This is possible because the A-chain is without effect on the 70S ribosomes of prokaryotic organisms. A total of 1 500 mg/l of culture has been obtained using such a system. The product is not glycosylated, which makes it attractive for the production of immunotoxins. Chain-A immunotoxins have been used *in vitro* to purge bone marrow of unwanted cells prior to transplantation. This has been done both in allogeneic and autologous bone marrow transplantations. In the allogeneic transplantations the immunotoxin has been used to destroy T-lymphocytes in the bone marrow from the donor, in order to reduce the incidence of graft-vs-host disease in patients receiving the transplant. In the autologous transplantations, a sample of the patient's own bone marrow is removed before destroying the remainder. Before re-introduction of the sample, malignant cells must be destroyed. This has been done with some success in the treatment of a variety of T-cell leukaemias and lymphomas.

The immunotoxins can also be used *in vivo*. Chain-A immunotoxins have been used successfully in the treatment of steroid-resistant, acute graft-vs-host disease. Problems in the treatment of solid tumours can arise due to poor access of the immunotoxin to the tumour, lack of specificity, tumour cell heterogeneity, antigen shedding, breakdown or rapid clearing of the immunotoxin and dose-limiting side effects. Repeated administration can be compromised by an immune response to both the antibody and the toxin component of the immunotoxin. Considerable effort has been and is being made to overcome these and other limitations.

Abrin

Abrus precatorius L. (*Fabaceae*) is a climbing shrub, growing in tropical and subtropical countries. The seeds are broadly ellipsoidal, 5–9 mm in diameter. The testa is black at the base and beautifully red at the top. The seeds are used in rosaries and jewellery and have caused poisoning as they contain a mixture of toxic proteins called *abrin*. Two abrins – abrin-a and abrin-b – have been isolated and characterized. Abrin-a binds to a Sepharose 4B column and agglutinates erythrocytes while abrin-b interacts very weakly with Sepharose 4B and does not agglutinate erythrocytes.

Both abrins are type II RIPs. The A-chain of abrin-a is composed of 250 amino acid residues, the sequence of which has been determined. The B-chains of both abrins have also been sequenced and consist of 268 amino acid residues and share 256 identical residues. Asparagine residues no. 101 and 141 in the B-chain of abrin-a are linked to sugar, involving several oligomannose-type and xylomannose-type sugar chains.

Lectins

Abrin and ricin are examples of *lectins*. A lectin is defined (H. Franz, *Naturwissenschaften* **77**, 103–109 (1990)) as a protein which is not an antibody or an enzyme, but which has the ability to attach itself to specific sugars. The binding is not covalent and the sugar can either be free or constitute part of a larger molecule, which may be present, for instance in a membrane. Lectins are found in plants – especially in seeds – but they have also been found to occur in fungi, bacteria, various marine organisms and in mammals. Seeds of plants belonging to the family *Fabaceae* are particularly rich in lectins.

Most lectins are glycoproteins. They are often produced as groups of several closely related compounds.

The biological function of lectins in plants is not clear, but in fungi and bacteria they enable these organisms to attach themselves to other cells. Thus, infections by pathogenic bacteria are effected via lectins. The formation of cancer metastases also seems to involve lectins which enable the tumour cells to attach themselves to the target organ.

Many plant lectins have the ability to agglutinate erythrocytes. The lectin attaches itself to specific carbohydrates on the erythrocyte membrane and one lectin molecule can bind more than one erythrocyte, thus forming aggregates which precipitate. Some lectins are unspecific in that they agglutinate erythrocytes of all blood groups, while others have a more specific action. For example, a lectin from Lima beans, *Phaseolus lunatus* L. (*Fabaceae*), agglutinates only group A erythrocytes and does not react with those of group B or O. Haemagglutinating lectins are used in the diagnosis of blood groups, to complement the diagnosis with mono- and polyclonal antibodies, e.g. a lectin from *Dolichos biflorus* L. (*Fabaceae*), is used for the differential diagnosis of groups A_1 and A_2 and one from *Vicia graminea* Sm. (*Fabaceae*) for the detection of group N.

Some lectins stimulate the mitosis of lymphocytes, e.g. concanavalin A from the jack bean *Canavalia ensiformis* DC. This lectin consists of four subunits, each containing 237 amino acids, as well as one Mn^{2+} ion and one Ca^{2+} ion. Concanavalin A acts only on T-lymphocytes.

Lectins can be used for the affinity chromatography of compounds containing specific sugars. Concanavalin A specifically binds α-D-glucose and α-D-mannose. Immobilization of concanavalin A on Sephadex® yields a medium which can be used for isolating substances in which these sugars are part of the structure. When a solution of the sample is passed through a column of the immobilized lectin, substances containing glucose or mannose moieties are adsorbed and, after washing the column, the compounds can be eluted with solutions of glucose or mannose, respectively.

Lectins are proteins and can therefore be marked by specific reagents. They can then be used to histochemically demonstrate the presence in tissues of compounds in which lectin-specific sugars are part of the structure.

Amanita toxins

The death cap, *Amanita phalloides* (Fr.) Link, and the less common "destroying angel", *Amanita virosa* (Fr.) Bertillon, often cause severe poisoning, which is sometimes fatal, because they are mistaken for the edible field mushroom, *Agaricus campestris* L. Both genera, *Amanita* and *Agaricus*, belong to the family *Agaricaceae* (agarics). The fruiting body consists of a stalk (stipe) and a cap (pileus), the underside of which bears numerous vertical, radiating gills (lamellae) carrying the basidiospores. Failure to distinguish field mushrooms and *Amanita* is most common with young specimens, but can easily be avoided if the following two diagnostic characters are noted:

1. *Amanita* always have white spores. If the pileus is broken, the gills and flesh of the pileus have almost the same white colour. Field mushrooms have blackish brown spores and for that reason, the gills are darker than the rest of the pileus.
2. *Amanita* have a universal veil, which breaks when the fungus develops. Its remnants are left at the base in the form of a sheath (volva). Field mushrooms lack this sheath.

The first symptoms of *Amanita* poisoning usually appear after 8–12 hours: stomach pains, diarrhoea and violent vomiting. Following hospital treatment, a deceptive period of improvement generally occurs, which may last 12–24 hours. Then signs of liver damage appear. In serious cases these progress rapidly and include an enlarged and pressure-sensitive liver, jaundice, stomach and intestinal bleeding, oliguria to anuria, and disturbances of consciousness. Death may intervene 4–7 days later during hepatic coma. This happens in up to 30 % of the cases.

The *Amanita* toxins comprise two groups of cyclic peptides: amatoxins and phallotoxins. The main representative of the amatoxin type is α-amanitin, and of the phallotoxin type phalloidin. A non-toxic peptide, antamanide, has also been isolated. This peptide is an antagonist of the phallotoxins and counteracts their effects if given together with these toxins.

Amatoxins As already mentioned, α-amanitin is the main representative of the first group of *Amanita* toxins, the amatoxins. The other members of this group differ from α-amanitin in amino acid 3, in which the δ-hydroxyl group may be missing, in amino acid 4, which may be tryptophan rather than 6-hydroxytryptophan, and in amino acid 1, which may be aspartic acid instead of asparagine.

α-Amanitine

γ,δ-Dihydroxyisoleucine

Hydroxyproline (Hpr)

6-Hydroxytryptophan residue

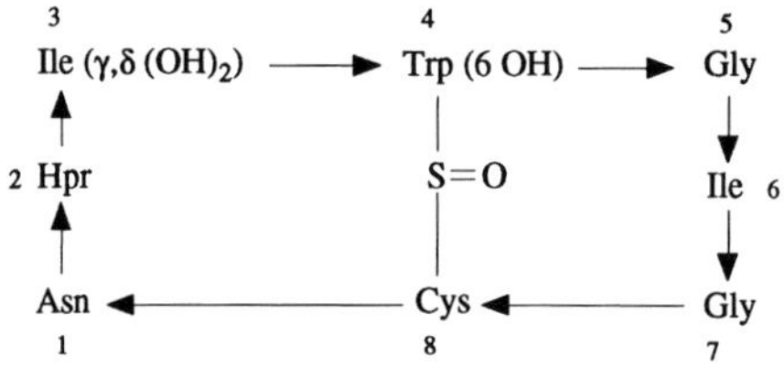

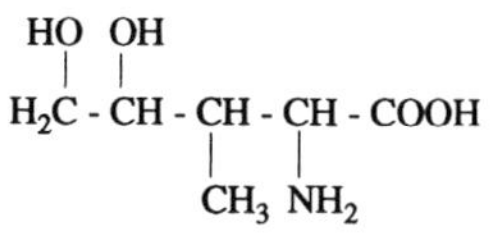

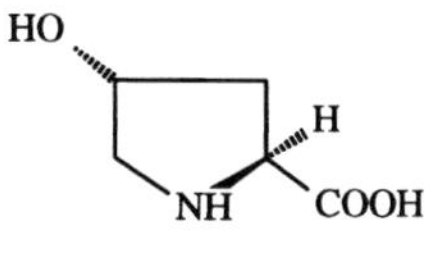

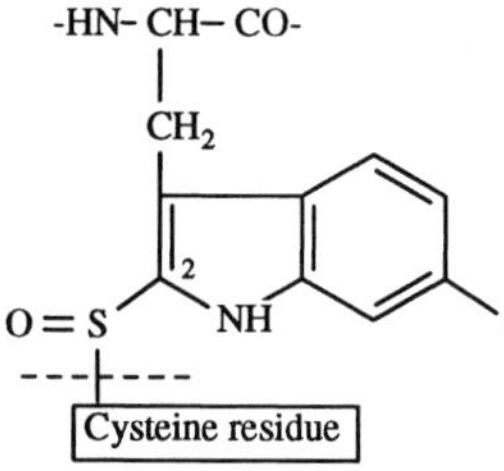

The LD_{50}, intraperitoneally in mice, of amanitin is 0.2–0.5 mg/kg. The effect of the poison is slow, with death occurring only after 2–6 days. The toxic effect arises both after injection and after ingestion and results in slow degeneration of the liver and kidneys. The toxin stops protein synthesis in eukaryotic cells by inhibiting the action of RNA polymerase B, thereby blocking the synthesis of messenger-

RNA. *In vivo*, the synthesis of ribosomal RNA is also stopped, although no blocking of RNA polymerase A is observed *in vitro*. The effect of amatoxins on RNA polymerase is limited to eukaryotic cells.

The toxic effect is prolonged, because amatoxins excreted in the bile reach the blood again via the enterohepatic circulation, thus renewing their assault on the liver. Part of the amatoxins may also be reabsorbed from the kidneys, thus likewise prolonging their destructive action.

Amanitin as a tool in biological research

Since inhibition of polymerase B in eukaryotic cells requires only nanomolar amounts of the toxin, amanitin is an ideal tool for investigating if complex biological processes are dependent on a step of transcription, namely *de novo*-synthesis of m-RNA. α-Aminitin has become as indispensable and specific a tool for the inhibition of transcription as cycloheximide is for the inhibition of translation. An important area is the exploration of hormones which frequently display their effect by stimulating the transcription of certain gene segments. As a consequence, certain hormones lose their effect in the presence of amanitin. Experiments with amanitin have demonstrated that not only many hormones, but also several vitamins, second messengers, growth factors and drugs act through induction of transcription. Amanitin is also employed in virus research. When the replication of a virus is inhibited by amanitin it is an indication that the development of the virus is dependent on the transcription apparatus of the host cell. Generally the rule holds true that viruses which multiply in the cytoplasm are resistant to amanitin whereas those that multiply in the nucleus are inhibited by amanitin.

Phallotoxins

Phalloidin is a bicyclic peptide, in which two uncommon amino acids, hydroxyproline and γ, δ-dihydroxyleucine, are integral parts. The structure is also characterized by an unusual sulphur bridge between cysteine and C-2 in the indole ring of tryptophan (see the

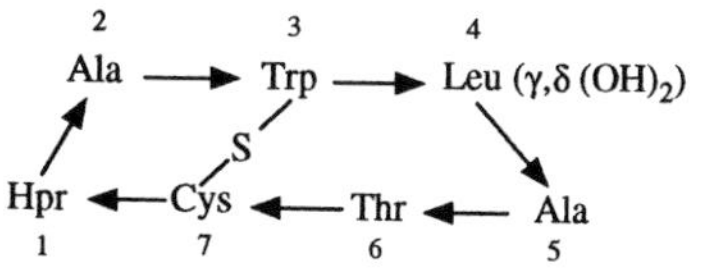

Phalloidin

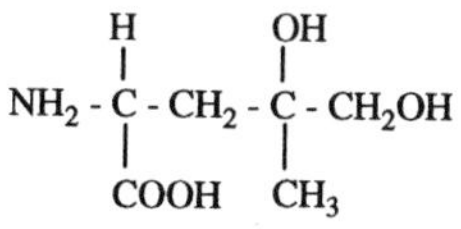

γ,δ-Dihydroxyleucine

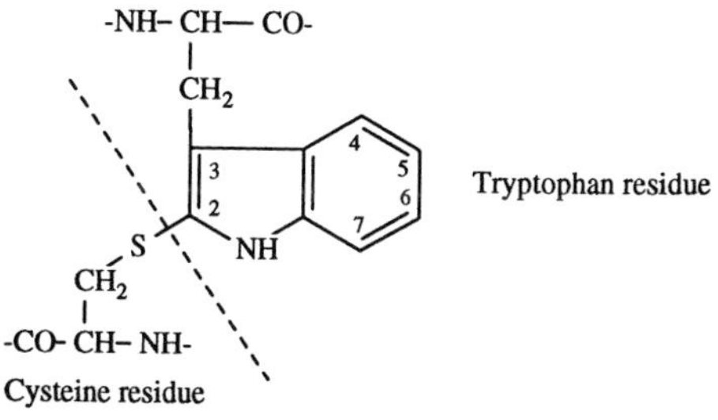

Tryptophan residue

Cysteine residue

structures of the amatoxins). Other peptides in the phallotoxin group differ from phalloidin in one of the amino-acid residues 4, 5 or 6. In amino acid 4, the hydroxyl group δ may be missing, or the second methyl group may also be hydroxylated.

The toxic effects of the phallotoxins appear fairly rapidly. LD_{50} intraperitoneally in mice is 1.5–2.5 mg/kg and death occurs after 2–3 hours. The phallotoxins act mainly on the liver. The plasma membranes of the liver cells are attacked and ions begin to leak from the cells – first calcium ions, then follows a massive outflow of potassium ions. The phallotoxins are also attached to membranes within the cell, e.g. the endoplasmic reticulum and lysosomes. When the membranes of the lysosomes give way, enzymes are set free and destroy the cell. These effects occur on parenteral administration of phallotoxins. However, these toxins are not likely to play a significant role in death-cap poisoning, as they are poorly absorbed in the intestines.

Pro - Phe - Phe - Val - Pro
Pro - Phe - Phe - Ala - Pro

Antamanide

Antamanide This peptide counteracts the effects of the phallotoxins if administered simultaneously. Limited protection against amatoxins can also be achieved by pre-treating the animal with antamanide for some days. Experiments with liver cells have indicated that the probable reason for the protection is that the antamanide stops the toxins from entering the cells.

Other toxins in Amanitas A high-molecular-weight toxin fraction – phallolysin – has also been isolated from the death cap. It has haemolytic and cytotoxic properties, and the lethal dose is 0.2 mg/kg for the mouse and 0.04 mg/kg for the rabbit. Phallolysin consists of two glycoproteins with molecular masses above 30 kDA. However, it is destroyed at 65 °C, which makes its contribution to accidental poisoning by mistakenly cooked death-cap mushrooms unlikely.

Antidotes against *Amanita* poisoning

Antamanide cannot be used as an antidote against poisoning, since it works only if given together with the toxins. A combination of penicillin and silybin (page 339) is considered the most effective treatment. Also intravenous administration of the iridoid aucubin (page 383) has in experiments on dogs been shown to have protective activity against *Amanita* poisoning.

Snake venoms

Poisonous snakes are kept in snake farms for serum production. One well-known snake farm is the Instituto Butantan in São Paulo, Brazil. The venom is obtained by "milking" the snake: it is made to bite through a rubber membrane, stretched over a glass container in which the venom is collected. The venom glands are gently pressed, to force as much venom as possible out through the fangs. The amount of venom produced by a snake varies from one drop to 2 ml, and the animal can be milked roughly every 3 weeks. The fresh venom is yellowish and transparent. It can be dried in a vacuum or lyophilized without losing its activity and it can be stored for years under refrigeration.

Serum against snake-bite is prepared by injecting increasing doses of the venoms into horses, which then produce antibodies against the poisons. It is customary to produce so-called polyvalent serum, by injecting several different venoms from related snake species. The content of antibodies in the blood of the horse is checked by biological determination of the antitoxic effect. When a sufficiently high titre has been attained, blood is withdrawn from the horse and the serum is separated and, if necessary, subjected to various purification procedures.

A snake venom is an exceedingly complex mixture of proteins with very different properties. Among the components is a large number of enzymes: hyaluronidases, phosphatases, cholinesterases, proteases, phospholipase A, esterases, exo- and endopeptidases, enzymes that affect blood coagulation, and kinin-liberating enzymes. In many snakes, the toxins proper are small basic proteins, of which there are two groups: neurotoxins and cardiotoxins. Other constituents are small peptides with enzyme-inhibiting effects.

Poisonous snakes are encountered in the following tribes of the families *Elapidae* and *Viperidae*:

Elapidae
　　Hydrophinae, sea snakes.
　　Elapinae, which includes very poisonous snakes
　　　　like cobras and mambas.
　　Acanthophiinae, very poisonous Australian snakes.

Viperidae
　　Viperinae, vipers, e.g. the European adder – *Vipera berus*.
　　Crotalinae, rattlesnakes.

There are great differences in the types of compounds that constitute the venoms from the snakes of these four tribes. Neurotoxins are prevalent in *Hydrophinae* and *Elapinae*. Cardiotoxins are found in the *Elapinae*. The venoms of *Crotalinae* and *Viperinae* contain proteolytic enzymes, which are rare in *Elapinae*.

Enzymes of snake venoms

Ancrod This is an enzyme (EC 3.4.21.74, accepted name *venombin A*) obtained from the venom of the Malayan pit viper *Agkistrodon rhodostoma* (*Crotalinae*). It is a glycoprotein with a molecular mass of 38 kDa. The carbohydrate moiety comprises 30 % of the molecule. The effect of ancrod is similar to that of thrombin, but ancrod releases only fibrinopeptide A from fibrinogen and does not activate the fibrin-stabilizing factor XIII. The enzyme is used medicinally to treat disorders of arterial circulation.

Many of the enzymes present in snake venoms are hyaluronidas-

es or proteases. They contribute to the overall toxic effect by fa
litating the distribution of the toxins in the tissues. Bites
rattlesnakes result in severe necrosis, due to the strong proteoly
activity of the venom.

Phospholipase A$_2$ (EC 3.1.1.4) causes haemolysis of red blo
corpuscles by splitting off lecithin to form lysolecithin which l
haemolytic activity. The enzyme also attacks phosphoglycerides
biological membranes thus damaging the membranes. Cardiotoxi
(page 525) potentiate the activity of the phospholipase. Phosphc
pase A$_2$ is considered to be one of the toxic components of sna
venom.

Neurotoxins

The most active components of the venom of *Hydrophinae* a
Elapinae are neurotoxins. Like curare, these neurotoxins act
blocking the transmission of nerve impulses in peripheral cholin
gic synapses. This causes muscular paralysis and death by asph
ia, when the muscles involved in respiration become affected.

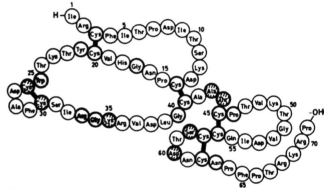

Neurotoxin siamensis 3.

The neurotoxins have been studied more extensively than a
other group of toxins from snake venom. Structurally they a
small, basic proteins, consisting of a chain of 60–62 (short neu
toxins) or 71–74 amino acids (long neurotoxins). In short neurotc
ins the chain of amino acids is crosslinked by four disulphi
bridges. Long neurotoxins contain an additional such bridge. T
way in which the toxin is built up is illustrated here by the amir
acid sequence of a long neurotoxin – siamensis 3 – from the Th
land cobra *Naja naja siamensis* (*Elapinae*).

As already mentioned, short and long neurotoxins act po
synaptically by blocking acetylcholine receptors so that acet
choline is no longer able to act. This property of neurotoxins l
been used to isolate acetylcholine receptors by affinity chromatc
raphy, using neurotoxins linked to an insoluble matrix.

Some snakes belonging to the family *Elapidae* have been fou to contain other neurotoxins, which act pre-synaptically, i.e. they block the liberation of acetylcholine. These toxins are 4–25 times as toxic as the short and long neurotoxins. The pre-synaptic neurotoxins have quite different structures to the post-synaptic ones. The Australian tiger snake *Notechis scutatus* ssp. *scutatus (Acanthophiinae, Elapidae)* contains the pre-synaptic neurotoxin notexin, whose amino acid sequence is known. This toxin consists of a chain of 119 amino acid residues, which is crosslinked through seven disulphide bridges. Besides its pre-synaptic neurotoxic effect, notexin also exerts a direct myotoxic effect, leading to complete degeneration of skeletal muscles. The most toxic pre-synaptic neurotoxin known is taipoxin from the Australian taipan *Oxyuranus scutellatus* spp. *scutellatus (Acanthophiinae, Elapidae)*. Its LD_{50} (i.v., mouse) is as low as 2 µg/kg, whereas the corresponding figure for post-synaptic neurotoxins is 50–100 µg/kg. Taipoxin is a glycoprotein with a molecular mass of 46 kDa, consisting of three sub-units: α, β and γ. Subunits α and β each consist of 120 amino acid residues, and each unit is crosslinked by seven disulphide bridges. Subunit γ contains 135 amino acid residues and eight disulphide bridges together with the carbohydrate moieties.

Cardiotoxins

Venoms from cobras (*Elapidae*) contain another group of toxins which depolarize cell membranes, so that their permeability to ions is increased. Skeletal muscles contract irreversibly and the toxins cause cardiac arrest in systole, an effect that has given rise to their name cardiotoxins. Cardiotoxins are less poisonous than neurotoxins. LD_{50} i.v. in the mouse is approximately 750 µg/kg. The cardiotoxins of cobras are built up from 60–62 amino acid residues with four disulphide bridges, the positions of which are the same as in the neurotoxins. A different cardiotoxin, molecular mass 22 kDa, and probably consisting of two sub-units, has been isolated from the Mojave rattlesnake, *Crotalus scutullatus (Crotalinae)*.

Lizard toxins

Exenatide Exenatide is a straight-chain polypeptide with 39 amino acids, which was first isolated from the venom glands of the Gila monster, *Heloderma suspectum*, and named exendin-4. Exenatide (trademark Byetta, Amylin Pharmaceuticals, Inc, USA) is a synthetic product with the same amino acid sequence as exendin-4, as illustrated below.

```
1                         10                              20
N-His-Gly-Glu-Gly-Thr-Phe-Thr-Ser-Asp-Leu-Ser-Lys-Gln-Met-Glu-Glu-Glu-Ala-Val-Arg-Leu-Phe-
                         30                              39
Ile-Glu-Trp-Leu-Lys-Asn-Gly-Gly-Pro-Ser-Ser-Gly-Ala-Pro-Pro-Pro-Ser-amide
```

The Gila monster (Heloderma suspectum). Photo IBL.

The Gila monster lives in the Gila River area in New Mexico an Arizona in the USA. It is about 50 cm long and has a short tail an beaded skin which is generally yellow and black in irregular blotc es or rings. The Gila monster, together with the Mexican bead lizard (*Heloderma horridum*) are the only known venomous lizard They do not have fangs, but secrete the venom from a chain of pe mandibular glands onto grooved teeth in the lower jaw, there enabling the venom to enter tissues of the prey bitten by the lizar The Gila monster venom contains several straight-chain polype tides, ranging from 35 to 39 amino acids in length. These peptid have a number of effects on mammalian tissues and recepto including increase in cellular cAMP and amylase release from di persed acini from guinea pig pancreas, indicating them to have an logues in the glucagon/vasoactive intestinal peptide (VIP)/secret superfamily of peptides.

Exenatide is used for treatment of type 2 diabetes in patients f which other oral antidiabetic drugs, such as metformin or sulph nylureas, do not give adequate glycaemic control.

Exenatide acts on receptors in the β-cells of the islets of Lange hans to potentiate glucose-stimulated insulin secretion, there mimicking the natural incretin hormone glucagon-like peptide (GLP1), which is released from cells in the gastrointestinal tract response to food intake. Failure of β-cells to secrete enough insul to compensate for peripheral tissue insulin resistance is a maj cause of hyperglycaemia in patients with type 2 diabetes. GLP1 rapidly degraded *in vivo* by dipeptidyl peptidase and is therefore n a suitable antidiabetic agent. Exenatide shows similar glucoregul tory activities to GLP1 but has a much longer half-life. It al reduces food intake and slows gastric emptying, thereby reduci the rate at which meal-derived glucose reaches the circulation. Du ing periods of hyperglycaemia exenatide lowers serum glucag concentrations.

Exenatide is administered by twice-daily injections, but longacti microsphere-based preparations have recently been developed.

Ziconotide

Ziconotide (omega-conotoxin MVIIA, Prialt®) is a synthetic ve sion of a peptide, consisting of 25 amino acids, which was origin ly isolated from the venom of the Indo-Pacific predatory mari cone snail *Conus magus* in 1979. The genus *Conus* contains abc 500 different species which are marine, venomous, fish-hunti predators.

Ziconotide, a highly polar and water-soluble molecule with molecular mass of 2 639 Da, consists of 25 L-amino acids, inclu ing three disulphide bonds, which determine the three-dimensio structure. The synthesized commercial product is identical to t natural peptide from the marine snail. It is noteworthy that no che ical modifications have been introduced.

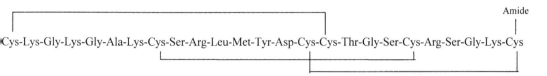

Fig.151. Amino acid sequence of ziconotide

Ziconotide is used for amelioration of chronic pain and is administered intrathecally (directly into the spinal canal) and approved for sale under the name Prialt by the FDA in USA in 2004 and the year after by EMEA in Europe.

Ziconotide is the first therapeutic agent in a new pharmacological class of "topically" active analgesics. The compound selectively targets neuron-specific (N-type) voltage-gated calcium channels and results in analgesia by interruption of calcium dependent, primary, afferent transmission of pain signals in the spinal cord.

Exploitation of peptides from cone snails is continuing and several other toxins are in development of drugs for use predominantly in the pain and CNS therapeutic areas.

Mistletoe toxins

The European mistletoe *Viscum album* L. (*Viscaceae*) is a semi-parasite, living on a number of woody (lignified) hosts. It absorbs water and nutritive salts from the host but undertakes its own photosynthesis and its constituents are partially different from those of the host. The mistletoe is sometimes considered a pest because it slowly destroys fruit trees and ornamental trees. The plant has a characteristic appearance, with evergreen leaves and pseudo-dichotomous branching.

The European mistletoe has been used in folk medicine over virtually the entire Europe for several thousand years and is believed to be effective against hypertension and cancer. A high-molecular-weight protein fraction with a good tumour-inhibiting effect *in vitro* has been isolated from fresh press-juice. This fraction is extremely heat-sensitive and loses its activity even on brief warming to 40°C. Moreover, it is toxic and hence useless as a drug. Crude mistletoe extract has a hypotensive effect on experimental animals when injected intravenously. This effect is due partly to the presence of choline and γ-aminobutyric acid in the extract and partly to small basic proteins, called *viscotoxins*, which consist of a chain of 46 amino-acid residues with three disulphide bridges. The three main viscotoxins are called viscotoxins A2, A3 and B. They exhibit a considerable degree of homology, with only minor differences in the amino-acid sequence. The primary structure of viscotoxin A3 is shown below.

Mistletoe toxins homologous with the viscotoxins have been found also in other mistletoes. The pharmacological effects of mistletoe toxins are very similar to those of cardiotoxins from cobra

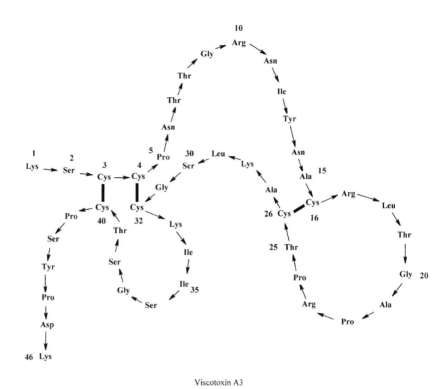

Viscotoxin A3

venoms, although only low amino-acid sequence homology has been shown. The viscotoxins and other mistletoe toxins are structurally highly homologous to purothionins from the endosperm of *Poaceae* plants, to crambin from *Crambe abyssinica* Hochst. ex R.E. Fries (*Cruciferae*) and to *Pyrularia* thionin, isolated from the parasitic plant *Pyrularia pubera* Michaux (*Santalaceae*). All these compounds are collectively named thionins after the Greek word for sulphur, describing one of the intrinsic properties of these proteins, namely the abundant presence of sulphur-containing cysteine residues. It has therefore been suggested that the viscotoxins should be renamed viscothionins and that the other mistletoe toxins should be termed mistletoe thionins.

V. album also contains toxic lectins (see page 518). Two compounds called VAA I (= Viscum-Lektin, Mistellektin I, Viscumin) and VAA II (= Viscumtoxin, Mistellektin II) have been isolated and characterized. Both are glycoproteins with the molecular masses 65 kDa and 58–60 kDa, respectively. Like the previously described plant toxins ricin and abrin (pages 513, 518), the mistletoe lectins are composed of two protein chains – A and B – which are joined via disulphide bridges. In VAA I the molecular masses (without the sugar) of these chains are 29 kDa (A) and 34 kDA (B). The corresponding figures for VAA II are 27 kDa (A) and 31 kDA (B). The mistletoe lectins specifically bind galactose and galactose amine residues, and agglutinate human erythrocytes in an unspecific manner. LD_{50} (mice) is in the range 30–50 µg/kg compared with

500 µg/kg for the viscotoxins. The mistletoe lectins are type II RIPs (page 513) in which the B-chain attaches the toxin to terminal galactose residues in the surface of the cell and triggers endocytosis, introducing the A-chain which acts as an *N*-glycosidase on 28S rRNA (see ricin page 513).

VAA I has been investigated as a possible cancer remedy. In pharmacological experiments anti-metastatic activity has been demonstrated as well as a decreased growth of tumours, but no curative effect has been obtained. Clinical tests with very low doses (1 ng/kg body weight, subcutaneous) have given ambiguous results. One report claims no effects while others say that the results are promising. Stimulation of the production of endorphins has been reported, an effect which might contribute to an enhanced well-being for the patients.

Cyclotides

Cyclotides are a family of plant proteins, whose *N*- and *C*-terminal amino acids are connected via an ordinary peptide bond, thus forming a circular structure. Three disulphide bonds stabilize the structure. Two of them, together with their connecting peptide chain form an embedded ring, which is penetrated by the third disulphide, thus forming a cyclic cystine knot (CCK) motif. This construction makes the cyclotides extremely stable against enzymatic, chemical and thermal degradation. The general structure is presented in Fig. 152. A three-dimensional structure of the cyclotide cycloviolacin O2 is shown on the book cover and highlights the disulphide bonds

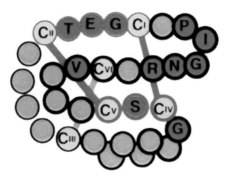

Fig. 152. General structure of cyclotides. The disulphide bonds connecting cysteine residues C_I–C_{IV} and C_V–C_{II}, together with the amino acid residues which are outlined in pink, form the embedded ring which is penetrated by the disulphide bond C_{VI}–C_{III}. Conserved amino acid residues are coloured green and variable residues blue. Those residues closer to the observer are outlined with a thicker black or red outline. From D.J. Craik, N.L. Daly, T. Bond and C. Waine "Plant cyclotides. A unique family of cyclic and knotted proteins that defines the cyclic cystine knot structural motif". Journal of Molecular Biology 294, 1327–1336 (1999). Reproduced with permission from the publisher, Elsevier B.V. (Academic Press)

(in yellow) and the elements of secondary structures present within the cyclic backbone (in blue). The β-sheet hairpin loop and the short α-helix, which is found in bracelet cyclotides, is also shown.

The cyclotides contain about 30 amino acids and they can be grouped in two subfamilies – bracelet and Möbius proteins. In the bracelet proteins the amino acid residues in the peptide chain are connected to form a regular ring whereas in the Möbius proteins the ring is twisted to form a Möbius strip. The twist is caused by a *cis* Pro peptide bond in the part of the peptide chain connecting cysteine residues CV and CVI.

The first discovery of a cyclotide was the result of independent observations made in 1965 by the Swedish professor of Pharmacognosy Finn Sandberg and in 1970 by the Norwegian Red Cross volunteer Lorens Gran, in the Central African Republic and Zaire respectively. In both countries women used a decoction obtained by boiling of the leaves of *Oldenlandia affinis* DC (*Rubiaceae*) to accelerate childbirth. In 1973 Gran isolated a peptide called Kalata B1, the structure of which (Fig. 153) was determined 20 years later and found to be the first cyclotide. Several other cyclotides have since then been isolated and characterized from this plant.

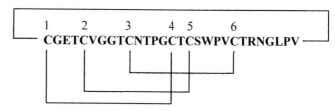

Fig. 153. Amino acid sequence of Kalata B1

So far cyclotides have been found only in two plant families – *Violaceae* and *Rubiaceae*. Over 100 different cyclotide sequences have been reported, but it has been suggested that *Violaceae* alone contains over 9 000 different cyclotides.

Besides uterotonic activity, cytotoxic, haemolytic, antibacterial and insecticidal activities have been attributed to cyclotides. They have also been reported to have anti-HIV activity and to have antifouling activities against barnacles. Their function in plants is not known, but they are suspected to be involved in host defence. The stability of the cyclotide structures together with their wide range of biological activities make them interesting candidates for pharmaceutical or agricultural applications. Chemical synthesis has been achieved and tolerance to residue substitutions has been demonstrated. Possible applications are exploitation of the biological activities of naturally occurring cyclotides and use of the CCK framework as a stable drug delivery vehicle, e.g. for instable small peptides that can be grafted to the stable cyclotide skeleton.

Biosynthesis

Biosynthetically the cyclotides are normal proteins, i.e. they are encoded by genes and synthesized on ribosomes. The cyclotides are produced by cleavage of a precursor protein, organized as illustrated in Fig. 154.

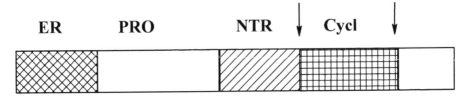

Fig. 154. Organization of the cyclotide precursor protein. ER = endoplasmic reticulum signal sequence region. PRO = proregion, NTR = N-terminal repeat region, Cycl = cyclotide region. NTR and Cycl can be repeated 1–3 times within one precursor protein, encoding different or identical cyclotides. Arrows indicate cleaving points for release of the cyclotide.

Release of the cyclotide is accomplished by cleavage at the arrows in Fig. 154. Cleavage on the N-terminal side of the cyclotide fragment is achieved prior to the sequence x-Leu-Pro, where x indicates an amino acid with small side-chain, e.g. Gly. On the C-terminal side, cleavage is performed following an Asn or Asp residue. The mechanism of the ring closure *in planta* is as yet unknown.

Non-ribosomal polypeptides

Most polypeptides and proteins are biosynthesized on ribosomes (see textbooks in biochemistry and molecular biology). Some microorganisms, however, biosynthesize polypeptides by a different mechanism, not involving mRNA, tRNA or ribosomes. Instead the amino acids are joined together on protein templates, called *Non-Ribosomal Peptide Synthases* (NRPSs). Like in polyketide biosynthesis (page 270) the procedure is a multistep process, directed by covalent catalysis, in which there are no free intermediates. Not only pure peptides are formed on protein templates, but also lipopeptides, depsipeptides and peptidolactones. The precursors are exceedingly diverse and include pseudo-, non-proteinogenic, hydroxy, *N*-methylated and D-amino acids. In contrast the nucleic acid-dependent ribosomal synthesis of peptides and proteins is restricted to encompass only the 21 proteinogenic amino acids (including selenocysteine).

The group includes a number of substances of medical and pharmaceutical interest such as antibiotics and compounds with antitumour, enzyme-inhibiting and immunosuppressive activity.

Biosynthesis

The NRPSs are very large proteins ranging in molecular mass from

61 kDa to 1600 kDa. They consist of *modules*, each of which co tains all the catalytic functions required for incorporation of o amino acid into the peptide chain including recognition, activatio modification and peptide bond formation. The number of modul in a synthetase is thus the same as the number of amino ac residues contained in the synthesized peptide and they are aligne in a sequence which is colinear with the amino acid sequence of t peptide. Each module contains a number of *domains* each of whic performs one special reaction required for the incorporation of t amino acid. The *apo*-enzyme resulting from translation of t genetic code for the synthetase is not active but must be subjecte to a posttranslational modification during which 4'-phosphopar tetheine (page 240) is covalently attached to the side-chain of a sp cial serine residue, located in the thiolation domain (see below) each module. This modification is performed by a special *4'-P transferase* which utilizes CoA as the source of 4'-phosphopantet eine (Fig. 155) yielding an active *holo-enzyme*.

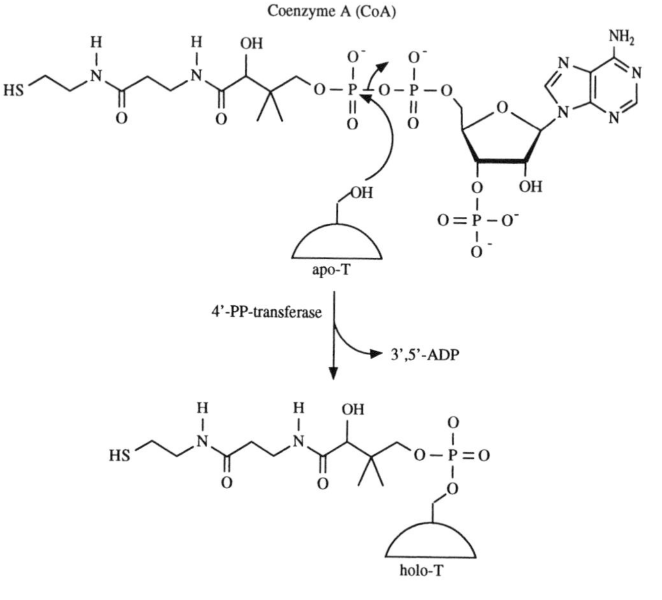

Fig. 155. Posttranslational phosphopantetheinylation of the thiolatio. domain (T) in a module of a peptide synthetase

The domains contained in a module are: the adenylation domain the thiolation domain, the condensation domain and possibly mod ification domains for epimerization or *N*-methylation.

The adenylation domain (the A domain)

In the process catalysed by the A domain, an amino acid is activat ed by reaction with Mg^{2+}-ATP yielding an aminoacyl adenylat according to Fig. 156.

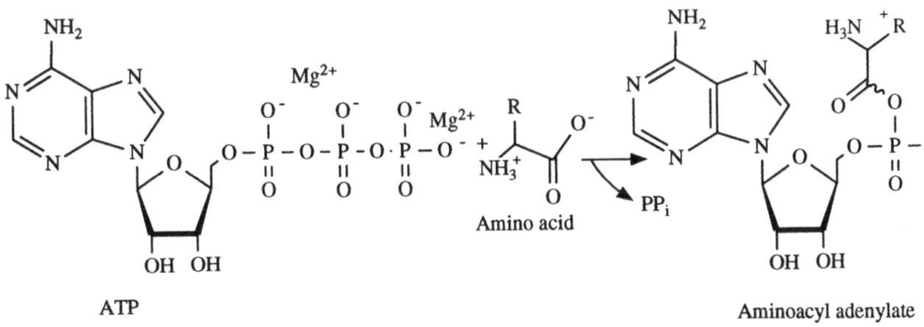

Fig. 156. Amino acid adenylation

The adenylate forming domains represent the non-ribosomal code as each domain in principle recognizes only one specific amino acid. The NRPSs are, however, considered less precise than their ribosomal counterparts as indicated by the existence of peptide iso-forms. Certain positions are nevertheless highly conserved. In cyclosporin (page 546) for instance, positions 2, 3 and 10 are strict-ly occupied by leucine while position 8 can contain leucine, valine or isoleucine.

In the NRPSs the highly conserved A domains are found as repet-itive blocks, the number of which is the same as the number of amino acids contained in the final product. The A domains have an average size of about 550 amino acids each. They share significant homology with acyl-CoA synthetases and luciferases, which also catalyse adenylation of carboxyl groups, and can thus be considered to be members of a superfamily of adenylate-forming enzymes. The 3D structure of the phenylalanine-activating A domain of *grami-cidin S₁ synthetase* 1 (see below) was in 1997 determined at a reso-lution of 1.9 Å. It is folded into two compact subdomains (in this context domain means a stable tertiary fold), a large N-terminal and a smaller C-terminal part that are connected with a short hinge. This achievement in connection with knowledge of the amino acid sequence (deduced from corresponding gene fragments) has allowed deductions about the mechanism of the adenylation reac-tion and the importance of highly conserved motifs in the sequence. However, this falls outside the scope of this book and the interested reader is referred to a review by Marahiel *et al.* in *Chemical Reviews* **97**, 2651–2673 (1997).

The thiolation domain (T domain)

A functional unit of about 100 amino acid residues is located direct-ly downstream of each A domain. This unit is the thiolation domain which carries 4'-phosphopantetheine (4'-PP) attached to the side-chain of a particular serine residue. The T domain is also called the Peptidyl Carrier Protein (PCP) and is an analogue to the acyl carrier proteins (ACP) of modular fatty acid and polyketide synthetases (pages 240, 274). If the module contains a modifying domain, such as a domain catalysing N-methylation, this may be located between the particular A and T domains. At the T domain the aminoacyl adenylate reacts with the thiol group of 4'-PP yielding a carboxy thioester (Fig. 157).

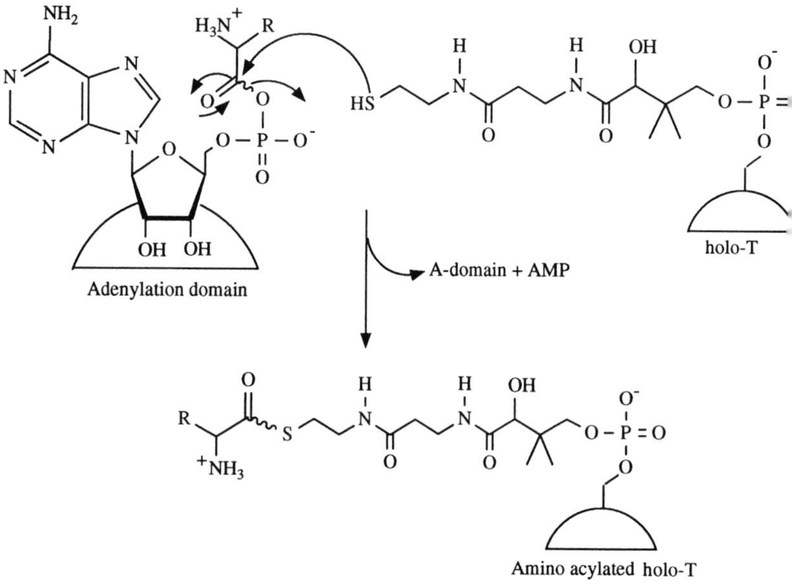

Fig. 157. Reaction of aminoacyl adenylate with 4'-PP

The formation of the relatively stable thioester is a prerequisite for peptide bond formation. The discovery of thioesters in peptide biosynthesis lent the name *thiotemplate mechanism* for the non-ribosomal peptide-forming systems.

The condensation domain (C domain)

The C domain is responsible for the condensation of two amino acids activated on adjacent modules, i.e. it catalyses elongation of the growing peptide chain. Within the polypeptide chains of peptide synthetases, the C domains are located in close proximity to the A and T domains (Fig. 159). The number of C domains is the same as the number of peptide bonds of the resulting peptide. A condensation domain is about 450 amino acids in length. At the C domain the nitrogen nucleophile of the thioester attached to one of the two T domains attacks the carbonyl of the thioester attached to the second T domain thus forming a peptide bond. A conserved histidine residue in the condensation domain is essential for the progress of the reaction which is accompanied by liberation of the second T domain (Fig. 158). When the two consecutive A and T domains are not located on the same enzyme and peptide formation thus has to be performed between amino acids activated on two synthetases, the C domain is found at the N-terminus of the amino acid-accepting synthetase (intermolecular amino acid transfer; as an example see the organization of the gramicidin S₁ synthetase complex and the tyrocidine A synthetase complex (Fig. 159).

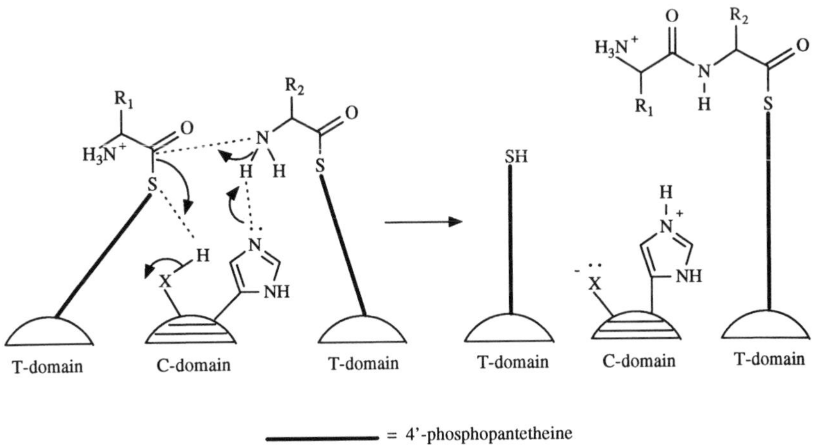

= 4'-phosphopantetheine

Fig. 158. Suggested mechanism for the formation of a peptide bond at the C domain. The imidazole ring indicates the second histamine in a conserved motif (HHxxxDG) in the condensation domains of peptide synthetases. This histamine residue is thought to act as the general base promoting the nucleophilic attack

Modification reactions

As mentioned above some non-ribosomal peptides contain modified amino acids such as D-amino acids or *N*-methylated amino acids. Modification of amino acid residues may occur at either of four stages: *a)* prior to activation, *b)* at the aminoacyl stage preceding elongation, *c)* after elongation at the peptidyl intermediate stage or *d)* following termination of peptide synthesis as a product transformation reaction.

In case *a* the modified amino acids are present in the cell and can serve as substrate for the respective activation domain. Transforming enzymes such as racemases can be detected and separated from the NRPSs. *N*-methylation and epimerization are the two best known reactions which take place during synthesis of the peptides. *N*-methylation occurs at the thioester stage prior to peptide bond formation and is thus an example of case *b* above. Particular domains for *N*-methylation have been found in the genes for *cyclosporin A synthetase* (Fig. 159) and *enniatin synthetase*. The sequence data revealed a novel type of module possessing an insertion of about 420 amino acids between the A and T domains. The occurrence of these insertions within the amino acid-activating modules coincides with the number of *N*-methylated residues in the corresponding peptides.

Epimerization can occur at various stages in the biosynthesis of non-ribosomal polypeptides. Gramicidin S_1 and tyrocidine A are cyclic peptides produced by *Bacillus brevis* (pages 538, 539). They contain D-phenylalanine and it has been shown that L-phenylalanine is the natural substrate and that epimerization occurs prior to condensation while the L-isomer is bound as a thioester (case *b* above). An epimerization domain is located between the thiolation domain and the following condensation domain of the genes coding for these peptides (Fig. 159). In the biosynthesis of δ-(L-α-aminoadipyl)-L-cysteinyl-D-valine (ACV, page 563), the precursor of isopenicillin N, epimerization takes place after formation of the peptide bond between L-cysteine and L-valine (case *c*). An epimerization domain is located at the C-terminal end of the gene *pcbAB* coding for ACV synthetase (Fig. 159). Cyclosporin (page 546) contains D-alanine which is provided as a substrate produced by non-integrated racemases (case *a*).

Termination reactions

When the growing peptide chain has reached the final module it is released either as a linear or as a cyclic product. Release of a linear peptide can be achieved either by hydrolysis (1), yielding the free enzyme and the peptide or by aminolysis (2) which yields an amide of the peptide and the free enzyme:

$$(1)\, ESA^nA^{n-1} \text{- - - -} A^1 \,+\, OH^- \longrightarrow A^nA^{n-1}\text{---}A^1 \,+\, ESH$$

$$(2)\, ESA^nA^{n-1} \text{- - - -} A^1 \,+\, RNH_2 \longrightarrow A^nA^{n-1}\text{---}A^1NHR + ESH$$

$$(ES = enzyme,\ A =\ amino\ acid)$$

Terminal amino groups or internal hydroxy or amino groups are involved in cyclizations which also can proceed by fragment condensations. Several NRPSs contain a thioesterase-like domain (see for example the gramicidin S_1 and tyrocidin A synthetase complexes, Fig. 159) at the C-terminal which is assumed to catalyse the release or cyclization of the peptide, but not all synthetases contain an integrated thioesterase.

The organization of modular peptide synthetases and the corresponding genes

NRPSs can consist either of a complex of several synthetases, each containing one or several modules, or of one single synthetase containing all the modules which are required for the complete biosynthesis. Examples of the first alternative are the bacterial peptide synthetase complexes for biosynthesis of gramicidin S_1 and tyrocidin A (page 537). The gramicidin S_1 synthetase complex consists of two integrated synthetases: *gramicidin S_1 synthetase 1* (GS₁ 1) and *gramicidin S_1 synthetase 2* (GS_1 2). GS_1 1 comprises one module for activation and epimerization of phenylalanine. This is the initiating module and does not contain a condensation domain. GS_1 2 has four modules which activate the other four amino acids L-Pro, L-Val, L-Orn and L-Leu and assemble the cyclic decapeptide product. The tyrocidin A synthetase complex comprises three integrated synthetases: *tyrocidine A synthetase 1* (TAS 1), *tyrocidine A synthetase 2* (TAS 2) and *tyrocidine A synthetase 3* (TAS 3). Like GS_1 1, TAS 1 has only one module for activation and epimerization of L-phenylalanine. This is also the initiating module and has no condensation domain. TAS 2 has three modules for activation of L-Pro and L-Phe as well as for activation and epimerization of a third L-Phe residue. The modules of TAS 2 also contain condensation domains and assemble the hitherto four activated amino acids to a peptide, which is further elongated by TAS 3, containing six more modules which complete the biosynthesis of tyrocidin A.

An example of the second alternative, mentioned above, is the fungal enzyme *cyclosporin A synthetase*, which is one single enzyme, comprising the 11 modules needed for the assembly of cyclosporin A (page 546).

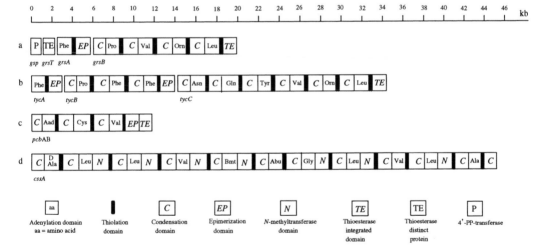

Fig. 159. Schematic diagram showing the modular organization of NRPSs on both the gene and the protein level. a) The gramicidin S_1 synthetase complex, encoded by the operon grs, which comprises the three genes grsT (encoding a thioesterase of unknown function), grsA (encoding gramicidin S_1 synthetase 1), and grsB encoding gramicidin S_1 synthetase 2). b) The tyrocidin A synthetase complex, encoded by the operon tyc with the genes tycA, (encoding tyrocidine A synthetase 1), tycB (encoding tyricidine A synthetase 2), and tycC (encoding tyrocidine A synthetase 3). c) ACV-synthetase encoded by the gene pcbAB. d) Cyclosporine A synthetase encoded by the gene cssA. The scale on top indicates the gene sizes in kb. Also shown is the gene gsp, encoding 4'-PP-transferase in the gramicidin S_1 synthetase complex. Aad = L-α-aminoadipic acid. Bmt = (4R)-4-[(E)-2-butenyl]-4-methyl-L-threonine. Abu = α-aminobutyric acid

A large number of bacterial operons and fungal genes which encode NRPSs have been cloned, sequenced and partially characterized. The complete DNA sequences of the bacterial operons *grs*, *tyc* and *bac* which encode NRPSs for the cyclic peptide antibiotics gramicidin S_1, tyrocidine A and bacitracin A (page 539) have been determined. These operons are 18–45 kb in size and the synthetases, as described above, comprise from one to six modules ranging in molecular size from 126 kDa for one-module enzymes to over 700 kDa for the six modules of tyrocidine synthetase 3.

The fungal gene *pcbAB*, which codes for ACV synthetase (page 564) has a size of 11.4 kb and the corresponding enzyme has a molecular mass of 426 kDa. The largest gene which so far has been characterized is *cssA* which encodes cyclosporin A synthetase. This gene is over 45 kb in length and encodes a polypeptide chain with over 15 000 amino acid residues with a molecular mass over 1 600 kDa. Fig. 159 is a schematic diagram showing the modular organization of these peptide synthetases on both the gene and the protein level.

Prospects for the construction of hybrid non-ribosomal peptides

The acquired knowledge of the genes and the organization of the NRPSs opens up a new field for combinatorial biosynthesis of peptide antibiotics and other bioactive peptides. Manipulation of the adenylation domains could allow the insertions of other amino acids in the primary sequence of a given peptide. It might also be possible to engineer complete biosynthetic systems (conveyor belts) for the synthesis of entirely new peptides. It has been shown that catalytic domains of the synthetases are also able to act as individual enzymes and can interact with other distinct domains, yielding a complex corresponding to a functional type II FAS or PKS (pages 240, 270). Construction of artificial type II NRPSs would increase the ability to generate manifold peptides with diverse structures.

Tyrothricin　　Tyrothricin is a mixture of peptide antibiotics which is produced by *Brevibacillus brevis*. The mixture consists of 20–30 % linear polypeptides, called gramicidins, and 70–90 % cyclic compounds called tyrocidines. These two groups can easily be separated.

Valine-Gramicidin A

HCO—L-Val —Gly—L-Ala —D-Leu—L-Ala—D-Val —L-Val —D-Val —L-TRP —
　　　　　　　1　　　2　　　3　　　4　　　5　　　6　　　7　　　8　　　9

D-Leu— L-Trp —D-Leu – L-Trp – D-Leu —L-Trp –NHCH$_2$CH$_2$OH
　10　　　11　　　12　　　13　　　14　　　15

Valine - Gramicidin A

Valine-Gramicidin A is the main component of the gramicidin fraction of tyrothricin. Its homologues have other amino acids at positions 1 and 11 in the chain. In the gramicidins the N-terminal amino group is formylated and the carboxy group forms an amide with ethanolamine. The gramicidin fraction is used as an ingredient in lozenges for the treatment of infections of the throat and topically for treating infections of the skin, the ear and the eye.

Tyrocidine A

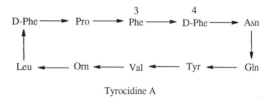

Tyrocidine A

Tyrocidine A is one component of the tyrocidin fraction of tyrothricin. Other tyrocidins differ in the amino acids at positions 3 and 4. Tyrocidins are about 50 times less potent as antibiotics than the gramicidins and are not used in therapy.

Gramicidin S₁

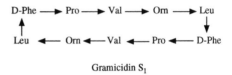

Gramicidin S₁

Gramicidin S is a mixture of cyclic antibiotic peptides which are produced by a different strain of *Bacillus brevis*. Its main component is gramicidin S₁. Gramicidin S is fairly toxic and is used only for topical applications.

Tyrocidine A and gramicidin S₁ are of interest because the modular organization of their NRPSs has been elucidated both at the gene and the protein level (see Fig. 159).

Gramicidines and tyrocidines act on bacterial membranes.

Bacitracin A

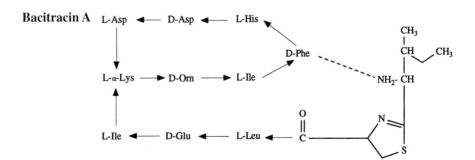

Bacitracin A

Various strains of *Bacillus subtilis* produce bacitracin, which is a mixture of cyclic polypeptides (bacitracins A, A', B, C, D, E, F, F_1, F_2, F_3 and G). About 70 % of this mixture is bacitracin A, which is used topically for treating infections of the skin, eyes, ears, nose and throat. The cyclic peptide part of the gramicidine molecule is formed by involvement of both the α- and the ε-amino groups of lysine. The amino-terminal part is formed by condensation of isoleucine and cysteine residues. Most of the bacitracin produced is used at subtherapeutic doses as an animal feed additive to increase feed efficiency. Therapeutic doses are used to treat various disorders in poultry and animals.

The operon encoding the bacitracin A synthetase comprises three genes, *bacA* encoding *bacitracin A synthetase 1* (5 modules), *bacB* encoding *bacitracin A synthetase 2* (2 modules), and *bacC* encoding *bacitracin synthetase 3* (5 modules). Epimerization domains catalyse the epimerization of glutamic acid, ornithine, phenylalanine and aspartic acid.

Polymyxin B and E Polymyxins A–E are antibiotic, cyclic polypeptides which originally were isolated from *Bacillus polymyxa*. Polymyxin E has also been isolated from *Bacillus colistinus* under the name colistin. Polymyxin B is a mixture of polymyxin B_1 and small amounts of polymyxin B_2. Also polymyxin E is a mixture and consists pre-

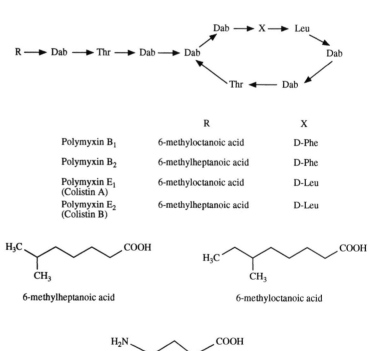

	R	X
Polymyxin B_1	6-methyloctanoic acid	D-Phe
Polymyxin B_2	6-methylheptanoic acid	D-Phe
Polymyxin E_1 (Colistin A)	6-methyloctanoic acid	D-Leu
Polymyxin E_2 (Colistin B)	6-methylheptanoic acid	D-Leu

6-methylheptanoic acid

6-methyloctanoic acid

Dab = L- α,γ - diaminobutyric acid

Fig. 160. Structures of polymyxins B_1, B_2, E_1 and E

dominantly of polymyxin E_1 (= colistin A) and minor amounts of polymyxin E_2 (= colistin B).

The polymyxins contain ten amino acid residues, six of which are L-α,γ-diaminobutyric acid (Dab). One Dab is N-terminal and a part of the molecule is cyclic due to an amide bond between the C-terminal threonine and the γ-amino group of one Dab. The α-amino group of the N-terminal Dab forms an amide bond with the carboxyl group of 6-methyloctanoic acid (polymyxin B_1 and E_1) or 6-methyl-heptanoic acid (polymyxin B_2 and E_2). Amino acid residue no. 6 is D-phenylalanine in the B series and D-leucine in the E series (Fig. 160). Due to the many free γ-amino groups of the Dab residues the polymyxins are strongly basic, which gives them detergent-like properties and causes them to bind to and damage bacterial membranes.

Polymyxin B and E are used topically for treatment of eye- or ear-infections and infected wounds, often in combination with other antibiotics. Internal use is avoided because of neurotoxic and nephrotoxic effects.

Daptomycin

Daptomycin is a cyclic polypeptide with a tail of decanoic acid, which is produced by fermentation of *Streptomyces roseosporus*. Normally this organism produces a complex of lipopeptides with varying lengths of the fatty acid tail, but by feeding decanoic acid during the fermentation, daptomycin is obtained as the main product. The peptide part of daptomycin is composed of 13 amino acids, 10 of which form the ring while 3 connect the ring to the fatty acid residue. Three amino acids (asparagine, alanine and serine) have D-configuration and two are the non-proteinogenic amino acids 3-methyl-L-glutamic acid and L-kynurenine. The N-terminal amino acid is L-tryptophan, the aminogroup of which forms an amide bond with the carboxyl group of decanoic acid. The ring is formed by an ester bond between the C-terminal kynurenine and the hydroxyl-group of threonine.

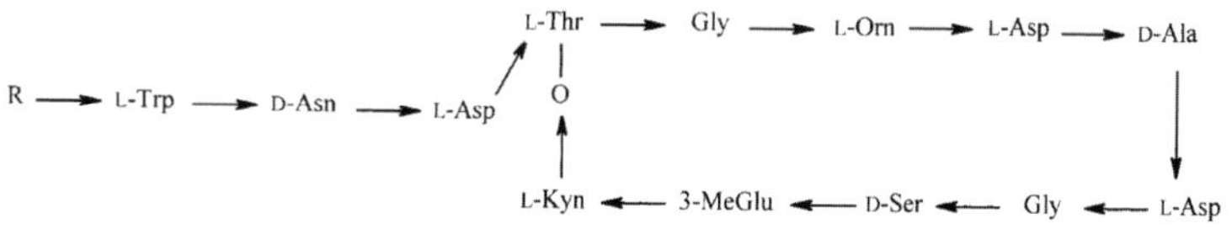

Daptomycin

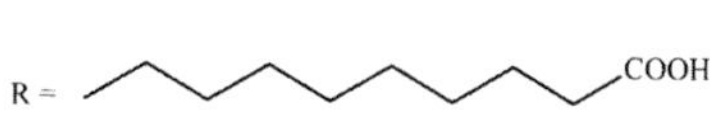

Decanoic acid

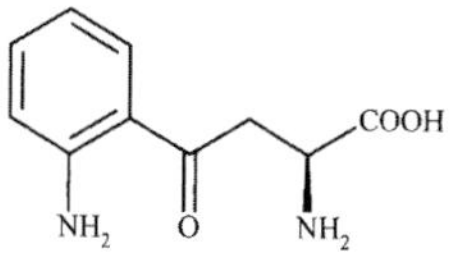

L-Kynurenine

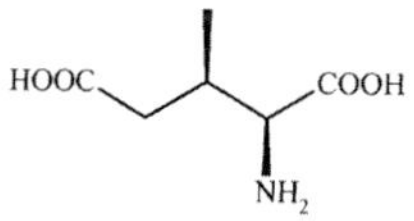

L-Threo-3-Methylglutamic acid

The gene cluster (*dpt*) coding for the NRPS of daptomycin has been characterized. Three large genes (*dptA, dptBC* and *dptD*) encode three proteins – DptA, DptBC and DptD – which catalyse formation of the peptide part of the molecule. The protein DptA contains modules for Trp-1, D-Asn-2, Asp-3, Thr-4 and Gly-5. Modules for Orn-6, Asp-7, D-Ala-8, Asp-9, Gly-10 and D-Ser-11 are contained in DptBC, while, finally, modules for mGlu-12 and Kyn-13 are found in DptD. The N-terminal C-domain of the module of DptA which codes for Trp-1 is different from the other C-domains and is proposed to have a role in coupling the decanoic acid to Trp-1. E and M domains, required for relevant amino acid modifications have been identified in DptA and DptBC. The module coding for Kyn-13 in DptD has a terminal Te domain, responsible for cyclization and release of the mature peptide from the enzymes.

In addition to these three large genes, genes for the supply of unusual amino acids and for regulation of the biosynthesis have also been identified in the *dpt* cluster as well as genes for fatty acid biosynthesis, acylation and ABC transporter components.

Daptomycin is used for treatment of skin and skin structure infections, caused by Gram-positive pathogens. It is also used for treatment of bacteraemia and endocarditis caused by *Streptococcus aureus* strains, including those resistant to methicillin. The drug is administered by i.v. infusion and its adverse effects appear comparable with those of vancomycin and semisynthetic penicillins.

Echinocandins

Echinocandin B is a lipopeptide with antifungal properties which was isolated in 1974 from *Aspergillus rugulosus*. In 1989 a new antifungal agent – pneumocandin B_0 – was obtained from a culture of the fungus *Glarea lozoyensis* (previously known as *Zalerion arboricola*), which was found in pond water from the valley of the Spanish river Lozoya. A third lipopeptide with similar structure and denoted as WF11899A, was isolated in 1994 from the fungus *Coleophoma empetri*. The structures of these three lipopeptides are presented in Fig. 161.

As seen from the figure, the natural echinocandins are cyclic hexapeptides connected to a long-chain fatty acid. In *echinocandin B* the peptide part is built of the amino acids 4,5-dihydroxy-L-ornithine (N-terminal), L-threonine, 4-hydroxy- L-proline, 3,4-dihydroxy- L-*homo*-tyrosine, L-threonine and 4-methyl-3-hydroxy L-proline (C-terminal). The ring is closed by a peptide bond between the carboxyl group of methyl-hydroxy-proline and the δ amino group of dihydroxy-ornithine. The side-chain is linoleic acid which forms an amide with the α-amino group of dihydroxy ornithine.

The structures of the other two lipopeptides are very similar. I *pneumocandin B_0* the side-chain consists of 10,12-dimethyl-myristic acid and amino acids 5 and 6 are 3-hydroxy- L-glutamine and 2 hydroxy- L-proline instead of L-threonine and 4-methyl-3-hydroxy L-proline. In *WF11899A* amino acid 5 is 3-hydroxy-L-glutamin

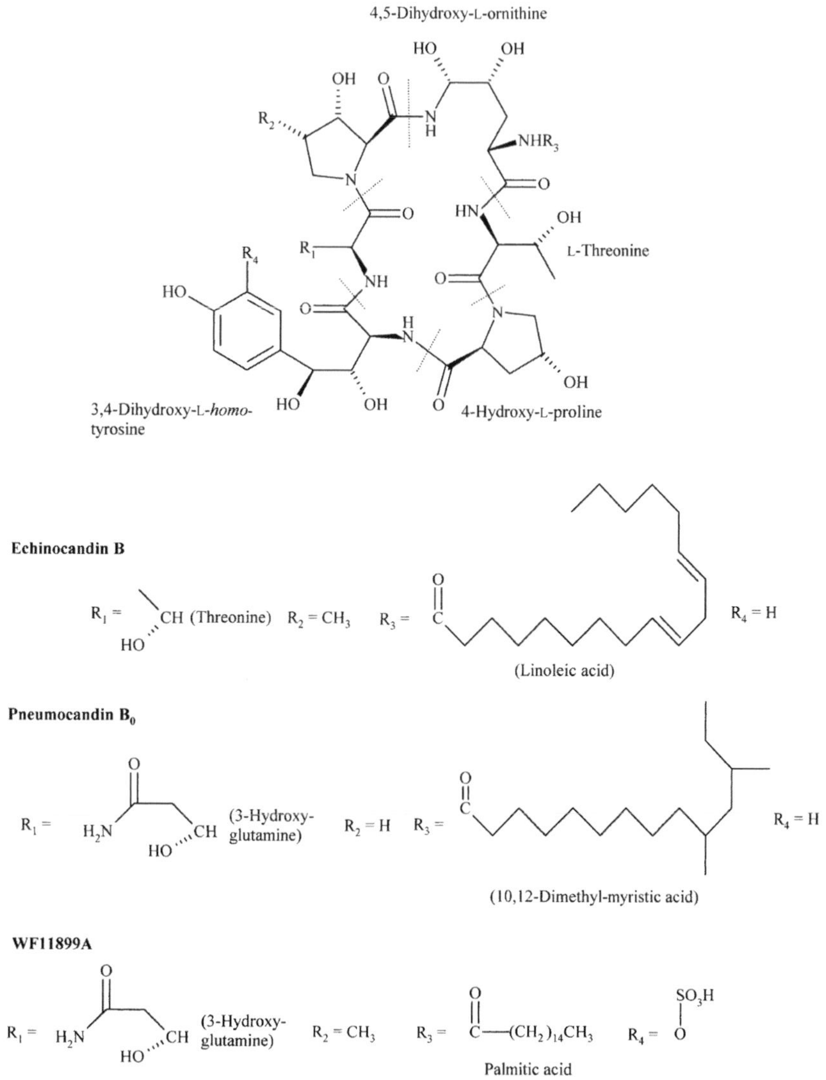

4,5-Dihydroxy-L-ornithine

3,4-Dihydroxy-L-*homo*-
tyrosine

L-Threonine

4-Hydroxy-L-proline

Echinocandin B

$R_1 =$ CH (Threonine) $R_2 = CH_3$ $R_3 =$ (Linoleic acid) $R_4 = H$

Pneumocandin B$_0$

$R_1 =$ (3-Hydroxy-glutamine) $R_2 = H$ $R_3 =$ (10,12-Dimethyl-myristic acid) $R_4 = H$

WF11899A

$R_1 =$ (3-Hydroxy-glutamine) $R_2 = CH_3$ $R_3 = C—(CH_2)_{14}CH_3$ Palmitic acid $R_4 = O—SO_3H$

Fig. 161. Structures of echinocandin B, pneumocandin B$_0$, and WF11899A

and the aromatic amino acid has an $O\text{-}SO_3H$ group connected to the ring. The side-chain is palmitic acid.

Very little is known about the biosynthesis of echinocandins. Tracer studies have shown that homotyrosine is formed from tyrosine by a chain elongation mechanism involving condensation with acetate. Only 4-hydroxy-proline derives from proline. The 4-methyl-3-hydroxy-proline residue is formed by cyclization of leucine.

None of the natural echinocandins is used as a drug. The first clinically used echinocandin was **Caspofungin**, which is a semi-synthetic derivative of pneumocandin, where amino acid 5 is 3-hydroxy-ornithine instead of 3-hydroxy-glutamine and the 5-

hydroxy group of dihydroxy-ornithine is replaced with a (2-aminoethyl)amino group.

The drug **Micafungin**, which was the second echinocandin in clinical use, is a derivative of WF11899A where palmitic acid is replaced with a complex aromatic side-chain. The peptide part of this molecule is the same as in WF11899A.

Attempts to obtain a useful drug from echinocandin failed at first, but finally a derivative where the linoleic acid was replaced with an alkoxytriphenyl side-chain could be introduced in the clinic as **Anidulafungin**. Also here the peptide part is the same as in the natural product.

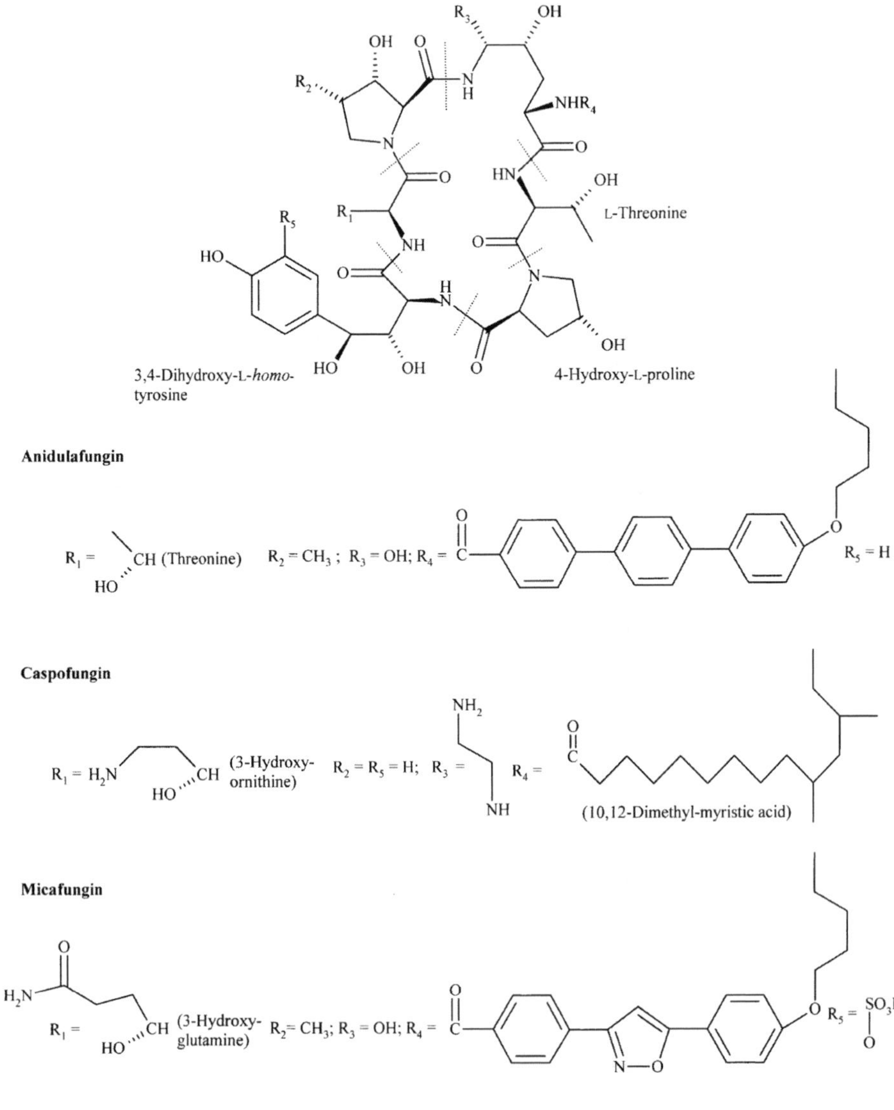

Fig. 162. Structures of Anidulafungin, Caspofungin and Micafungin

The echinocandins are used as antifungal remedies, particularly against candida infections. They have low oral bioavailability and are therefore given intravenously. The echinocandins have low toxicity and do not affect the cytochrome P450 system or P-glycoprotein and the risk of drug interactions is therefore minimal. The mechanism of action is inhibition of β-(1,3)-glucan synthesis which damages the fungal cell walls. *In vitro* and *in vivo* the echinocandins are fungicidal against *Candida* spp. and fungistatic against *Aspergillus* spp.

Dactinomycin Dactinomycin (actinomycin D) is a peptide antibiotic which is too toxic to be used as such but which is used in therapy as a cancer remedy. The compound is produced by *Streptomyces parvullus*. The structure comprises a phenoxazinone dicarboxylic acid (acti-nocin) to which two identical pentapeptides are attached by amide bonds to the amino groups of their N-terminal threonine residues. The peptides are cyclic due to lactone bonds between the hydroxy groups of the N-terminal threonine residues and the C-terminal *N*-methylated valines. Also a glycine residue in each peptide is *N*-methylated.

Biosynthetically the phenoxazinone ring system is derived from L-tryptophan via 3-hydroxyanthranilic acid which is methylated to form 3-hydroxy-4-methyl-anthranilic acid (3-HMA). The enzyme complex responsible for biosynthesis of the pentapeptide lactone comprises three enzymes. *Actinomycin synthetase I* (ACMS I, mol-

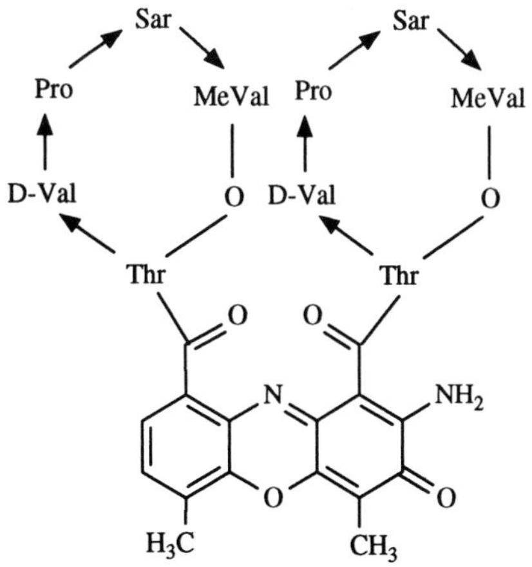

Dactinomycin
(Actinomycin D)

ecular size: 45 kDa) activates 3-HMA, forming adenylyl-3-HMA from 3-HMA and ATP. *Actinomycin synthetase II* (ACMS II) is a 280 kDa multienzyme that receives 3-HMA adenylate and attaches it to L-threonine. L-Valine is then added and epimerized to D-valine.

A third enzyme, *Actinomycin synthetase III* (ACMS III) is a 48 kDa polypeptide chain which has 3 modules for incorporation of ⬛ Pro, L-Gly and L-Val as well as for *N*-methylation of glycine an⬛ valine. Finally, the lactone ring between the C-terminal Me-Val an⬛ the OH-group of threonine is formed and one molecule of the ⬛ HMA pentapeptide lactone reacts with a second molecule, formin⬛ dactinomycin in an oxidative amino phenol condensation (Fi⬛ 163).

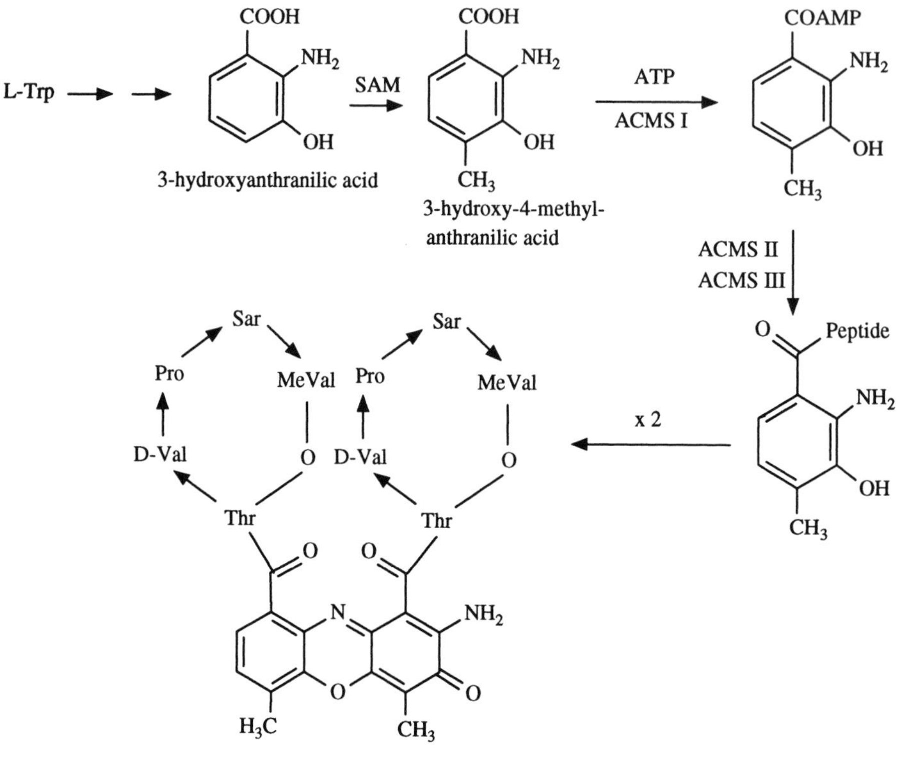

Dactinomycin

Fig. 163. Biosynthesis of dactinomycin

Dactinomycin is used against Wilm's tumour of the kidney an⬛ for treatment of carcinomas in the testicles and the uterus. Howeve⬛ it has serious side effects and must be used with great care and onl⬛ for short periods. Its activity is due to the planar phenoxazinone rin⬛ which intercalates with double-stranded DNA and inhibits DNA dependent RNA polymerases.

Cyclosporin   Cyclosporin A is a cyclic polypeptide built up from 11 amino acids some of which have a unique structure. Cyclosporin A is produce⬛ by the fungi *Cylindrocarpon lucidum* and *Tolypocladium inflatum* which are fungi belonging to the group of fungi imperfecti.

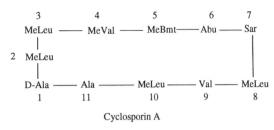

```
        3           4           5          6         7
      MeLeu ——— MeVal ——— MeBmt——— Abu ——— Sar
        |                                            |
  2   MeLeu                                           |
        |                                            |
      D-Ala ——— Ala ——— MeLeu ——— Val ——— MeLeu
        1        11          10         9        8
```

Cyclosporin A

MeBmt = (4R)-4-[(E)-2-butenyl]-4,*N*-dimethyl-L-threonine;
Abu = α-Aminobutyric acid; MeLeu = *N*-Methylleucine;
Sar = Sarcosine (= *N*-methylglycine).

MeBmt

Cyclosporin A is produced industrially by *Tolypocladium* grown in submerged culture. In normal fermentation broths it is the main component. Addition of D,L-α-aminobutyric acid almost completely suppresses the production of unwanted cyclosporin analogues.

Cyclosporin A is biosynthesized by *cyclosporin A synthetase*, a multifunctional NRPS (page 531). The enzyme contains domains for *N*-methylation of amino acids, but two amino acids must be supplied by separate pathways: D-alanine which is formed from L-alanine by a specific racemase, and the unusual amino acid (4R)-4-[(E)-2-butenyl]-4-methyl-L-threonine (= Bmt). Bmt originates from the acetate pathway and is in fact a *C*-methylated C_8 polyketide derivative into which the amino group has been introduced by transamination. During the biosynthesis of cyclosporin A, Bmt is *N*-methylated to form MeBmt.

Cyclosporin is an immunosuppressant which acts by binding to calcineurin, thereby inhibiting the activation of T-lymphocytes. It also has anti-inflammatory activity and does not affect bone marrow, but its nephrotoxic effect necessitates careful monitoring of kidney function. Cyclosporin is used to prevent allograft rejection in organ transplants and its introduction into medicine has made possible the transplantation of most organs, including hearts.

Glycopeptide antibiotics

Vancomycin Vancomycin is a glycopeptide antibiotic which is produced by the actinomycete *Amycolatopsis orientalis* (= *Streptomyces orientalis*). The structure is complicated and comprises the unusual amino acids 3,5-dihydroxyphenylglycine (C-terminal), 4-hydroxyphenylglycine (two residues), and β-hydroxy-chloro tyrosine (two residues). The N-terminal leucine residue is *N*-methylated. An asparagine residue is also contained in the structure. The tricyclic structure is generated by three phenolic oxidative coupling reactions involving the aromatic amino acid derivatives. The sugar part of vancomycin is composed of an unusual aminosugar – vancosamine – and glucose and is attached to the molecule via the hydroxy group of one of the two 4-hydroxyphenylglycine residues.

Vancomycin is active against Gram-positive bacteria and is used

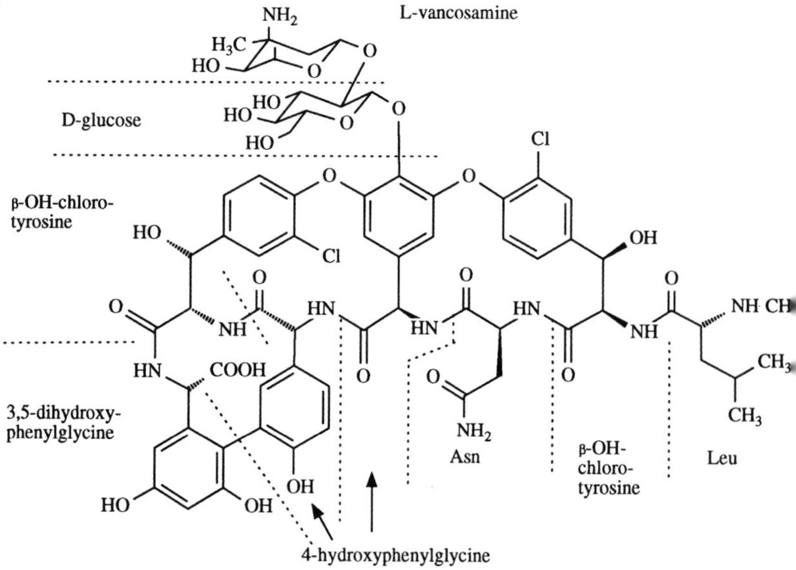

Vancomycin

for treatment of severe infections which are resistant to other antib
otics, particularly infections caused by *Staphylococci*. It is used pa
enterally but can be given orally for treatment of pseudomembran
ous colitis caused by *Clostridium difficile*, an infection which ca
result from treatment with other antibiotics.

Biosynthesis   Vancomycin contains a backbone of seven amino acids. Numbere
in order from the N-terminal they are: **1** = leucine, **2** = β-hydroxy
chloro tyrosine, **3** = asparagine, **4** = **5** = 4-hydroxyphenylglycine,

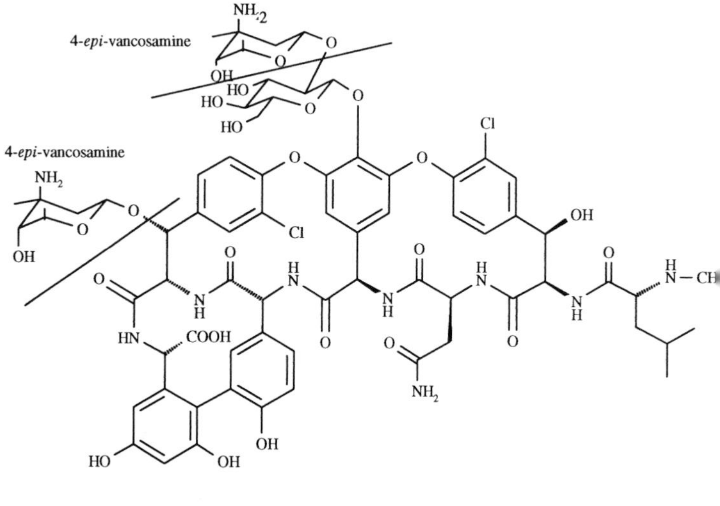

Chloroeremomycin

548

= β-hydroxy-chloro tyrosine and **7** = 3,5-dihydroxyphenylglycine, Studies with labelled precursors have shown that **7** is derived from four units of acetate and thus requires the activity of a polyketide synthase. Residues **2** and **6** as well as **4** and **5** derive from tyrosine.

Work on identification of the genes that are responsible for the biosynthesis of vancomycin has been performed on a related glycopeptide – chloroeremomycin – which is produced by the same organism as vancomycin and has the same peptide backbone but differs in the sugars attached to the peptide.

A cell-free extract from a chloroeremomycin-producing strain was fractionated and a fraction identified that was able to catalyse the addition of glucose to residue **4** of the aglucon of chloroeremomycin. A part of the amino acid sequence of the isolated polypeptide was determined and cDNA probes corresponding to this sequence were synthesized and used to locate the biosynthetic cluster of genes for the production of chloroeremomycin. The sequence of 72 kb of genomic DNA led to the location of 39 putative genes (Fig. 164). About half of the anticipated protein products of these genes have clearly implicated roles in the biosynthesis of the antibiotic.

Three large genes – 3, 4 and 5 – encode three proteins – CepA, CepB and CepC – which catalyse the formation of the peptide backbone. CepA is responsible for the biosynthesis of the N-terminal

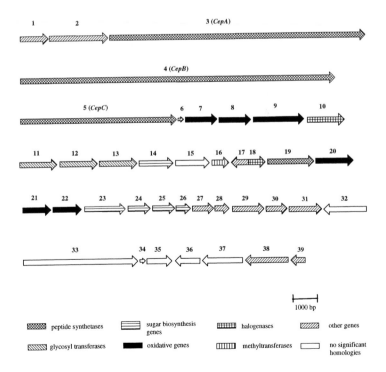

Fig. 164. Organization of the genes and putative enzymatic roles for their protein products in the biosynthesis of chloroeremomycin

tripeptide. The condensation domain that catalyses the condensation of the tripeptide to the next residue in the chain is located at the beginning of CepB, which elongates the tripeptide to a hexapeptide. Finally CepC adds amino acid **7** and cleaves the heptapeptide from the biosynthetic apparatus by the action of a thioesterase. The functional organization of the peptide synthetases is presented in Fig. 165.

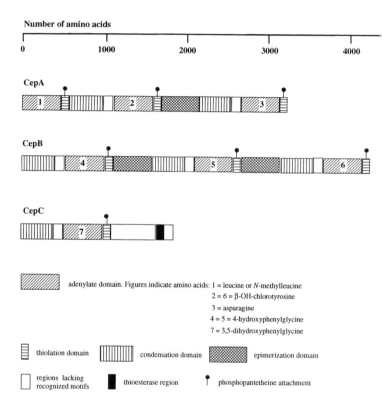

Fig. 165. Functional organization of the three peptide synthetases CepA, CepB and CepC which are responsible for construction of the heptapeptide backbone of chloroeremomycin (and vancomycin)

CepA contains three modules with domains for attachment of (*R*)-leucine (or its *N*-methyl analogue), (*S*)-tyrosine and (*S*)-asparagine, respectively. In CepB there are three modules with domains for two residues of (*S*)-*p*-hydroxyphenylglycine and one residue of (*S*)-tyrosine. Finally CepC contains only one module with a domain for attachment of the C-terminal amino acid – 3,5-dihydroxyphenylglycine. The tyrosine residues might or might not be hydroxylated and/or chlorinated prior to incorporation. The stereochemistry at the α-C centres of the peptide backbone in chloroeremomycin and vancomycin is *R, R, S, R, R, S, S* and it might therefore be expected that modules 1, 2, 4 and 5 would contain a domain for inverting the stereochemistry from *S* to *R*. This is true for modules 2, 4 and 5 but not for module 1, which implies that (*R*)-leucine or its *N*-methyl analogue may be directly loaded on to the biosynthetic apparatus.

Conversion of the linear heptapeptide into the aglucone of chloroeremomycin (and vancomycin) requires seven oxidative processes. These are the introduction of β-hydroxy groups on to tyrosine residues **2** and **6**, coupling of rings in the amino acids **5** to **7**, **4** to **6** and **4** to **2** and introduction of chlorine atoms on to the rings in amino acids **2** and **6**. Genes 7–10, 18 and 20–22 might encode enzymes involved in oxidative processes. Of these genes, 10 and 18 show homology to non-haem haloperoxidases and therefore are likely to be involved in the introduction of the chlorine atoms. Genes which encode glycosyl transferases responsible for the addition of the three sugars to the heptapeptide of chloroeremomycin have also been identified.

Teicoplanin Teicoplanin, which is produced by the bacterium *Actinoplanes teichomyceticus* (*Actinomycetales*), is a mixture of five related compounds whose structures are very similar to that of vancomycin. They are called teicoplanin T-A2-1 to T-A2-5 and differ only in the nature and length of a fatty acid chain which is attached to the sugar D-glucosamine. The amino group of the N-terminal 4-hydroxyphenylglycine is methylated. The C-terminal is 3,5-dihydroxyphenylglycine. The third amino acid is also 3,5-dihydroxyphenylglycine and its presence has allowed a fourth oxidative coupling, generating a fourth ring system. The teicoplanins also contain two

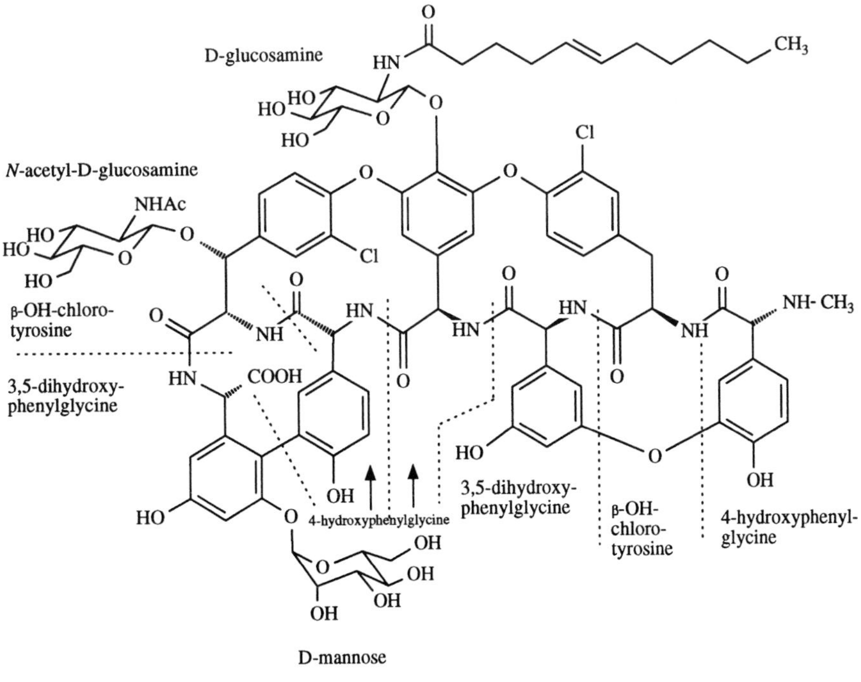

Teicoplanin T-A2-1

β-hydroxy-chlorotyrosine residues and two more 4-hydroxyphenyl-glycine residues. Besides the already mentioned D-glucosamine residue (attached via amino acid residue 4) the molecule also contains N-acetyl-D-glucosamine (attached via the β-hydroxy group of a β-hydroxy-chlorotyrosine residue) and a D-mannose residue, attached via a hydroxy group in the C-terminal 3,5-dihydroxy-phenylglycine residue.

Teicoplanin has similar antibacterial activity to that of vancomycin and is administered by intramuscular or intravenous injection.

Mode of action of glycopeptides

The antibacterial action of glycopeptides is due to their binding to the peptidoglycan layer of the bacterial cell wall. This layer is composed of a cross-linked polymeric network made from disaccharide units and peptide chains. The disaccharide module consists of one unit of N-acetylglucosamine (N-AcGlc) and one unit of N-acetylmuramic acid (N-AcMur) connected with a $(1 \rightarrow 4)$-β-glycosidic linkage. This disaccharide unit is repeated up to one hundred times and cross-linked to other oligosaccharide strands to introduce rigidity through a peptide framework. The peptide unit consists of an L-Ala-L-Glu-L-Lys-D-Ala-D-Ala fragment that is attached to the carboxylate group of the muramic acid through the N-terminus of L-Ala. The free amino group of the L-lysine residue provides a dense cross-linking of the peptidoglycan by a pentapeptide bridge to the D-Ala of a second peptide chain. The glycopeptides bind reversibly to the -L-Lys-D-Ala-D-Ala fragment of the peptidoglycan monomer, thereby inhibiting transglycosidation and transpeptidation from occurring. This interference leads to the collapse of the peptidoglycan by decisively shifting its dynamic equilibrium towards de-assembly, which precipitates cell lysis and death of the bacterium.

The potency of vancomycin and a number of other glycopeptides is enhanced by their ability to dimerize in solution, thereby providing two binding sites for -L-Lys-D-Ala-D-Ala units. Two hypotheses have been put forward to account for the enhanced potency of the dimers. The first of these says that the dimer has higher activity because half of the dimer binds to the substrate in the usual way while the other half finds its target through essentially intramolecular binding. This cooperative effect decreases the entropy factor for binding. The second hypothesis ascribes the increased potency to allosteric effects. Thus the hydrogen bonding within the dimer enhances the ability of the binding pocket to bind the ligand by polarizing the amide bonds. Teicoplanin cannot dimerize but gains increased potency by anchoring into the cell's phospholipid bilayer through the long hydrocarbon chain which is attached to D-glucosamine.

Bleomycin   Bleomycin is a mixture of glycopeptides produced by the bacterium *Streptomyces verticillus*. It has antibiotic properties but is used as an anti-cancer drug. The main component (55–70 %) of the mixture is bleomycin A_2 which has a complicated structure containing a tripeptide of β-hydroxy-histidine, 4-amino-3-hydroxy-2-methyl-valeric acid and threonine. The amino group of the N-terminal amino acid (β-OH-His) forms an amide bond with a complicated acid containing a pyrimidine ring and the carboxy group of threonine is amide-bound to a planar dithiazole ring system. A disaccharide composed of D-mannose and L-gulose is attached to the molecule via a glycosidic bond involving the hydroxy group of β-hydroxy-histidine. Besides bleomycin A_2, bleomycin also contains about 30 % bleomycin B_2 and minor amounts of the parent compound bleomycinic acid. The latter is inactive.

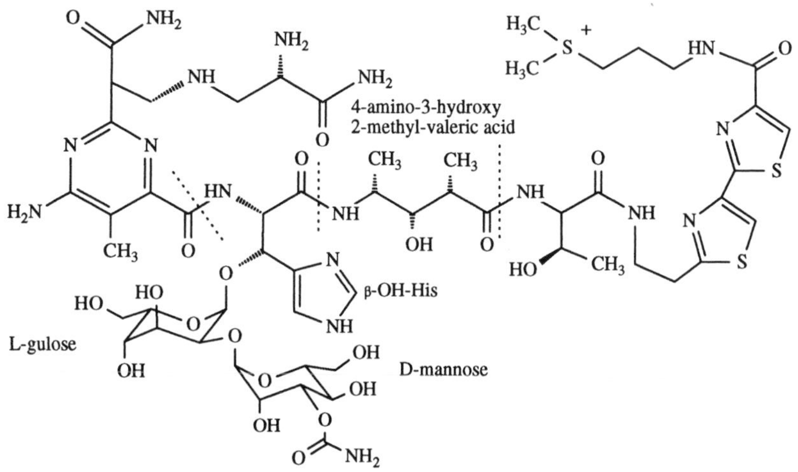

Bleomycin A_2

Bleomycin is used as a complement to treatment with other anti-cancer drugs against squamous cell carcinomas of various organs, Hodgkin's disease and tumours of the testicles. In contrast to other antitumour antibiotics it produces very little bone-marrow suppression.

Streptogramin antibiotics

Members of a group of *Streptomyces* produce antibiotics – **streptogramins** – with unique properties. The group consists of two sub-groups – A and B – which are produced by the same organism in the ratio 70:30. Group A are cyclic macrolactones of a hybrid peptide/polyketide structure. The ring is closed by an internal ester bond between the carboxyl group of the C-terminal amino acid (generally Pro) and an internal hydroxyl group of the polyketide. A representative of group A is *pristinamycin-IIA*, which is produced by *Streptomyces pristinaespiralis*.

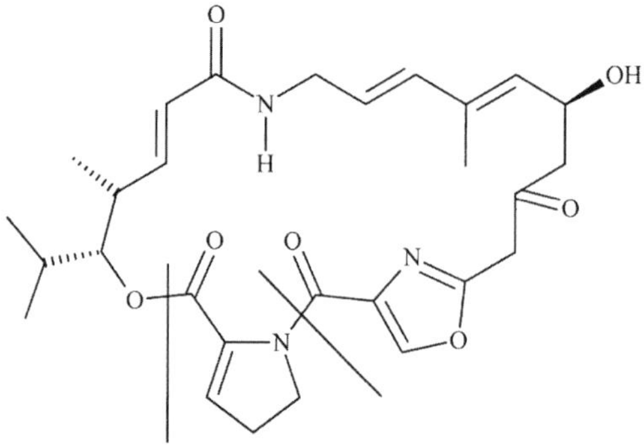

Dehydroproline

Pristinamycin IIa

Structural variations involve desaturation of the proline residue (as in pristinamycin II-A) and its substitution for alanine or cysteine.

Group B streptogramins are cyclic hepta- or hexa-depsipeptides in the structure of which uncommon amino acids are incorporated. An example is *pristinamycin-IA* which is produced by the same organism as pristinamycin-IIA. Pristinamycin I-A is composed of L-threonine, D-aminobutyric acid, L-proline, 4-*N*,*N*-(dimethyl-amino)-L-phenylalanine, L-4-oxo-pipecolic acid and L-phenyl-glycine. The peptide bond between proline and 4-*N*,*N*-(dimethyl-amino)-L-phenylalanine is *N*-methylated. The *N*-terminal threonine is *N*-acylated with 3-hydroxypicolinic acid and the ring is closed by an ester bond between the carboxygroup of the C-terminal phenyl-glycine and the secondary hydroxyl group of threonine. In other group-B-streptogramin antibiotics D-aminobutyric acid can be replaced by D-alanine, 4-*N*,*N*-(dimethylamino)-L-phenylalanine with L-phenylalanine and 4-oxopipecolic acid with 4-hydroxyp-ipecolic acid or aspartic acid or proline. The *N*-terminal L-threonine L-proline and the C-terminal L-phenylglycine are invariant.

Group A and B streptogramins are bacteriostatic but when a com-pound of group A is administered together with a compound of group B, the mixture acts synergistically and becomes bactericidal Streptogramins were discovered as early as in the 1950s but found only marginal clinical use in the following decades. They have mainly been used as feed additives in agriculture. The emergence of vancomycin-resistant enterococci, however, spurred renewed inter-est in the streptogramins which has resulted in development of semisynthetic, water-soluble derivatives of pristinamycin-IA and –IIA. These compounds are denoted as **dalfopristin** (from pristi-namycin-IIA) and **quinipristin** (from pristinamycin-IA). They are combined in the ratio 7:3 in the antibiotic **Synercid**, which is used for intravenous treatment of serious infections caused by antibiotic-resistant Gram-positive bacteria such as vancomycin-resistant *Ente-*

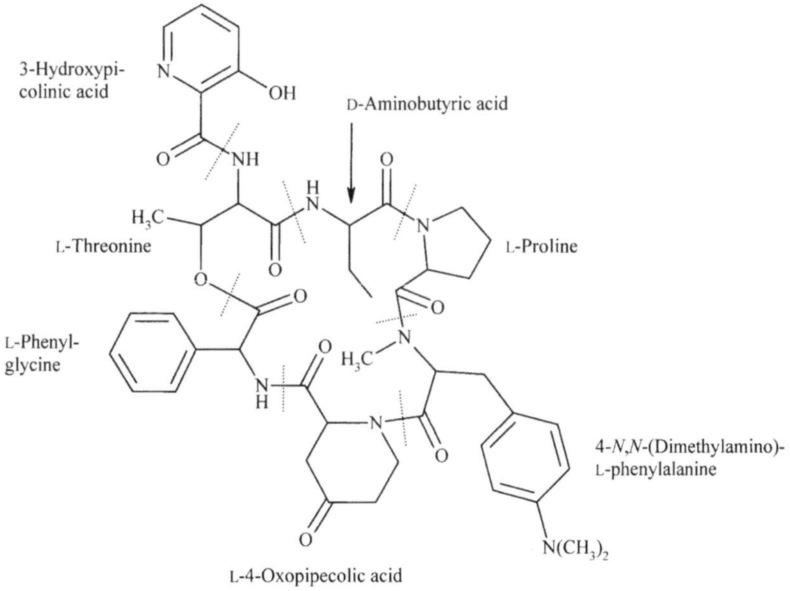

3-Hydroxypi-
colinic acid

D-Aminobutyric acid

H₃C

L-Threonine

L-Proline

L-Phenyl-
glycine

4-*N*,*N*-(Dimethylamino)-
L-phenylalanine

N(CH₃)₂

L-4-Oxopipecolic acid

Pristinamycin Ia

rococcus faecium (VREF), methicillin- and multidrug-resistant strains of *Staphylococcus aureus* and *Staphylococcus epidermidis* as well as penicillin- and macrolide-resistant *Streptococcus pneumoniae.*

Biosynthesis

Group A streptogramins

The C-terminal dehydroproline residue in pristinamycin-IIA is derived from proline. A two-enzyme system comprised of an FMN (FlavinMonoNucleotide)-dependent monooxygenase and an FMN

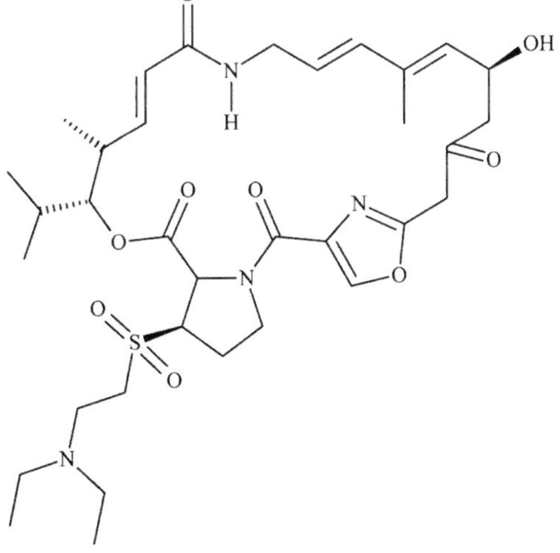

Dalfopristin

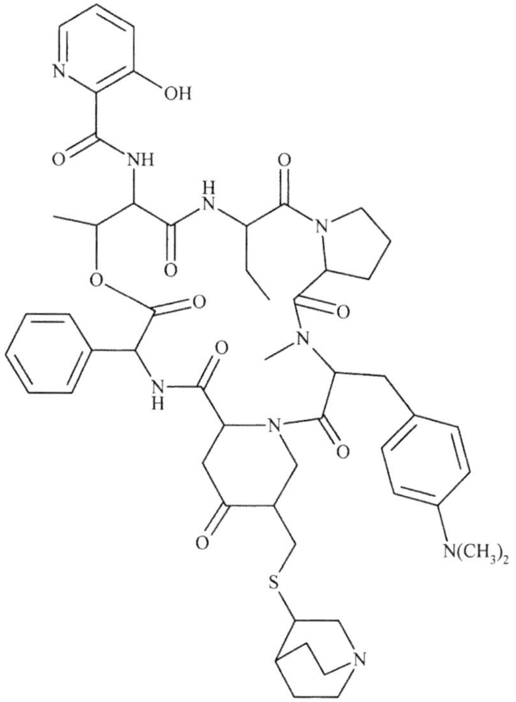

Quinipristin

reductase that catalyses proline oxidation have been purified and characterized from *Streptomyces pristinaespiralis*.

As shown in Fig. 166. the oxazole ring derives from serine which cyclizes in a manner involving a preceding carboxyl group, derived from acetate in the polyketide synthesis which forms most of the molecule.

The isopropyl group originates in valine and carbons 9 and 10 (and most probably the amide nitrogen) derive from glycine. The methyl groups 32 and 33 have different origin. Methyl 32 comes from methionine whereas 33 is derived from acetate, in an uncharacterized decarboxylation recation.

Group B streptogramins
The biosynthesis of the uncommon amino acids in the group B streptogramins has not been extensively studied. From *Streptomyces pristinaespiralis* a four-gene system for the production of 4-N,N-(dimethylamino)-L-phenylalanine has been cloned and partially characterized. Fig. 167 illustrates the biosynthetic reactions. The gene *papA* codes for a chorismate aminotransferase, papB for a mutase and papC for a dehydrogenase which catalyses formation of 4-aminophenylpyruvate. An unidentified transaminase converts the ketone to 4-aminophenylalanine. The final steps are two successive methylations of the 4-amino group, which are catalysed by an enzyme coded for by the gene *papM*.

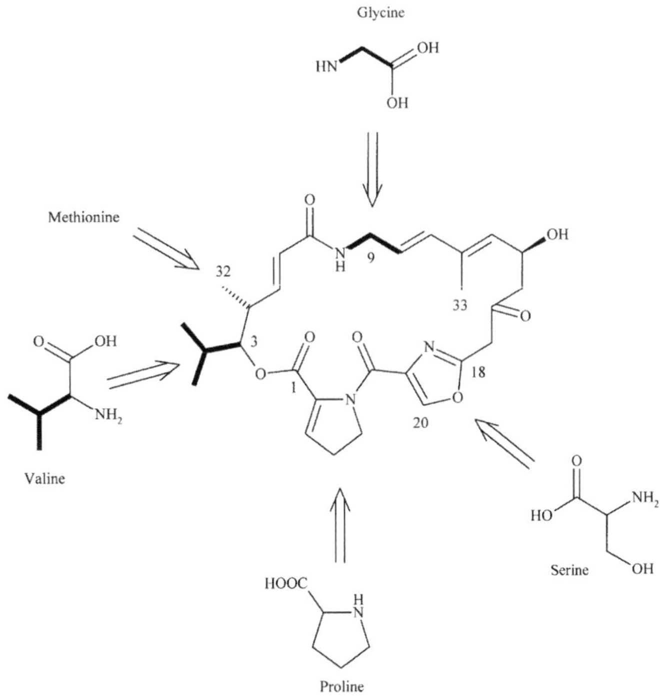

Fig.166. Biosynthetic origin of various parts of the structure of streptogramin group A antibiotics, illustrated for pristinamycin-IIA

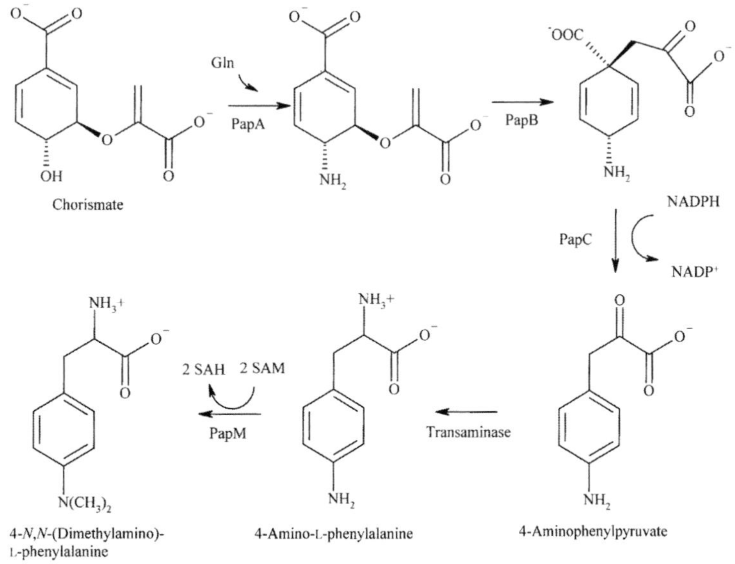

Fig.167. Biosynthesis of 4-N,N-(dimethylamino)-L-phenylalanine. PapA,
PapB, PapC and PapM are enzymes coded for by the genes PapA, PapB,
PapC and PapM, respectively

This biosynthesis is a variation of the biosynthesis of phenylalanine which also starts from chorismate (page 202). The two uncommon amino acids 3-hydroxypicolinic acid and 4-oxopipecolic acid have been found to be derived from lysine. Phenylalanine has been shown to be the source of phenylglycine.

Group B streptogramins are synthesized by non-ribosomal peptide synthetases (NRPSs). Three genes, responsible for biosynthesis of pristinamycin-1A, have been purified and cloned from *Streptomyces pristinaespiralis*. The first gene – *snbA* – encodes a 3 hydroxypicolinic acid:AMP ligase (SnbA, PI synthetase 1), a polypeptide of 582 amino acids with a predicted molecular size of 61.4 kDa, which contains only one adenylation domain which activates 3-hydroxypicolinic acid (3-HPA). No thiolation domain is present in SnbA. The second gene – *snbC* – codes for SnbC (PI synthetase 2), a protein with molecular size 240 kDa. SnbC contains two adenylation domains, two thiolation domains, two condensation domains and one C-terminal epimerization domain (Fig. 168) and activates and binds L-threonine and L-aminobutyric acid, which is converted to D-aminobutyric acid by the epimerization domain. The mechanism for formation of the depside bond between 3-hydroxy pipecolinic and threonine is proposed to be the following: The 3 HPA-adenylate, formed by SnbA is charged to the phosphopantheteinyl arm of the first thiol domain in SnbC and is then transferred to a waiting position in the first condensating domain of SnbC. Threonine is then adenylated by the first adenylation domain in SnbC, charged to the now free first thiol domain and then transferred to the second condensating domain where condensation takes place, forming the depsipeptide 3-HPA-Thr. In the third step L-aminobutyric acid is adenylated by the second adenylation domain of SnbC, the product is charged to the second thiol domain and then transferred to the second condensation domain where formation of 3-HPA-Thr-L-aminobutyric acid takes place. Finally L-aminobutyric acid is epimerized to D-aminobutyric acid by the epimerization domain. The concept of waiting positions in the condensation domains is analogous to the mechanism of FAS (page 245) in the biosynthesis of fatty acids. A good candidate for this waiting position is a serine residue in the condensation domain.

The third gene – *snbDE* – encodes SnbDE (PI synthetase 3), a protein composed of 4 849 amino acid residues with a predicted molecular mass of 521.8 kDa. This synthetase contains 4 adenylation domains, 4 thiolation domains, 4 condensation domains, one *N*-methylation domain and one C-terminal thioesterase domain (Fig 168) which allow biosynthesis of the rest of the pristinamycin-IA structure by connecting 3-HPA-Thr-L-aminobutyric to successively L-proline, 4-*N*,*N*-(dimethylamino)-L-phenylalanine, L-4-oxo-pipecolinic acid and L-phenylglycine. The *N*-methylase methylates the peptide bond between proline and dimethylamino-phenylalanine and the C-thioesterase domain forms the lactone bridge between phenylglycine and the second hydroxyl group of threonine.

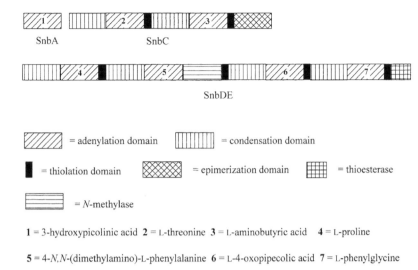

Fig. 168. Functional organization of the enzymes SnbA, SnbC and SnbDE *which catalyse the biosynthesis of pristinamycin-IA*

Mode of action and synergy

Like macrolides (page 271) and lincomycin (page 150) streptogramins act by inhibiting protein synthesis and bind the P site of the 50S subunit of bacterial ribosome. Group A streptogramins bind to the peptydiltransferase complex (PTC) only in the absence of aminoacyl-t-RNAs and block substrate attachment to the donor and acceptor sites, preventing the early event in elongation. In addition they cause a conformational change in the conformation of the ribosome that increases the activity of the B streptogramins by a factor of 100. Group B streptogramins prevent the extension of protein chains, cause the release of incomplete peptides and can bind to the ribosomes at any step of protein synthesis. A complex of dalfopristin and quinipristin, bound to the 50 S ribosomal subunit of the bacterium *Deinococcus radiodurans,* has been further studied and shed new light on the basis of synergy for these antibiotics. This, however, falls outside the scope of this book. For a review see T.A. Mukhtar and G.D. Wright in *Chemical Reviews,* **105**, 529–542 (2005).

Future prospects

A drawback of Synercid is that it is not orally active and has to be administered intravenously. Efforts are therefore being made to obtain new, orally available streptogramins and several patents have been filed. An example is RPR-106972, a 2:1 molar mixture of pristinamycin-IB with pristinamycin-IIB. The difference between pristinamycin-IB and pristinamycin-IA is that IB contains a 4-*N*-

(methylamino)-phenylalanine residue instead of the 4-*N*,*N*-(dimethylamino)-phenylalanine of pristinamycin-IA. In pristinamycin-IIB the dehydroproline residue of pristinamycin-IIA is replaced with proline.

β-Lactam antibiotics

This group of antibiotics comprises some of the most important and widely used antibiotics: penicillins and cephalosporins. Cephamycins are also members of the group. These compounds are peptides but they have most unusual structures, characterized by the presence of a 4-membered cyclic amide (β-lactam) ring which is condensed with a thiazolidine ring in the penicillins or a dihydrothiazine ring in the cephalosporins.

The β-lactams exert their antibacterial effects by interfering with the synthesis of the bacterial cell wall. The wall is built up on the outside of the cell in two steps. First mucopeptides (disaccharide units

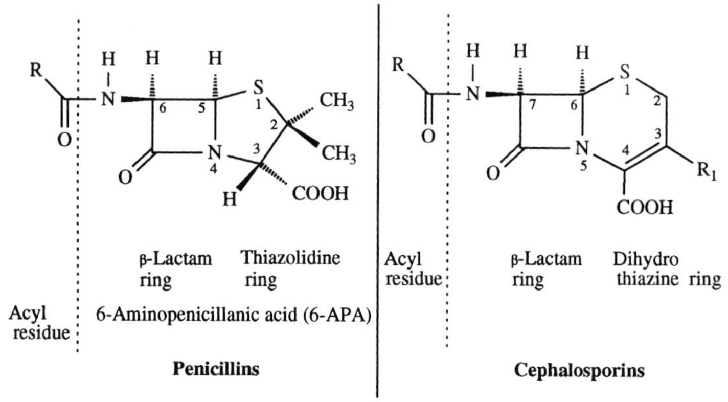

| β-Lactam ring | Thiazolidine ring | | β-Lactam ring | Dihydro thiazine ring |

Acyl residue 6-Aminopenicillanic acid (6-APA)

Acyl residue

Penicillins **Cephalosporins**

with pendant peptides) are exported from the cytoplasm to the outside of the cell where they are joined together by a transglycolase enzyme. Second these long polysaccharide chains are linked together through their peptide chains by a transpeptidase, thereby conferring mechanical strength to the construction. The transpeptidase recognizes the sequence D-alanyl-D-alanine at the end of the peptide chain, cleaves off the terminal alanine and joins the remainder to a peptide chain from an adjacent polysaccharide. Penicillin is believed to act by mimicking the structure of the transition state for the reaction of the D-alanyl-D-alanine sequence when it is bound to the transpeptidase enzyme. Because of this structural similarity, the enzyme binds penicillin and when it does so, the strained and reactive peptide bond of the β-lactam ring is opened by the hydroxyl of a serine residue in the binding pocket of the enzyme. This results in the penicillin being covalently bound to the active site of the enzyme thereby rendering it inactive for further cell wall synthesis. The bacteria die as a result of the cessation of the cell wall synthesis.

Penicillins Penicillins are produced by fungi of the genera *Penicillium* and *Aspergillus* (*Ascomycota, Trichocomaceae*). The discovery of the antibiotic properties of fermentation products from *Penicillium notatum* Westling, (nowadays known as *Penicillium chrysogenum*) and the subsequent identification of **benzylpenicillin** (penicillin G), as the active compound has been described previously (page 89). High-producing strains of *Penicillium chrysogenum* have been developed (page 93) and are used today for the production of benzylpenicillin. The fermentation is performed in two stages: the growth phase and the production phase. In the growth phase the medium contains inorganic salts, corn-steep liquor, lactose and growth stimulants. During

Natural penicillins this phase the mycelium grows but production of penicillin is very low. In the production phase a suitable carboxylic acid is added to the medium. This acid determines the substituent, R, in the formula above. Thus, addition of phenylacetic acid gives **benzylpenicillin** (R = $C_6H_5CH_2$) and if phenoxyacetic acid is added, phenoxymethylpenicillin (penicillin V, R = C_6H_5-O-CH_2) is obtained. Only monosubstituted acetic acids can be used in production of these so-called "biosynthetic" penicillins because of the specificity of the enzymes involved in the activation of the acids to their CoA esters (see below). The fermentation proceeds for 5–8 days at a temperature of 24–27 °C. The content of penicillin in the medium is then about 0.7 %. The mycelium is removed by filtration and pressing and penicillin is extracted from the filtrate with amyl acetate. Finally the penicillin is precipitated as its K-, Na- or Ca-salt.

Semisynthetic penicillins There are also semisynthetic penicillins on the market. These are produced from natural benzylpenicillin by enzymic degradation of the side-chain (R-CO-) yielding 6-aminopenicillanic acid which can be chemically acylated. The introduction of the biosynthetic and semisynthetic penicillins has overcome one drawback of the natural benzylpenicillin, namely instability at the low pH in the gastric juice. Extensive use of penicillin and its analogues has caused many bacterial strains to become resistant to these antibiotics. The mechanism of the resistance is that these bacteria have developed the ability to produce destructive β-lactamases (penicillases) that hydrolyse the penicillins thus making them inactive. Some semi-synthetic penicillins are not sensitive to those enzymes. **Phenoxymethylpenicillin** (penicillin V) was the first acid-resistant penicillin brought on the market. **Ampicillin** is an example of an acid-resistant, broad-spectrum penicillin and **cloxacillin** is resistant both to β-lactamase and acid. **Methicillin**, with 2,6-dimethoxy benzene as the side-chain, was the first commercial β-lactamase-resistant penicillin. It was acid sensitive and is no longer used. It is name, however, is still used in the term **methicillin-resistant *Staphylococcus aureus* strain (MRSA)**, denoting a strain of *S.a.* that is resistant to all penicillins.

Penicillins are active against Gram-positive aerobic and anaerobic bacteria and are used very frequently for a number of infections. Benzyl- and phenoxymethylpenicillin have low activity

561

Ampicillin

Cloxacillin

against Gram-negative bacteria but several semisynthetic penicillins, containing very polar groups on the α-carbon of the side-chain, are active also against these bacteria (broad-spectrum penicillins). As the cell membranes of mammalian cells are not attacked by penicillins these are very specific and safe antibiotics. However, development of allergy towards penicillins is not unusual. The responses range from a mild rash to fatal anaphylactic shock.

Cephalosporins The first cephalosporins were detected in the fungus *Acremonium chrysogenum (Ascomycota, Hypocreaceae)* in 1953. These compounds are also produced by fungi of other genera, e.g. *Emericellopsis* and *Paecilomyces*. Cephalosporin C is a natural product, which is used as a starting material for the production of semisynthetic cephalosporins, differing from each other in the substituents R and R_1 (see structure on page 560). The side-chain cannot be varied by addition of other carboxylic acids during the fermentation, as is done for the penicillins (see above). Only chemical modification is possible. A great number of semisynthetic cephalosporins are available. They are classified by dividing them into "generations", based primarily on their antibacterial spectrum but also related to the year of their introduction. **Cephalexin** is an example of a first generation cephalosporin. It is active against Gram-positive cocci – including penicillase-producing staphylococci – bacillae and Gram-negative enterobacteria. **Cefuroxime** is a second generation, broad-spectrum cephalosporin that is highly resistant against β-lactamases. An example of a third generation cephalosporin is **ceftriaxone** which is active against many Gram-negative organisms and has a longer half-life than other cephalosporins.

The cephalosporins are active against many types of microorganisms. They are not attacked by most β-lactamases and are comparatively acid-resistant. However, some bacteria have developed β-lactamases which attack cephalosporins as well. It is also possible in this group of antibiotics to influence the stability and antibacterial spectrum by alterations to the side-chain. The risk for development of allergy is lower with the cephalosporins than with the penicillins.

Cephalosporin C

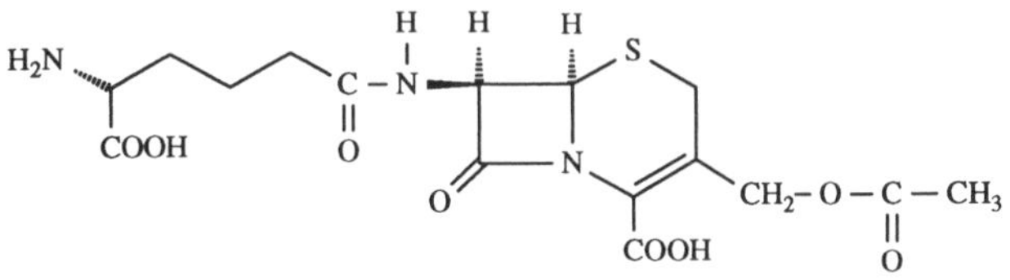

Cephalexin

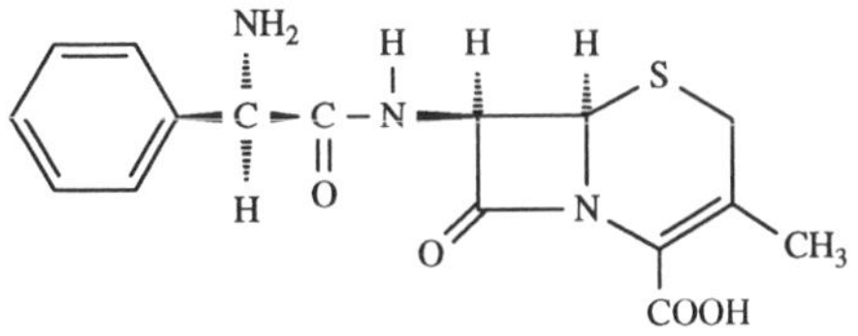

Cefuroxime

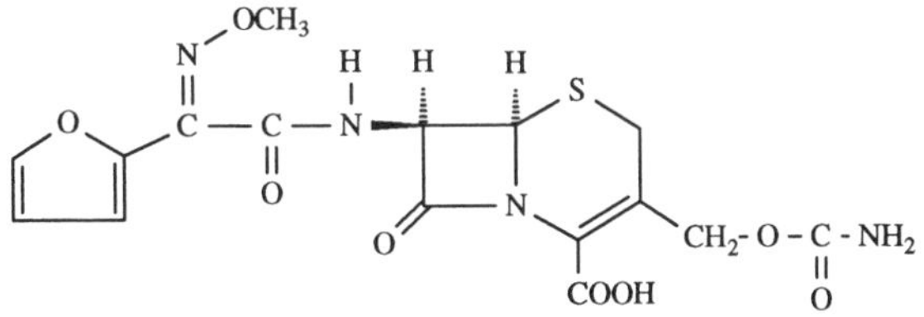

Ceftriaxone

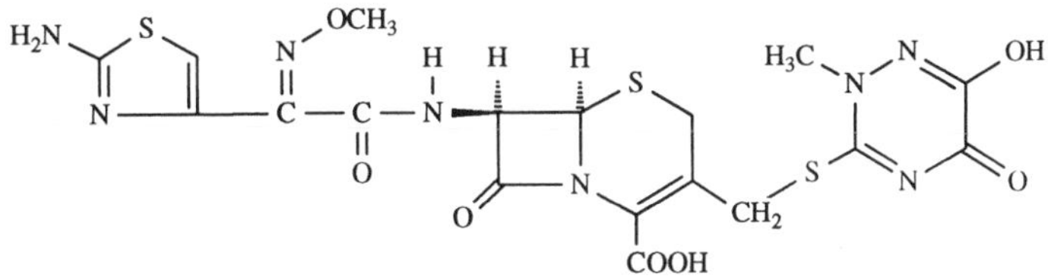

Cephamycins

Cephamycins are β-lactams present in bacteria such as *Streptomyces* species. **Cephamycin C** is produced by *Streptomyces clavuligerus* and is the most widely studied of the natural cephamycins. Structurally it is related to cephalosporin C but has an additional α-methoxy group in the C(7) position. Cephamycin C is resistant to β-lactamases and is active against Gram-negative but not against Gram-positive bacteria. A semisynthetic derivative – **cefoxitin** – is also active against Gram-positive bacteria, particularly against bowel flora including *Bacteroides fragilis*, and is used in the treatment of perotinitis.

Biosynthesis of β-lactams

The biosynthetic pathway to penicillins, cephalosporins and cephamycins starts with formation of *isopenicillin N,* which is the precursor from which all three groups of antibiotics are biosynthesized. The first step in the biosynthesis of this compound is the formation of the tripeptide δ-(L-α-Aminoadipyl)-L-Cysteinyl-D-Valine (ACV) from the amino acids L-α-aminoadipic acid (L-α-

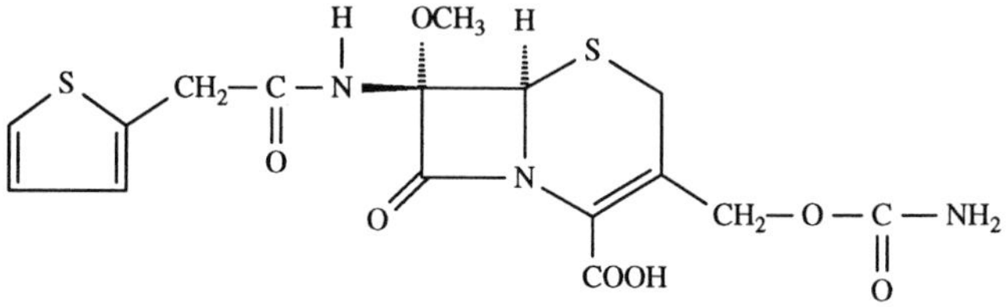

AAA), L-valine and L-cysteine. L-α-Aminoadipic acid is formed in the biosynthesis of lysine in fungi (page 494). In β-lactam-producing fungi there is thus a competition for this amino acid between the pathway for biosynthesis of lysine and that of β-lactams.

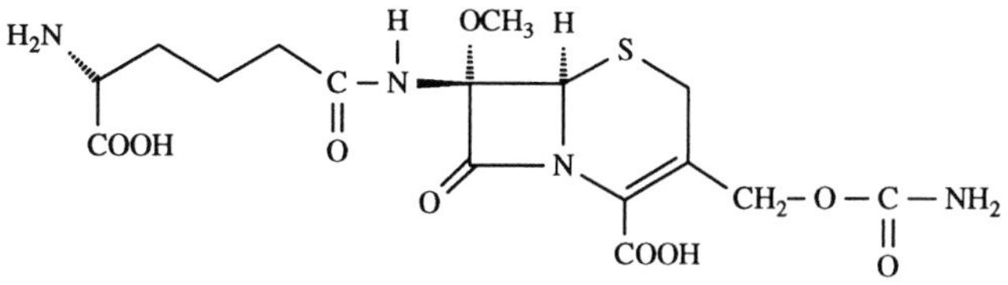

Cephamycins are produced by bacteria, which produce lysine via a different pathway, not involving α-aminoadipic acid (page 493). The cephamycin producers obtain their α-aminoadipic acid by degradation of lysine in two steps. The enzyme *L-lysine 6-transaminase* (EC 2.6.1.36) converts lysine to α-aminoadipic acid 6-semialdehyde, which is cyclized to form piperideine-6-carboxylic acid. The enzyme *piperideine-6-carboxylate dehydrogenase* then oxidizes this compound to α-aminoadipic acid. Piperideine-6-carboxylate dehydrogenase is found in cephamycin producers but not in actinomycetes that do not produce β-lactams.

The synthesis of ACV does not involve ribosomes but is catalysed by the multifunctional enzyme N-*(5-amino-5-carboxypentanoyl)-L-cysteinyl-D-valine synthase* (EC 6.3.2.26), also known as ACV synthetase), which has been purified from, among others, *Aspergillus nidulans* and *Acremonium chrysogenum*. The enzyme has been shown to catalyse the ATP-dependent activation of each of the L-amino acids, the binding of the activated amino acids as thioesters, the polymerization of the amino acids and the epimer-

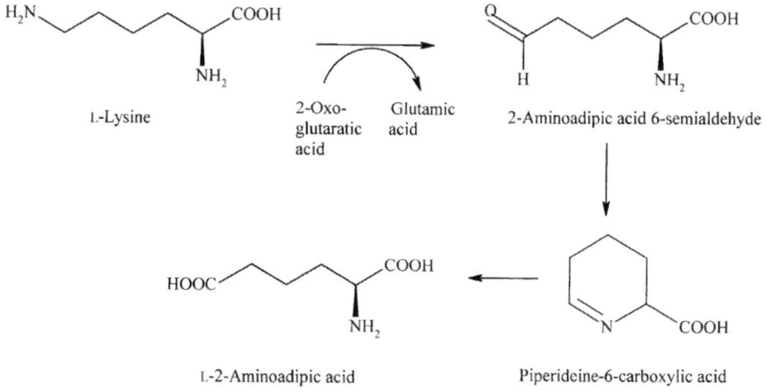

Fig. 169. Biosynthesis of α-aminoadipic acid from lysine

izativ n of L-valine to D-valine which seems to take place at the peptide stage. ACV synthetase is structurally and functionally related to
other large multifunctional enzymes known to catalyse non-ribosomal peptide bond formation (see page 537).

The ACV synthetase is encoded by a gene called *pcb*AB. The
designation *pcb* stands for genes involved in penicillin and
cephalosporin biosynthesis. The letters AB indicate that it was first
supposed that two genetic loci were responsible for the synthesis,
making first the dipeptide α-aminoadipyl-cysteine followed by the
epimerization and addition of valine to complete the tripeptide
ACV. However, genetic evidence indicates that a single gene
encodes all the activities needed for the production of ACV. *pcb*AB
has been cloned from *Penicillium chrysogenum*. It is an unusually
large gene with an open reading frame of 11 376 bp encoding a protein of 3 792 amino acids with a molecular mass of 426 kDa (see
also page 537). The *pcb*AB gene has also been cloned from the fungi *Acremonium chrysogenum* and *Aspergillus nidulans* and from β-
lactam-producing bacteria such as *Amycolatopsis lactamdurans*.
Sequence similarity is high among the *pcb*AB genes. A comparison
of the nucleotide sequences encoding the three repeated domains
shows that parallel interfungal domains share on average about
71 % nucleotide sequence identity, whereas the average sequence
identity for parallel fungal–bacterial domains is about 48 %.

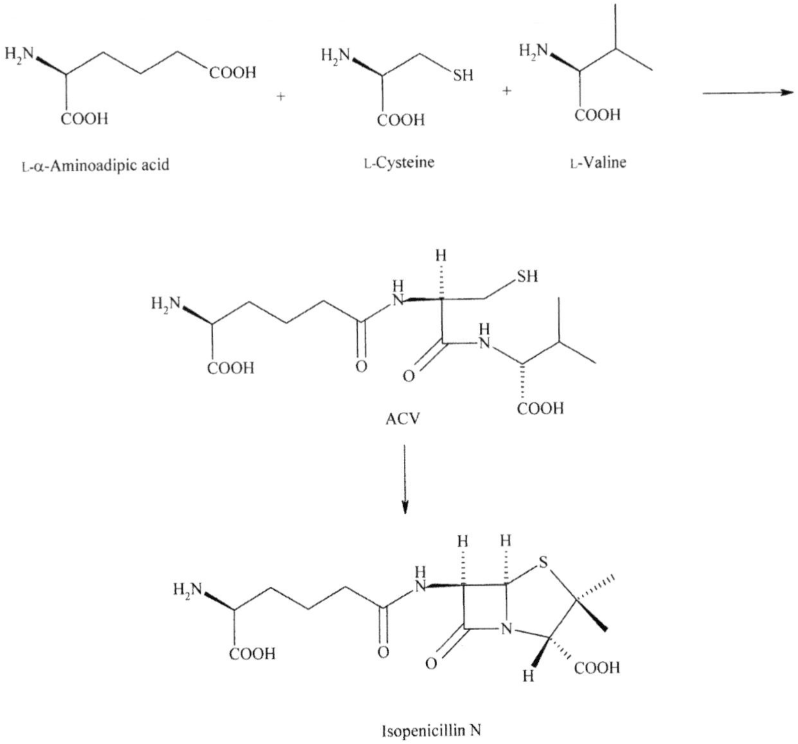

L-α-Aminoadipic acid   L-Cysteine   L-Valine

ACV

Isopenicillin N

Fig. 170. Biosynthesis of isopenicillin N

Following formation of ACV, the enzyme *isopenicillin N synthase* (EC 1.21.3.1, also called cyclase and abbreviated as IPNS) catalyses oxidation (desaturation) of the linear tripeptide to form the bicyclic compound isopenicillin N (Fig.170).

The process involves loss of four hydrogen atoms which are transformed to one molecule of dioxygen (O_2) to form two molecules of water. The process can be described as "desaturative cyclization" in that the loss of hydrogen is coupled to cyclization of ring structures. In addition to molecular oxygen, the reaction requires ferrous iron and an electron donor, e.g. ascorbate. The following mechanism has been proposed (Fig. 171): In the Fe^{2+}-IPNS complex the iron is attached to the active centre of IPNS via ligation to two histidines and one aspartic acid. One molecule of water is also attached to the iron atom. ACV forms a non-covalent bond to the iron via its cysteinyl thiol and is also bound to the enzyme via hydrogen bonds involving the carboxyl groups of valine and L-aminoadipic acid. Approach of one molecule of dioxygen results in a species in which the oxygen is bound to the iron with the distal oxygen atom having an unpaired electron (B in Fig.171). This species is highly reactive and removes a hydrogen from the position α to the sulphur in ACV, resulting in formation of compound C in Fig. 171. Subsequent cleavage of the hydroperoxide, with concomitant deprotonation of the amide N-H allows simultaneous β-lactam closure and ferryl formation (D → E). In the process, the isopropyl group is rotated (C → D) to relieve its steric interactions with the sulphur ligand. This rotation directs the valine β-hydrogen towards the ferryl-oxo species with which it reacts (E) allowing closure of the second ring (F). Finally, the Fe^{2+}-IPNS complex is split off and the formed isopenicillin N is liberated.

IPNS has been isolated from *Acremonium chrysogenum* and purified to homogeneity. The corresponding gene – *pcb*C – has also been identified, cloned and expressed in *Escherichia coli*, yielding a recombinant enzyme consisting of a single polypeptide of 338 amino acids with a deduced molecular mass of 38 kDa. *pcb*C has also been cloned from eight other fungal and bacterial species. As expected the sequence homology is high. Most of the fungal and bacterial IPNS genes have been expressed in *E. coli* and very high yields of the enzyme have sometimes been obtained. The recombinant proteins are indistinguishable in their biochemical properties from those of the parent organisms, except that the protein from *Acremonium chrysogenum* lacks the amino-terminal methionine and glycine residues that are predicted from the open reading frame of the gene.

IPNS is not very discriminating with respect to substrate. A whole range of unusual bicyclic β-lactams have been obtained by synthetic modification of the structural features of the tripeptide and by employing IPNS as a "synthesizing machine".

566

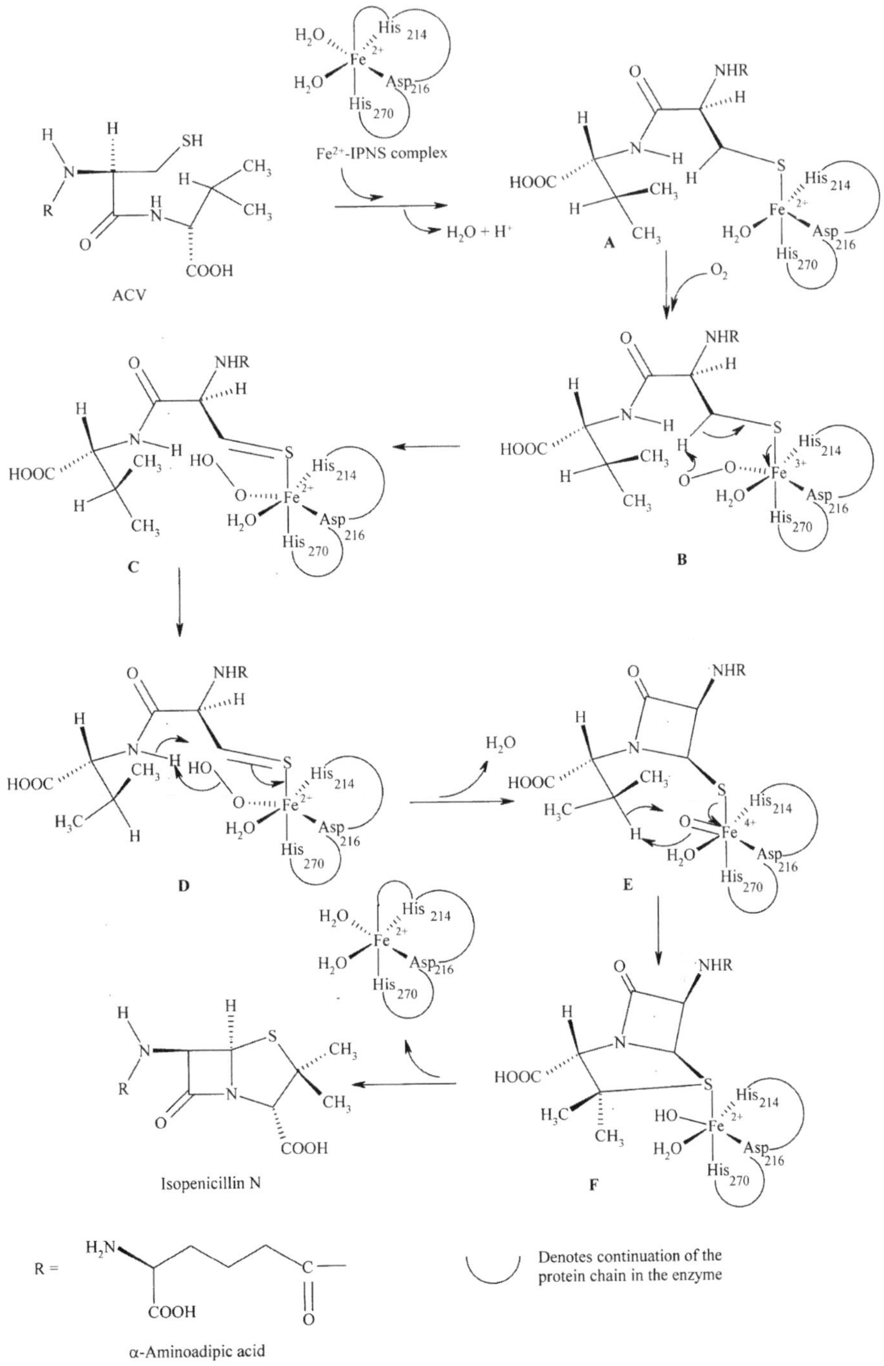

Fig. 171. Mechanism for the cyclization of ACV to isopenicillin N

The penicillin pathway

Following biosynthesis of isopenicillin N, the pathways to penicillin and to cephalosporin/cephalomycin diverge. In the penicillin

567

pathway the L-α-aminoadipyl side-chain of isopenicillin N is replaced by an intracellular or exogenously supplied carboxylic acid. Following activation of the acid by transformation to the corresponding acyl-CoA derivative, the exchange is catalysed by the enzyme *isopenicillin N N-acyltransferase* (EC 2.3.1.164, abbreviated as IAT). This protein has been isolated and shown to consist of two subunits with approximate molecular masses of 11 kDa and 29 kDa, respectively. The enzyme has dual activity. The L-α-aminoadipyl side-chain is first cleaved off to yield 6-aminopenicillanic acid (6-APA), which is then acylated with the new acid. For formation of benzylpenicillin, 6-APA is acylated with the phenylacetyl moiety of phenylacetyl-CoA, and the acid involved in biosynthesis of penicillin V (phenoxymethylpenicillin) is phenoxyacetic acid.

The gene which codes for IAT (named *pen*DE, where *pen* stands for penicillin) has been identified, cloned and expressed in *Escherichia coli*. The open reading frame corresponds to a protein with a deduced molecular mass of 39 kDa in perfect agreement with the molecular mass found for the isolated heterodimer. *pen*DE has been cloned from *Aspergillus nidulans* and *Penicillum chrysogenum* and both these fungal genes have three introns present in the amino-terminal half. This is the first example of a fungal β-lactam gene containing introns.

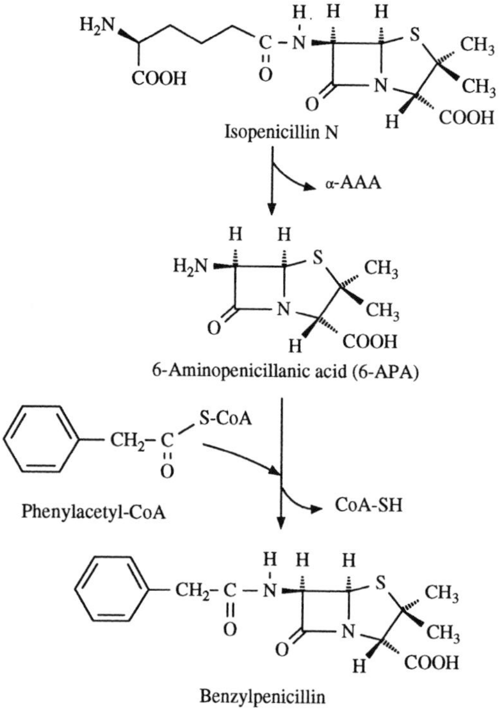

Fig. 172. Biosynthesis of benzylpenicillin

The cephalosporin/cephamycin pathway

In the biosynthesis of cephalosporin the first reaction after formation of isopenicillin N is isomerization of the L-α–aminoadipyl side-chain to the D-enantiomer, yielding penicillin N which is the substrate for the following ring-expansion enzymes. In *Acremonium chrysogenum* the epimerase activity is very labile, but it was found to be more stable in the bacterium *Streptomyces clavuligerus* from which organism the enzyme could be isolated. The isolated *isopenicillin N epimerase* (EC 5.1.1.17) has a molecular mass of approximately 47 kDa and is a racemase which catalyses a reversible isomerization between isopenicillin N and penicillin N. The corresponding gene, denoted *cef*D, was identified and found to have a single open reading frame coding for a protein of 398 amino acids corresponding to a molecular mass of 43 497 D.

The following two reactions in the biosynthesis of cephalosporin C involve ring expansion to yield deacetoxycephalosporin C which is hydroxylated to deacetylcephalosporin C. Ring expansion is catalysed by *deacetoxycephalosporin C synthase* (EC 1.14.20.1), which converts the five-membered thiazolidine ring of penicillins into the six-membered dihydrothiazine ring of cephalosporins. This enzyme requires Fe^{2+} and 2-oxoglutarate. The product – deacetoxy-cephalo-sporin C – is hydroxylated to form deacetylcephalosporin C by *deacetoxycephalosporin C hydroxylase* (EC 1.14.11.26). This enzyme, which requires 2-oxoglutarate, incorporates an oxygen atom from O_2 into the exocyclic methyl group of deacetoxy-cephalosporin C. Genes encoding these two enzymes – *cef*E and *cef*F, respectively – have been cloned from several cephalosporin-producing organisms and expressed in *Escherichia coli*, yielding the corresponding enzymes. In *Acremonium chrysogenum*, however, these two enzymes are located on a single protein, encoded by the gene *cef*EF.

Deacetylcephalosporin C serves as a precursor for both cephalosporin C and cephamycin C. The enzyme *deacetylcephalosporin C acetyltransferase* (EC 2.3.1.175) catalyses acetylation of deacetylcephalosporin C with acetyl-CoA to yield cephalosporin C. The gene coding for this enzyme – *cef*G – has been cloned and expressed. It contains two introns and encodes a protein of 444 amino acids with a molecular mass of 49 269 D.

The last steps in the biosynthesis of cephamycin C are less well known. Deacetylcephalosporin C is converted into *O*-carbamoyl-deacetyl cephalosporin C by an *O*-carbamoyltransferase that transfers a carbamoyl group from carbamoylphosphate. The methoxyl group at C(7) derives from molecular oxygen and methionine by the action of a monooxygenase and methyltransferase.

Organization of the genes involved in the biosynthesis of the β-lactams

Antibiotic biosynthetic genes tend to exist in gene clusters. The

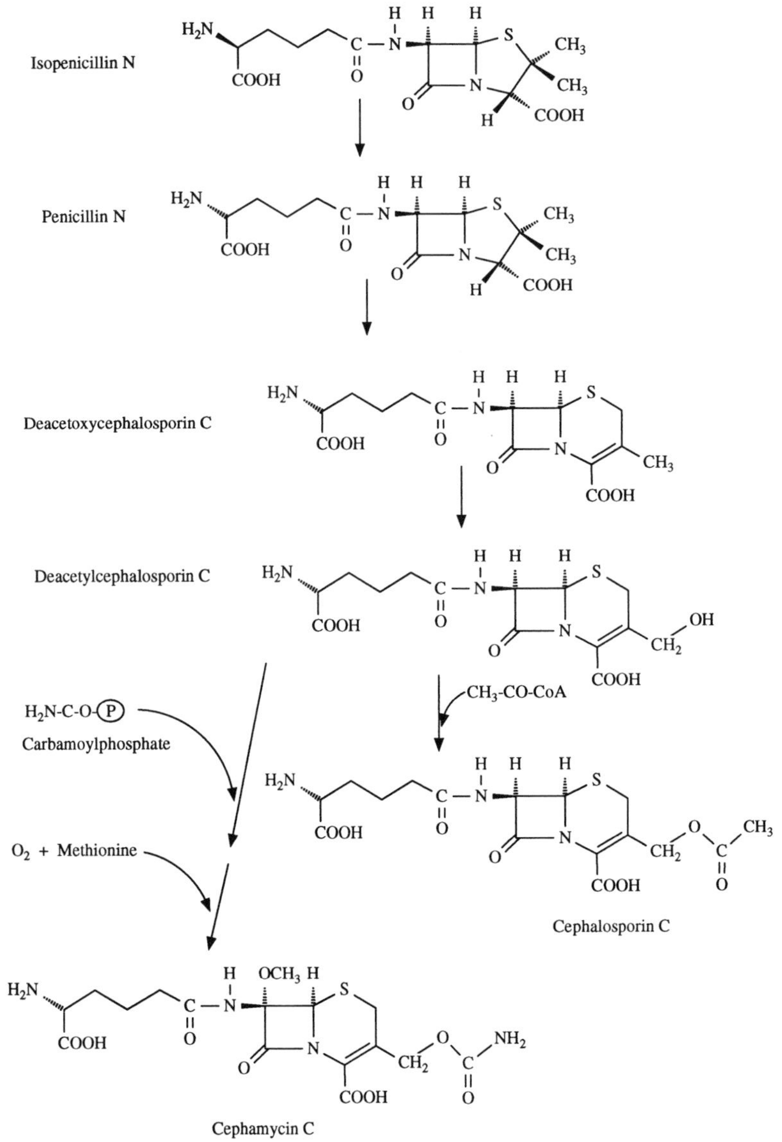

Fig. 173. Biosynthesis of cephalosporin C and cephamycin C

three genes responsible for penicillin biosynthesis in *Penicillium chrysogenum* and *Aspergillus nidulans* are all located together. *pcb*AB is separated from *pcb*C and *pen*DE by about 1 100 bp in *P. chrysogenum* and by 872 bp in *A. nidulans*. In both organisms *pcb*AB is transcribed in opposite orientation compared with the two other genes (Fig. 174). In *Acremonium chrysogenum*, *pcb*AB and *pcb*C are located together on chromosome VI in analogy with the penicillin producing genes. They are separated by 1 233 bp, and

Penicillium chrysogenum

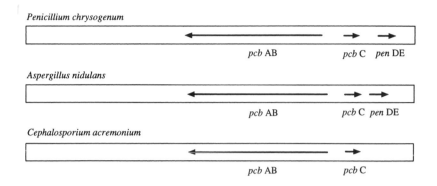

Aspergillus nidulans

Cephalosporium acremonium

Fig. 174. Organization of the genes involved in biosynthesis of β-lactams in Penicillium chrysogenum, Aspergillus nidulans *and* Acremonium chrysogenum *(previously known as* Cephalosporium acremonium*). Arrows indicate orientation of transcription*

transcribed in opposite orientations. The later genes *cef*EF and *cef*G, on the other hand, have been mapped to chromosome II. *cef*G is linked to *cef*EF but is expressed in opposite orientation to *cef*EF.

Other β-lactams

In the penicillins, cephalosporins and cephamycins the β-lactam ring is fused to a sulphur-containing 5-membered or 6-membered ring. There are also compounds in Nature where the β-lactam is fused to other types of rings. One such example is the antibiotic clavulanic acid where an oxygen-containing 5-membered ring is substituted for the sulphur-containing ring of the penicillins.

Clavulanic acid

Clavulanic acid is produced by *Streptomyces clavuligerus*, which also produces cephamycin (see above, page 563). Although these two compounds are produced by the same organism their biosynthetic pathways are not at all related. The precursors of clavulanic acid are arginine and a C_3-unit which provide all the carbons. The β-lactam ring originates from the C_3-unit and the α-amino group of arginine while the rest of the molecule is derived from the residual arginine. The precursor of the C_3-unit has not yet been identified and various candidates such as glycerol, glycerate, acrylate, malonic semialdehyde, D-lactate, β-hydroxipropionate and pyruvate have been suggested. Following formation of the β-lactam ring from the C_3-unit and arginine the arginine moiety is cyclized in three reactions, catalysed by one enzyme – *clavaminate synthase* (EC 1.14.11.21)

Clavulanic acid has no use as an antibiotic but it is able to bind irreversibly to β-lactamase enzymes, thereby rendering them inactive. It is therefore combined with β-lactamase-susceptible penicillins which thus become effective against a wider range of organisms.

Clavulanic acid

571

Thienamycin Thienamycin is an example of a β-lactam derivative where the 5-membered ring does not contain a heteroatom. Two of the carbons of the β-lactam ring are derived from acetate and glutamic acid forms the rest of the ring system. The side-chain containing the amino group is provided by cysteine and the hydroxyethyl side-chain is constructed via two C-methylations involving S-adenosyl-methionine (SAM, page 497). Thienamycin is produced by *Streptomyces cattleya* but the yields are insufficient for commercial production. The compound is therefore made by total synthesis.

Imipenem Thienamycin is unstable and is therefore converted to an N-formimidoyl derivative, imipenem, which has a broad spectrum of activity against aerobic and anaerobic Gram-positive and Gram-negative bacteria. It is not attacked by β-lactamases and even has β-lactamase inhibitory activity. A drawback in the use of imipenem is its sensitivity towards a dehydropeptidase in the kidney. Preparations of imipenem must therefore be combined with an inhibitor of this enzyme – cilastatin.

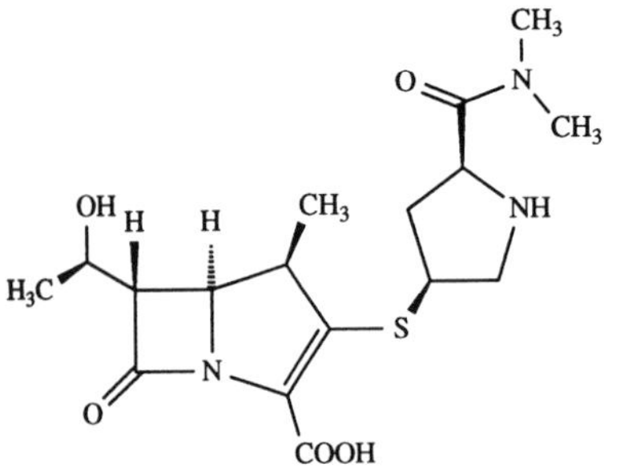

Meropenem

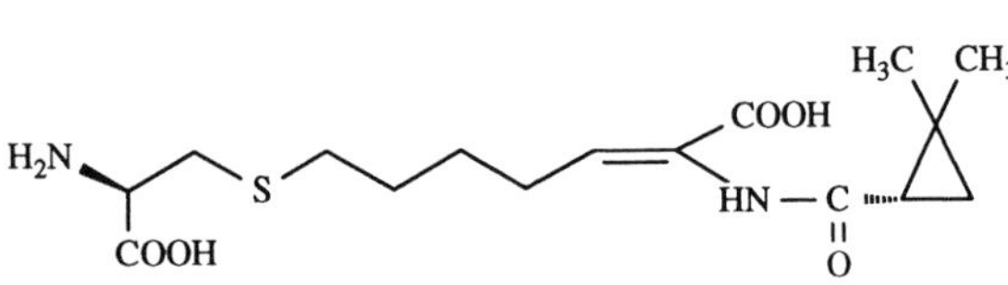

Cilastatin

Meropenem A more recent derivative is meropenem which is resistant to the dehydropeptidase.
Imipenem and meropenem act like other β-lactams by interfering with the formation of the bacterial cell wall. They are both broad-spectrum antibiotics and are used in treatment of serious bacterial infections such as meningitis and pneumonia.

Aztreonam Aztreonam is a synthetic, monocyclic β-lactam antibiotic, originally isolated from *Chromobacterium violaceum*.
Aztreonam is used parenterally for treatment of infections caused by Gram-negative organisms such as bone and joint infections, gonorrhoea, intra-abdominal and pelvic infections, lower respiratory-tract infections in patients with cystic fibrosis, meningitis

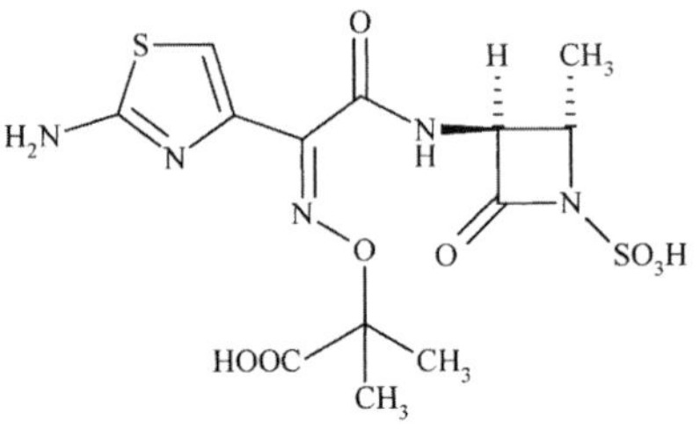

Aztreonam

572

septicaemia, skin and soft-tissue infections and urinry-tract infections.

Vitamins derived from amino acids

**Pantothenic acid
(Vitamin B$_5$)**

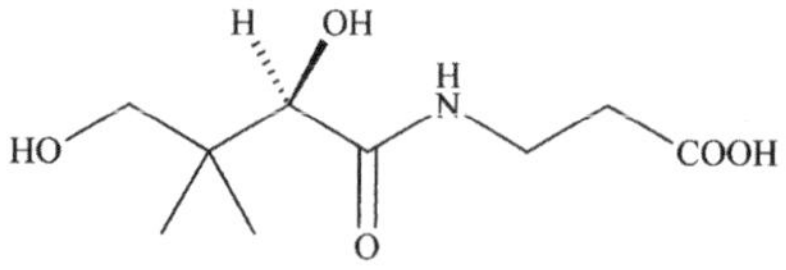

D-Pantothenic acid

The dextrorotary (D) isomer of pantothenic acid is essential to all forms of life since it is needed for biosynthesis of the important cofactor coenzyme A (CoA, see page 720) and of acyl carrier protein (page 240). Plants and microorganisms can biosynthesize pantothenic acid, but mammals, including man, must obtain it in their diet. Most foods contain small quantities of pantothenic acid and whole grain and eggs are particularly rich in this vitamin. Pantothenic acid in vitamin preparations is synthetic. About 4 000 tonnes are produced anually in a process that requires high optical resolution to separate the inactive L-isomer from the biologically active D-isomer.

The daily requirement of pantothenic acid ranges from 2 mg for infants to 5 mg for adults. Pregnant and breastfeeding women need 6, respectively 7 mg/day. Pantothenic acid deficiency is exceptionally rare. The symptoms are similar to other vitamin B deficiencies and include fatigue, allergies, nausea and abdominal pain. In serious cases adrenal insufficiency and hepatic encephalopathy are reported.

Biosynthesis   The pathways to pantothenic acid are similar in plants and bacteria but there are important differences both in the enzymes involved and in the overall organization of the pathway. The bacterial pathway is best known and is illustrated in Fig. 175. The biosynthesis consists of two branches, one starting with valine and the other with aspartic acid. The first reaction in the valine branch is deamination of valine to form 3-methyl-2-oxobutanoate (α-ketoisovalerate), catalysed by the enzyme *branched-chain-amino-acid transaminase* (EC 2.6.1.42). In the next step α-ketoisovalerate is methylated with formation of 2-dehydropantoate by *3-methyl-2-oxobutanoate hydroxymethyltransferase* (EC 2.1.2.11) using 5,10-methylene tetrahydrofolate as a cofactor. 2-Dehydropantoate is then reduced to (*R*)-pantoate in a reaction catalysed by *2-dehydropantoate 2-reductase* (EC 1.1.1.169).

In the aspartic acid branch of the pathway, aspartate is decarboxylated to form β-alanine by the enzyme *aspartate 1-decarboxy-*

lase (EC 4.1.1.11). In the final step of the pantothenic acid biosynthesis, β-alanine is condensed with (*R*)-pantoate in a reaction catalysed by *pantoate-β-alanine ligase* (EC 6.3.2.1) yielding (*R*)-pan-
tothenate (= D-pantothenic acid). Genes for all four enzymes of the
pathway have been cloned, the recombinant proteins overexpressed
and purified and their 3D structures determined.

Biosynthesis of pantothenic acid in plants is less well known
Radio-labelling experiments have shown that the pathway is similar
to that of bacteria but the enzymes involved have not been clarified
to full extent. Molecular biology approaches have shown the presence of two genes encoding 3-methyl-2-oxobutanoate hydroxymethyltransferase. The reason for the presence of two isoforms of
that enzyme is not yet known. cDNA encoding the last enzyme in
the pathway – pantoate-β-alanine ligase – has been cloned from
several species of plants. Expression of the enzyme, however, gave
a protein which catalysed the reaction at pantotate concentrations
below 0.5 mM, whereas higher concentrations inhibited the reaction. A gene encoding 2-dehydropantoate 2-reductase has not yet
been found in plants. Activity corresponding to this enzyme is clearly present in plants so it is possible that the enzyme is quite different from the bacterial enzyme. Little is known about the source of
β-alanine in plants. No evidence of the presence in plants of an

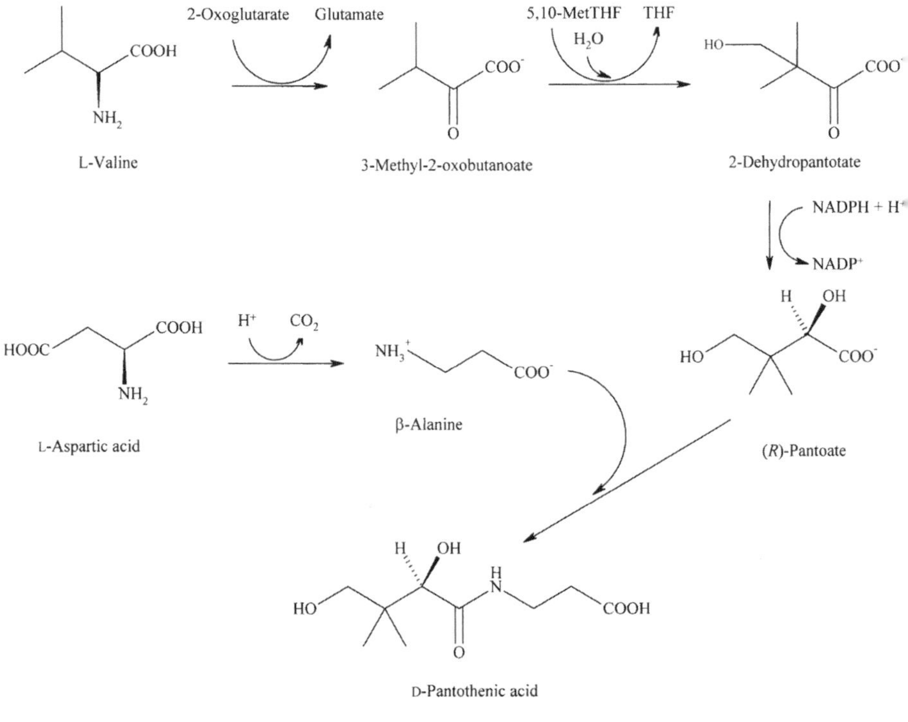

Fig. 175. Biosynthesis of pantothenic acid in bacteria. 5,10-MetTHF =
5,10-methylenetetrahydrofolate. THF = tetrahydrofolate

enzyme corresponding to aspartate 1-decarboxylase has been found. One possibility is that the β-alanine required for biosynthesis of pantothenic acid in plants derives from a three-step degradation of uracil.

Pyridoxalphosphate and vitamin B$_6$

Pyridoxalphosphate is the active form of vitamin B$_6$ and acts as a cofactor for approximately 100 enzymes that catalyse essential chemical reactions in the human body, such as amino acid metabolism, niacin biosynthesis, nucleic acid synthesis, thiamin biosynthesis and glucose metabolism. It is also a potent antioxidant. Vitamin B$_6$ is implicated in more bodily functions than any other single nutrient and is associated with nervous system function (serotonin synthesis), red blood cell formation (haem biosynthesis), immune system function, hormone function and ion transport modulation. It is also of importance for prevention of cardiovascular disease, relief of depression and as an anti-tumour agent. Three compounds – pyridoxine, pyridoxal and pyridoxamine – are considered as vitamin B6 because they are transformed to pyridoxal phosphate in the human body.

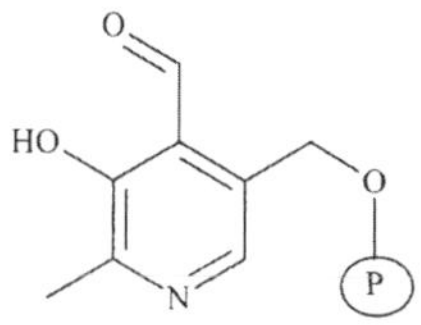

Pyridoxal 5´-phosphate 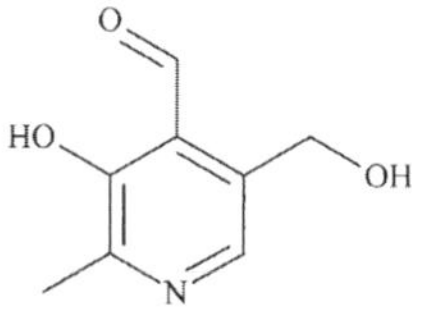   Pyridoxal 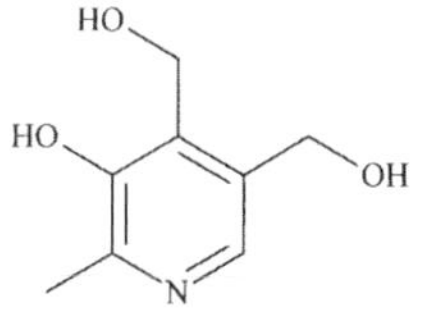   Pyridoxine 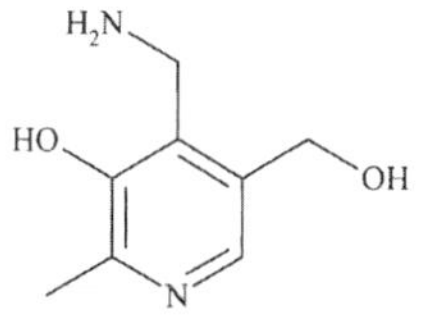   Pyridoxamine

Severe deficiency of vitamin B$_6$ is not common. Alcoholics can be at risk due to low dietary intakes and impaired metabolism of the vitamin. Neurologic symptoms in severe deficiency include irratibility, depression and confusion. Inflammation of the tongue and mouth ulcers may also occurr.

Diet sources, comparatively rich in vitamin B6 are banana, salmon, turkey, chicken, field salad, spinach, potatoes with skin, green beans, yeast and hazelnuts. The recommended daily intake is about 2 mg which is just about reached in a normal diet. Strict vegetarians and old people might need to increase their vitamin B$_6$ intake by eating foods rich in the vitamin or by taking supplements. Drugs can interfere with the metabolism of B$_6$ and might cause deficiency. Known examples are tuberculosis medications, including isoniazid and cycloserine, penicillamine and antiparkinsonian drugs including L-Dopa.

Biosynthesis

Only microorganisms, fungi and plants can synthesize vitamin B$_6$. The biosynthesis has been extensively studied in *Escherichia coli* and starts (Fig.176) with oxidation of erythrose 4-phosphate, catalysed by *erythrose-4-phosphate dehydrogenase* (EC 1.2.1.72) to 4-phosphoerythronate, which is further oxidized to (3R)-3-hydroxy-

2-oxo-4-phosphonooxybutanoate by *4-phosphoerythronate dehydrogenase* (EC1.1.1.290). *Phosphoserine transaminase* (EC 2.6.1.52) then converts this compound to 4-phosphonooxy-L-threonine, which in the next reaction, catalysed by *4-hydroxythreonine-4-phosphate dehydrogenase* (EC 1.1.1.262), forms 2-amino-3-oxo-4-phosphonooxybutanoate which spontaneously decarboxylates to 3-amino-2-oxopropylphosphate. This compound is condensed with 1-deoxy-D-xylulose 5 phosphate (DXP) to form pyridoxine 5′-phosphate in a reaction catalysed by *pyridoxine 5′-phosphate synthase* (EC 2.6.99.2). DXP is formed from pyruvate and glyceraldehyde 3-phosphate and pyruvate in the start of the non-mevalonate pathway for the biosynthesis of isoprenoids (page 369). Finally *pyridoxal 5′-phosphate synthase* (EC 1.4.3.5) oxidizes pyridoxine 5′-phosphate to pyridoxal 5′-phosphate.

This pathway has long been assumed to be ubiquitous in all vitamin B_6 synthesizing organisms but since 1999 genetic evidence has accumulated that the majority of organisms must have a path-

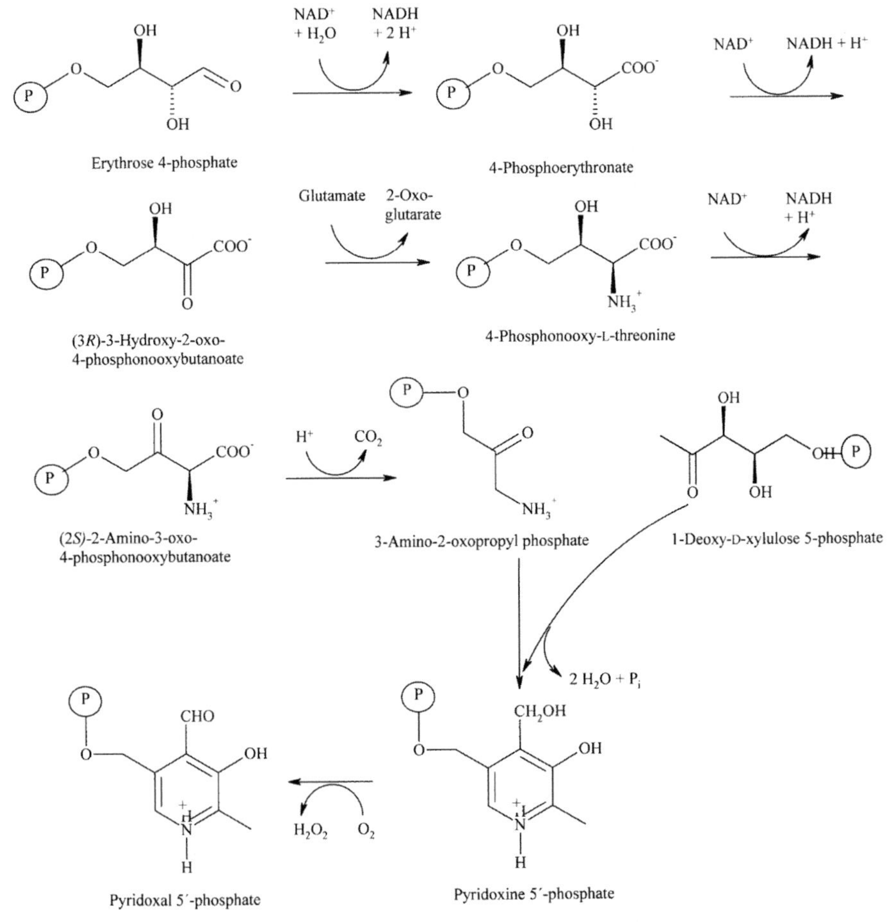

Fig. 176. Biosynthesis of pyridoxal 5′-phosphate in Escherichia coli

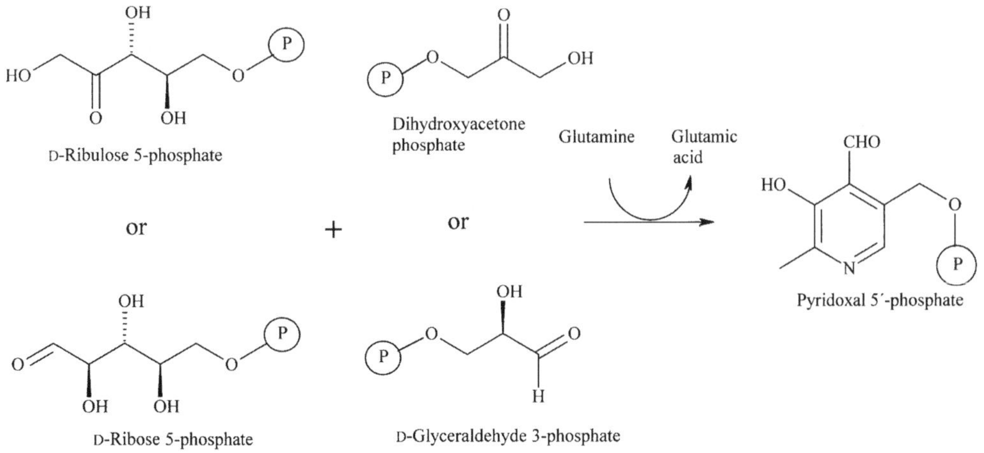

Fig. 177. The DXP-independent pathway of pyridoxal 5´-phosphate biosynthesis

way of vitamin B6 biosynthesis completely different from the one established for *Escherichia coli*. Two genes – *Pdx1* and *Pdx2* – have been found to be widely distributed and are found in all archaea, fungi, plants and most bacteria. These genes code for two proteins – Pdx1 and Pdx2 – which function together as a glutamine amidotransferase, which transfers an aminogroup from glutamine to either ribose 5-phosphate or ribulose 5-phosphate, which then in combination with glyceraldehyde 3-phosphate or dihydroxyacetone phosphate form pyridoxal 5´-phosphate. It is therefore apparent that a single amidotransferase is capable of synthesizing pyridoxal 5´-phosphate in most organisms. This pathway has been termed the DXP-independent pathway of pyridoxal 5´-phosphate biosynthesis.

Crude drug **Dried yeast** is obtained as a by-product of the brewery industry and consists of bottom yeast (under yeast, sediment yeast), the industrial term for the portion of yeast that settles to the bottom of the fermentation vat. Dried yeast is prepared by washing the yeast, first with a 1 % soda solution to remove bitter substances, then with water and finally drying. Dried yeast contains the entire vitamin-B complex and is an ingredient in vitamin preparations.

Biotin (vitamin H) Biotin is a cofactor for some carboxylases, i.e. enzymes which catalyse incorporation of CO_2 in biochemical reactions. An example is *acetyl-CoA carboxylase* (EC 6.4.1.2) which catalyses the formation of malonyl-CoA in the biosynthesis of fatty acids (page 239). Biotin is synthesized by bacteria, yeasts, moulds, algae and some plant species. Humans cannot make biotin but might absorb biotin that has been produced by bacteria in the colon. Biotin deficiency is very rare. A daily intake of 30 µg is recommended. It is supposed to strengthen hair and nails and is therefore an ingredient in cosmetic products.

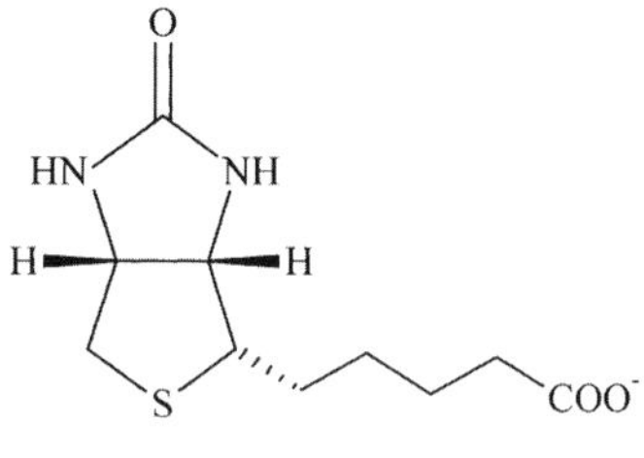

Biotin

Biosynthesis   The enzyme *6-carboxyhexanoate -CoA ligase* (EC 6.2.1.14) cataly ses the first step in the biosynthesis of biotin, where CoA is attached to pimelic acid forming 6-carboxyhexanoyl-CoA (= pimeloyl CoA). In the next reaction, catalysed by *8-amino-7-oxononanoate synthase* (EC 2.3.1.47), L-alanine is condensed with pimeloyl-CoA yielding 7-keto-8-aminopelargonic acid (=KAPA = 8-amino-7 oxononanoate). KAPA then reacts with SAM to form 7,8 diaminopelargonic acid (= DAPA = 7,8-diaminononaoate). In this reaction SAM acts as an amino group donor. The reaction is catal ysed by *adenosylmethionine-8-amino-7-oxononanoate trans-ami nase* (= DAPA synthase, EC 2.6.1.62). The reduced imidazole ring of biotin is formed by incorporation of CO_2 between the two amine

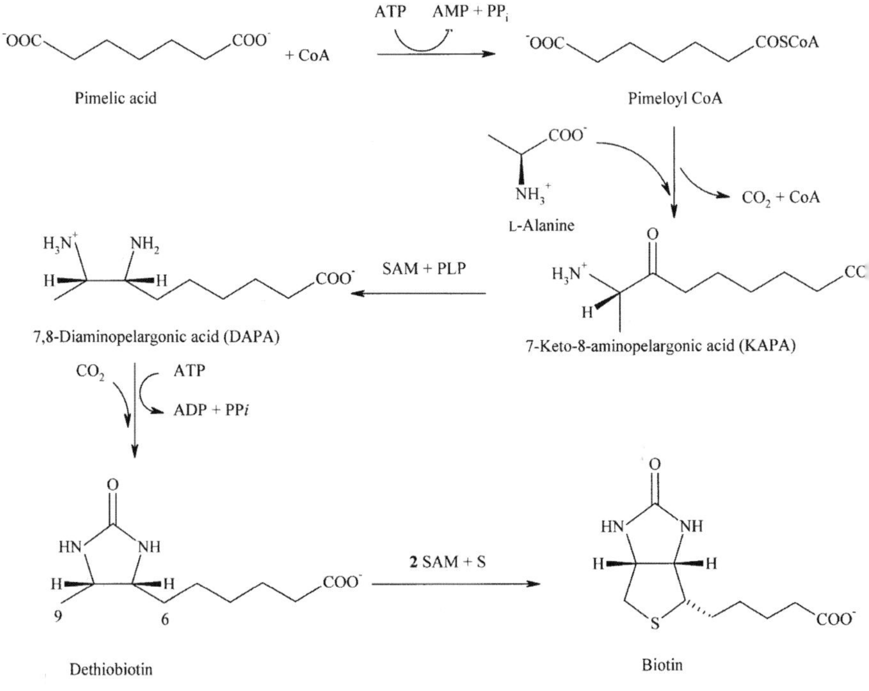

Fig. 178. Biosynthesis of biotin

groups of DAPA to form dethiobiotin in a reaction catalysed by *dethiobiotin synthase* (EC 6.3.3.3). Finally, incorporation of a sulphur atom between C-6 and C-9 of dethiobiotin, catalysed by *biotin synthase* (EC 2.8.1.6) yields biotin. The sulphur donor in this reaction is not definitely identified. It is neither elemental sulphur nor does it originate from SAM. Evidence has been provided that it might be donated from an Fe-S cluster in the enzyme itself.

The enzymes 8-amino-7-oxononanoate synthase, adenosylmethionine-8-amino-7-oxononanoate transaminase and dethiobiotin synthase have been fully characterized and their three-dimensional structure elucidated. The *bioB* gene of *Escherichia coli* codes for a protein which has been referred to as biotin synthase, but does not sustain biotin formation *in vitro* at a catalytic rate.

The four steps from pimeloyl CoA to biotin are common to most bacteria and plants.

Vitamins B$_{12}$ Vitamins B$_{12}$ are very complicated molecules based on a *corrin* ring, a structure which like the porphyrin ring of haem, cytochromes and chlorophyll (page 118) contains four pyrrole rings. In contrast to the porphyrin structure, two of the pyrrole rings in corrin are directly bonded to each other (Fig. 179). Vitamin B$_{12}$ contains cobalt which is coordinatively bonded to nitrogen in the four pyrrole rings and to a dimethylbenzimidazole moiety in the rest of the structure. A sixth coordinative bond is variable and involves a cyanide group in vitamin B$_{12}$ (cyanocobalamin) and a hydroxyl group in vitamin B$_{12a}$ (hydroxycobalamin), the two forms of the vitamin which are used in medicine. Cyanocobalamin is in fact an artefact which was formed during the original isolation of B$_{12}$. The cyanide group is derived from CN$^-$ ions which were present in the animal charcoal used in the purification process. In other varieties of vitamin B$_{12}$ the sixth coordinative bond connects the cobalt with water (vitamin B$_{12b}$, aquocobalamin), a nitro group (vitamin B$_{12c}$, nitritocobalamin) or a methyl group (methyl vitamin B$_{12}$, methylcobalamin). None of these cobalamin derivatives is physiologically active. In the body they are transformed to coenzyme B$_{12}$ (5'-deoxyadenosylcobalamin), by substitution of a 5'-deoxyadenosyl residue for the groups forming the sixth coordinative bond (Fig. 179).

Vitamin B$_{12}$ occurs in meat, liver, kidneys and seafood. Bacteria in the intestine of the animals produce B$_{12}$ which is taken up and stored in the liver which contains particularly large amounts of the vitamin. Vegetables contain practically no vitamin B$_{12}$ and vegetarians thus are at risk of obtaining insufficient amounts of this vitamin which might lead to development of *pernicious anaemia*, a disease the symptoms of which are nervous disturbances and low production of red blood cells and which is fatal if not treated.

Commercially, vitamin B$_{12}$ is produced from cultures of *Streptomyces griseus*, *Propionibacterium freudenreichii* spp. *shermanii* and *Pseudomonas denitrificans*. A total cobalamin extract is pre-

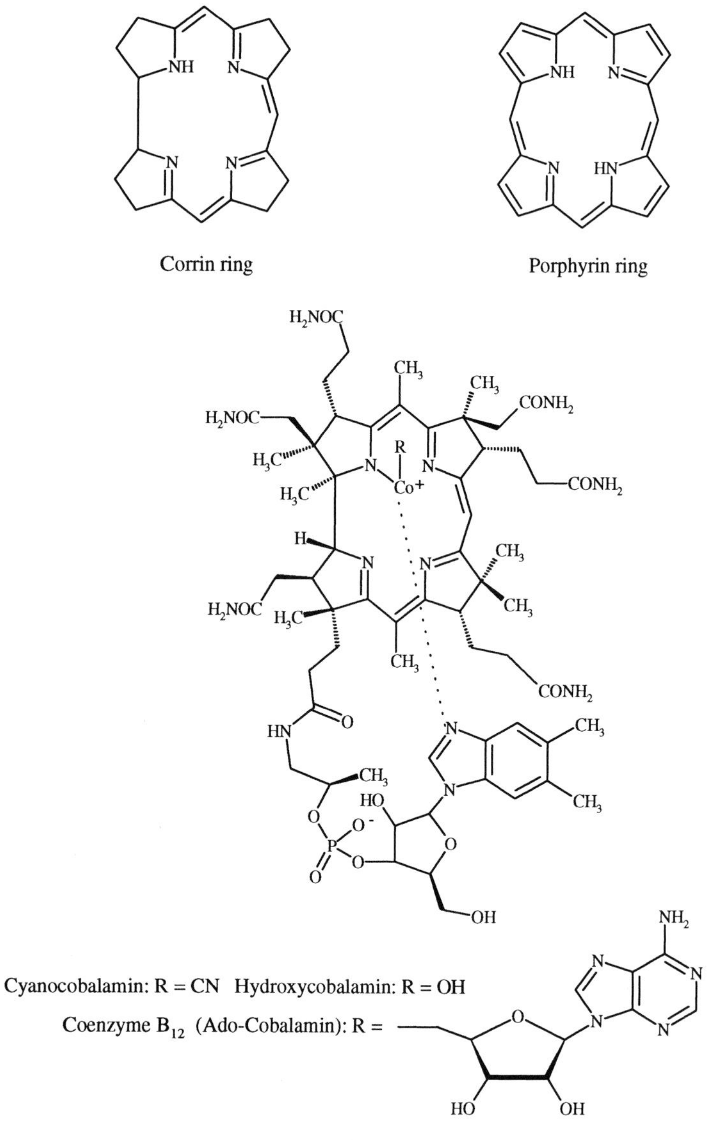

Corrin ring   Porphyrin ring

Cyanocobalamin: R = CN   Hydroxycobalamin: R = OH

Coenzyme B$_{12}$ (Ado-Cobalamin): R =

Fig. 179. Structures of corrin, porphyrin, cyanocobalamin, hydroxycobal
amin and coenzyme B$_{12}$

pared, and the mixture of cobalamins is converted to hydroxycobal
amin or cyanocobalamin.

Biosynthesis   The biosynthesis of vitamin B$_{12}$ is extremely complicated and it
elucidation has been characterized as an achievement comparabl
with the climbing of Mount Everest. The starting point (Fig. 180) i
5-aminolaevulinic acid (ALA), which can be formed either from
succinylCoA and glycine or, more commonly, from glutamic acid
In the latter case, the biosynthesis starts with charging of a gluta
mate-accepting tRNA with glutamate, a reaction catalysed by *glu*

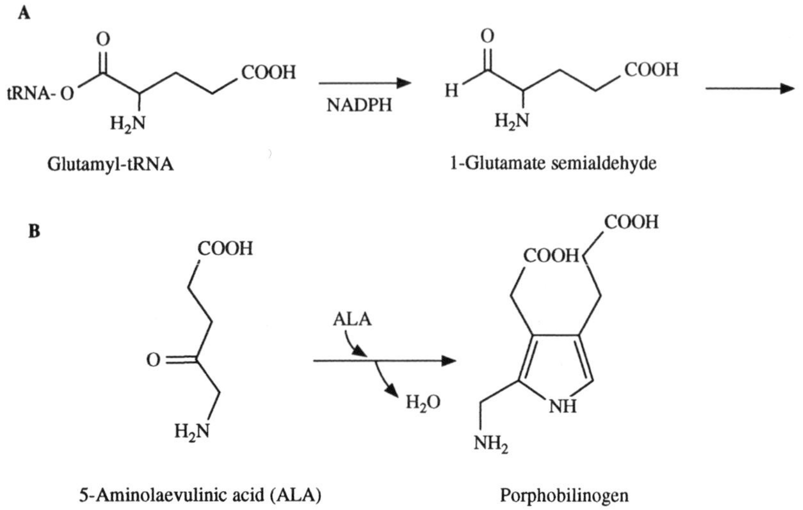

A

Glutamyl-tRNA   1-Glutamate semialdehyde

B

5-Aminolaevulinic acid (ALA)   Porphobilinogen

Fig. 180. Biosynthesis of hydrogenobyrinic acid. A: Formation of ALA. B: Self-condensation of ALA to form porphobilinogen B (reaction 1)

tamate-tRNA ligase (EC 6.1.1.17) which is an ATP requiring enzyme. Glutamyl-tRNA is reduced to L-glutamate semialdehyde in a reaction mediated by *glutamate-tRNA reductase* (EC 1.2.1.70), an enzyme which requires NADPH. Finally ALA is formed in a transamination reaction catalysed by *glutamate-1-semialdehyde 2,1-aminomutase* (EC 5.4.3.8). The enzymes catalysing these reactions are encoded by two genes called *hemA* and *hemL*, respectively.

Thirteen complicated reactions transfer ALA to hydrogenobyrinic acid, which is a cobalt-free precursor of vitamin B_{12}. *The first* of these reactions (Fig. 180) is a dehydrative self-condensation of ALA to porphobilinogen (PBG), which is catalysed by *porphobilinogen synthase* (EC 4.2.1.24), an enzyme encoded by the gene *hemB*.

In the *second reaction* (Fig. 182) *hydroxymethylbilane synthase* (EC 2.5.1.61, also known as PBG deaminase) (encoded by *hemC*) catalyses tetramerization of PBG to *hydroxymethylbilane* (*HMB*, also called preuro'gen). This tetramerization is a remarkable process which has been extensively studied employing PBG deaminase obtained by expression of the *hemC* gene in *Escherichia coli*. PBG deaminase contains a dipyrromethane cofactor, covalently attached to one of the four cysteine residues in the enzyme. This cofactor is derived from the substrate and the enzyme is unique in that the active holoenzyme is formed by automatic transformation of two PBG units to the apoenzyme without the intervention of another enzyme (Fig. 181).

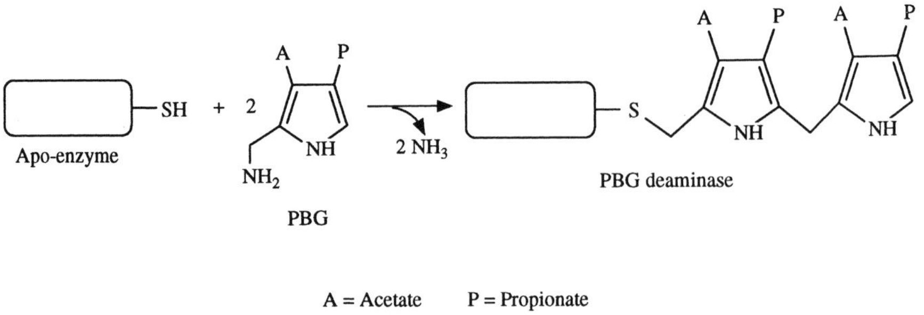

A = Acetate   P = Propionate

Fig. 181. Biosynthesis of hydrogenobyrinic acid. Formation of active PBG deaminase

The tetramerization of PBG is initiated by the attachment (with loss of NH_3) of one molecule of PBG to the free α-pyrrole position of the dipyrromethane complex of the enzyme. This process is repeated until four PBG units have been assembled (rings A to D in Fig. 182). Site-specific cleavage of the hexapyrrole chain releases an azafulvene which is stereospecifically hydrated to yield HMB.

The ***third reaction*** (Fig. 183) is catalysed by *uroporphyrinogen-III synthase* (EC 4.2.1.75, also known as uro'gen III synthase) and involves the transformation of HMB into *uro'gen III*, (uroporphyrinogen III). Uroporphyrinogen-III synthase has been produced by cloning of the genes *hemC* and *hemD* together and overexpression in *Escherichia coli*. The reaction involves twisting of ring D (Fig. 182) and the mechanism for this procedure is not yet determined, although two different hypotheses have been proposed. Uro'gen III is precursor not only for vitamin B_{12} but also for other "pigments of life": haems, sirohaem, chlorophylls and factor F_{430}.

The genes and enzymes required for reactions 1–3 have been identified from *Escherichia coli*, which, however, does not synthesize vitamin B_{12}. The genes required for the continuation of the pathway have been identified from two different organisms: *Salmonella typhimurium* and a strain of *Pseudomonas denitrificans* used for commercial production of Vitamin B_{12}. The biosynthetic pathways of these two organisms show some differences. *Pseudomonas denitrificans* requires molecular oxygen (reaction 7 below) and adds cobalt to the ring at a comparatively late stage (reaction 15 below), while *Salmonella typhimurium* is independent of oxygen and adds cobalt early (at the level of precorrin-2, see below) in the biosynthetic sequence. The complete step-by-step biosynthesis of coenzyme B_{12} has so far only been elucidated for *Pseudomonas denitrificans* and this pathway will be described here.

In ***reaction 4 and 5*** (Fig. 183), uro'gen III is methylated at positions 2 and 7 to form *precorrin-2*. S-adenosylmethionine (SAM) delivers the methyl groups required for this and subsequent methylations. The reaction is catalysed by *uroporphyrin III C-methyl-*

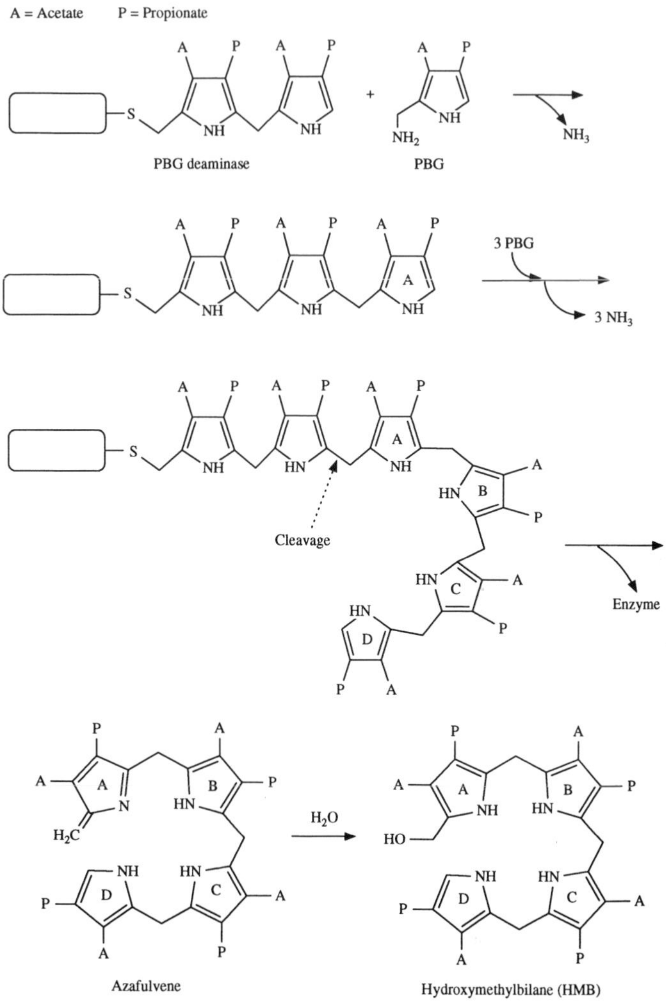

A = Acetate   P = Propionate

Fig. 182. Biosynthesis of hydrogenobyrinic acid. Formation of HMB from PBG (reaction 2)

transferase (EC 2.1.1.107) which is encoded by the gene *cobA* in *Pseudomonas denitrificans*. *Escherchia coli* also produces precorrin-2 but uses it for synthesis of sirohaem. In *Pseudomonas denitrificans* eight clustered genes, labelled *cobF–cobM* code for the enzymes required for the reactions that transform precorrin-2 into hydrogenobyrinic acid. Homologues for many of these genes are part of a large cobalamin operon in *Salmonella typhimurium*.

In reaction 6 (Fig. 183) *precorrin-2, C^{20}-methyltransferase* (EC 2.1.1.130) (gene *cobI*) methylates precorrin-2 at position 20, yielding *precorrin-3A* which in *reaction 7* (Fig. 183) is oxidized to form *precorrin-3B*. The catalysing enzyme is *precorrin-3B synthase* (EC 1.14.13.83) (gene *cobG*) which requires molecular oxygen for activity.

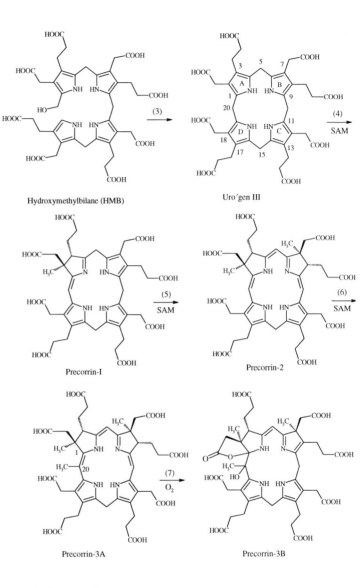

Fig. 183. Biosynthesis of hydrogenobyrinic acid. Transformation of HMB into precorrin-3B (reactions 3–7)

In reaction 8 (Fig. 184) *precorrin-3B C¹⁷-methyltransferase* (EC 2.1.1.131) (gene *cobJ*) catalyses a ring contraction and methylation at C-17, yielding *precorrin-4*.

In reaction 9 (Fig. 184) further methylation at C-11, catalysed by *precorrin-4 C¹¹-methyltransferase* (EC 2.1.1.133) (gene *cobM*) forms *precorrin-5*, which in **reaction 10** (Fig. 184) loses acetic acid in a combined methylation reaction, introducing methyl at C-1, with formation of *precorrin-6A*. This reaction is catalysed by *precorrin-5 C¹-methyltransferase (deacetylating)* (EC 2.1.1.152) (gene *cobF*).

In reaction 11 (Fig. 184) the double bond between C-18 and C-

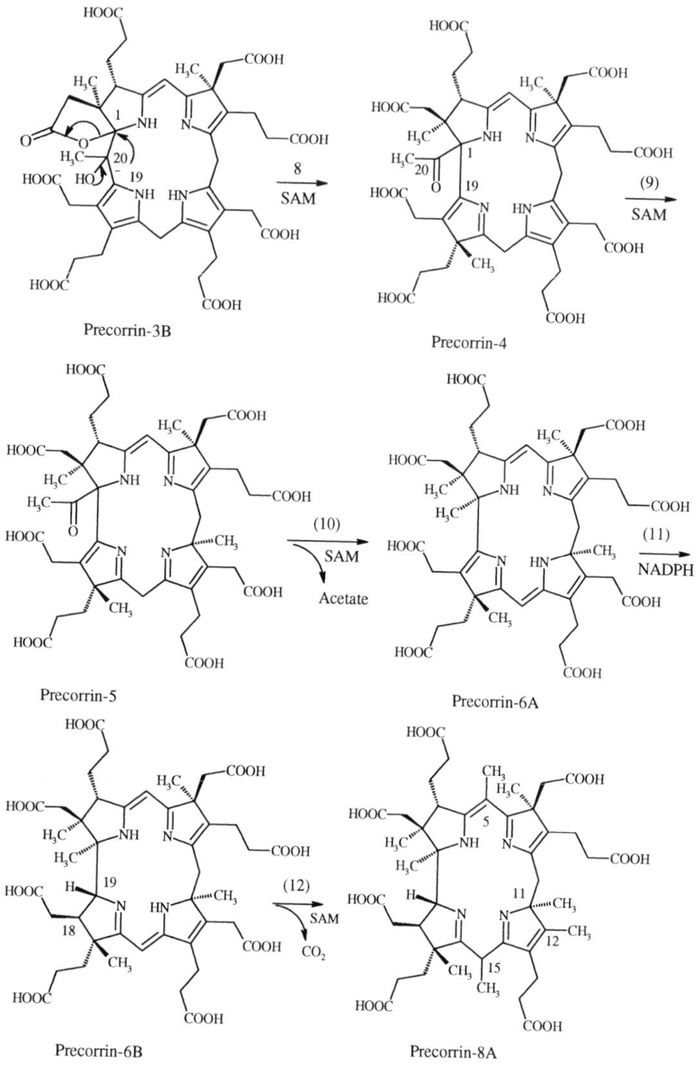

Fig. 184. Biosynthesis of hydrogenobyrinic acid. Formation of precorrin-8A from precorrin 3-B (reactions 8–12)

19 is reduced in an NADPH-dependent reaction involving *precorrin-6X reductase* (EC 1.3.1.54) (gene *cobK*) causing the formation of *precorrin-6B*.

In reaction 12 (Fig. 184) *precorrin-6Y $C^{5,15}$-methyltransferase (decarboxylating)* (EC 2.1.1.132) and *methyltransferase* (an enzyme encoded by the gene *cobL*) catalyse introduction of methyl groups at C-5 and C-15, accompanied by rearrangement of neighbouring double bonds, and decarboxylation of the acetyl group at C-12. The product formed is termed *precorrin-8X*.

Reaction 13 (Fig. 185) is the last reaction in the biosynthesis of *hydrogenobyrinic acid* and involves a methyl shift from C-11 to C-12, catalysed by *precorrin-8X methylmutase* (EC 5.4.1.2) (gene: *cobH*).

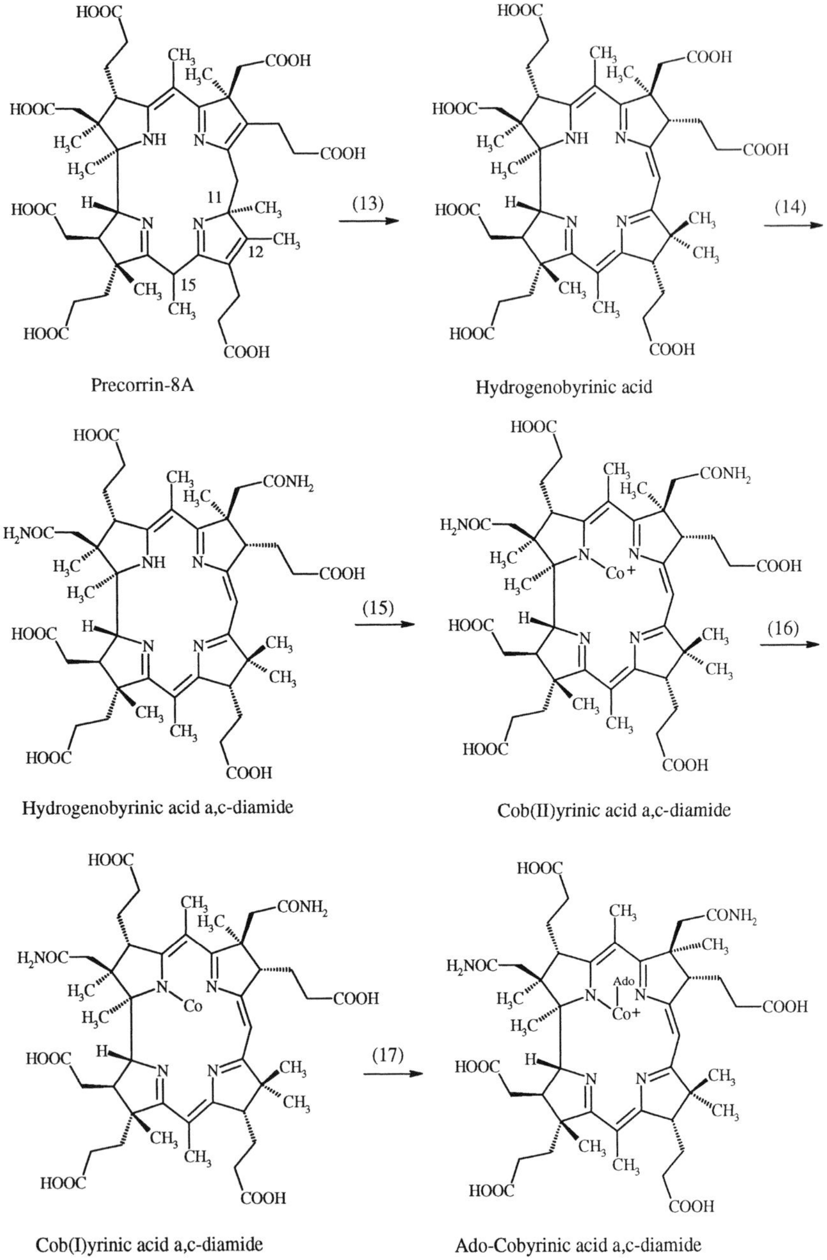

Fig. 185. Completion of the biosynthesis of hydrogenobyrinic acid (reaction 13) and continuation of the biosynthesis of coenzyme B_{12}. Reactions 14 –17).

The elucidation of the pathway described above has made it possible to perform a "one pot" synthesis of hydrogenobyrinic acid from ALA. All the 12 enzymes involved are available in large quantities by expression of the corresponding genes in *Escherichia coli*, and thus it was possible to mix ALA, SAM and all enzymes in a reaction which yielded hydrogenobyrinic acid in an overall yield of 20 %. Hydrogenobyrinic acid can be converted to cyanocobalamin by chemical means, and the complete synthesis of vitamin B_{12} from its starting material ALA is thus now possible. This example shows that combination of applied biotechnology and chemical synthesis is possible in the synthesis of natural products that are currently inaccessible by conventional chemical synthesis, a technique which in the future can be expected to be of great significance for the production of important natural products, including drugs.

The completion of the pathway from hydrogenobyrinic acid to coenzyme B_{12} involves eight more reactions, the genes and enzymes of which have been characterized (Figs 185 and 186). The enzyme *hydrogenobyrinic acid* a,c-*diamide synthase (glutamine-hydrolysing)* (EC 6.3.5.9) encoded by the gene *CobB*, amidates the acetate side-chains in rings A and B using glutamine as the amido donor, yielding *hydrogenobyrinic acid* a,c-*diamide* (**reaction 14** Fig. 185). In **reaction 15** (Fig. 185) a polypeptide complex (*cobaltochelase*, EC 6.6.1.2) encoded by genes *cobN, cobS* and *cobT* inserts Co^{2+} in hydrogeno-byrinic acid *a,c-diamide* with formation of *cob(II)yrinic acid* a,c-*diamide*. In **reaction 16** (Fig. 185) the flavin-requiring enzyme *cob(II)yrinic acid* a,c-*diamide reductase* (EC 1.16.8.1) catalyses reduction to *cob(I)yrinic acid* a,c-*diamide*, which is adenosylated (**reaction 17** Fig. 185) to *adocob(I)yrinic acid* a,c-*diamide*. The enzyme for this reaction *(cob(I)yrinic acid* a,c-*diamide adenosyltransferase*, EC 2.5.1.17) is encoded by the gene *cobO*. Amidation of four side-chains, catalysed by the enzyme encoded by the gene *cobQ*), *adenosylcobyric acid synthase (glutamine hydrolysing)* (EC 6.3.5.10) yields *ado-cobyric acid* (**reaction 18** Fig. 186). Also here glutamine is the source of the amide groups. In **reaction 19** (Fig. 186) an aminopropanol residue is attached to the remaining acetate side-chain in ring D, yielding *ado-cobinamide*. The enzyme catalysing this reaction is *adenosylcobinamide phosphate synthase* (EC 6.3.1.10). Ado-cobinamide is phosphorylated to form *ado-cobinamide phosphate* (**reaction 20** Fig. 186). The reaction involves ATP or GTP and is catalysed by *adenosylcobinamide kinase* (EC 2.7.1.156). Ado-cobinamide phosphate reacts with guanosine triphosphate yielding *ado-GDP-cobinamide* (**reaction 21** Fig. 187). The enzyme, catalysing this reaction is *adenosylcobinamide phosphate guanylyltransferase* (EC 2.7.7.62). *Adocobalamine (coenzyme B_{12})* is finally formed in a reaction (**22** Fig. 187) involving exchange of the guanosinediphosphate part of ado-GDP-cobinamide with α-ribazole, catalysed by *adenosylcobinamide-GDP ribazoletransferase* (EC 2.7.8.26).

α-Ribazole originates from riboflavin, which via flavinmononu-

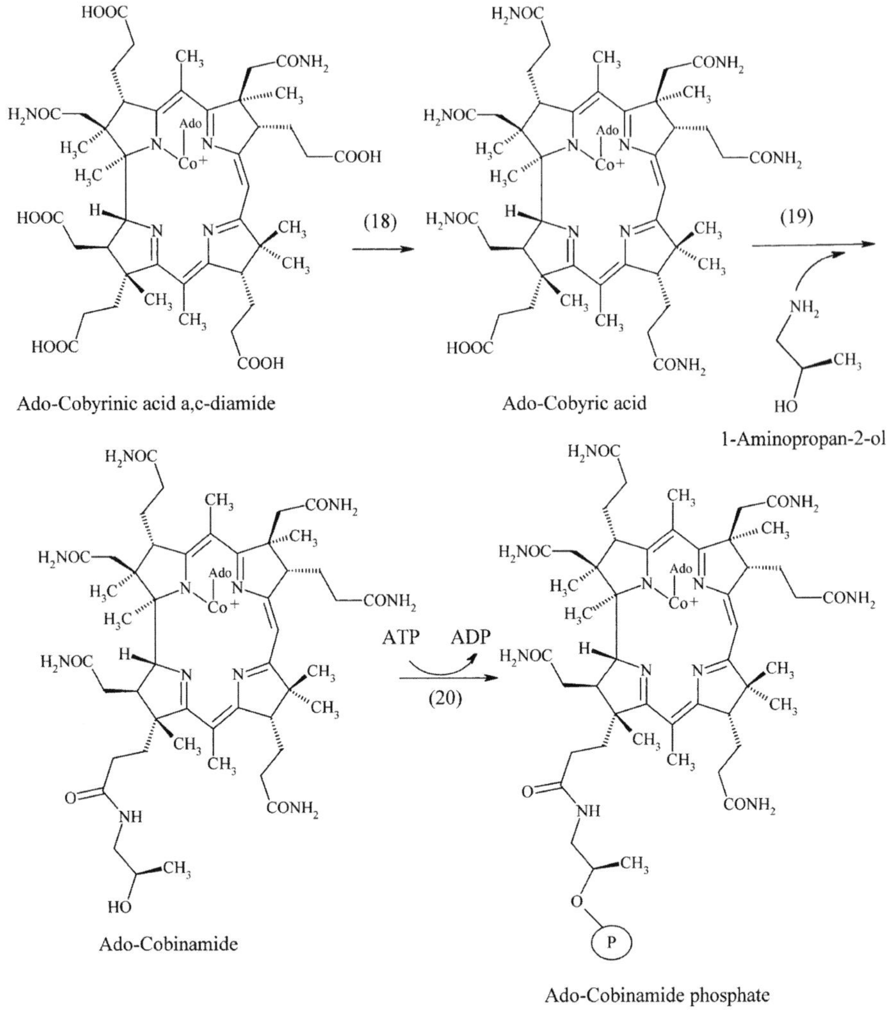

Fig. 186. Biosynthesis of coenzyme B_{12}. Reactions 18 – 20

cleotide is transformed to 5,6-dimethylbenzimidazole (Fig. 188). The enzyme *nicotinate-nucleotide – dimethylbenzimidazole phosphoribosyltransferase* (EC 2.4.2.21) catalyses a reaction between this compound and β-nicotinate D-ribonucleotide, resulting in formation of α-ribazole-5′-phosphate from which the phosphate-group is split off by *α-ribazole phosphatase* (EC 3.1.3.73).

Cyanogenic glycosides

Many plants contain glycosides that form hydrogen cyanide on hydrolysis and are therefore known collectively as cyanogenic gly-

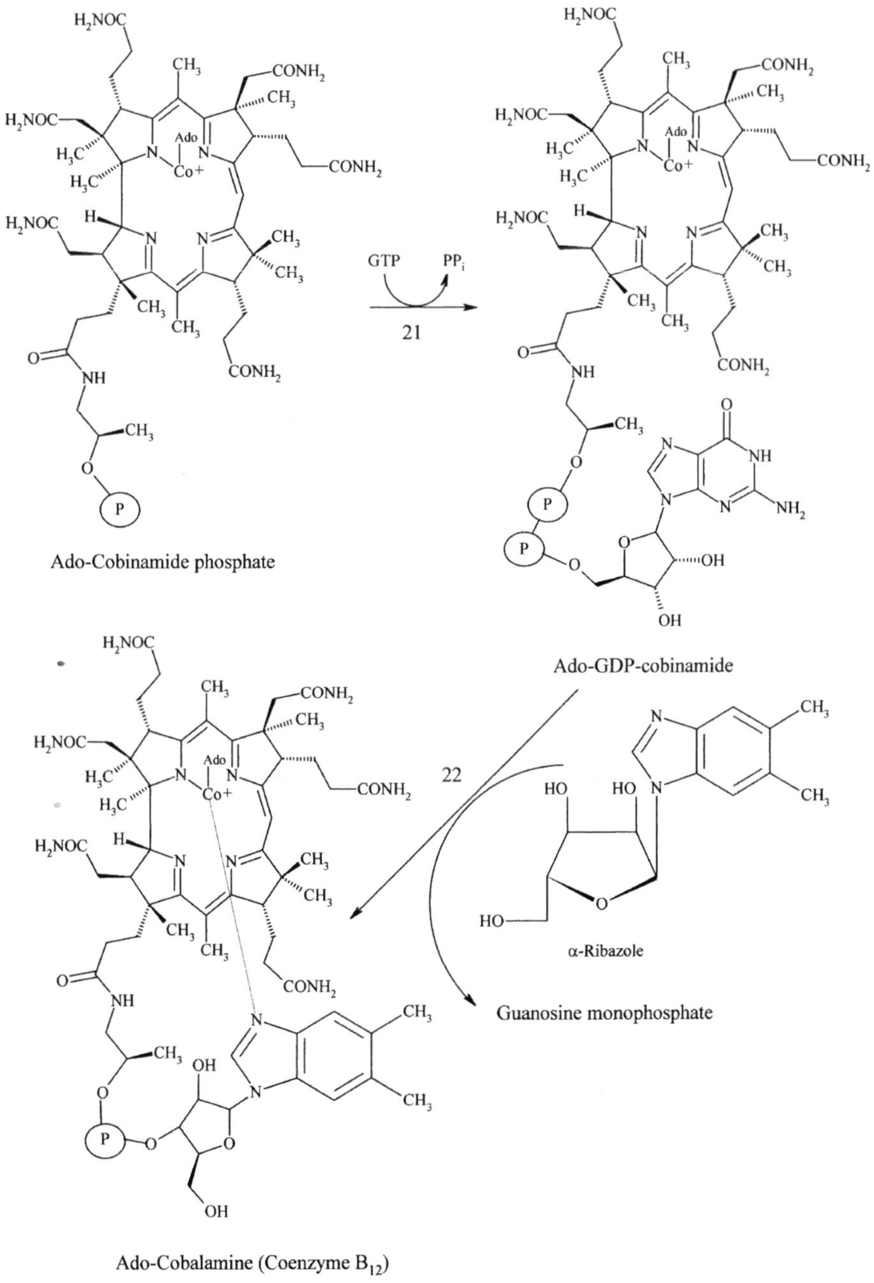

Fig. 187. Biosynthesis of coenzyme B_{12}. Reactions 21 + 22

cosides. The first one to be isolated was amygdalin, from bitter almonds. On hydrolysis, amygdalin yields glucose, hydrogen cyanide and benzaldehyde:

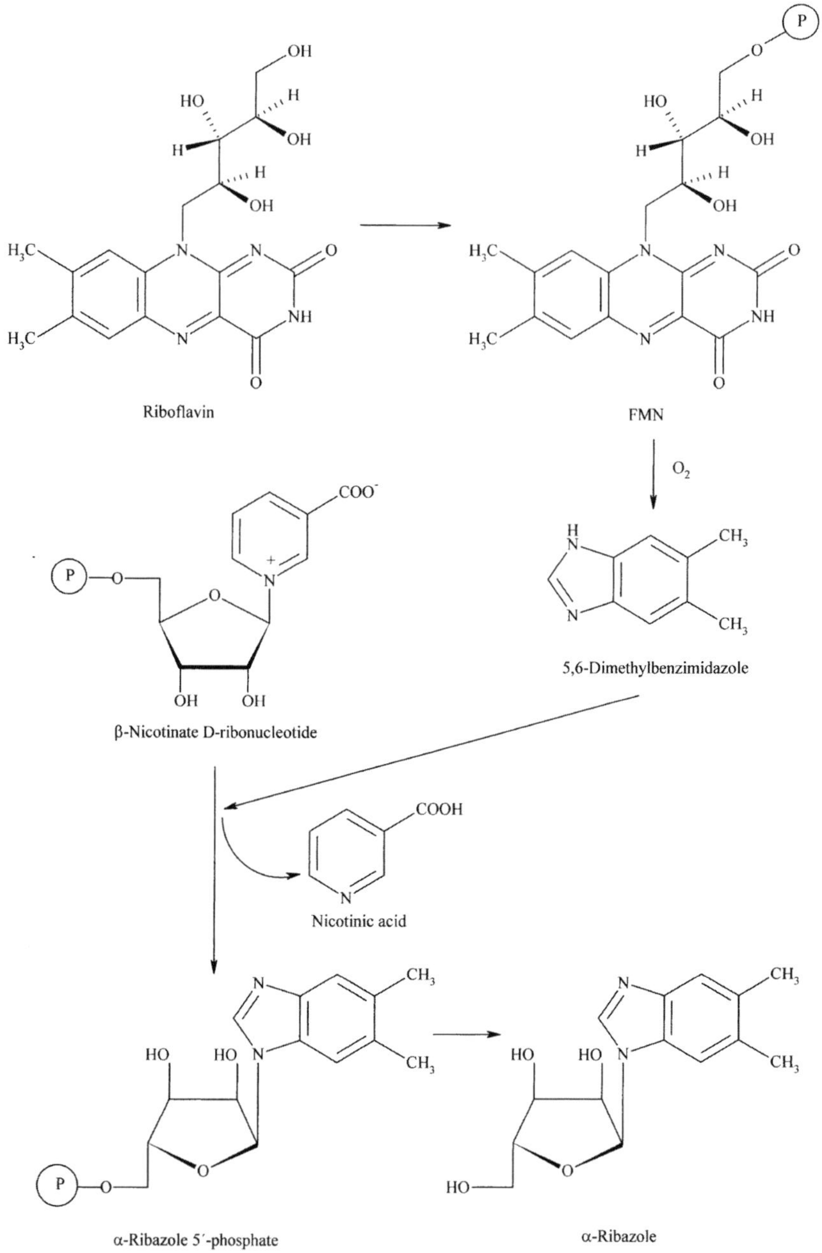

Fig. 188. Biosynthesis of ribazole from riboflavin

Almonds contain a mixture of enzymes called *emulsin*, which consists of *amygdalin hydrolase*, *prunasin hydrolase* and *hydroxy nitrile lyase*. When bitter almonds are crushed in the presence of water, emulsin catalyses the above series of reactions.

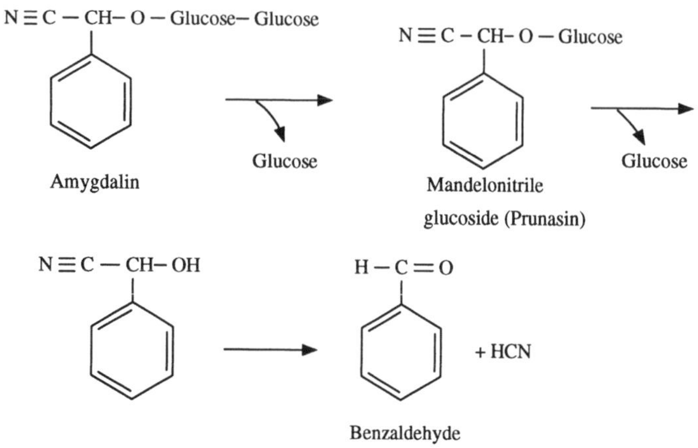

Fig. 189. Hydrolysis of amygdalin

Cyanogenic glycosides are found throughout the plant kingdom, and in at least 70 families. Besides amygdalin, which occurs in species of *Rosaceae*, about 15 other glycosides have been isolated and structurally elucidated. They can be divided into four structural groups:

1. Glycosides derived from 2-hydroxy-2-acetonitrile or its derivatives: the *amygdalin* type. See Fig. 189.
2. Glycosides with saturated aliphatic aglycones: the *linamarin* type.
3. Glycosides in which the aglycone contains a double bond in conjugation with the nitrile group: the *acacipetalin* type.
4. Glycosides with an unsaturated alicyclic aglycone: the *gynocardin* type.

Biosynthesis of cyanogenic glycosides

Biosynthetically, the aglycones of cyanogenic glycosides are derived from L-amino acids. The biosynthesis can be exemplified by the formation of *dhurrin*, a glycoside from the important cereal sorghum (*Sorghum vulgare* Pers., *Graminae*, also called great millet). The precursor of dhurrin is L-tyrosine which is oxidized to *N*-hydroxytyrosine. *N,N*-dihydroxytyrosine is formed in a second oxidation reaction and reduction of this compound yields(*E*)-*p*-

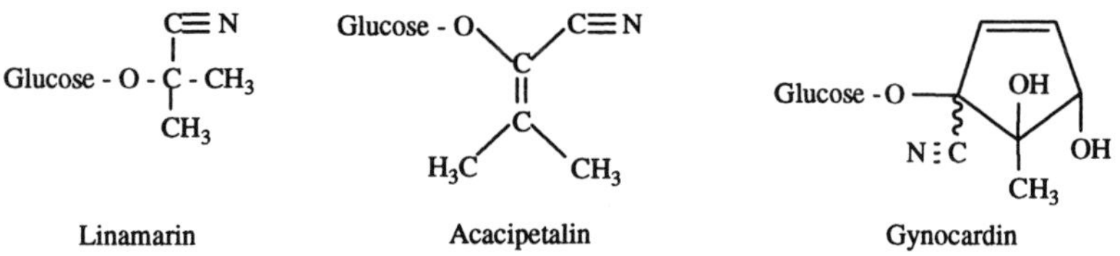

Linamarin

Acacipetalin

Gynocardin

hydroxy-phenylacetaldoxime, which is epimerized to (Z)-p-hydroxy-phenylacetaldoxime. All these reactions are catalysed by a multifunctional, cytochrome P450 dependent enzyme, designated a *CYP 79 (cytochrome P450_{TYR})*. A full length cDNA clone encoding this enzyme has been isolated and sequenced. The open reading frame encodes a protein with a molecular mass of 61 887 Da. This cDNA has been expressed in *Escherichia coli* yielding an active enzyme with high substrate specificity.

A second multifunctional cytochrome P450 enzyme called *CYP71E1 (P450ox)* catalyses dehydration of the Z-aldoxime to *p*-hydroxyphenylacetonitrile which is oxidized to *p*-hydroxymandelonitrile. Both the dehydration and the oxidation are dependent on the presence of NADPH. The mechanism by which CYP71E1 catalyses both a dehydration and an oxidation is at present not understood. CYP71E1 is an unstable enzyme which has been isolated from microsomal membranes of etiolated sorghum seedlings. The

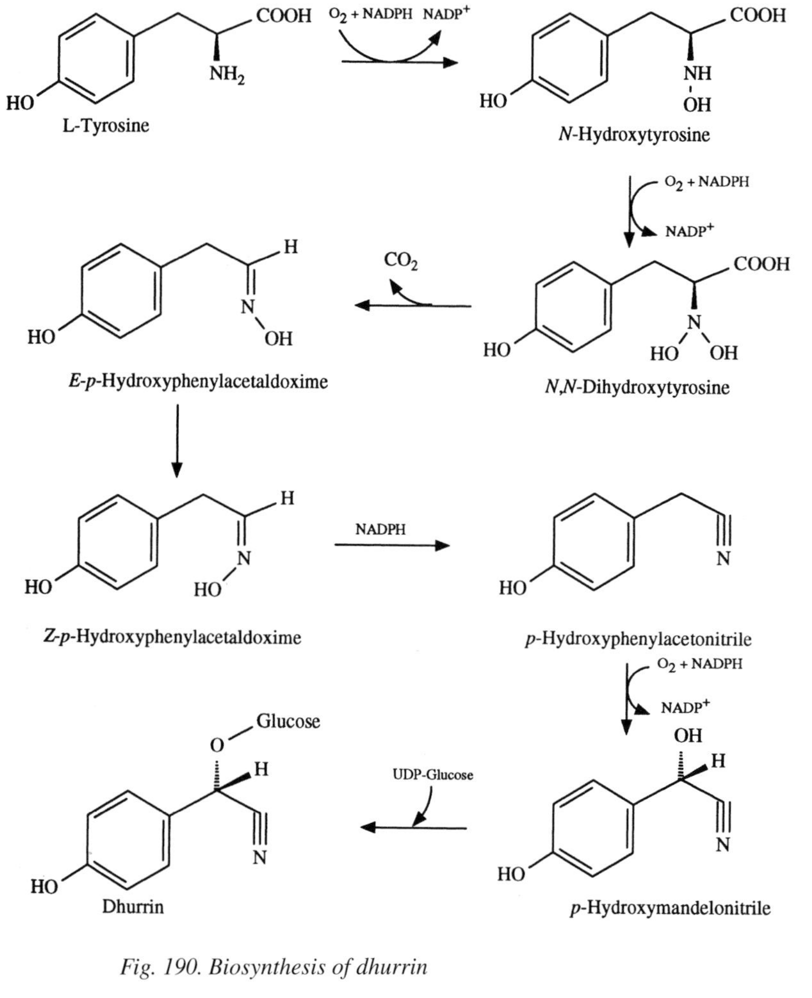

Fig. 190. Biosynthesis of dhurrin

sequence of 14 N-terminal amino acids of the native enzyme has been determined. CYP71E1 has been cloned and expressed in *Escherichia coli*. The deduced amino acid sequence comprises 532 amino acids.

The final step in the biosynthesis of dhurrin is glucosidation of *p*-hydroxymandelonitrile, a reaction catalysed by a soluble UDPG glucosyltransferase that has been partially purified. Mixing tyrosine, recombinant CYP79, the isolated CYP71E1, partially purified UDPG glucosyltransferase and the required cofactors permitted biosynthesis of dhurrin *in vitro*.

Other cyanogenic glycosides are formed in a similar way with other amino acids as the ultimate precursors. Amygdalin is thus biosynthesized from phenylalanine and linamarin from valine.

Crude drugs **Bitter almond** (Amygdali amarae semen) and **Bitter almond oil** (Amygdali amarae aetheroleum). Bitter almond is the seed of the bitter variety of the almond tree *Prunus dulcis* (Mill.) D. A. Webb (*Rosaceae*). There are two physiological varieties: one with bitter seeds, bitter almond, and one with sweet seeds, sweet almond. The varieties differ only in this respect. Morphologically, they are identical. The almond tree is 3–5 metres high, and native to western Central Asia. It is cultivated in Mediterranean countries as well as in other regions with a similar climate. Bitter almond seed contains 1–3 % of the cyanogenic glycoside amygdalin and also the enzyme mixture emulsin, which, as already seen, in the presence of water splits the amygdalin into the end products hydrogen cyanide, benzaldehyde and glucose. The seed also contains about 45 % of fixed oil, which can be obtained by expression (see further page 251). The residue left after expressing the fixed oil gives genuine bitter almond oil on steam distillation. Bitter almond oil consists mainly of benzaldehyde, but it also contains 2–4 % hydrogen cyanide. Bitter almond and bitter almond oil are used as spices, but this use is not entirely safe because of the hydrogen cyanide content. About 60 bitter almonds constitute a lethal dose for an adult person, and 6–10 almonds pose a great risk for a child. Nowadays, genuine bitter almond oil has for the most part been replaced by synthetic benzaldehyde.

Cyanogenic glycosides as a toxic hazard in foodstuffs

Some foodstuffs, used mainly in developing countries, contain cyanogenic glycosides and can cause poisoning if not properly handled. One of these is *cassava root* (tapioca root), which is obtained from *Manihot utilissima* Pohl., (*Euphorbiaceae*) and which can contain up to 0.4 % linamarin. The root is used as a source of starch and after crushing must be carefully washed with water to remove the cyanogenic glycoside. Otherwise, if eaten, the glycoside is split in the gastrointestinal tract by enzymes of the bacterial flora, causing poisoning by the hydrogen cyanide liberated.

Another foodstuff containing cyanogenic glycosides is the *Lima bean* which is obtained from *Phaseolus lunatus* L. (*Fabaceae*). Also this product must be crushed and washed carefully before any food can be prepared from it.

Glucosinolates (mustard-oil glucosides)

Glucosinolates or mustard-oil glucosides are a group of glucosides with the following general structure:

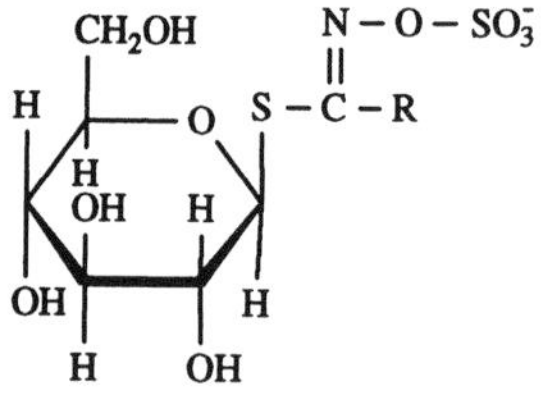

The sugar moiety in these β-glycosides is always glucose, which is linked to the aglycone via a sulphur atom (S-glycoside). The aglycone is an anion, in the plant forming a salt with an inorganic or an organic cation. R can be either aliphatic or aromatic. Glucosinolates are hydrolysed by *thioglucosidase* (EC 3.2.1.147, also known as myrosinase) according to the following reaction scheme:

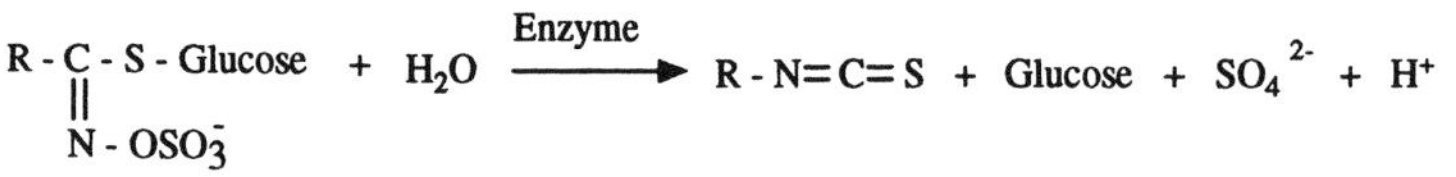

$$R - \underset{\underset{N - OSO_3^-}{\|}}{C} - S - Glucose \; + \; H_2O \; \xrightarrow{\text{Enzyme}} \; R - N = C = S \; + \; Glucose \; + \; SO_4^{2-} \; + \; H^+$$

As shown, the hydrolysis is associated with a rearrangement of the aglycone to an isothiocyanate. Glucosinolates are tasteless and odourless compounds, whereas the isothiocyanates, formed on hydrolysis, are liquids with a sharp smell and taste (= mustard oils). Glucosinolates have been encountered in 11 plant families. Among these, the family *Brassicaceae* (mustard-oil plants) is outstanding, as all its genera contain glucosinolates.

Myrosinase is a group of isoenzymes which consist of glycosylated dimeric proteins with subunit masses from 62 to 77 kDa. Isoenzymes have been isolated and characterized from some species and shown to have varying degrees of glycosylation, ascorbic acid activation and hydrolysis rates on different glucosinolates. The myrosinase isoenzymes are encoded by a multigene family consisting of two subgroups. Two nuclear genes representing each of these two subgroups have been cloned and sequenced from *Brassica campestris* L. These genes are known as *Myr1.Bn1* and *Myr2.Bn1*. *Myr1,Bn1* has a 19 amino acid signal peptide and consists of 11 exons of sizes ranging from 54 to 256 bp and 10 introns of sizes from 75 to 229 bp. The *Myr2.Bn1* gene has a 20 amino acid signal peptide and consists of 12 exons ranging in size from 35 to 262 bp and 11 introns of sizes from 81 to 131 bp. *Myr1.Bn1* has been shown to encode seed storage myrosinase and the *Myr2.Bn1* gene encodes myrosinase present in vegetative tissue. The calculated molecular masses of the myrosinases encoded by the *Myr1* and *Myr2* genes are 58 497 Da and 60 384 Da, respectively.

Myrosinase is found in a specialized type of idioblast called *myrosin cells*. Myrosin cells are dispersed throughout the tissue, but in the radicle a large proportion is present in the cortex area. In the cotyledons the myrosin cells are found in the outermost part. The enzyme–substrate system is believed to act as a defence system and the localization of the myrosin cells suggests that they can act as a toxic mine or tripwire when the tissue is destroyed.

Biosynthesis of glucosinolates

Biosynthetically, glucosinolates are derived from amino acids. The course of the biosynthesis is analogous to that of the aglycones in cyanogenic glycosides in that the end product contains both the nitrogen atom and the entire carbon skeleton of the amino acid except for the carboxyl group. The biosyntheses of these two types of glycosides are closely related and aldoximes play a key role in the early stages.

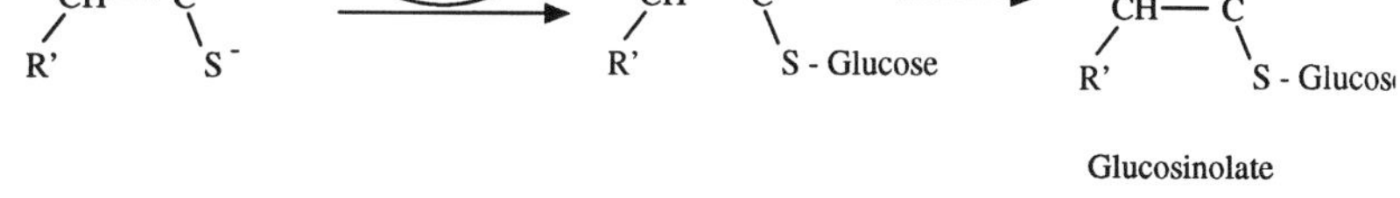

Fig. 191. Biosynthesis of glucosinolates

Many glucosinolates have carbon skeletons derived from uncommon amino acids. It has been shown that these acids are formed from the amino acids methionine and phenylalanine by elongation of the chain in one or more steps. The sulphur atom linking the aglycone to the sugar part comes from sulphur-containing amino acids, especially cysteine. The sulphate group is introduced by a sulphotransferase, with 3'-phosphoadenosine 5'-phosphosulphate (PAPS) as the donor.

Crude drugs   **Black mustard** (Sinapis nigri semen) is the seed of *Brassica nigra*

Koch (*Brassicaceae*), a herb grown in Europe, Asia and the United States. The seed is small, 1–1.6 mm in diameter, and has a dark testa. Black mustard contains sinigrin, which is split by myrosinase into allyl isothiocyanate, glucose and potassium hydrogen sulphate:

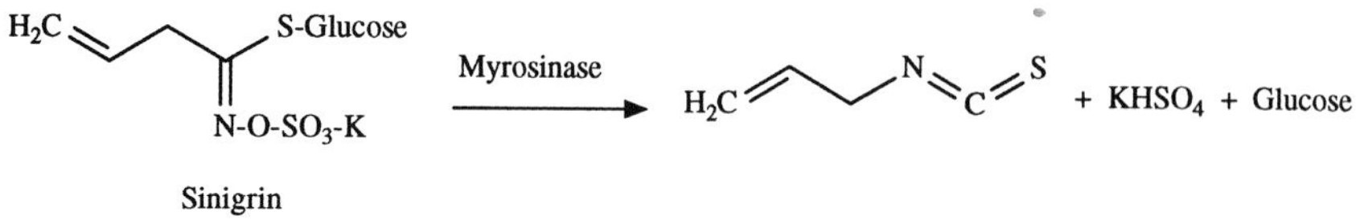

Sinigrin

The drug is used as a condiment because of the sharp taste of the allyl isothiocyanate. The seed also contains 30 % fatty oil.

White mustard (Sinapis albae semen) is the seed of *Brassica alba* Boiss. (*Brassicaceae*). This herb is cultivated in Europe, southwestern Asia, and the United States. The seed is slightly larger than that of the black mustard, 1.5–2.5 mm in diameter and has a yellowish white coat. The glucosinolate of white mustard is sinalbin, which is hydrolysed by myrosinase to *p*-hydroxybenzyl isothiocyanate, sinapin hydrogen sulphate and glucose.

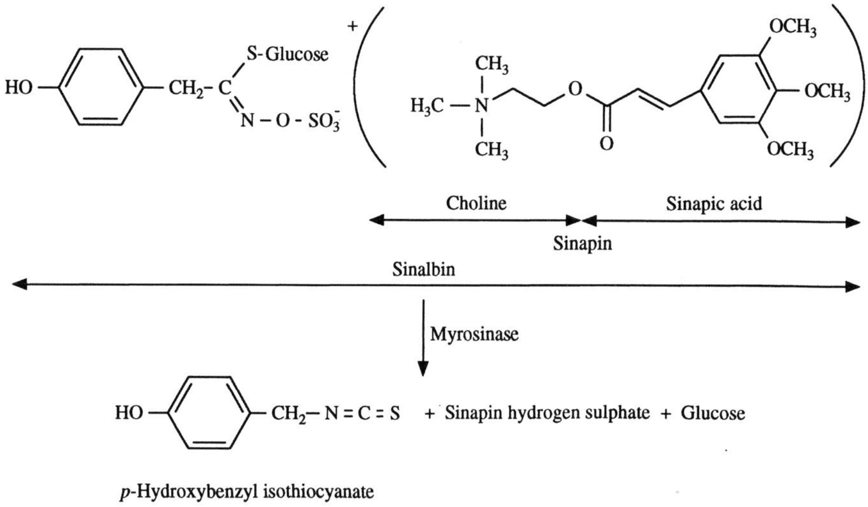

p-Hydroxybenzyl isothiocyanate

The seed also contains about 30 % fixed oil. White mustard is used as a spice. Powdered white, and black, mustard can be stirred into water and taken as an emetic.

Further reading

Bromelain

Harrach, T., Eckert, K., Schulze-Forster, K., Nuck, R., Grunow, D. and Maurer, H.R. Isolation and partial characterization of basic proteinases from stem bromelain. *Journal of Protein Chemistry*, **14** (1), 41–52 (1995).

Maurer, H.R. Bromelain: biochemistry, pharmacology and medical use. *Cellular and Molecular Life Sciences*, **58** (9), 1234–1245 (2001).

Rowan, A.D., Buttle, D.J. and Barrett, A.J. The cysteine proteinases of the pineapple plant. *Biochemical Journal*, **266**, 869–875 (1990).

RIP toxins

Chen, Y.-L., Chow, L.-P., Tsugita, A. and Lin, J.-Y. The complete primary structure of abrin-a B chain. *FEBS Letters*, **309** (2), 115–118 (1992).

Kimura, Y., Hase, S., Ikenaka, T. and Funatsu, G. Structures of sugar chains of abrin a obtained from *Abrus precatorius* seeds. *Biochimica et Biophysica Acta*, **966**, 150–159 (1988).

Kimura, M., Sumizawa, T. and Funatsu, G. The complete amino acid sequences of the B-chains of abrin-a and abrin-b, toxic proteins from the seeds of *Abrus precatorius. Bioscience, Biotechnology and Biochemistry*, **57** (1), 166–169 (1993).

Lord, J.M., Roberts, L.M. and Robertus, J.D. Ricin: Structure, mode of action, and some current applications. *The FASEB Journal*, **8**, 201–208 (1994).

Olsnes, S. and Kozlov, J.V. Ricin. *Toxicon*, **39** (11), 1723–1728 (2001).

Pastan, I., Chaudhary, V. and FitzGerald, D.J. Recombinant toxins as novel therapeutic agents. *Annual Review of Biochemistry*, **61**, 331–354 (1992).

Robertus, J. The structure and action of ricin, a cytotoxic *N*-glycosidase. *Seminars in Cell Biology*, **2**, 23–30 (1991).

Lectins

Scott, M.L. and Bird, G.W.G. Some contributions of the plant kingdom to transfusion medicine – lectins and plant enzymes. *Transfusion Medicine Reviews*, **6** (2), 103–115 (1992).

Amanita mushrooms

Chang, I.-M. and Yamaura, Y. Aucubin, a new antidote for poisonous *Amanita* mushrooms. *Phytotherapy Research*, **7**, 53–56 (1993).

Wieland, T. The toxic peptides from *Amanita* mushrooms. *International Journal of Peptide and Protein Research*, **22**, 257–276 (1983).

Wieland, T. and Faulstich, H. Fifty years of amanitin. *Experientia*, **47**, 1186–1193 (1991).

Lizard toxins

Brubaker, P.L. Incretin-based therapies: mimetics versus protease inhibitors. *Trends in Endocrinology and Metabolism*, **18** (6), 240–245 (2007).

Davidson, M.B., Bate, G. and Kirkpatrick, P. Exenatide. *Nature reviews/Drug discovery*, **4**, 713–714 (2005).

Raufman, J.-P. Bioactive peptides from lizard venoms. *Regulatory Peptides*, **61**, 1–18 (1996).

Ziconotide

Klotz, U. Ziconotide – a novel neuron-specific calcium channel blocker for the intrathecal treatment of severe chronic pain – a short review. *International Journal of Clinical Pharmacology and Therapeutics*, **44** (10), 478–483 (2006).

Mistletoe toxins and thionins

Florack, D.E.A. and Stiekema, W.J. Thionins: properties, possible biological roles and mechanism of action. *Plant Molecular Biology*, **26**, 25–37 (1994).

Debreczeni, J.E., Girmann, B., Zeeck, A., Krätzner, R. and Sheldrick, G.M. Structure of viscotoxin A3: disulfide location from weak SAD data. *Acta Crystallografica*, **D59**, 2125–2132 (2003).

Cyclotides

Craik, D.J., Daly, N.L., Bond, T. And Waine, C. Plant cyclotides: A unique family of cyclic and knotted proteins that defines the cyclic cystine knot structural motif. *Journal of Molecular Biology*, **294**, 1327–1336 (1999).

Daly, N.L., Rosengren, K.J. and Craik, D.J. Discovery, structure and biological activities of cyclotides. *Advanced Drug Delivery Reviews*, **61**, 918–930 (2009).

Jennings, C., West, J., Waine, C., Craik, D. and Anderson, M. Biosynthesis and insecticidal properties of plant cyclotides: The cyclic knotted proteins from *Oldenlandia affinis*. *Proceedings of the National Academy of Sciences*, **98** (19), 10614–10619 (2001).

β-Lactam antibiotics

Aharonowitz, Y., Cohen, G. and Martin, J.F. Penicillin and cephalosporin biosynthetic genes: structure, organization, regulation, and evolution. *Annual Review of Microbiology*, **46**, 461–495 (1992).

Baldwin, J.E. The biosynthesis of penicillins and cephalosporins. *Journal of Heterocyclic Chemistry*, **27** (1), 71–78 (1990).

Byford, M.F., Baldwin, J.E., Shiau, C.-Y. and Schofield, C.J. The mechanism of ACV synthetase. *Chemical Reviews*, **97**, 2631–2649 (1997).

Cooper, R.D.G. The enzymes involved in biosynthesis of penicillin and cephalosporin; their structure and function. *Bioorganic & Medicinal Chemistry*, **1** (1), 1–17 (1993).

Lendenfeld, T., Ghali, D., Wolschek, M., Kubicek-Pranz, E.M. and Kubicek, C.P. Subcellular compartmentation of penicillin biosynthesis in *Penicillium chrysogenum. The Journal of Biological Chemistry*, **268** (1), 665–671 (1993).

Martín, J.F., Gutiérrez, S., Fernández, F.J., Velasco, J., Fierro, F., Marcos, A.T. and Kosalkova, K. Expression of genes and processing of enzymes for the biosynthesis of penicillins and cephalosporins. *Antonie van Leeuwenhoek*, **65**, 227–243 (1994).

Martín, J.F. and Gutiérrez, S. Genes for β-lactam antibiotic synthesis. *Antonie van Leeuwenhoek*, **67**, 181–200 (1995).

Martín, J.F. New aspects of genes and enzymes for β-lactam antibiotic biosynthesis. *Applied Microbiology and Biotechnology*, **50**, 1–15 (1998).

Roach, P.L., Clifton, I.J., Hensgens, C.M.H., Shibata, N., Schofield, C.J., Hajdu, J. and Baldwin, J.E. Structure of isopenicillin N synthase complexed with substrate and the mechanism of penicillin formation. *Nature*, **387**, 827–830 (1997).

Weil, J., Miramonti, J. and Ladisch, M.R. Cephalosporin C: mode of action and biosynthetic pathway. *Enzyme and Microbial Technology*, **17**, 85–87 (1995).

Weil, J., Miramonti, J. and Ladisch, M.R. Biosynthesis of cephalosporin C: regulation and recombinant technology. *Enzyme and Microbial Technology*, **17**, 88–90 (1995).

Clavulanic acid

Baggaley, K.H., Brown, A.G. and Schofield, C.J. Chemistry and biosynthesis of clavulanic acid and other clavams. *Natural Products Reports*, **14**, 309–333 (1997).

Non-ribosomal polypeptides

von Döhren, H., Keller, U., Vater, J. and Zocher, R, Multifunctional peptide synthetases. *Chemical Reviews*, **97**, 2675–2705 (1997).

Mankelow D.P. and Neilan, B.A. Non-ribosomal peptide antibiotics *Expert Opinion on Therapeutic Patents*, **10** (10), 1583–1591 (2000).

Marahiel, M.A., Stachelhaus, T. and Mootz, H.D. Modular peptide synthetases involved in nonribosomal peptide synthesis. *Chemical Reviews*, **97**, 2651–2673 (1997).

Schwarzer, D., Finking, R. and Marahiel, M.A. Nonribosomal peptides: from genes to products. *Natural Products Reports*, **20**, 275–287 (2003).

Daptomycin

Baltz, R.H., Miao, V. and Wrigley, S. K. Natural products to drugs: daptomycin and related lipopeptide antibiotics. *Natural Products Reports*, **22**, 717–741 (2005).

Baltz, R.H. Biosynthesis and genetic engineering of lipopeptide antibiotics related to Daptomycin. *Current Topics in Medicinal Chemistry*, **8**, 618–638 (2008).

Schriever, C.A., Fernández, C., Rodvold, K.A. and Danziger, L.H. Daptomycin: A novel cyclic lipopeptide antimicrobial. *American Journal of Health-System Pharmacy*, **62** (11), 1145–1158 (2005).

Echinocandins

Bills, G.F., Platas, G., Peláez, F. and Masurekar, P. Reclassification of a pneumocandin-producing anamorph, *Glarea lozoyensis* gen. et sp. Nov., previously identified as *Zalerion arboricola*. *Mycological Researchs*, **103** (2), 179–192 (1999).

Denning, D.W. Echinocandin antifungal drugs. *The Lancet*, **362**, 1142–1151 (2003).

Hensens, O.D., Liesch, J.M., Zink, D.L., Smith, J.L., Wichmann, C.F. and Schwarz, R.E. Pneumocandins from *Zalerion arboricola*. III. Structure elucidation. *The Journal of Antibiotics*, **45** (12), 1875–1885 (1992).

Iwamoto, T., Fujie, A., Sakamoto, K., Tsurumi, Y., Shigematsu, N., Yamashita, M., Hashimoto, S., Okuhara, M. and Kohsaka, M. WF11899A, B and C, novel antifungal lipopeptides. 1. Taxonomy, fermentation, isolation and physico-chemical properties. *Journal of Antibiotics*, **47** (10), 1084–1091 (1994).

Keller-Juslén, C., Kuhn, M., Loosli, H.R., Petcher, T.J., Weber, H.P. and von Wartburg, A. Struktur des cyclopeptid-antibiotikums SL 7810 (= echinocandin B). *Tetrahedron Letters*, **46**, 4147–4150 (1976).

Cyclosporin A

Offenzeller, M., Su, Z., Santer, G., Moser, H., Traber, R., Memmert, K. and Schneider-Scherzer, E. Biosynthesis of the unusual amino acid (4*R*)-4-[(E)-2-butenyl]-4-methyl-L-threonine of cyclosporin A. *The Journal of Biological Chemistry*, **268** (35), 26127–26134 (1993).

Vancomycin

Nicolaou, K.C., Boddy, N.C., Bräse, S. and Winssinger, N. Chemistry, biology, and medicine of the glycopeptide antibiotics. *Angewandte Chemie. Int. Ed.*, **38**, 2096–2152 (1999).

Williams, D.H. The glycopeptide story – How to kill the deadly "superbugs". *Natural Products Reports*, **13**, 469–477 (1996).

Williams, D.H. and Bardsley, B. The vancomycin group of antibiotics and the fight against resistant bacteria. *Angewandte Chemie. Int. Ed.*, **38,** 1172–1193 (1999).

Streptogramin antibiotics

de Crécy-Lagard, V., Blanc, V., Gil, P., Naudin, L., Lorenzon, S., Famechon, A., Bamas-Jacques, N., Crouzet, J. and Thibaut, D. Pristinamycin I biosynthesis in *Streptomyces pristinaespiralis*: Molecular characterization of the first two structural peptide synthetase genes. *Journal of Bacteriology*, **179** (3), 705–713 (1997).

de Crécy-Lagard, V., Saurin, W., Thibaut, D., Gil, P., Naudin, L., Crouzet, J. and Blanc, V. Streptogramin B biosynthesis in *Streptomyces pristinaespiralis* and *Streptomyces virginiae*: Molecular characterization of the last structural peptide synthetase gene. *Antimicrobial Agents and Chemotherapy*, **41** (9), 1904–1909 (1997).

Mukhtar, T.A. and Wright, G.D. Streptogramins, oxazolidones and other inhibitors of bacterial protein synthesis. *Chemical Reviews*, **105**, 529–542 (2005).

Cyanogenic glycosides

Bak, S., Kahn, R.A., Nielsen, H.L., Möller, B.L. and Halkier, B.A. Cloning of three A-type cytochromes P450, CYP71E1, CYP98, and CYP99 from *Sorghum bicolor* (L.) Moench by a PCR approach and identification by expression in *Escherichia coli* of CYP71E1 as a multifunctional cytochrome P450 in the biosynthesis of the cyanogenic glucoside dhurrin. *Plant Molecular Biology*, **36**, 393–405 (1998).

Kahn, A.R., Bak, S., Svendsen, I., Halkier, B.A. and Möller, B.L. Isolation and reconstitution of cytochrome P450ox and in vitro reconstitution of the entire biosynthetic pathway of the cyanogenic glucoside dhurrin from sorghum. *Plant Physiology*, **115**, 1661–1670 (1997).

Vetter, J. Plant cyanogenic glycosides. *Toxicon*, **38**, 11–36 (2000)

Pantothenic acid

Chakauya, E., Coxon, K.M., Whitney, H.M., Ashurst, J.L., Abell, C. and Smith, A.G. Pantothenate biosynthesis in higher plants: advances and challenges. *Physiologia Plantarum*, **126**, 319–329 (2006).

Webb, M.E., Smith, A.G. and Abell, C. Biosynthesis of pantothenate. *Natural Product Reports*, **21** (6), 695–721 (2004).

Pyridoxalphosphate

Fitzpatrick, T.B., Amrhein, N., Kappes, B., Macheroux, P., Tews, I. and Raschle, T. Two independent routes of *de novo* vitamin B_6 biosynthesis: not that different after all. *Biochemical Journal*, **407**, 1–13 (2007).

Pyridoxal biosynthesis. www.chem.qmul.ac.uk/iubmb/enzyme/reaction/misc/pyridoxal.html.

Roje, S., Vitamin B biosynthesis in plants. *Phytochemistry*, **68**, 1904–1921 (2007).

Glucosinolates

Mithen, R. Glucosinolates – biochemistry, genetics and biological activity. *Plant Growth Regulation*, **34**, 91–103 (2001).

Thangstad, O.P., Winge, P., Husebye, H. and Bones, A. The myrosinase (thioglucoside glucohydrolase) gene family in *Brassicaceae*. *Plant Molecular Biology*, **23**, 511–524 (1993).

Biotin

Schneider, G. and Lindqvist, Y. Structural enzymology of biotin biosynthesis. *FEBS Letters*. **495**, 7–11 (2001).

Vitamin B_{12}

Battersby, A.R. Biosynthesis of vitamin B_{12}. *Accounts of Chemical Research*, **26**, 15–21 (1993).

Battersby, A.R. Tetrapyrroles: The pigments of life. *Natural Products Reports*, **17**, 507–526 (2000).

Crouzet, J., Levy-Schil, S., Cameron, B., Cauchois, L., Rigault, S., Rouyez, M.C., Blanche, F., Debussche, L. and Thibaut, D. Nucleotide sequence and genetic analysis of a 13.1 kilobase pair *Pseudomonas denitrificans* DNA fragment containing five *cob* genes and identification of structural genes encoding cob(I)alamin adenosyltransferase, cobyric acid synthase, and bifunctional cobinamide kinase-cobinamide phosphate guanylyltransferase. *Journal of Bacteriology*, **173** (19), 6074–6087 (1991).

Lingens, B., Schild, T.A., Vogler, B. and Renz, P. Biosynthesis of vitamin B12. Transformation of riboflavin 2H-labeled in the 1´R position or 1´S position into 5,6-dimethylbenzimidazole. *European Journal of Biochemistry*, **207**, 981–985 (1992).

Raux, E., Schubert, H.L. and Warren, M.J. Biosynthesis of cobalamin (vitamin B12): a bacterial conundrum. *CMLS, Cellular and Molecular Life Sciences*, **57**, 1880–1893 (2000).

Roessner, C.A. and Scott, A.I. Genetically engineered synthesis of natural products: from alkaloids to corrins. *Annual Review of Microbiology*, **50**, 467–490 (1996).

Scott, A.I. How Nature synthesizes vitamin B_{12} – a survey of the last four billion years. *Angewandte Chemie. International Edition in English*, **32** (9), 1223–1243 (1993).

Warren, M.J., Raux, E.R., Schubert, H.L. and Escalante-Semerena, J.C. The biosynthesis of adenosylcobalamin (vitamin B_{12}. *Natural Products Reports*, **19**, 390–412 (2002)

10. Alkaloids

General aspects

Alkaloids are formed from amino acids, but other precursors, e.g. terpenes or steroids, are often also built into the final alkaloid skeleton. For this reason, the biosynthesis of alkaloids cannot be accommodated in any one of the main biosynthetic pathways described earlier, but frequently involves a combination of products from two or more of these routes. We shall, therefore, classify the alkaloids by the ring systems that constitute the main part of their structures, treating the biosynthesis of these ring systems in connection with each group of alkaloids.

What precisely is meant by the term alkaloid is not easy to define, but the term is normally used for nitrogenous plant constituents with basic properties and having physiological effects in animals or man. There are many borderline cases, however. Colchicine, for example, is regarded as an alkaloid, although it is not basic, and a number of nitrogenous, physiologically active compounds of animal origin have also been termed alkaloids. To avoid these problems, the following definition has been suggested[1]: *"An alkaloid is a cyclic organic compound containing nitrogen in a negative oxidation state which is of limited distribution among living organisms"*. This definition evidently encompasses all compounds which have hitherto been regarded as alkaloids, but excludes such nitrogenous compounds as simple amines, amino acids, peptides, proteins, nucleic acids, nucleotides, porphyrins, vitamins and nitro and nitroso compounds.

Occurrence in the plant kingdom

Alkaloids are not found in all plant families. They are relatively uncommon in bacteria, algae, fungi and lichens, and they are not often encountered in ferns and conifers. According to a review by Cordell *et al.* (see *Further reading* below) the total number of alkaloids found in monocots, dicots and gymnosperms is 21 120. Alkaloids are distributed in 7 231 species of higher plants in 1 730 genera (approx. 14.2 %) within 186 plant families. In 35 families alkaloids have been detected but not isolated and in another 20 plant families no alkaloids have as yet been detected. Among the monocotyledons, the families *Amaryllidaceae*, *Liliaceae* and *Poaceae* are rich in alkaloids (based on the number of isolated alkaloids) and

---

[1] S. William Pelletier. The Nature and Definition of an Alkaloid. In S. William Pelletier (Ed.), *Alkaloids. Chemical and Biological Perspectives*. John Wiley, New York, pp. 1–31, 1983.

many orchids have now been found to contain alkaloids, although earlier regarded as being devoid of this type of compound. Most alkaloids occur in dicotyledons, especially in the following families: *Annonaceae, Apocynaceae, Asteraceae, Berberidaceae, Boraginaceae, Buxaceae, Celastraceae, Fabaceae, Lauraceae, Loganiaceae, Menispermaceae, Papaveraceae, Piperaceae, Ranunculaceae, Rubiaceae, Rutaceae* and *Solanaceae*. The highest number of alkaloids have been isolated and characterized from plants of the following five families (the figures within parentheses are the number of alkaloids isolated): *Apocynaceae s. lat.* (2 844), *Rutaceae* (1 730), *Ranunculaceae* (1 559), *Fabaceae* (1 452) and *Papaveraceae* (1 309).

In some large families, such as the *Lamiaceae* and *Rosaceae*, only a limited number of species have been found to contain alkaloids. However, a large number of plants remain to be investigated, and the list just given may well have to be extended in future.

Location in the plant

Alkaloids may occur in every part of a plant, but usually one or more organs have a higher content than others. Barks, leaves and fruits are often rich in alkaloids, which may occur in particular tissues or cells. Thus, opium alkaloids are encountered specifically in the latex vessels of the opium poppy, and the tropane alkaloids of *Datura* species are contained in the leaf petiole and in the tissue adjacent to the midrib of the leaf. In the individual cell, the alkaloids usually occur in the vacuoles and not in the protoplasm or the cell wall.

The organ with the highest alkaloid content is not necessarily the place where the alkaloids are formed. An active transport of alkaloids from one organ to another has been observed in many plants. Such processes can be studied by means of grafting experiments and plants belonging to the family *Solanaceae* have received particular attention: The aerial parts of a tomato plant were grafted on to the root of the deadly nightshade (*Atropa belladonna*), a plant producing tropane alkaloids. After some time, tropane alkaloids could be isolated from the tomato graft, where they do not normally occur. Conversely, when the aerial parts of *Atropa* were grafted on to the root of the tomato, the tropane alkaloids gradually disappeared from the *Atropa* graft. These experiments show that the tropane alkaloids of *A. belladonna* are formed in the root and transported to the leaves for storage. Analogous experiments have demonstrated that the reverse is true for lupin alkaloids. They are synthesized in the stem and then transported to the roots. Further work, using radioactively labelled precursors of the alkaloids, indicates that the situation can be still more complicated. Thus, most of the alkaloid nicotine is formed in the root of the tobacco plant and transferred to the leaves, but a small amount is synthesized in the stem. The alkaloid anabasine, an isomer of nicotine, is formed as a secondary alkaloid in the roots of the tobacco species *Nicotiana tabacum* and *N. rustica,* but as the main alkaloid in the stem of *N. glauca*.

The transport of alkaloids synthesized in the roots seems to take place mainly via the vessels, because alkaloids have been found both in the vessels and in the fluid dripping from leaf tips when water transport in the plant is very active. Less is known regarding the transport of those alkaloids that are formed in leaves and stems. Transport via the phloem would seem likely, though alkaloids have never been found there.

Role in the plant

There has been much speculation about the possible advantage that a plant may derive from its alkaloid content, but little is actually known about this. Moreover, the same is true about most other secondary metabolites in plants. The term secondary metabolites is used to denote substances that are formed in plants but that do not participate in those metabolic processes which are necessary for the life and development of the plant. Most of the plant constituents that are used medicinally are secondary metabolites.

As regards alkaloids, attention has been called to the fact that the presence of alkaloids may protect plants from being eaten by grazing cattle. Most alkaloids have a bitter taste and it has been shown that animals learn by experience to avoid these plants. In a heavily grazed area those plants which contain alkaloids will often be found to be untouched, whilst almost everything else has been consumed. However, it is impossible to tell cause from effect in this case. The plants containing alkaloids survive in an overgrazed area because they contain alkaloids, but this is obviously not the reason why the plant has developed resources for alkaloid synthesis in the first place, as these resources were developed many millions of years before the evolutionary arrival of cattle. If the presence of alkaloids, or other highly specialized groups of compounds such as sesquiterpenelactones or iridoids, were a determining factor in the struggle for existence, plants void of these resources would long ago have been overwhelmed by those employing them. Applying an evolutionary perspective on the occurrence of many of these secondary metabolites, there are obvious co-evolutionary patterns between plants and grazers, a continuing race of arms where the "decision" to explore a new set of "weapons" is at the same time a possibility and a risk. The possibility is to gain an advantage, the risk implies spending too many resources in a field where the opposition already have developed their countermeasures.

Other hypotheses propose that alkaloids are substances formed when the organism is detoxifying poisonous substances arising from its normal metabolism, or that alkaloids might constitute reserve nutrition for the plant, storing nitrogen for example. However, these hypotheses hardly seem to be compatible with the complicated biochemical systems which plants have developed for the synthesis of alkaloids.

Although nothing definite can be said about the importance of secondary metabolites for plants, most scientists seem to accept the idea that they have a role to play as defence substances and hence are of importance for the ability of plants to survive and thus to compete in evolutionary terms with other plants.

Detection and isolation of alkaloids

Most alkaloids are present in the plant as salts. As such they are usually relatively soluble in water or aqueous ethanol but not in organic solvents like chloroform, ether or toluene. As bases their solubility is the opposite. Alkaloid bases are insoluble in water, moderately soluble in ethanol but easily soluble in organic solvents. This difference in solubility is utilized both for detection of alkaloids in plants and for their isolation. A test for the presence of alkaloids in a plant comprises making an aqueous or ethanolic extract, adding a base such as ammonia or NaOH and extracting with an immiscible organic solvent like chloroform. Following evaporation of the solvent the residue is dissolved in a dilute acid and a few drops of an alkaloid-precipitating reagent like *Mayer's reagent* (potassiomercuric iodide solution) is added. Formation of a precipitate indicates the presence of alkaloids in the extract. It is also possible to spot the organic extract on a thin-layer plate, which after development in a suitable solvent mixture is dried and sprayed with *Dragendorff's reagent* (solution of potassium bismuth iodide). Alkaloids are detected as orange spots on the plate. Purine bases like caffeine are not precipitated by the alkaloid-precipitating reagents. They must be detected by other means such as the *murexide test* in which the base is mixed with a very small amount of potassium chlorate and a drop of hydrochloric acid and the mixture evaporated to dryness. The residue is then exposed to ammonia vapour, which produces a purple colour.

Alkaloids containing a quaternary nitrogen can usually not be extracted with organic solvents from an aqueous solution even at alkaline pH. Such alkaloids cannot be detected by the above described procedure, but must be directly precipitated from the aqueous or ethanolic extract with Mayers' reagent or ammonium reineckate. It should, however, be observed that false results can be obtained if the extract contains proteins which also give a precipitate. Care must therefore be taken to remove proteins, e.g. by heat denaturation, before performing the test. Most plant extracts contain choline which is also precipitated by these reagents.

That alkaloids as a group are relatively easy to extract and purify is undoubtedly one of the reasons why alkaloids have been studied more than other groups of natural products that can be isolated from plants. Isolation of alkaloids (except those with quaternary nitrogen) is performed according to the same principle as described above for detection of the compounds: The powdered plant material is moistened with water and mixed with calcium hydroxide (slaked lime), which decomposes the alkaloid salts in the plant material and converts the alkaloids to the free bases. At the same time, the lime combines with organic acids, tannins and other unwanted constituents. The alkaloids are then extracted from the mixture with an organic solvent. The extract is purified by shaking with dilute acid, into which the alkaloids pass in the form of salts. Further purification can be achieved by addition of excess alkali to the aqueous solution and renewed extraction with an organic solvent. The extract leaves a crude alkaloid mixture on evaporation of the solvent. Ammonia or sodium carbonate can be used instead of slaked lime in this procedure. The choice of organic solvent is influenced by several factors. Good solvents for alkaloids are, for example, chloroform, ether, ethyl acetate, benzene, toluene and light petroleum. For industrial extractions, the price of the solvent and the possibility of recovering it are important factors. Chloroform is an excellent solvent for most alkaloids, but it also dissolves many unwanted compounds and is relatively expensive and like other chlorinated solvents it is also a hazard to human health. Hence, it is often preferable to choose a less efficient solvent and compensate by applying a continuous cyclic extraction process: The basified plant material is extracted with the organic solvent. The extract is treated with dilute acid (often in counter-current), whereby the alkaloids pass into the aqueous phase. The solvent is returned to the raw material and dissolves a further portion of alkaloid, and so on. In this manner, a large amount of alkaloid can be extracted with a small quantity of solvent.

The next step is to separate the individual components of the crude alkaloid mixture obtained. This can be very difficult, because the mixture is often very complex. Working on an industrial scale, the main constituents can sometimes be isolated by fractional crystallization of either the bases or their salts. Frequently, however, it is necessary to resort to chromatographic separation methods.

Besides alkaloids with a quaternary nitrogen, phenolic alkaloids and alkaloid N-oxides cannot be isolated by the standard procedure. At high pH phenolic alkaloids are present as phenolates which cannot be extracted with organic solvents. Following their liberation from the salts by addition of a strong base, the pH of the extract must therefore be lowered to a value at which the phenolates are broken, before extraction with an organic solvent can be undertaken. Alkaloid N-oxides are more polar than the corresponding alkaloids and are therefore more difficult to extract with an organic solvent. They can be extracted with ethyl acetate or chloroform but not with less polar solvents like diethyl ether.

Alkaloids in medicine and their potential in drug discovery

Most alkaloids have pharmacological effects. They display a wide range of biological responses and can have extremely high potency (in the nM range). Many alkaloids are in current use as drugs, among them very important remedies such as the painkiller morphine, and the cancer remedies vincristine and vinblastine. Some alkaloids are potent toxins, for example aconitine and the pyrrolizidine alkaloids, and some have an illicit use as narcotics, for example cocaine and heroin. Plant alkaloids comprise about 15.6 % of the known natural products but they constitute almost 50 % of the plant-derived natural products of pharmaceutical and biological significance. Alkaloids represent a vast reservoir for development of new drugs but there has been very little cooperation between pharmacologists and chemists, the latter being mainly interested in the isolation and structure determination of alkaloids. Thus many alkaloids have been isolated and their structure determined but they have never been subjected to investigation of their biological and pharmacological properties. According to the previously (page 604) cited review by Cordell *et al.*, 16 132 (= 76.4 %) of the hitherto isolated alkaloids have not been tested for pharmacological effects in any *in vitro* or *in vivo* bioassay; 2 291 have been tested in one single assay, 1 995 in 2–5 assays and only 167 have been subjected to 20 or more assays. Among these 167 most frequently tested alkaloids over one-third (35.9 %) are pharmaceutically significant. Compared with acetogenins, terpenoids and phenylpropanoids, alkaloids show more substantial skeletal and functional group diversity. There are 1 872 different structural skeletons in the total of 21 120 known alkaloids. The total number of plant-derived natural products is 135 500 and there are about 5 750 different structural skeletons among these. Thus alkaloids, which represent only 15.6 % of the total number of known compounds from plants, possess 32.5 % of the total number of skeletons. The range of structures of alkaloids, from linear chains to planar, multi-ring systems, to globular molecules, represents a vast amount of conformational rigidity and flexibility. Most alkaloids have single or multiple chiral centres and they rarely occur in racemic mixtures but are usually obtained with a high degree of optical purity of one enantiomeric form. There are thus many possibilities for interactions with alkaloids and enzymes and a renewed focus on alkaloids and alkaloid-containing plants might well give a substantial pay-off in the drug discovery process.

Amino alkaloids

The nitrogen atom of an amino alkaloid is located in an amino group and is not a member of a heterocyclic ring, a common feature in other alkaloids.

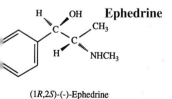

(1*R*,2*S*)-(-)-Ephedrine

Ephedrine

Ephedrine is obtained from *Ephedra sinica* and related species. The natural product is L-ephedrine, which can also be synthesized from 1-hydroxy-1-phenylacetone and methylamine. 1-Hydroxy-1-phenylacetone is prepared by fermentation of benzaldehyde with brewer's yeast.

Ephedrine is a sympathomimetic, which means that its effects

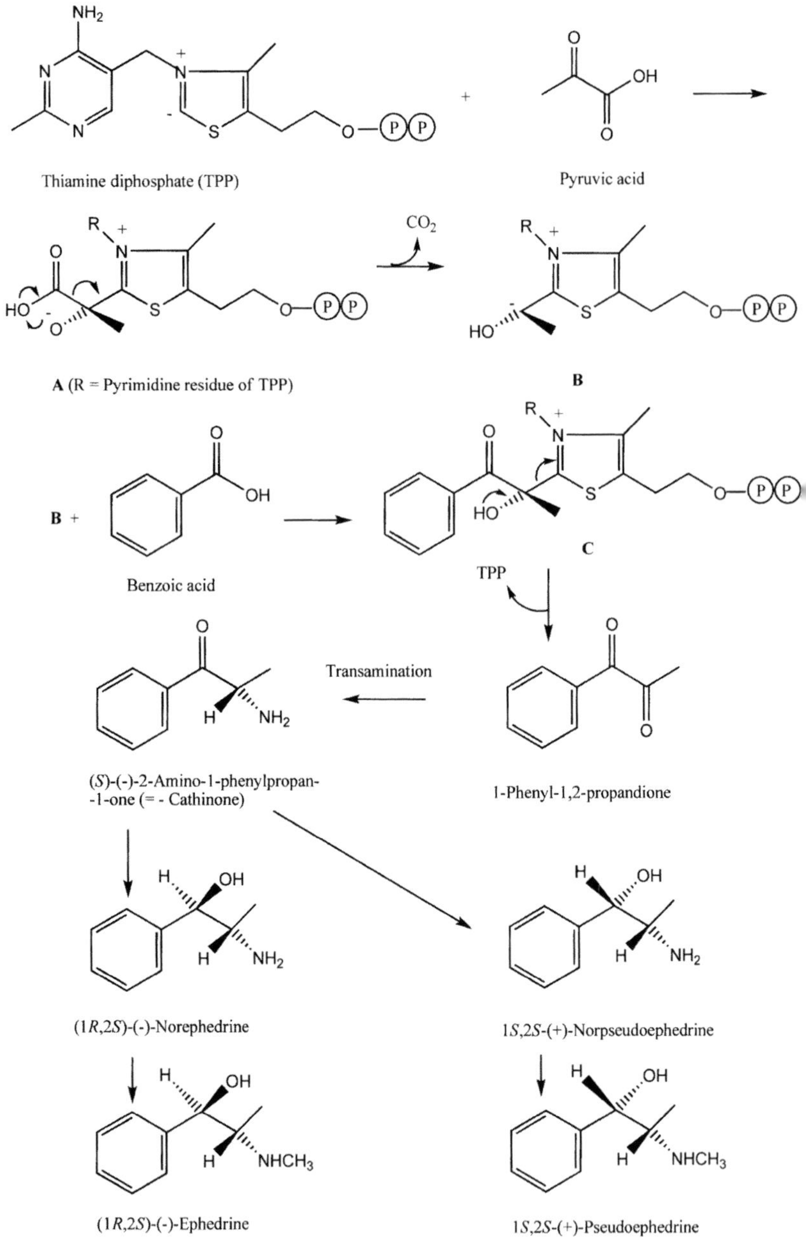

Fig. 192. Biosynthesis of ephedrine and pseudoephedrine. R denotes the pyrimidine residue of thiamine diphosphate

are similar to those which arise on stimulation of sympathetic nerves. Ephedrine causes increased blood pressure and pulse, contraction of blood vessels, and dilation of the bronchi. It also stimulates the central nervous system in a way similar to that of amphetamine, but less strongly. Ephedrine sulphate and chloride are used as bronchodilators in the treatment of asthma and colds.

Biosynthesis Biosynthetically, ephedrine is derived from phenylalanine, which contributes the benzene ring and the adjacent (the first) carbon atom of the side-chain. The second and third carbon atoms of the side-chain come from pyruvate, and nitrogen is introduced by a transamination reaction at a late stage of the biosynthesis, which is presented in Fig. 192. The incorporation of pyruvate is mediated by thiamine diphosphate, the thiazole ring of which has an acidic hydrogen and is thus capable of yielding a carbanion which acts as a nucleophile towards the carbonyl group of pyruvic acid yielding compound A. Decarboxylation of A results in a product (B) which reacts with benzoic acid (derived from phenylalanine) resulting in compound C. Splitting off the thiamine residue results in formation of 1-phenyl-1,2-propandione, which by transamination yields cathinone, a compound which is present in minor amounts in *Ephedra* species, but which is one of the main alkaloids in khat from *Catha edulis* (see below). Reduction of cathinone yields norephedrine and norpseudoephedrine which are *N*-methylated to ephedrine and pseudoephedrine. The methyl group emanates from methionine. The nor-alkaloids and pseudoephedrine have been isolated from *Ephedra sinica* and other *Ephedra* species.

Crude drugs **Ephedra** (Ephedrae herba) consists of the dried aerial parts of *Ephedra sinica* Stapf (*Gnetaceae*). The plant is a shrub, 60–90 cm high, with scale-like leaves that clasp the stem. *E. sinica* comes from China, where it has been used for thousands of years under the name "ma huang", among others, for treatment of the symptoms of the common cold. The main sources of the plant are nowadays India and Pakistan. The drug is collected in the autumn when the alkaloid content, which can vary between 0.5 and 2 %, is at its highest; 30–90 % of the alkaloid mixture consists of ephedrine.

Another constituent is pseudoephedrine, a stereoisomer of ephedrine. Pseudoephedrine has α-adrenergic activity and is used to treat rhinitis of allergic or vasomotor origin, because of its ability to decrease swelling of the mucous membranes.

Other *Ephedra* species containing ephedrine are *E. equisetina* Bunge and *E. distachya* L. (sometimes also known as *E. gerardiana* Wall.) growing in southern France, India and Pakistan and *E. nebrodensis* Tieno (previously known as *E. major* Host), a Spanish species. Ephedra is used exclusively for the extraction of ephedrine.

Khat or **Abyssinian tea** is the leaf of *Catha edulis* Forssk. (*Celastraceae*), a shrub or small tree native to East Africa. This plant is cultivated in the highlands of Yemen, Ethiopia and Kenya. It grows also on Madagascar and in South Africa. The leaves are either chewed or used for preparing tea, which is drunk as a stimulant. Fresh leaves are preferred to dried ones. Khat counteracts fatigue, facilitates strenuous muscular work, and causes a light elation with talkativeness and sociability. Particularly in Yemen, the chewing of khat is a socially important event which is often practised in a ceremonial fashion, predominantly by men. Khat-chewing by women is much less common and is also less formal. The chewers meet in some private house where the khat session takes place in a specially arranged room. The participants arrange their sitting according to their social status. Three phases of the ceremony can be recognized. In the first part each participant chews about 100–200 g of fresh leaves. The leaves are chewed one by one, the juice is swallowed and the leaf residues are retained in a large quid inside the cheek, which is stretched considerably, finally reaching the size of a tennis-ball in experienced chewers. After about half an hour the stimulant phase appears during which the participants become elated and talkative. Social and personal themes are discussed and minor conflicts are conciliated in a relaxed and happy atmosphere. This status lasts for about two hours following which silence prevails and each participant feels as if she is the centre of the world. During the third phase reality returns and a feeling of unrest appears which is counteracted by the drinking of black tea.

The effects described are due to the cathinone ((S)-$(-)$-2-amino-1-phenylpropan-1-one) present in the leaves. Fresh or well preserved leaves contain 0.05–0.15 % cathinone. The active constituent was earlier believed to be cathine ($(1S,2S)$-$(+)$-norpseudoephedrine), but this substance is probably a transformation product of cathinone. The effect of cathinone is similar to that of amphetamine (note the structural similarity!). The leaves also contain small amounts of a

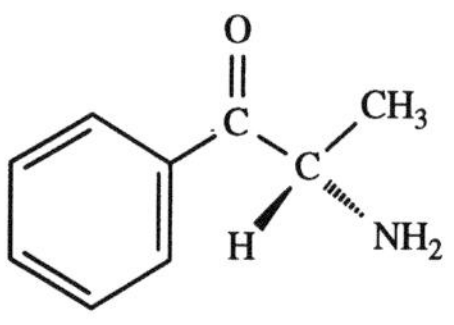

(-)-Cathinone

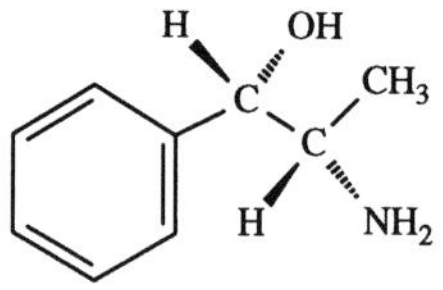

(+)-Cathine

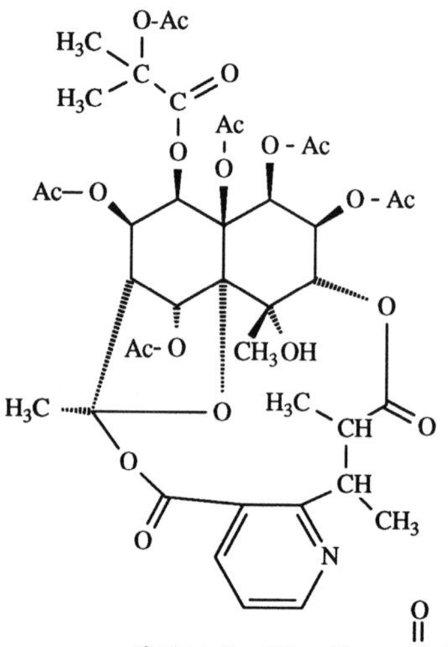

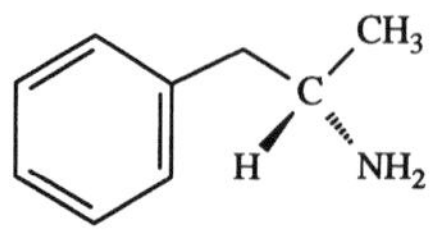

(+)-Amphetamine

Catheduline K1   (Ac = - C - CH$_3$)

large number of alkaloids – cathedulines – with complex structures. Their possible contribution to the effects of khat is not yet known.

Although originally restricted to Arabia and East Africa, khat-chewing has now spread also to many other countries, often introduced by refugees from the civil wars in Ethiopia and Somalia. Air transports make the fresh drug available. The chewing of khat has social consequences. The drug is expensive and late khat sessions make it difficult to work the following day. Although abstinence symptoms are not particularly severe, some khat-chewers develop a craving for the stimulant causing them to spend much time and money in procuring the drug. As khat also produces anorexia these people often do not eat enough food thus achieving a deficient nutritional state which favours infections. In Europe the legal status of khat differs. The drug is prohibited in Sweden, France and Switzerland while it is tolerated in Great Britain and the Netherlands. Although the effects of khat are very much the same as those of amphetamine, there is a security factor in that the maximal dose that can be ingested is determined by the volume of the material. Therefore, because cathinone is metabolized rather rapidly, there is probably a certain limit to the plasma concentrations that can be attained by chewing the leaves.

Mescaline

Mescaline is a hallucinogenic alkaloid which is obtained from the cactus *Lophophora williamsii* (*Cactaceae*). It also occurs in *Trichocereus pachanoi* and in other *Trichocereus* species (*Cactaceae*). Besides mescaline, these cacti contain other phenethylamines as well as tetrahydroisoquinoline derivatives, e.g. anhalonidine.

Oral doses of 100–400 mg of mescaline cause euphoria, changes in the conception of time, and brilliant, coloured hallucinations. Mescaline has been used to induce model psychoses in studying the aetiology of mental diseases.

Mescaline

Biosynthesis

Biosynthetically, these alkaloids come from the amino acid tyrosine (1) (Fig. 193). Decarboxylation to tyramine (2) is followed by oxidation to dopamine (3), which is then methylated to compound (4). Oxidation to 3,4-dihydroxy-5-methoxyphenethylamine (5) is the next step. This substance is the precursor of both mescaline and the tetrahydroisoquinoline alkaloids present in *L. williamsii*. The next methylation step determines the type of alkaloid formed: mescaline is produced if the hydroxyl group in position 3 is methylated first (6, 7); tetrahydroisoquinoline alkaloids are formed if the hydroxyl group in position 4 is the first to be methylated (8, 9, 10).

The one- and two-carbon fragments needed for the ring closures to form compounds (9) and (10) derive from glyoxalate and pyruvate, respectively.

Anhalonidine

Crude drug

Mescal buttons are dried slices of the trunk and tap-root of the peyote cactus *Lophophora williamsii* Coult. (conserved against the older name *Anhalonium williamsii* Lem) (*Cactaceae*), which grows in

Plate 19.
Lophophora williamsii Coult. The stem is sliced to yield mescal buttons which contain the hallucinogenic alkaloid mescaline (page 613). Photograph by Gunnar Samuelsson.

Plate 20.
Amanita muscaria (L.) Pers. Fly agaric. This mushroom contains hallucinogenic compounds such as ibotenic acid and muscimole (page 616). Photograph by Gunnar Samuelsson.

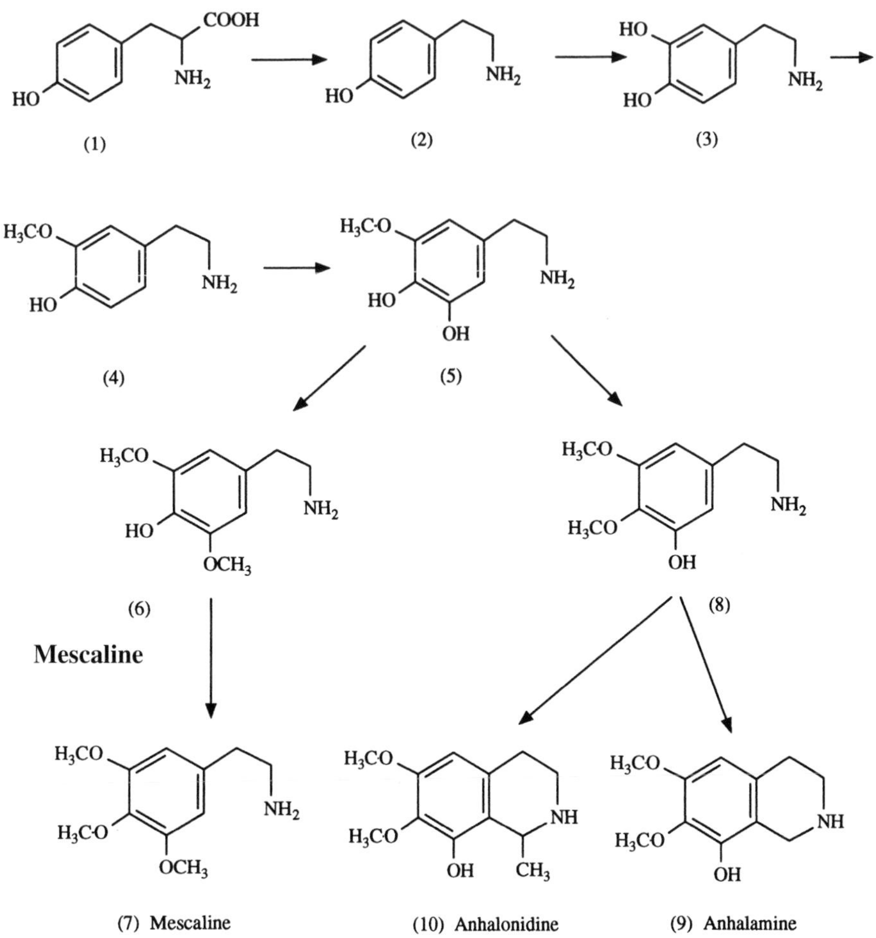

Fig. 193. Biosynthesis of mescaline and tetrahydroisoquinoline alkaloids

Southern United States and Northern Mexico (Plate 19, page 614). This is a small cactus, consisting essentially of trunk extending into a tap-root, in all 6–20 cm high and 4–12 cm in diameter. The centre, as seen from above, is rounded to concave. The colour is bluish green, mainly due to the thick wax cuticula, and the top of the plant has white tufts of hair. In the literature, this cactus has a confused taxonomy and has been described under various names, e.g. *Anhalonium williamsii* (Salm-Dyck) Rümpler, *A. lewinii* Hennings and *L. lewinii* Rusby. In areas where this plant grows, Indians have long used it as a means of inducing hallucinations during religious ceremonies.

Trichocereus pachanoi Britton et Rose (*Cactaceae*) is a columnar cactus which is used by the Indians of Peru to produce the hallucinogenic drink "cimora".

$$HO - \bigcirc - CH_2 - NH - \overset{\overset{\textstyle O}{\|}}{C} - (CH_2)_4 - CH = CH - CH \overset{\textstyle CH_3}{\underset{\textstyle CH_3}{<}}$$

$$OCH_3$$

Capsaicin

Capsaicin   Capsaicin is a pungent substance occurring in the fruits of certain *Capsicum* species (*Solanaceae*) which causes irritation of the skin. It can be regarded as the amide of vanillylamine (4-hydroxy-3-methoxybenzylamine) with an unsaturated, branched C_{10} fatty acid. Biosynthetically, the vanillylamine moiety originates from phenyl-alanine. In the fatty-acid part, the four terminal carbon atoms come from the amino acid valine and the other six from three acetate residues.

Capsaicin is used as a rubefacient in liniments for the sympto-matic treatment of rheumatic pains. It increases the blood circula-tion in the skin, giving a feeling of warmth, but in contrast to can-tharidin, for example, it does not give rise to blisters (page 412). A cream containing capsaicin can be used to counteract postherpetic neuralgia, i.e. pains which remain after all other symptoms of shing-les have disappeared. The mechanism for this effect is that capsaicin affects the pain receptors, making them less sensitive.

Crude drug   **Capsicum, chilli** (Capsici fructus) is the dried fruit of *Capsicum annuum* L. var. *longum* Sendter (*Solanaceae*). This is an annual herb from the tropical parts of America, now cultivated in all tropi-cal and subtropical countries. The capsaicin is located in the fruit wall, especially in the ridges carrying the seeds. There are many varieties of *C. annuum,* with widely different capsaicin contents. Strains which lack capsaicin, but which have a high content of ascorbic acid (vitamin C), are used as a vegetable under the name of sweet pepper. The variant *longum* contains 0.1–0.5 % capsaicin, together with a number of other structurally related pungent com-pounds, which differ as regards chain length, branching, and the presence or absence of a double bond in the fatty-acid moiety.

Capsicum is used as a spice and for the extraction of capsaicin.

Cayenne pepper (Capsici fructus acer) is the dried fruit of *Capsicum frutescens* L. (*Solanaceae*), a perennial shrub with much smaller fruits than *C. annuum* var. *longum.* It is cultivated in Africa, India, America and Japan. Cayenne pepper has a higher capsaicin content (0.6–0.9 %) than chillies, but is likewise used as a spice and for the extraction of capsaicin.

Muscarine, ibotenic acid,   These compounds are found in the toadstool or fly agaric, *Amanita*
muscimole and muscazone   *muscaria* (L) Pers. (Plate 20, page 614). Muscarine is the only one that belongs to the amino alkaloids. The others contain another nitrogen atom, located in an isoxazole or oxazole ring. Muscimole

N(CH₃)₃

(+)-Muscarine

COOH
H₂N – CH

OH

Ibotenic acid

NH₂
CH₂

OH

Muscimole

COOH
H₂N – CH

NH

Muscazone

is formed through decarboxylation of ibotenic acid. Both ibotenic acid and muscazone can also be regarded as amino acids.

In certain remote parts of eastern Siberia the inhabitants have used the fly agaric as an intoxicant. The mushroom was usually eaten fresh or partially dried. The effects appear after 15–20 minutes. A small dose (1–4 mushrooms) causes dizziness, nausea, sleepiness, a feeling of weightlessness, and coloured hallucinations. A larger dose (5–10 mushrooms) gives more pronounced symptoms of poisoning, with spasms and more vivid hallucinations. A higher dose than 10 mushrooms may be lethal. The hallucinogens are excreted in the urine, so that several people could intoxicate themselves with the same lot of mushrooms by drinking each other's urine.

Of the four constituents mentioned, muscarine has a strong cholinergic but no psychotropic effect. The muscarine content of the fly agaric is low (2–3 mg per kg of fresh mushrooms), so that muscarine is not involved in producing the symptoms of poisoning just described. In contrast, muscarine is responsible for poisoning by certain species of *Inocybe*, which may contain up to ca. 1 % of muscarine. This compound also occurs in some representatives of the genus *Clitocybe*. The symptoms of muscarine poisoning appear 15–30 minutes after ingestion and include increased salivation, perspiration, floods of tears, stomach pains, severe nausea and diarrhoea. The pulse is low, the breathing is asthmatic, and the pupils are contracted. Poisoning may lead to death through heart failure or respiratory paralysis, but this is uncommon. No psychotropic effects have been observed.

Ibotenic acid and muscimole are responsible for most of the symptoms described in cases of fly agaric poisoning. Muscimole has the strongest effect. Doses of 10–15 mg give an adult all the symptoms with the possible exception of hallucinations, which do not seem to be as typical as on poisoning with the intact mushroom. The effects of muscazone are similar to those of ibotenic acid and muscimole, but a good deal weaker. Possibly, other hallucinogenic constituents remain to be discovered in the fly agaric.

Colchicine

H
NH– C – CH₃
O

H₃CO

OCH₃

Colchicine is an alkaloid, which can be obtained from the autumn crocus (the meadow-saffron), *Colchicum autumnale*, and from *Gloriosa superba* (both of the family *Colchicaceae*). The molecule consists of an aromatic ring (A) with three methoxyl groups, a 7-membered ring (B) carrying an acetylated amino group, and a tropolone ring (C) whose hydroxyl group is methylated. As the nitrogen atom is part of an amide function, colchicine is non-basic.

In medicine, colchicine is used as a remedy against gout, a disease caused by the deposition of uric acid in the joints. Colchicine is highly poisonous, however, and the treatment must be carefully supervised. The substance inhibits cell division and is used in plant breeding (page 54) to produce polyploidy, as it does not prevent chromosome division. Colchicine also inhibits the division of ani-

Plate 21.
Colchicum autumnale L. Autumn crocus or meadow saffron. Colchicine (page 617) is extracted from the seeds and corms of this plant. The Botanical Garden, Uppsala. Photograph by Gunnar Samuelsson.

Plate 22.
Nicotiana tabacum L. The leaves are used for the production of tobacco (page 625). Photograph by Gunnar Samuelsson.

mal cells, but it is too poisonous to be used to arrest tumour growth. *Demecolcine* (Fig. 194, structure 1) has a wider margin of safety and is used for the treatment of chronic myelogenic leukaemia and malignant lymphomata. Both colchicine and demecolcine are microtubule inhibitors (page 219).

Biosynthesis   Biosynthetically, ring A is derived from phenylalanine, which also contributes carbon atoms 5, 6 and 7 of ring B. The tropolone ring is formed from tyrosine by ring expansion involving the β-carbon atom of tyrosine which becomes C-12 of the tropolone ring. The nitrogen atom in colchicine is also tyrosine-derived. The methyl groups of the four methoxyl substituents come from methionine or methanol. A phenethylisoquinoline derivative [(S)-autumnaline] is postulated to be an important intermediate in the biosynthesis, which is summarized in Fig. 194. Phenylalanine is deaminated and reduced to cinnamaldehyde which is oxidized to 4-hydroxycin-namaldehyde. Tyrosine is decarboxylated and oxidized to dopamine which is condensed with 4-hydroxycinnamaldehyde to yield compound A. Ring closure, oxidation and methylation reactions lead to (S)-autumnaline, which can be rewritten as indicated in the figure. (S)-autumnaline is transformed by ring closure, oxidation and methylation into O-methylandrocybine, in which one of the seven-membered rings of colchicine appears for the first time. This compound has been isolated from *Colchicum autumnale*. A tropolone ring is formed by expansion of ring C of O-methylandrocybine by incorporation of C-12 (originally the β-carbon atom of tyrosine). C-13 becomes the formyl group of the resulting compound – N-formyldemecolcine. This ring expansion is a unique process of considerable mechanistic interest and an attempt to explain it has been published (P. W, Sheldrake *et al. J. Chem. Soc. Perkin Trans.* I, 3003–3009 [1998]). Final conclusions, however, have to await elucidation of the enzymology involved.

In the following reactions N-formyldemecolcine loses the formyl group and yields demecolcine (1), which is oxidized to N-formyl-N-deacetylcolchicine (2). Loss of the formyl group produces deacetylcolchicine (3) which is acetylated to colchicine.

Crude drug   **Colchicum seed** (Colchici semen) is the seed of the autumn crocus or meadow-saffron, *Colchicum autumnale* L. (*Colchicaceae*) (Plate 21, page 618). This is a bulbous plant with an unusual flowering period, which has given rise to its name in various languages (German: "Zeitlose"). In the late summer, a young corm can be seen developing in an axil in the outer part of the old corm. In August or September, 2–6 flowers, which are 10–12 cm long, grow out from the young corm. They have six stamens and a perianth consisting of six lilac-coloured segments joined together to form a long tube. The superior ovary lies at the bottom of this tube. More than half the length of the flower is underground, so that the corm and the surrounding soil protect the fruit during the winter. The fruit, which is a three-lobed capsule, is carried up above ground in the spring by the growing leaves and can be harvested in July or August, when the seeds are ripe. During these events, the new corm grows at the cost of the old one, which gradually dies, but, before doing so, may in its second spring first produce a few more small daughter corms, thus providing for the vegetative propagation of the plant.

Colchicum autumnale is a European plant and colchicum seed is produced in Poland, Czechoslovakia, Yugoslavia and the Netherlands. They contain 0.6–1.2 % colchicine and smaller amounts of demecolcine and related alkaloids which differ essentially as regards the substituents to the aromatic ring. The crude drug, as well as the corm, are used for extraction of colchicine.

Colchicine is present in other species of *Colchicum* and in other

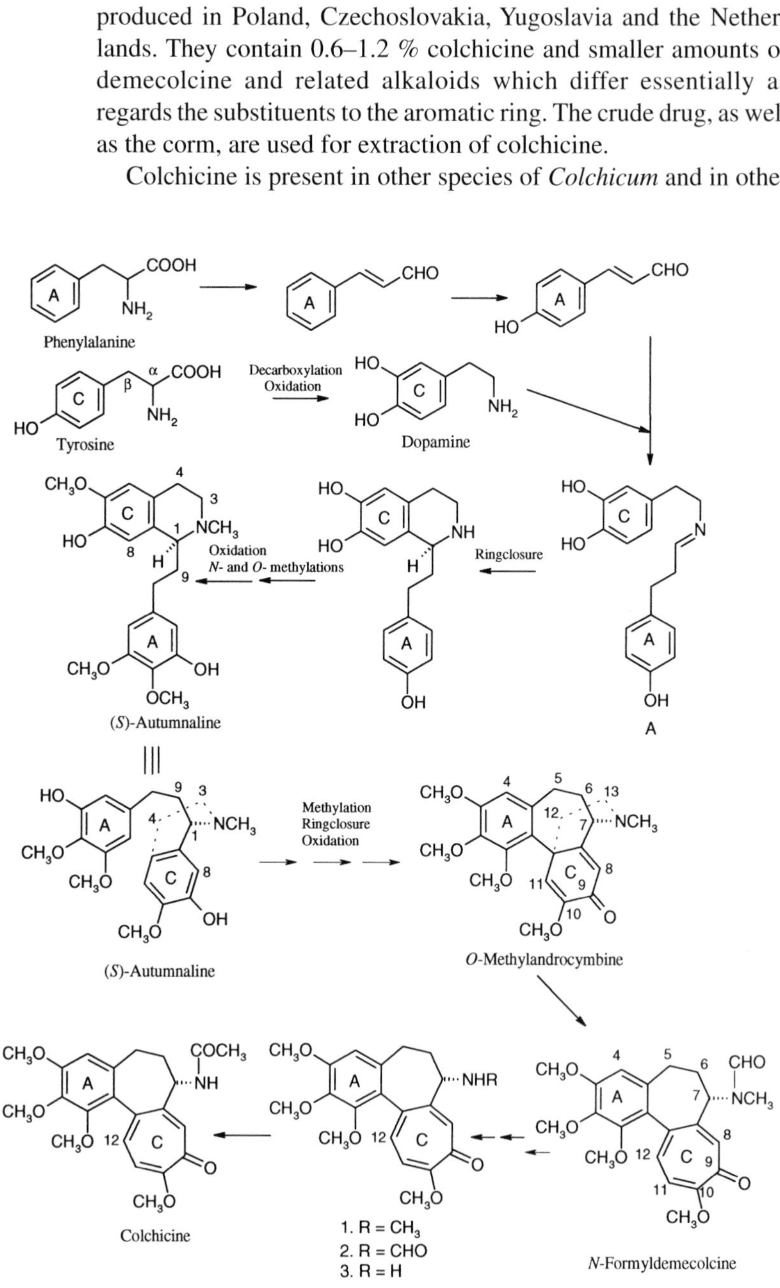

Fig. 194. Biosynthesis of colchicine

genera of the *Colchicaceae*. *Gloriosa superba* L., a plant grown in India, is another commercial source of colchicine.

Erythrophleum alkaloids

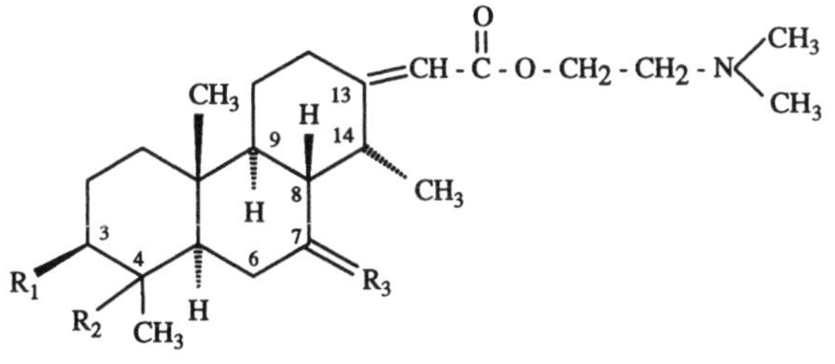

	R₁	R₂	R₃
Cassaine	OH	CH₃	O
Cassaidine	OH	CH₃	H, OH (β)
Ivorine	$-O-\overset{O}{\overset{\|}{C}}-CH=C\overset{CH_3}{\underset{CH_3}{<}}$	CH₃	O
Erythrophlamine	OH	$-\overset{O}{\overset{\|}{C}}-O-CH_3$	O
Cassamine	H	$-\overset{O}{\overset{\|}{C}}-O-CH_3$	O

These alkaloids are encountered in *Erythrophleum* species (*Fabaceae*), which are trees native to Africa, East and South-East Asia, and Australia. The alkaloids are esters of *N*-monomethyl-ethanolamine or *N,N*-dimethylethanolamine and various derivatives of cassenic acid that differ as regards their substituents at C-3, C-4 and C-7. The table above lists a few examples of the 32 or so *Erythrophleum* alkaloids isolated up to 2002. The cassenic acid derivatives are diterpenes, biosynthesized via the mevalonic-acid pathway through cyclization of geranylgeranyl diphosphate (page 413). Ethanolamine can be formed by decarboxylation of serine or through a transamination reaction between glycolaldehyde and glutamic acid.

Erythrophleum alkaloids have the same effect as the cardiac glycosides, i.e. they slow the pulse and increase the force of contraction of the heart muscle. However, their effect is short-lived and their margin of safety is narrow. Hence, their medicinal use is limited. Besides their cardiotonic effect, these alkaloids act as local analgesics.

Crude drug **Sassy bark** (Erythrophlei cortex) is the bark of *Erythrophleum suaveolens* (Guill. et Perrot.) Brenan (*Fabaceae*) or the closely related species *E. guineense* G. Don. It is an African tree. Its bark contains alkaloids of the type described above, cassaine being the main component. Other alkaloids present in sassy bark are cassai dine, erythrophlamine and cassamine (see the table). Ivorine has been isolated from *E. ivorense* A. Chevalier.

Sassy bark has been used in West Africa as an ordeal poison. A person under suspicion of a crime was made to drink a decoction of the bark. If he survived, this was taken as a sign of his innocence. The bark is used for preparing *Erythrophleum* alkaloids, especially cassaine.

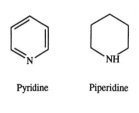

Pyridine   Piperidine

Pyridine and piperidine alkaloids

This heading embraces alkaloids that contain either a pyridine or a piperidine ring. Other *N*-heterocyclic rings may also occur in their structures. Pyridine and piperidine rings may be formed biosyn thetically in various ways.

Coniine

Coniine occurs in the poison hemlock, *Conium maculatum* L. (*Api aceae*) where it is the main alkaloid representing 90 % of the alka loids contained in the plant. Other alkaloids present are γ-coniceine, *N*-methylconiine, conhydrine and pseudoconhydrine. Unripe fruits have a higher alkaloid content (2 %) than other parts of the plant. On ripening, the content decreases to less than half the previous val ue (0.7 %). Coniine is relatively unstable, which leads to a very low alkaloid content in fruits that have been stored for some time. Fresh leaves and flowers contain 0.2 % alkaloids.

Coniine, which is an oil, is highly toxic. A dose of 6–8 poison hemlock leaves is said to be able to kill a human being. The cause of death is paralysis of the respiratory organs. Contamination of aniseed with the similar-looking hemlock fruits, a potential source of poisoning, used to be mentioned in the older literature, but rarely occurs nowadays. Hemlock is of historical interest too, because it has generally been assumed to have been in the potion that Socrates drank when he was condemned to death.

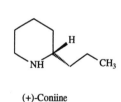

(+)-Coniine

Biosynthesis

Coniine is formed from four acetate residues (Fig. 195). These give a polyketo acid (A), which is reduced to 5-keto-octanoic acid (B) and 5 keto-octanal (C). The latter compound undergoes a transamination reaction, for instance with alanine, to form 5-keto-octylamine (D). Ring closure then gives γ-coniceine (E), which is finally reduced to coniine (F).

The biosynthesis of coniine is thus intimately connected with the acylpolymalonate pathway (page 237), and is an interesting variant of the numerous biosynthetic routes leading to alkaloids.

Arecoline

Arecoline is a liquid alkaloid occurring in the seed of the betel-nut palm, *Areca catechu*. Arecoline (in the hydrochloride form) is used as a vermifuge in veterinary medicine. It works on the worm, on the

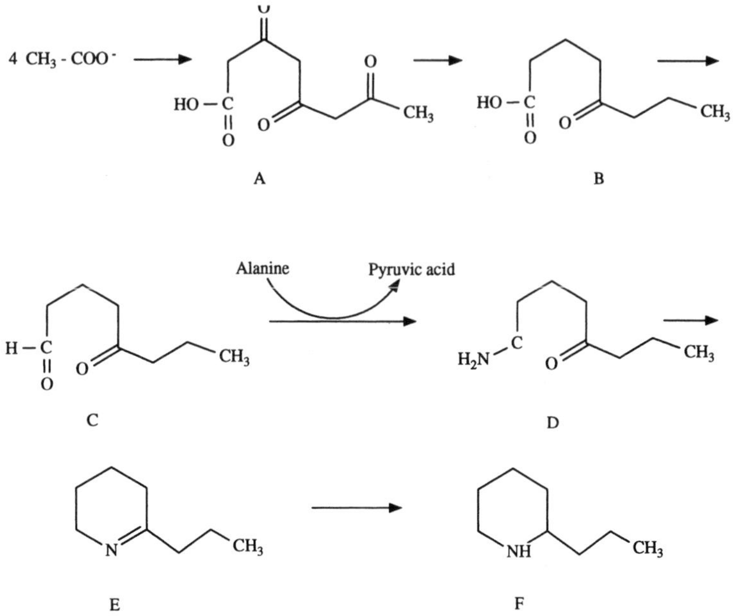

Fig. 195. Biosynthesis of coniine

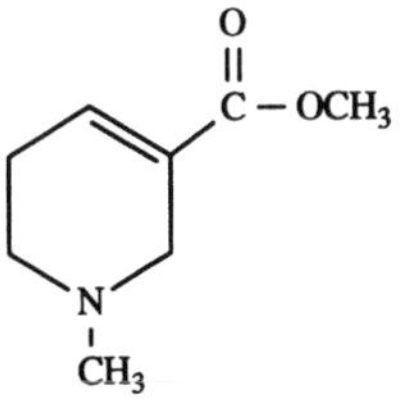

Arecoline

one hand, by damaging its muscles, and on the host, on the other hand, by stimulating peristalsis, so that the worm is more easily expelled.

Arecaidine

Biosynthesis Arecoline is formed from nicotinic acid, which in turn derives from aspartic acid and glyceraldehyde-3-phosphate (formed during the dark reactions in photosynthesis, page 128):

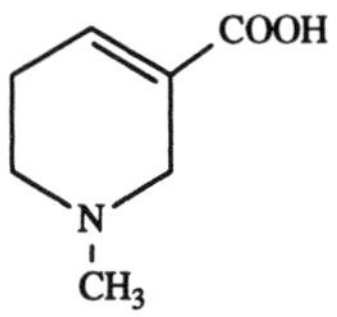

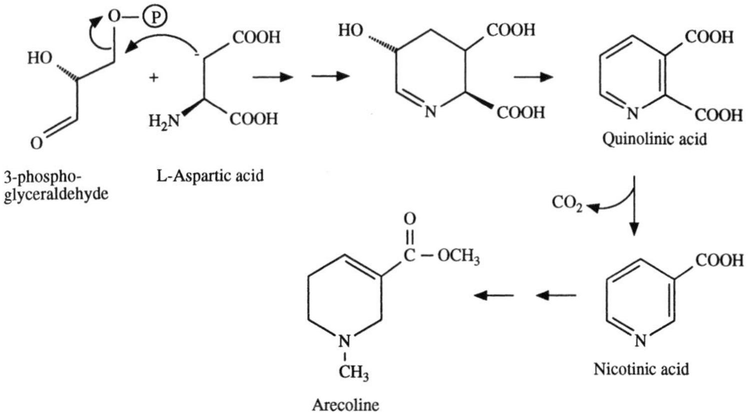

Fig. 196. Biosynthesis of arecoline

This biosynthesis is another example of a natural pathway to pyridine and piperidine rings.

Crude drug

Guvacine

Areca nut (Arecae semen) is the dried, ripe seed of *Areca catechu* L. (*Arecaceae*). This is a palm, 15–17 m high, growing in India, Sri Lanka, Indonesia and East Africa. To prepare the drug, the fruit (a nut) is crushed, and the seed is boiled in water containing lime and dried. The drug is used for the extraction of arecoline. It also contains arecaidine and the nor-derivatives guvacine and guvacoline.

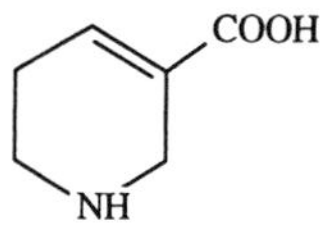

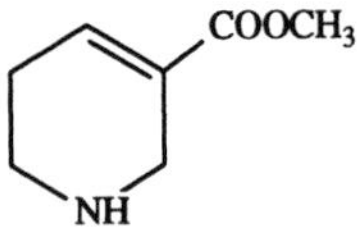

Areca nut is used in many countries for betel chewing. Fresh leaves of betel (*Piper betle* L., *Piperaceae*) are mixed with areca nut, slaked lime and tannin (gambir, catechu) and the mixture is chewed. The effect is refreshing and relaxing. The active constituent is arecaidine, which is formed from arecoline through ester hydrolysis under the influence of the lime present in the quid. Betel chewing has been reported to be associated with an increased risk of oesophageal cancer, possibly arising from the presence of arecoline.

Guvacoline

Lobeline

Lobeline occurs in *Lobelia inflata*. Lobeline stimulates respiration, but has to be administered intravenously, as it is quickly degraded in the body, and the resultant effect is of short duration. Oral administration requires very high doses for there to be any action at all. As regards other effects, lobeline is very similar to nicotine, and the substance has been tried as a cure for smoking.

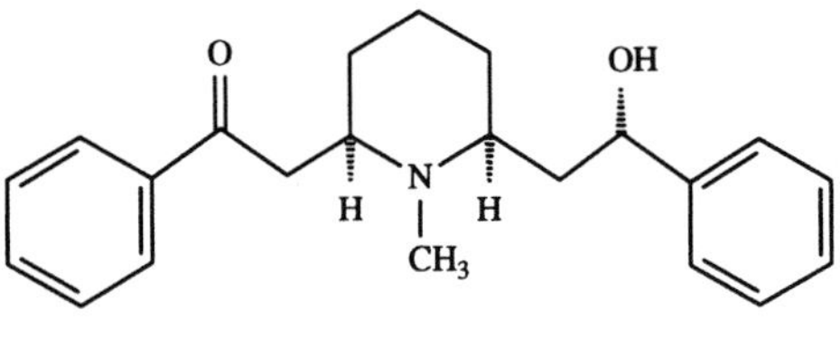

Lobeline

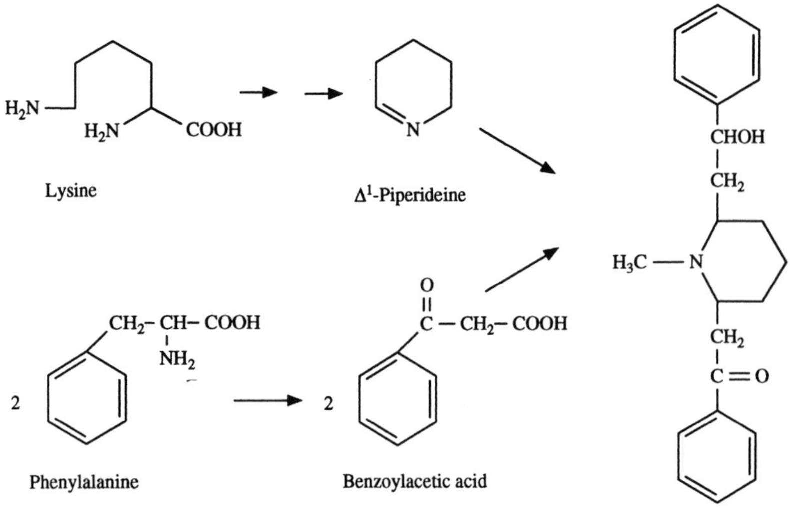

Lysine

Δ^1-Piperideine

Phenylalanine

Benzoylacetic acid

Lobeline

Fig. 197. Biosynthesis of lobeline

Biosynthesis Biosynthetically, the piperidine ring derives from lysine. This is the third route by which piperidine rings can arise. Δ^1-Piperideine is an intermediate. Formation of the aromatic rings follows the shikimic-acid pathway via phenylalanine.

Crude drug **Lobelia** (Lobeliae herba) consists of the dried aerial parts of *Lobelia inflata* L. (*Campanulaceae*). This is an annual herb of the eastern and central parts of the United States. The drug was used by the North American Indians both as a medicament and as a substitute for tobacco. It contains a large number of piperidine alkaloids, the main one being lobeline; the total alkaloid content is 0.3–0.4%. Galenical preparations of lobelia have been used as expectorants, but their effect cannot be attributed to lobeline because of the instability of this compound (see above). The active constituent is probably a different alkaloid, isolobinine, which strongly irritates the mucous membranes on oral administration. The anti-asthmatic effect of isolobinine is regarded as a reflex action due to the irritation of the mucous membranes of the stomach.

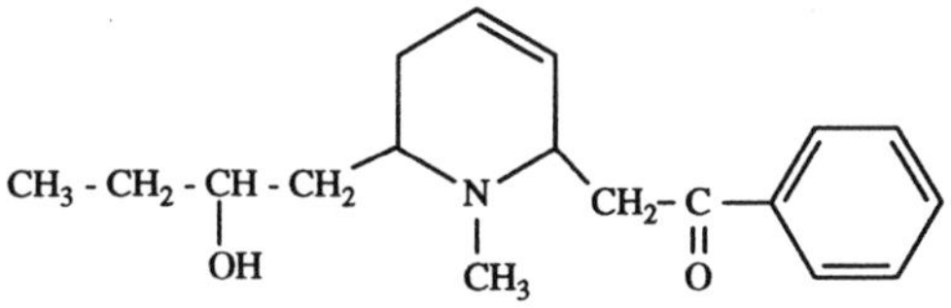

$$CH_3 - CH_2 - \underset{\underset{OH}{|}}{CH} - CH_2$$

Isolobinine

Nicotine Nicotine is the main alkaloid of the tobacco plant *Nicotiana tabacum* and is the compound responsible for the habit of smoking tobacco. Nicotine, which is an oil, has many pharmacological effects. In low doses, such as those inhaled in smoking, nicotine causes hypertension, respiratory stimulation and stimulation of secretion from several glands. It also has a CNS-stimulating effect. Nicotine is a strong poison which interacts with the nicotinic acetylcholine receptors. The lethal dose in man is 50–100 mg which approximately corresponds to the nicotine content of 5 cigarettes. However, in smoking most of the nicotine is destroyed by the heat or is distributed into the air. In contrast to the effect of low doses, toxic doses cause hypotension and death occurs as a result of respiratory arrest.

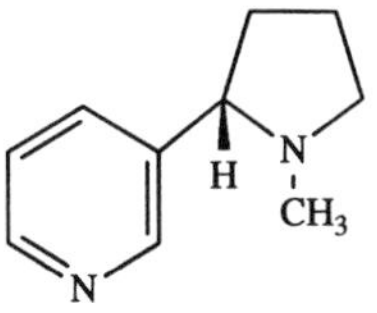

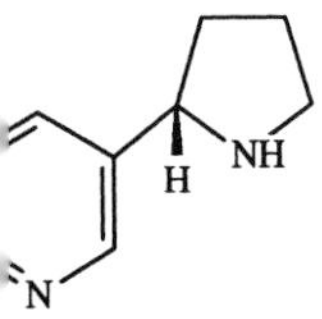

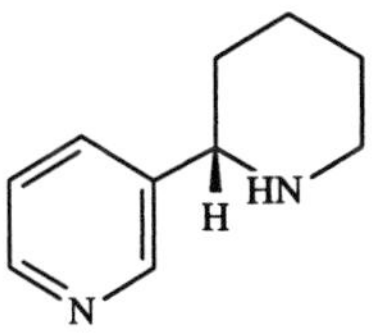

Nicotine

Nornicotine

Nicotine is used as an insecticide, especially in gardening, and is prepared by isolation from tobacco waste. For medical purposes nicotine is available in the form of chewing gum, nasal spray and nicotine-impregnated patches to be used by smokers who want to stop their habit.

Anabasine

625

Biosynthesis   The pyridine ring, like the ring in arecoline (page 623) is derived from glyceraldehyde-3-phosphate and aspartic acid. The pyrrolidine ring is formed from the amino acid ornithine through decarboxylation, methylation, oxidative deamination and ring closure (Fig. 198). The methyl group comes from methionine. The decarboxylation of ornithine is catalysed by *ornithine decarboxylase* (EC 4.1.1.17). The product of ornithine decarboxylation – putrescine – is a precursor of the polyamines spermidine and spermine, but in the biosynthesis of nicotine and the tropanalkaloids (page 630) it is methylated, a reaction which is catalysed by *putrescine N-methyltransferase* (EC 2.1.1.53). A cDNA encoding this enzyme has been isolated from *Nicotiana tabacum* and expressed in *Escherichia coli*. Oxidative deamination of *N*-methylputrescine, catalysed by a diamine oxidase, yields *N*-methyl-γ-aminobutanal, which spontaneously cyclizes to the 1-methyl-Δ^1-pyrrolinium cation, which reacts with 1,2-dihydropyridine (from nicotinic acid) to yield a product which is oxidized to nicotine.

For the biosynthesis of nicotinic acid see page 623.

Crude drug   **Tobacco** (Nicotianae folium) consists of the dried and fermented leaves of *Nicotiana tabacum* L. (*Solanaceae*, Plate 22, page 618) and related species, e.g. *N. rustica* L. The *Nicotiana* species used in making tobacco come from North America, where the Indians knew how to use their leaves as stimulants. After the discovery of the New World, the use of tobacco spread to the rest of the world and the plant is now grown widely in tropical as well as subtropical and

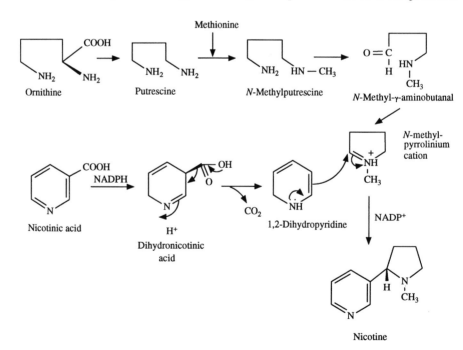

Fig. 198. Biosynthesis of nicotine

temperate regions. Tobacco is prepared by "curing", which involves slow drying and fermentation to give the drug its aroma and brown colour. The quality of the product depends partly on the botanical origin (species, variety, hybrid) and partly on the way in which the drying and fermentation have been carried out. The nicotine content of tobacco varies within very wide limits (0.05–9 %). Smoking tobacco is preferably prepared from strains low in nicotine. Besides nicotine, tobacco contains nornicotine and anabasine. The species *N. glauca* R. Grah. is relatively rich in anabasine, the piperidine ring of which is formed biosynthetically from lysine.

Smoking tobacco involves a considerable risk to the health. Nicotine causes arteriosclerosis and diseases of the heart. The smoke contains more than 3 000 compounds, some of which are strongly carcinogenic, and smokers run a pronounced risk of developing cancer, particularly cancer of the lungs.

Histrionicotoxins

The skin of South American frogs of the family *Dendrobatidae* contains toxic compounds which have been identified as alkaloids. One group, contained in the skin of frogs of the genus *Phyllobates* has a steroid skeleton and will be discussed later (page 701). Another large group of frog alkaloids contains bicyclic compounds, consisting of either a spiropiperidinol ring system (histrionicotoxins) or an indolizidine ring system (pumiliotoxins).

Histrionicotoxins have been found mainly in frogs of the genera *Dendrobates* and *Epipedobates* and were first detected in the species *Dendrobates histrionicus* of western Colombia and north-western Ecuador. About 16 different histrionicotoxins are known.

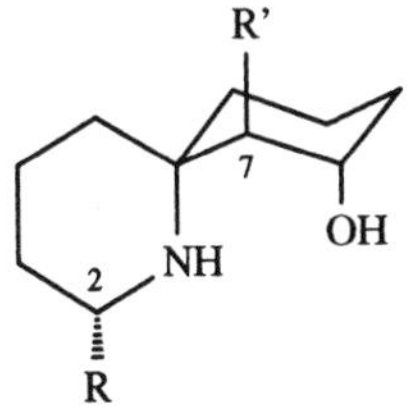

Histrionicotoxin: R = *cis*-CH₂CH=CHC≡CH
R' = *cis*-CH=CHC≡CH

The structural differences between these compounds are located in two carbon chains connected at C-2 and C-7 of the ring system. The length of the chains at C-2 varies from 3 to 5 carbon atoms and those at C-7 can be 2–4 carbon atoms long. The chains can be saturated, but they can also contain double bonds and triple bonds. Histrionicotoxin (below) is an example of an alkaloid with unsaturated chains at both C-2 (five carbons) and C-7 (four carbons). A few of the compounds have been given trivial names but most are labelled by a code developed for alkaloids from dendrobatid frogs. This code consists of a combination of figures and letters where the figures denote the molecular weight and the letters are used to differentiate between compounds with the same molecular weight. Isomers are characterized by prefixes (e.g., *cis*, *trans*, epi, iso) and primes (e.g. A', A", etc.). Thus histrionicotoxin is dendrobatid alkaloid 283A in this system.

Histrionicotoxins have relatively low toxicities. In a mouse, a subcutaneous dose of 1 mg causes locomotor difficulties and prostration. The compounds exert their pharmacological activity by affecting at least three classes of channels in nerve and muscle. In *receptor-regulated channels*, particularly the nicotinic acetylcholine receptor channel, histrionicotoxins block the channel conductance in a time- and stimulus-dependent manner. They also accelerate the desensitization or inactivation of the channel. In *voltage-dependent sodium channels* the compounds reduce conductances in a manner reminiscent of local anaesthetics. The third class are *voltage-dependent potassium channels*, where the histrionicotoxins reduce conductances in a time- and stimulus-dependent manner. Radioactively labelled histrionicotoxin is used as a pharmacological tool for the study of these channels.

Pumiliotoxins   A great number of pumiliotoxins have been isolated from frogs of the family *Dendrobatidae*. They have also been found in amphibians of the genera *Pseudophryne* (*Myobatrachidae*), *Mantella* (*Ranidae*) and *Melanophryniscus* (*Bufonidae*). These alkaloids have an indolizidine ring system and can be divided into two groups: pumiliotoxins and allopumiliotoxins (Fig. 199). The two groups differ in the substitution at C-7 where the allopumiliotoxins have a hydroxyl group. The individual alkaloids differ in the length and structure of the side-chain at C-6.

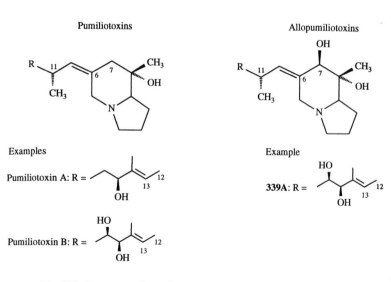

Fig. 199. Structures of pumiliotoxins and allopumiliotoxins

Pumiliotoxin A and B were found in *Dendrobates pumilio* from the Bocas archipelago of Panama, and are the first detected representatives of this group. Pumiliotoxins and allopumiliotoxins often occur together. In analogy with the histrionicotoxins only a few have been given trivial names, whereas most are denoted by the code system described above in connection with the histrionicotoxins. In this system pumiliotoxins A and B are dendrobatid alkaloids 307A and 323A, respectively. The pumiliotoxins are not synthesized by the frogs but originate from ants which are part of the frogs's food. Thus the two alkaloids pumiliotoxin A and B have been found in ants of the genera *Brachymyrmex* and *Paratrechina* and species of these genera have also been found in the stomach contents of *Dendrobates pumilio*. It is very probable that all alkaloids found in frogs derive from their diet.

Pumiliotoxin A and B are comparatively toxic compounds with subcutaneous lethal doses for mice of 50 μg and 20 μg, respectively. Allopumiliotoxins are less toxic. Pumiliotoxin B interacts with voltage-dependent sodium channels to elicit an increased influx of sodium ions and, in brain and heart preparations, a stimulation of phosphoinositide breakdown. The latter effect can, via inositol triphosphate, cause release of calcium from internal storage sites. Pumiliotoxin has both cardiotonic and myotonic activity in isolated atrial or rat phrenic nerve diaphragm preparations. The cardiotonic activity correlates well with the stimulation of phosphoinositide breakdown. The effect on neuromuscular preparations is interpreted as due primarily to effects on sodium channels but an additional direct effect on calcium mobilization is also possible.

Pumiliotoxin B and congeners are valuable research tools, and perhaps models, for the development of new myotonic or cardiotonic agents.

Epibatidine   Epibatidine was isolated in very small amounts from the Ecuadoran frog *Epipedobates tricolor* (*Dendrobatidae*). Following determination of the structure, total synthesis has been accomplished.

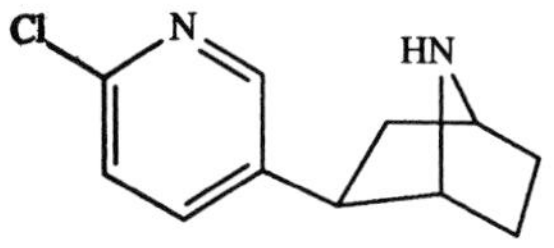

Epibatidine

Epibatidine is 200-fold more potent than morphine as an analgesic and unlike morphine the analgesic activity is not blocked by the opioid receptor antagonist naloxone. The target of activity is the nicotinic acetylcholine receptor. Epibatidine has a marked analgesic activity at a dose of 0.01 μmol/kg but is quite toxic at only slightly higher doses. In animals, tolerance to the analgesic effects of epibatidine are minimal to modest and hence the compound could have a potential for long-term treatment of chronic neuropathic, arthritic or cancer-elicited pain. However, a major obstacle to the clinical utility is its very limited therapeutic index and its lack of marked selectivity towards the multiple subtypes of the nicotinic acetylcholine receptor.

The commercial availability of racemic epibatidine and both enantiomers has made the compound a ligand of choice for the investigation of nicotine acetylcholine receptors. The discovery of the mechanism for the activity of epibatidine has also triggered extensive research with the aim of finding analgesics with the same potency but with a better therapeutic index and better selectivity for the subtypes of the nicotinic acetylcholine receptor

Tropane alkaloids

The most important representatives of this group of alkaloids are hyoscyamine, scopolamine (=hyoscine) and cocaine (see below). The tropane skeleton consists of a pyrrolidine and a piperidine ring, which share the nitrogen atom and two carbon atoms.

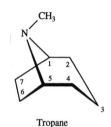

Tropane

Biosynthesis of the tropane skeleton

The pyrrolidine part of this structure is biosynthesized from the N-methyl-Δ^1-pyrrolinium cation, which is formed from ornithine in the same way as in the biosynthesis of nicotine (Fig. 198, page 626). Previously it was thought that the rest of the tropane skeleton was biosynthesized differently in hyoscyamine and scopolamine on the one hand and in cocaine on the other. However, it has been shown that the skeleton is formed chiefly in the same way in all hitherto

N-Methylpyrrolium cation   Acetoacetat

A

Carboxytropinone

Fig. 200. Biosynthesis of the tropane skeleton

Plate 23.
Atropa belladonna L. The leaves contain the alkaloid (-)-hyoscyamine (page 632). The Botanical Garden, Uppsala. Photograph by Gunnar Samuelsson.

Plate 24.
Datura stramonium L. The leaves contain the alkaloids (-)-hyoscyamine and scopolamine (page 632). The Botanical Garden, Uppsala. Photograph by Gunnar Samuelsson.

studied tropane alkaloids (Fig. 200). The formation of the skeleton from the N-methyl-Δ^1-pyrrolinium cation requires addition of three carbon atoms, and 4-(1-methyl-2-pyrrolidinyl)-3-oxobutanoate (A in Fig. 200) has unequivocally been shown to be a precursor. A is formed by the reaction of C-4 in acetoacetate with N-methyl-Δ^1-pyrrolinium salt. Cyclization of A yields carboxytropinone which is the first product containing the tropane skeleton.

Hyoscyamine and scopolamine

Hyoscyamine and scopolamine are found in several genera of the family *Solanaceae* (see below under "crude drugs"). The alkaloids are esters of tropane-3α-ol (respectively its 6–7 epoxide) and an unique acid – tropic acid.

The α-carbon atom of tropic acid is asymmetric, so that two stereoisomers are possible. The natural form, the one occurring in pharmacologically active hyoscyamine, is the (-)-form. The alkaloid is easily racemized, yielding atropine, which has the same pharmacological effects as (-)-hyoscyamine, but must be used in doses that are twice as great, because the (+)-component is practically inactive.

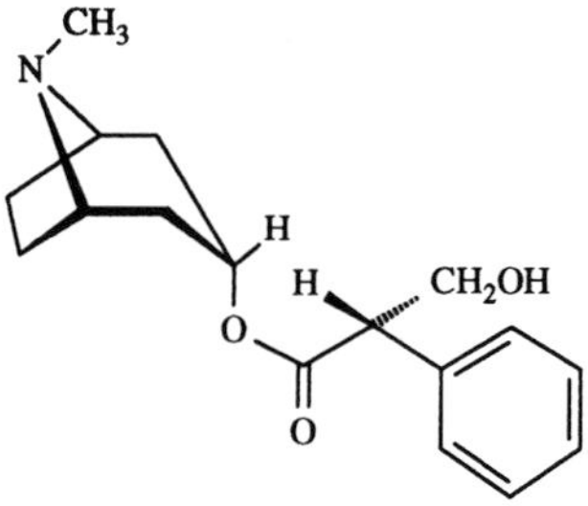

(-)-Hyoscyamine 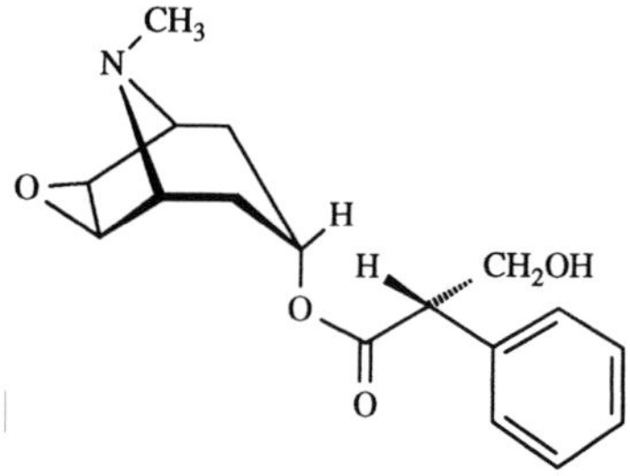   Scopolamine

Hyoscyamine has an anticholinergic (parasympatholytic) effect. Hyoscyamine or atropine is used as a remedy against spasms in skeletal muscles and for the suppression of secretions, e.g. an overproduction of hydrochloric acid in the stomach. These alkaloids are also used as antidotes in poisoning by cholinesterase inhibitors, e.g. physostigmine or organic phosphoric-acid insecticides. Among the side-effects are xerostomia, due to decreased saliva secretion, and dilation of the pupils of the eyes, leading to problems on exposure to strong light. Hallucinations may occur as a toxic effect of overdoses of hyoscyamine. This explains why plants, containing the alkaloid, are used by primitive peoples in religious rites.

Like hyoscyamine, scopolamine has an anticholinergic effect, but unlike hyoscyamine, it is a central nervous system depressant. It is used prophylactically against motion sickness (sea sickness, car sickness, etc.) and, often in combination with morphine, for premedication before operations.

Semisynthetic derivatives of hyoscyamine and scopolamine

Ipratropium bromide

This compound is a derivative of hyoscyamine, in which the nitrogen atom has been quaternized by introduction of an isopropyl group. The substance is used as a bronchodilator in chronic bronchitis and bronchial asthma.

Homatropine methobromide

This substance is a quaternized hyoscyamine derivative in which an additional methyl group has been introduced and the tropic acid moiety replaced by mandelic acid. The substance has antispasmodic properties.

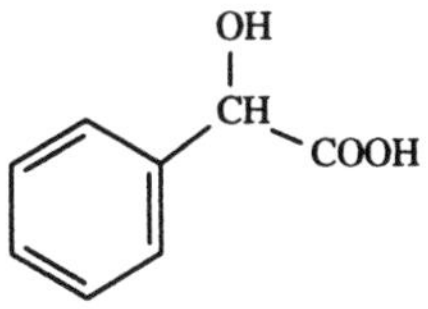

OH
|
CH
COOH

Mandelic acid

Butylscopolamine and methylscopolamine

These are quaternary derivatives of scopolamine containing a butyl and a methyl group, respectively. The compounds have less pronounced central effects than scopolamine and are used as antispasmodics.

Biosynthesis of hyoscyamine and scopolamine

The formation of the tropine part of these alkaloids could involve decarboxylation of carboxytropinone to tropinone, a reaction which has not yet been proved. It is also possible that tropinone is formed directly from 4-(1-methyl-2-pyrrolidinyl)-3-oxobutanoate (A in Fig. 200) by a concomitant cyclization and decarboxylation reaction. Tropinone is reduced to the corresponding alcohol – tropine (tropane-3α-ol) – by an enzyme *tropinone reductase I* (EC 1.1.1.206). A cDNA encoding this enzyme has been isolated from *Datura stramonium* and expressed in *Escherichia coli*. A cDNA for a second tropinone reductase (*tropinone reductase II*, EC 1.1.1.236) has also been cloned and expressed. This reductase forms pseudotropine, which is the 3β epimer of tropine. A 120-amino acid residue peptide at the C-terminus of each reductase determines the stereospecificity of the reaction it catalyses. Tropine forms an ester with tropic acid (via phenyl-lactic acid, see below) yielding hyoscyamine which is oxidized to 6β-hydroxyhyoscyamine. Scopolamine is eventually formed through direct attack by the 6β–hydroxyl group, leading to displacement of the 7β-hydrogen atom.

The oxidation of hyoscyamine to 6-β-hydroxyhyoscyamine and the following epoxide formation is catalysed by *hyoscyamine 6S-dioxygenase* (EC 1.14.11.11) an enzyme which has been cloned and expressed in *E. coli*. The enzyme requires 2-oxoglutarate, ferrous ion, ascorbate and molecular oxygen for catalysis and therefore belongs

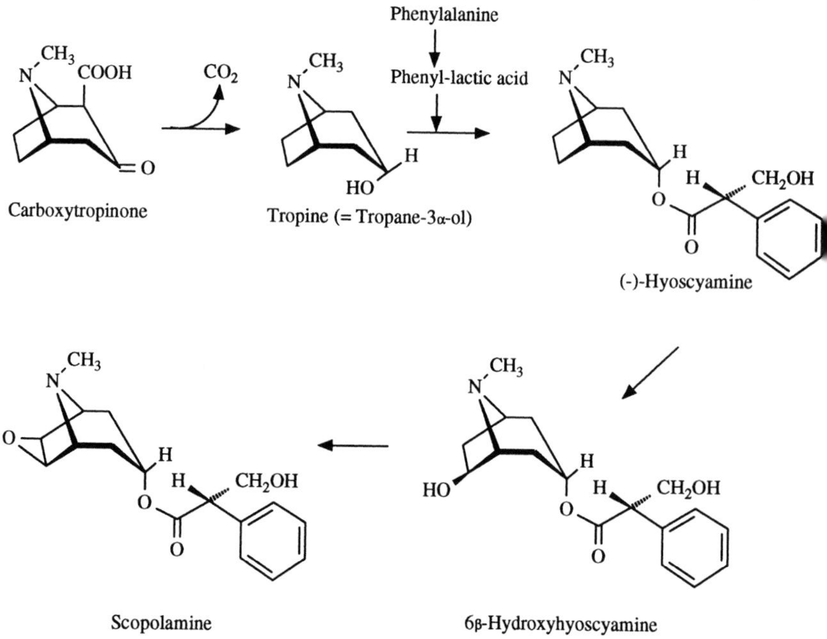

Fig. 201. Biosynthesis of hyoscyamine and scopolamine from carboxytropinone

to the 2-oxoacid-dependent oxygenase family. The gene for this enzyme has been found only in scopolamine-producing solanaceous species. The enzyme is localized to the root pericycle in scopolamine-producing species of *Hyoscyamus*, *Atropa* and *Duboisia*. The cloning of the cDNA coding for hyoscyamine 6S-dioxygenase has permitted production of transgenic plants. Thus *Atropa belladonna* produces hyoscyamine as the main alkaloid but transgenic plants of the species, containing this gene, produce elevated levels of scopolamine. As scopolamine is of higher economic interest than hyoscyamine, this finding might be of commercial interest.

(S)-(-)-Tropic acid is a compound which is known only as an acid moiety of tropane alkaloid esters. Tropic acid derives from phenyl-alanine but the formation requires an intramolecular rearrangement of the phenylalanine skeleton.

The formation of tropic acid in connection with the biosynthesis of hyoscyamine is illustrated in Fig. 202. Transamination converts phenylalanine to phenylpyruvic acid which is reduced to phenyl-lactic acid. This acid forms an ester with tropine yielding the alka-loid littorine, which occurs together with hyoscyamine in tropane alkaloid-producing solanaceous plants. Isotope labelling studies have indicated that C-1′ and C-3′ of littorine form the phenylacetate moiety of hyoscyamine and that C-2′ forms the hydroxymethyl group in a stereochemically controlled reaction. Several hypotheses have been proposed for the enzymology of this rearrangement.

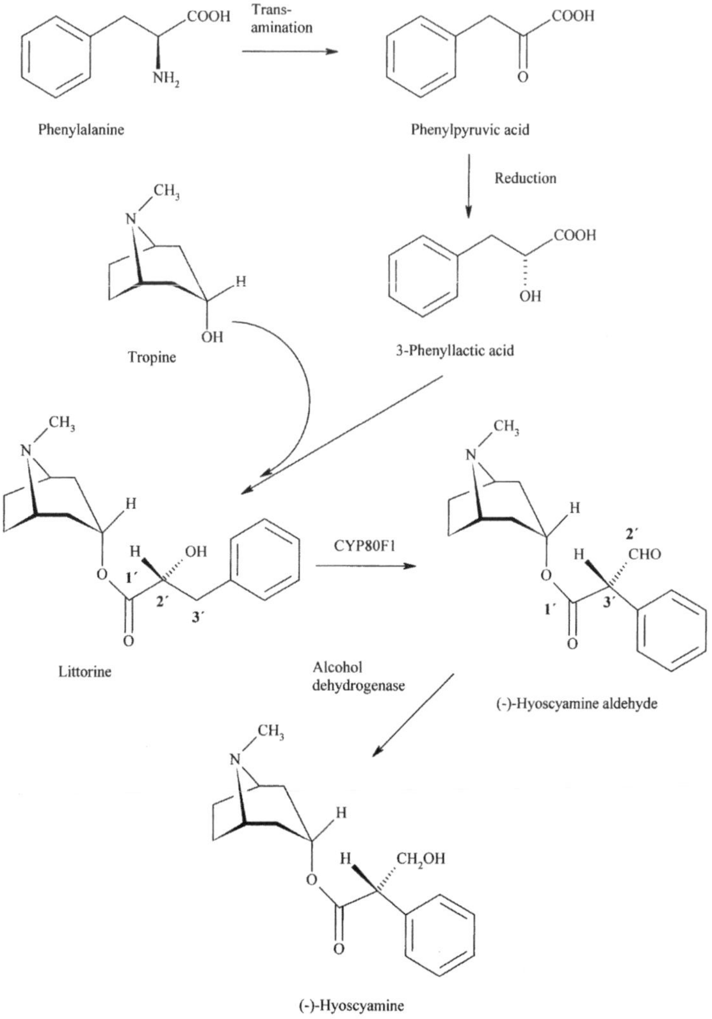

Fig. 202. Biosynthesis of (-)-hyoscyamine involving formation of (S)-(-)-tropic acid

Strong evidence has now been presented that a cytochrome P450 enzyme and an alcohol dehydrogenase are involved. This proposition is supported by the isolation of a cDNA coding for the cytochrome P45 – CYP80F1 – and the subsequent cloning of the cDNA and expression of the enzyme in yeast. CYP80F1 was found to catalyse the oxidation of (R)-littorine with rearrangement to form hyoscyamine aldehyde which is then reduced to hyoscyamine by an alcohol dehydrogenase.

The rearrangement thus occurs following formation of the ester and not, as previously thought, in the free acid. Further modifications of the tropane skeleton also occur on the ester and not on the free alcohol.

Crude drugs **Belladonna leaf** (Belladonnae folium) are the dried leaves of *Atropa belladonna* L. (*Solanaceae*, Plate 23, page 631). This is a perennial herb, up to 1.5 metre high. It is native to central and southern Europe and is cultivated in many other countries, including India and the United States. Belladonna leaves have a total alkaloid content of 0.2–0.6 %. *The European Pharmacopoeia* requires at least 0.35 %. About 75 % of the total alkaloids is (-)-hyoscyamine, and the rest is atropine, which is formed during the drying process. Fresh leaves contain almost exclusively (-)-hyoscyamine. The scopolamine content is very low. Extracts of belladonna leaves are used for the same indications as (-)-hyoscyamine.

A. belladonna produces black glossy berries, about the size of a cherry, which contain hyoscyamine and which have caused poisoning in children who have eaten them. The name "belladonna" is Italian and means "beautiful woman". It refers to the fact that juice from the berries was used in Italy in olden times as a beauty preparation. It was dropped into the eyes and the hyoscyamine enlarged the pupils, which was considered to enhance the beauty of the woman.

Stramonium (Stramonii folium) is the dried leaf of the thorn-apple, *Datura stramonium* L. (*Solanaceae*, Plate 24, page 631). The plant is an annual herb, up to 2 metres in height, native to the region around the Caspian Sea but occurring also in other parts of Europe and in the United States; it is cultivated in South America. The name thorn-apple refers to the fruit, which is a capsule, 3–4 cm long, covered with prickles. Stramonium has a total alkaloid content of 0.2–0.5 %. About two-thirds of this material consists of hyoscyamine, partially racemized to atropine, and the rest is scopolamine. An extract of the drug is used to a limited extent for the same indications as (-)-hyoscyamine. Powdered stramonium used to be employed in the form of cigarettes as a remedy for asthma, the smoke containing hyoscyamine which has an antispasmodic effect on the bronchi.

Hyoscyamus (Hyoscyami folium) consists of the dried leaves of henbane, *Hyoscyamus niger* L. (*Solanaceae*). This plant is a herb, annual or biennial, native to Europe but now spread all over the world. The alkaloid content is 0.02–0.15 %, three-quarters of which is hyoscyamine and the rest is scopolamine. Hyoscyamus may be employed in the form of an extract, just like belladonna leaves and stramonium, but nowadays only small amounts are used.

Egyptian henbane consists of the dried leaves and flowering tops of *Hyoscyamus muticus* L. (*Solanaceae*), a perennial herb occurring in deserts in Egypt and the Middle East. It is also cultivated in the United States. The drug contains 0.5–1.5 % alkaloids, essentially (-)-hyoscyamine, and is used for the extraction of the alkaloid.

Duboisia is the dried leaf of *Duboisia myoporoides* R. Br.

(*Solanaceae*), a perennial shrub, growing along the east coast of Australia. The leaves can be repeatedly harvested from the same shrub and the season at which they are collected determines the proportions of the alkaloids they contain. The total alkaloid content may be up to 3 %. Hyoscyamine is the main alkaloid in leaves harvested in October, whereas scopolamine dominates in leaves collected in April. Besides these two constituents, the drug contains other tropane alkaloids, which are either esters of tropine with acids other than tropic acid or esters of Ψ-tropine (tropane-3β-ol). The species *D. myoporoides* has been shown to comprise several chemical races, which exhibit constant differences in their alkaloid content. Duboisia is used for the extraction of hyoscyamine and scopolamine. The leaves of another species, *D. leichhardtii* F. Muell., which occurs in a limited area in south-eastern Australia, can also be used. The dominant alkaloid is hyoscyamine.

Cocaine   Cocaine occurs in *Erythroxylum coca*. The alkaloid is an ester of ecgonine, which is a derivative of Ψ-tropine (tropane-3β-ol) carrying a carboxyl group at C-2. In cocaine, this carboxyl group is esterified with methanol and the hydroxyl group is esterified with benzoic acid. Cocaine is used as a local anaesthetic, especially in ophthalmic practice, but its application is limited because the compound is also a central stimulant, which has led to its misuse as a narcotic. However, the compound has played an important part in the development of other local anaesthetics, which have been synthesized using cocaine as a model, e.g. procaine.

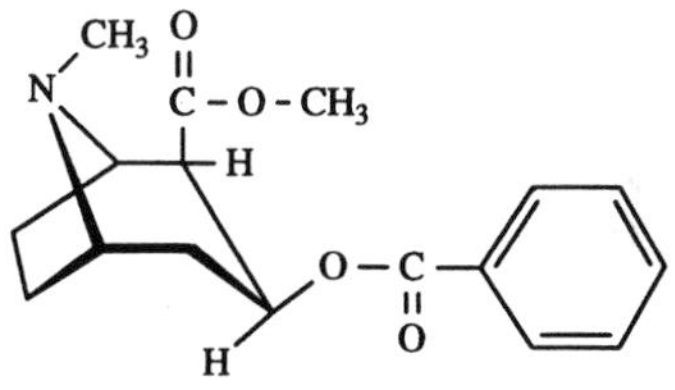

Cocaine

Biosynthesis   The biosynthesis of the tropane skeleton of tropane alkaloids has been outlined previously (page 630, Fig. 200). In the biosynthesis of cocaine (Fig. 203) carboxytropinone is methylated and reduced to methyl ecgonine, the β-hydroxy group of which is esterified with benzoic acid.

Crude drug   **Coca leaf** (Cocae folium) is the dried leaf of *Erythroxylum coca* Lam. or *E. novogranatense* Hieron. (*Erythroxylaceae*). These plants are shrubs, reaching a height of 5 metres; they come from the eastern slopes of the Andes in Peru and Bolivia. The species are only known in cultivation and the wild-growing forms from which they have come are no longer known. They require a climate with an even temperature in the range 15–25 °C. Three harvests per annum are collected.

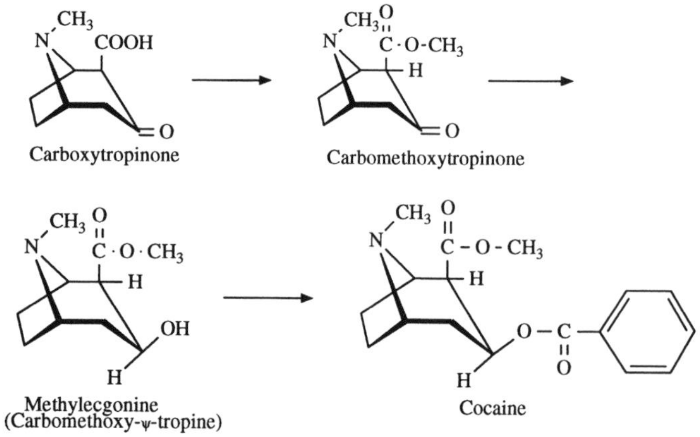

Fig. 203. Biosynthesis of cocaine

Trujillo coca or Peruvian coca is the leaves of *E. truxillense* Rusby (sometimes known as *E. novogranatense* var. *truxillense* (Rusby) Plowman). It is grown in Peru, near the city of Trujillo, and in the Maraon valley. A cocaine-free extract of Trujillo coca is an ingredient of Coca-Cola and contributes to the flavour of this drink. Trujillo coca has a higher content of methyl salicylate than other kinds of coca.

Coca chewing   Coca leaves are used as a stimulant among the Indians in Bolivia, Peru and Colombia. This practice is ancient and was in use long before the discovery of the New World. The dried leaves are mixed with an alkaline mineral called *llipta* and the mixture is kept in the mouth, where it becomes thoroughly soaked with saliva. The material is moved from time to time in the mouth rather than chewed, and the saliva is swallowed. The Indians of the Amazon basin use the leaves of *E. coca* var. *ipadu* Plowman, which has a lower cocaine content than ordinary coca leaves. The leaves are toasted and mixed with the ash of *Cecropia sciadophylla* Mart. The mixture is pounded to a very fine powder, a portion of which is placed in the cheek and worked with the tongue until thoroughly moistened with saliva. The quid is allowed gradually to dissolve and is not ejected as in ordinary coca chewing. This method of use probably ensures an optimal utilization of the low cocaine content. The coca chewer can stand great hardship under the influence of the alkaloids thus extracted. Feelings of fatigue and hunger are alleviated and he can, for instance, carry heavy burdens in rough terrain. The damaging effect of coca chewing is much debated. It seems that the crude drug is not as liable to cause addiction and personality destruction as is the abuse of pure cocaine that occurs in the United States and Europe. For that reason, it has been suggested that cocaine is not the active principle in coca chewing. However, it has been shown that when coca is chewed, cocaine appears in the blood about 30 minutes after taking the drug. The concentration of cocaine in the blood reaches its maximum after 1–2 hours and falls to zero within 5–7 hours and these changes in the cocaine concentration agree well with the degree of stimulation experienced. The dose of cocaine normally ingested (15–50 mg) is, however, low compared with the doses consumed by cocaine addicts (up to 10 g per day). There is also a great difference in the time during which the cocaine is absorbed by the body. In the case of the addict, there is a sudden surge of cocaine into the bloodstream, producing a "rush and crash" effect. In contrast, the chewing of coca results in a slow release of cocaine which is gradually absorbed from the gastrointestinal tract. This might explain why chewing coca is less harmful than cocaine abuse.

Alkaloid content   Coca leaves contain other ecgonine esters besides cocaine, such as cinnamoylcocaine, which has a cinnamoyl rather than a benzoyl group, and α- and β-truxillines, in which the benzoic-acid residue has been replaced by α- and β-truxillic-acid (dimers of cinnamic acid) residues, respectively. Other alkaloids present are tropacocaine (a derivative of Ψ-tropine), and cuscohygrine and hygroline (derivatives of hygrine).

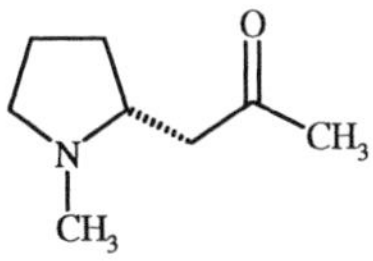

| (+)-Hygrine | Hygroline | Cuscohygrine |

| α-Truxillic acid | Tropacocaine | Cinnamoylcocaine |

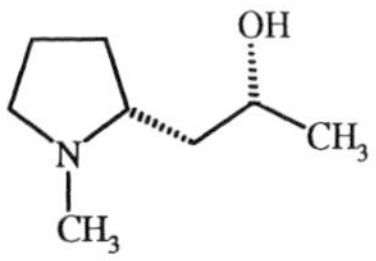

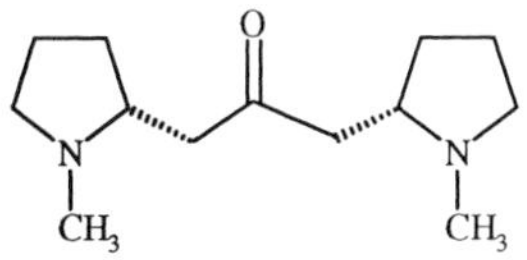

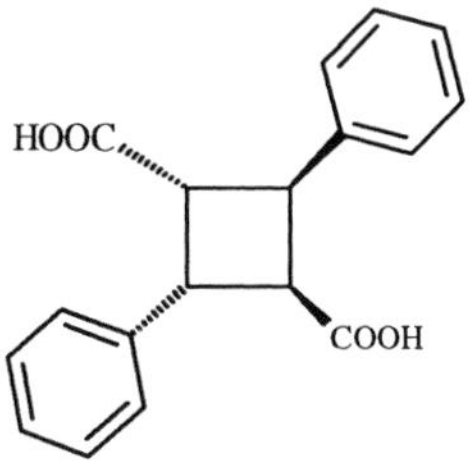

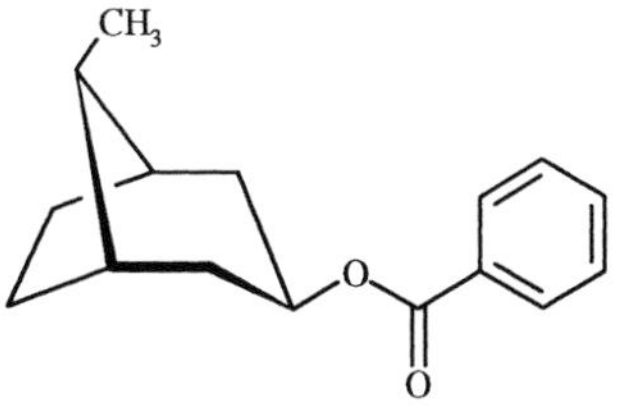

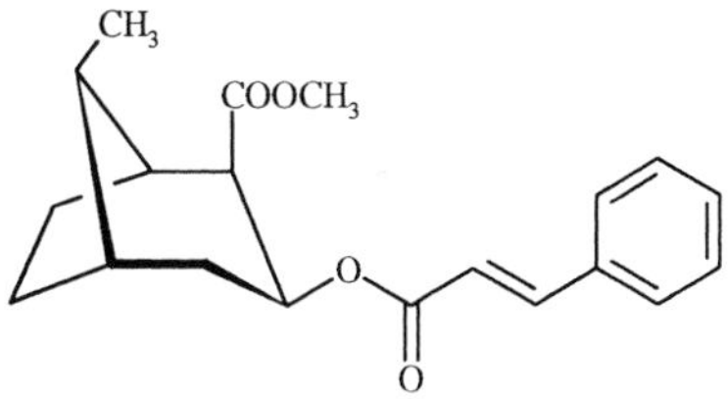

Production of cocaine   The ecgonine esters comprise 0.5–2 % of the coca leaves, the proportion of cocaine varying from 30 % to almost 100 %, depending on the origin of the crude drug. Cocaine is prepared by extraction of the leaves, followed by hydrolysis of the crude alkaloid mixture obtained. Ecgonine is isolated from the hydrolysate, and then benzoylated and methylated to yield cocaine.

Pyrrolizidine alkaloids (*Senecio* alkaloids)

Pyrrolizidine alkaloids constitute another group of alkaloids biosynthetically derived from the amino acid ornithine. The structures of some typical pyrrolizidine alkaloids are as follows:

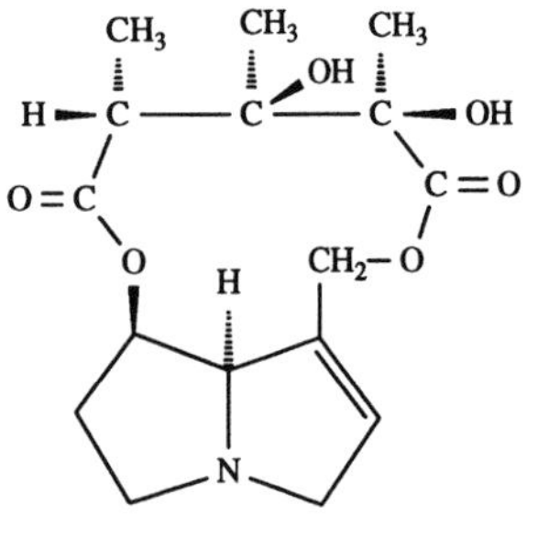

Monocrotaline

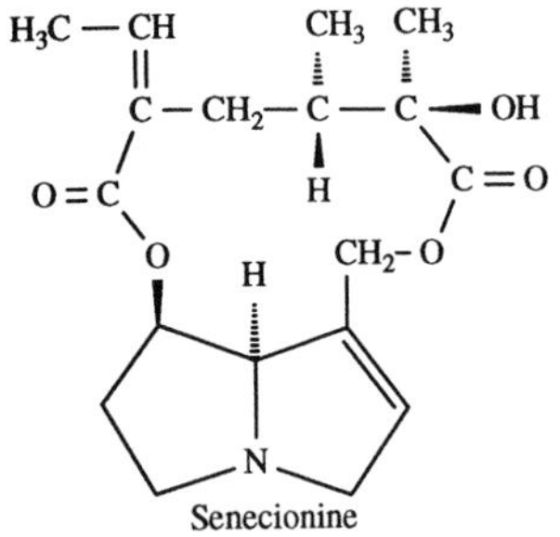

Senecionine

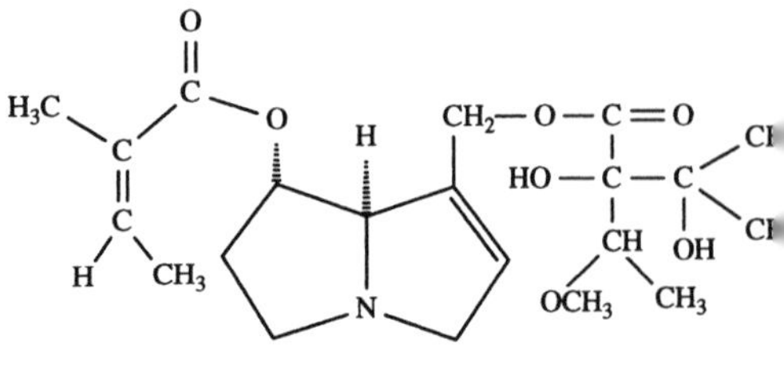

Lasiocarpine

In these alkaloids, the nitrogen atom is located in a pyrrolizidine structure, called the necic base, which is esterified with structurally complex acids (necic acids)

Pyrrolizidine alkaloids occur in plants belonging to the *Asteraceae*, *Boraginaceae* and *Fabaceae*. As regards the substances shown above, monocrotaline is found in *Crotalaria* species (*Fabaceae*), lasiocarpine in *Heliotropium* species (*Boraginaceae*) and senecionine in *Senecio* species (*Asteraceae*). The latter genus, especially, is rich in pyrrolizidine alkaloids, and this group of alkaloids is also often called *Senecio* alkaloids.

Pyrrolizidine alkaloids are potent poisons which often give rise to poisoning in grazing livestock. Poisoning in humans is reported from developing countries following the use of pyrrolizidine alkaloid-containing plants in traditional medicine. The main acute effect is liver damage, and it appears that the alkaloids are metabolized to related pyrrole derivatives which cause the damage. Pyrrolizidine alkaloids are also carcinogenic and formation of tumours can occur long after the ingestion of the alkaloids.

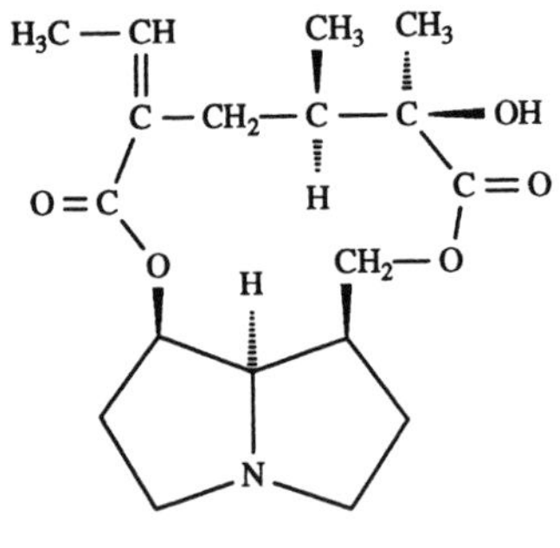

Platyphylline is a pyrrolizidine alkaloid found in several *Senecio* species and does not have hepatotoxic activity. This is ascribed to the fact that there is no double bond in the necic base. The alkaloid is used in the former Soviet Union for the treatment of hypertension.

Platyphylline

Biosynthesis of pyrrolizidine alkaloids

The biosynthesis of pyrrolizidine alkaloids is depicted below. The necic base of the pyrrolizidine alkaloids derives from homospermidine, which is formed from one molecule each of the primary metabolites spermidine and putrescine which are derived from arginine and thus, ultimately from ornithine (page 489).

The reaction is catalysed by *homospermidine synthase (spermidine specific)* (EC 2.5.1.45). The gene encoding this enzyme in *Senecio vernalis* has been cloned, sequenced and overexpressed in *Escherichia coli*. As illustrated in Fig. 204 spermidine is dehydrogenated to yield dehydrospermidine from which a 4-aminobutyli-

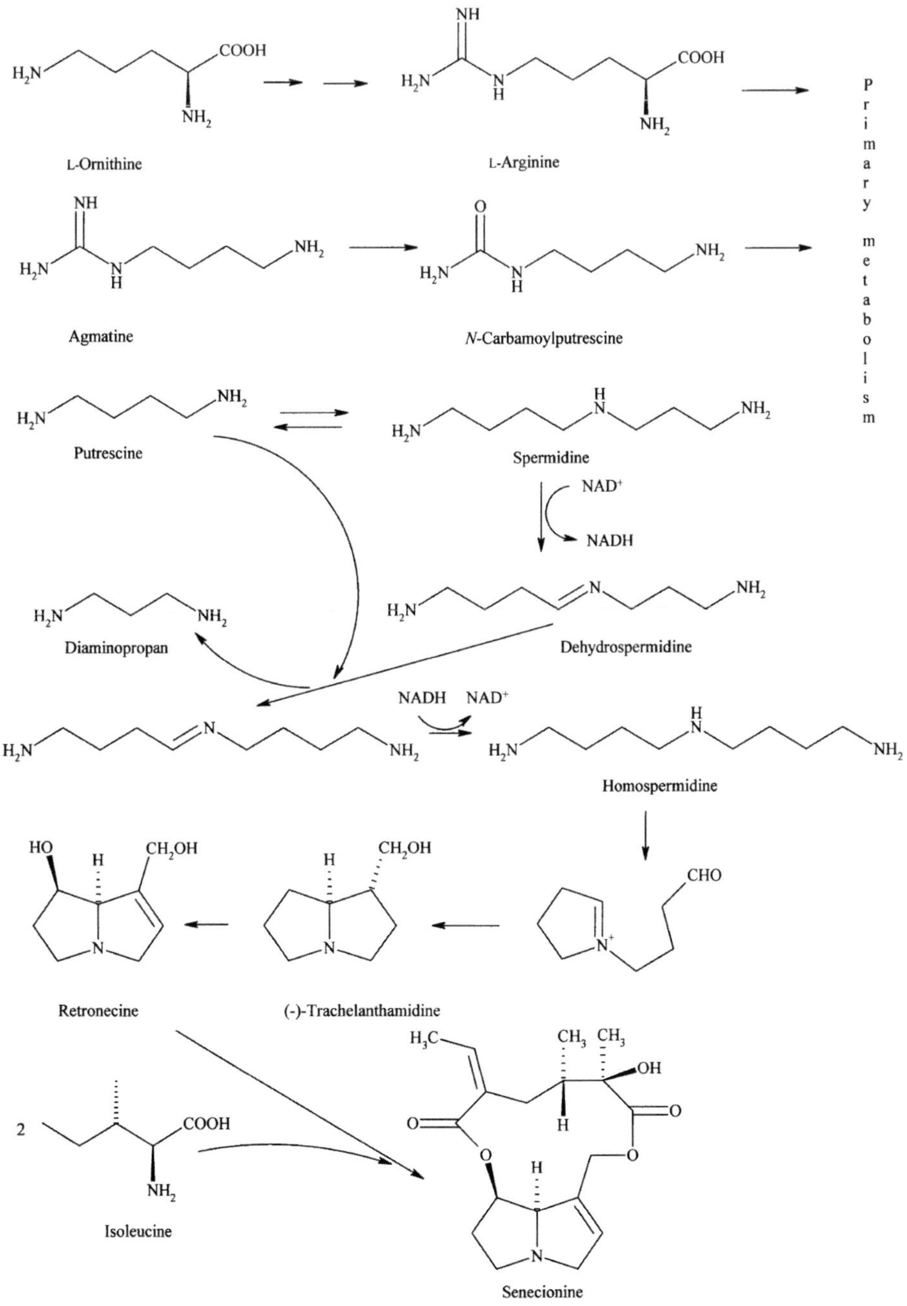

Fig. 204. Biosynthesis of pyrrolizidine alkaloids

dine group is transformed to putrescine with formation of an imine intermediate which is reduced to homospermidine. The following steps in the synthesis of the necine base moiety have not yet been characterized on an enzymatic level. Most likely an intermediate iminium ion is formed, the cyclization of which may lead to four stereoisomers of 1-hydroxymethylpyrrolizidine, all of which have been identified as necines in naturally occurring pyrrolizidine alkaloids. One of those is (-)-trachelanthamidine which is the specific precursor of retronecine, the by far most widespread necine base.

The necic acids are formed from amino acids such as isoleucine, leucine, valine and threonine. Combination of a necic acid with a necic base gives the final pyrrolizidine alkaloid.

Quinolizidine alkaloids (*Lupinus* alkaloids)

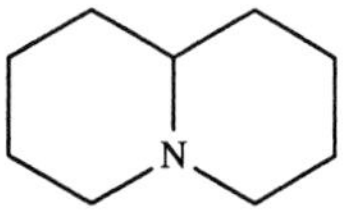

Quinolizidine

The basic skeleton of this group of alkaloids is quinolizidine (norlupinane), which consists of two saturated six-membered rings sharing a nitrogen atom. A number of quinolizidine alkaloids, e.g. sparteine (see below), are tetracyclic and consist of two norlupinane skeletons joined together.

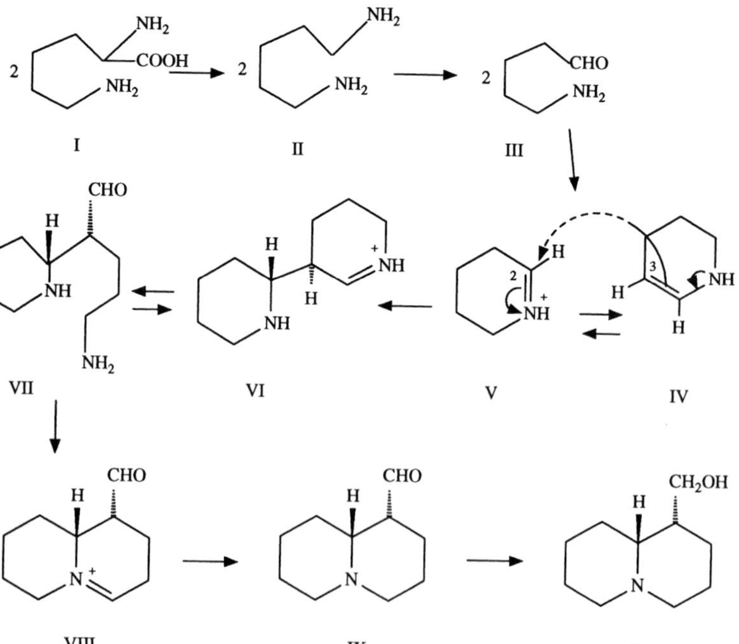

Fig. 205. Biosynthesis of quinolizidine alkaloids

Biosynthesis of quinolizidine alkaloids

The norlupinane ring system is formed from the amino acid lysine. Two lysine molecules (I) are decarboxylated to two cadaverine molecules (II), which undergo oxidative deamination to yield the corresponding aldehydes (III). The aldehydes then undergo ring closure forming either (IV) or (V), which are in equilibrium with each other. The next step is a stereospecific dimerization, as indicated in Fig. 205. In the product (VI), the ring containing the double bond is then opened by oxidation, leading to formation of the aldehyde (VII). Deamination and ring closure yields (VIII), which is reduced to (IX). Finally, the aldehyde function is reduced to form lupinine (X).

In forming the tetracyclic quinolizidine alkaloids (Fig. 206), the intermediate (VI) (Fig. 205) reacts with another molecule of intermediate (V) yielding compound (XI), which subsequently affords

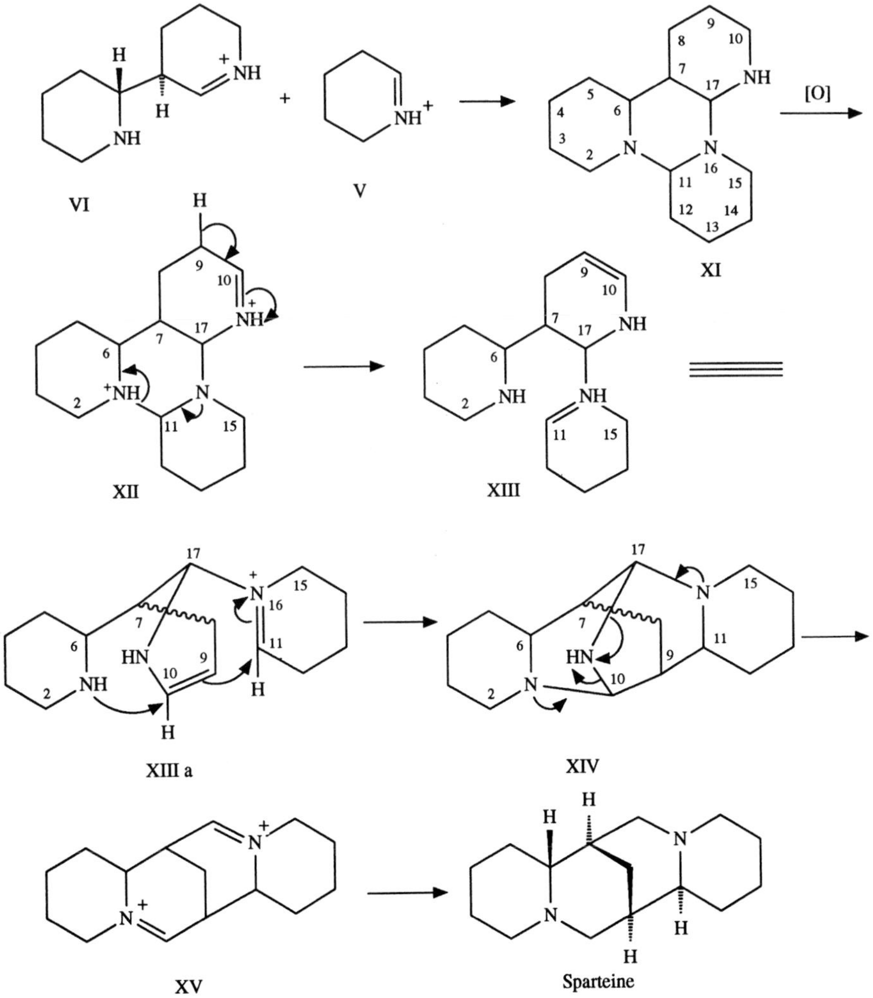

Fig. 206. Biosynthesis of tetracyclic quinolizidine alkaloids. The numbering of the atoms in the intermediate compounds is like that of the final product

643

(XII) and (XIII). To understand the later reactions, it is convenient to rewrite formula (XIII) as (XIIIa). Ring closure giving (XIV), is followed by loss of a nitrogen atom with the formation of (XV), which is reduced to sparteine.

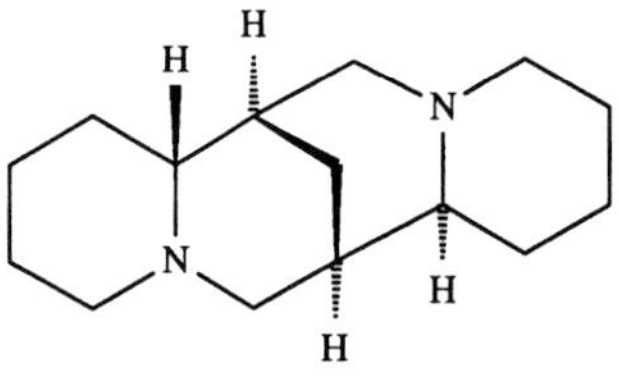

Sparteine

Sparteine is found in the common broom – *Cytisus scoparius* Link – as well as in species of broom belonging to the genus *Genista*; all these plants belong to the *Fabaceae*. The alkaloid has also been encountered in other plant families. The common broom is a shrub, growing in Europe and western Asia. The sparteine content in its leaves and branch tops is at its highest in the month of May and decreases after flowering, which takes place in June.

Sparteine has an effect on atrioventricular conduction which resembles that of quinidine, and it causes contraction of the uterus. It has a limited use in treating cardiac arrythmias and in bringing about contractions of the uterus during delivery.

Sparteine

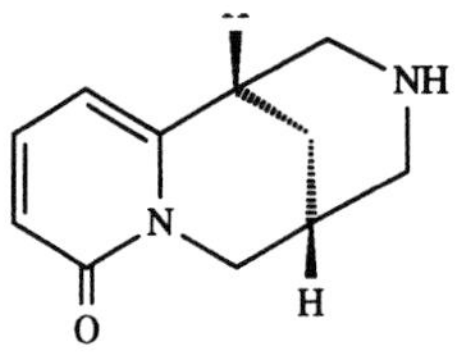

Cytisine

Cytisine occurs in the ornamental trees *Laburnum anagyroides* Medic. and *L. alpinum* J. Presl. (*Fabaceae*). In the spring, they produce yellow flowers in large hanging racemes and the fruit is a legume similar to a pea pod. All parts of the plant contain cytisine. The content is especially high in the seeds – up to 3 %. The pharmacological effects of cytisine resemble those of nicotine. The compound is not used medicinally, but it sometimes causes poisoning in children who have eaten laburnum fruits or sucked pieces of the bark.

Cytisine

Isoquinoline alkaloids

The isoquinoline skeleton is present in a large number of alkaloids, occurring in several only remotely related plant families.

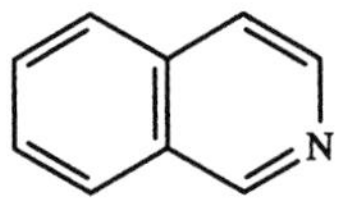

Isoquinoline

The amino acid tyrosine is the biosynthetic precursor of isoquinoline alkaloids. The biosynthesis of the tetrahydroisoquinoline alkaloids occurring in the peyote cactus, *Lophophora williamsii* has already been described (page 613) in connection with the biosynthesis of mescaline. These alkaloids contain a reduced isoquinoline skeleton, whereas other alkaloids in the group have a more complex composition and contain additional rings.

Protoberberine alkaloids

Berberine

Berberine is a yellow alkaloid which is found in several genera within the families *Berberidaceae* (*Berberis, Hydrastis*) and *Ranunculaceae* (*Coptis, Thalictrum* and other genera). Berberine has anti-amoebic, antibacterial and anti-inflammatory properties. It is widely used in Asia as a drug, usually in the form of crude plant extracts. In Japan the alkaloid is produced in cell cultures of *Coptis orientalis*.

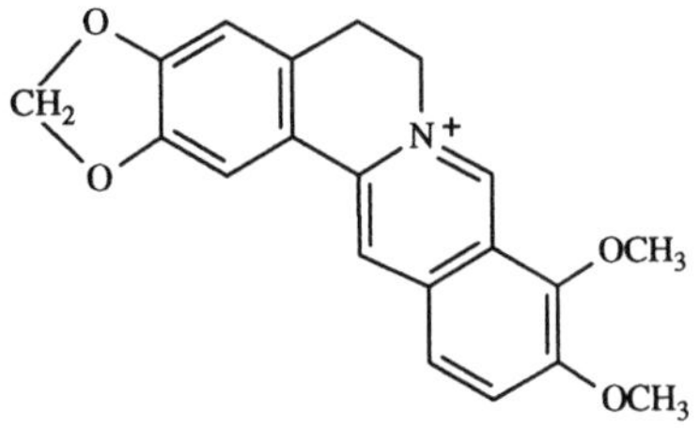

Berberine

Biosynthesis

Berberine is of great interest from a biosynthetic point of view, as all 13 enzymes required for its formation have been isolated and characterized. The first 9 steps of the pathway, resulting in formation of (*S*)-(+)-reticuline, are also present in the biosynthetic pathways to benzylisoquinoline alkaloids such as the medicinally important alkaloids morphine, codeine and noscapine (page 648) and to benzophenanthridine alkaloids like sanguinarine (page 665).

The studies on the biosynthesis of berberine have been performed on cell cultures mainly of *Coptis orientalis* and species of *Thalictrum*. Tyrosine is the amino acid from which these groups of alkaloids originate. The pathway from tyrosine to berberine is illustrated in Fig. 207.

The enzyme *tyrosine 3-monooxygenase* (EC 1.14.16.2) catalyses hydroxylation of tyrosine to L-dopa, which in the next reaction is decarboxylated to dopamine by *aromatic-L-amino-acid decarboxylase* (EC 4.1.1.28). *Tyrosine transaminase* (EC 2.6.1.5) catalyses the reaction between a second molecule of tyrosine and 2-oxoglutaric acid yielding 4-hydroxyphenylpyruvic acid which is decarboxylated by *4-hydroxyphenylpyruvate decarboxylase* (EC 4.1.1.80) resulting in the formation of 4-hydroxyphenylacetaldehyde. *(S)-norcoclaurine synthase* (EC 4.2.1.78) catalyses a stereoselective condensation of dopamine and 4-hydroxy-phenylacetaldehyde to (*S*)-norcoclaurine. Methylation of the 6-OH of norcoclaurine is catalysed by *(R,S)-norcoclaurine 6-O-methyltransferase (=6-OMT)*, EC 2.1.1.128) yielding (*S*)-coclaurine. *N*-methylation of coclaurine is mediated by *(S)-coclaurine-N-methyltransferase* (EC 2.1.1.140) resulting in the formation of (*S*)-*N*-methylcoclaurine which is oxidized by N-*methylcoclaurine 3´-monooxygenase* (EC 1.14.13.71) to yield 3'-hydroxy-*N*-methylcoclaurine. Methylation of 4'-OH in this compound, catalysed by *3'-hydroxy-*N-*methyl-(S)-coclaurine-4'-O-methyltransferase* (=4'-OMT, EC 2.1.1.116) yields (*S*)-(+)-reticuline. This alkaloid is a key-compound in the biosynthesis also of other benzylisoquinoline alkaloids, e.g. the opium alkaloids (see below). 4'-OMT was obtained together with 6-OMT from cell cultures of *Berberis koehneana*. The two enzymes could not be separated by conventional means but the 6-OMT can be inactivated by thermal denaturation. At -20 °C, 6-OMT loses all activity in 3 months while after 1 year the 4'-OMT retains 95 % of its original activity. In both reactions the methyl group derives from *S*-adenosyl-L-methionine (SAM). 4'-OMT is highly specific. Only the (*S*)-enantiomer of the substrate is recognized and the enzyme is regiospecific for only one hydroxyl in ring C.

Hitherto all reactions have occurred in the cytosol, but before the next reaction takes place reticuline has to enter a specific vesicle in which *reticuline oxidase* (EC 1.21.3.3., also known as the berberine bridge enzyme) catalyses oxidative cyclization of the *N*-methyl group resulting in formation of (*S*)-scoulerine, which is the parent compound of a multitude of tetrahydroprotoberberine alkaloids and derivatives thereof. The formation of (*S*)-scoulerine requires consumption of one mole of O_2 and is accompanied by formation of one mole of H_2O_2. It is inhibited in the presence of metal chelators. The reaction is unusual and cannot be mimicked by current chemical techniques and does not have an equivalent in Nature. The exact mechanism has not yet been been elucidated. The berberine bridge enzyme was originally purified from cell-suspension cultures of *Berberis beaniana* and characterized as a single polypeptide with a molecular mass of 52 ± 4 kDa. The enzyme has also been isolated from elicited cell-suspension cultures of the California poppy, *Eschscholzia californica*. The cDNA for this enzyme has been cloned and found to contain the entire reading frame for the berberine bridge enzyme, coding for 538 amino acids. The first 22 amino acids from the amino terminus constitute the putative signal peptide that should direct the preprotein into the specific vesicle where the enzyme is accumulated. These 22 amino acids are absent from the mature protein which has a molecular weight of 57 352, excluding carbohydrate, as calculated from the translation of the nucleotide sequence. Three carbohydrate consensus sequence sites were found in the sequence. The cDNA was expressed in *Saccharomyces cerevisiae* and gave an active enzyme with all properties identical to those obtained for the enzyme previously isolated from the cell cultures of *E. californica*.

The next enzyme in the pathway to berberine is *(S)-scoulerine-9-O-methyltransferase* (EC 2.1.1.117) which is a cytosolic enzyme, implying that the newly formed (*S*)-scoulerine has to leave the vesicle before methylation of the hydroxyl group at C-9 yielding (*S*)-tetrahydrocolumbamine. This alkaloid re-enters the vesicle where *(S)-canadine synthase* (EC 1.14.21.5) mediates the formation of the methylenedioxy bridge of canadine. Finally berberine is formed by

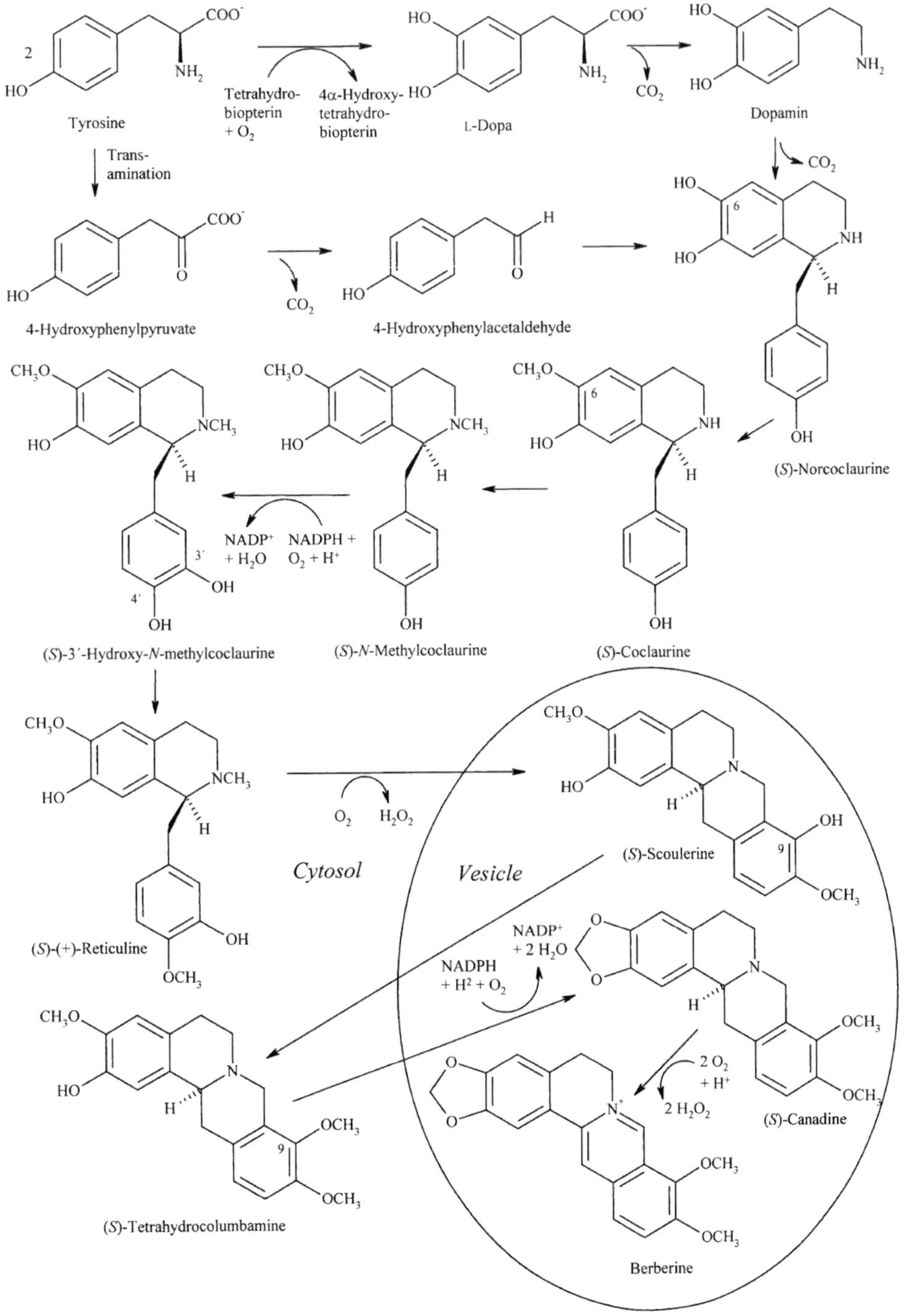

Fig. 207. Biosynthesis of berberine

aromatization of the C-ring, a reaction which is catalysed by *tetrahydroberberine oxidase* (EC 1.3.3.8). The reaction requires the presence of molecular oxygen and is accompanied by formation of two moles of hydrogen peroxide. The enzyme has a molecular mass of 58 kDa and consists of two identical subunits.

Benzylisoquinoline alkaloids

Opium alkaloids

The opium poppy, *Papaver somniferum*, is the source of the crude drug opium (page 657), which contains a large number of alkaloids. Four of them have attained wide medicinal use: the benzylisoquinoline derivative papaverine, the phenanthrene derivatives morphine and codeine, and the phthalide-isoquinoline derivative noscapine (narcotine).

Papaverine

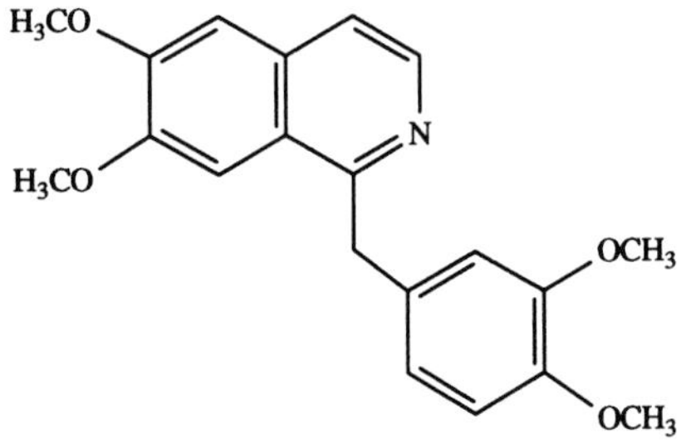

Papaverine

Papaverine is structurally one of the simplest opium alkaloids and is typical of the benzylisoquinoline group of alkaloids. The alkaloid can be isolated from opium, which contains 1 %, and it is used in medicine as an antispasmodic.

Morphine

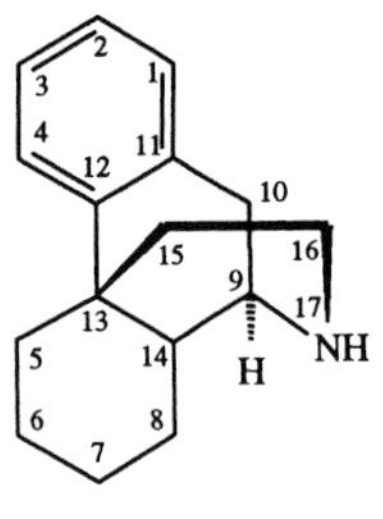

Morphinan

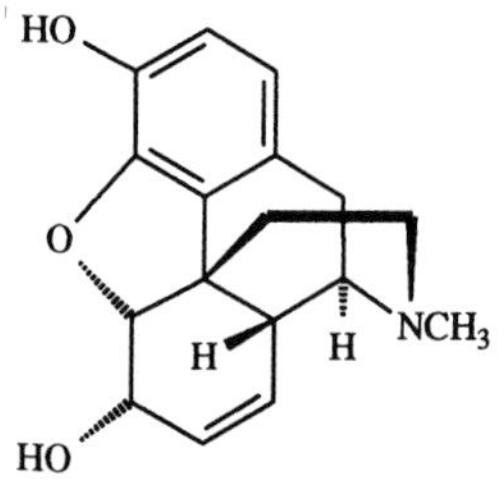

Morphine

From a structural point of view, morphine can be regarded as a phenanthrene derivative, but studies of its biosynthesis have shown that the alkaloid is formed from tyrosine via the benzylisoquinoline alkaloid reticuline (see below). The 4-ring system contained in morphine is called *morphinan* and is numbered as indicated above. Alkaloids containing this ring system are often referred to as *morphinan alkaloids*. Morphine is the main alkaloid of opium (10–20 %) and, because of its phenolic properties, can be separated relatively easily from the other alkaloids. Morphine is used as a very effective analgesic, but it must be handled with great care because of the risk of addiction. Morphine is classified as a narcotic. In addition to the analgesic effect, it causes euphoria, i.e. a strong sense of wellbeing. On repeated use, increased doses are required to bring about this effect, and in the end the patient cannot do without the drug. If it is not provided, the body reacts intensely and the victim has the feeling that he must have morphine at any price, in order to be able to live.

Codeine   Codeine is another constituent of opium (0.8–2.5 %), but it is mostly prepared by the methylation of morphine. Codeine is also an analgesic, but a much weaker one than morphine. It enhances the effects
of other analgesics, such as acetyl-salicylic acid, and large amounts
of codeine are used for this purpose. Codeine is also an antitussive.
The risk of addiction is much smaller than with morphine.

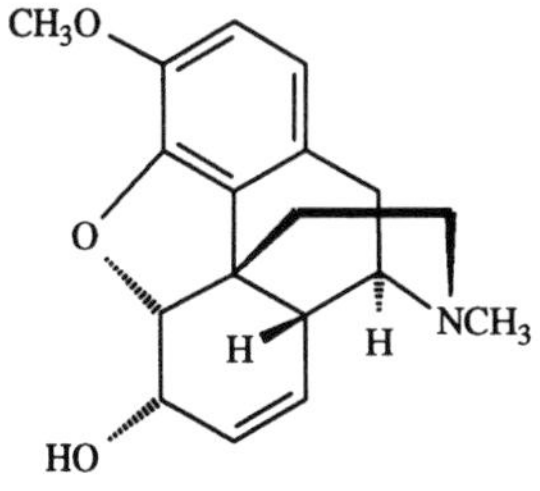

Codeine

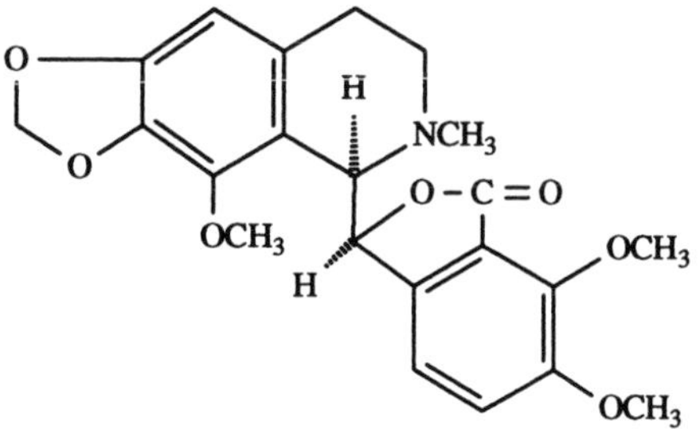

Noscapine

Noscapine   Opium contains 4–8 % of noscapine. This alkaloid subdues the
cough reflex and is used to alleviate a hacking cough. Noscapine
has no euphoric effect.

Thebaine   Thebaine has no use in medicine but some derivatives are important
drugs (see below). Thebaine is also a key intermediate in the
biosynthesis of morphine and codeine (Fig. 210).

Semisynthetic and synthetic derivatives of morphine and thebaine

Diamorphine (heroin)   Diamorphine is the diacetyl derivative of morphine (Fig. 208)
which is very easy to synthesize. It was marketed in the beginning
of the 20th century as a safer drug than morphine for relieving pain,
because it did not cause respiratory depression, a not infrequent
cause of death when patients with severe injuries were treated with
high doses of morphine. The name heroin was given to the drug
because it was thought to be a heroic drug in being a safe painkiller.
It also became very popular as a cough remedy and was thought to
be as safe as codeine. However, it soon became apparent that heroin also caused euphoria and was even more addictive than morphine
and in 1912 a report appeared in *Journal of the American Medical
Association* which warned of its abuse by addicts and called for legislation to prevent its free sale. Today heroin is one of the worst illicit drugs, the addiction to which it is very difficult to get rid of. Most
of the illicitly produced opium is used for production of heroin.
Most addicts inject the drug, often using the same syringe for several persons, a habit which has contributed to the spreading of
AIDS. There is also a preparation available for smoking. The med-

ical use of heroin is banned in the USA but it is used in the UK in terminal care, particularly for cancer patients.

Ethylmorphine

Ethylmorphine is the monoethyl-ether of morphine (Fig. 208). The compound is used in mixtures as a cough-relieving remedy.

Dextromethorphan

Dextromethorphan is a synthetic compound which retains much of the structure of morphine (Fig. 208). The stereochemistry of the nitrogen-containing bridge is, however, the opposite of that in morphine and the ether bridge is missing. The cyclohexene ring is saturated and has no substituents. The substance is completely devoid of euphoric activities and does not relieve pain. It is used as a cough remedy.

Pentazocine

Pentazocine is another synthetic compound, based on the structure of morphine (Fig. 208). In comparison with morphine the cyclohexene ring is removed as well as the ether bridge and a 3-methyl-2-butenyl group is substituted for the methyl group at the nitrogen. Pentazocine is used as an analgesic. It has a slight euphoric effect but the risk for addiction is much less pronounced. In some patients it can cause hallucinations when injected. These can be counteracted by the morphine-antagonist naloxone (see below).

Pethidine (mepiridine)

It is difficult to recognize any structural relationship between pethidine and morphine. Pethidine was not synthesized as a morphine analogue but as an atropine analogue in the hope of finding an antispasmodic drug for use in the treatment of bladder or intestinal spasm. However, it was discovered that it causes the so-called Straub effect on the mouse tail which consists of the tail flipping back in a rigid S-shape. This effect is typical of narcotic analgesics and indeed the compound was found to have about one-tenth the potency of morphine with rapid onset of analgesia and a shorter duration of action. At the beginning it was thought that it would not have the ability to induce tolerance and addiction because of the seeming lack of structural relation to morphine, but it was soon found that it resembled morphine in these effects as well. These results show that the essential structural elements for the pharmacological activities of morphine are the aromatic ring and the piperidine ring. Pethidine is used against severe pain.

Etorphine

As shown in Fig. 209 thebaine is used for semisynthesis of several drugs. Etorphine is synthesized using the conjugated diene structure in thebaine for a Diels–Alder cyclo-addition reaction. The comparatively long side-chain makes the compound more lipophilic permitting greater penetration into the central nervous system. The compound is several thousand times more potent than morphine but is not used clinically as it is not more safe than morphine and has no other desirable feature. It is, however, used for immobilization of large animals because only small doses are required and can easily be injected by darts.

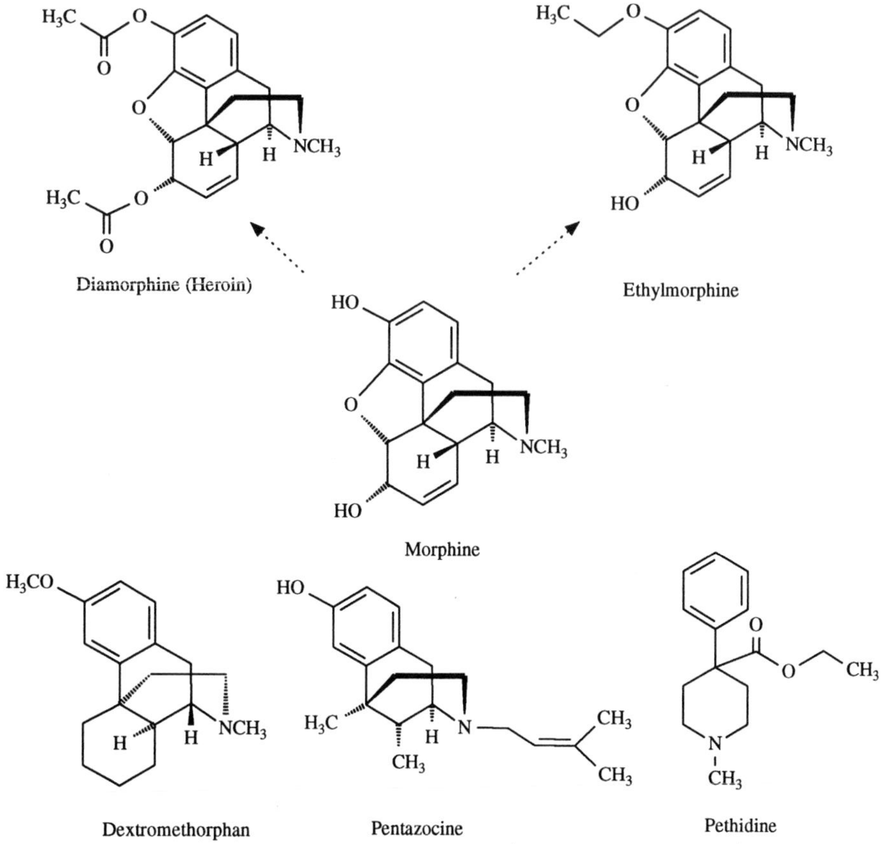

Fig. 208. Semisynthetic and synthetic derivatives of morphine

Buprenorphine Buprenorphine shows a close structural resemblance to etorphine, the main difference being the *N*-cyclopropylmethyl substituent. Buprenorphine combines morphine agonist and antagonist properties. As an analgesic it is about 100 times as potent as morphine, and the activity lasts for 6–8 hours. It has to be administered by injection or by sublingual application because of rapid hepatic metabolism. Buprenorphine is used for relief of postoperative pain. Because of its antagonistic properties patients addicted to other opiates can suffer withdrawal symptoms when treated with buprenorphine.

Naloxone Naloxone is a powerful opiate antagonist which is made semisynthetically from thebaine. Medically it is used to treat opiate poisoning including children born to heroin addicts. Naloxone is also a powerful research tool as it binds to all opioid receptors with a particularly high affinity for the μ-receptor which mediates analgesia, respiratory depression and addiction.

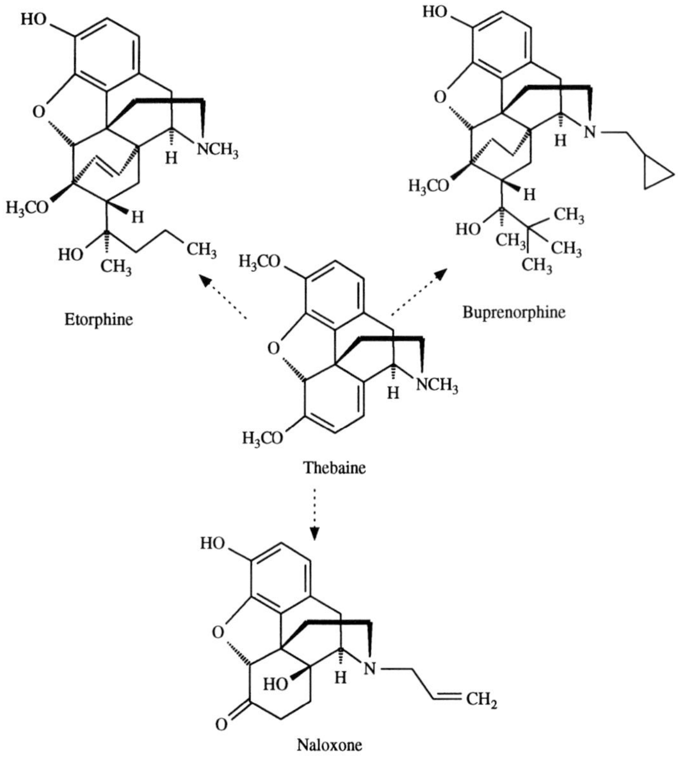

Etorphine

Buprenorphine

Thebaine

Naloxone

Fig. 209. Semisynthetic derivatives of thebaine

Biosynthesis of the opium alkaloids

Codeine and morphine The first 9 steps of the biosynthetic pathway to codeine and morphine, resulting in the formation of (*S*)-(+)-reticuline, are the same as in the formation of this compound in the berberine pathway (page 647, Fig. 207). In order to understand the continuation of the biosynthesis of codeine and morphine from reticuline, the structure for (*S*)-reticuline can be rewritten as shown on the following page.

The molecule as written in Fig. 207 (page 647) is first turned upside down in the plane of the paper (A). Then the tetrahydro-isoquinoline ring is rotated 180° about the bond to the methylene group (B). Finally the *N*-methyl group and carbon *a* are retained in their position while the rest of the tetrahydro-isoquinoline ring system is rotated 180° about the dotted line (C).

Reticuline occurs in two stereoisomeric forms: *S*-(+)- and *R*-(-)-reticuline. The two isomers are interconvertible via the 1,2-dehydroreticulinium ion, an intermediate which has been found to be naturally occurring in *Papaver somniferum*. An enzyme, *1,2-dehydroreticuline reductase* (EC 1.5.1.27), has been isolated from seedlings of *Papaver somniferum*. This enzyme stereospecifically reduces 1,2-dehydroreticuline to (*R*)-(-)-reticuline. The isolated enzyme is cytosolic, NADPH dependent and constitutes a single

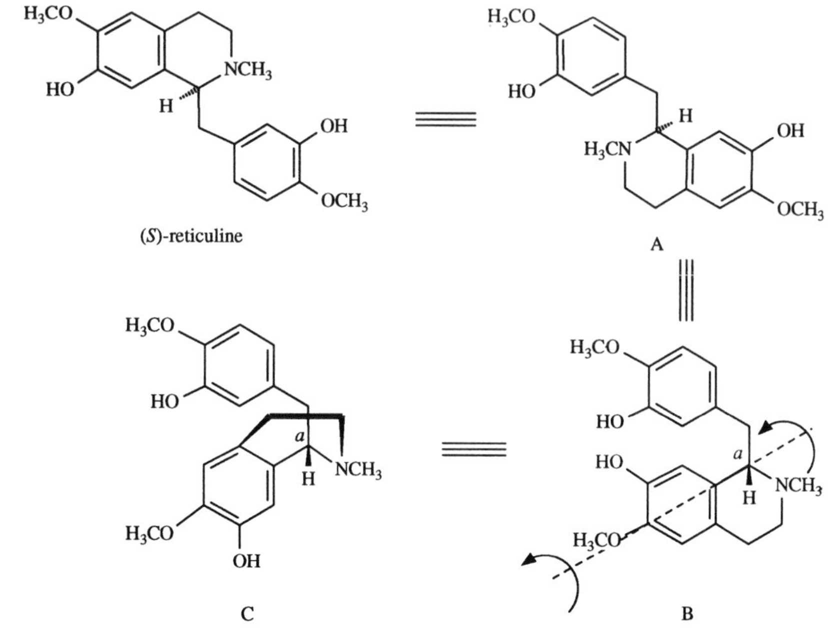

(S)-reticuline

A

C

B

polypeptide with a molecular mass of 30 kDa. It is highly substrate specific and has been found only in morphinan alkaloid-containing plants.

The biosynthesis of codeine and morphine from R-(-)-reticuline is illustrated in Fig. 210. (R)-(-)-reticuline is first transformed to the dienone salutaridine by regioselective *para-ortho* oxidative coupling, catalysed by *salutaridine synthase* (EC 1.14.21.4) a microsomal NADPH-dependent cytochrome P450 enzyme. Unlike most of the enzymes of this class salutaridine synthase functions as an oxidase rather than as a monooxygenase – there is no associated incorporation of oxygen into the substrate during the course of reaction.

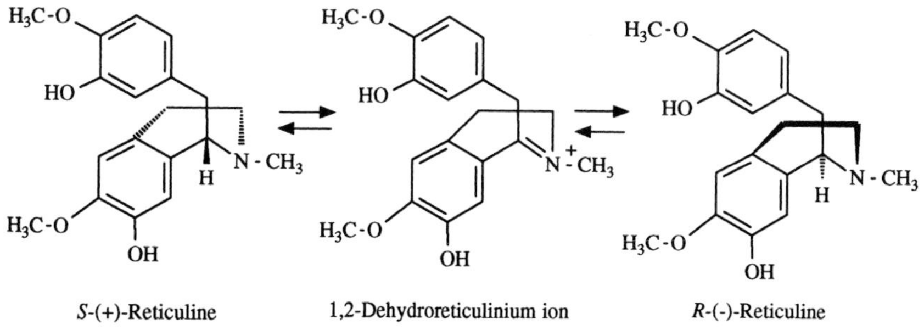

S-(+)-Reticuline   1,2-Dehydroreticulinium ion   R-(-)-Reticuline

Salutaridine is reduced to salutaridinol (7S) by *salutaridine reductase* (EC 1.1.1.248, also known as salutaridine: NADPH 7-oxidoreductase). This reaction proceeds at pH 6.0–6.5. The enzyme can also catalyse the reverse reaction – salutaridinol to salutaridine – but the pH optimum for that reaction is much higher: 9.0–9.5. The enzyme is a single polypeptide of molecular mass 52 ± 3 kDa which is absolutely dependent on NADPH/NADP as pyridine nucleotide cofactors. It was originally isolated from cell cultures of *Papaver somniferum* but has also been shown to be present in capsules and seedlings of the plant. This enzyme has not been found in any other *Papaver* species except the thebaine-producing *P. bracteatum* (page 659). It is thus highly specific for plants producing the morphinandienone skeleton.

The oxide bridge in codeine and morphine is closed in a reaction catalysed by the enzyme *salutaridinol 7-O-acetyltransferase* (EC 2.3.1.150) which transfers an acetyl group from acetyl-CoA to the hydroxy group at C-7 of salutaridinol. The product – salutaridinol-7-O-acetate – undergoes a spontaneous allylic elimination yielding the alkaloid thebaine. The spontaneous elimination reaction has been demonstrated to occur *in vitro* at pH 8–9 and no enzyme catalysing this reaction has been found. It is therefore assumed that the elimination reaction occurs spontaneously also *in vivo*. Salutaridinol-7-O-acetyltransferase was isolated from a cell culture of *Papaver somniferum* and has a molecular mass of 50 ± 1 kDa. The enzyme is highly substrate specific. The epimer of salutaridinol – 7-*epi*-salutaridinol – is not acetylated by the enzyme, thus confirming the previous finding that this compound is not a precursor for thebaine.

Thebaine is demethylated to yield neopinone which spontaneously rearranges to codeinone. Codeinone is reduced to codeine by a highly substrate-specific and stereoselective enzyme called *codeinone reductase (NADPH)* (EC 1.1.1.247). Codeinone reductase has been purified and found to be a monomeric protein with the molecular mass 35 kDa. Four cDNAs encoding codeinone reductase have been isolated from *Papaver somniferum* and have been expressed in *Escherichia coli*. The four isoforms have very similar properties and substrate specificity.

Finally, demethylation of codeine yields morphine. The two demethylation reactions are unusual in a biosynthetic pathway. It has been suggested that the methyl groups serve as protective groups during the biosynthesis, thereby avoiding other possible reactions in which the hydroxy groups in question could participate.

Noscapine   The precursor of noscapine is (*S*)-(+)-reticuline. The lactone carbonyl group is derived from the *N*-methyl group of reticuline. It has been assumed that a derivative of protoberberine, e.g. scoulerine, is formed and then oxidized and methylated to noscapine (see Fig. 211, page 656).

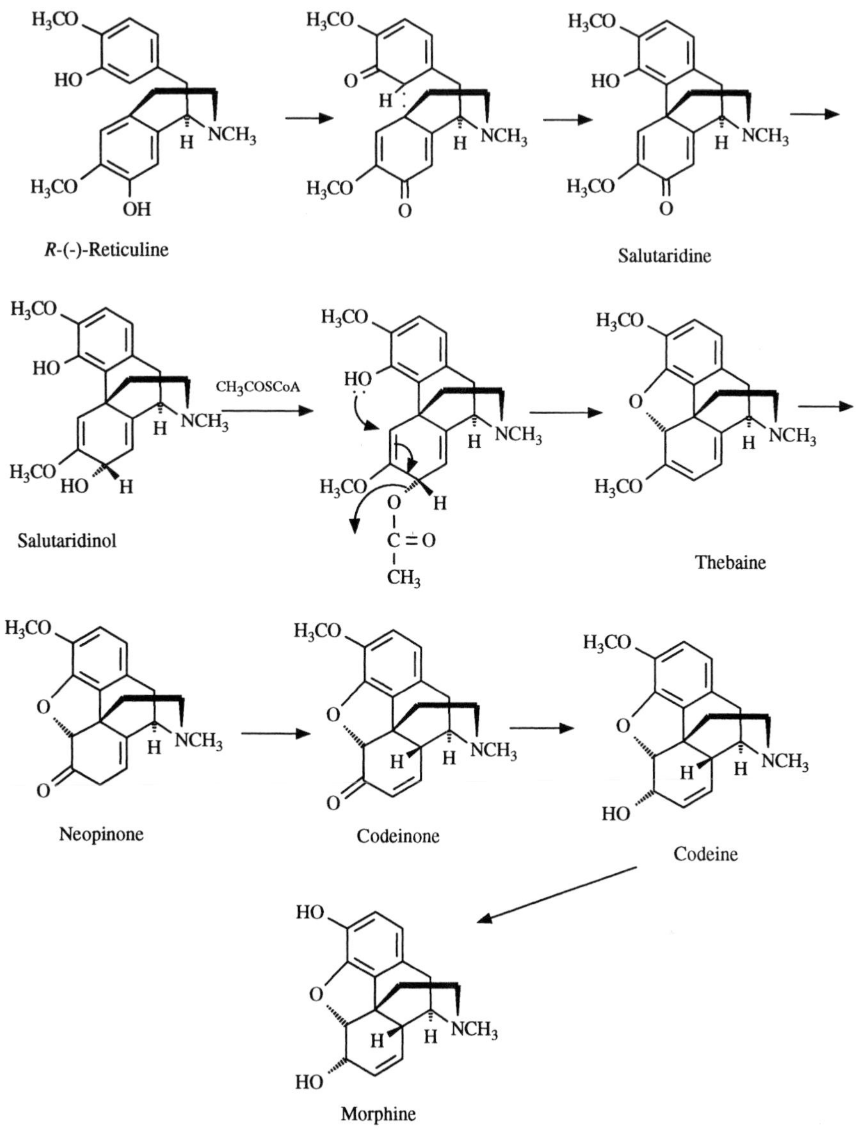

Fig. 210. Biosynthesis of codeine and morphine

Papaverine   The precursor of papaverine is norcoclaurine which is biosynthesized from dopamine and 4-hydroxyphenylacetaldehyde as previously described (page 647, Fig. 207). Norcoclaurine is methylated
in several steps, the first poduct being *N*-nor-reticuline. Further
methylation yields tetrahydropapaverine, which is finally dehydrogenated to papaverine (Fig. 212).

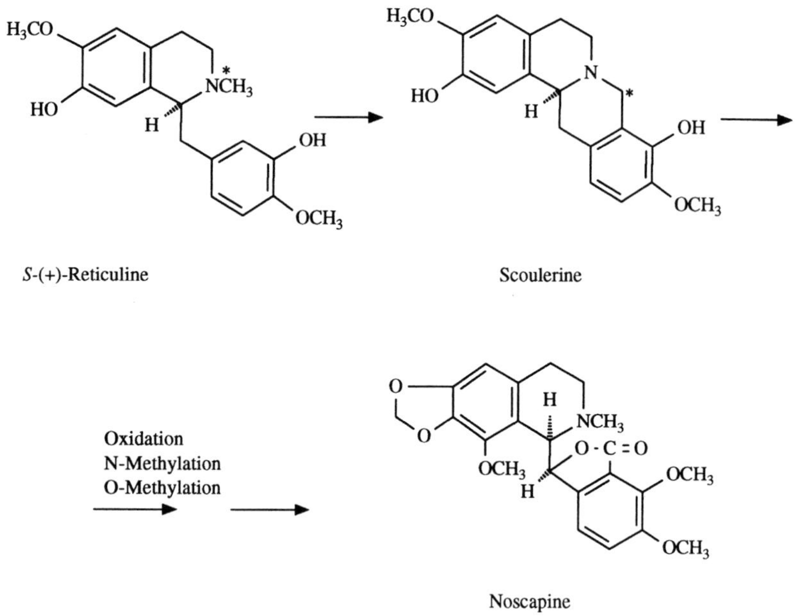

S-(+)-Reticuline

Scoulerine

Oxidation
N-Methylation
O-Methylation

Noscapine

Fig. 211. Biosynthesis of noscapine

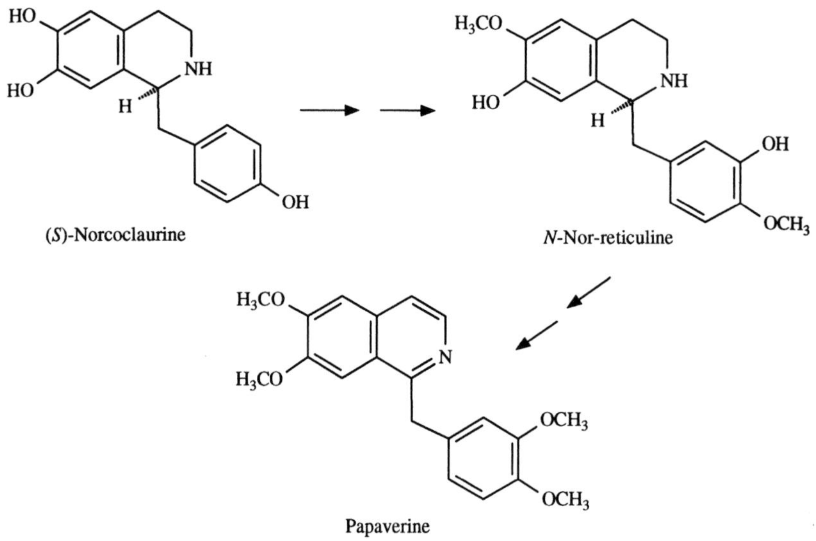

(*S*)-Norcoclaurine

N-Nor-reticuline

Papaverine

Fig. 212. Biosynthesis of papaverine

Crude drug   **Opium** is the inspissated latex from unripe fruit capsules of *Papaver somniferum* L. (*Papaveraceae*), the opium poppy. This plant was known as an analgesic and a sedative more than a thousand years before Christ, and the custom of obtaining opium from the plant was practised at least as early as the 3rd or 4th century BC. *Papaver somniferum* is an annual herb, up to 1–1.5 metres in height, with elongated ovate leaves and white, violet or purple flowers. Current opinion favours the western Mediterranean as the region where the opium originally came from. It is grown in Bulgaria, Greece, Iran, Turkey, Yugoslavia, the former Soviet Union, India, China, Vietnam and Mexico. The cultivation of opium poppy for the production of opium and alkaloids is controlled by the International Narcotics Control Board of the United Nations. Turkey was formerly an important producer of opium, but nowadays the opium poppy is grown there only for the production of alkaloids from the capsules (poppy straw). The illegal production of opium occurs, however, in many countries, particularly in the border region where Burma, Thailand and Laos meet, the so-called "Golden triangle" and in Afghanistan. In the year 2000 the global production of opium and poppy straw concentrate was estimated at 8 700 tons, 54 % of which were illicit. The area occupied for this production was over 300 000 hectares of which 71 % were used for illicit purposes [Blakemore, P.R. and White, J.D. Morphine, the Proteus of organic molecules. *Chemical Communications*, 1159–1168 (2002)] *P. somniferum* can also be grown in colder regions like Scandinavia.

The unripe fruit contains latex in an anastomosing system of laticifers. If the surface of the capsule is damaged, the latex oozes out in the form of white droplets which quickly turn brown and become semi-solid. The product is opium. To obtain opium, the capsules are incised just before ripening, using special knives or spikes which penetrate only half-way through the fruit wall. Care must be taken to avoid cutting right through the capsule wall, as this will lead to losses through leakage of the latex into the inside of the capsule. The incision, or incisions, can be made either horizontally or vertically. In principle, one cut is sufficient, as the latex vessels are all interconnected. The incision is made in the afternoon, and next day the opium is scraped off with a special tool. When carrying out the incision, the emerging latex, which is liquid at first, must not be shaken off. The weather should be dry and calm. Rain in the period between making the incision and collecting the opium will spoil the harvest. The opium scraped off the capsules is collected and left to dry further in the air. It is then made into lumps and sent to the factory for initial processing. Here, the morphine content of the various batches is determined, and batches with different contents are mixed so that crude opium with a reasonably constant content of morphine is obtained. Today, India is the only exporter of opium, which is marketed in the form of rounded, somewhat flattened cakes, weighing 5 kg. The morphine content is 9–12 %.

Plate 25.
Papaver bracteatum Lindley. In this species, the biosynthesis of opium alkaloids does not proceed beyond thebaine (page 659). Jakobslund, Östergötland, Sweden. Photograph by Gunnar Samuelsson.

Plate 26.
Catharanthus roseus (L.) G. Don. The plant is the raw material for the extraction of vincristine and vinblastine (page 681). Joowhar, Somalia. Photograph by Gunnar Samuelsson.

Opium has a very characteristic smell and can be identified microscopically by the presence of typical fragments of the capsule wall.

Opium is used as a raw material for preparing morphine, codeine and noscapine, but these alkaloids are nowadays to a large extent produced by direct extraction from the capsules and so-called poppy straw (see below) and not from opium.

Opium for smoking. In the Orient, people smoke opium in order to become intoxicated. The opium to be used for this purpose is called *chandu* and is prepared in the following manner:

Opium is mixed with twice the amount of water and the water is evaporated while stirring until the residue is a thick mass. Any foam formed is removed during the evaporation process. The thick residue is roasted over a charcoal fire at about 200 °C and the mass is dissolved in water and filtered. The filtrate is boiled down to a syrup. After cooling, the extract is vigorously stirred to produce a foam and then left to ferment for several months. A product with a fine and characteristic aroma is obtained, which is smoked in special pipes. Opium smoking is habit-forming and subverts the personality, like morphinism.

Production of morphine from poppy straw

The production of opium is very laborious and can, therefore, be practised only in countries with cheap labour. For this reason, methods have been developed for preparing morphine directly from the capsules and other aerial parts of *P. somniferum*. These methods can be combined with the cultivation of the opium poppy for seed production. The seeds contain an excellent fixed oil, which is very suitable for margarine production. The opium poppy is grown in several European countries for its seeds, and morphine is extracted from poppy straw in France, Switzerland, Hungary and as mentioned above, Turkey. Most of the morphine obtained is methylated to codeine, which is used in much larger amounts than morphine.

The risks attached to morphine have led to the search for other sources of codeine. Another poppy species, *P. bracteatum* Lindley (Plate 25, page 658), native to Iran, lacks an enzyme for the demethylation of thebaine and, consequently, it produces this alkaloid instead of morphine and codeine. The thebaine content can reach 3.5 % in the dried capsules, which could be sufficient for economic production on a large scale. Chemical demethylation of thebaine to codeine is relatively simple. *P. bracteatum* is a perennial, so that the plant can be harvested every year. Cultivation on a pre-production scale has been backed by the United Nations but has not been successful because of technical difficulties. Among other things, it has proved difficult to develop an economic method of weed control. The project appears to have been abandoned.

Endogenous morphine

Receptors for morphine and codeine have been detected in the brain. This discovery resulted in the isolation of the enkephalins, endorphins and dynorphins, which are small peptides produced primarily but not exclusively in the pituitary gland. These peptides interact with the opioid receptors inducing analgesia and depressing respiratory function and several other processes much in the same way as morphine, and their activity is also counteracted by opiate antagonists like the semisynthetic thebaine derivative naloxone (page 651). They are also implicated in analgesia produced by the old Chinese practice of acupuncture. It is now well established that these peptides play several important roles in the central nervous system in addition to their role in the process of analgesia.

Tyr-Gly-Gly-Phe-Met-Thr-Ser-Glu-Lys-Ser-Gln-Thr-Pro-Leu-Val-Thr-

Leu-Phe-Lys-Asn-Ala-Ile-Val-Lys-Asn-Ala-His-Lys-Lys-Gly-Gln

β-Endorphin

Tyr-Gly-Gly-Phe-Met	*Tyr-Gly-Gly-Phe*-Leu
Met-enkephalin	Leu-enkephalin

Tyr-Gly-Gly-Phe-Leu-Arg-Arg-Ile-Arg-Pro-Lys-Leu-Lys-Trp-Asp-Asn-Gln

Dynorphin A

Fig. 213. The amino acid sequence of endogenous opiod peptides. Italics indicate an amino acid sequence common to all four peptides

Interestingly enough it has also been demonstrated that codeine and morphine are present in mammalian tissue, particularly in brain tissue and the feeding of (*R*)-(-)-reticuline to the tissues resulted in production of salutaridine. These findings have led to the suggestion that the morphinane alkaloids might be biosynthesized in mammalian tissues by a pathway similar to that in *Papaver somniferum*.

The concentration of morphine is very low (toad skin: 3 pmol/g tissue; bovine cerebral cortex: 0.05 pmol/g tissue), concentrations which are much below the concentration of morphine required for an effective interaction with μ-opioid receptors (about 1 nmol). However, in rats with adjuvant-induced arthritis the spinal cord level of morphine was observed to be increased four times in comparison with normal animals, and there was also a considerable increase in urinary excretion of morphine. It is possible that endogenous morphine may be involved in the regulation of some kinds of pain.

It has also been observed that morphinan alkaloids are normal constituents of human urine although in very low concentrations (codeine: 3 pmol/ml; morphine: 2 pmol/ml).

Bisbenzylisoquinoline alkaloids

The bisbenzylisoquinoline class of alkaloids are dimers of tetrahydrobenzylisoquinolines connected by one to three ether linkages formed by phenol coupling. Over 270 members of this class of alkaloids are known. Many of them are pharmacologically active agents such as skeletal muscle relaxants, antihypertensives, antimalarials and cytotoxins. The structural variations include substitutions on the phenyl rings, the regiospecificity of the ether linkages and the stereospecificity of the isoquinoline moieties.

Biosynthesis of bisbenzylisoquinoline alkaloids

An initial insight into the biosynthesis of these alkaloids has been obtained through studies on cell suspension cultures of *Berberis stolonifera* which produce the alkaloids berbamunine and guattegaumerine. Berbamunine is formed from one molecule of (S)-N-methylcoclaurine and one molecule of (R)-N-methylcoclaurine which are connected by an ether linkage, the formation of which is catalysed by *berbamunine synthase* (EC 1.14.21.3) (Fig. 214). This enzyme also couples two molecules of (R)-N-methylcoclaurine to form guattegaumerine. The enzyme has been isolated and part of the N-terminal amino acid sequence determined. The cDNA encoding berbamunine synthase has been isolated and expressed in an insect cell culture. It is a cytochrome P450-dependent oxidase which is highly regio- and stereoselective. The reaction catalysed by this enzyme is novel for cytochromes P450 in that it proceeds

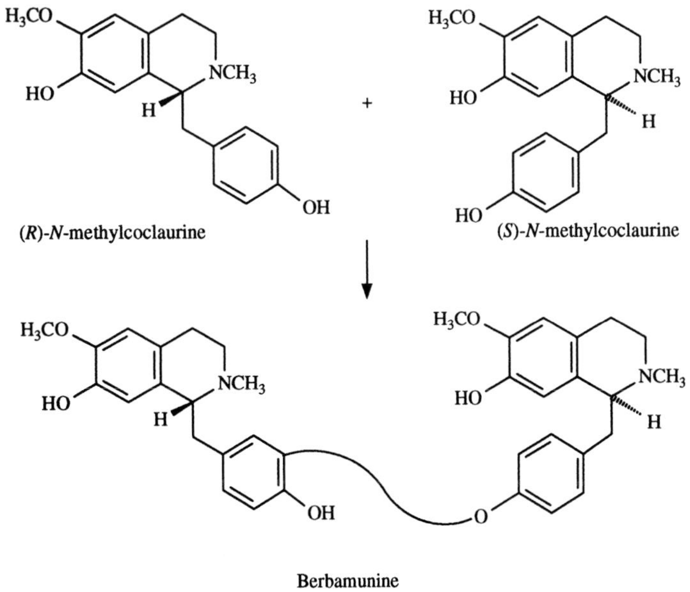

(R)-N-methylcoclaurine

(S)-N-methylcoclaurine

Berbamunine

Fig. 214. Biosynthesis of berbamunine

without incorporation of oxygen into the product, i.e. it functions as an oxidase rather than as a monooxygenase.

Tubocurarine A medicinally useful alkaloid belonging to this group, tubocurarine, can be isolated from the South American dart poison curare.

One of the two nitrogen atoms in the isoquinoline ring systems in tubocurarine is quaternary. The alkaloid causes relaxation of the skeletal muscles through blockade of the acetylcholine receptor on

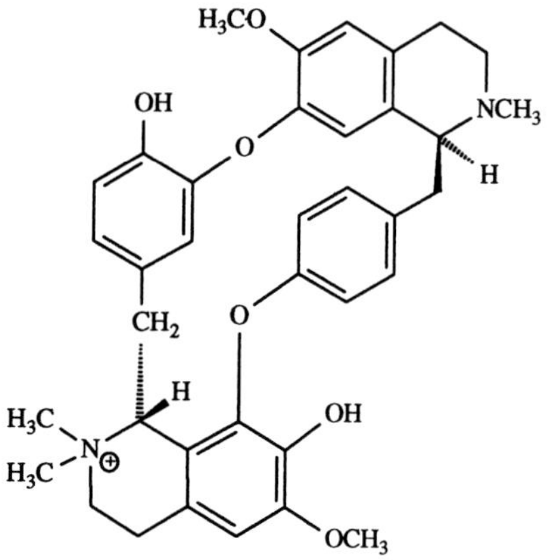

Tubocurarine

the motor endplates. The effect is of limited duration and the substance is used to relax the muscles during surgical operations. Tubocurarine was the first compound to be used for this purpose. Intensive studies on the relation between structure and pharmacological effects have led to the synthesis of several other substances having the same effect. One of these is **Atracurium** – a bis-quaternary compound comprising two papaverinium residues (see page 648) joined by a chain of 13 atoms which includes two ester groupings. Tubocurarine is a fine example of the structure of a natural product of ethnic origin being used as a model for the synthesis of drugs with improved properties.

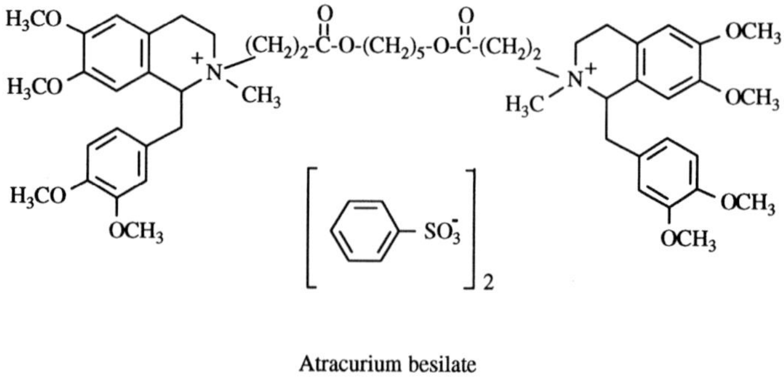

Atracurium besilate

Crude drugs   **Curare** is a generic term for the muscle-relaxant dart and arrow poisons that are prepared and used by the Indians of the Amazon region of South America. Curare is produced from several plants, not all of which contain toxic substances. Formerly, the different kinds of curare were grouped according to the type of container: bamboo tubes, unglazed earthenware pots, or small calabashes. Nowadays, curares may be put into the nearest handy container available and this classification is no longer appropriate. It is best replaced by a chemical one: (a) curares with bisbenzylisoquinoline alkaloids as principal active constituents (from *Chondrodendron, Menispermaceae*); (b) curares containing both bisbenzylisoquinoline alkaloids (from *Chondrodendron* and related genera) and dimeric alkaloids (from *Strychnos, Loganiaceae*) as active principles; and (c) curares with dimeric alkaloids (from *Strychnos*) as the chief essential constituents. The dimeric indole alkaloids will be discussed in the chapter on indole alkaloids (page 691).

***Chondrodendron*-based curare.** This type of curare is obtained from *Chondrodendron tomentosum* Ruiz et Pav. which is the only species known to contain the muscle-relaxant alkaloid (+)-tubocurarine. The curare is prepared on a semi-industrial scale in the Upper Amazon (Río Huallaga) region of Peru and is usually exported in tins containing about 1 kg. It serves as the raw material for the isolation of the alkaloid.

Amaryllidaceae alkaloids

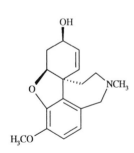

Galanthamine

As the name implies these alkaloids occur in plants of the family *Amaryllidaceae*, many of which are ornamental plants like *Narcissus* sp. (daffodils) and *Galanthus* sp. (snowdrops). Over one hundred alkaloids have been isolated from plants of this family. These alkaloids are toxic and only one – galanthamine – has found use in medicine.

Galanthamine was originally isolated from bulbs of snowdrops and has also been found in other genera of the *Amaryllidaceae* family. It is extracted from bulbs of *Leucojum aestivum* L. (snowflake) which contain up to 2 % (dry weight). Also daffodils (*Narcissus* sp.) are used as a commercial source of galanthamine. A patent for an industrial process for synthesis with an overall yield of 18 % was filed in 1997 but does not yet seem to have resulted in a commercial synthetic product.

Biosynthesis   The structure of galanthamine indicates that the molecule is the product of an intramolecular oxidative coupling within a precursor of the C_6C_2-N-C_1C_6 type. As seen from Fig. 215 the C_6C_1- unit comes from a molecule of phenylalanine while tyrosine delivers the -N-C_2C_6 unit. Phenylalanine is transformed into caffeic acid via cinnamic and *p*-coumaric acid. Loss of two carbon atoms results in

3,4-dihydroxybenzaldehyde which reacts with tyrosine in a reaction, also involving a methylation step, yielding 4´-O-methylnorbelladine. This is the C_6C_2-N-C_1C_6 unit which undergoes *ortho-para*-coupling, resulting in formation of a dienone (A in Fig. 215), which in a spontaneous reaction cyclizes to form compound B. Finally, B is reduced to N-demethylgalanthamine which is N-methylated to yield galanthamine.

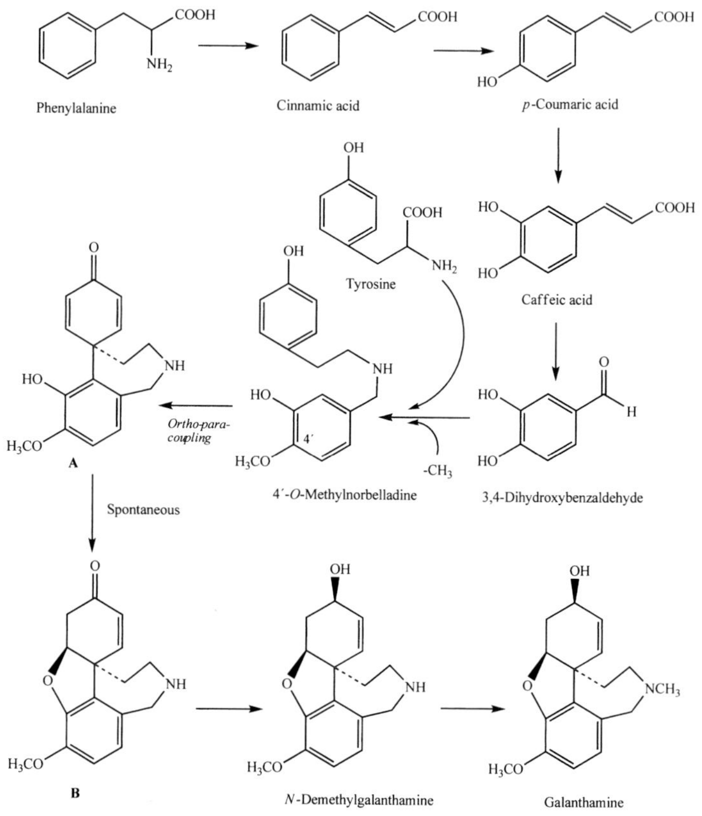

Fig. 215. Biosynthesis of galanthamine

Galanthamine is an acetylcholinecholinesterase inhibitor which is used for treatment of Alzheimer's disease (AD), an age-dependent neurodegenerative disorder characterized by multiple cognitive deficits, including worsening of memory, judgement and comprehension. The cognitive deficits are thought to be primarily related to the degeneration of cholinergic neurons in the cortex and hippocampus, resulting in deficits of cholinergic transmission and reduced levels of acetylcholine in these patients. Patients may, however, also have a loss of other neurotransmitters. There is no cure for patients suffering from AD and symptomatic treatment focuses on the restoration of cholinergic function. Of the available methods, only inhibition of acetylcholinesterase has been shown to significantly improve cognition in AD patients. Galanthamine is a reversible, competitive inhibitor of acetylcholinesterase that potentially enhances cholinergic function in the brain through two mechanisms of action: the inhibition of the enzyme and the potentiation of the effects of acetylcholine at nicotinic acetylcholine receptors.

In Eastern Europe, galanthamine is used as a reversal agent in anaesthetic practice because it neutralizes the neuromuscular blockade induced by tubocurarine. Besides this, galanthamine acts as a mild analeptic and shows analgesic power as strong as that of morphine. Applied in eye drops it reduces the intraocular pressure.

Benzophenanthridine alkaloids

Sanguinarine Sanguinarine is an alkaloid with antibacterial properties, which is used in toothpastes and other oral hygiene products to counteract the formation of plaque. Sanguinarine can be produced in cell cultures of *Papaver somniferum*, *P. bracteatum* and *Eschscholzia californica*.

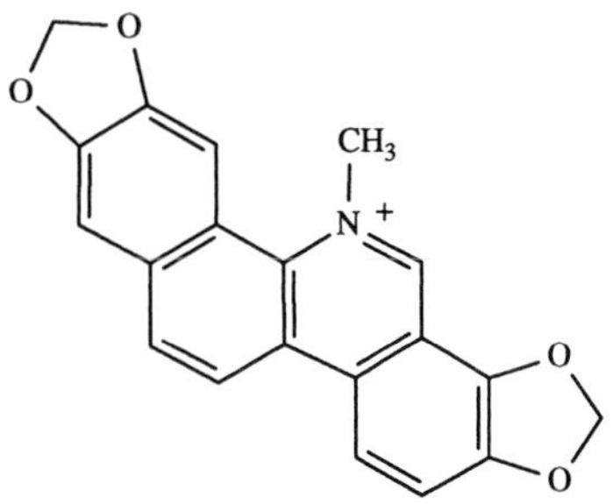

Sanguinarine

Biosynthesis The biosynthesis of sanguinarine starts with tyrosine and the first 10 steps are the same as in the biosynthesis of berberine (page 645) resulting in the formation of (*S*)-scoulerine. The enzyme (*S*)-*cheilantifoline synthase* (EC 1.14.21.2) converts (*S*)-scoulerine to (*S*)-cheilantifoline, which is transformed to (*S*)-stylopine by (*S*)-*stylopine synthase* (EC 1.14.21.1). Both enzymes are highly substrate-specific and stereoselective cytochrome P450-containing enzyme complexes. The reactions require the presence of molecular oxygen and NADPH. (*S*)-*Tetrahydroprotoberberine* N-*methyltransferase* (EC 2.1.1.122) transfers a methylgroup from S-adenosylmethionine to the nitrogen of (*S*)-stylopine resulting in the formation of (*S*)-*cis*-N-methylstylopine. This enzyme is highly stereoselective and only methylates tetrahydroberberine alkaloids with the (*S*)-configuration. The substitution pattern at rings A and D are also important. (*S*)-scoulerine is not methylated by this enzyme while (*S*)-stylopine shows maximal activity as a substrate. In the next step the microsomal cytochrome P450-containing, NADPH and O_2-dependent

enzyme complex *methyltetrahydroprotoberberine 14-monooxyge*
nase (EC 1.14.13.37) hydroxylates carbon 14 stereo- and
regiospecifically, thereby breaking the bond to the nitrogen to yield
protopine. The 6-position of protopine is specifically hydroxylated
by *protopine 6-monooxygenase* (EC 1.14.13.55) in the presence of

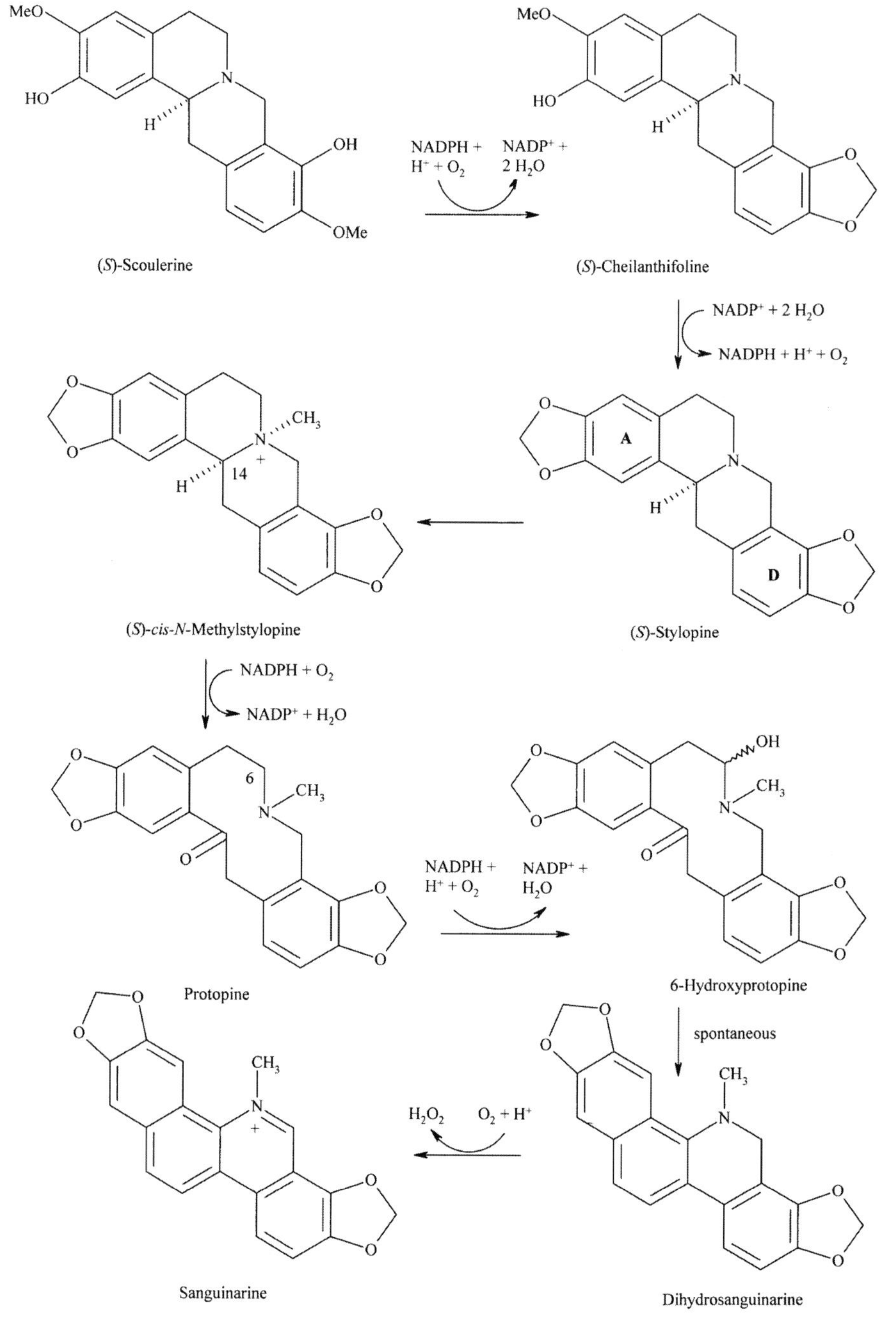

Fig. 216. Biosynthesis of sanguinarine

O_2 and NADPH yielding 6-hydroxyprotopine which spontaneously rearranges to form dihydrosanguinarine. In the final step the cytosolic enzyme *dihydrobenzophenantridine oxidase* (EC 1.5.3.12) transforms dihydrosanguinarine to sanguinarine in the presence of O_2. The complete pathway from tyrosine to sanguinarine thus involves 16 enzymes which have all been characterized.

Terpenoid tetrahydroisoquinoline alkaloids

In these alkaloids, which are encountered in the genus *Cephaëlis* of the family *Rubiaceae*, a terpenoid fragment joins two tetrahydroisoquinoline ring systems.

Emetine and cephaeline occur in roots of *Cephaëlis ipecacuanha* and *C. acuminata*.

Emetine Emetine is used as a specific remedy for the treatment of amoebic dysentery, a tropical disease caused by the protozoa *Entamoeba histolytica*. The treatment, however, suffers from serious side effects and is nowadays little used. Both emetine and cephaeline are emetics.

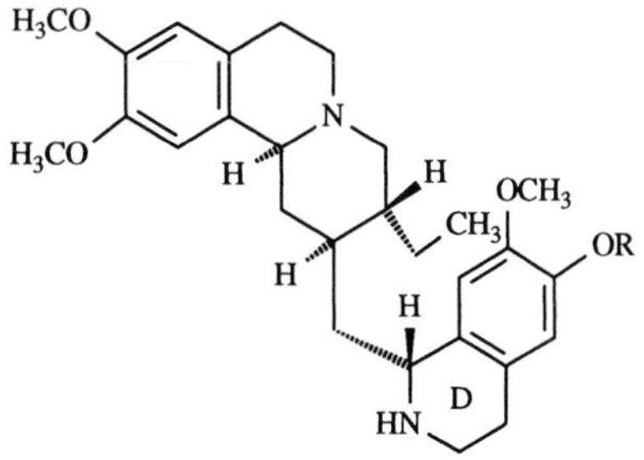

Emetine: R = CH₃
Cephaeline: R = H

Biosynthesis The monoterpene which is contained in the structures of these alkaloids has been shown to be the iridoid loganin (I, Fig. 217), which is formed via the non-mevalonate pathway (page 384). It is transformed into secologanin (II), which condenses with dopamine to form deacetylisoipecoside (III). Rearrangement of the terpene residue and condensation with another dopamine molecule gives emetine and cephaeline.

Crude drug **Ipecacuanha root** (Ipecacuanhae radix) is the root of *Cephaëlis ipecacuanha* Rich. (also described as *Psycotria ipecacuanha* Stokes or *C. acuminata* Karst.) (*Rubiaceae*). Both plants are small shrubs. *C. ipecacuanha* grows in Brazil, especially in the State of Matto Grosso. The roots are collected all the year round, except dur-

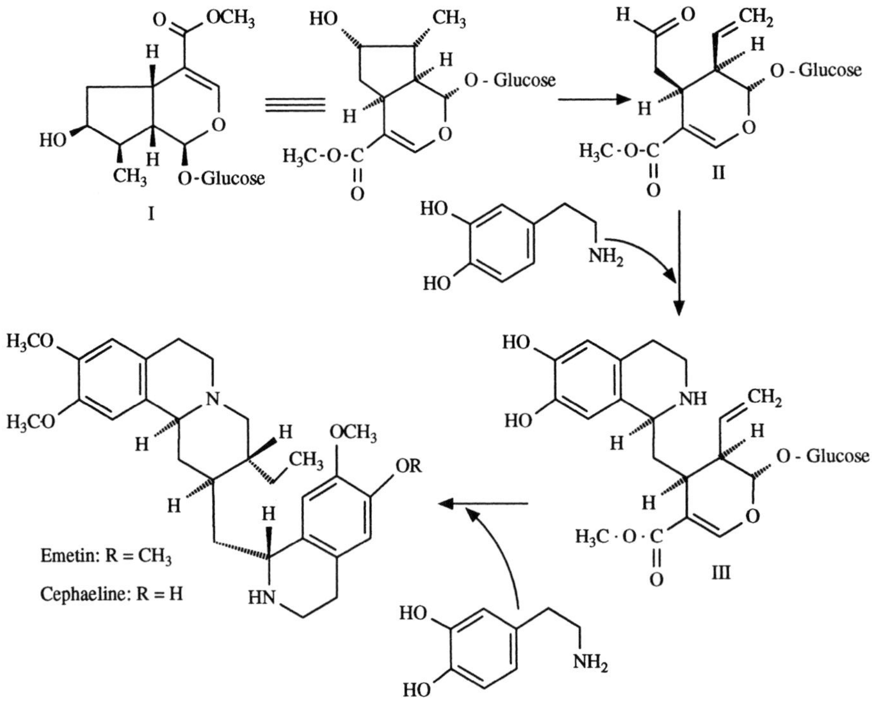

Fig. 217. Biosynthesis of emetine and cephaeline

ing the rains. The drug is best collected from 3–4-year-old plants; these have the highest alkaloid content. The drug derived from *C. ipecacuanha* is often called *Rio ipecacuanha*, after the port from which it is shipped. The plant is nowadays also grown in India.

Drug from *C. ipecacuanha* grown in the north of Colombia, in Nicaragua and in Panama is called *Cartagena ipecacuanha*, again after one of the ports from which it is shipped. The root of this variety is thicker and lighter than that of *C. ipecacuanha*.

Both drugs contain emetine and cephaeline as the main alkaloids. They also contain psychotrine, an analogue of cephaeline in which the ring-D nitrogen is present as an imino function. Rio ipecacuanha has a total alkaloid content of about 2 % and the ratio of emetine to cephaeline is 2:1. The corresponding figures for Cartagena ipecacuanha are 2.2–2.5 % and 1:2.

Ipecacuanha is used for the extraction of emetine. An extract of the crude drug is used as an emetic and an expectorant.

Indole alkaloids

This group of alkaloids is characterized by the indole skeleton, derived biosynthetically from tryptophan. Most indole alkaloids also contain other complicated ring systems, which are biosynthe-sized via mevalonic acid and include a C_5 unit (ergot alkaloids) or a C_{9-10} monoterpene unit (*Rauwolfia*, *Strychnos* and *Catharanthus* alkaloids).

Simple indole alkaloids

Psilocin and psilocybin

From time immemorial the Indian population in Central America has used certain mushrooms to create a state of intoxication, enabling them, as they believe, to communicate with the gods and learn, for example, the reasons why a person has become ill and

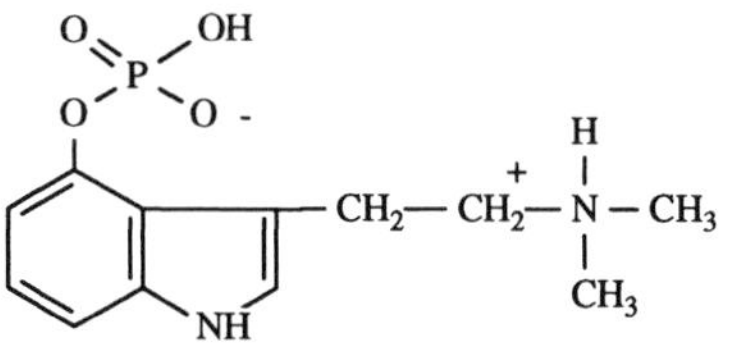

Psilocybin

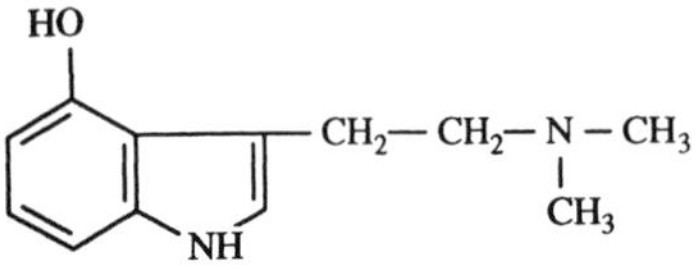

Psilocin

how the illness should be cured. The mushrooms are also used at festivals and religious ceremonies. The important role played by these hallucinogenic mushrooms in religion is evident from certain 3 000-year-old finds in Guatemala. These are "stone mushrooms", shaped like toadstools, on the stalks of which a god or a demon has been carved. In some remote areas in the south of Mexico, the Indi-ans still use these mushrooms for magic rites. Their name for these mushrooms is "Teonanácatl" (divine fungus). In the 1950s, the ancient Mexican mushroom cult was studied by the American eth-nomycologists R. Gordon Wasson and his wife Valentina Pavlovna. They made several expeditions to southern Mexico where the mush-room cult still persists. In the summer of 1955, Wasson was able, for the first time, to take part in a secret mushroom ceremony in Huaut-la de Jiménez, Oaxaca, and he was probably the first white man to ingest the holy mushrooms. The Wassons invited the French mycol-ogist Roger Heim to collaborate with them to establish the identity of the mushrooms and in 1956 he concluded that they constituted previously unknown species of the genus *Psilocybe*, as well as a species of the genus *Stropharia* and a species of the genus *Conocybe*. Heim was able to cultivate some of the *Psilocybe* species in the laboratory and one in particular, *P. mexicana* Heim, gave a very good crop which made it possible to send about 100 grams to Albert Hofmann at Sandoz AG for chemical investigation. In 1958, Hof-mann succeeded in isolating the hallucinogenic components and found them to be indole alkaloids. They were named psilocybin and psilocin.

The isolation was hampered by the fact that the hallucinogenic effects could not be demonstrated in experiments with animals.

Hofmann and his collaborators had to test the various fractions by experimenting on themselves. Hofmann described the first dramatic experiment, when on July 1, 1957, he ate 32 dried toadstools with a total weight of 2.4 g, as follows:

"Thirty minutes after my taking the mushrooms, the exterior world began to undergo a strange transformation. Everything assumed a Mexican character. As I was perfectly well aware that my knowledge of the Mexican origin of the mushroom would lead me to imagine only Mexican scenery, I tried deliberately to look on my environment as I knew it normally. But all voluntary efforts to look at things in their customary forms and colors proved ineffective. Whether my eyes were closed or open, I saw only Mexican motifs and colors. When the doctor supervising the experiment bent over me to check my blood pressure, he was transformed into an Aztec priest and I would not have been astonished if he had drawn an obsidian knife. In spite of the seriousness of the situation, it amused me to see how the Germanic face of my colleague had acquired a purely Indian expression. At the peak of the intoxication, about 1 1/2 hours after ingestion of the mushrooms, the rush of interior pictures, mostly abstract motifs rapidly changing in shape and color, reached such an alarming degree that I feared that I would be torn into this whirlpool of form and colors and would dissolve. After about six hours the dream had come to an end. Subjectively, I had no idea how long this condition had lasted. I felt my return to everyday reality to be a happy return from a strange, fantastic but quite real world to an old and familiar home."[2]

Psilocybin and psilocin have the same pharmacological effects, but psilocybin is chemically more stable than psilocin. These alkaloids are not used as drugs, but, like LSD (page 675), psilocybin has been tested therapeutically in psychiatry, as well as in attempts to induce model psychoses for research purposes.

Physostigmine   Physostigmine is obtained from the seed of *Physostigma venenosum*. The alkaloid is unstable and needs to be kept in airtight containers with the exclusion of light. Physostigmine is a reversible inhibitor of the enzyme cholinesterase and is used in combination with pilocarpine (page 699) for the treatment of glaucoma. It is

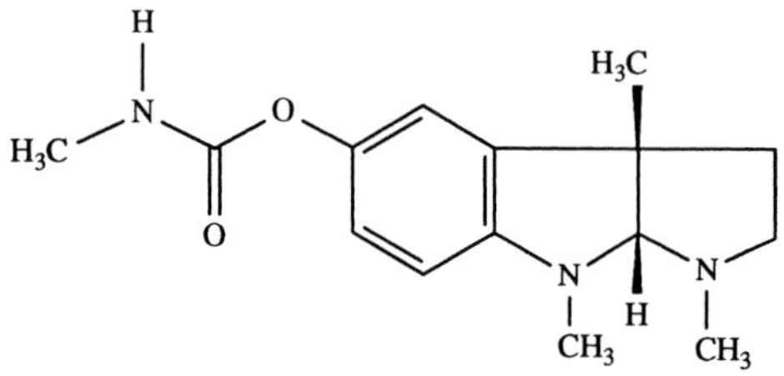

---

[2] From "History of the basic chemical investigations on the sacred mushrooms of Mexico", by Albert Hofmann, in "*Teonanácatl, Hallucinogenic Mushrooms of North America*". J. Ott and J. Bigwood (Eds.), Madrona Publishers, Inc., Seattle, U.S.A. (1978).

also used as an antidote to anticholinergic poisons such as hyoscyamine/atropine and scopolamine. Experimentally it has been used in the treatment of Alzheimer's disease but with limited success.

Biosynthesis   Physostigmine is biosynthesized (Fig. 218) from 5-hydroxytryptophan (1), which is decarboxylated to 5-hydroxytryptamine (2) which, after rearrangement (3), is methylated at C-3 (4), and the amino group adds on to the double bond (5). Methylation at both nitrogen atoms gives eseroline (6), and subsequent formation of the urethan grouping leads to physostigmine (7).

Crude drug   **Calabar bean** (Physostigmatis semen, Calabar semen) is the seed of *Physostigma venenosum* Balf. (*Fabaceae*). This is an approximately 15 m long climber, growing in the countries along the coast of the Gulf of Guinea in West Africa. The fruit is a pod with 1–3 kidney-shaped seeds, which are 2–3 cm long. The calabar bean contains 0.5 % alkaloids, mainly physostigmine. The drug is used for the extraction of this alkaloid.

Formerly, the native population used the calabar bean as an ordeal poison. A person suspected of being guilty of a crime was forced to eat the crushed beans or drink an aqueous extract of them. If he survived, he was considered innocent. This usage was reported from remote areas in the then Belgian Congo as late as 1959.

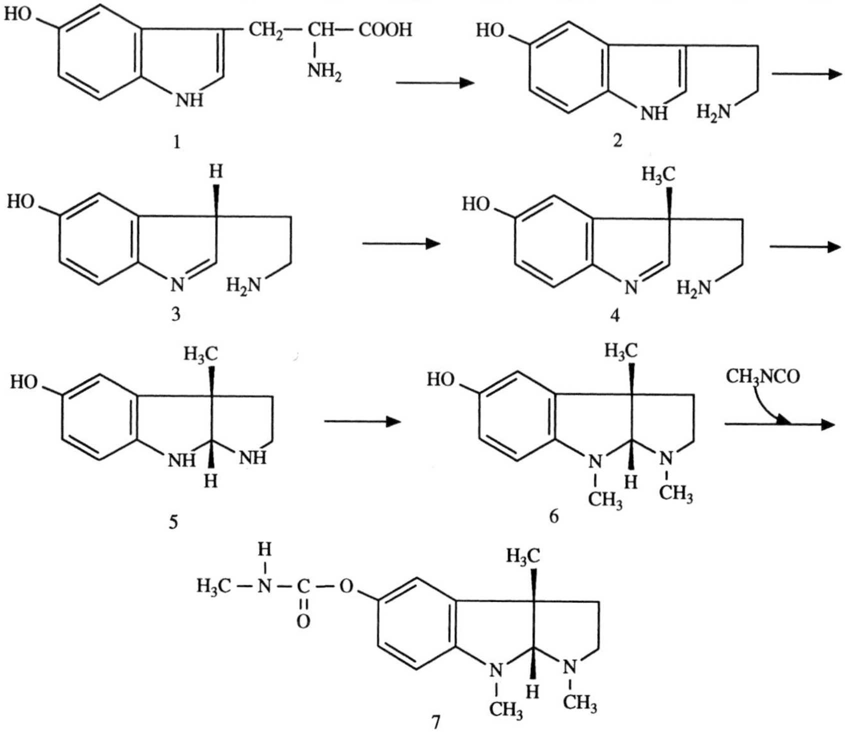

Fig. 218. Biosynthesis of physostigmine

Terpenoid indole alkaloids

Ergot alkaloids

Ergot, which is the sclerotium of the fungus *Claviceps purpurea* (see below), contains a number of alkaloids (more than 50 have been isolated) which are amides of the indole derivative D-lysergic acid. This acid is readily isomerized to D-isolysergic acid, so that the fungus contains two series of alkaloids, derived from D-lysergic acid and D-isolysergic acid, respectively. Lysergic-acid alkaloids have also been isolated from other species of the genus *Claviceps* and from members of the family *Convolvulaceae*, e.g. *Ipomoea violacea* (morning glory) and *Ipomoea burmanni* (also known as *Rivea corymbosa*. Seeds of these species have been misused as narcotics due to their content of lysergic acid amide (ergine) which is hallucinogenic like LSD (see page 675) but is less potent.

The complicated structures that are linked via an amide function to lysergic acid in the alkaloids ergotamine, ergocristine, α- and β-ergocryptine, and ergocornine are cyclic tripeptides containing the following amino acids:

Ergotamine:	α-hydroxyalanine, proline, phenylalanine
Ergocristine:	α-hydroxyvaline, proline, phenylalanine
α-Ergocryptine:	α-hydroxyvaline, proline, leucine
β-Ergocryptine:	α-hydroxyvaline, proline, isoleucine
Ergocornine:	α-hydroxyvaline, proline, valine

In the structures in Fig. 219 the positions of the three amino acids are indicated by means of dotted lines. The α-amino group of the α-hydroxyamino acid is attached to the carboxyl group of the lysergic acid, while their carboxyl group is linked via a normal peptide bond with the amino group of the variable amino acid (phenylalanine, leucine, isoleucine, valine). The carboxyl group of this acid is joined in the normal way to the amino group of the proline moiety with formation of a peptide bond. The carboxyl group of the proline not only produces a lactone with the hydroxyl group of the α-hydroxyamino acid, but also a lactam with the amino group of the variable amino acid. This can be explained on the basis of the ortho-form of the carboxyl group in proline:

One hydroxyl group is involved in forming the lactone bond, the second one in producing the lactam, and the third one is free.

In addition to the peptide alkaloids discussed so far, other peptide alkaloids have also been isolated. They are, however, of no medicinal value.

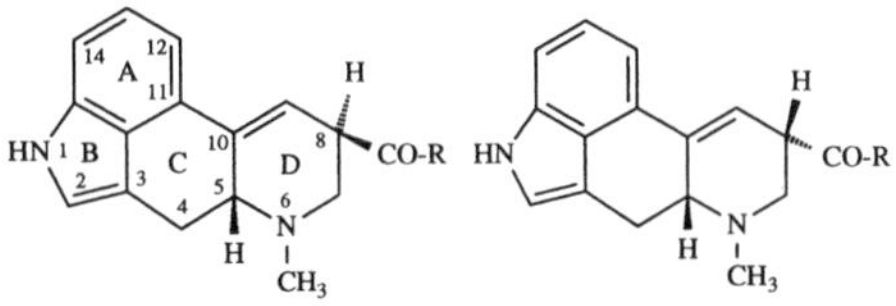

R = OH	D-Lysergic acid	D-Isolysergic acid
R = NH₂	Lysergic acid amide (Ergine)	Isolysergic acid amide (Erginine)

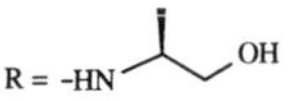

R = -HN⟍⟍OH — Ergometrine (Ergobasine Ergonovine) — Ergometrinine

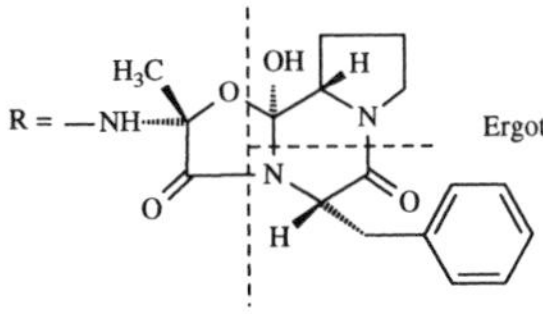

R = —NH⟍ ... Ergotamine — Ergotaminine

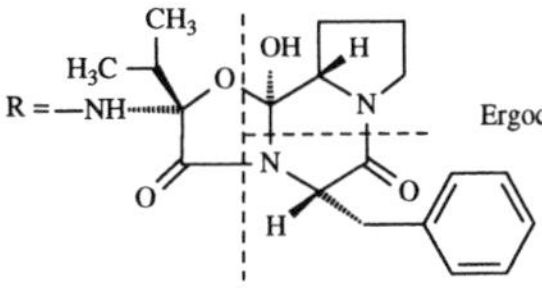

R = —NH⟍ ... Ergocristine — Ergocristinine

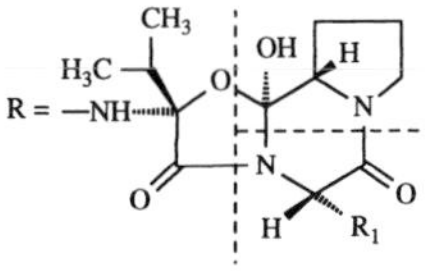

R = —NH⟍ ...

R₁ = - CH₂ - CH - CH₃   α-Ergocryptine   α-Ergocryptinine
         CH₃

R₁ = - CH - C₂H₅   β-Ergocryptine   β-Ergocryptinine
     CH₃

R = — N ... Ergocornine — Ergocorninine

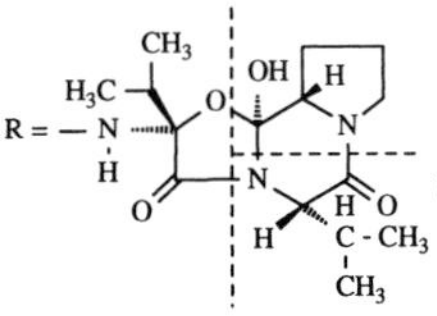

Semisynthetic derivatives, not occurring in Nature:

R = - N(C₂H₅)(C₂H₅)   Lysergic acid diethylamide (LSD)

R = - NH - CH - CH₂ - CH₃   Methylergometrine
       CH₂OH

R = - NH - CH - CH₂ - CH₃
       CH₂OH
and methylated indole nitrogen   Methysergide

Fig. 219. Structures of ergot alkaloids

Ergot also contains a group of alkaloids having the same ring system (ergoline) as lysergic acid, but lacking the carboxyl group. These are called clavine alkaloids. They are without medicinal interest, though they are important as intermediates in the biosynthesis of the lysergic-acid alkaloids (page 676). Many clavine alkaloids have been isolated from *Claviceps purpurea* and related species; some of their structures are given here:

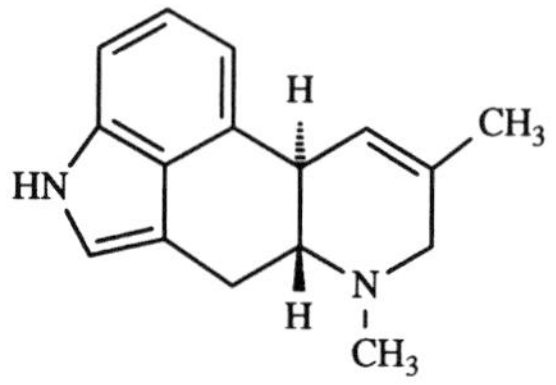

Agroclavine

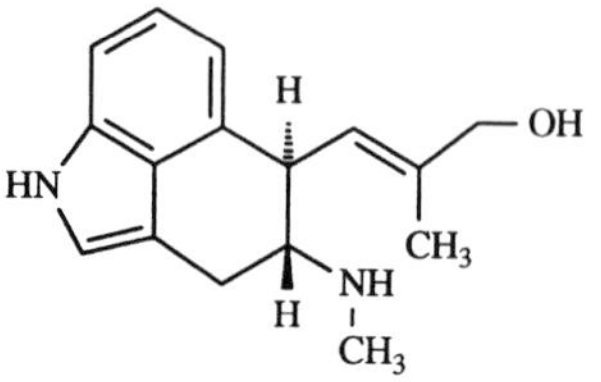

Chanoclavine I

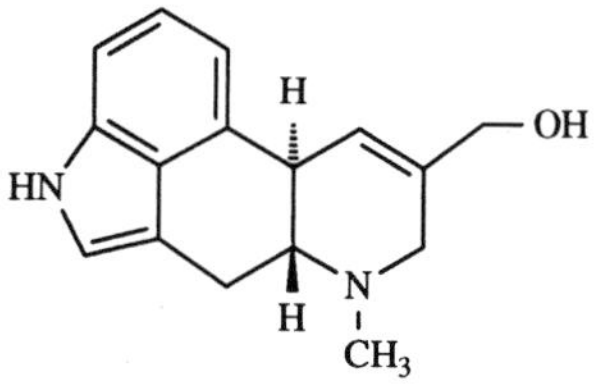

Elymoclavine

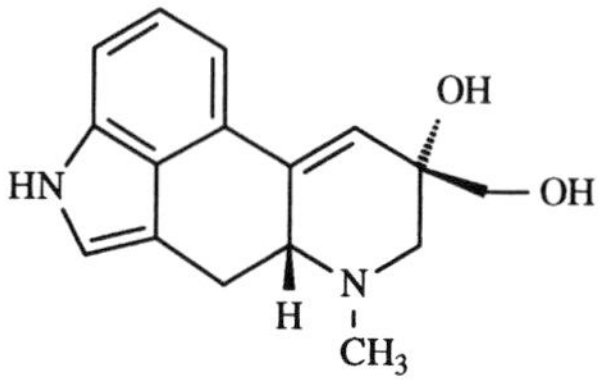

Penniclavine

The effects and medicinal use of ergot alkaloids

Ergometrine

Ergometrine causes contractions of the uterus and injections of the maleate are used to reduce postpartum haemorrhage. They can also initiate delivery, but nowadays the natural hormone oxytocin is preferred for this purpose.

The semisynthetic derivative methylergometrine, in which the side-chain consists of a 2-aminobutanol rather than 2-aminopropanol residue, has a more prolonged effect than the natural alkaloid.

Ergotamine

Ergotamine also has oxytocic activity, but it is used medicinally mainly as an effective sympatholytic agent. Ergotamine tartrate is employed, often in combination with caffeine, for the treatment of migraine. The alkaloid is also used both prophylactically and therapeutically against hypotension in spinal anaesthesia – a use which is based on constriction of the peripheral blood vessels.

Hydrogenation of the 9,10-double bond in the lysergic-acid moiety yields dihydroergotamine, which is also employed as a remedy for migraine and is often better tolerated by the patients. The dihydro compound is a weaker oxytocic and vasoconstrictor than the parent alkaloid.

Ergotoxine

Ergotoxine was one of the first crystalline alkaloid preparations to be isolated from ergot. It was long believed to be a single substance but later it was found to be a mixture of the four alkaloids: ergocristine, α- and β-ergocryptine, and ergocornine. A mixture of equal amounts of the methanesulphonates of the 9,10-dihydro derivatives of these alkaloids is used under the name Hydergine® as a vasodilator for disturbances in peripheral blood circulation, i.e. its effect is the opposite of that of ergotamine.

2-Bromo-ergocryptine 2-Bromo-ergocryptine (bromocriptine), which does not occur in Nature, is a dopamine agonist, i.e. it has the same effect as dopamine, and is used against Parkinsonism. The substance also reduces the secretion of the hypophysial hormone prolactin. For this reason it is used to inhibit or stop the production of milk. The disease *acromegaly*, which is characterized by abnormal growth of the hands, feet, head and chest, can also be treated with the drug, because bromocriptine reduces growth-hormone secretion.

Lysergic acid diethylamide Lysergic acid diethylamide (LSD) is a synthetic derivative of lysergic acid, and does not occur in ergot. Its effect was discovered accidentally in 1943 by Albert Hofmann (page 669). He had synthesized the compound in order to test its oxytocic effect and happened to become affected by a small quantity of the product. He described the experience as follows:

"Last Friday, the 16th of April, I had to leave my work in the laboratory and go home because I felt strangely restless and dizzy. Once there, I lay down and sank into a not unpleasant delirium which was marked by an extreme degree of fantasy. In a sort of trance with closed eyes (I found the daylight unpleasantly glaring) fantastic visions of extraordinary vividness accompanied by a kaleidoscopic play of intense coloration continuously swirled around me. After two hours this condition subsided."[3]

In order to corroborate his suspicion that it was LSD that had caused these effects, the following week Hofmann took 0.25 mg of the preparation, which he believed to be a small dose. In fact, this dose was 5 times greater than what was later to be regarded as an average dose (30–50 μg). All the symptoms reappeared, but greatly intensified because of the higher dose. This time, the state of intoxication lasted for more than 6 hours, and Hofmann mentions that particularly during the latter part of this period, all aural sensory impressions were transformed into colour sensations of rare intensity.

Like psilocybin (page 669), LSD has been tested both for diagnostic and therapeutic use in psychiatry, but it has not fulfilled the initial expectations of the compound.

LSD is used illegally as a hallucinogenic drug. Under its influence the personality disintegrates and people have also killed themselves by jumping out of high buildings, etc., believing that they could sail through the air.

---

[3] From: Sidney Cohen, "*The Beyond Within. The LSD Story*", Atheneum, New York, 1965.

Methysergide   Methysergide is a serotonin antagonist. For this reason, it is used prophylactically against migraine, but it is not suitable for the treatment of acute cases. For these, ergotamine or dihydroergotamine is the drug of choice (see above).

Biosynthesis of ergot alkaloids

The terpene part of the structure of ergot alkaloids is formed via the mevalonate pathway. As illustrated in Fig. 220 the amino acid tryp-

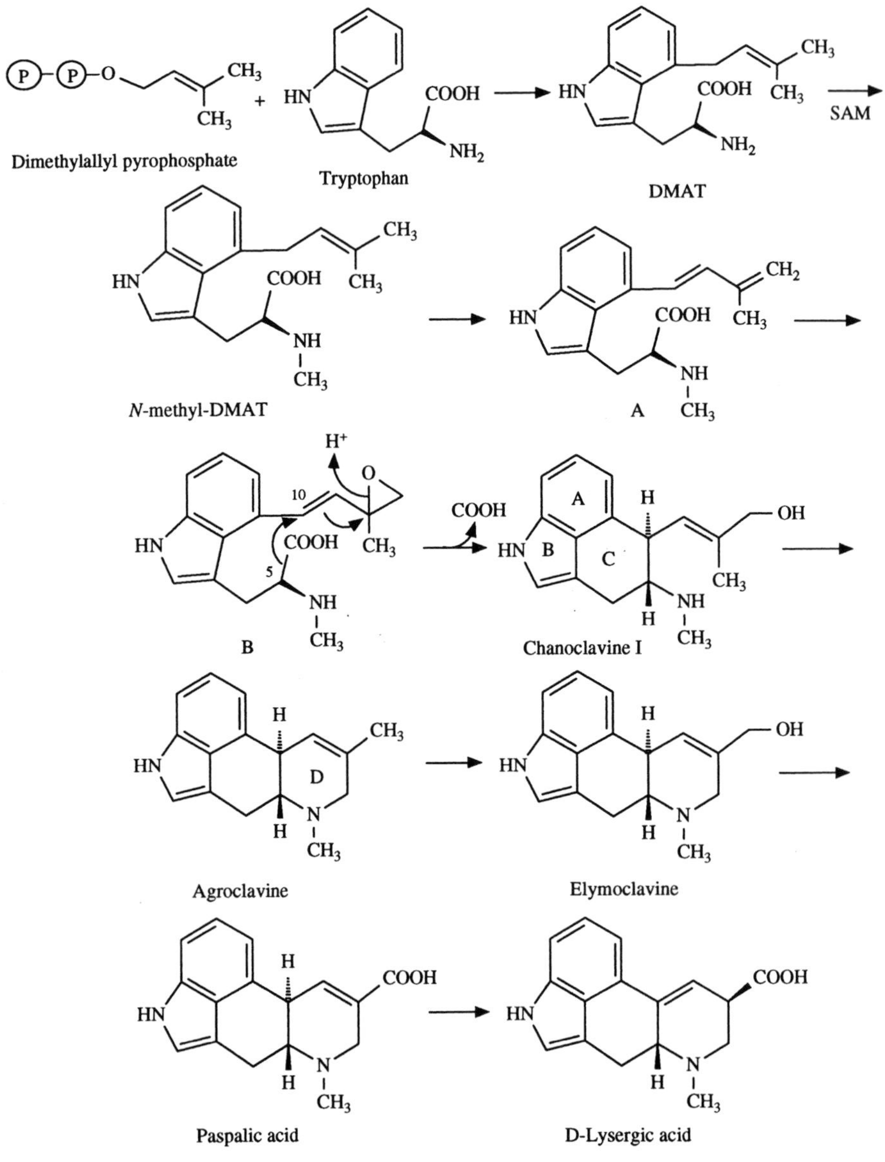

Fig. 220. Biosynthesis of lysergic acid

tophan and dimethylallyl diphosphate, formed from isopentenyl diphosphate (page 400), are precursors of the ergoline ring system that occurs in lysergic acid and clavine alkaloids. The two precursors react to form dimethylallyltryptophan (DMAT). Formation of DMAT is the rate-limiting step in ergot alkaloid biosynthesis. The enzyme, *tryptophan dimethylallyltransferase* (EC 2.5.1.34, also known as dimethylallyltryptophan synthase or DMAT synthase), which catalyses this reaction, has been isolated. Also the gene, *dmaW*, coding for the enzyme has been identified and expressed in yeast. The gene codes for a protein with a predicted molecular mass of 52 kDa consisting of 455 amino acid residues.

In the next reaction DMAT is methylated to form *N*-methyl-DMAT which is transformed to a diene derivative (A in Fig. 220). In a reaction involving molecular oxygen the diene is then oxidized to form an epoxide (B). Decarboxylation followed by an attack of the resulting C-5 anion on C-10 and subsequent opening of the epoxide ring gives chanoclavine I, which is a clavine alkaloid containing ring C of the ergoline ring system (page 673). The next step involves the formation of ring D, which probably occurs via chanoclavine I aldehyde, resulting in the clavine alkaloid agroclavine. Agroclavine is then oxidized to elymoclavine in a process requiring molecular oxygen and probably catalysed by a cytochrome P450 monooxygenase. Further oxydation of elymoclavine finally yields lysergic acid.

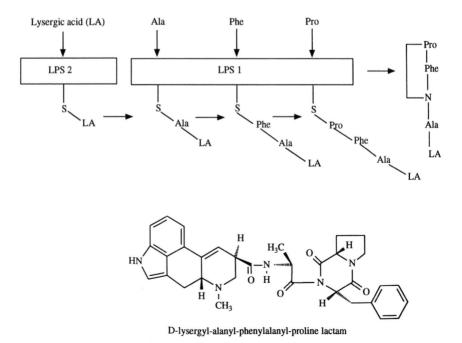

D-lysergyl-alanyl-phenylalanyl-proline lactam

Fig. 221. Biosynthesis of D-lysergyl-alanyl-phenylalanyl-proline lactam

In analogy with the biosynthesis of non-ribosomal polypeptides (page 531) the ergot peptide alkaloids are synthesized by a multi-functional enzyme complex that harbours all catalytic activities required to assemble D-lysergic acid and the amino acids. This complex has been isolated and termed D-*lysergyl peptide synthase (LPS)*. LPS consists of two polypeptide chains of 370 kDa (LPS 1) and 140 kDa (LPS 2), respectively. LPS 1 activates the amino acids as adenylates and subsequently as covalent thioesters. LPS 2 only activates D-lysergic acid as a thioester. As in the other non-ribosomal polypeptidases 4'-phosphopantetheine is an important cofactor for covalent binding of amino acids and D-lysergic acid. Fig. 221 illustrates the formation of D-lysergyl-alanyl-phenylalanyl-proline lactam, which is the immediate precursor of ergotamine. The process starts with activation of D-lysergic acid by LPS 2. The activated lysergic acid is transferred to LPS 1 where it reacts with alanine, which is the first amino acid in the peptide chain. Phenylalanine and proline are then added in the following two reactions and finally lactam formation takes place and the ergot alkaloid is released. The isolation of LPS has shed new light on the biosynthesis of the peptide ergot alkaloids. According to previous theories the biosynthesis was supposed to start with proline, leading to formation of a tripeptide which was then supposed to be coupled to lysergic acid. As shown above, the sequence of events is in fact the opposite.

Two genes – *cpps1* and *cpss2* – which code for LPS 1 and LPS 2, respectively, have been identified as well as a gene which might encode chanoclavine cyclase. Future functional analysis of all genes involved in ergoline ring synthesis and identification of the corresponding steps by enzymatic analysis of the gene products will allow development of new drugs by enzyme engineering and by biocombinatorial approaches.

Crude drug **Ergot** (Secale cornutum) is the sclerotium (the resting stage) of the fungus *Claviceps purpurea* (Fr.) Tul. (*Clavicipitaceae*), a parasite on rye (*Secale cereale* L., *Poaceae*). Ergot can also grow on other cereals and grasses. The life cycle of the fungus begins in the spring, when the sclerotium germinates and forms perithecia (fruiting bodies), in which asci develop and produce ascospores. These are carried by the wind to the ears of rye (or other grass), where they penetrate the ovary and germinate. The resulting hyphae (elongated, thread-like fungal cells) soon form a tissue, which penetrates and destroys the ovary. This tissue produces another type of spore – conidiospores – and a sweet secretion, called honeydew. The honeydew attracts insects, which carry the conidiospores to other ears of rye, thus spreading the infection. Later, in the vegetation period, the hyphae become more compact and dark in colour. They enter the resting stage as the sclerotium, which is pointed, cylindrical, 1–5 cm long, and about 5 mm in diameter. In the autumn, the sclerotium falls to the ground, where it hibernates. In the following spring it germinates and the cycle is completed.

The presence of ergot in rye involves the risk of ergot poisoning (ergotism). For this reason, every flour mill has to be equipped with a device that removes the ergots from the corn before it is milled. In olden times, this was unknown and ergot poisoning was common, especially after wet and cold summers, which favour the development of the fungus. The effects of these poisonings were very severe and were characterized by gangrene (necrosis) of the extremities, which could lead to the loss of hands and feet. If combined with lack of vitamin A, ergot poisoning could also give rise to spasms. The reason why the ingestion of ergot causes gangrene is the powerful vasoconstricting activity of its alkaloids, especially ergotamine, on the peripheral blood vessels. One of the symptoms in the early stages of ergot poisoning is a burning sensation in the extremities, often called St. Anthony's fire. Since the middle of the 17th century, when the correlation between ergot and St. Anthony's fire was discovered, epidemics of ergotism have decreased, but even in recent times poisoning has occurred, due to the use of inefficient devices for separating the ergots.

Production of ergot alkaloids

Wild ergot (collected at flour mills) can be used as a raw material for the extraction of ergot alkaloids for medicinal use. The wild drug is nowadays obtained mainly from Spain and Portugal, but the amount available is insufficient to meet the demand. Ergot can be grown on rye, starting with sclerotia, which are left to grow on a nutrient medium. They form a mycelium which produces conidiospores. These are then injected into the ears of rye by means of special machines pulled by tractors. In these machines the ears of rye move on to a felt pad, where they are punctured with pointed needles with the simultaneous injection of a suspension of conidiospores. This procedure gives a high yield of sclerotia (about 75–250 kg per hectare). Ergot is cultivated in some European countries (Czech republic, Hungary, Germany, Switzerland), but to a decreasing extent, because nowadays ergot alkaloids can be produced by growing the fungus in cultures in tanks, similar to those used for the production of antibiotics (page 90). The first ergot species that could be induced to produce lysergic-acid alkaloids under such conditions was *Claviceps paspali,* a species found on the grass *Paspalum distichum* L. It produces simple lysergic acid amides, from which medicinally useful alkaloids can be synthesized chemically. Strains of *Claviceps purpurea* have now been developed, which, in tank cultures, yield both ergometrine and ergotamine in amounts sufficient for a financially sound industrial operation.

Terpenoid indole alkaloids derived from strictosidine

This is a very large group of alkaloids comprising over a thousand substances. Their structures contain a C_{9-10} terpene residue and they can usually be classified as members of three types, called the *Corynanthe-Strychnos*, *Aspidosperma* and *Iboga* types. The names indicate the genera in which the respective types are found most frequently.

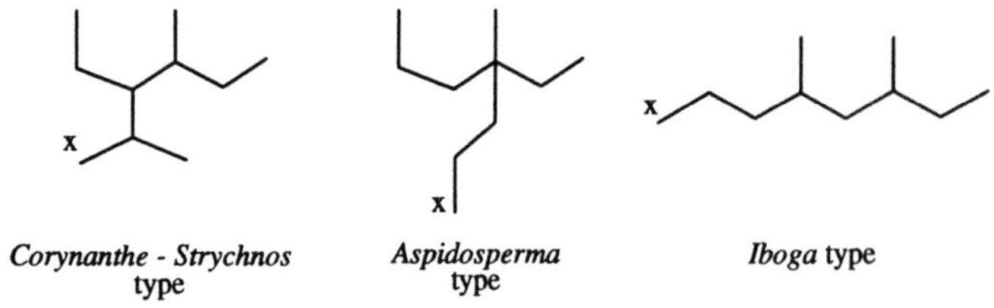

Corynanthe - Strychnos
type

Aspidosperma
type

Iboga type

Those alkaloids which have a C_9 terpene moiety lack the carbon atom marked with an x in the above structures.

Biosynthesis of strictosidine-derived terpenoid indole alkaloids

Extensive biosynthetic studies have shown that the three terpene skeletons have a common precursor, the secoiridoid secologanin, which plays the same role in the biosynthesis of the monoterpenoid indole alkaloids as in the biosynthesis of emetine (page 667). Secologanin condenses with tryptamine (from tryptophan) to form $3\alpha(S)$-strictosidine (Fig. 222), which is the immediate precursor of this group of indole alkaloids and also of the cinchona alkaloids

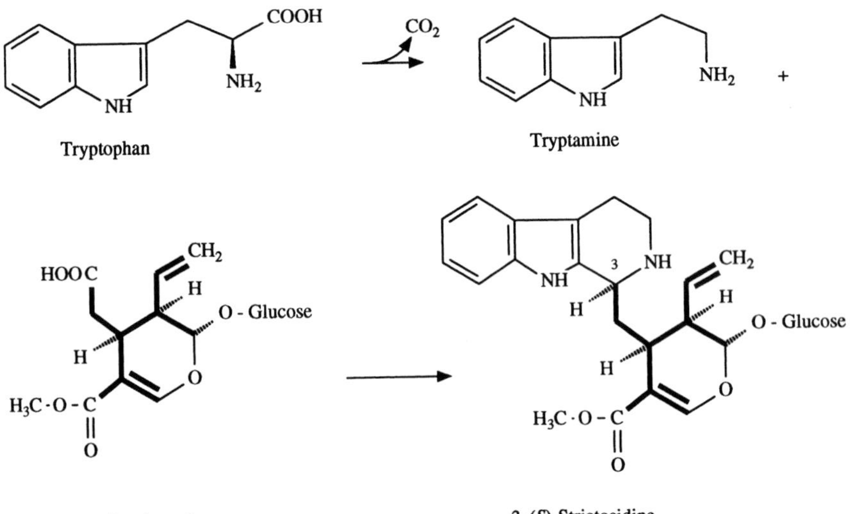

Tryptophan

Tryptamine

Secologanin

$3\alpha(S)$-Strictosidine

Fig. 222. Biosynthesis of $3\alpha(S)$-strictosidine

(page 693) where, however, the indole nucleus has been rearranged into a quinoline system. Only the 3α(S)-isomer can function as a precursor in these biosynthetic pathways. Strictosidine itself contains the 10-carbon skeleton of the *Corynanthe*-group and the two other types arise by rearrangement of this framework in the biosynthetic steps in Fig. 222.

The biosynthesis of secologanin via 10-hydroxygeraniol and loganin has already been outlined in the chapter on the iridoids and secoiridoids (page 384). Tryptamine is formed by decarboxylation of tryptophan, a reaction which is catalysed by *aromatic-L-amino-acid decarboxylase* (EC 4.1.1.28), also known as tryptophan decarboxylase). This enzyme, which has a high substrate specificity, has been isolated from cell suspension cultures of *Vinca rosea* (previously known as *Cataranthus roseus*) and has been purified to homogeneity. Its molecular mass is 115 kDa and it consists of two identical subunits. Corresponding cDNAs have been isolated and sequenced.

Condensation of tryptamine and secologanine to form strictosidine is mediated by *strictosidine synthase* (EC 4.3.3.2). The reaction is stereospecific and yields only 3α(S)-strictosidine. The enzyme has been isolated and purified to homogeneity from cell suspension cultures of *Rauwolfia serpentina* and *Vinca rosea*. Both plants produce medicinally important alkaloids (see below). Strictosidine synthase from *R. serpentina* is a single polypeptide with a molecular mass of 30 kDa and contains 5.3 % carbohydrate. It is a stable enzyme which can easily be immobilized on CNBr-activated Sepharose for preparative production of strictosidine. Gram quantities of this compound are readily produced by this method. Also the enzyme from *V. rosea* is a glycoprotein. The strictosidine synthase from *V. rosea* has been found to comprise six isoforms whereas only one isoform of the enzyme from *R. serpentina* has been identified. Strictosidine synthase has also been identified in many other species representing different genera within the family *Apocynaceae*.

cDNA sequences have been reported for strictosidine synthase from both *R. serpentina* and *V. rosea*. A single gene encodes the enzyme in both species which indicates that the isoforms from *V. rosea* are results of post-translational modifications of a single precursor. Extensive investigations have shown that they are not artefacts resulting from the isolation procedure. The physiological significance of the isoforms has not yet been explained.

Vinblastine and vincristine Structurally these alkaloids are heterodimers of vindoline and an oxidation product of catharanthine. Vindoline and catharanthine are terpenoid indole alkaloids which are also present in the plant and at a higher percentage than the heterodimers. They are effective against certain types of cancer. Vinblastine is used against Hodgkin's disease and also against choriocarcinomas that are resistant to other treatments. Vincristine is employed against leukaemia, lymphomas and small cell lung cancer. The tumour inhibiting effect of these compounds is due to their activity as microtubule inhibitors. They act by binding to tubulin and inhibit formation of protofilaments and microtubules (page 219).

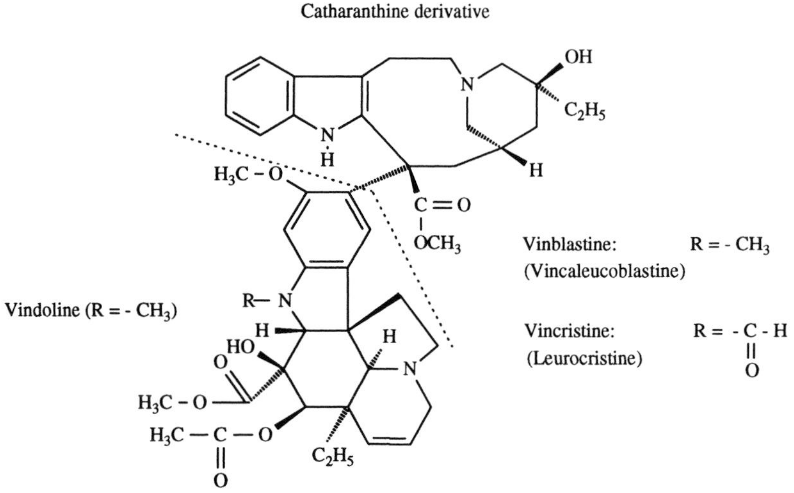

Catharanthine derivative

Vindoline (R = - CH₃)

Vinblastine: R = - CH₃
(Vincaleucoblastine)

Vincristine: R = - C - H
(Leurocristine)   ‖
                        O

Vinblastine and vincristine are obtained from the aerial parts of the Madagascar periwinkle *Vinca rosea* L. (also known as *Catharanthus roseus* (L.) G. Don and *Lochnera rosea* (L.) Reichb.) (*Apocynaceae*) (Plate 26, page 658). This species is a permanently flowering perennial herb. Its origin is Madagascar, and it is grown as an ornamental plant in tropical countries and also as a common potted plant. It is 40–80 cm high and is woody at the base. The flowers are pink or white. The plant has been used in folk medicine as a remedy for diabetes and scientific investigation gave interesting and unexpected results. It was found to contain a very large number of alkaloids, about 100 of which have been isolated so far. Some of these do indeed have a hypoglycaemic effect, confirming the popular use of the plant, but they have proved unsuitable for clinical application. On the other hand, it was discovered that several of the alkaloids had anti-neoplastic activity and this ultimately led to the introduction into clinical use of vinblastine and vincristine. Most of this work was carried out at Eli Lilly, an American pharmaceutical company, under the guidance of Gordon H. Svoboda.

The vinblastine and vincristine content in *Vinca rosea* is very low and the demand for the compounds is hard to satisfy in spite of the large-scale cultivation in many tropical countries. Over 500 kg of the plant is needed for production of 1 g of vincristine and the extraction and purification process is very complicated. The content of vinblastine is higher, but it is vincristine that is in the highest demand. However, as mentioned earlier, the content of catharanthine and vindoline in the plant is much higher than that of the heterodimers. It is now possible to synthesize vinblastine from these alkaloids in about 40 % yield. Vincristine can be obtained from vinblastine by controlled chromic acid oxidation or via microbiological *N*-demethylation by *Streptomyces albogriseolus*. Much effort has been invested into attempts to produce vinblastine and vincristine from cell cultures but so far these have not resulted in a commercially profitable process.

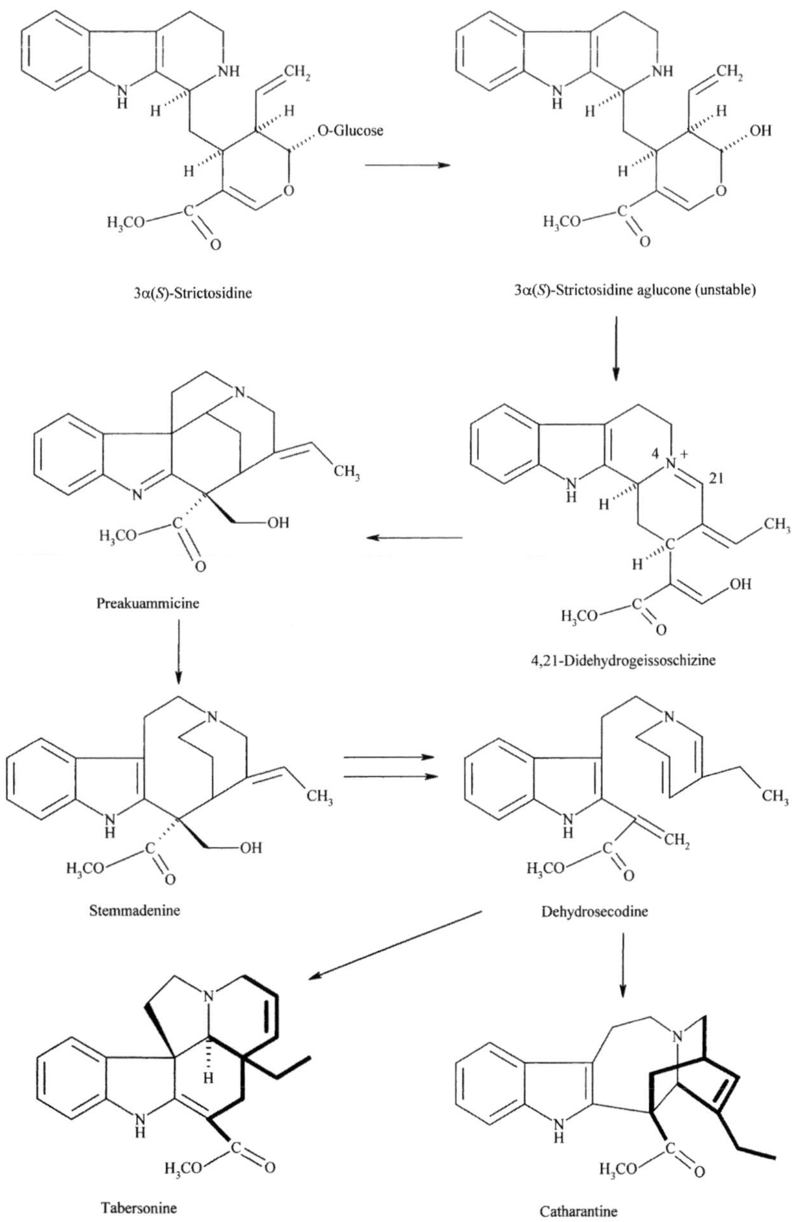

Fig. 223. Biosynthesis of catharantine (Iboga-*type) and tabersonine* (Aspidosperma-*type) from 3 (S)-strictosidine. Highlighted bonds indicate the C$_{10}$ -terpene skeletons*

Biosynthesis The biosynthetic pathways from strictosidine to alkaloids representing the three types of monoterpenoid indole alkaloid have been
most thoroughly studied in cell cultures of *Vinca rosea.* The many
alkaloids from this plant belong to all three types.

The first reaction is removal of the glucose moiety of strictosidine to form an unstable aglucone. The enzyme catalysing this reaction is *strictosidine β-glucosidase* (EC 3.2.1.105) which has two
isozymic forms I and II with molecular masses 230 and 450 kDa,
respectively. These isozymes are specific and hydrolyse the 3β-iso-
mer of strictosidine at only 10 % of the rate with which the α-iso-
mer is hydrolysed. The unstable aglucone is converted to another
unstable intermediate – 4, 21-didehydrogeissoschizine. This derivative is the precursor of three different pathways leading to representatives of the three different types of terpenoid indole alkaloids.
The continuation of the paths to tabersonine (*Aspidosperma*-type,
C_{10}) and to catharanthine (*Iboga* type, C_{10}) is illustrated in Fig 223.
The path leading to ajmalicine, another indole alkaloid of medicinal
interest and which represents the *Corynanthe*-type will be described
later (page 687, Fig. 226). The enzymes which catalyse these reactions are not yet known.

The pathway leading to vindoline from tabersonine involves six
steps (Fig. 224). The enzymes involved in these reactions (three
hydroxylations, *O*-methylation, *N*-methylation and an *O*-acetyla-
tion) have all been identified and characterized and have all been
shown to possess high selectivity and affinity for their respective
substrates.

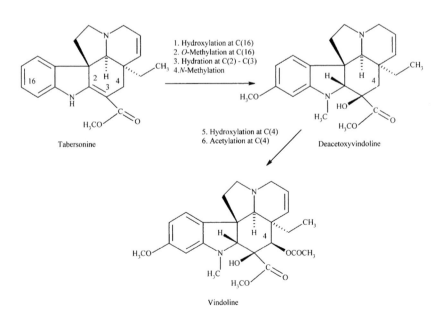

Fig. 224. Biosynthesis of vindoline from tabersonine

The coupling of catharantine and vindoline to form vinblastine is illustrated in Fig. 225. Catharantine is oxidized to form a peroxide which subsequently loses its peroxide group, permitting a nucleophilic attack by vindoline on to the conjugated iminium system, yielding an unstable dihydropyridinium derivative which is reduced to form anhydrovinblastine. A final oxidative reaction yields vinblastine. Vincristine is formed from vinblastine by oxidation of the *N*-methyl group.

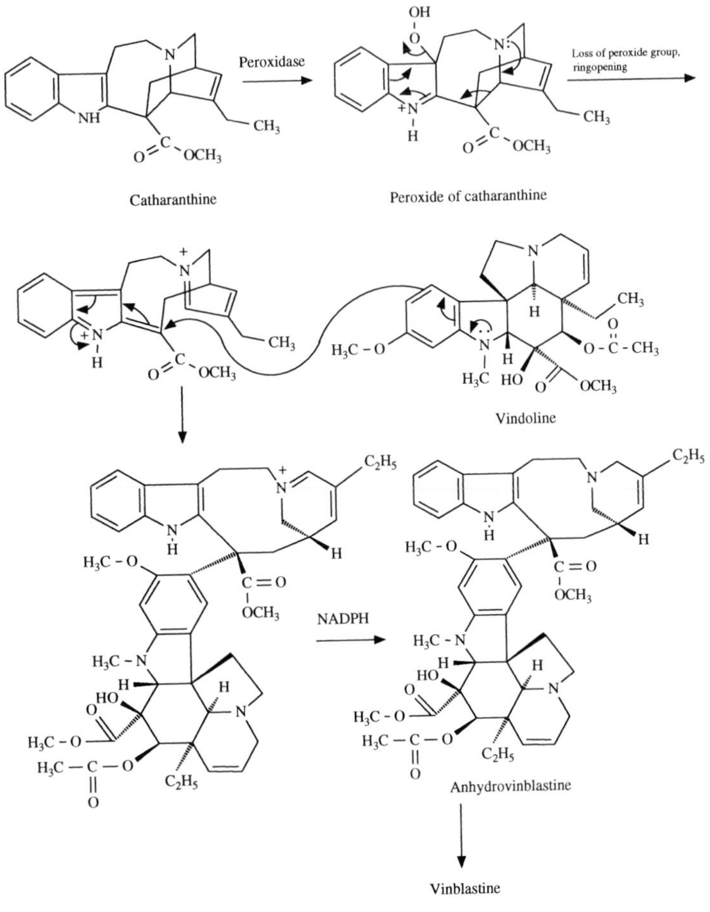

Fig. 225. Biosynthesis of vinblastine from catharanthine and vindoline

685

Vindesine

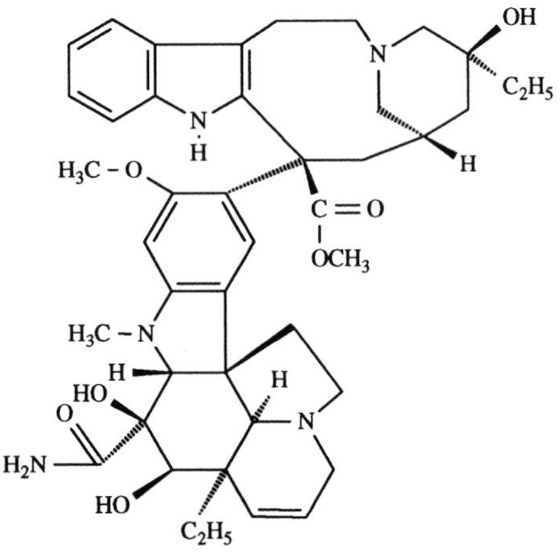

Vindesine

Vindesine is a derivative of vinblastine where the acetate substituent in the vindoline part of the molecule has been hydrolysed and the neighbouring methylester has been converted to an amide. It is used against acute lymphoid leukaemia in children and is also active against non-small-cell lung cancer and malignant melanoma.

Vinorelbine Vinorelbine is 3',4'-didehydro-4'-deoxi-8'-norvincaleukoblastine and is obtained from anhydrovinblastine which in turn is synthesized from vindoline and catharanthine by electrochemical oxidation. Shortening of the C_2N-bridge in anhydrovinblastine by one carbon yields vinorelbine. The compound is used against non-small-cell lung cancer.

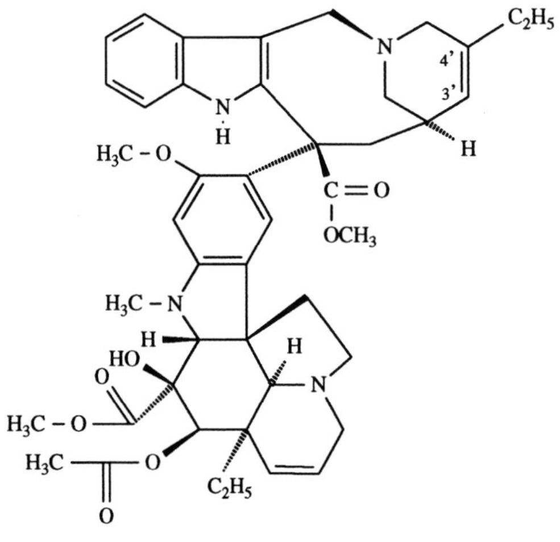

Vinorelbine

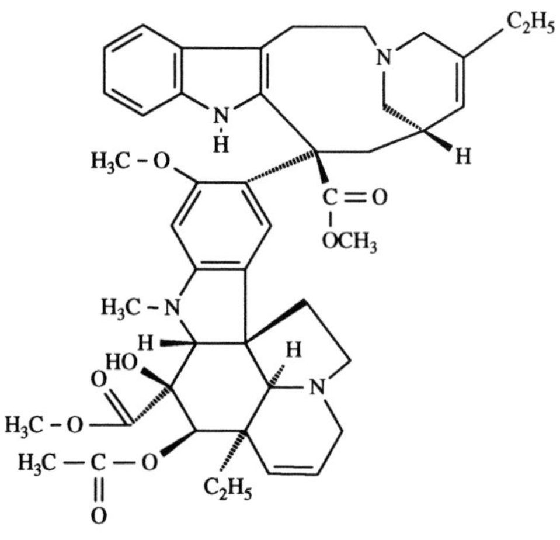

Anhydrovinblastine

Ajmalicine

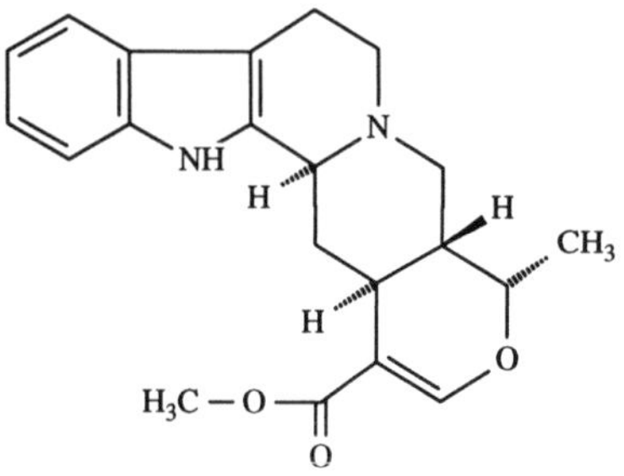

Ajmalicine

Ajmalicine can be extracted from roots of *Vinca rosea* and is also available from the roots of *Rauwolfia* species (page 688). It can also be obtained by hydrogenation of its oxidation product – serpentine – which can be isolated from *Rauwolfia* species. Ajmalicine is used as an antihypertensive and to increase the blood flow in the brain and peripheral parts of the body.

Biosynthesis   As illustrated in Fig. 226, the unstable intermediate didehydrogeissoschizine (page 684) is converted into cathenamine by the enzyme

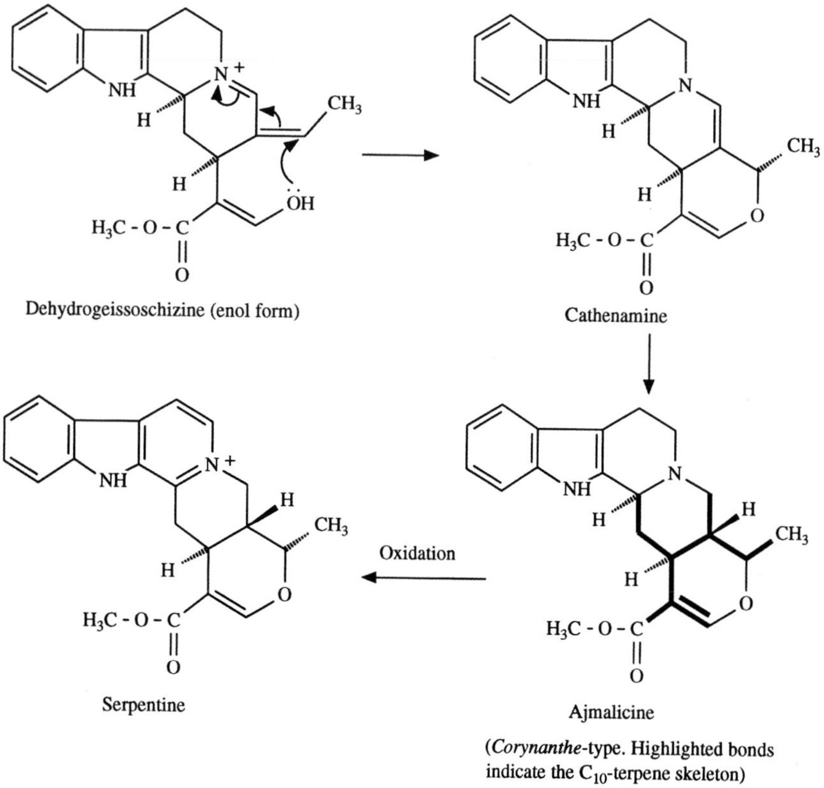

Dehydrogeissoschizine (enol form)

Cathenamine

Serpentine

Ajmalicine

Oxidation

(*Corynanthe*-type. Highlighted bonds indicate the C_{10}-terpene skeleton)

Fig. 226. Biosynthesis of ajmalicine and serpentine

cathenamine synthase. The enzyme *cathenamine reductase* then transforms cathenamine into ajmalicine. Ajmalicine can be converted to serpentine by a peroxidase. These three enzymes have not yet been completely characterized. In cell cultures this reaction takes place in light whereas dark grown cultures accumulate ajmalicine.

Reserpine

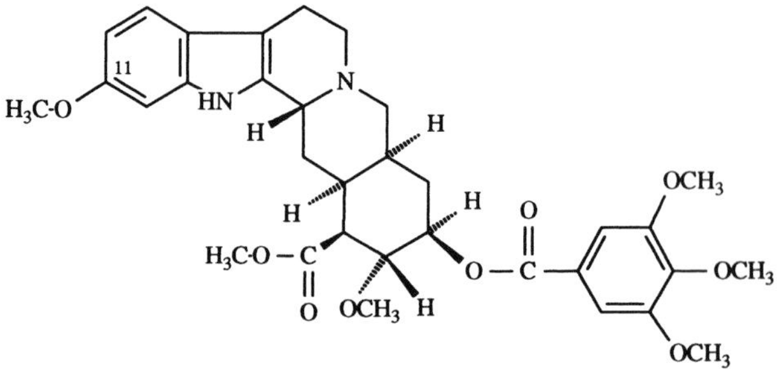

Reserpine

Plants belonging to the genus *Rauwolfia* (*Apocynaceae*) contain numerous alkaloids, three of which – reserpine, ajmaline and ajmalicine – are of medicinal importance. Reserpine was once an important drug for treatment of hypertension but also has a tranquilizing effect and high doses could cause mental fatigue and depression. Investigations of the mechanism of the tranquillizing effect led to the discovery that reserpine powerfully decreased the levels of the neurotransmitters 5-hydroxytryptamine (5-HT, serotonin) and noradrenaline in the brain and that the depression observed as a result of high doses was the result of exhaustion of the brain's stores of these compounds. This discovery was followed by intense studies of the functions of the monoaminergic systems and how these could be influenced by drugs. This research resulted in the introduction of the psychopharmacological drugs which revolutionized the treatment of psychoses and mental illness. Today reserpine has practically no use as a hypotensive drug or a tranquillizer. It has been superseded by more effective synthetic compounds.

Reserpine was first isolated from the root of *Rauwolfia serpentina* Benth. ex Kurz, an Indian shrub, the root of which had been used for centuries in popular medicine for the treatment of many diseases, including mental conditions. The discovery of the tranquillizing effect of reserpine gave a scientific explanation to this old tradition.

Ajmaline

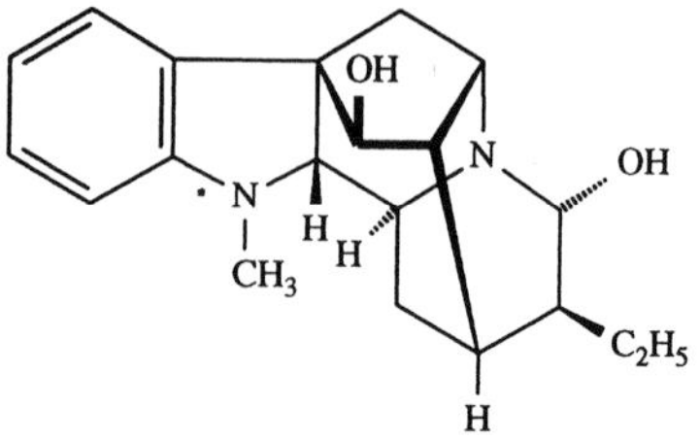

Ajmaline

This alkaloid is extracted from the roots of *R. serpentina*, *R. vomitoria* Afzel. and other *Rauwolfia* species. Ajmaline is used as a remedy for heart arrhythmias.

Biosynthesis   Ajmaline is biosynthesized from 4,21-didehydrogeissoschizine, which is derived from strictosidine as described on page 684. Didehydrogeissoschizine is transformed to polyneuridine aldehyde by a *sarpagan bridge enzyme*. The mechanism for this process has not yet been elucidated. The enzyme *polyneuridine aldehyde esterase* (EC 3.1.1.78) catalyses hydrolysis and decarboxylation of the methylester of polyneuridine aldehyde, yielding 16-*epi*-vil-

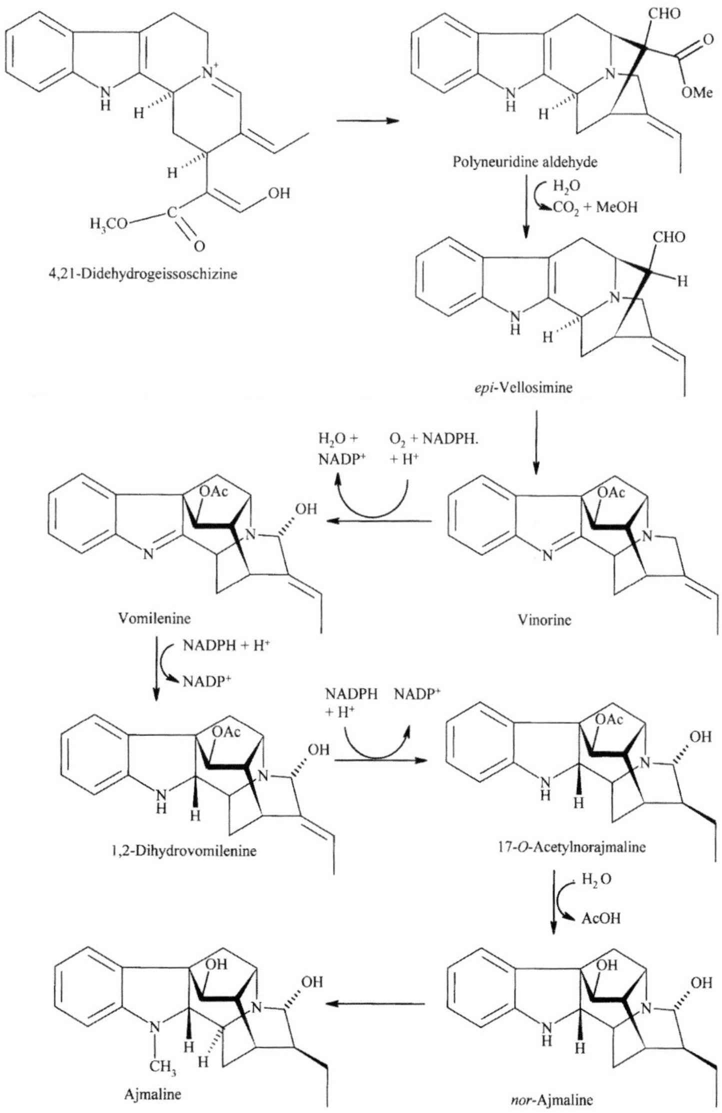

Fig. 227. Biosynthesis of ajmaline

losimine, which is transformed to vinorine by *vinorine synthase* (EC 2.3.1.160). Vinorine is then hydroxylated to vomilenine by *vinorine hydroxylase* (EC 1.14.13.75). In the next step the indolenine bond of vomilenine is reduced by *vomilenine reductase* (EC 1.5.1.32) with formation of 1,2-dihydrovomilenine, which is further reduced by *1,2-dihydrovomilinene reductase* (EC 1.3.1.73) to yield 17-*O*-acetylnorajmaline. *Acetylajmaline esterase* (EC 3.1.1.80) hydrolyses the acetyl linkage of acetylnorajmaline, yielding norajmaline, which is finally methylated at the indole nitrogen to form ajmaline.

The enzymes participating in this biosynthesis have been partly characterized. The sarpagan bridge enzyme has not yet been isolated, but might be a cytochrome P450 enzyme, which is dependent on NADPH and molecular oxygen. Polyneuridine aldehyde esterase has been purified from *Rauwolfia* cell cultures and sequencing of protein fragments enabled a clone of the enzyme to be isolated from a cDNA library. The enzyme has been overexpressed in *E. coli* and subjected to detailed mechanistic studies. Vinorine synthase has also been purified from *Rauwolfia* cell culture, sequenced and cloned from a cDNA library. The enzyme has been expressed in *E. coli*, crystallized and subjected to site-directed mutagenesis studies, which have led to a proposed mechanism of catalysis. Vinorine hydroxylase is a P450 enzyme which has not been purified from plant material in the active form. The two reductases have been purified and partially sequenced but no active clones have yet been obtained. Acetylajmaline esterase has been purified, partially sequenced and a full length clone isolated and expressed in tobacco leaves, yielding a protein with the expected enzymatic activity. The *N*-methyl transferase that catalyses the final reaction in the biosynthesis of ajmaline has not been further characterized.

Strychnine and brucine

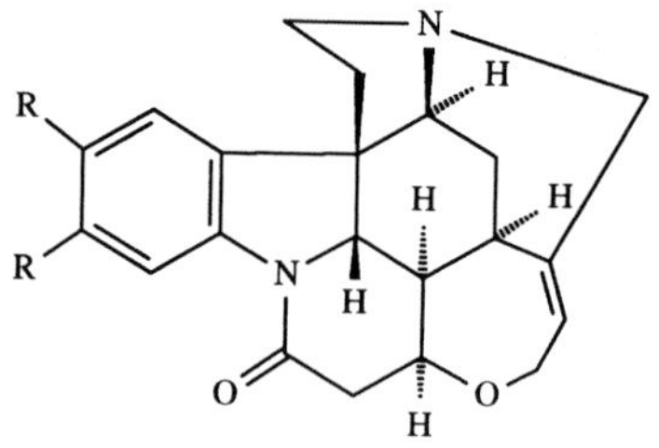

Strychnine: R = H

Brucine: R = OCH₃

Strychnine and brucine are obtained from the seed of *Strychnos nux-vomica*. Strychnine stimulates the central nervous system, greatly increasing the reflex excitability. Very small doses can give a subjective feeling of stimulation. For this reason, strychnine was formerly used as an ingredient in so-called tonics, i.e. invigorating drugs used during convalescence and in conditions of debility. In larger doses (30–100 mg), strychnine is a deadly poison. The increase in reflex excitability leads to violent muscular convulsions. All the muscles contract, forcing the patient into a position, with the back arched and resting only on the head and heels.

The effects of brucine are much weaker (about 1/50th of those of strychnine). The alkaloid has a very bitter taste and can be used as a standard substance for the determination of bitterness values (page 388).

Crude drug **Nux vomica** (Strychni semen) is the seed of *Strychnos nux-vomica* L. (*Loganiaceae*). This plant is a tree, about 12 m high, native to the south-west of India (the Malabar coast) but growing also in Bangladesh, Sri Lanka, Thailand, Laos, Cambodia and Vietnam. The fruit is a berry with 3–5 seeds which, after drying, constitute the crude drug. Nux vomica is used for the extraction of strychnine and brucine.

Alkaloids of Strychnos-based curares

As mentioned earlier (page 663), the South American dart and arrow poison curare derives partly from plants of the family *Menisperma-ceae* and partly from plants of the family *Loganiaceae*. Strychnos-based curares are prepared from the bark of several species, especially *S. guianensis* Baill., *S. toxifera* Schomb. ex Lindl., and *S. castelnaei* Wedd. Numerous alkaloids have been detected in and isolated from these three species, many of which are indole alkaloids with quaternary nitrogen. The most active bases contain two quaternary nitrogen atoms and their skeleton comprises 38 carbon atoms. An example is toxiferine. They can be regarded as dimers of C_{19} alkaloids related to strychnine, which have also been isolated from samples of curare. Similar dimeric quaternary alkaloids have been isolated from an African *Strychnos* species.

N,N^1-Diallylbisnortoxiferine is a semisynthetic derivative of toxiferine which is used as a short-lasting muscle relaxant in surgical operations.

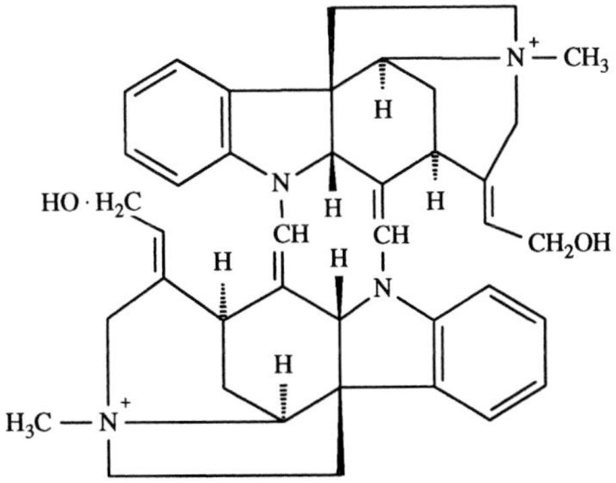

Toxiferine

Quinoline alkaloids

Quinoline

The medicinally important *Cinchona* alkaloids, quinine and quinidine, belong to the group of alkaloids containing a quinoline nucleus in their skeleton. Biosynthetically, they are alkaloids derived from the amino acid tryptophan and a monoterpenoid skeleton of the *Corynanthe* type.

Cinchona alkaloids The barks of trees belonging to the genus *Cinchona* contains some 25 related alkaloids. Quinine and quinidine are stereoisomers. There are also two other stereoisomers, cinchonine and cinchonidine, which lack the methoxyl group at C-6' in the quinoline nucleus. These two alkaloids are not used medicinally, although they show effects similar to those of quinine and quinidine. The structural formula given below can be used if it is not required to indicate the configuration. The stereoisomers in these pairs differ as regards the configurations at C-8 and C-9 (see Fig. 228).

$R = OCH_3$: Quinine and quinidine

$R = H$: Cinchonine and cinchonidine

Quinine is used as a remedy for malaria; quinidine is also active against malaria, but is mainly employed against cardiac arrhythmias. Large amounts of quinine are also used as a bitter flavouring in soft drinks ("tonic water").

Malaria This disease is found mostly in tropical areas, where it constitutes a major medical problem. It is characterized by attacks of high fever, which recur at regular intervals (every second or third day). There is also a severe form with irregular attacks of fever. The disease is caused by a parasitic protozoon of the genus *Plasmodium*, which uses mosquitos of the genus *Anopheles* as an intermediary host. When an infected mosquito bites a person, *sporozoites* enter the blood, where they penetrate the red blood corpuscles and divide to form *merozoites*. When this process is complete, the blood corpuscles burst open and the merozoites enter the blood plasma. The fever occurs at this point.

Some of the merozoites infect new blood corpuscles, while others develop into the sexual form, *gametes*. The gametes can pass to a healthy mosquito when it bites a person suffering from malaria. The gametes conjugate in the mosquito, forming sporozoites and the circle is completed. Quinine acts mainly on the merozoites.

Quinine has largely been replaced by synthetic antimalarials (chloroquine, etc.), which have fewer side-effects. In recent years, however, there has been a resurgence in the use of quinine for treating malaria, as forms of the disease have appeared which are resistant to the synthetic drugs, but which still respond to quinine. See also artemisinin on page 408.

Biosynthesis Cinchona alkaloids are biosynthesized from tryptophan and loganin in the same way as the monoterperoid indole alkaloids. However,

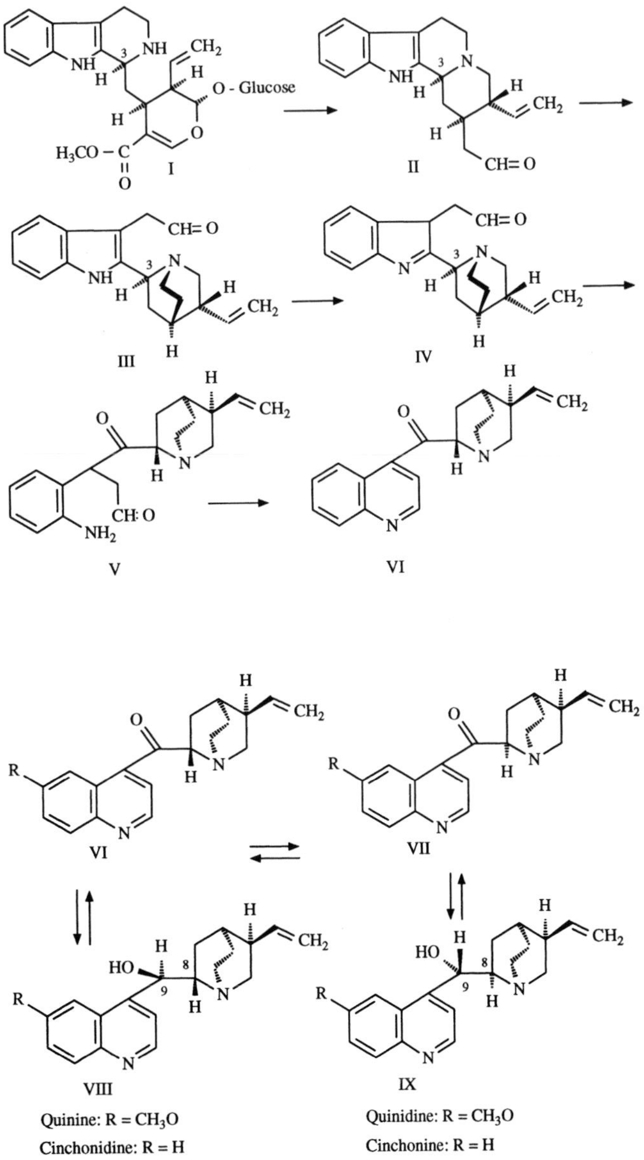

Fig. 228. Biosynthesis of Cinchona alkaloids

the pathway is more complicated and includes rearrangements both in the indole nucleus and in the terpenoid part of the structure.

Tryptophan and secologanin form strictosidine (I, Fig. 228,) as previously explained (page 680), which then gives corynantheal (II). Rearrangements in the terpene moiety (II–IV), followed by rearrangements in the indole structure (IV–VI), lead to the ketone cinchonidinone (VI), from which the two stereoisomeric groups of *Cinchona* alkaloids can be derived (VI–IX).

Crude drug **Cinchona** (Cinchonae cortex) is obtained from *Cinchona cordifolia* Mutis ex Humb., (sometimes known as *Cinchona pubescens* Vahl.) (*Rubiaceae*). The alkaloids can be isolated from this species but also from other species, e.g. *C. succirubra* Pavon ex Klotzsch, *C. ledgeriana* Moens ex Trimen (Plate 27, page 695), *C. calisaya* Wedd. and hybrids. *Cinchona* species are 15–20 m high trees native to Colombia, Ecuador, Peru and Bolivia, where they grow in the Andes at altitudes of 1 000–3 000 m. *Cinchona* is cultivated especially in Indonesia, but also in India, and many other parts of the world (Africa, Central America and South-East Asia).

The effect of cinchona against malaria became known following the Spanish conquest of Peru at the beginning of the 17th century, though how the discovery was made is uncertain. A legend tells that the wife of the Spanish viceroy, the Countess de Chinchon, was cured of malaria by means of cinchona and that this event made the drug known in Europe. There is no truth in this legend. When Linnaeus named the genus, he intended to honour the Countess de Chinchon, but he happened to make a spelling mistake, so that the genus received the name *Cinchona* instead of the intended name Chinchona.

Up to the end of the 19th century, cinchona was obtained from wild trees in South America. The producers there tried to monopolize cinchona production, but in the middle of the 19th century seeds of *C. calisaya* and of *C. cordifolia* were smuggled out. The former were planted in Java, the latter in India and Ceylon. However, the alkaloid content of the bark from these trees was too low, and only when a new species became available – *C. ledgeriana,* smuggled out of Bolivia – did the Dutch succeed in establishing extensive plantations of high-grade *Cinchona* trees in Java. These plantations soon came to dominate the world production of cinchona and at the time of the Second World War the Dutch had an almost complete monopoly of cinchona and Cinchona alkaloids.

Java was occupied by the Japanese in 1942, which led to a severe shortage of quinine on the side of the Allies. Attempts were made to overcome this by going back to collecting the bark from wild trees in South America, and also by developing and producing synthetic antimalarials. The latter attempts were successful and, after the war, the synthetic drugs completely dominated the treatment of malaria. However, as quinidine was still in demand, the plantations on Java continued to operate. During the war the area planted fell from

Plate 27.
Stripping the bark of *Cinchona ledgeriana* Bern. Moens ex Trimen, India. The bark is used for the extraction of quinine and quinidine (page 694). Photograph by F. Sandberg.

Plate 28.
Veratrum album L. The root is used for extraction of Protoveratrines A and B (page 701). Pirin mountains, Bulgaria. Photograph by Gunnar Samuelsson.

14 400 ha to 13 600 ha, but otherwise things were largely unaffected. After the war, export of the bark was greatly reduced, but production of cinchona alkaloid salts rose considerably. New plantations have been started in other tropical countries, especially in India where good results have been obtained with a hybrid of *C. ledgeriana* and *C. calisaya.* These developments have been helped by the fact that, as mentioned above, quinine has come back into use as an antimalarial because forms of malaria resistant to the synthetic drugs have appeared.

Cinchona species require a warm and humid tropical climate and grow best at an elevation of 1 000–2 000 metres. The average temperature should be 15–20 °C with little variation (20–30 °C in the day, 10–15 °C in the night), and the air humidity 75–95 %. The rainfall should be 1 500–2 000 mm per year, evenly distributed throughout the year. In plantations, the trees are allowed to become about 10 years old. They are then pulled up and the bark is collected from the entire tree.

Cinchona bark has a total alkaloid content of 3–15 %, and pharmaceutical bark must contain at least 6 %. The quinine content varies with the species. It is highest in *C. ledgeriana,* where it constitutes about 80 % of the total alkaloids. In *C. pubescens,* the quinine content is a good deal lower; cinchonine and cinchonidine often dominate and comprise up to 50 % of the total alkaloids present. The quinidine content is low in all species – about 1 % of the total alkaloids.

Camptothecin

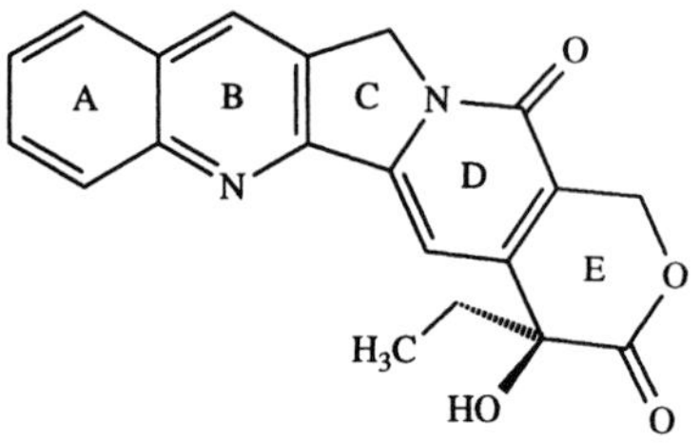

Camptothecin

Camptothecin is a cytotoxic quinoline alkaloid which was first discovered in the stem wood and bark of *Camptotheca acuminata* Decne. *(Nyssaceae)*, a tree native to China and Tibet. Later on it has also been found in other plants such as *Mappia tomentosa* Miers [also known under the names *Stemonurus foetida* Wight and *Nothapodytes foetida* (Wight) Sleumer *(Olacaceae)*], *Stemonurus megacarpus* Hemsl. [also known as *Merriliodendron megacarpum* Kaneh. *(Olacaceae)*], *Ophiorrhiza mungos* L. *(Rubiaceae)* and *Tabernaemontana heyneana* Wall *(Apocynaceae)*. Like paclitaxel (page 426) its discovery was the result of a systematic screening of natural products for anticancer drugs performed by the National Institute of Health (USA). In preliminary clinical tests camptothecin showed a broad-spectrum anticancer activity but also a high toxicity. Another problem was the poor solubility of the compound. A more soluble derivative is the sodium salt, the formation of which includes opening of the lactone ring E. The salt, however, has only 1/10 the activity of the parent compound. The anticancer activity of camptothecin is due to inhibition of topoisomerase I, an enzyme implicated in various DNA transactions, such as replication, transcription and recombination. Camptothecin binds to a complex formed by DNA with topoisomerase I, thereby inhibiting protein synthesis and cell division.

The low solubility of camptothecin and its serious side effects have hampered clinical research and efforts have been made to find more suitable derivatives. **10-Hydroxy camptothecin** (Fig. 229) is an alkaloid which is also present in *Camptotheca acuminata* although in very low concentration. It is more active than camptothecin and is used in China against cancers of the neck and head. Another derivative is **9-amino-camptothecin**, which is also poorly soluble but which is active at much lower doses than camptothecin (Fig. 229).

Total synthesis has yielded derivatives with better properties. One of these – **Irinotecan** (7-ethyl-10-[4-(1-piperidino)]carbonyl-oxy-camptothecin, Fig. 229) – has been approved in Japan and France as a remedy against colon cancer. Irinotecan is a prodrug which is rapidly metabolized by carboxyesterases to the active compound 7-ethyl-10-hydroxycamptothecin. Another synthetic compound is **Topotecan** (Fig. 229) which is currently tested clinically against several tumours, such as brain tumours in children.

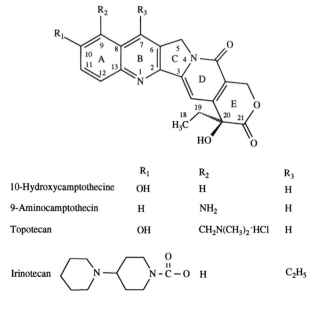

	R_1	R_2	R_3
10-Hydroxycamptothecine	OH	H	H
9-Aminocamptothecin	H	NH_2	H
Topotecan	OH	$CH_2N(CH_3)_2 \cdot HCl$	H
Irinotecan		H	C_2H_5

Fig. 229. Derivatives of camptothecin

697

Biosynthesis   3α(*S*)-Strictosidine is also the precursor for camptothecin. Ring D is formed by hydrolysis of the methylester and subsequent formation of an amide with the nitrogen in ring C yielding strictosamide (Fig. 230). In the next reaction the double bond in ring B is broken by oxidation forming two carbonyl groups. An aldol-type condensation yields pumiloside which undergoes allylic isomerization, reduction and oxidation to yield camptothecin.

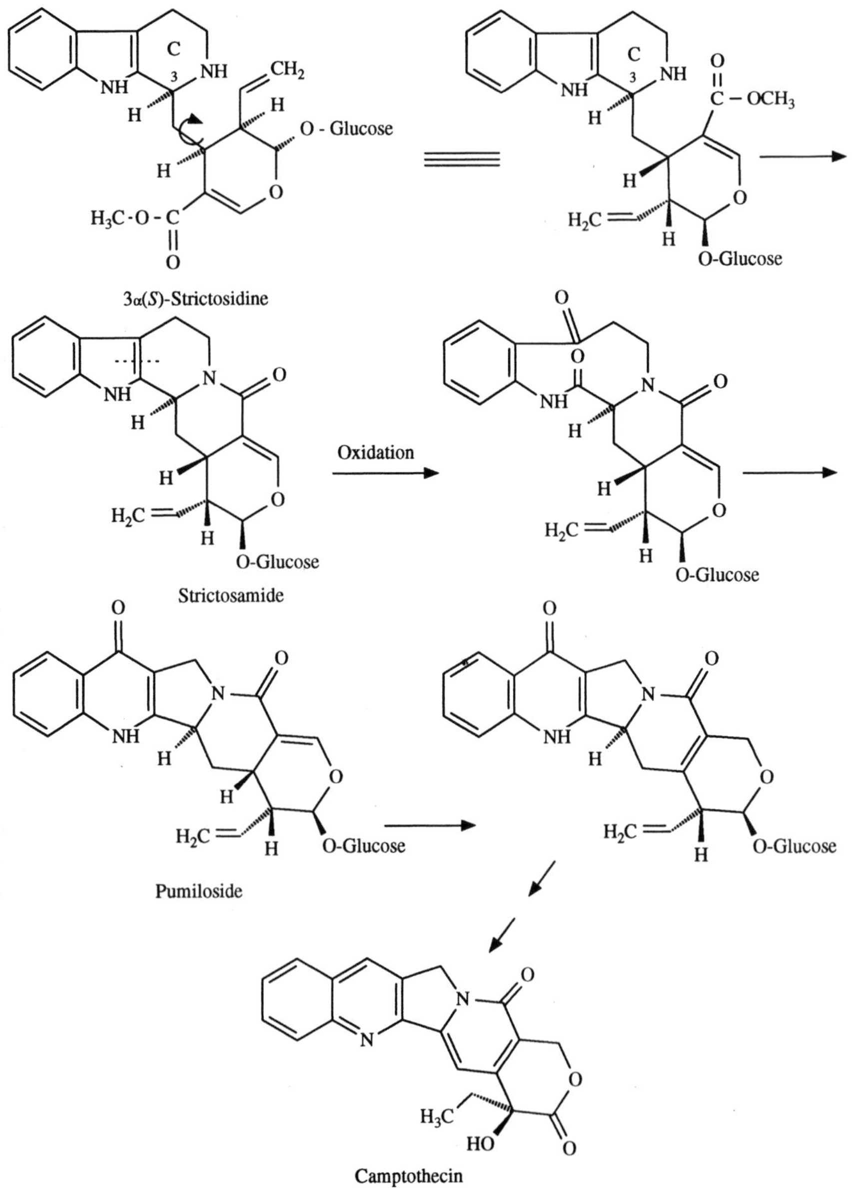

Fig. 230. Biosynthesis of camptothecin

Imidazole alkaloids

Pilocarpine Pilocarpine is found in the leaves of *Pilocarpus* species. It is a cholinergic substance which in the form of the hydrochloride or the nitrate is used in eye drops for the treatment of glaucoma. The biosynthesis of pilocarpine is unknown, but it is likely that the imidazole structure is derived from the amino acid histidine.

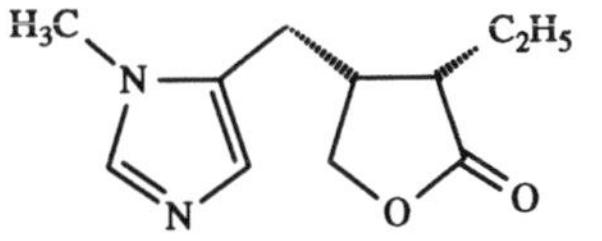

Pilocarpine

Crude drug **Jaborandi** (Jaborandi folium) is the leaflets of *Pilocarpus* species (*Rutaceae*). These are shrubs or trees, native to South and Central America and to the West Indies. The most important species at present is *P. microphyllus* Stapf, which grows in Brazil and which gives a drug sold under the name of Maranham jaborandi. The drug contains about 0.7–0.8 % pilocarpine and is used as a source of this alkaloid. Other kinds of jaborandi, with larger leaves, are also available on the market, but their alkaloid content is lower, 0.5–0.7 %. Pernambuco jaborandi comes from *P. jaborandi* Holmes and Paraguay jaborandi from *P. pennatifolius* Lem.

Steroidal alkaloids

Veratrum alkaloids *Veratrum album* and *Veratrum viride* contain a large number of alkaloids. Some of these have a normal steroid skeleton, while others have a modified one in which rings C and D of the steroid nucleus have changed places (C-nor-D homo-steroids). The latter alkaloids are divided into three groups: jervanine, veratramine and cevanine alkaloids. Rubijervine is one of the alkaloids with a normal steroid skeleton; jervine is a representative of the jervanine group; and veratramine belongs to the veratramine group.

The alkaloids belonging to these groups occur both as free alkamines, e.g. rubijervine, jervine and veratramine, and as esters or glycosides of the alkamines. None of the members of these three groups of alkaloids has been used in medicine.

Cevanine alkaloids The most important alkaloids of the cevanine group are protoveratrines A and B. These are tetra-esters of the alkamine protoverine and have a hypotensive effect. They are used for the treatment of severe cases of hypertension, but only in hospital and under strict control, because the difference between the therapeutic and toxic doses is small.

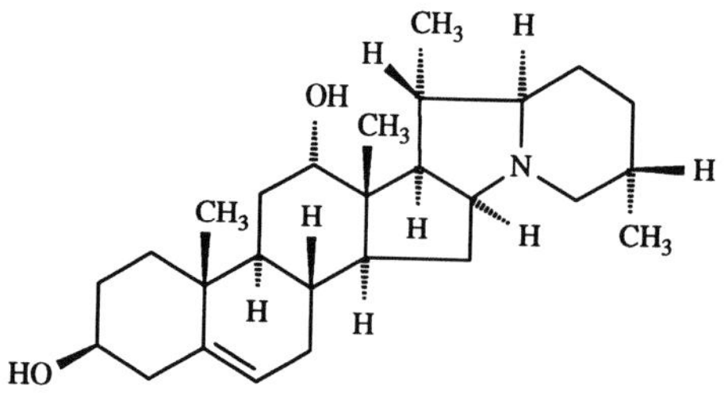

Rubijervine

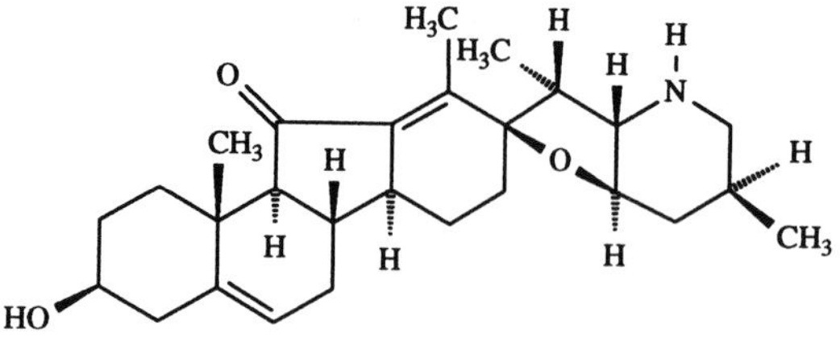

Jervine

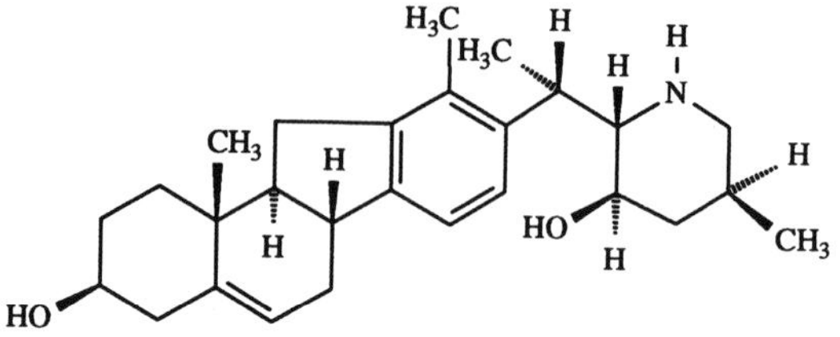

Veratramine

Within this group, there are a large number of alkaloids which differ as regards the number and nature of the ester-bound acids. There are also esters of the alkamine germine, which differs from protoverine in lacking a hydroxyl group at C-6. Besides protoveratrines A and B, other esters of protoverine and germine also have a hypotensive effect, the strength of which depends on the number of ester bonds. Alkamines and monoesters are weak, whereas tri- and tetraesters are strong hypotensives.

The biosynthesis of the veratrum alkaloids is not known in detail, but experiments appear to show that a substance with a normal steroidal skeleton is the precursor of the C-nor-D-homo-steroids.

Crude drug **European Veratrum** (White Hellebore, Veratri rhizoma) is the rhizome of *Veratrum album* L. (*Liliaceae*). The plant is a perennial herb growing in mountainous districts in Europe and northern Asia (Plate 28, page 695). The rhizome is harvested in the autumn and sliced to facilitate drying. The total alkaloid content of the drug is 1–1.5 %. Decoctions of white hellebore were formerly used as insecticides. Nowadays, the drug is not used medicinally as such, but it is the raw material from which protoveratrines A and B are extracted.

American Veratrum (Green Hellebore, Veratri viridae rhizoma) consists of the rhizome of *V. viride* Ait. (*Liliaceae*). The plant is similar to *V. album* and grows in North America, where it is used for the preparation of a standardized alkaloid mixture for the treatment

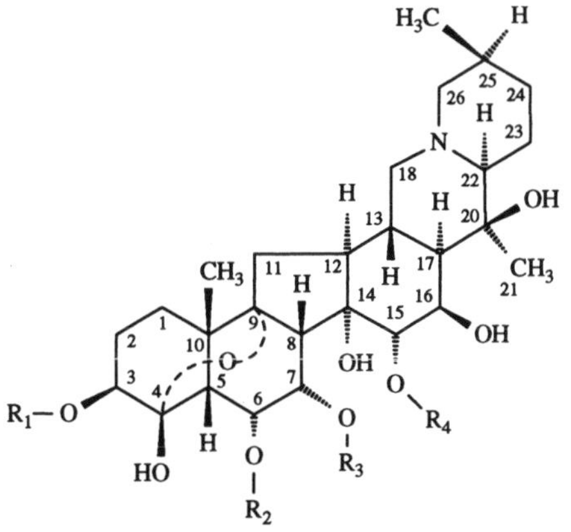

Protoverine: $R_1 = R_2 = R_3 = R_4 = H$

Protoveratrine A:

$$R_1 = -\overset{O}{\underset{\|}{C}} - \overset{OH}{\underset{CH_3}{C}} - CH_2 - CH_3$$

$$R_2 = R_3 = -\overset{O}{\underset{\|}{C}} - CH_3$$

$$R_4 = -\overset{O}{\underset{\|}{C}} - \overset{CH_3}{\underset{|}{CH}} - CH_2 - CH_3$$

Protoveratrine B:

$$R_1 = -\overset{O}{\underset{\|}{C}} - \overset{HO}{\underset{CH_3}{C}} - \overset{OH}{CH} - CH_3$$

$$R_2 = R_3 = -\overset{O}{\underset{\|}{C}} - CH_3$$

$$R_4 = -\overset{O}{\underset{\|}{C}} - \overset{CH_3}{\underset{|}{CH}} - CH_2 - CH_3$$

of hypertension. The most important alkaloids in this mixture are germidine and germitrine. Both are esters of the alkamine germine.

Cevadilla seed (Sabadillae semen) is the seed of *Schoenocaulon officinale* A. Gray (*Liliaceae*). This is a bulbous plant, coming from Mexico and the West Indies. The seed contains about 1 % veratrum alkaloids. The most important one is cevadine, a cevanine alkaloid. Extracts of the drug are insecticidal and have been used against head-lice.

Veratrine is a mixture of alkaloids from cevadilla seed, which is not only a potent insecticide, but a strong poison as well.

Batrachotoxins   Skin secretions of certain brightly coloured frogs of the genus *Phyllobates* (family *Dendrobatidae*), native to rainforests west of the Andes in Colombia, are sufficiently toxic to be used by the Indians for preparation of poison blow darts. The toxins of these skin secre-

701

tions have been identified as steroidal alkaloids called batrachotoxins. Batrachotoxin, homobatrachotoxin and batrachotoxinin A are the main components. 4β-Hydroxy-batrachotoxin and 4β-hydroxy-homobatrachotoxin are present in minor amounts.

Batrachotoxins have been found only in five species of the genus *Phyllobates*. Only three of these – *P. terribilis*, *P. bicolor* and *P. aurotaenia* – have high enough levels of batrachotoxins in their skin to allow preparation of dart poisons. *P. terribilis* has the highest content: about 500 µg batrachotoxin, 300 µg homobatrachotoxin and 200 µg batrachoxinin A per frog skin. This frog is so toxic that the darts can be poisoned by simply scraping the grooved tips across the frog's back. The other two species are less poisonous. Their skin contains roughly 20 µg batrachotoxin, 10 µg homobatrachotoxin and 50 µg batrachotoxinin A per skin. To obtain the dart poison these frogs are impaled on sticks. This treatment provokes a profuse secretion of the liquid which can then be collected.

Batrachotoxin is extremely toxic. The LD_{50} subcutaneously in mice is about 40 µg. Homobatrachotoxinin is slightly less toxic whereas the LD_{50} of batrachotoxinin A is 500 times higher. The toxicity is due to interaction with a site on voltage-dependent sodium channels of nerve and muscle, causing stabilization of the channel in an open formation. This causes massive influx of sodium ions and depolarization of nerve and muscle.

Batrachotoxin has no use as a drug but is an important tool for mechanistic studies of the function of voltage-dependent sodium channels and for investigation of the role of depolarization and/or influx of sodium ions on physiological functions.

Batrachotoxin-containing frogs as well as frogs containing histrionicotoxins (page 627), pumiliotoxins (page 628) and epibatidine (page 629) completely lack skin alkaloids when raised in cap-

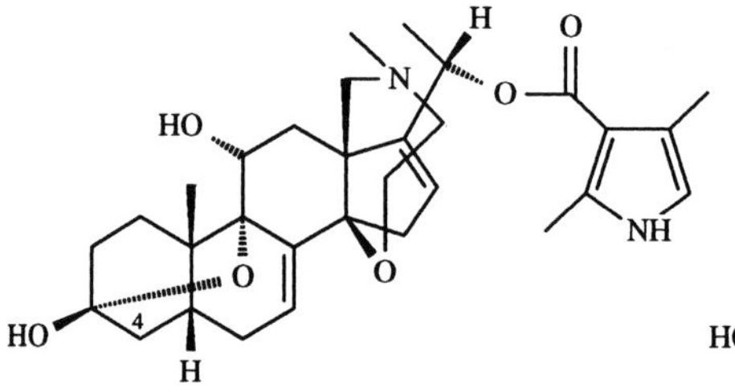

Batrachotoxin

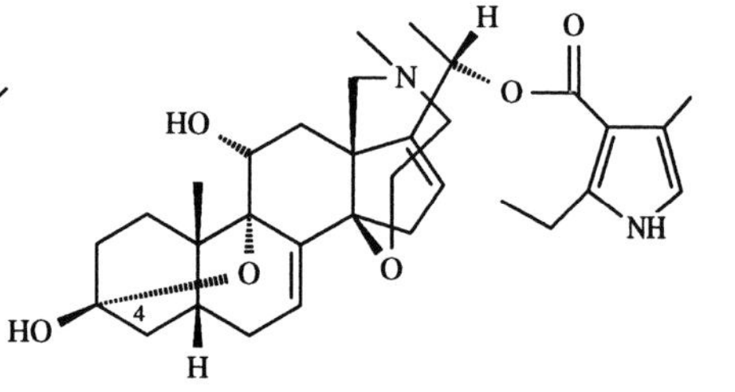

Homobatrachotoxin

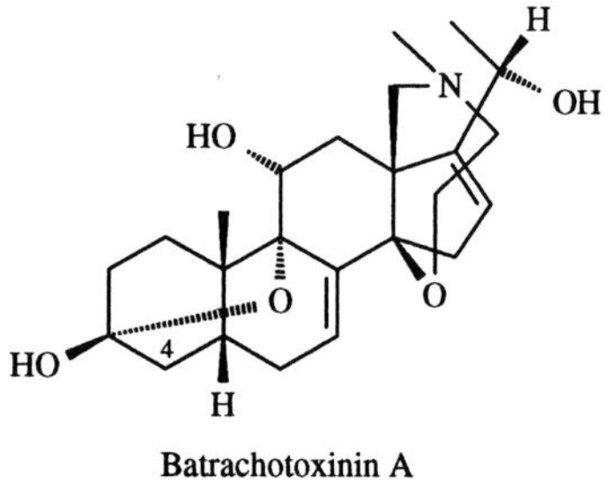

Batrachotoxinin A

Plate 29.
Aconitum napellus L. The plant contains aconitine (page 704). Jakobslund, Östergötland, Sweden. Photograph by Gunnar Samuelsson.

Plate 30.
Coffea canephora Pierre ex Froehner. Unripe fruits. Coffee is prepared from the ripe fruit (page 732). The Botanical Garden, Funchal, Madeira. Photograph by Gunnar Samuelsson.

tivity. A variety of enviromental manipulations do not trigger alkaloid production, but frogs fed alkaloid-dusted fruit flies efficiently accumulate unchanged such alkaloids in their skin. This indicates that the alkaloids are not biosynthesized by the frogs but that they probably are of dietary origin.

Aconitum alkaloids

Alkaloids with a diterpene structure are found in the genera *Aconitum* and *Delphinium* (*Ranunculaceae*), *Garrya* (*Garryaceae*) and *Inula* (*Asteraceae*). As regards their structure, these alkaloids can be formally derived from phyllocladene by linking the nitrogen atom of 2-aminoethanol, ethylamine or methylamine to C-16 and C-17.

Aconitine

Only one alkaloid in this group is of medicinal interest, aconitine from the aconite (monk's hood, wolfsbane) *Aconitum napellus* (*Ranunculaceae*). Aconitine is an ester alkaloid in which the alkamine aconine is esterified with acetic and benzoic acids. Aconitine is highly toxic. The lethal dose for man is 3–6 mg. The alkaloid affects the central, as well as the peripheral, nervous system, being at first stimulating and later paralysing. The latter effect is the reason why aconitine or liniments of aconite root have been used as external remedies for trigeminal neuralgia (see below). Toxic doses reveal their effects after only a few minutes. The first reaction is numbness of the tongue and buccal cavity, followed by nausea and vomiting. Then follow formication (a sensation as of ants crawling over the skin) and a burning sensation, followed by a feeling of intense cold. The body temperature goes down and severe pain arises in various parts of the body. The patient is conscious throughout and dies of respiratory paralysis or, after high doses, of heart failure in $1/2$–3 hours. The effect of aconitine depends on the ester bonds being intact. Hydrolysis of one or both ester bonds reduces the pharmacological activity drastically.

Crude drug

Aconite root (Aconiti tuber) is the storage root of wolfsbane, *Aconitum napellus* L. (*Ranunculaceae*), a perennial herb with blue,

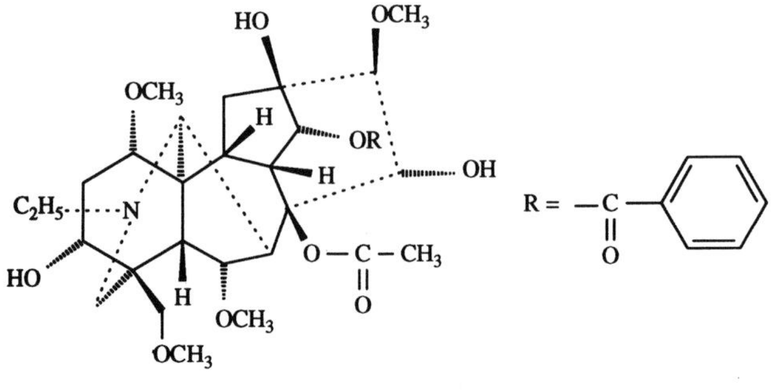

Aconitine

helmet-shaped flowers (Plate 29, page 703). It grows to a height of 0.5–2 m. It is native to mountainous regions in Europe, but is also a common ornamental garden plant. During the flowering period, the plant produces one or several new tubers, which are also used as a drug. Aconite root contains 0.5–1.5 % alkaloids, of which aconitine is the most important. The aconitine content of the drug varies a great deal, because the alkaloid is easily hydrolysed to the weakly active compounds benzoylaconine and aconine. The drug is used for aconitine extraction. In some countries, a tincture of aconite root is used as an external analgesic for neuralgia, especially trigeminal neuralgia.

Being a common ornamental plant, the risk of poisoning associated with wolfsbane should be borne in mind. Aconitine is present in all parts of the plant.

Guanidinium alkaloids

Tetrodotoxin

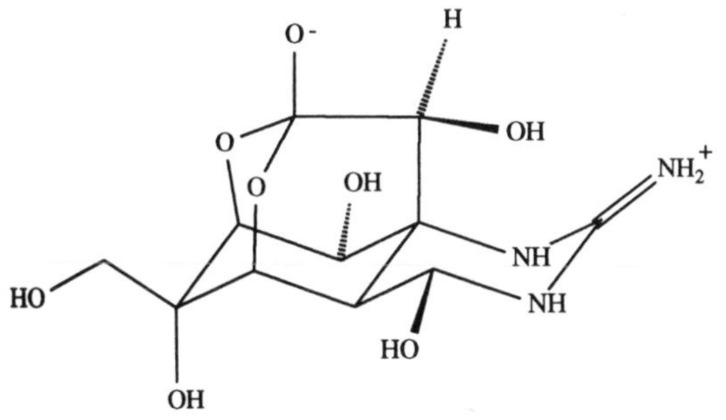

Tetrodotoxin

Tetrodotoxin was first isolated from the Japanese puffer fish, *Takifugu* sp. (*Tetraodontidae*), where it occurs in liver, ovaries and testes. These organs have to be carefully removed before consumption of the fish, which is considered a delicacy (called fugu) in Japan. Specially trained cooks are employed for cooking fugu, and eating this delicacy is a considerable risk, as failure to completely remove all the tetrodotoxin-containing organs results in serious, often fatal, poisoning. Tetrodotoxin has also been found in many other marine organisms, e.g. crabs, and in amphibians such as salamanders, newts, frogs and toads. Tetrodotoxins are of bacterial origin and are not produced by the organisms containing them. The biosynthesis of tetrodotoxin is not known, but the amino acid arginine is a precursor and the isopentenyl diphoshate pathway is also involved.

Tetrodotoxin is an extremely toxic compound. The LD_{50} on intraperitoneal injection in mice is about 0.2 µg. The substance causes paralysis of the neuromuscular system and acts by binding to the sodium channels, thereby blocking the influx of sodium ions

through excitable nerve membranes. Tetrodotoxin is a valuabl
pharmacological tool for the study of sodium channels.

Saxitoxin   Saxitoxin is another very potent neurotoxin that originates from
dinoflagellates which are consumed by many shellfish and mol
luscs, such as scallops, oysters and mussels which accumulate th
toxin and thereby become poisonous. These dinoflagellates live i
tropical and temperate waters and producers of shellfish in thes
areas must therefore monitor their product continuously for toxici
ty. The risk of shellfish becoming poisonous is particularly relate
to periods when the causative dinoflagellates multiply profoundly
called red tide as the amount of dinoflagellates colour the sea.

Saxitoxin

The biosynthesis of saxitoxin is not known in detail. The mole
cule contains a reduced purine ring system but the biosynthesis i
not related to that of the purines. The amino acid L-arginine i
known to supply most of the ring system, and acetate and methion
ine are also involved.

Saxitoxin is a very toxic compound with approximately the same
potency as tetrodotoxin, and acts in the same way as an inhibitor o
nerve conduction by blocking of the sodium channels. Also saxi
toxin is used as a pharmacological tool.

Further reading

General review

The Alkaloids, a continuing review covering mainly chemistry bu
    also pharmacological aspects, starting with Vol. I, 1950. Aca
    demic Press, London. Latest volume **63** (2006)

Cordell, G.A., Quinn-Beattie, M.L. and Farnsworth, N.R. The
    potential of alkaloids in drug discovery. *Phytotherapy
    Research*, **15**, 183–205 (2001).

Biosynthesis

Hashimoto, T. and Yamada, Y. Alkaloid biogenesis: molecular aspects. *Annual Review of Plant Physiology and Plant Molecular Biology*, **45**, 257–285 (1994).

Herbert, R.B. The biosynthesis of plant alkaloids and nitrogenous microbial metabolites. *Natural Products Reports*, **20**, 494–508 (2003); **18**, 50–65 (2001); **14**, 359–372 (1997); **13**, 45–58 (1996); **12**, 445–464 (1995) and older reviews in the same journal.

Kutchan, T.M. Alkaloid biosynthesis – the basis for metabolic engineering of medicinal plants. *The Plant Cell*, **7**, 1059–1070 (1995).

Amino alkaloids

Al-Bekairi, A.M., Abulaban, F.S., Qureshi, S. and Shah, A.H. The toxicity of *Catha edulis* (khat). A review. *Fitoterapia*, **62** (4), 291–300 (1991).

Barker, A.C. *et al.* Biosynthesis. Part 28. Colchicine: definition of intermediates between O-methylandrocymbine and colchicine and studies on speciosine. *Journal of the Chemical Society, Perkin Transactions I*, 2989–2994 (1998).

Grue-Sørensen, G. and Spenser, I.D. The biosynthesis of ephedrine. *Canadian Journal of Chemistry*, **67**, 998–1009 (1989).

Grue-Sørensen, G. and Spenser, I.D. Biosynthetic route to the *Ephedra* alkaloids. *Journal of the American Chemical Society*, **116**, 6195–6200 (1994).

Kalix, P. *Catha edulis*, a plant that has amphetamine effects. *Pharmacy World & Science*, **18** (2), 69–73 (1996).

McDonald, E. *et al.* Biosynthesis. Part 27. Colchicine: studies of the phenolic oxidative coupling and ring-expansion processes based on incorporation of multiply labelled 1-phenethylisoquinolines. *Journal of the Chemical Society, Perkin Transactions I*, 2979–2987 (1998).

Sheldrake, P.W. *et al.* Biosynthesis. Part 30. Colchicine: studies on the ring expansion step focusing on the fate of the hydrogens at C-4 of autumnaline. *Journal of the Chemical Society, Perkin Transactions I*, 3003–3009 (1998).

Woodhouse, R.N. *et al.* Biosynthesis. Part 29. Colchicine: studies on the ring expansion step focusing on the fate of the hydrogens at C-3 of autumnaline. *Journal of the Chemical Society, Perkin Transactions I*, 2995–3001 (1998).

Tropane alkaloids

Abraham, T.W. and Leete, E. New intermediate in the biosynthesis of the tropane alkaloids in *Datura innoxia*. *Journal of the American Chemical Society*, **117**, 8100–8105 (1995).

Chesters, N.C.J.E., O'Hagan, D. and Robins, R.J. The biosynthesis of tropic acid in plants: evidence for the direct rearrangement of 3-phenyllactate to tropate. *Journal of the Chemical Society, Perkins Transactions I*, 1159–1162 (1994).

Chesters, N.C.J.E., Walker, K., O'Hagan, D. and Floss, H.G. The biosynthesis of tropic acid: a reevaluation of the stereochemical course of the conversion of phenyllactate to tropate in *Datura strammonium*. *Journal of the American Chemical Society*, **118**, 925–926 (1996).

Hashimoto, T., Yun, D.-J. and Yamada, Y. Production of tropane alkaloids in genetically engineered root cultures. *Phytochemistry*, **32** (3), 713–718 (1993).

Huang, M.N., Abraham, T.W., Kim, S.H. and Leete, E. 1-Methylpyrrolidine-2-acetic acid is not a precursor of tropane alkaloids. *Phytochemistry*, **41** (3), 767–773 (1996).

Humphrey, A. and O'Hagan, D. Tropane alkaloid biosynthesis. A century old problem unresolved. *Natural Products Reports*, **18**, 494–502 (2001).

Li, R., Reed, D.W., Liu, E., Nowak, J., Pelcher, L., Page, J.E. and Covello, P.S. Functional genomic analysis of alkaloid biosynthesis in *Hyoscyamus niger* reveals a cytochrome P450 involved in littorine rearrangement. *Chemistry & Biology*, **13**, 513–520 (2006).

Robins, R.J., Chesters, N.C.J.E., O'Hagan, D., Parr, A.J., Walton, N.J. and Woolley, J.G. The biosynthesis of hyoscyamine: the process by which littorine rearranges to hyoscyamine. *Journal of the Chemical Society, Perkin Transactions I*, 481–485 (1995).

Robins, R.J., Abraham, T.W., Parr, A.J., Eagles, J. and Walton, N.J. The biosynthesis of tropane alkaloids in *Datura stramonium*: The identity of the intermediates between *N*-methylpyrrolinium salt and tropinone. *Journal of the American Chemical Society*, **119**, 10929–10934 (1997).

Pyrrolizidine alkaloids

Hartmann, T. and Ober, D. Biosynthesis and metabolism of pyrrolizidine alkaloids in plants and specialized insect herbivores. *Topics in Current Chemistry*, **209**, 207–243 (2000).

Ober, D. and Hartmann, T. Homospermidine synthase, the first pathway-specific enzyme of pyrrolizidine alkaloid biosynthesis, evolved from deoxyhypusine synthase. *Proceedings of the National Academy of Sciences*, **96** (26), 14777–14782 (1999).

Robins, D.J. Biosynthesis of pyrrolizidine and quinolizidine alkaloids. *The Alkaloids*, **46**, pp. 1–61, Academic Press Inc. (1995).

Roeder, E. Medicinal plants in Europe containing pyrrolizidine alkaloids. *Pharmazie*, **50**, 83–98 (1995).

Berberine

Dittrich, H. and Kutchan, T.M. Molecular cloning, expression, and induction of berberine bridge enzyme, an enzyme essential to the formation of benzophenanthridine alkaloids in the response of plants to pathogenic attack. *Biochemistry*, **88**, 9969–9973 (1991).

Frenzel, T. and Zenk, M.H. S-adenosyl-L-methionine:3'-hydroxy-N-methyl-(S)-coclaurine-4'-O-methyl transferase, a regio- and stereoselective enzyme of the S-reticuline pathway. *Phytochemistry*, **29** (11), 3305–3511 (1990).

Galneder, E. and Zenk, M.H. Enzymology of alkaloid production in plant cell cultures. in H.J.J. Nijkamp, L.H.W. Van Der Plas and J. Van Aartrijk (Eds), *Progress in Plant Cellular and Molecular Biology*, **9**, 754–762, Kluwer Academic Publishers, Dordrecht, The Netherlands (1990).

Okada, N., Shinmyo, A., Okada, H. and Yamada, Y. Purification and characterization of (S)-tetrahydroberberine oxidase from cultured *Coptis japonica* cells. *Phytochemistry*, **27** (4), 979–982 (1988).

Rueffer, M. and Zenk, M.H. Distant precursors of benzylisoquinoline alkaloids and their enzymatic formation. *Zeitschrift für Naturforschung*, **42c**, 319–332 (1987).

Steffens, P., Nagakura, N. and Zenk, M.H. Purification and characterization of the berberine bridge enzyme from *Berberis beaniana* cell cultures. *Phytochemistry*, **24** (11), 2577–2583 (1985).

Morphine

De-Eknamkul, W. and Zenk, M.H. Purification and properties of 1,2-dehydroreticuline reductase from *Papaver somniferum* seedlings. *Phytochemistry*, **31** (3), 813–821 (1992).

Gerardy, R. and Zenk, M.H. Purification and characterization of salutaridine:NADPH 7-oxidoreductase from *Papaver somniferum*. *Phytochemistry*, **34** (1), 125–132 (1993).

Gollwitzer, J., Lenz, R., Hampp, N. and Zenk, M.H. The transformation of neopinone to codeinone in morphine biosynthesis proceeds non-enzymatically. *Tetrahedron Letters*, **34** (36), 5703–5706 (1993).

Lenz, R. and Zenk, M.H. Closure of the oxide bridge in morphine biosynthesis. *Tetrahedron Letters*, **35** (23), 3897–3900 (1994).

Lotter, H., Gollwitzer, J. and Zenk, M.H. Revision of the configuration at C-7 of salutaridinol-I, the natural intermediate in morphine biosynthesis. *Tetrahedron Letters*, **33** (18), 2443–2446 (1992).

Novak, B.H. *et al*. Morphine synthesis and biosynthesis – an update. *Current Organic Chemistry*, **4**, 343–362 (2000).

Zenk, M.H., Gerardy, R. and Stadler, R. Phenol oxidative coupling of benzylisoquinoline alkaloids is catalysed by regio- and stereo-selective cytochrome P450 linked plant enzymes: salutaridine and berbamunine. *Journal of the Chemical Society, Chemical Communications*, 1725–1727 (1989).

Endogenous morphine

Hosztafi, S. and Fürst, Z. Endogenous morphine. *Pharmacological Research*, **32** (1), 15–20 (1995).

Sanguinarine

Kutchan, T.M., Dittrich, H., Bracher, D. and Zenk, M.H. Enzymology and molecular biology of alkaloid biosynthesis. *Tetrahedron*, **47** (31), 5945–5954 (1991).

Galanthamine

Eichhorn, J., Takada, T., Kita, Y. and Zenk, M.H. Biosynthesis of the *Amaryllidaceae* alkaloid galanthamine. *Phytochemistry*, **49**, 1037–1047 (1998).

Scott, L.J. and Goa, K.L. Galantamine. A review of its use in Alzheimer's disease. *Drugs*, **60** (5), 1095–1122 (2000).

Ergot alkaloids

von Döhren, H., Keller, U., Vater, J. and Zocher, R. Multifunctional peptide synthetases. *Chemical Reviews*, **97**, 2675–2705 (1997).

Kozikowski, A.P., Chen, C., Wu, J.-P., Shibuya, M., Kim, C.-G. and Floss, H.G. Probing ergot alkaloid biosynthesis: intermediates in the formation of ring C. *Journal of the American Chemical Society*, **115**, 2482–2488 (1993).

Tsai, H.-F., Wang, H., Gebler, J.C., Poulter, C.D. and Schardl, C.L. The *Claviceps purpurea* gene encoding dimethylallyltryptophan synthase, the commited step for ergot alkaloid biosynthesis. *Biochemical and Biophysical Research Communications*, **216** (1), 119–125 (1995).

Tudzynski, P., Correia, T. and Keller, U. Biotechnology and genetics of ergot alkaloids. *Applied Micobiology and Biotechnology*, **57**, 593–605 (2001).

Terpenoid indole alkaloids

De Waal, A., Meijer, A.H. and Verpoorte, R. Strictosidine synthase from *Catharanthus roseus:* purification and characterization of multiple forms. *Biochemical Journal*, **306**, 571–580 (1995).

Kutchan, T.M. Strictosidine: from alkaloid to enzyme to gene. *Phytochemistry*, **32** (3), 493–506 (1993).

Kutney, J.P. Biosynthesis and synthesis of indole and bisindole alkaloids in plant cell cultures: a personal overview. *Natural Products Reports*, **7**, 85–103 (1990).

Liu, D.-H., Jin, H.-B., Chen, Y.-H., Cui, L.-J., Ren, W.-W., Gong, Y.-F. and Tang, K.-X. Terpenoid indole alkaloids biosynthesis and metabolic engineering in *Catharanthus roseus. Journal of Integrative Plant Biology*, **49** (7), 961–974 (2007).

Misra, N., Luthra, R. and Kumar, S. Enzymology of indole alkaloid biosynthesis in *Catharanthus roseus*. *Indian Journal of Biochemistry & Biophysics*, **33**, 261–273 (1996).

O'Connor, S.E. and Maresh, J.J. Chemistry and biology of monoterpene indole alkaloid biosynthesis. *Natural Products Reports*, **23**, 532–547 (2006).

Camptothecin

Torck, M. and Pinkas, M. Camptothécine et derivés: une nouvelle classe d'agents antitumoraux. *Journal de Pharmacie de Belgigue*, **51**, 200–207 (1996).

Wall, M.E. and Wani, M.C. Camptothecin and taxol: discovery to clinic – thirteenth Bruce F. Cain Memorial Award Lecture. *Cancer Research*, **55**, 753–760 (1995).

Amphibian alkaloids

Daly, J.W., Garaffo, H.M. and Spande, T.F. Amphibian alkaloids in Cordell, G.A. (Ed.). *The Alkaloids, Chemistry and Pharmacology*, **43**, pp. 185-288, Academic Press, San Diego (1993).

Daly, J.W. Thirty years of discovering arthropod alkaloids in amphibian skin. *Journal of Natural Products*, **61**, 162–172 (1988).

Daly, J.W. *et al*. Alkaloids from frog skin: the discovery of epibatidine and the potential for developing novel non-opioid analgesics. *Natural Products Reports*, **17**, 131–135 (2000).

Saporito, R.A., Garraffo, H.M., Donnelly, M.A., Edwards, A.L., Longino, J.T. and Daly, J.W. Formicine ants: An arthropod source for the pumiliotoxin alkaloids of dendrobatid poison frogs. *Proceedings of the National Academy of Sciences USA*, **101** (21), 8045–8050 (2004).

Smith, S.Q. and Jones, T.H. Tracking the cryptic pumiliotoxins. *Proceedings of the National Academy of Sciences USA*, **101** (21), 7841–7842 (2004).

11. Purines and Pyrimidines

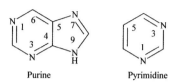

Purine   Pyrimidine

Purine and pyrimidine are heterocyclic compounds, derivatives of which are structural elements in the nucleic acids RNA and DNA, which are natural products of vital importance for all living organisms. Purine- and pyrimidine derivatives are also enzyme cofactors as well as important energy sources and activators in biosynthetic processes. The purine skeleton is contained in the drugs caffeine, theobromine and theophylline. Such compounds will be treated in this chapter, whereas for information about DNA, which constitutes the genome, and RNA which is responsible for transciption of the genetic information, the reader is referred to textbooks of biochemistry and molecular biology.

Purine derivatives

Biosynthesis

In purine derivatives the purine skeleton is attached to a sugar moiety (ribose or 2-deoxyribose) which can be phosphorylated (usually in the 5´-position). The purine ring system is not synthesized separately but is assembled on ribose phosphate as illustrated in Fig. 231. Deoxyribose derivatives are synthesized by reduction of the ribose moiety. The basic biochemical and molecular analysis of purine biosynthesis has been done in microorganisms and animals, but recent studies have shown that plants use similar reactions. The process starts with formation of 5-phospho-β-D-ribosylamine from ribose-5-phosphate (page 129) and ammonia, catalysed by *ribose-5-phosphate-ammonia ligase* (EC 6.3.4.7). This substance can also be formed via 5-phospho-α-D-ribose 1-diphosphate, which is obtained from ribose-5-phosphate by phosphorylation with ATP, catalysed by *ribose-phosphate diphosphokinase* (EC 2.7.6.1). The amino group is then introduced from glutamine by *amidophosphoribosyltransferase* (EC 2.4.2.14). These alternative reactions are not shown in Fig. 231. In the next reaction the carboxyl group of glycine is phosphorylated and reacts with the amino-group of phosphoribosylamine, forming N[1]-(5-phospho-D-ribosyl)glycinamide.

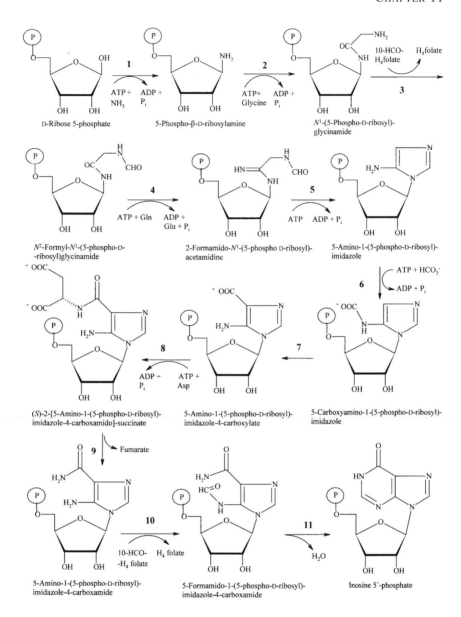

Fig. 231. Biosynthesis of inosine 5′-phosphate.

Participating enzymes; 1. ribose-5-phosphate-ammonia ligase *(EC 6.3.4.7), 2.* phosphoribosylamine-glycine ligase *(EC 6.3.4.13), 3.* phosphoribosylglycinamide formyltransferase *(EC 2.1.2.2), 4.* phosphoribosylformylglycinamidine synthase *(EC 6.3.5.3), 5.* phosphoribosylformylglycinamidine cyclo-ligase *(EC 6.3.3.1), 6.* 5-(carboxyamino)imidazole ribonucleotide synthase *(EC 6.3.4.18), 7.* 5-(carboxyamino) imidazole ribonucleotide mutase *(EC 5.4.99.18), 8.* phosphoribosylaminoimidazole-succinocarboxamide synthase *(EC 6.3.2.6), 9.* adenylosuccinate lyase *(EC 4.3.2.2), 10.* phosphoribosylaminoimidazolecarboxamide formyltransferase *(EC 2.1.2.3), 11.* IMPcyclohydrolase *(EC 3.5.4.10)*

Activated formate from N^{10}-formyltetrahydrofolate is added to the original glycine-amino group with formation of N^2-formyl-N^1-(5-phospho-D-ribosyl)glycinamide. Amination of the carbonyl group of this compound yields 2-formamido-N^1-(5-phospho-D-ribosyl)acetamidine. The amino-group derives from glutamine in an ATP-dependent reaction. The five-membered imidazole ring of purines is then formed by cyclization, with consumption of ATP, forming 5-amino-1-(5-phospho-D-ribosyl)imidazole.

In bacteria and yeasts the next step is phosphorylation of bicarbonate followed by reaction with the exocyclic amino-group to yield 5-carboxyamino-1-(5-phospho-D-ribosyl)imidazole which is rearranged to 5-amino-1-(5-phospho-D-ribosyl)imidazole-4-carboxylate. In vertebrates and plants this compound is formed directly from 5-amino-1-(5-phosphoribosyl)imidazole and CO_2 in a reaction catalysed by *phosphoribosylaminoimidazole carboxylase* (EC 4.1.1.21). This reaction is not shown in Fig. 231. Phosphorylation of the imidazole carboxylate group followed by displacement of the phosphate group by the amino group of aspartic acid yields (*S*)-2-[5-amino-1-(5-phospho-D-ribosyl)imidazole-4-carboxamido]-succinate. In the next reaction fumarate leaves the aspartic residue allowing the amino group to form an amide with the carboxyl carbon of the imidazole ring, yielding 5-amino-1-(5-phospho-D-ribosyl)imidazole-4-carboxamide. A formyl group from N^{10}-formyltetrahydrofolic acid is added to the amino nitrogen, yielding 5-formamido-1-(5-phospho-D-ribosyl)imidazole-4-carboxamide. In the final reaction loss of water and cyclization yields the purine nucleotide inosine 5´-phosphate (IMP) which is the starting material for AMP and GMP as illustrated in Fig. 232.

For formation of adenosine 5´-phosphate (AMP), *adenylosuccinate synthase* (EC 6.3.4.4) catalyses a reation with inosine 5´-phosphate, guanosine triphosphate (GTP) and aspartic acid to form N^6-[(*S*)-1,2-dicarboxyethyl]-AMP. This compound loses fumarate in a reaction catalysed by *adenylosuccinate lyase* (EC 4.3.2.2) yielding AMP.

For formation of guanosine 5´-phosphate (GMP), IMP is oxidized to xanthosine 5´-phosphate by *IMP dehydrogenase* (EC 1.1.1.205) and NAD^+. The carbonyl group formed, participates in a transamination reaction with glutamine or ammonia yielding guanosine 5´-phosphate (GMP). Catalysing enzymes are *GMP synthase* (EC 6.3.4.1) in the reaction with NH_3 or *GMP synthase (glutamine hydrolysing)* (EC 6.3.5.2) in the reaction with glutamine.

AMP and GMP are building blocks of RNA and are also precursors of a number of other purine derivatives. The corresponding deoxyribonucleotides deoxyadenosine 5´-monophosphate (dAMP) and deoxyguanosine 5´-phosphate (dGMP), which are building blocks of DNA, are formed from the ribonucleotides by reduction of the ribose moiety.

The genes coding for the pathway to purines have been identified both in microorganisms, mammals and plants. Prokaryotes have

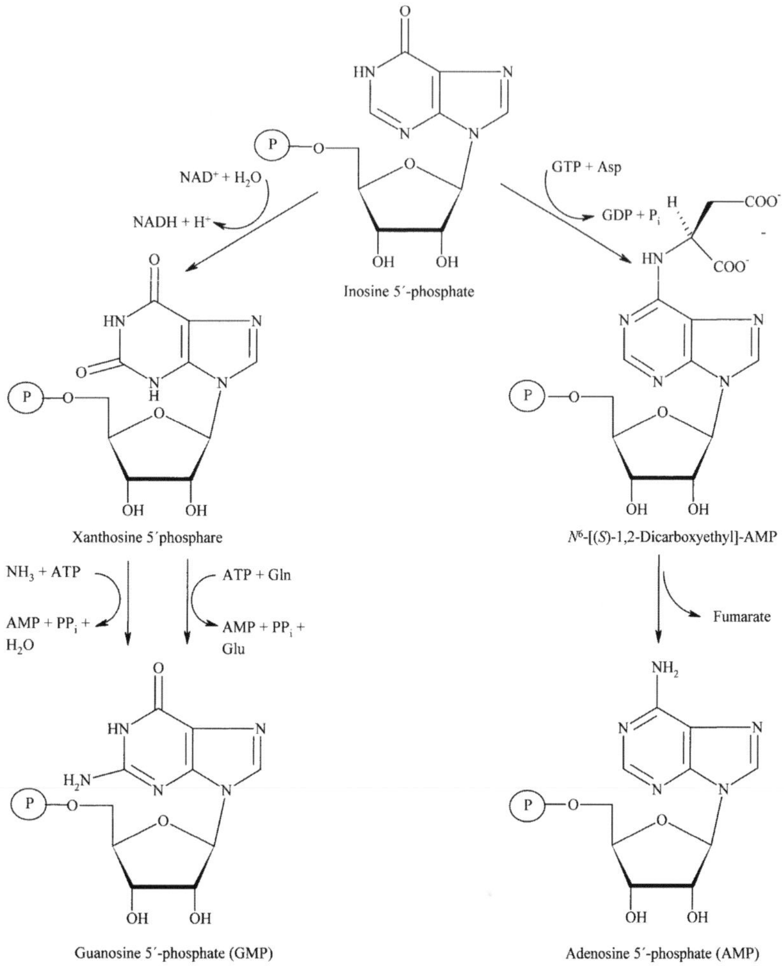

Fig. 232. Biosynthesis of AMP and guanosine 5´-phosphate (GMP) from IMP

single genes encoding monofunctional proteins whereas higher eukaryotes contain single genes encoding mono-, bi- and trifunctional proteins. In mammals these are probably organized as enzyme complexes. Plants are more similar to prokaryotes, with mono-functional proteins.

Besides the here described *de novo* biosynthesis of purines, these can also be formed from degradation products of purine derivatives in a salvage pathway, which requires less energy than the *de novo* biosynthesis.

Adenosine 5´-triphosphate (ATP)

ATP is the primary cellular energy carrier. Because of its content of two high-energy phospho-anhydride bonds it can serve as a source of energy in many biochemical processes. Hydrolysis of ATP to

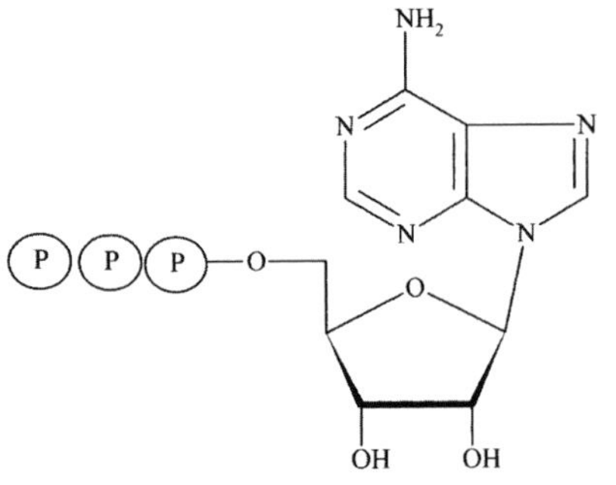

Adenosine triphosphate

form adenosine diphosphate (ADP) and inorganic phosphate or adenosine monophosphate (AMP) and inorganic diphosphate yields 7.3 kcal/mole of free energy and in this book there are many examples of biosynthetic reactions which are powered by hydrolysis of phosphate bonds in ATP. ATP is involved in the biosynthesis of nucleic acids and is a source of the purine skeleton in other biosynthetic pathways.

ATP is synthesized by phosphorylation of ADP in the light reactions of photosynthesis (Page 127), during glycolysis, in the citric acid cycle and in mitochondria. The ADP required can be formed *de novo* from AMP by phosphorylation catalysed by *adenylate kinase* (EC2.7.4.3). It can also originate from hydrolysis of ATP in energy requiring reactions.

Nicotinamide adenine dinucleotide (NAD+) and nicotinamide adenine dinucleotide phosphate (NADP+)

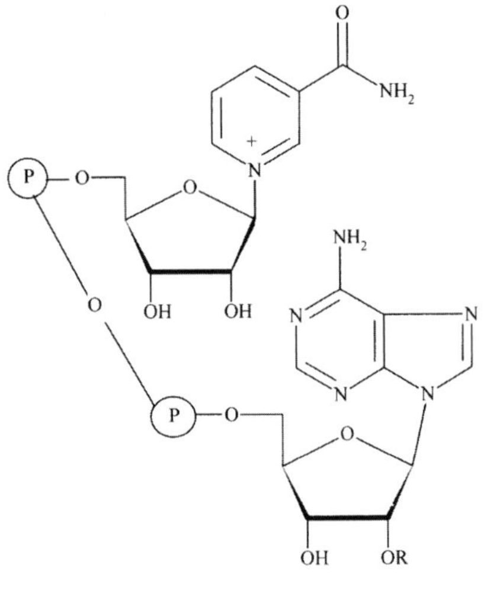

NAD$^+$: R = H   NADP$^+$: R = —(P)

Nicotinamide adenine dinucleotide (NAD$^+$) and nicotinamide adenine dinucleotide phosphate (NADP$^+$) are important coenzymes which participate in redox reactions where they accept or donate electrons. In oxidation reactions two protons (one hydride ion and one proton) are removed from the substrate (M in Fig. 233). The proton is released into solution and from the hydride electron pair one electron is transferred to the positively charged nitrogen of the nicotinamide group of NAD$^+$ or NADP$^+$ and the hydrogen attaches to the carbon atom opposite to that nitrogen, yielding NADH and NADPH, respectively. The process is reversed in reduction reactions.

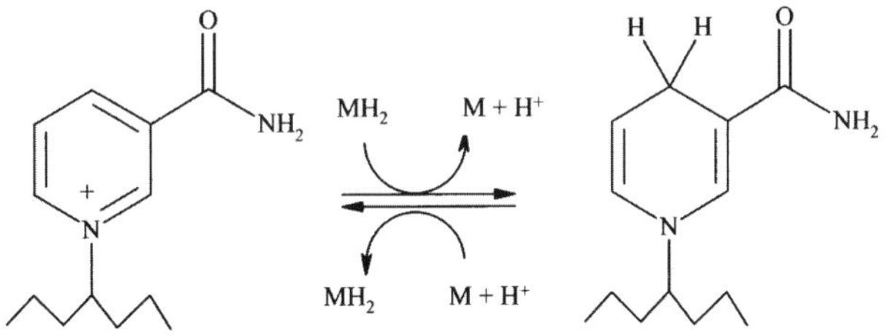

Fig. 233. Redox reactions involving NAD$^+$ and NADP$^+$. M = metabolite. The zigzag line represents the rest of the NAD molecule

Biosynthesis

The nicotinamide part of the NAD$^+$ and NADP$^+$ molecules are derived from the amino acids aspartic acid or tryptophan. The aspartic acid pathway is used by most prokaryotes and many plants (e.g. *Arabidopsis thaliana*), whereas fungi (e.g. yeast) and mammals use the tryptophan pathway. Both pathways result in formation of the NAD$^+$-precursor quinolinic acid (= pyridine-2,3-dicarboxylate, Fig. 234).

In the aspartic acid pathway the enzyme L-*aspartate oxidase* (EC 1.4.3.16) oxidizes L-aspartate to form α-iminosuccinic acid, which in a reaction catalysed by *quinolate synthase* (2.5.1.72), is condensed with dihydroxyacetone phosphate and cyclized to yield quinolinic acid.

The tryptophan pathway to quinolinic acid, which comprises five steps, starts with a reaction between tryptophan and molecular oxygen, catalysed by the enzymes *tryptophan 2,3-dioxygenase* (TDO, EC 1,13,11,11) or *indoleamine 2,3-dioxygenase* (IDO, EC 1,13,11,52), resulting in the formation of *N*-formyl-kynurenine. Both TDO and IDO are utilized by mammals, whereas only one of these enzymes is used by the other organisms who employ the tryptophan pathway. Thus *Saccharomyces cerevisiae* (yeast) uses IDO while *Drosophila melanogaster* (fruit fly) uses TDO.

717

N-formyl-kynurenine is hydrolysed by *arylformamidase* (EC 3.5.1.9) to L-kynurenine which is hydroxylated by *kynurenine 3-monooxygenase* (EC 1.14,13,9) to form 3-hydroxy-L-kynurenine. In the fourth step 3-hydroxy-L-kynurenine loses the side-chain with formation of 3-hydroxy-anthranilic acid in a reaction catalysed by *kynureninase* (EC 3.7.1.3). The enzyme *3-hydroxyanthranilate 3,4-dioxygenase* (EC 1,13,11,6) catalyses the final two stage reaction with initial formation of 2-amino-3-carboxymuconate semialdehyde, which spontaneously is converted to quinolinic acid.

The enzyme *nicotinate-nucleotide diphosphorylase (carboxylating)* (also known as quinolinic acid phosphoribosyltransferase, EC 2.4.2.19) catalyses formation of nicotinic acid mononucleotide from quinolinic acid and 5-phosphoribosyl-1-diphosphate (Fig. 235) and in the following reaction, catalysed by *nicotinate-nucleotide-adenylyltransferase* (EC 2.7.7.18), this compound reacts with ATP to form deamido-NAD⁺, which finally is amidated

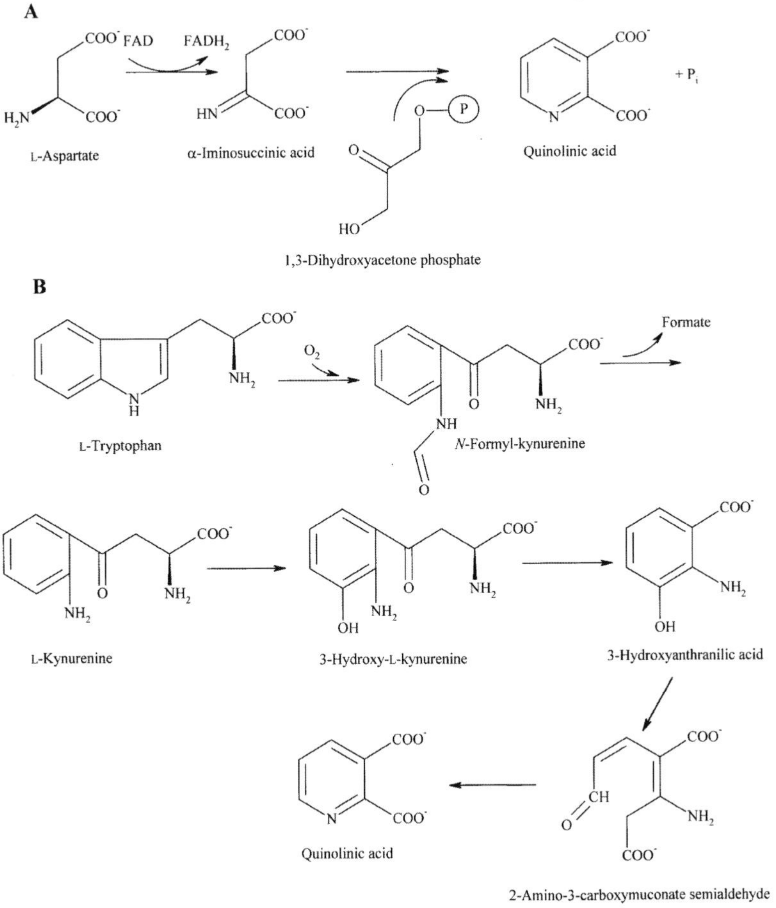

Fig. 234. Biosynthesis of the precursor quinolinic acid in the biosynthesis of NAD⁺ and NADP⁺. A: The aspartic acid pathway. B. The tryptophan pathway

to NAD⁺ in a reaction catalysed by *NAD⁺ synthase (glutamine-hydrolysing)* (EC 6.3.5.1).

NADP⁺ is produced from NAD⁺ by *NAD⁺ kinase* (EC 2.7.1.23), which phosphorylates NAD⁺ using ATP.

In mammals the above-described tryptophan pathway for NAD⁺ biosynthesis is less efficient and formation of NAD⁺ and NADP⁺ proceeds mainly via two alternative pathways, one from nicoti-

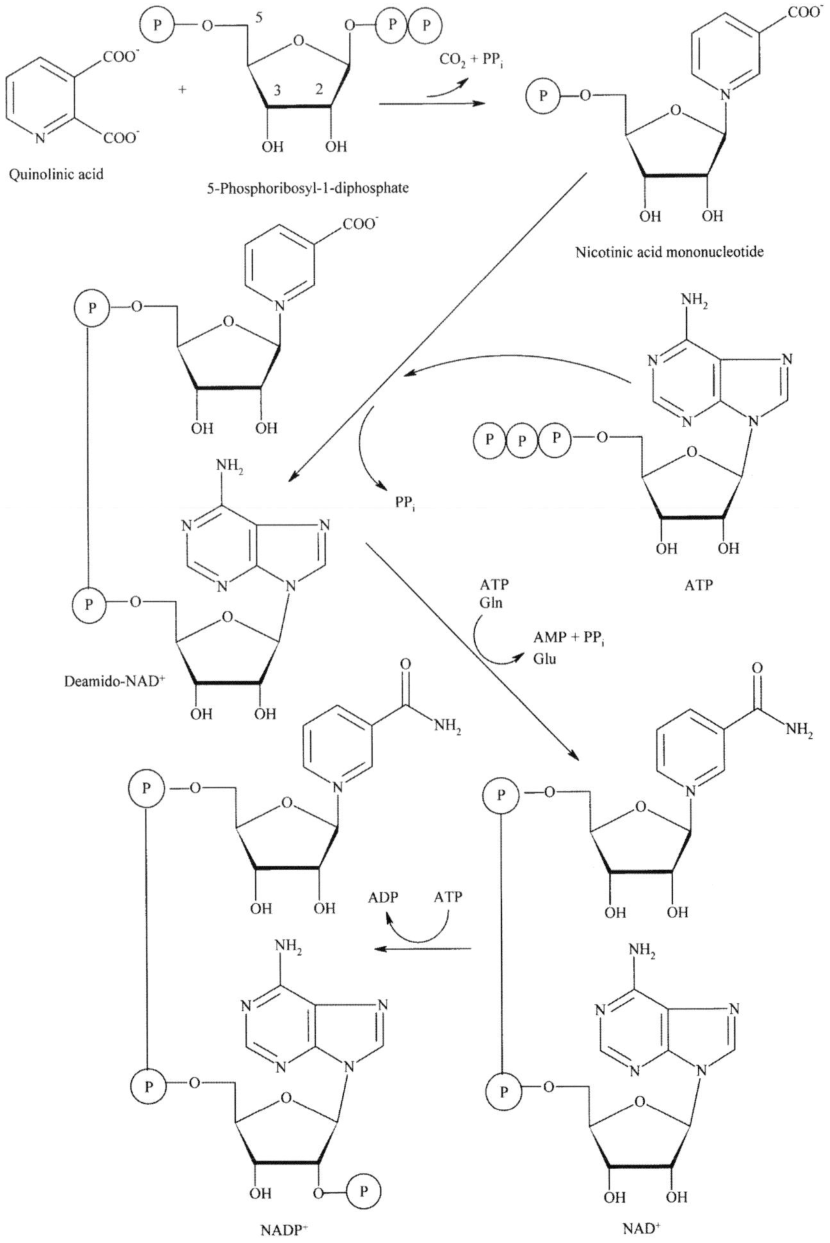

Fig. 235. Biosyntheis of NAD⁺ and NADP⁺ from quinolinic acid

namide via nicotinic acid, nicotinic acid mononucleotide and deamido NAD^+ and the other from nicotinamide via nicotinamide mononucleotide. Mammals cannot synthesize tryptophan *de novo* and insufficient dietary supply of this amino acid and of nicotinic acid or nicotine amide therefore causes insufficient biosynthesis of NAD^+ and $NADP^+$. In man, severe nicotinic acid deficiency results in development of the disease pellagra, characterized by diarrhoea, dementia, dermatitis and finally death. This disease can be cured by the supply of nicotinic acid in the diet. Nicotinic acid is thus a vitamin, usually called **vitamin B₃ (niacin)**.

In plants nicotinic acid is biosynthesized from aspartic acid via quinolinic acid (see page 623). Good dietary sources of nicotinic acid are yeast, meat, poultry, red fish (e.g. salmon), cereals, legumes and seeds. In maize, nicotinic acid is bound as a glycoside which cannot be split in the digestive system, thus decreasing the bioavailability. Treatment of the maize with lime releases nicotinic acid, which was known to the early settlers in USA who depended heavily on maize as a food. In the beginning of 1900 this had been forgotten in the southern USA, which caused a severe outbreak of pellagra among the poor population there.

Coenzyme A (CoA)

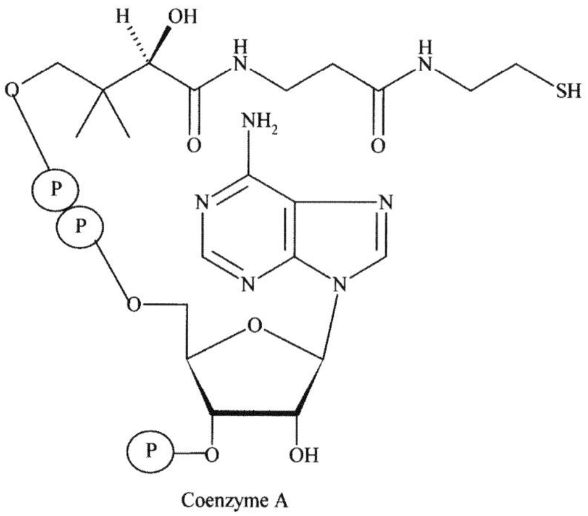

Coenzyme A

Coenzyme A is a very important cofactor in all living organisms. It functions as an acyl group carrier and a carbonyl-activating group and we have already come across it in many biosynthetic processes, particularly as the source of 4´-phosphopantetheine as the prosthetic group of fatty acid- and polyketide synthases as well as of non-ribosomal peptide synthases (pages 240, 270 and 531).

Biosynthesis   The biosynthesis of CoA proceeds in five steps and is illustrated in Fig. 236. The starting material is pantothenic acid (page 573) which is phosphorylated to form 4′-phosphopantothenate in a reaction catalysed by *pantothenate kinase* (EC 2.7.1.33). The enzyme *phosphopantothenate-cysteine ligase* (EC6.3.2.5) catalyses condensation of 4′-phosphopantothenate with cysteine, forming 4′-phosphopantothenoylcysteine which is decarboxylated by *phosphopantothenoylcysteine decarboxylase* (EC 4.1.1.36) to yield pantetheine 4′-phosphate. In bacteria these two enzyme activities are located on a bifunctional protein which uses cytidine triphosphate (CTP, page 739) for activation of the condensation step. In plants and mammals the condensation and decarboxylation steps are catalysed by separate enzymes and the condensing enzyme uses ATP for activation. In the next step AMP is added with formation of dephospho-CoA, a reaction catalysed by *pantetheine-phosphate adenylyltransferase*

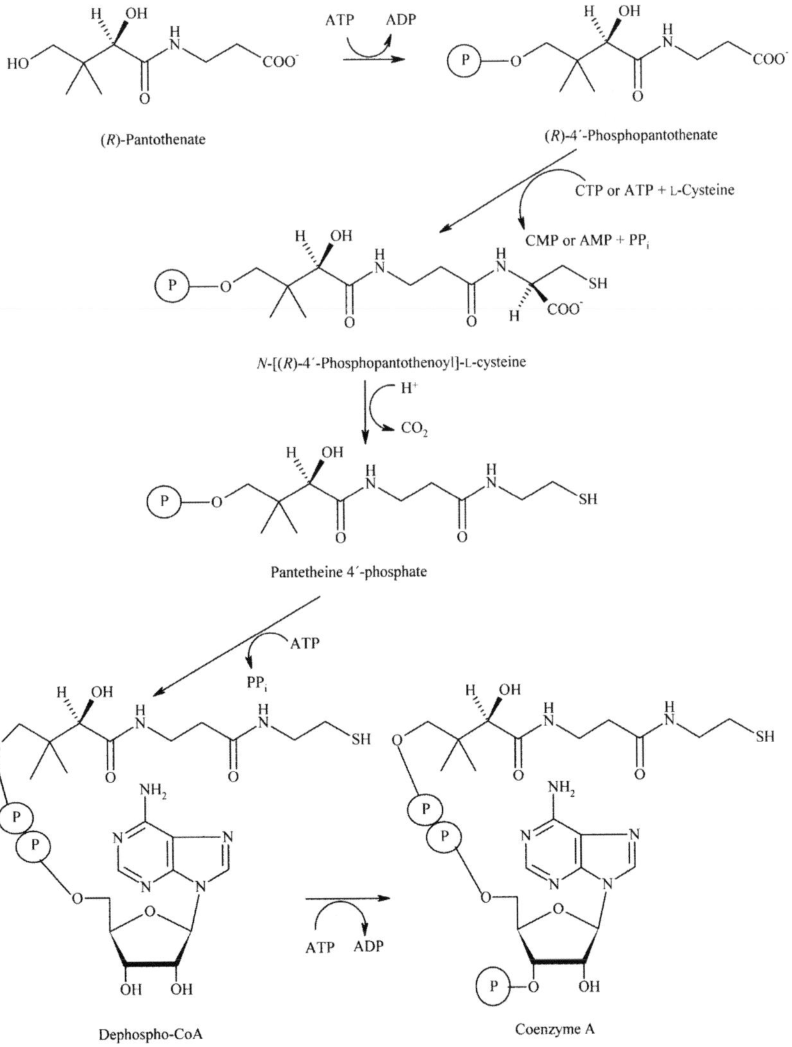

Fig. 236. Biosynthesis of coenzyme A

721

(EC 2.7.7.3). Finally dephospho-CoA is phosphorylated on the 3´-hydroxyl yielding CoA. This reaction is catalysed by *dephospho-CoA kinase* (EC 2.7.1.24). In bacteria and plants the last two reactions are catalysed by separate enzymes, whereas in mammals these two enzymatic activities are located on a bifunctional protein.

Most of the genes, coding for the enzymes involved, are known from bacteria, plants and mammals.

Guanosine 5´-triphosphate (GTP)

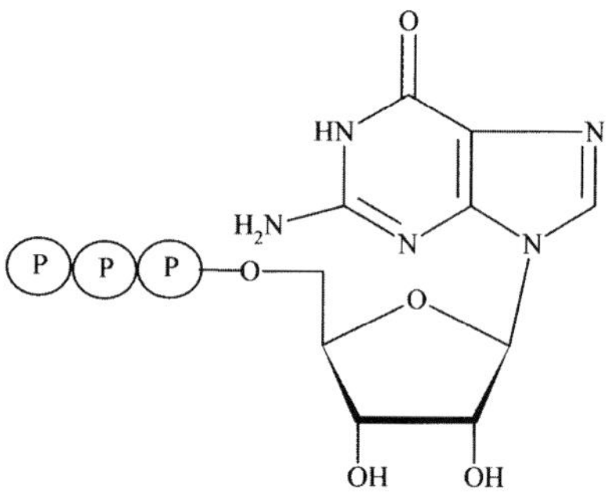

Guanosine triphosphate

Phosphorylation of guanosine 5´-monophosphate (GMP) yields GTP which is an alternative source of energy in biosynthetic reactions and a precursor for riboflavin (vitamin B_2), flavin mononucleotide (FMN), flavin-adenin dinucleotide (FAD) and tetrahydrofolate (Vitamin B_9). It is also involved in the biosynthesis of nucleic acids.

Riboflavin (Vitamin B_2)

Riboflavin is not a purine derivative, but is biosynthetically derived from the purine-nucleotide guanosine triphosphate. Vitamin B_2 is available in foods such as milk, liver, kidney, dairy products, eggs, meat and fresh vegetables. Yeast is a particularly rich source of this vitamin. Riboflavin may be produced synthetically or by fermentation, using the fungus *Ashbya gossypii, Candida* yeasts or *Bacillus subtilis*. The fermentation methods are progressively replacing synthetic preparation. Riboflavin is required for the metabolism of fats, carbohydrates and proteins, is an important antioxidant and activates vitamin B_3, B_6 and folic acid. It is also important for liver metabolism of poisons and drugs. Dietary defiency is manifested by problems of the skin and eyes.

Biosynthesis   The imidazole ring of guanosine 5´-triphosphate is opened in a reaction catalysed by *GTP cyclohydrolase II* (EC 3.5.4.25) with release of formate and diphosphate to yield 2,5-diamino-6-ribosyl-

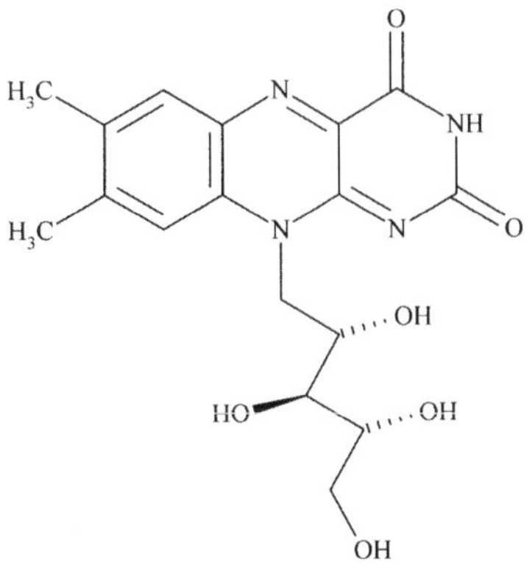

Riboflavin

amino-4(3*H*)-pyrimidone 5´phosphate (2 in Fig. 237). This compound is converted to 5-amino-6-ribitylamino-2,4(1*H*,3*H*)-pyrimidinedione 5´phosphate (5) in a two-stage reaction involving hydrolytic cleavage of the position 2 amino group of the heterocyclic ring and the reduction of the ribosylside chain. This proceeds differently in bacteria and yeast. In bacteria a bifunctional enzyme catalyses deamination of 2 to yield 5-amino-6-ribosylamino-2,4(1*H*,3*H*)-pyrimidinedione 5´phosphate (3), followed by reduction of the phosphoribosyl side-chain. In yeast the first reaction is reduction of the phosphoribosyl side-chain, yielding ribitylaminopyrimidine (4), followed by deamination. Compound 5 is dephosphorylated to yield 5-amino-6-ribitylamino-2,4(1*H*,3*H*)-pyrimidinedione (6) which is condensed with 3,4-dihydroxybutanone 4-phosphate (7, formed from ribulose 5-phosphate) with formation of 6,7-dimethyl-8-ribityllumazine (8). This reaction is catalysed by *6,7-dimethyl-8-ribityllumazine synthase* (lumazine synthase). The final step in the biosynthesis of riboflavin is dismutation of 6,7-dimethyl-8-ribityllumazine (8) catalysed by *riboflavin synthase* (EC 2.5.1.9). In this reaction two molecules of 8 yield riboflavin and one molecule of 5-amino-6-ribitylamino-2,4 (1*H*, 3*H*)-pyrimidinedione (6), which is thus brought back into the process.

Crude drug   **Dried yeast** is an ingredient in vitamin preparations which is rich in vitamin B_2 but also contains vitamins B_1 (page 741), B_3 (page 720), B_5 (page 573) and B_6 (page 575). Dried yeast is obtained as a byproduct of the brewery industry and consists of bottom yeast (under yeast, sediment yeast), the industrial term for the portion of yeast that settles to the bottom of the fermentation vat. Dried yeast is prepared by washing the yeast, first with a 1 % soda solution to remove bitter substances and then with water, and finally drying.

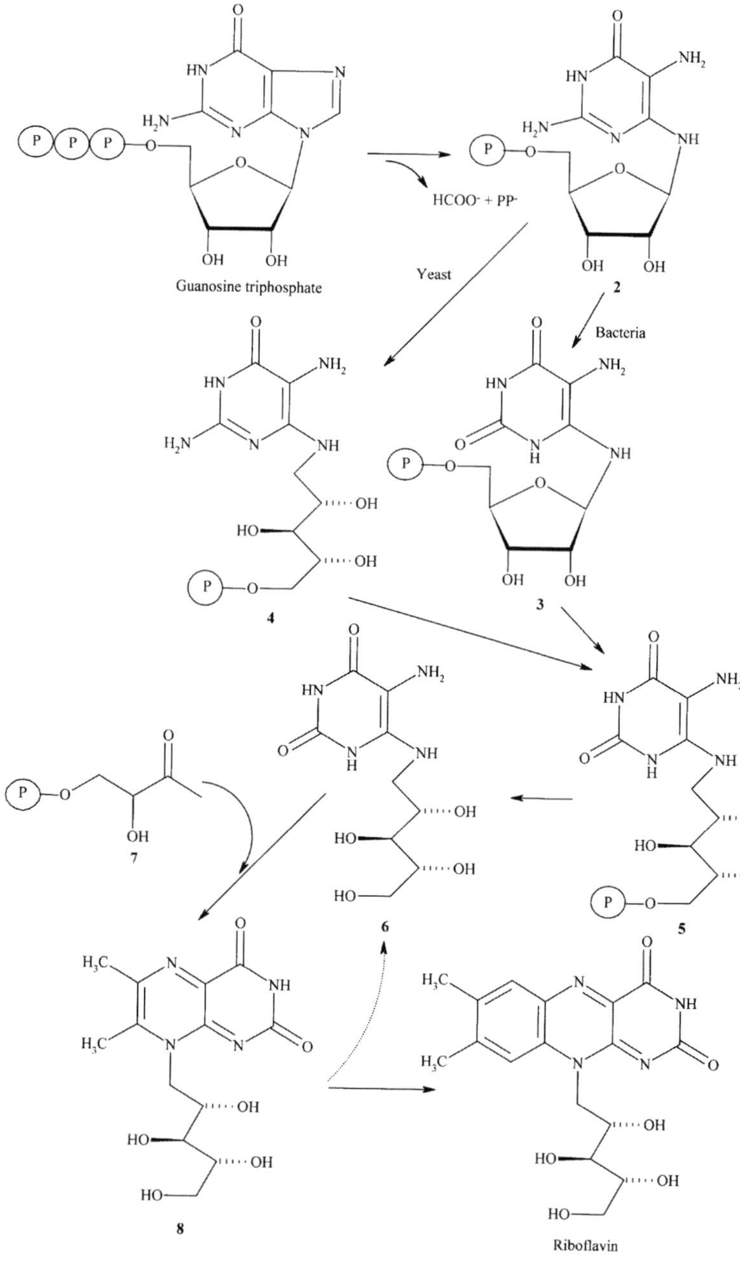

Fig. 237. Biosynthesis of riboflavin

Flavin mononucleotide (FMN) and flavin adenine dinucleotide (FAD)

FMN and FAD are cofactors for many enzymes. Mitochondrial electron transport, photosynthesis, fatty acid oxidation and metabo-

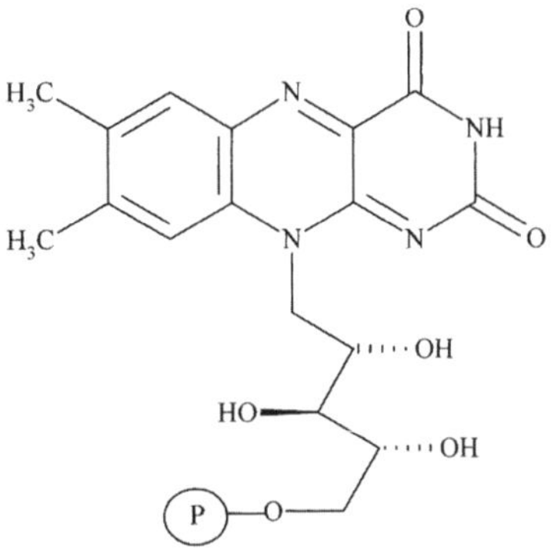

Flavin mononucleotide

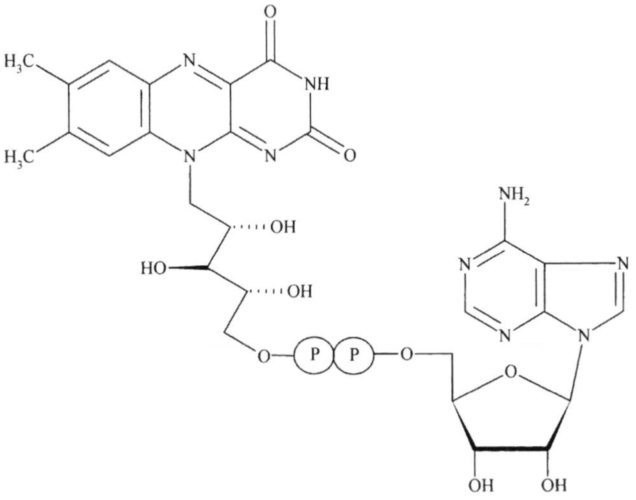

Flavin adenine dinucleotide

lism of vitamins B_6, B_{12} and folates are among the vital processes in which these two flavins participate. Biosynthetically they derive from riboflavin and are formed by phosphorylation and adenylation respectively, of this compound.

5,6,7,8-Tetrahydrofolate (THF, Vitamin B$_9$)

THF is not a purine derivative but its biosynthetic origin is the purine guanosine 5′-triphosphate (GTP). Folic acid is a synthetic compound, usually referred to as vitamin B_9 and contained in vitamin preparations. In the body the synthetic folic acid is reduced to form tetrahydrofolate, which is the active compound. Naturally occurring vitamin B_9, also called folate, is a complex containing THF but also derivatives of THF with an additional number of glutamic acid residues (up to 6) in the peptide chain. Natural vitamin B_9 can be found in vegetables such as spinach and turnip greens. It

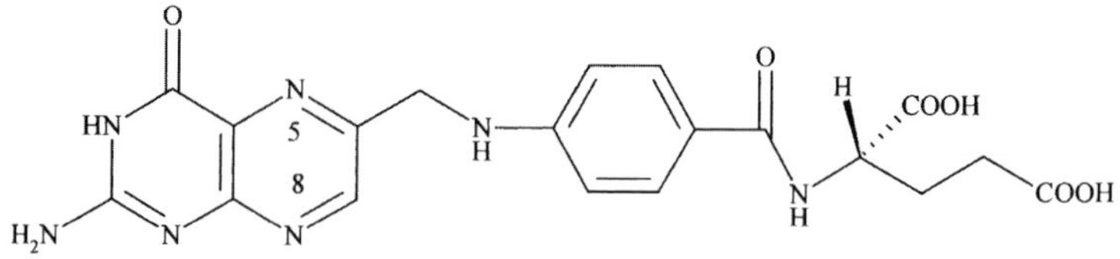

Folic acid

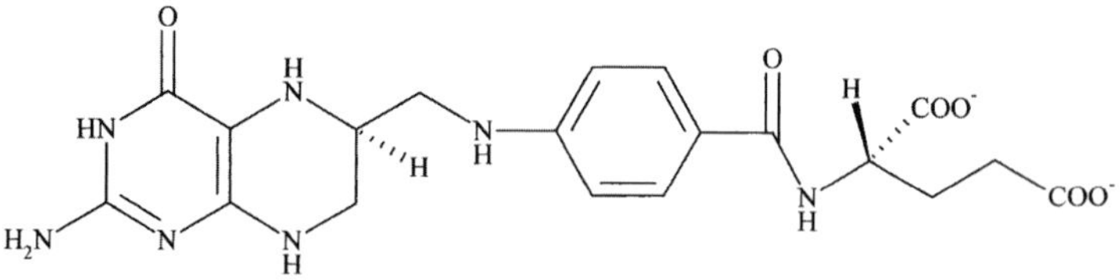

5,6,7,8-Tetrahydrofolate

is also present in yeast, beans, peas and liver. Cooking may destroy up to 90 % of the vitamin B_9 contained in the natural sources.

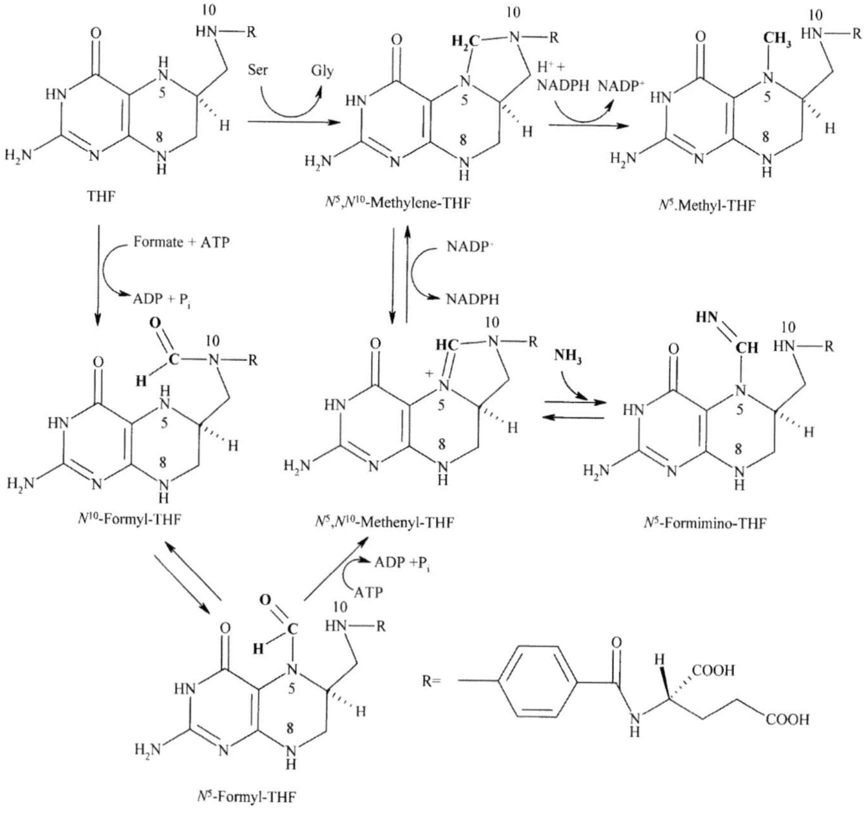

Fig. 238. Conversions of one-carbon units attached to THF

In biochemical processes THF functions as a carrier of one-carbon groups such as methyl-, methylene-, methenyl-, formimino- or formyl groups. We have already encountered this in several biosynthetic pathways such as the biosynthesis of methionine (page 497) and of the purine skeleton (page 712). The one-carbon group carried by THF is bonded to the N-5 or N-10 nitrogen (denoted as N^5 and N^{10}) or to both. The one-carbon units carried by THF are interconvertible as illustrated in Fig. 238.

Vitamin B_9 (folate) is necessary for the production and maintenance of new cells and is needed for the replication of DNA. This is particularly important during pregnancy and infancy when cell growth is extremely rapid. Both adults and children need vitamin B_9 to make normal red blood cells and to prevent anaemia. Folate deficiency is particularly dangerous for pregnant women and can cause malformations in the child including neural tube defects, resulting in malformations of the spine (spina bifida), skull and brain. Adequate folate intake during the periconceptional period and the first month following conception is very important. An intake corresponding to 400 μg/day of synthetic folic acid has been suggested for the whole pregnancy period. Breastfeeding mothers need 280 μg/day, and non-pregnant women as well as men 200 μg/day.

Biosynthesis

Plants, yeast and bacteria synthesize THF from GTP and chorismic acid.

The biosynthesis proceeds first along two parallell pathways, one from GTP to 6-hydroxymethyl-7,8-dihydropterin diphosphate (Fig. 239) and the other from chorismate to p-aminobenzoate. These two compounds are then combined to form 7,8-dihydropteroate from which two reactions lead to formation of THF (Fig. 240).

A. Formation of 6-hydroxy-methyl-7,8-dihydropterin diphosphate (Fig. 239)

The biosynthesis starts with formation of 7,8-dihydroneopterin 3′-triphosphate from GTP in a reaction catalysed by *GTP cyclohydrolase I* (EC 3.5.4.16). The phosphate groups are removed from this compound in two reactions catalysed by *7,8-dihydroneopterin triphosphate pyrophosphohydrolase* and a non-specific phosphatase with formation of 7,8-dihydroneopterin. The enzyme *dihydroneopterin aldolase* (EC 4.1.2.25) then catalyses removal of 2-hydroxyacetaldehyde yielding 6-hydroxymethyl-7,8-dihydropterin, which is diphosphorylated to 6-hydroxymethyl-7,8-dihydropterin diphosphate in a reaction catalysed by *2-amino-4-hydroxy-6-hydroxymethyldihydropteridine diphosphokinase* (EC 2.7.6.3).

B. Formation of 7,8-dihydropteroate and THF (Fig.240)

The enzyme *aminodeoxychorismate synthase* (EC 2.6.1.85) catalyses substitution of an amino group from glutamine for the 4-hydroxy group of chorismate with formation of 4-amino-4-deoxy-chorismate, which loses pyruvate to yield 4-aminobenzoate in a reaction catalysed by *aminodeoxychorismate lyase* (EC 4.1.3.38). *Dihydropteroate synthase* (EC 2.5.1.15) then catalyses the combi-

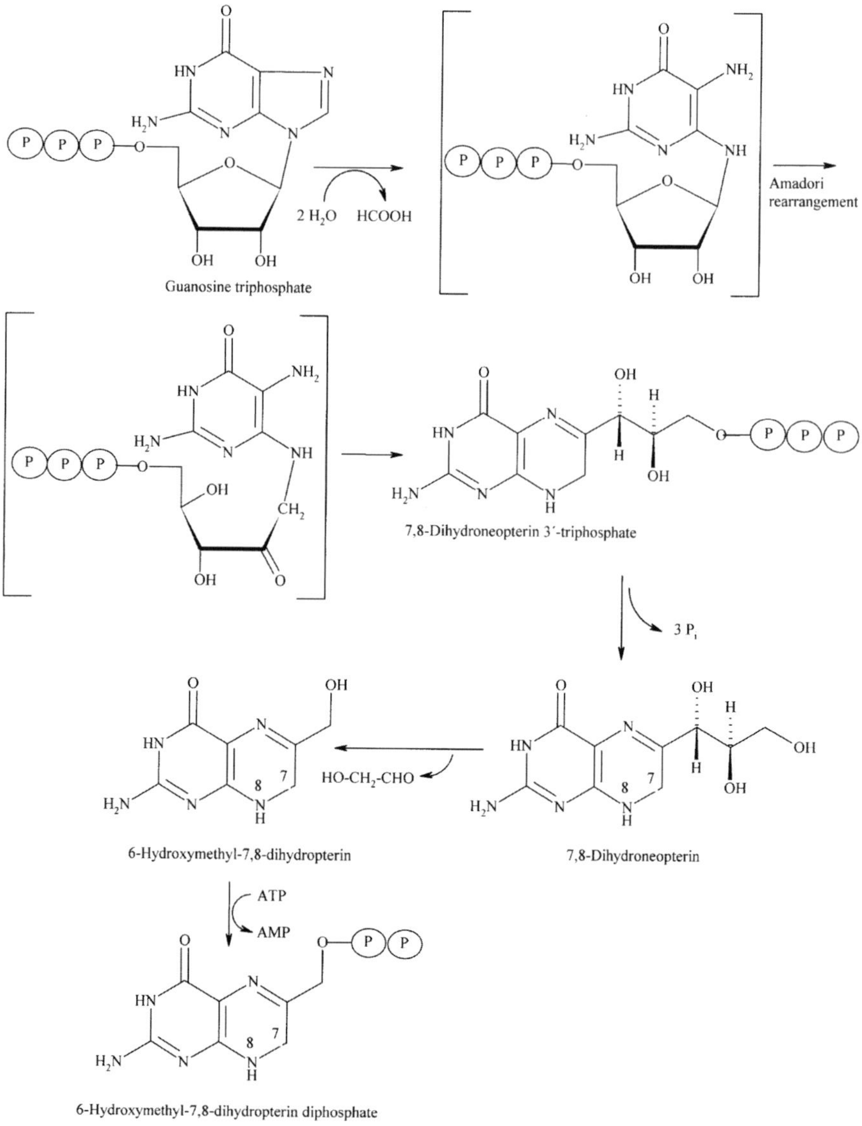

Fig. 239. Biosyntheis of 6-hydroxymethyl-7,8-dihydropterin diphosphate

nation of 4-aminobenzoate and 6-hydroxymethyl-7,8-dihydropterin diphosphate with formation of 7,8-dihydropteroate. A peptide bond is then formed between the carboxyl group of dihydropteroate and the amino group of glutamate yielding 7,8-dihydropteroylglutamate (= 7,8-dihydrofolic acid). This ATP-depending reaction is catalysed by *dihydrofolate synthase* (EC 6.3.2.12). The final reaction in the biosynthesis of THF is reduction of the double bond between atoms 5 and 6 by *dihydrofolate reductase* (EC 1.5.1.3).

Other members of the vitamin B$_9$ complex are formed by elongation of the peptide chain with up to six glutamic acid residues in reactions catalysed by the enzyme *tetrahydrofolate synthase* (EC 6.3.2.17).

Caffeine, theobromine and theophylline

The purine derivatives caffeine, theobromine and theophylline are practically devoid of basic properties, but they are still usually classified as alkaloids because they contain nitrogen and are pharmacologically active. Caffeine is the most common of the three compounds and is distributed throughout the plant kingdom, occurring in genera which are not closely related systematically. The most important caffeine-containing plants are: coffee, species of the genus *Coffea* (*Rubiaceae*); tea, *Camellia thea* Link. previously known as *Thea sinensis* L. or *Camellia sinensis* Kuntze (*Theaceae*); guarana, *Paullinia cupana* H.B. & K. (*Sapindaceae*); kola, *Cola* species (*Sterculiaceae*); and maté, *Ilex paraguensis* A. St.Hil. and a few related species (*Aquifoliaceae*). Theobromine is found particularly in the seeds of *Theobroma cacao* L (*Sterculiaceae*). Theophylline occurs in small amounts in tea. In the living plants, the purines are bound to sugars, phenols and tannins, but they are set free during the fermentation and roasting processes which are involved in the production of purine drugs.

Caffeine is extracted from tea waste. It can also be obtained on roasting coffee in which process a considerable amount of caffeine sublimes and can be collected. Caffeine stimulates the central nervous system, increases the activity of the heart, and acts as a diuretic. The substance is used medicinally as a stimulant.

Theobromine is extracted from cocoa-bean husks, a waste product from the production of cocoa (see below). Theobromine does not stimulate the central nervous system, but it is diuretic and has a relaxing effect on smooth muscle. The compound is employed as a diuretic and for the treatment of asthma attacks.

Theophylline is present in tea, but in amounts too small to be worth isolating. Instead, the compound is prepared either by demethylation of caffeine or by total synthesis. Theophylline has a diuretic effect and an antispasmodic effect on smooth muscle. Like theobromine, it is used as a diuretic and for the treatment of asthma attacks.

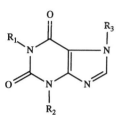

	R$_1$	R$_2$	R$_3$
Caffeine	CH$_3$	CH$_3$	CH$_3$
Theobromine	H	CH$_3$	CH$_3$
Theophylline	CH$_3$	CH$_3$	H

Biosynthesis Caffeine and theobromine are biosynthesized from the nucleotide xanthosine 5′-phosphate, which is an intermediate in the biosynthesis of guanosine 5′-phosphate (Fig. 232). As illustrated in Fig. 241 the enzyme *5′-nucleotidase* (EC 3.1.3.5) removes the phosphate group with formation of xanthosine which is methylated by *7-methylxanthosine synthase* (EC 2.1.1.158) forming 7-methylxanthosine. The ribose group is then removed by N-*methyl nucleosidase* (EC 3.2.2.25) yielding 7-methylxanthine, which is methylated

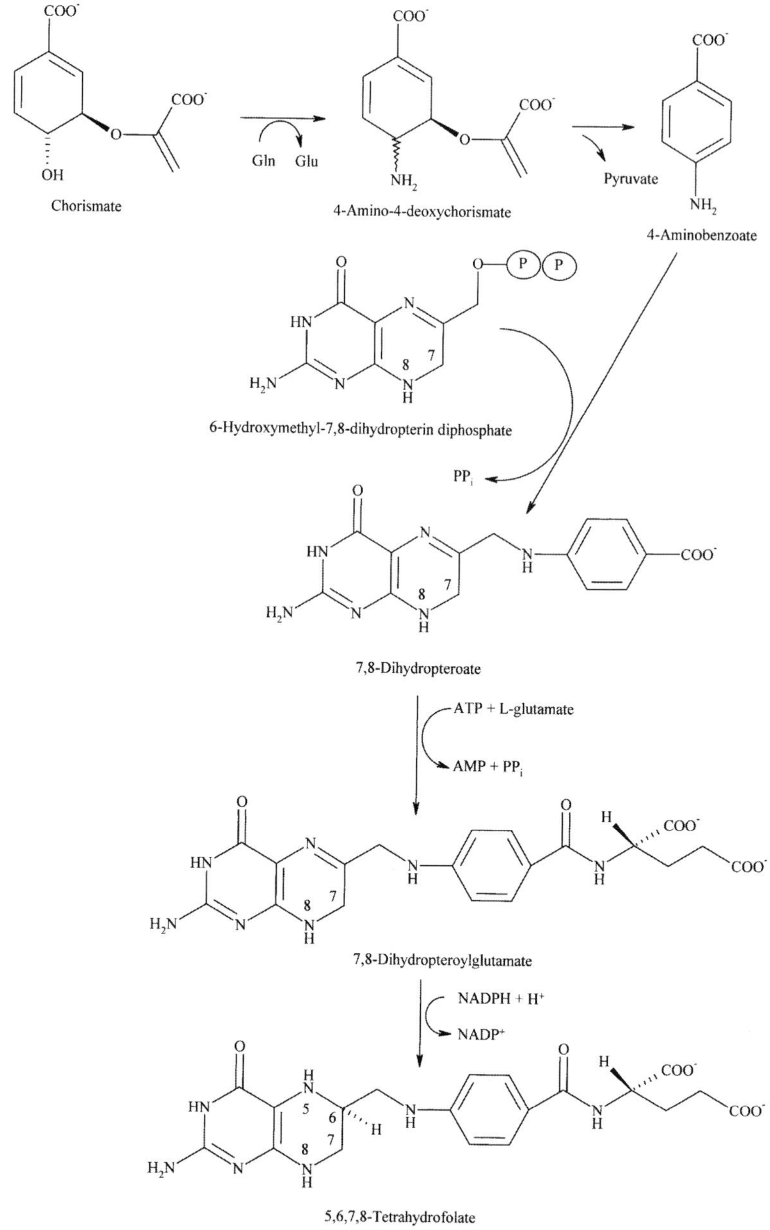

Fig. 240. Biosynthesis of 5,6,7,8-Tetrahydrofolate (THF)

by *theobromine synthase* (EC 2.1.1.159) to theobromine. Finally *caffeine synthase* (EC 2.1.1.160) methylates theobromine to caffeine.

The biosynthesis of theophylline has been suggested to proceed as illustrated in the upper part of Fig. 241. Xanthosine is transformed to xanthine by loss of the ribose moiety and xanthine is then methylated in two steps to produce theophylline.

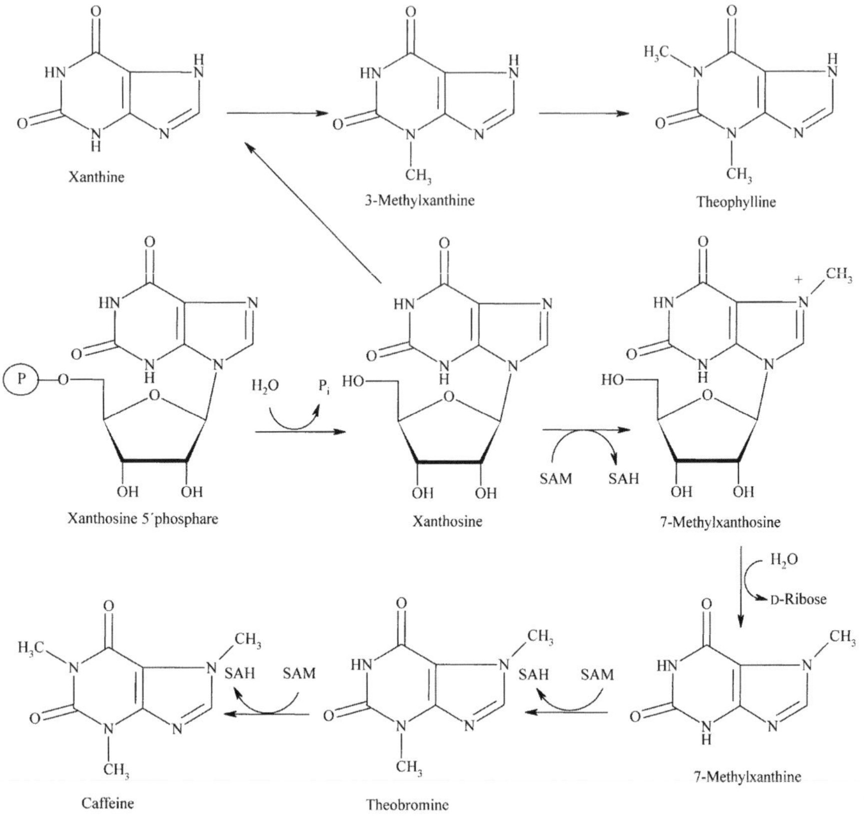

Fig. 241. Biosynthesis of theobromine, caffeine and theophylline. SAM = S-adenosyl-L-methionine, SAH = S-adenosyl-L-homocysteine

The biosynthesis of caffeine has been studied most extensively in coffea fruits and tea leaves. Genes encoding the enzymes of the main pathway have been cloned and expressed. It has been found that the pathway via xanthosine also operates in *Theobroma cacao* and that the accumulation of theobromine in cacao fruits is explained by the very slow conversion of theobromine to caffeine in this plant.

As illustrated in Fig. 241 caffeine formation is closely associated with the *S*-adenosyl- L-methionine (SAM) cycle. SAM is used as the methyl donor in three methylation steps and in the process it is converted to *S*-adenosyl-L-homocysteine (SAH), which is then hydrolysed to L-homocysteine and adenosine. Adenosine can be metabolized to xanthosine via the pathway adenosine → AMP → IMP → xanthosine-5′-phosphate → xanthosine, which enters the main pathway to caffeine. Homocysteine is recycled to replenish SAM levels. Three moles of SAH are produced for each mole of caffeine that is synthesized and it has been suggested that the SAM cycle is the sole source of both the purine skeleton and the methyl groups needed for caffeine and theobromine biosynthesis.

Crude drugs   **Coffee** (Coffeae semen) is the roasted seed of *Coffea* species (*Rubiaceae*). The most important species is *Coffea arabica* L., which is native to Sudan and the highlands of Ethiopia, and cultivated all over the tropics as several subspecies and varieties. Other species are grown, such as *Coffea liberica* Hiern, from Liberia and Sierra Leone, and *Coffea canephora* Pierre ex Froehner and *Coffea robusta* Linden, (Plate 30, page 703), the origin of which is the Congo region (Republic of Central Africa, Zaïre). *C. liberica* is more resistant to disease than *C. arabica,* but gives a lower quality coffee. *C. canephora* is also less easily attacked by disease.

The annual world production of coffee is of the order of 4 million tons. Most of it comes from South America, where Brazil is the greatest producer (ca. 1.5 million tons), followed by Colombia (ca. 0.5 million tons). Africa produces approximately a quarter of the world production and Asia about 5 %. *C. arabica* and its subspecies and varieties are grown in South America, whereas *C. canephora* dominates in Africa and Asia.

Coffea species are shrubs or small trees with entire, opposite, coriaceous (leathery) leaves, stipules and white flowers. The fruit is a berry, the size of a cherry, turning red when ripe. The endocarp of the fruit is parchment-like and inside it are two seeds, each covered with a thin testa, that is also parchment-like or papery. The seed consists mainly of endosperm and the embryo is small.

The coffee is harvested when the fruit is completely ripe. Unripe fruits must be avoided, because they will reduce the quality of the coffee. It is therefore necessary to pick coffee by hand. After the harvest, the seed is separated from the fruit pulp and the endocarp. There are wet and dry methods of doing this. In the wet process, the fruits are first made to swell in water and most of the fruit flesh is then removed in a roller mill. The remaining fruit flesh is destroyed in a fermentation process, lasting 14–48 hours. This process must be monitored carefully, because it influences the aroma of the coffee. After washing and drying, the papery testa is removed in special machines. The rest of the seed – the coffee-bean – is then ready for roasting.

In the dry process, the fruits are dried in the sun or in hot air, and this is followed by mechanical removal of the fruit pulp, endocarp and testa in a single operation.

The crude coffee, obtained by either process, is finally roasted at 200–250 °C, a procedure that gives the coffee its brown colour and

Chlorogenic acid

its aroma. Unroasted coffee contains 1–2.5 % caffeine, which is combined with chlorogenic acid.

In the roasting process, the caffeine is set free and many substances are formed to which the coffee owes its special aroma. More than 200 different components have been isolated so far, including volatile acids, alkaloids, phenols, aldehydes (especially furfural), ketones, amines, pyrrole, pyridine, furfuryl mercaptan, etc. Coffee is used for the preparation of a stimulating drink, the effects of which are due to its caffeine content. As already mentioned, part of the caffeine sublimes during the roasting process.

Black tea (Theae folium) is the fermented and dried leaf of *Camellia thea* Link. (*Theaceae*). When growing wild, the plant is a 5–10 m tall tree, but in cultivation it is maintained as a shrub, through constant picking of the top shoots and young leaves (Plate 31, page 735). The tea plant comes from South-East Asia and is today grown for commercial purposes in India, Sri Lanka, Indonesia, China and Japan. There are also plantations in Africa, mainly Kenya and South America.

Black tea is produced from the top shoot and the two uppermost leaves of each branch of the shrub. The picking can begin when the tea plant is 3 years old and can continue for 25–50 years; the yield is highest when the plant is about 10 years old. The leaves are quickly transported to the tea factory, which is usually located next to the plantation. Here, they are first of all dried, which may take 15–36 hours, depending on the temperature and humidity of the air. In modern factories, air is blown through the mass of leaves to accelerate the drying process. The leaves are then crushed (rolled) in special machines. During this process, enzymes (especially polyphenol oxidases) come into contact with polyphenols in the cell sap and oxidation takes place, giving the end product its dark colour. The crushed leaf material is then fermented at 30 °C for 0.5–2 hours, by which time the enzymic reactions that started during the crushing process are completed.

After the fermentation, the material is heated to 115–120 °C, so that the water content is reduced to 4–5 % and a stable product is obtained. Conveyor belts slowly carry the tea, spread on trays, through ovens, in which the desired temperature is reached. The black product is finally sifted, whereby different grades with respect to the particle size are obtained.

Black tea contains 2.5–4.5 % caffeine, 0.02–0.04 % theophylline and ca. 0.05 % theobromine. Other constituents include tannins of the catechin type (page 200), as well as oxidized polyphenols, these latter giving the tea its colour. The aroma is due to small amounts of a volatile oil with a very complex composition.

Black tea is the type of tea used in the West to prepare a stimulating drink. The waste material from the production – fine powder, petioles, etc. – is used for caffeine extraction.

In eastern Asia, the main type of tea used is **Green tea**. Its preparation differs from that of black tea in that the fresh leaves are immediately heated. This destroys the enzymes and the tea keeps its green colour during the subsequent rolling and drying processes.

Green tea is reported to have a cardioprotective effect, which is attributed to its content of flavonoids, particularly (-)-*epi*gallocatechin 3-*O*-gallate, the content of which is higher in green tea than in black tea. This cardioprotective effect of the flavonoids depends on their antioxidant, antithrombogenic and anti-inflammatory properties and they also improve the coronary flow velocity reserve.

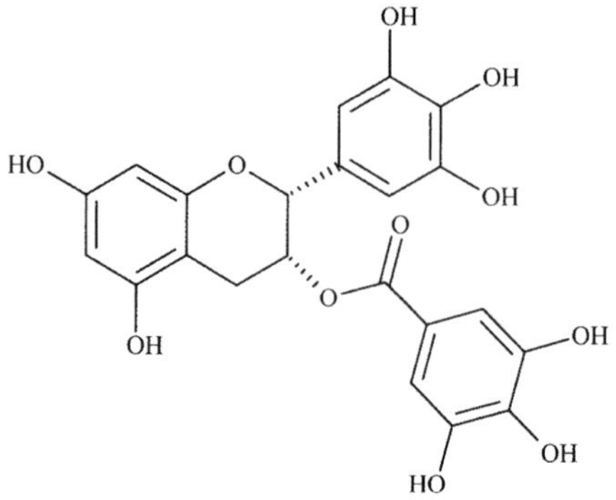

(-)-*Epi*gallocatechin 3-*O*-gallate

Maté leaf (Maté folium) is the dried leaf of *Ilex paraguensis* A. St.Hil. (*Aquifoliaceae*). Other *Ilex* species can also give this drug.

Ilex paraguensis is a 6–12 m high tree, native to southern Brazil, Paraguay and northern Argentina. In cultivation, the plant is kept as a shrub by pruning. For maté production, the ends of the branches are cut off and pulled through fire so that the enzymes of the leaves are inactivated and the green colour of the leaves is retained in the subsequent drying process. The dried and crushed leaves are used in South America for the preparation of a stimulating drink. Maté leaf contains about 1 % caffeine.

Kola seed (Colae semen) is the seed of *Cola acuminata* Schott et Endl. (*Sterculiaceae*). The drug consists of the two big cotyledons of the embryo. *Cola* species are trees, 15–25 metres high, endemic to the tropical part of West Africa, but also cultivated in other tropical countries, e.g. India, Brazil and Jamaica. Fresh kola seeds are eaten as a stimulant in their home countries. Dried Kola seed is of some medicinal use as a stimulating ingredient in so-called tonics. The drink Coca-Cola contains an extract of Kola seed. The drug has a caffeine content of 1–3 % and it also contains about 5 % tannins of the catechin type (page 200).

Guarana (Pasta guarana) is prepared from the seeds of *Paullinia*

Plate 31.
Camellia thea Link. Tea is prepared from the leaves (page 733). Plantation in Sri Lanka. Photograph by Gunnar Samuelsson.

Plate 32.
Fruit of *Theobroma cacao* L. Sri Lanka. (page 736) Photograph by Gunnar Samuelsson.

cupana Kunth (*Sapindaceae*). This plant is a liana of the Amazon basin, where it is also cultivated. The fruits, the size of hazel nuts, usually contain only one seed, which consists essentially of the cotyledons of the embryo. The seed is roasted, crushed and stirred into water to make a paste, which is moulded into sticks or other forms, such as animal figures, and dried. Guarana contains 4–8 % caffeine and ca. 8 % catechin-type tannins (page 200). It is used in Brazil to make a stimulating drink and has recently become a popular ingredient in so-called 'energy drinks'.

Cocoa seed (Cacao semen) is obtained from *Theobroma cacao* L. (*Sterculiaceae*). This is a tree, 4–6 metres in height, which originates from Central South America and the equatorial forests of South America and is cultivated in most tropical countries. About two-thirds of the world's production of cocoa seed comes from Africa, particularly the Ivory Coast, Ghana, Nigeria and Cameroon.

The flowers are borne in very large numbers directly on the trunk and on the thicker branches, but only some of them develop into fruits. The fruits are berries, 15–20 cm long, with a thick and hard pericarp turning orange when ripe (Plate 32, page 735). They are attached to the trunk or a branch through a short peduncle and contain 50–60 light-coloured seeds, embedded in a fruit flesh. The seeds are taken out of the fruits and piled in bins for several days. This starts a fermentation process during which the brown colour and the typical aroma develop. Following fermentation the seeds are dried and shipped to factories for further preparation. The first step is roasting in large roasting machines, during which the shells of the seeds become brittle and can be removed in the following cracking and fanning step. The remaining seed kernel contains about 50 % fat, which melts during the subsequent grinding process resulting in a dark-brown liquid called *chocolate liquor*. This is poured into moulds where it solidifies. *Cocoa powder* for preparation of drinking cocoa is produced by pumping the molten chocolate liquor into a filter press which removes most of the fat – *cocoa butter* (page 254). On the filter cloth remains a light-brown cake of solid particles which is ground and sifted to form cocoa powder. In making various forms of confectionary, chocolate liquor is mixed with cocoa butter, different aroma substances and sugar. Addition of milk gives milk chocolate.

Cocoa seed contains theobromine. During the roasting process much of the theobromine is transferred to the shells which thus are a source for extraction of theobromine. It should be noted that theobromine is poisonous for dogs, which metabolize the compound very slowly. The lethal dose is estimated to be 90–250 mg/kg of body weight, while a dose of 12 mg/kg is reputed not to cause any toxic effects. The content of theobromine in the various chocolate products is difficult to estimate and depends on the content of cocoa powder or chocolate liquor. The content in cocoa powder is 500–2 600 mg/100 g and in cooking or plain chocolate it is estimated to

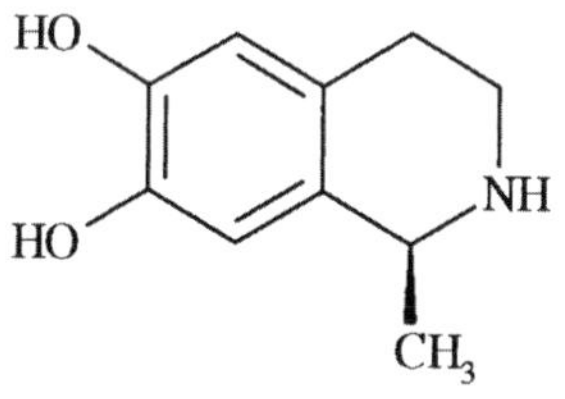

(S)-Salsolinol

be 450–1 600 mg/100 g. A medium-sized dog (20 kg) could thus die from eating about 70 g of cocoa powder and only 9 g of the powder can be considered to be safe. For cooking chocolate the corresponding lethal dose is 113 g.

Cocoa seed also contains a small amount (0.2-0.5 %) of caffeine and a tetrahydroisoquinoline alkaloid – (S)-salsolinol in an amount of up to 25 μg/g.

Salsolinol has dopaminergic activity and binds strongly to the D3 receptor which is thought to be involved in the reward system and consequently in the pathogenesis of addiction induced by different drugs or stimulants. The content of salsolinol in cocoa products is high enough to cause an interaction with the D3 receptor when eating greater amounts. This might explain the craving for and addiction to chocolate which is observed in some persons.

Pyrimidine derivatives

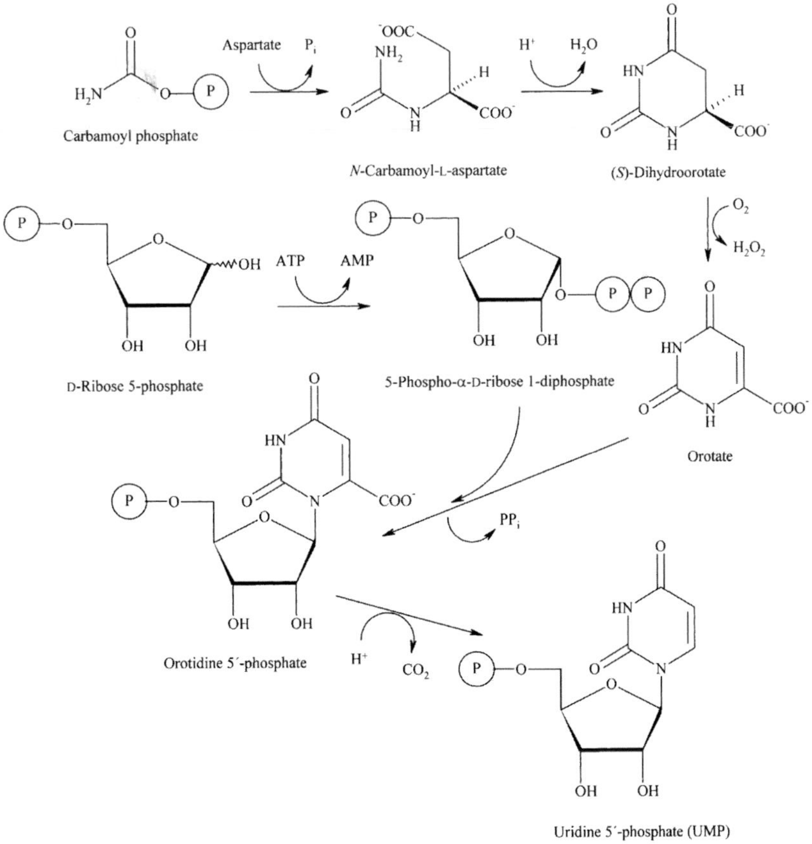

Fig. 242. Biosynthesis of uridine 5´-phosphate (UMP)

Biosynthesis The pyrimidine ring is biosynthesized from carbamoyl phosphate (page 490) and aspartic acid as illustrated in Fig. 242. In a reaction catalysed by *aspartate carbamoyltransferase* (EC 2.1.3.2), carbamoyl phosphate reacts with aspartate under formation of *N*-carbamoyl-L-aspartate. Ring-closure, catalysed by *dihydroorotase* (EC 3.5.2.3), yields (*S*)-dihydroorotate, which is dehydrogenated by *dihydroorotate oxidase* (EC 1.3.3.1) to form orotate, which is the starting material for biosynthesis of pyrimidine nucleosides and nucleotides. For introduction of the ribosyl moiety D-ribose 5-phosphate is diphosphorylated forming 5-phospho-α-D-ribose 1-diphosphate in a reaction catalysed by *ribose-phosphate diphosphokinase* (EC 2.7.6.1). *Orotate phosphoribosyltransferase* (EC 2.4.2.10) catalyses the next reaction in which orotate and 5-phospho-α-D-ribose 1-diphosphate are condensed to form orotidine 5′-phosphate which is decarboxylated to uridine 5′-phosphate by *orotidine-5′-phosphate decarboxylase* (EC 4.1.1.23).

The genes coding for the enzymes involved in the biosynthesis of pyrimidines are known both for prokaryotes and eukaryotes. In bacteria each enzymatic step is catalysed by a different protein that is encoded by a specific gene. In plants, the first three steps are catalysed by three different proteins, encoded by specific genes, whereas in mammals all three steps are catalysed by a single multifunctional protein. In both mammals and plants the last two steps are catalysed by distinct domains of a bi-functional protein. The genes for the two enzymatic functions are, however, separate.

Also for pyrimidines a salvage pathway permits biosynthesis from breakdown products of pyrimidines requiring less energy than *de novo* biosynthesis.

Uridine 5′-phosphate (UMP) is a building block of RNA, and is a precursor of thymidine monophosphate (TMP), contained in DNA. It is also the starting point for formation of the important pyrimidine nucleotides uridine triphosphate (UTP) and cytidine triphosphate (CTP).

Uridine triphosphate (UTP)

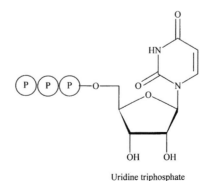

Uridine triphosphate

The main role of uridine triphosphate (UTP) is as a substrate for the synthesis of RNA, but it also acts as a source of energy or an activator in metabolic reactions in analogy with ATP although it is more specific. Thus UTP reacts with glucose-1-phosphate with formation of UDP-glucose and inorganic phosphate in the biosynthesis of sucrose (page 150).

UTP is formed by phosphorylation of UMP in two stages. First uridine diphosphate (UDP) is formed in a reaction mediated by ATP and the enzymes *UMP kinase* (EC 2.7.4.22) in prokaryotes and *cytidylate kinase* (EC 2.7.4.14) in eukaryotes. UTP is then formed by phosphorylation of UDP by *nucleoside-diphosphate kinase* (EC 2.7.4.6) and ATP. In contrast to the monophosphate kinases this enzyme has a broad specificity and can phosphorylate several different nucleotide diphosphates.

Cytidine triphosphate (CTP)

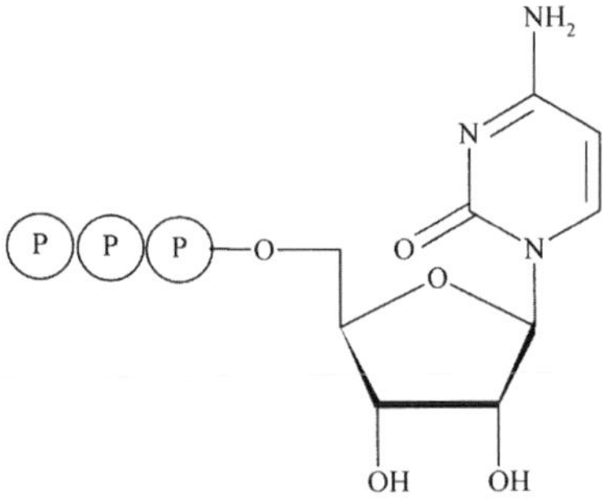

Cytidine triphosphate

As for UTP the main role for CTP is in the biosynthesis of RNA. It can also act as a specialized source of energy or an activator in metabolic reactions. Thus CTP is used as an activator in the bacterial biosynthesis of CoA (page 721). CTP is formed from UTP by the replacement of a carbonyl group by an amino group in a reaction catalysed by *CTP synthase* (EC 6.3.4.2). The aminogroup originates from glutamine. Fig. 243 illustrates the process.

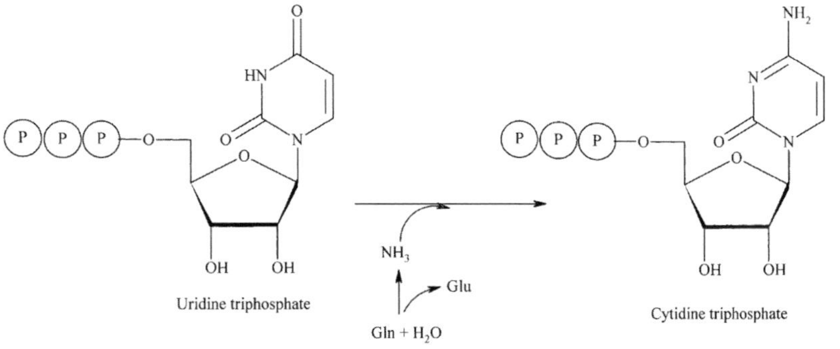

Fig. 243. Formation of CTP from UTP

Deoxycytidine monophosphate (DMP), which is a constituent of DNA, is formed from CTP by dephosphorylation to CDP followed by reduction of the ribose moiety.

Cytarabine (Ara-C)

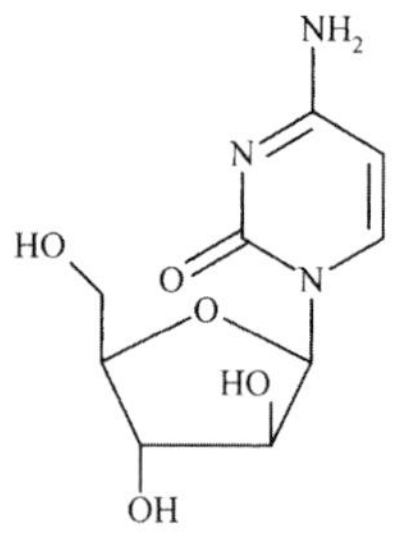

Cytarabine

In the 1950s the American chemist Werner Bergmann and coworkers studied marine sponges. This resulted in isolation and characterization of two unusual nucleosides named spongothymidine and spongouridine from the Caribbean sponge *Cryptotethya crypta*. These nucleosides were identified as 1-β-D-arabofuranosylthymine and 1-β-D-arabofuranosyluracil, respectively. The spongonucleosides were found in free state in the sponge and were extractable by acetone. At this time there was a great interest in nucleosides and nucleic acids owing to the discoveries reported by Watson and Crick about the function and structure of DNA, and anti-cancer research was focused on designing compounds that could interfere with the replication of DNA and stop fast-dividing cancer cells. The first efforts to design anti-cancer drugs based on nucleosides focused on modifications of the base unit with an intact sugar part. As the discovered sponge nucleosides were unusual DNA and RNA building blocks with a modified sugar part, another research direction was initiated. Now molecules were synthesized with a modified sugar moiety, which led to cytosine arabinoside, which is an analogue of the sponge nucleosides with cytosine instead of thymidine or uracil. This compound was called Ara-C.

In the 1960s several research groups studied the biological effects of Ara-C in both bacterial and mammalian systems. Ara-C was found to have a potent effect against leukaemia in mice induced by the mammalian cancer cell line L-1210. The studies resulted in the marketing of Ara-C under the name cytarabine as a remedy against acute leukaemia in children. Recently, other analogues of nucleosides have been designed and synthesized, to overcome the mechanisms of resistance, and developed to commercial drugs against acute leukaemia.

Thiamin (Vitamin B$_1$)

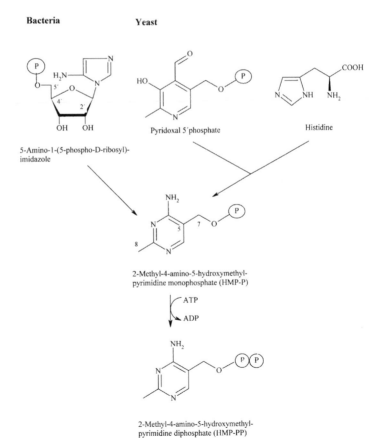

Thiamin

Thiamin is the precursor of the cofactor thiamin diphosphate which is required by numerous enzymes that participate in the metabolism of carbohydrates and amino acids. Humans cannot synthesize thiamin *de novo* and this compound must therefore be obtained from the diet. Severe deficiency of thiamin causes beriberi, which is characterized by impairment of the nerves and heart. The symptoms are loss of appetite, weakness and pain in the limbs, emotional disturbances, impaired sensory perception and periods of irregular heart rate. The disease appears in three forms: wet beriberi, dry beriberi

Bacteria   **Yeast**

5-Amino-1-(5-phospho-D-ribosyl)-imidazole

Pyridoxal 5′phosphate

Histidine

2-Methyl-4-amino-5-hydroxymethyl-pyrimidine monophosphate (HMP-P)

ATP
ADP

2-Methyl-4-amino-5-hydroxymethyl-pyrimidine diphosphate (HMP-PP)

Fig. 244. Biosynthesis of 2-methyl-4-amino-5-hydroxymethylpyrimidine diphosphate (HMP-PP)

and cerebral beriberi. In wet beriberi the heart symptoms are dominating with oedema resulting from cardiac failure and poor circulation. In dry beriberi the long nerves deteriorate, first in the legs and then in the arms with atrophy of muscles and loss of reflexes. The symptoms are mainly the result of impaired metabolism of carbohydrates. In the absence of the cofactor thiamin diphosphate, pyruvic acid and lactic acid accumulate in the tissues and are believed to be responsible for most of the neurological and cardiac manifestations. Cerebral beriberi is the result of thiamin deficiency affecting the central nervous system and may lead to Wernicke's encephalopathy and Korsakoff's psychosis. These diseases are particularly manifested in alcoholics.

Prior to the end of the 19th century beriberi was common in wealthy people in Asia, whose diet consisted mainly of polished white rice. Poor people, who consumed unpolished rice, were not affected by the disease. These observations led to the discovery in

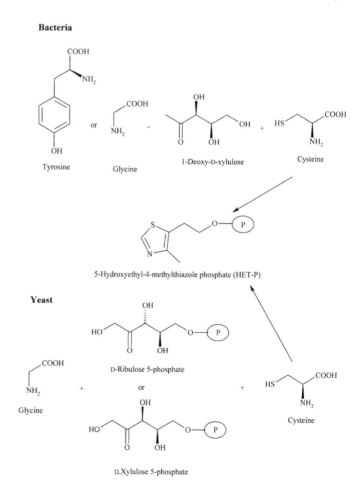

Fig. 245. Biosynthesis of 5-hydroxyethyl-4-methylthiazole phosphate (HET-P)

rice bran of a factor that cured beriberi. This was crystallized in 1926 and its structure was determined in 1935 when it was given the name thiamin. Nowadays beriberi is encountered mainly in cases of chronic alcoholism.

In Nature thiamin occurs as free thiamin and as mono-di- and triphosphate. Besides rice bran, thiamin occurs in green peas, spinach, liver, beef, pork, nuts, bananas, soyabeans, whole grains, bread and yeast. The recommended average daily intake is 1.1–1.4 mg for adults. Children under 3 years of age require 0.2–0.5 mg, whereas older children and youngsters should have 0.6–1.0 mg/day.

Biosynthesis

The structure of thiamin contains a thiazole ring and a pyrimidine ring. These are formed separately and then coupled to form thiamine phosphate as outlined in Fig. 246.

Formation of the pyrimidine structure

Two pathways are known. for formation of 2-methyl-4-amino-5-hydroxymethylpyrimidine diphosphate (HMP-PP), which is the unit used in the subsequent condensation process. In bacteria and probably also in plants HMP-PP is formed from 5-amino-1-(5-phospho-D-ribosyl)-imidazole (= AIR), which is an intermediate in *de novo* biosynthesis of purine (see page 712 and Fig. 231). This is transformed to 2-methyl-4-amino-5-hydroxymethylpyrimidine monophosphate (HMP-P) in a complex rearrangement reaction where C5, C7 and C8 of the molecule are derived from the 4′ (or 5′), 5′ (or 4′) and 2′ ribose carbons. HMP-P is then phosphorylated to HMP-PP by the enzyme *phosphomethylpyrimidine kinase* (EC 2.7.4.7). These reactions are illustrated in Fig. 244 yeast cells make HMP-P from pyridoxine and L-histidine. These two pathways to PMP-PP have been elucidated by incorporation experiments but the genes and enzymes involved are incompletely known.

Formation of the thiazole structure

The compound which is condensed with HMP-PP in the formation of thiamine phosphate is 5-hydroxyethyl-4-methylthiazole phos-

2-Methyl-4-amino-5-hydroxymethyl-pyrimidine diphosphate (HMP-PP)

5-Hydroxyethyl-4-methylthiazole phosphate (HET-P)

Thiamin phosphate

Thiamin

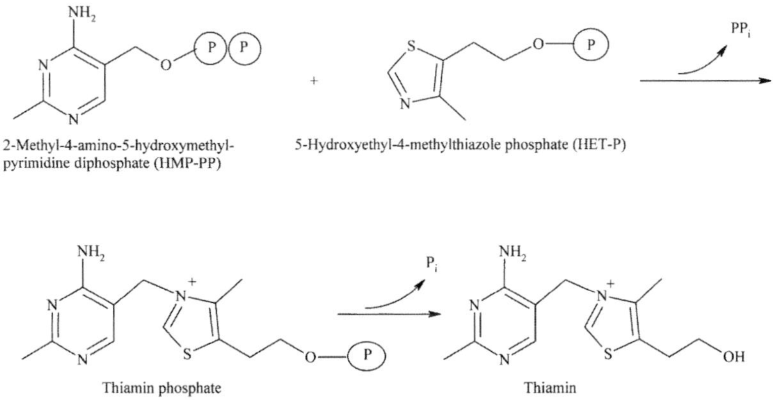

Fig. 246. The final stage in the thiamin biosynthesis pathway

phate (HET-P). Also here the pathways for biosynthesis are different in yeast and bacteria (Fig. 245). In bacteria HETP-P is formed from tyrosine or glycine, deoxy-D-xylulose and cysteine, whereas in yeast the starting materials are glycine, D-ribulose 5-phosphate or D-xylulose 5-phosphate and cysteine. Also here the knowledge of the starting materials is based on incorporation experiments and the genes and enzymes involved are only partly known. In bacteria, six gene products, probably involved in a multi-enzyme complex, have been shown to participate in the complex biosynthetic process. In yeast and plants only one gene (THI4) has been identified, the transcript of which is one of the most abundant proteins in the cell under thiamin depleted conditions.

The knowledge of the biosynthesis of HET-P in plants is controversial. There is evidence from experiments with spinach chloroplasts that the bacterial pathway is used, but there is also evidence that the biosynthesis involves L-cysteine, glycine and an unidentified metabolite, which possibly is NAD+ and that homologues of THI4 are at work, thus suggesting that the yeast pathway is used in plants.

As illustrated in Fig. 246. the enzyme *thiamin phosphate synthase* catalyses the condensation of HMP-PP and HET-P to thiamin phosphate, which is then dephosphorylated to thiamin. In the thiamin-producing organisms and also in the human body, thiamin is diphosphorylated to thiamin diphosphate, which is the working cofactor. Humans can utilize thiamin mono- and diphosphate as the intestinal flora contains enzymes which hydrolyse these compounds to thiamin which can then be diphosphorylated.

Further reading

Purine biosynthesis, AMP and GMP biosynthesis

International Union of Biochemistry and Molecular biology. Database http://www.chem.qmul.ac.uk/iubmb/enzyme/reaction/misc/purine1.html

NAD⁺ and NADP⁺

Katoh, A. and Hashimoto, T. Molecular biology of pyridine nucleotide and nicotine biosynthesis. *Frontiers in Bioscience*, **9**, 1577–1586 (2004).

Katoh, A., Uenohara, K., Akita, M. and Hashimoto, T. Early steps in the biosynthesis of NAD in Arabidopsis start with aspartate and occur in the plastid. *Plant Physiology*, **141**, 851–857 (2006).

Rongvaux, A., Andris, F., Van Gool, F. and Leo O. Reconstructing eukaryotic NAD metabolism. *BioEssays*, **25**, 683–690 (2003).

CoA

International Union of Biochemistry and Molecular biology. Database http://www.chem.qmul.ac.uk/iubmb/enzyme/reaction/misc/CoA1.html

Leonardi, R., Zhang, Y.-M., Rock, C.O. and Jackowski, S. Coenzyme A: Back in action. *Progress in Lipid Research*, **44**, 125–153 (2005).

Riboflavin

Bacher, A., Eberhardt, S., Fisher, M., Kis, K. and Richter, G. Biosynthesis of Vitamin B2 (Riboflavin). *Annual Review of Nutrition*, **20**, 153–167 (2000).

Roje, S. Vitamin B biosynthesis in plants. *Phytochemistry*, **68**, 1904–1921 (2007).

Folate (Vitamin B₉)

International Union of Biochemistry and Molecular biology. Database http://www.enzyme-database.org/reaction/misc/folate1.html

Roje, S. Vitamin B biosynthesis in plants. *Phytochemistry*, **68**, 1904–1921 (2007)

Caffeine, theobromine, tea

Anaya A.L., Cruz-Ortega, R. and Waller, G.R. Metabolism and ecology of purine alkaloids. *Frontiers in Bioscience*, **11**, 2354–2370 (2006).

Ashihara, H. and Suzuki, T. Distribution and biosynthesis of caffeine in plants. *Frontiers in Bioscience*, **9**, 1864–1876 (2004).

International Union of Biochemistry and Molecular biology. Database http://www.chem.qmul.ac.uk/iubmb/enzyme/reaction/misc/caffeine.html

Cheng, T.O. All teas are not created equal. The Chinese green tea and cardiovascular health. *International Journal of Cardiology*, **108**, 301–308 (2006).

Melzig, M.F. *et al*. In vitro pharmacological activity of the tetrahydroisoquinoline salsolinol present in products from *Theobroma cacao* L. like cocoa and chocolate. *Journal of Ethnopharmacology*, **73**, 153–159 (2000).

Biosynthesis of pyrimidine nucleotides

Huang, M. and Graves, L.M. De novo synthesis of pyrimidine nucleotides: emerging interfaces with signal transduction pathways. *Cellular and Molecular Life Sciences*, **60**, 321–336 (2002).

International Union of Biochemistry and Molecular biology. Database http://www.chem.qmul.ac.uk/iubmb/enzyme/reaction/misc/pyrimid.html

Zrenner, R., Stitt, M., Sonnewald, U. and Boldt, R. Pyrimidine and purine biosynthesis and degradation in plants. *Annual Review of Plant Biology*, **57**, 805–836 (2006).

Cytarabine

Bergmann, W. and Burke, D.C. Contributions to the study of marine products XXXIX. The nucleosides of sponges III. Spongothymidine and Spongouridine. *Journal of Organic Chemistry*, **20**, 1501–1507 (1955).

Howard, J.P, Cevik, N. and Murphy, M.L. Cytosine arabinoside (NSC-63878) in acute leukemia in children. *Cancer Chemotherapy Report*, **50** (5), 287–291 (1966).

Thiamine

Begley, T.P., Downs, D.M., Ealick, S.E., McLafferty, F.W., Van Loon, A.P.G.M., Taylor, S., Campobasso, N., Chiu, H.-J., Kinsland, C., Reddick, J.J. and Xi, J. Thiamin biosynthesis in prokaryotes. *Archives of Microbiology*, **171**, 293–300 (1999).

Nosaka, K. Recent progress in understanding thiamin biosynthesis and its genetic regulation in *Saccharomyces cerevisiae*. *Applied Microbiology and Biotechnology*, **72**, 30–40 (2006)

Roje, S. Vitamin B biosynthesis in plants. *Phytochemistry*, **68**, 1904–1921 (2007).

Textbooks of biochemistry and molecular biology

Berg, J.M., Tymoczko, J.L. and Stryer, L. *Biochemistry*, 6th Edn. W.H. Freeman and Company, New York, 2006.

Lodish, H., Berk, A., Kaiser, C.A., Krieger, M., Scott, P., Bretscher, A., Ploegh, H. and Matsudaira, P. *Molecular Cell Biology*. 6th Edn. W.H. Freeman and Company, New York (2008).

Appendix I

Plant-derived crude drugs, used for preparation of herbal remedies

English name	Pharmaceutical name	WHO-Approved name (See page 47)	Page
Aloe vera	-	*Aloe vera* (L.) Burm. f., leaf mucilage	319
American ginseng	-	-	448
Aniseed	Anisi fructus	*Pimpinella anisum* L., fruit	213
Barbados aloes	Aloe barbadensis	*Aloe vera* (L.) Burm. f., dry leaf juice	319
Benzoin	Siam benzoin (Benzoe tonkinensis) (Gum benzoin)	*Styrax tonkinensis* (Pierre) Craib ex. Harwich, resin	234
Bitter fennel	Foeniculi amari fructus	*Foeniculum vulgare* Mill., fruit	214
Bitter-orange peel	Aurantii pericarpium	*Citrus aurantium* L., peel	380
Buckbean leaf	Menyanthis folium	*Menyanthes trifoliata* L., leaf	387
Camphor	Camphora	*Cinnamomum camphora* (L.) Presl., gum	381
Cape aloes	Aloe capensis	*Aloe ferox Mill.*, dry leaf juice	319
Capsicum	Capsici fructus	*Capsicum annuum* L., fruit	616
Caraway fruit	Carvi fructus	*Carum carvi* L. fruit	381
Cascara	Rhamni purshianae cortex	*Rhamnus purshiana* DC, bark	318
Cinnamon	Cinnamomi cortex	*Cinnamomum verum* J. Presl, bark	210
Devil´s claw root	Harpagophyti radix	*Harpagophytum procumbens* DC ex Meisn., root	389
Eleutherococcus	Eleutherococci radix	*Eleutherococcus senticosus* (Rupr. & Maxim.) Maxim, root	222
Feverfew	Chrysanthemi parthenii folium	*Tanacetum parthenium* (L.) Sch. Bip., leaf	408
Frangula bark	Frangulae cortex	*Rhamnus frangula* L., bark	317
Gentian root	Gentianae radix	*Gentiana lutea* L., root	388
Ginger	Zingiberis rhizoma	*Zingiber offizinale* Roscoe, rhizome	382
Ginkgo	Ginkgo folium	*Ginkgo biloba* L., leaf	423
Ginseng	Ginseng radix	*Panax ginseng* C.A. Mey., root	447
Hamamelis leaf	Hamamelidis folium	*Hamamelis virginiana* L., leaf	202
Holy thistle	Cardui benedicti herba	*Cnicus benedictus* L., herb	404
Horse chestnut seeds	Hippocastani semen	*Aesculus hippocastanum* L., seed	445
Hypericum	Hyperici herba	*Hypericum perforatum* L., herb	325
Indian psyllium	Plantaginis ovatae semen	*Plantago ovata* Forssk., seed	174
Ispaghula husk	Plantaginis ovatae seminis tegumentum	*Plantago ovata* Forssk., seed husk	174
Ispaghula seed	Plantaginis ovatae semen	*Plantago ovata* Forssk., seed	174
Kava	Kava-Kava rhizoma	*Piper methysticum* Forester, rhizome	337
Linseed	Lini semen	*Linum usitatissimum* L., seed	174

English name	Pharmaceutical name	WHO-Approved name (See page 47)	Page
Liquorice extract	Glycyrrhizae extractum crudum	*Glycyrrhiza glabra* L., root-extract	442
Liquorice root	Liquiritae radix	*Glycyrrhiza glabra* L., root	442
Manna	Manna	*Fraxinus ornus* L., juice	176
Marshmallow root	Althaeae radix	*Althaea officinalis* L., root	174
Matricaria flower	Matricarie flos	*Matricaria recutita* (L.) Rausch., flower	405
Oil of turpentine	Terebinthinae aetheroleum	*Pinus pinaster* Ait., essential oil	375
Peppermint leaf	Menthae piperitae folium	*Mentha x piperita* L., leaf	378
Peppermint oil	Menthae piperitae aetheroleum	*Mentha x piperita* L., essential oil	378
Peru balsam	Balsamum peruvianum	*Myroxylon balsamum* var. *pereirae* (Royle) Harms, balsam	209
Quassia wood	Quassiae lignum	*Quassia amara* L., wood and *Picrasma excelsa* (Sw.) Planch., wood	449
Quillaja bark	Quillajae cortex	*Quillaja saponaria* Molina, bark	444
Rhodiola	Arctic root	*Rhodiola rosea* L., root	216
Rhubarb	Rhei radix	*Rheum palmatum* L., root	319
Roman chamomile flower	Chamomillae romanae flos	*Chamaemelum nobile* (L.) All., flower	406
Rose hips	Cynosbati fructus	*Rosa canina* L.	179
Rosemary leaf	Rosmarini folium	*Rosmarinus officinalis* L., leaf	380
Saffron	Croci stigma	*Crocus sativus* L., stigma	477
Schisandra		*Schisandra chinensis* (Turcz.) Baill., fruit	223
Senega root	Senega radix (Polygala radix)	*Polygala senega* L., root	445
Senna leaf	Sennae folium	*Senna alexandrina* Mill., leaf	321
Senna pods, Alexandrian	Sennae fructus acutifoliae	*Senna alexandrina* Mill., fruit	322
Senna pods, Tinnevelley	Sennae fructus angustifoliae	*Senna alexandrina* Mill., fruit	322
St. Mary Thistle	Cardui mariae fructus	*Silybum marianum* (L.) Gaertn.	339
Thyme	Thymi herba	*Thymus vulgaris* L., herb	381
Valerian root	Valerianae radix	*Valeriana officinalis* L., root	386
Wormwood	Absinthii herba	*Artemisia absinthium* L., herb	404

750

751

753

766

770